Common Conversio

To Convert	To	Do This
Permeability in m^2	Darcy	multiply by 10^{12}
Permeability in m^2	Hydraulic Conductivity in $m\text{-}s^{-1}$ (Pure Water at 20 °C)	multiply by 9.77×10^6
Permeability in m^2	Hydraulic Conductivity in $m\text{-}year^{-1}$ (Pure Water at 20 °C)	multiply by 3.08×10^{14}
Temperature in °C	°K	add 273.16
Temperature in °C	°F	multiply by 1.8 and add 32
Pressure in psi (pound per square inch)	Pascals	multiply by 6894.74
Pressure in atm (atmospheres)	Pascals	multiply by 101325.
Pressure in bars	Pascals	multiply by 10^5
Length in Meters	Feet	multiply by 3.28084
Energy in Calories	Joules	multiply by 4.184
Mass in Kilograms	Mass in Pounds	multiply by 2.20462
Density in $g\text{-}cm^{-3}$	Density in $kg\text{-}m^{-3}$	multiply by 1000
Volume in Gallons (US liquid)	Liters	multiply by 3.7854
Volume in Liters	m^3	multiply by 10^{-3}
Length in Miles	Feet	multiply by 5280
Area in Acres	Square Feet	multiply by 43560

INTRODUCTION TO HYDROGEOLOGY

Introduction to Hydrogeology

David Deming
University of Oklahoma

Boston Burr Ridge, IL Dubuque, IA Madison, WI New York San Francisco St. Louis
Bangkok Bogotá Caracas Kuala Lumpur Lisbon London Madrid Mexico City
Milan Montreal New Delhi Santiago Seoul Singapore Sydney Taipei Toronto

McGraw-Hill Higher Education

A Division of The **McGraw-Hill** *Companies*

INTRODUCTION TO HYDROGEOLOGY

Published by McGraw-Hill, a business unit of The McGraw-Hill Companies, Inc., 1221 Avenue of the Americas, New York, NY 10020.

This book is printed on acid-free paper.

1 2 3 4 5 6 7 8 9 0 DOC/DOC 0 9 8 7 6 5 4 3 2 1

ISBN 0–07–232622–0

Publisher: *Margaret J. Kemp*
Developmental editor: *Renee Russian*
Associate marketing manager: *Tami Petsche*
Senior project manager: *Kay J. Brimeyer*
Senior production supervisor: *Laura Fuller*
Coordinator of freelance design: *Rick D. Noel*
Cover designer: *Sheilah Barrett*
Cover image: *"Artesian Well at Prairie Du Chien, Wisconsin," from the Fifth Annual Report of the US Geological Survey, 1885.*
Senior photo research coordinator: *Carrie K. Burger*
Media technology producer: *Judi David*
Compositor: *GAC / Indianapolis*
Typeface: *10/12 Times Roman*
Printer: *R. R. Donnelley & Sons Company/Crawfordsville, IN*

Library of Congress Cataloging-in-Publication Data

Deming, David, 1954–
Introduction to hydrogeology / David Deming. — 1st ed.
p. cm.
Includes bibliographical references.
ISBN 0–07–232622–0
1. Hydrogeology. I. Title.

GB1003.2 .D46 2002
551.49—dc21

2001044509
CIP

www.mhhe.com

This book is dedicated to students, everywhere.

CONTENTS

CHAPTER 4

PROPERTIES OF GEOLOGIC FLUIDS *92*

CHAPTER 5

TRANSIENT FLOW *115*

CHAPTER 6

NEAR SURFACE FLOW *143*

CHAPTER 7

DRIVING FORCES AND MECHANISMS OF FLUID FLOW *187*

CHAPTER 11

HEAT TRANSPORT *340*

CHAPTER 12

EARTHQUAKES, STRESS, AND FLUIDS *363*

CHAPTER 13

FLUIDS IN THE OCEANIC CRUST *387*

CHAPTER 14

FLUIDS AND ORE DEPOSITS *410*

Preface

This book grew out of my hydrogeology lectures at the University of Oklahoma from 1992 onward. At the time I began teaching hydrogeology, I found there was no single textbook that covered the topic in the manner that I wanted. There were, and remain, many excellent books on the subject, but none of them presented hydrogeology quite in the way that I thought appropriate for an introductory class consisting primarily of geology majors.

This book is intended—as titled—to be an introduction to hydrogeology, not a comprehensive treatise. It has been my experience that about two-thirds of the material contained herein is sufficient to cover a semester. My preference is to teach the first seven chapters, which cover what I consider to be the basic physics and geology of fluid flow in the crust, and then cover perhaps half of the remaining chapters according to student interest or my own whims.

My primary goal in writing this text was to emphasize the geologic aspects of hydrogeology. The text is not designed solely for those who desire to specialize in hydrogeology. Fluids are intimately involved in nearly all geologic processes, and this text is designed to introduce the average geology student to that world. Hopefully, this book can work on several different levels. The mathematical level is such that the material is accessible to anyone with a first-year knowledge of basic differential and integral calculus. I have also tried whenever possible to derive mathematical expressions without skipping steps or pulling results out of a hat. The references and recommended reading are included so as to allow professionals and graduate students to delve more deeply into the specific topics that interest them.

There is some material in this book that is absent from other texts in the genre. In accordance with the promotion of a broad intellectual perspective, I have included a number of short essays that explore the history and folklore of topics related to groundwater. The biographies contained herein are intended to convey more than the mere facts relating to the development of the science of hydrogeology. In addition to the chronology of ideas, there is a revealed human drama that celebrates the vision, dedication, and moral courage exhibited by the founders of our science.

I know there are some who will find this book anomalous in its brief treatment of areas such as the pumping of water wells, along with the inclusion of topics such as oil migration and fluids in the oceanic crust. The choice of material reflects the natural evolution of our science; it was never my intention to write a book that perpetuated the existing state of affairs. Rather my motive was to change how hydrogeology is taught to geologists and help hydrogeology take its place as a truly geologic science.

David Deming
University of Oklahoma

ACKNOWLEDGEMENTS

This book could not have been completed without the assistance of Claren Kidd and the staff of the Laurence S. Youngblood Energy Library at the University of Oklahoma. I also received research assistance from Dr. Kerry Magruder, the librarian of the University of Oklahoma History of Science Collection. I found the GeoRef database developed by the American Geological Institute to be an invaluable resource.

I am indebted to numerous colleagues. Steve Ingebritsen of the U.S. Geological Survey answered incessant questions on varied topics. John Bredehoeft contributed nearly all of the material that formed the basis for the history of groundwater research at the U.S. Geological Survey. The following individuals made direct contributions to the book by writing biographies: John Bredehoeft, Ben Rostron of the University of Alberta, Lenny Konikow and Paul Hsieh of the U.S. Geological Survey, Russell G. Slayback, and Fred F. Meissner. David Stearns and Fred Meissner helped me with the biography of M. K. Hubbert by discussing their personal recollections and assessments of Hubbert with me.

Many people either helped me locate photographs or donated photos from their personal collections. These include Dorcas Hendershott of the School of Earth Sciences at Stanford University, Dr. Richard E. Goodman, Dr. Louis A. Frank of the University of Iowa, Dr. Dale Lightfoot at Oklahoma State University, and Devin Galloway of the U.S. Geological Survey.

Finally, I thank my wife, Jerry A. Deming, for her patience and faith.

FLUIDS IN THE CRUST:
The Science of Hydrogeology

1.1 DEFINITIONS

Hydrology is the study of terrestrial water. **Hydrogeology** is the branch of hydrology that studies underground fluids and their interaction with solid geologic materials. Traditionally, hydrology has not included the study of the oceans, the water within them, or the processes therein. The study of ocean waters is considered to be the realm of oceanography.

The word "hydrology" as commonly used today refers to water in the atmosphere and on the surface of the continents and the processes that affect it. Some of the topics with which hydrology concerns itself are evaporation, streamflow, erosion caused by water, floods, rivers, lakes, and wetlands. The terminology may become complex. Meinzer (1942, p. 4) divided hydrology into the subdisciplines of **potamology** (study of surface streams), **limnology** (study of lakes), and **cryology** (study of ice and snow), and geohydrology. The term **geohydrology** as used by Meinzer (1942) and others is essentially the same as hydrogeology. Underground processes such as ore formation tend to lie exclusively in the domain of hydrogeology. But, some geologic processes such as streamflow depend upon both underground flow and overland flow. In such cases, the distinction between hydrogeology and hydrology becomes blurred. Terminology and classification are not ends in and of themselves; excessive reliance upon them may obscure rather than reveal knowledge.

Although most underground fluids are aqueous, the study of hydrogeology encompasses the interaction of other fluids (e.g., petroleum) with groundwater. What is groundwater? The most common definition of groundwater dates back to Meinzer (1923), who defined groundwater to be water below the water table, differentiating it from water in the unsaturated or vadose zone that he termed "vadose water." Meinzer's rationale was that "the water that supplies springs and wells is groundwater." The definition was thus centered on the fact that in 1923 hydrogeology was almost exclusively preoccupied with the flow of water to wells and factors relating to the development of groundwater as a resource. There is no physical or chemical distinction between water found in the unsaturated zone and that found below the water table. Indeed, most water in the saturated zone originates in the unsaturated zone. Hydrogeology today is no longer concerned solely with wells and the development and management of groundwater as a human resource. Therefore, it is both natural and appropriate to define **groundwater** as any subsurface aqueous fluid, either saline or fresh.

Nearly all groundwater in the Earth's crust is found in the pore spaces of rocks. Rocks and unconsolidated sediments (e.g., gravel, sand, silt) are **porous media.** A porous medium is, literally, a material with holes. In rocks, these holes may be present as fractures or as interstitial spaces between solid grains. O. E. Meinzer (1923) wrote:

> The rocks that form the crust of the Earth are in few places, if anywhere, solid throughout. They contain numerous open spaces, called voids or interstices, and these spaces are the receptacles that hold the water that is found below the surface of the land and is recovered in part through springs and wells.

The solid part of a rock is the **matrix.** Nearly all rocks and sediments have appreciable permeability, as well as the ability to store and release water. **Permeability** is the capacity of a porous medium that allows it to transmit fluids in the presence of a potential energy gradient. Porous media in the Earth (rocks and sediments) may not only transmit fluids, but also store and release them. The latter processes are analogous to the uptake and release of water by a household sponge. A sponge, of course, is more compressible than a rock, but the same physical principles apply.

Our interest in hydrogeology is motivated by the fact that fluids play an important role in virtually all geologic processes. Groundwater is an important source of drinking and agricultural water, as well as a potential mechanism for the transport of contaminants and pollutants. Moving fluids are important in the formation of certain ore deposits, including zinc, lead, and gold. Our technological civilization depends largely upon petroleum and natural gas as energy sources, and understanding their movement in the subsurface is often a key to economic exploitation. Fluid pressure affects the state of stress within the Earth, and thus the occurrence of earthquakes. The ability of moving subsurface fluids to transport both heat and mass implies that nearly all rocks have been diagenetically altered through interactions with these fluids. **Diagenesis** is any post-depositional change that takes place in a sediment or rock.

Our primary concern is to understand fluids as they occur in the Earth's crust, especially as they pertain to geologic problems. Fluids are an important component of the Earth's crust, they affect the mineralogical, mechanical, and thermal evolution of the rocks in the crust. In the first seven chapters of this book, the following basic questions are addressed: Where are fluids found in the Earth's crust? What is their composition? What physical properties of Earth materials (rocks and sediments) affect fluid movement? What forces drive fluids in the Earth's crust? The second half of the book (chapters 7 through 14) is devoted to a discussion of how fluids interact and affect specific geologic phenomena and processes.

1.2. A Brief History of Hydrogeology.

The study of underground water by man extends into the indefinite past. Our oldest written records and archaeological excavations indicate ancient man was interested in phenomena such as springs and the source of the water that fed them.

1.2.1. The Ancient World: Speculation and Empiricism.

Water is essential for survival; many ancient civilizations were located on the banks of rivers. These include the Tigris and Euphrates rivers in Iraq, the Nile in Egypt, and the Indus in Pakistan. Some of the earliest hydrological studies were done with the goal of understanding and controlling river flow. The first systematic hydrological studies may have been done by the ancient Egyptians (3500–3000 BC), who studied the Nile River and developed a system by which downstream communities could be warned of advancing floods. Water levels of the Nile were carefully monitored and compared with previous years. If levels rose precipitously fast

Figure 1.1 The hydrologic cycle as it was once understood. Ocean water was thought to travel through subterranean tunnels into the continents where it collected in caverns that fed springs. The springs in turn supplied water to rivers.
(From Kircher, 1678)

at upstream locations, the swiftest rowers would propel their boats rapidly downstream, outpacing the advancing flood and warning of its approach (Biswas, 1970).

In Europe, prehistoric man exploited salt springs as a commercial source of salt. The remains of Neolithic and Bronze Age saltworks have been discovered at numerous locations throughout what are now Austria, Germany, and France. The development of salt-processing facilities may have occurred as part of a shift from hunting-gathering to farming as a means of sustenance. With less meat in their diet, agrarian societies may have needed a supplemental source of salt. Salt was used not only as a spice, but also for food preservation. The salt industry was an important part of the economy of Prehistoric Europe. Salt-producing communities tended to be prosperous, and were often important centers of trade (Hanor, 1987).

The chief hydrological mystery with which the ancients concerned themselves was the nature of the hydrologic cycle (Figure 1.1). Water ran off the continents through rivers into the sea, but what supplied the rivers? Similarly, the source of water from springs was unknown. Clearly, there had to be some recirculation of oceanic waters back to the continents, but how this occurred was a mystery. Some of the earliest speculations on the nature of the return mechanism were made by the ancient Greeks. Thales (about 650 BC) stated that springs and streams are supplied by ocean water that is driven into the rocks of the continents by winds and then elevated into the mountains by the pressure of the rocks. Plato (427–347 BC) in his dialogue titled

Figure 1.2 The Roman aqueduct at Pont du Gard ("Bridge of the Gard"), France. This combination bridge-aqueduct was constructed around 19 BC to carry water to the city of Nîmes over the Gard River in southern France. The structure has three tiers, the highest of which rises to a height of 47 meters. The Pont du Gard is built of limestone blocks, which weigh as much as six tons each and are fitted together without the use of mortar.

Phaedon offered the opinion that the waters that form the seas, lakes, rivers, and springs come from a vast cavern named Tartarus, and that all waters return to Tartarus through various routes. One of the difficulties in invoking the oceans as a source of the fresh waters found on the continents is that ocean water is salty. Recirculation to the continents thus not only required an energy source, but a process of either filtration or distillation to remove salt. Aristotle (384–322 BC) authored a treatise titled *Meteorologica* in which he recognized the processes of evaporation and distillation. He wrote that the rays of the sun turned water into air, and that when the air became cold it turned back into water and fell as rain. Aristotle believed, however, that most river flow was not supplied by rainfall, but rather by subterranean condensation that fed springs, which in turn supplied water to rivers.

The Romans contributed little to the ideas originated by the Greeks, but were great engineers who built a system of aqueducts to supply Rome with water. An **aqueduct** is a pipe or channel designed to transport water from a remote source, usually by gravity (Figure 1.2). Altogether, the Romans built eleven aqueducts that evolved over 500 years to meet their needs. The first aqueduct was built in 312 BC, the last completed in AD 226. By AD 97, nine aqueducts brought about 322 million liters (85 million gallons) of water a day into Rome. The longest of these was the *Aqua Marcia*, which was 91 km in length. To deliver water to Rome, Roman engineers had to overcome the problem of uneven topography while maintaining a smooth and continuous downhill slope over the course of the aqueduct. They had no external power source to drive water flow, all flow had to be downhill. Two books on the water supply of Rome were written by the water commissioner of the city, Julius Frontinus, in AD 97. There is little indication in Frontinus' book that he understood any of the essential physics of fluid flow. The Roman aqueducts, however, worked and did so extremely well.

1.2.2. The Sixteenth Century: The Importance of Observation.

From about AD 200 to 1500, little scientific progress was made in any field of study, and hydrology was no exception. The earliest signs of a resurgent intellectual interest in hydrological problems are found in the writings of the Florentine artist and scientist, Leonardo da Vinci (1452–1519). Meinzer (1942) concluded that da Vinci's writings showed a clear understanding of the hydrologic cycle and the origin of rivers and streams by the infiltration of rainwater. Leonardo was the first to apply the principles and methods of modern science to hydrologic problems. That is, he based his conclusions not on speculation, but observation. He even conducted experiments and used models to study the flow of water in channels. Although Leonardo is generally better known as a Renaissance scientist, more substantive contributions to hydrology were made by Bernard Palissy (1509–1589). Palissy was also the inventor of enameled pottery and a pioneer in the field of paleontology. Because he was born poor, Palissy lacked a classical education and never learned Greek or Latin. In sixteenth-century Western Europe, all serious scientific and philosophical works were written in Latin. Thus Palissy lacked access to virtually all recorded knowledge and was forced to rely upon his own observations. Unencumbered by existing paradigms, Palissy was able to make a number of original and significant contributions. He was the first to categorically endorse the theory of infiltration. The **theory of infiltration** is the theory that states the source of the water that discharges from springs and streams is precipitation, which has infiltrated slowly into a porous Earth. With great skill and logic, Palissy argued that rivers have no other sources of water other than rainfall. He also made contributions to the understanding of artesian wells, the recharge of wells from nearby rivers, and advocated forestation for the prevention of soil erosion.

1.2.3. The Seventeenth and Eighteenth Centuries: The Importance of Measurement.

The beginning of modern hydrology is in the seventeenth century, with the introduction of quantitative measurement. Three individuals from this period of time made noteworthy contributions. They were the French physicists Pierre Perrault (1608–1680) and Edmé Mariotté (1620–1684), and the English astronomer Edmund Halley (1656–1742). Perrault measured rainfall in the drainage basin of the Seine River and also the discharge of the Seine. His measurements showed that the total annual precipitation in the basin was about six times the cumulative discharge from the Seine River. Thus, Perrault showed the fallacy of the age-old assumption that rainfall alone was inadequate to account for the discharge of water by springs and streams. Mariotté made measurements that essentially verified Perrault's results, but were more precise. Mariotté vigorously defended the infiltration theory. He maintained that the water from rain and snow infiltrates into the "pores of the Earth," percolating downward until it reached impermeable rock. Mariotté then envisioned groundwater moving laterally, eventually discharging in springs or streams. This conception is essentially correct. The English astronomer, Halley, studied the other part of the hydrologic cycle, and proved that enough water evaporated from the oceans to provide for the rainfall that replenished continental waters.

The science of hydrology continued to develop in the eighteenth century. Instrumentation was invented and quantitative investigations continued. In 1715, Antonio Vallisnieri, president of the University of Padua, Italy, published a book that gave a correct explanation of the source of artesian water and the mechanism of its flow. His book was illustrated with some of the first geologic cross-sections ever drawn. In 1738, Daniel Bernoulli (1700–1782) published a mathematical formula now known as Bernoulli's Law. **Bernoulli's Law** relates pressure changes to velocity and elevation changes in a moving fluid (see chapter 2). In 1791, La Métherie investigated the permeability of different types of rocks and correctly divided precipitation into three parts: runoff, infiltration, and evaporation.

1.2.4. The Nineteenth Century: Hydrogeology Emerges as a Distinct Science.

In the nineteenth century, the science of hydrology grew very rapidly and hydrogeology began to emerge as a distinct science in its own right. The most significant accomplishments in the nineteenth century were made by French scientists. In 1856, the French engineer Henry Darcy (1803–1858) published the mathematical law that is the basis for understanding all groundwater flow, Darcy's Law (see chapter 2). Another French civil engineer, Arsene Dupuit (1804–1866) analyzed flow in unconfined aquifers (see chapter 6). Dupuit's work was later refined by the Austrian, Philipp Forcheimer (1852–1933) who introduced complex mathematical analyses to the solution of groundwater problems. Significant contributions were also made by the German Adolph Thiem (1836–1908) (see chapter 6). Thiem introduced field methods for making tests of groundwater flow and studied the flow of groundwater to wells. Under Thiem's influence, Germany became Europe's leading country in the utilization of groundwater as a resource.

Most of the advances that were made in the nineteenth century were pragmatic studies, made with the goal of obtaining a better understanding of how to use groundwater as a resource. In the United States, the study of groundwater was boosted by the founding of the U.S. Geological Survey in 1879. The latter years of the nineteenth century and the first decades of the twentieth were largely devoted to groundwater resource studies; most of these were undertaken by personnel of the U.S. Geological Survey. A noteworthy accomplishment was the 1885 publication of T. C. Chamberlin's classic monograph, *The Requisite and Qualifying Conditions of Artesian Wells,* in the Fifth Annual Report of the U.S. Geological Survey. Chamberlin's work showed that the conditions under which groundwater resources formed, and the degree to which they could be exploited, were first and foremost geological problems.

1.2.5. The Twentieth Century: Hydrogeology Becomes a Geological Science.

Most hydrogeological research during the first part of the twentieth century was devoted to groundwater utilization. In the United States, the U.S. Geological Survey placed particular emphasis on the discovery and appraisal of groundwater resources for the western states. Two of the more significant contributions were made by N. H. Darton and O. E. Meinzer. In 1905, Darton published a remarkable monograph titled *Preliminary Report on the Geology and Underground Water Resources of the Central Great Plains.* The work was 433 pages long and included 72 plates. It was one of eleven papers that Darton published in 1905. In 1923, the Chief of the Ground Water Division of the U.S. Geological Survey, O. E. Meinzer, published two works that constituted the first systematic and complete textbooks in hydrogeology. These were *The Occurrence of Ground Water in the United States, with a Discussion of Principles,* and *Outline of Ground-Water Hydrology, with Definitions.* Domenico and Schwarz (1998, p. 2) identified the publication of Meinzer's treatises as the end of the era of exploration for water resources in the United States. Further work on water resources in the United States would be devoted to development, inventory, and management.

Two significant developments mark the maturation of hydrogeology in the twentieth century. The first of these was the expansion of hydrogeology beyond the search for water resources. Hydrogeology was born from the search to procure groundwater for drinking, irrigation, and sanitation. As water resources were discovered and developed and major aquifer systems and the hydraulics of wells became better understood, hydrogeologists began to turn their attention to the role of fluids in geologic processes. The second major development in twentieth century hydrogeology was the advances that were made in understanding and mathematically describing the storage and movement of fluids in the Earth's crust.

In 1928, O. E. Meinzer demonstrated through his studies of the Dakota Sandstone (see chapter 3) that rocks as porous media are compressible, and that most groundwater in the crust is stored and released as a result of aquifer compression/expansion. In 1935, another member of the U.S. Geological Survey, C. V. Theis, developed the mathematical formulation that allowed the determination of aquifer characteristics (see chapter 6).

In 1940, M. K. Hubbert (1903–1989) published a lengthy paper in *The Journal of Geology* titled simply "The Theory of Groundwater Motion." In this paper, Hubbert laid out in precise terms the fundamental physics of groundwater flow, demonstrating that fluid flow is governed not by pressure gradients, but by potential energy gradients. Hubbert subsequently set the tone for the science of hydrogeology in the twentieth century by demonstrating that fluids played important roles in diverse geological problems. Two of Hubbert's most significant contributions were his development of the hydrodynamic theory of oil migration (Hubbert, 1953), and, with W. W. Rubey, his exposition of the importance of fluid pressures in faulting (Hubbert and Rubey, 1959). In the same year that Hubbert published "The Theory of Groundwater Motion", 1940, another paper of equal significance appeared by C. E. Jacob. Jacob derived the differential equation that described the transient movement of groundwater and the change of potential energy within a fluid. The mathematical basis of hydrogeology was then complete. Darcy (1856) had shown that groundwater flow occurs in response to potential energy gradients. Hubbert (1940) made a complete exposition of fluid potential energy, specifying the respective roles of fluid pressure and elevation. Finally, Jacob (1940) derived an equation that made it possible to predict how potential energy gradients and thus groundwater flow would change and evolve within a porous medium.

Just as the need to develop groundwater resources in the nineteenth century motivated much of the development of hydrogeologic science, other needs in the latter half of the twentieth century moved hydrogeology into new fields. As the water resources of the more developed nations of the world became well explored and developed, the emphasis in applied hydrogeological studies shifted from resource development to underground wastes and contamination. Groundwater contamination is now an important field of study, and many hydrogeologists find employment in the industry that deals with the remediation of environmental problems. Most importantly, we are now obtaining an appreciation for the ubiquitous presence of fluids in the Earth's crust and their role in virtually all geologic processes. In its modern form, hydrogeology is a science that deals with all of the geologic ramifications of fluids in the Earth's crust. Some of the important questions that concern hydrogeologists today are the following: What role do fluids play in diagenesis and metamorphism? What forces drive fluids through the crust to form ore deposits? How does petroleum migrate? How does the circulation of ocean water through the young, hot rocks at mid-ocean spreading centers alter the composition of both the rocks and the oceans? How does fluid pressure affect faulting and seismicity in the crust?

1.3. THE DYNAMIC PARADIGM.

With the advent of plate tectonics, the conception of a static Earth was replaced by a new picture of the Earth's crust as a domain caught up in a process of continual evolutionary change. An increasing knowledge of the importance of fluids in the Earth's crust is now invalidating the old conception of the crust as being composed of unchanging bodies of solid rock. Aqueous fluids are active agents in the mechanical, chemical, and thermal processes that determine the nature of the crust. As geologic fluids move through the crust they enter into chemical reactions with solid components of the crust, mobilize and transport both heat and mass, alter the geologic and geophysical properties of crustal rocks, and create economic concentrations of important ore minerals (Figure 1.3). Fluids play an important role in nearly all geologic processes and are ubiquitous in the continental crust to depths of at least 10 to 15 km (see Table 1.1 and Nesbitt and Muehlenbachs, 1989, 1991; Zoback et al., 1988; Bredehoeft et al., 1990, Oliver, 1986; 1992). In the new paradigm, the continental crust is viewed as a two-component system: a solid framework that evolves through interactions with geologic fluids (Deming, 1994b).

A convenient division may be made between hydrologic regimes in the upper and lower continental crust. In the upper continental crust, the permeability of sedimentary and crystalline rocks

Groundwater at the U.S. Geological Survey, 1879–2000[a]

The U.S. Geological Survey (USGS) was established by an act of the United States in 1879 to provide a permanent Federal agency to conduct the systematic and scientific "classification of the public lands, and examination of the geological structure, mineral resources, and products of national domain." From its inception, the USGS has concerned itself with the study of the nation's water resources. Much of the early hydrological work was concerned with the study of surface water resources to evaluate their possible development for use in irrigating crops in the arid western states.

The first groundwater publication of the USGS was T. C. Chamberlin's classic exposition of the fundamental principles of artesian flow, *The Requisite and Qualifying Conditions of Artesian Wells,* that appeared in the Fifth Annual Report of the USGS in 1885. Chamberlin was perhaps the first to unequivocally state the maxim of the dynamic school of hydrogeology: "no stratum is entirely impervious."

In addition to field work, the USGS also financed theoretical investigations of groundwater flow by University of Wisconsin mathematics professor, Charles Sumner Slichter (1864–1946). Slichter was the first to apply potential field theory to groundwater flow and demonstrate that the flow of groundwater was mathematically analogous to the flow of electricity or heat. His most important contribution was *Theoretical Investigation of the Motion of Ground Waters,* published in 1899 in the 19th Annual Report of the USGS.

From about 1904 through 1908, Walter Curran Mendenhall (1871–1957) studied groundwater resources in the semi-arid region of Southern California and published his results in ten USGS Water-Supply Papers. Mendenhall summarized his conclusions in a classic paper published in the journal *Economic*

O. E. Meinzer (1876–1948) examining an alkali flat near Wilcox, as part of his study of the groundwater resources in Sulphur Spring Valley, Arizona, during the fall of 1910.
(Photograph courtesy of the U.S. Geological Survey Photographic Library, Denver, Colorado.)

Geology in 1909, "A Phase of Ground Water Problems in the West." Mendenhall went on to a distinguished career at the USGS, becoming Director of the Survey from 1930 through 1943.

The appraisal and mapping of much of the groundwater resources of the western states was done by Nelson Horatio Darton (1865–1948) in the late 19th century and early twentieth century. Darton systematically mapped the Dakota Aquifer discharge area in eastern South Dakota and the recharge areas in the Black Hills, and along the Rocky Mountain front in Colorado and Wyoming. Most of our modern ideas about artesian aquifers date back to the work by Darton. Classic works by Darton include *Preliminary Report on the Geology and Underground Water Resources of the Central Great Plains* that was published as USGS Professional Paper 32 in 1905, and *Geology and Underground Waters of South Dakota,* published as USGS Water-Supply Paper 227 in 1909.

Oscar E. Meinzer (1876–1948) was made acting chief of the Division of Ground Water in 1912 (see photograph). A year later, he became the permanent head of Ground Water, a position he held until 1946. Meinzer's contributions to hydrogeology were substantial. His exposition of the principles of groundwater occurrence as published in USGS Water-Supply Paper 489 in 1923, *The Occurrence of Ground Water in the United States, with a Discussion of the Principles,* was a standard reference in the field for decades. Another classic from Meinzer's more than 120 publications was his explanation of the storage capacity of porous media, "Compressibility and Elasticity of Artesian Aquifers," published in the journal *Economic Geology* in 1928.

In the early part of the twentieth century the working schedule for geologists at the USGS was to go to the field in summers and return to Washington, DC in the winter. Most of the Survey geologists belonged to the Geological Society of Washington (DC) that met at the Cosmos Club in Washington—it still meets there. At one of the meetings the Chairperson tried to introduce O. E. Meinzer as the *Father of Ground Water*; when asked to stand, N. H. Darton, who was also in the audience stood. This gives some idea of the personality of these men; Darton probably the most prolific geologic mapper of his day, perhaps of all time, had a big ego. Nevertheless, Darton's work on the Dakota Sandstone stands out as a pioneering effort—from our perspective Darton may well have deserved the accolade—*Father of Ground Water.*

As Chief of the Ground Water Branch, Meinzer set about hiring groundwater hydrologists. He hired two types of individuals—engineers and geologists. Meinzer understood that groundwater studies involved advanced mathematical theory; he hired engineers to fill this need. By 1930, there were approximately 20 professionals in the Ground Water Branch at the USGS. At the same time (1930) there were perhaps 30 to 40 individuals seriously studying groundwater in the western world. There was a small group working with Thiem in Austria/Germany; there was a group with the geochemist Henri J. Schoeller in France; and a small group at Stanford working with Professor Cyrus Fisher Tolman (1873–1942). Nevertheless, 50 to 70% of these individuals worked at the USGS; the USGS was the center of groundwater studies in the western world. It would stay the center for groundwater studies until the 1970s.

Groundwater flow theory in the 1920s was steady state flow. The principal theory for flow to a well was the Thiem equation (see chapter 6). By the late 1920s, the profession realized that the steady state analysis was inadequate; the steady state theory did not describe the observations of aquifer response to pumping or other stresses. In 1928 and 1929, there was an enlightening series of discussions in the journal *Economic Geology* on the source of groundwater to the Dakota Aquifer. Both O. E. Meinzer and Karl Terzaghi (1883–1963), among others, contributed; both men discussed the idea of aquifer elasticity. Clearly, the search was on at the Ground Water Branch for a new non-equilibrium theory. Charles Vernon Theis (1900–1987) provided that new theory in 1935, publishing his solution in the *Transactions of the American Geophysical Union,* and following up with expositions in the journals *Economic Geology* in 1938 and *Civil Engineering* in 1940. Theis' new theory of transient flow was derived by applying theoretical results from the study of heat flow to the flow of groundwater. The new theory was met with skepticism, especially by O. E. Meinzer. Theis had both an undergraduate engineering degree and a Ph.D. in geology from the University of Cincinnati—perhaps the best Geology Department of its day. Meinzer had expected that the new theory would come from one of his engineers whose background was more theoretical. Much to Theis' dismay Meinzer, his boss, remained skeptical until Charles Edward Jacob (1914–1970) showed the analogy to be correct. The Theis non-equilibrium equation became the cornerstone on which the modern theory of groundwater hydraulics is built.

In the 1940s, C. E. Jacob was the preeminent theoretician within the Ground Water Branch, and in the groundwater profession. Jacob started work with the USGS on Long Island, New York. There he ran experiments on aquifer elasticity by observing in a well the nearby movement of a railroad locomotive. In 1940, Jacob derived the groundwater flow equation from first principles; he showed that the equation has the same form as the equation of heat flow—thus vindicating Theis. The critical part of the analysis was Jacob's derivation of the storage coefficient; Jacob's result has stood the test of time, including attacks on its validity. Jacob published his result in the *Transactions of the American Geophysical Union* in 1939 ("Fluctuations in Artesian Pressure Produced by Passing Railroad Trains as Shown in a Well on Long Island"), and 1940 ("On the Flow of Water in an Elastic Artesian Aquifer").

The 1940s were a time in which the new non-equilibrium theory was applied to various problems in well hydraulics. Jacob was the leader in the application; there were a host of new well-hydraulic solutions. Perhaps the most important conceptual advance was the recognition that the confining layers for artesian aquifers were not impermeable. Darton had speculated on leakage to and from the Dakota Aquifer in his 1909 Water-Supply Paper. There is a story that Hilton H. Cooper Jr. used to tell on himself. He was advising, on behalf of the USGS, on the construction of a water-supply well for a paper mill at Fernandina Beach, Florida, during World War II. He had estimated the pumping water-level in the well using the Theis equation and the fact that the recharge area was in central Florida, some distance inland. He made the mistake of giving his estimate to the driller—an old timer in the area. The driller told him "no way sonny (Hilton was young at the time), the water level will never be that deep." The driller was right; the Theis theory neglected leakage through the confining layer. Quickly, Jacob and Mahdi S. Hantush (1921–1984) revised the theory—first the leaky aquifer theory, and then the modified leaky-aquifer theory that included storage in the confining layers.

The 1940s were a period in which the USGS was the only groundwater game in town. Almost every well-known hydrogeologist or groundwater engineer of the day had done an apprenticeship at the USGS. By 1950, the Ground Water Branch at the USGS had approximately 1000 employees—perhaps 50 times more than any other groundwater group in the western world.

At the same time that well hydraulics were being revised there was recognition that using well hydraulics alone one could not analyze an aquifer system. Robert Raymond Bennett, working in Maryland, was a leader in the realization that one had to analyze an entire aquifer system if one was to evaluate its potential yield. Bennett did a flow net analysis of the Baltimore area to illustrate his ideas. Working with Herbert E. Skibitzke in Arizona, they developed the electric analog model; the analog model was designed to analyze an entire aquifer system. Electric analog analyzers had been used before to solve flow problems—there is an analogy between the flow of electricity, the flow of heat, and the flow of groundwater. Skibitzke, a mathematician by training, introduced new ideas into the electric analog model. He realized you could represent the permeability/transmissivity field using a finite-difference grid of low precision resistors. Similarly, the aquifer storage coefficient could be represented by a finite-difference network of low precision capacitors. The low precision components made it feasible to build large networks of resistors and capacitors. The network was pulsed using continuous electric function generators and read out using oscilloscopes; pulsing was also Skibitzke's innovation. By the late 1950s, the USGS Ground Water Branch had built an analog computer laboratory in Phoenix, Arizona, where they were building large-scale analog models on a production basis for the analysis of groundwater problems throughout the United States. Bill Walton and Tom Prickett built a smaller but similar analog model facility at the Illinois Water Survey. These were the first groundwater computer models of aquifer systems as we know them today; they were the predecessor of the digital computer models of today. Herb Skibitzke was the genius behind their development.

There was also a recognition at the USGS that groundwater geochemistry was important. John D. Hem pioneered in the development of analytical methods for groundwater chemistry. In 1990, Hem won the Meinzer award of the Geological Society of America for USGS Water-Supply Paper 1473, *Study and interpretation of the chemical characteristics of natural water.* Arthur M. Piper (1898–1989) invented a graphical method (the "Piper diagram") for displaying water chemistry data that became widely used by hydrogeologists. William Back pioneered in the use of geochemistry in the study of groundwater flow systems. These are a few of the geochemical achievements of the early group at the USGS.

Things changed in the 1960s. The country became aware of water. The Congress passed a bill that provided funding for a water-resources institute in every state. Quickly, state universities throughout the country created water-resources institutes. These institutes hired staffs, many of them groundwater professionals. This created the first serious competition for groundwater professionals; the USGS monopoly in groundwater was effectively ended with the creation of the water-resources institutes. Luna Leopold, the Chief Hydrologist at the USGS, and his colleague Walter Langbein, were instrumental in convincing the Congress to pass the bill creating the water-resources institutes. Earlier, Luna Leopold had created the Water Resources Division at the USGS by combining the Ground Water, Surface Water, and Quality of Water Branches.

Gradually, groundwater expertise became widely distributed in North America. Individuals at the USGS were to play an important role in developing digital groundwater models. George F. Pinder (1942 –) and John D. Bredehoeft (1933 –) developed the first widely used groundwater digital models. In 1969, the American Geophysical Union awarded the Robert E. Horton award to Pinder and Bredehoeft for their paper, "Application of the Digital Computer for Aquifer Evaluation", that had been published in the journal *Water Resources Research* in 1968. The digital models were widely adopted at the USGS in its Districts where there was already the tradition of the analog models. By the mid-1970s the digital models replaced the analog model. Bredehoeft and Pinder also introduced contaminant transport models; these built on the digital flow models. In 1976, the Geological Society of America presented the O. E. Meinzer award to Bredehoeft and Pinder for their paper "Mass Transport in Flowing Groundwater" published in the journal *Water Resources Research* in 1973. The transport models were not widely used until the 1980s when society became concerned with widespread industrial contamination. MODFLOW, the USGS groundwater flow simulation code, is still the standard flow code of the profession.

Groundwater investigations are still a cornerstone of the work of the Water Resources Division at the USGS, however, the glory days when the USGS dominated the groundwater profession are a bygone era. To a large extent the discipline of groundwater as we know it today was invented at the U.S. Geological Survey. At the close of the twentieth century, the USGS continues to be the largest organization conducting geology and hydrologic investigations in the United States.

[a]This article is largely based on a personal communication from John D. Bredehoeft to the author, August 2000.

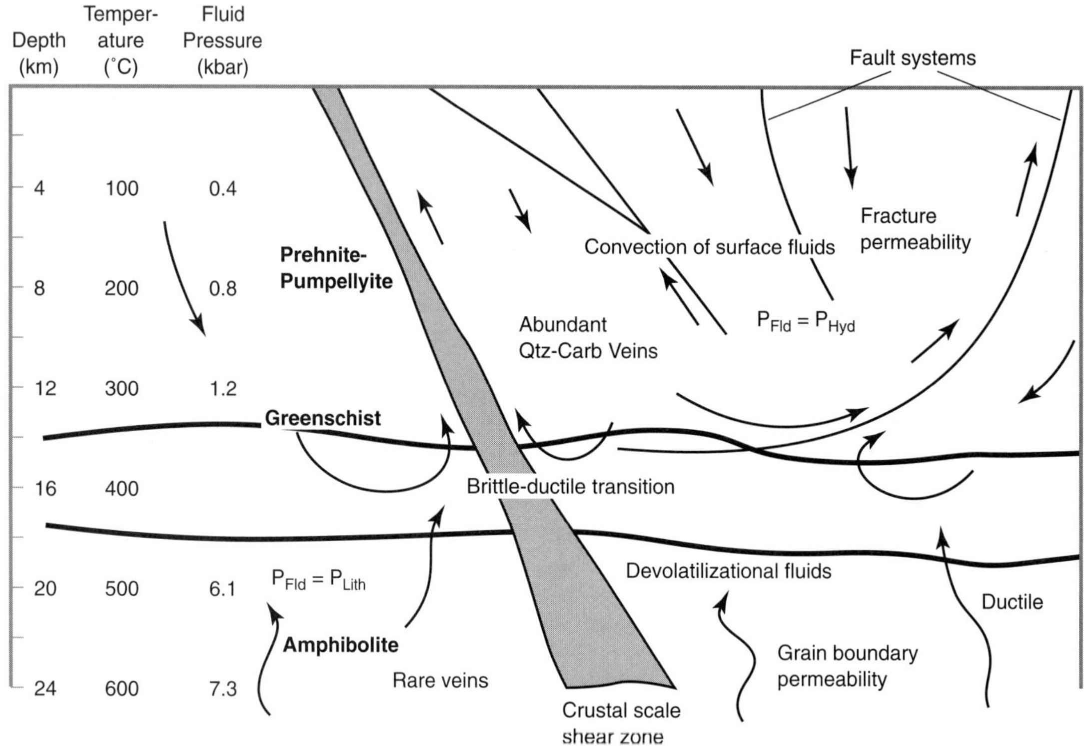

Figure 1.3 Generalized hydrogeological model for the continental crust. Fluids flow freely through a permeable upper crust; in the lower crust, fluids are released from metamorphic reactions and processes. Arrows indicate flow directions.

(From Nesbitt and Muehlenbachs, 1995, p. 1049.)

TABLE 1.1 Possible Indicators of Free Water in the Crust *(After Bredehoeft et. al., 1990, p. 5).*

Indicator	Depth Range
Water table	0 – 2 km
Deep wells	0 – 12 km
Reservoir induced seismicity	0 – 12 km
Crustal low velocity zones	7 – 12 km
Crustal electrical conductivity zones	10 – 12 km
Oxygen isotopes	0 – 12 km
Metamorphism	> 20 km
Formation of hydrothermal ore deposits	> 5 km
Crustal seismic attenuation zones	7 – 12 km
Low stress on faults	0 – 10 km
Silicic volcanism	near surface

may be sufficiently high to allow for the existence of continuous flow systems that may conceivably persist for tens of millions of years. Below depths of approximately 10 to 20 km, however, the continental crust is thought to be essentially impermeable. At these depths, temperatures and pressures are so high that rocks flow, and open fractures cannot exist for appreciable lengths of geologic time. Nevertheless, there is considerable geochemical evidence from examination of exhumed terrains that large amounts of fluid have circulated through the lower crust (Walther, 1990). The mechanism(s) by which fluid flow in the lower crust takes place remains a mystery.

Fluid flow is also important in the oceanic crust. The oceanic crust is much thinner than the continental crust (6–8 km versus 40 km), and its composition is different (Bott, 1982; Christensen and Mooney, 1995). In general, the rocks of the continental crust, especially, the upper crust, tend to be richer in aluminum and silicon, while the rocks of the oceanic crust tend to have relatively more iron and magnesium. The composition of the continental crust is exemplified by the rock granodiorite, while the oceanic crust is composed largely of basalt and gabbro. In the highly fractured and permeable pillow basalts that make up much of the upper oceanic crust near spreading centers, the circulation of seawater is believed to be pervasive, and responsible for transporting much of the Earth's background heat flow. Seawater circulation in the young, hot oceanic rocks that are formed at mid-ocean spreading centers is quite vigorous, but diminishes as the permeability of the oceanic crust is reduced by fractures and void spaces filling up with minerals. Permeability may also reduce the development of a surficial sediment blanket.

The precise volume of water that circulates through the oceanic crust is difficult to estimate, but it is enormous. Davis and Chapman (1996) estimated that about 50 ocean volumes circulate through young oceanic crust before it is sealed by chemical alteration or sediment blanketing. A corollary is that fluid circulation in the oceanic crust may be a major factor in controlling the composition of the Earth's oceans.

The modern dynamic paradigm embraces the following axioms:

1. Crustal fluids are present everywhere. The uppermost few meters or tens of meters of the Earth's crust are partially saturated with fluids. The thickness of this **unsaturated** or **vadose zone** depends upon the local geology and climate. Below the unsaturated zone, void spaces in rocks and sediments are completely filled with aqueous fluids.
2. Crustal fluids are in motion everywhere, they simply move faster in some locations than others.
3. With very few exceptions, all rocks are permeable. Some examples of exceptional rocks that are virtually impermeable include rock salt, unfractured crystalline rocks, and permafrost.
4. The Earth's crust is a solid-fluid system. The universal presence of moving fluids in the crust with the capability of transporting both mass and energy, implies that the crust cannot be viewed as an unchanging body of solid rock. Rather, it is conceptually more useful to view the crust as a two-component system: a solid rock matrix that continually evolves through mass and energy interchanges with moving fluids.
5. Fluid flow occurs on all scales. Fluid flow systems occur on a scale from centimeters to hundreds of kilometers. The long distance migration of groundwater over distances of up to 1000 kilometers in the continental crust has been documented by geochemical studies. Banner et al. (1989) showed that groundwater discharging from natural springs and artesian wells in central Missouri originated as meteoric recharge in the Front Range of Colorado, about 1000 kilometers away. Similarly, groundwater in the Great Artesian Basin in Australia (Figure 1.4) moves distances up to 1000 km in its westward migration (Cathles, 1990).

1.4. THE HYDROLOGIC CYCLE.

The **hydrologic cycle** is the circulation of water between the continents, the oceans, and the atmosphere (Figures 1.5 and 1.6). The nature of the hydrologic cycle was one of the chief mysteries that ancient man pondered. How could the rivers continually run into the sea while the sea appeared never to rise and the rivers never run dry? As the Bible states in Ecclesiastes (Chapter 1, verse 7):

> All the rivers run into the sea; yet the sea is not full; unto the place from whence the rivers come, thither they return again.

As described above, the first clear understanding of the nature of the hydrologic cycle dates from Bernard Palissy (1509–1589), who advocated the

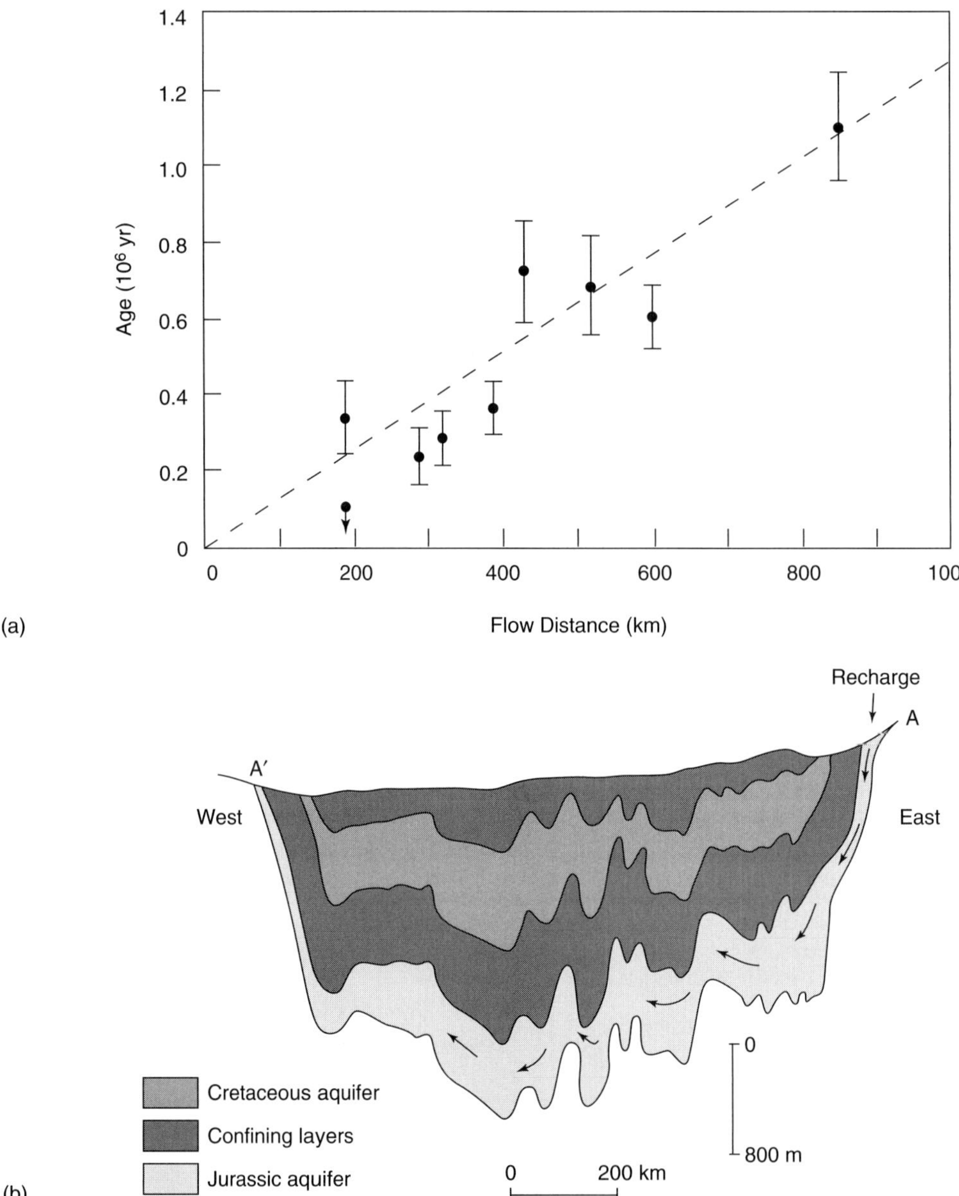

Figure 1.4 (a) Age of water in the Jurassic-age aquifer of the Great Artesian Basin in Australia increases with increasing distance from the recharge region in the Great Divide Range. Dividing the age of the water by the distance traveled suggests an average linear velocity on the order of 1 meter per year. (b) Generalized geologic cross-section A-A′ through the Great Artesian Basin in Australia showing the Jurassic-age aquifer, overlying confining layer, and recharge area.

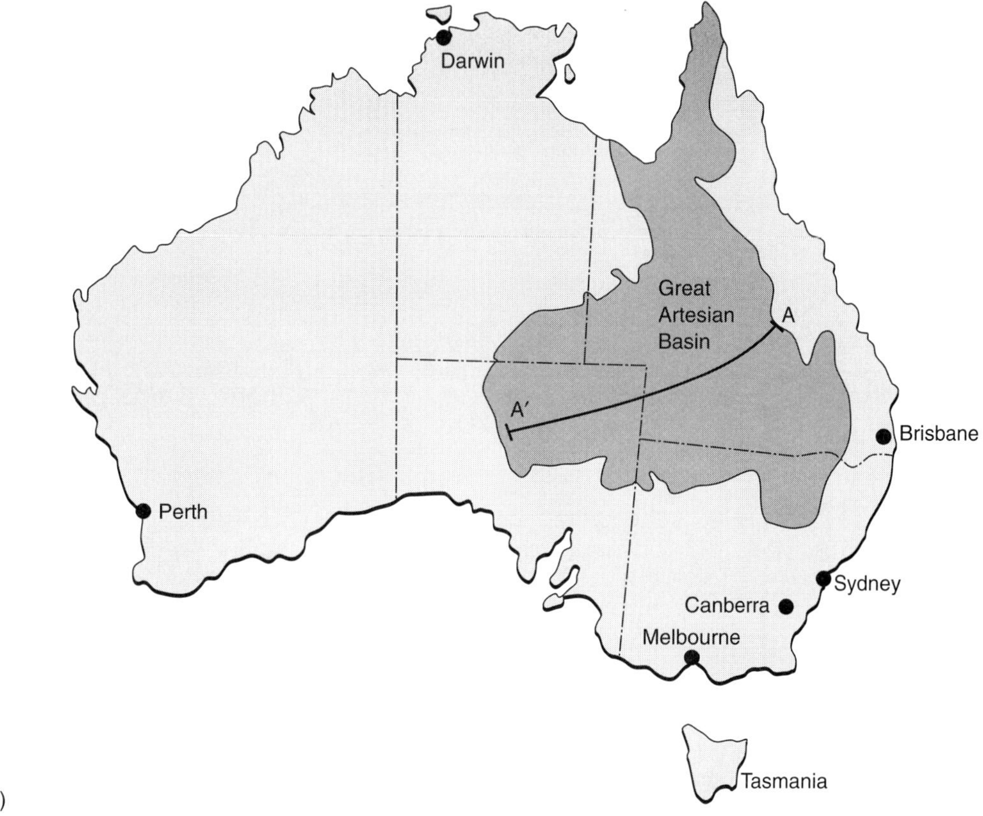

Figure 1.4 (continued) (c) Map, showing location of Great Australian Basin and cross-section A-A′. (After Bentley et al., 1986.)

theory of infiltration. Water enters the continents by precipitation and recharges streams by both overland flow and subterranean flow. Underground flow is maintained by infiltration.

1.4.1. Water Inventory and Residence Times.

The total amount of water found near and on the Earth's surface has a volume of approximately 1.4 $\times$ 10^{18} m^3 and a corresponding mass of 1.4 $\times$ 10^{21} kg (Oki, 1999, p. 10). About 94% of the world's water is found in the oceans, both as liquid water and as sea ice. About 4% of the Earth's water is present in the subsurface as groundwater and approximately 2% of the Earth's water is found in glaciers and ice caps. Lakes, swamps, rivers, and other reservoirs including the atmosphere each contain less than 0.01% of the Earth's total water inventory.

In the context of the hydrologic cycle, the continents, oceans, and atmosphere each constitute a reservoir that holds water for a certain period of time. The average amount of time a water molecule spends in a reservoir is termed the **residence time.** The residence time of water in a reservoir is calculated by dividing the total volume of water in that

The Hydrologic Cycle

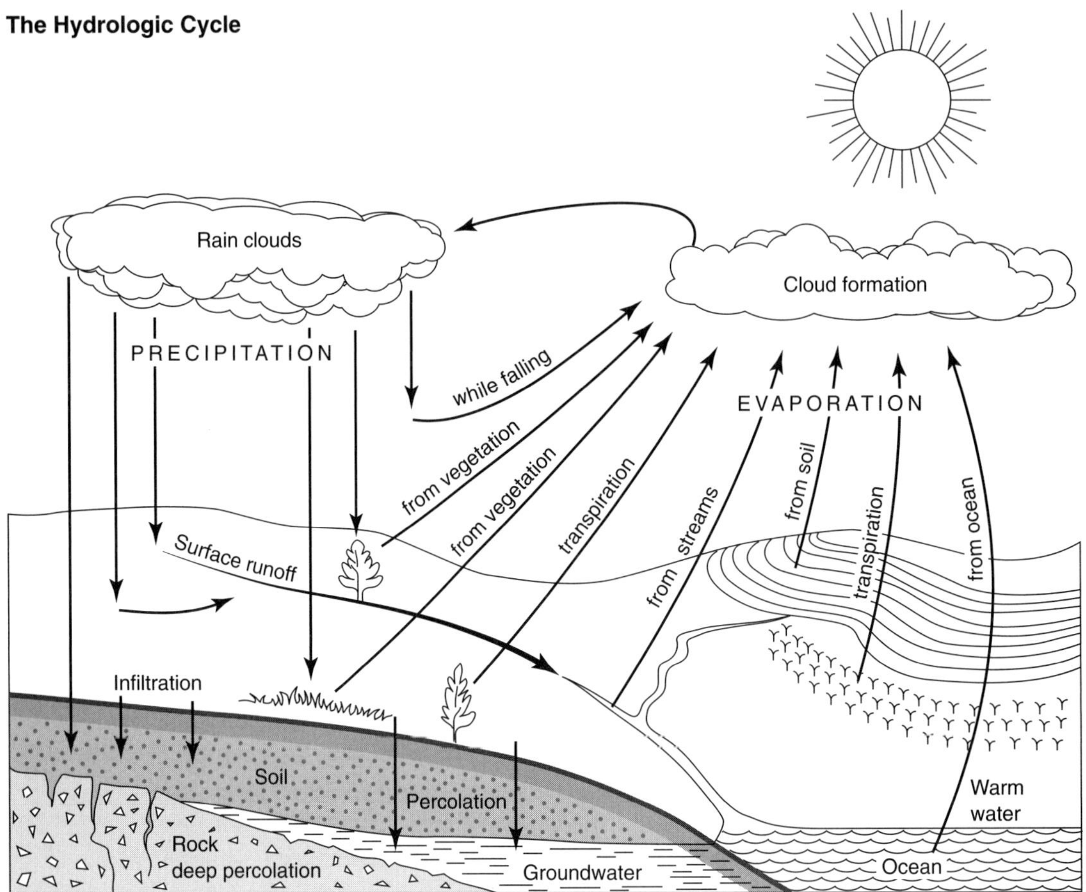

Figure 1.5 Processes and reservoirs in the hydrological cycle.
(From Ackermann et al., 1955, p. 42.)

reservoir by the rate at which water is added or subtracted from it. The residence time of water in the oceans is about 40,000 years (Gross, 1993; Berner and Berner, 1996). The residence time of water in the atmosphere is only 11 days. Residence times for reservoirs such as lakes and rivers vary widely because of the wide variability of circumstances. Freeze and Cherry (1979) give an average residence time for rivers of about two weeks. The residence time of groundwater may vary from literally minutes to hundreds of millions of years. During an intense rainstorm, water may enter the ground for a very short period of time and then leave again as it runs downhill. In contrast, the stable craton of the Canadian Shield contains salty, dense brines that may have been present for a significant portion of Earth's total history.

The relative amount of water in different terrestrial reservoirs has not been constant over geologic time. During the Pleistocene Ice Ages of the last million years, substantial amounts of water were tied up in glaciers and continental ice sheets. At the end of the last Ice Age, sea level rose a total of 120 meters, which is about 3% of an average ocean depth of 3800 meters. The slow transfer of water from ice to the oceans took several thousand years, starting 19,000 years ago and reaching what are approximately present-day levels around 5000

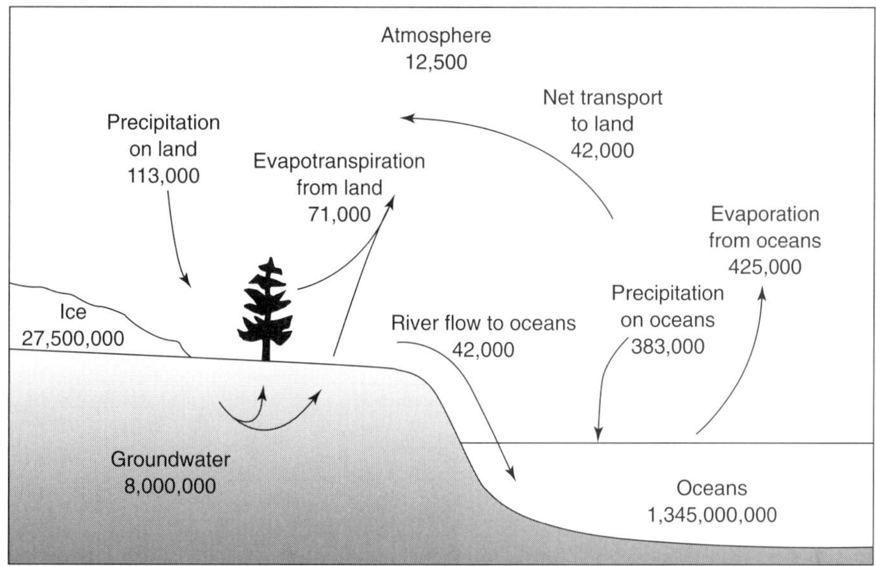

Figure 1.6 Quantitative representation of processes in the hydrological cycle showing transfer rates between reservoirs in units of cubic kilometers per year.
(From Winter et al., 1998, p. 2.)

years before present (Winograd et al., 1997). On longer time scales, ocean levels relative to land are known to have undergone eustatic cycles on three major orders (Vail et al., 1977). A **eustatic** sea level change is one that is worldwide, not merely a local change due to land subsidence. It is very difficult to quantify the amounts by which sea level has fallen and risen over geologic time. Nevertheless, it is clear from the geologic record that global changes in sea level are a major feature of the geologic record throughout Phanerozoic time. The **Phanerozoic** time period consists of the last 570 million years. The cause of these changes in eustatic sea level are unknown. One hypothesis is that the volume of the ocean basins has changed due to changes in the rate of sea-floor spreading.

1.4.2. Energy Sources and Climate.

Water is the most abundant and important substance on the surface of the Earth. The Earth is unique among the planets of the solar system for possessing large amounts of water on its surface. The range of temperature and pressure conditions on and near the surface of the Earth allow the simultaneous presence of all three water phases: liquid, solid (ice), and gas (steam or water vapor). Water is necessary for life, and water is an essential part of Earth's climate system. Some of the ways in which the hydrologic cycle controls and affects climate are through ocean currents, clouds, and the greenhouse effect.

Ocean currents moderate climate by redistributing heat. In the Atlantic Ocean, the Gulf Stream current warms the countries of Iceland and Great Britain during the winter months. In the Pacific Ocean, the California Current brings the chilly water of the North Pacific down the coast of California, lowering summer temperatures. The American author, Mark Twain, is reputed to have said that the coldest winter he ever spent was "a summer in San Francisco." On average, about 60% of the Earth's surface is covered by clouds at any given time (Oki, 1999, p. 10). Clouds cool surface temperatures by blocking and reflecting sunlight.

The concentration of water vapor in the atmosphere controls the mean planetary temperature through the greenhouse effect. The **greenhouse effect** refers to the tendency of certain gases in the atmosphere to retain heat and thus warm the Earth. During the day, the atmosphere and solid Earth are warmed by the absorption of visible light. Warmed substances emit heat in the form of infrared radiation. Certain gases such as water vapor and carbon dioxide are transparent to visible light, but are opaque to infrared radiation. Thus, greenhouse gases admit heat to the Earth in the form of visible light, but prevent the same heat in the form of infrared radiation from escaping. A common experience is to walk near the south wall of a brick building after a hot, sunny summer day. As the sun sets, one can "feel" the heat pouring off the brick wall. This is infrared radiation; it is invisible to the unaided human eye. There is a natural greenhouse effect that keeps the mean planetary temperature of Earth about 33°C warmer than it would otherwise be. At least 90% of the natural greenhouse warming is due to water vapor in the atmosphere (Dickinson, 2000). The role of water vapor in controlling Earth's climate is so important that Broecker (1997) concluded that for the Earth to have gone through the dramatic coolings of the Pleistocene Ice Ages, the water content of the atmosphere must have been significantly lower.

Two energy sources drive fluid circulation in the Earth's crust and atmosphere: solar insolation and terrestrial heat flow from the Earth's interior. **Insolation** is solar radiation incident upon the surface of the Earth. The physical units of both solar insolation and terrestrial heat flow are W-m^{-2} (Watts per square meter). The **Watt** is a unit of power, or energy per unit time. The standard metric unit of energy is the **Joule.** A Watt is defined as a 1 Joule per second. Thus "W-m^{-2}" refers to energy per unit area per unit time. Averaged over latitude and 24 hours, solar insolation is about 235 W-m^{-2}. Each year, solar radiation evaporates about 423×10^3 km^3 of water from the world's oceans. About 9% of this water precipitates over continents and returns to the oceans by overland flow or through the discharge of groundwater. Most return flow to the oceans (~94 to 99%) is by way of surface flow through rivers; groundwater discharge accounts for 1 to 6% of the total return flow (Cathles, 1990, p. 323).

Terrestrial heat flow provides the ultimate driving force for plate tectonics, and thus is indirectly responsible for providing a driving force for fluid flow on continents by creating topographic gradients. Terrestrial heat flow also impels fluid flow by free convection, thermal expansion, and temperature-dependent diagenetic and metamorphic changes involving fluid release or uptake. The worldwide average heat flow is 0.087 W-m^{-2}. The average oceanic heat flow is 0.101 W-m^{-2}, the average continental heat flow is 0.065 W-m^{-2} (Pollack et al., 1993, p. 273). In general, the insolation flux from the sun and the near-surface hydrologic flow that it generates overwhelms fluid circulation in the deeper crust driven by the flow of heat from the Earth's interior. But, over geologic time the integrated flux from deeper flow systems can transport considerable amounts of mass and heat, resulting in phenomena such as ore deposits.

1.5. Surface Water and Groundwater as a Resource.

A human being requires about 0.5 gallons (1.9 liters) of water daily to maintain normal metabolic functions, however, water is also required for preparing food, washing, and sanitation. In agricultural societies, the daily per capita use is about 3 to 5 gallons (11 to 19 liters) (Frank, 1955, p. 4). The water needs of industrial and technological societies are much greater. From 1900 to 1990, per capita water use in the United States increased by a factor of 5 to 8. Nearly all of the increase in water usage on a per capita basis took place from 1900 through 1975. Two reasons for this rapid increase were the expansion of irrigation agriculture in western states, and increased use of water for cooling thermal electric power plants (Thompson, 1999, p. 124). In the last 25 years of the twentieth century, per capita water usage in the U.S. declined. In 1975, per capita water usage in the U.S. peaked at 1944 gallons (7359 liters) per day. By 1995, per capita water use in the U.S. had declined to 1500 gallons (5678 liters) (Solley et al., 2000). Reductions in water usage were achieved through more efficient technologies for irrigation and for industrial applications. Water quality laws and regulations adopted pursuant to these laws forced industries to adopt water recycling technologies. Between 1954 and 1985, water recycling increased by a factor of four (Anderson, 1995, p. 268).

Surface and groundwater use can be divided into instream and offstream use. **Instream use** is water used within a stream for purposes such as transportation, hydroelectric power, and recreation. **Offstream use** refers to surface water withdrawn from a lake, stream, or other body of water, or groundwater withdrawn from the solid Earth for human use. Offstream use (Figure 1.7) can be divided into four broad categories: (1) domestic and commercial use, (2) industrial and mining, (3) thermoelectric (cooling of electric power plants), and (4) agriculture (irrigation and livestock). In the U.S. as of 1995, 85% of water withdrawals were freshwater, obtained from both surface water sources (78% of freshwater withdrawals) and groundwater (22% of freshwater withdrawals). The single largest category of freshwater use was irrigation and livestock (41%); a close second was thermoelectric (39%). Domestic and commercial use accounted for 12% of freshwater use, with industrial and mining activities accounting for the remaining 8%. Most saline water that is withdrawn is used for cooling of electric power plants in coastal locations.

On a planet that is three-fourths covered with water, it should come as no surprise that water is not a scarce resource, even if most of the planet's water is not fresh. The per capita water availability as of the year 2000 was projected to range from a low of 2387 gallons (9036 liters) per day for Asia, to a high of 20,471 gallons (77,491 liters) per day for South America (Thompson, 1999, p. 121). Nevertheless, water shortages can occur at specific locations at specific times; arid countries are especially vulnerable. In Israel, there are no untapped sources of freshwater; all new demand must be met by treating and reusing wastewater. Jordan, Israel, Algeria, Egypt, Tunisia, and the countries of the Arabian Peninsula use nearly 100% of their available water supply. In contrast, Canada annually uses only 1% of its available freshwater.

Groundwater is a renewable resource, but not necessarily on a human time scale. **Mining groundwater** refers to extraction of groundwater faster than it can be recharged. Several aquifers in the U.S. have been subjected to groundwater mining; the classic example is the High Plains Aquifer, also known as the Ogalalla Aquifer after its chief water-bearing formation. The **High Plains Aquifer** (Figure 1.8) is a water-bearing stratum in the central U.S. that consists primarily of sands and gravels deposited by streams draining the Rocky Mountains. It extends from the state of South Dakota south to Texas, underlying 174,000 square miles (450,658 square kilometers).

The area underlain by the High Plains Aquifer was settled in the late nineteenth century and became one of the most important agricultural areas in the U.S. The drought of the 1930s provided an impetus for increased use of groundwater for irrigation, and a rapid increase in exploitation of the High Plains Aquifer began around 1940. Approximately 20% of the irrigated land in the United States is in the High Plains, and about 30% of the groundwater used for irrigation in the U.S. is pumped from the High Plains Aquifer. Intense pumping of groundwater has resulted in large water-level declines and significant decreases in the thickness of the saturated zone in some areas. By 1980, water levels in the High Plains Aquifer in parts of Texas, New Mexico, and southwestern Kansas had declined more than 100 feet (30 meters).

Irrigation withdrawals from the High Plains Aquifer in 1990 were greater than 14 billion gallons (53 billion liters) per day. This withdrawal rate is 78 times higher than the natural recharge rate of 180 million gallons (681 million liters per day). The water budget for the High Plains Aquifer (Figure 1.9) has subsequently been modified, with the natural recharge increasing by a factor of 21 due to irrigation, and the natural discharge from seeps and

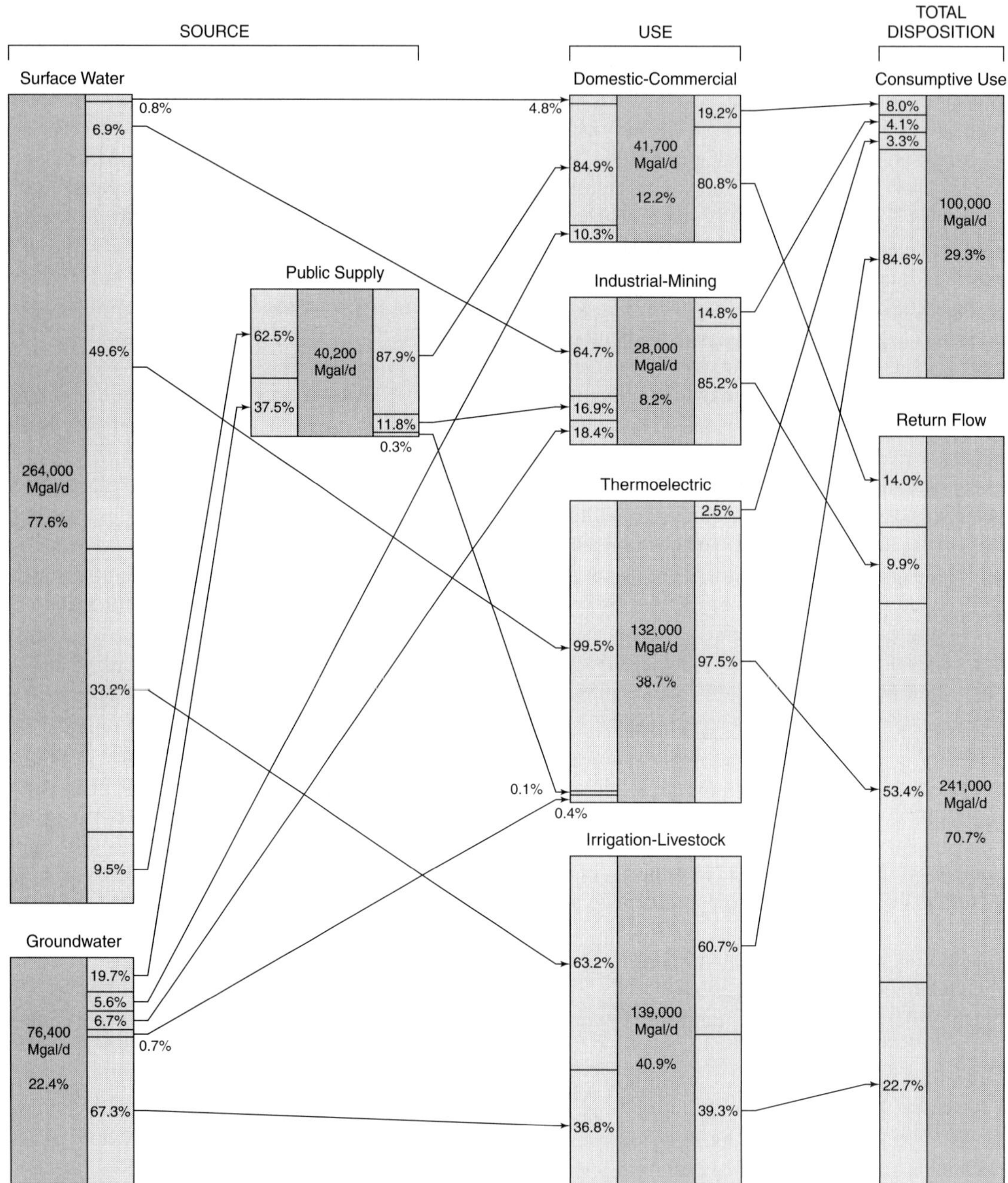

Figure 1.7 Source, use, and disposition of freshwater in the United States, 1995. (From Solley et al., 2000.)

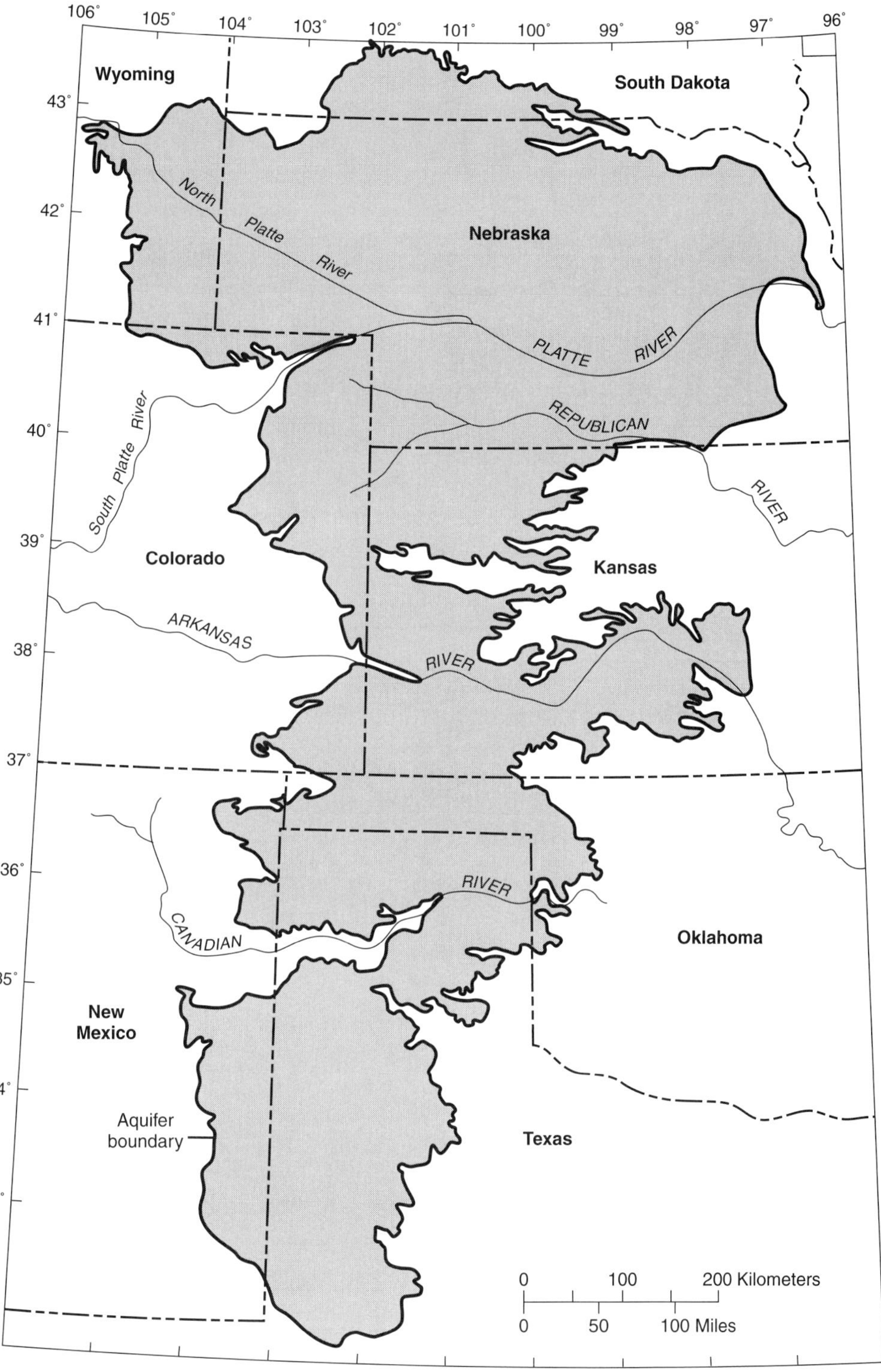

Figure 1.8 Location of the High Plains Aquifer.
(From Thelin and Heimes, 1987, p. C-2.)

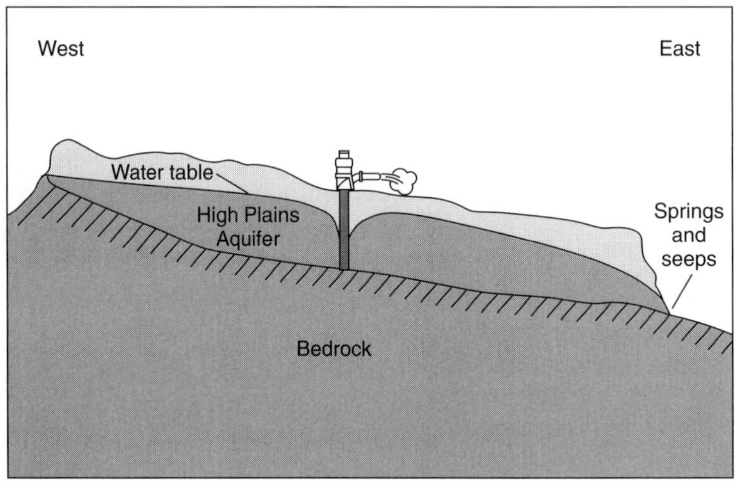

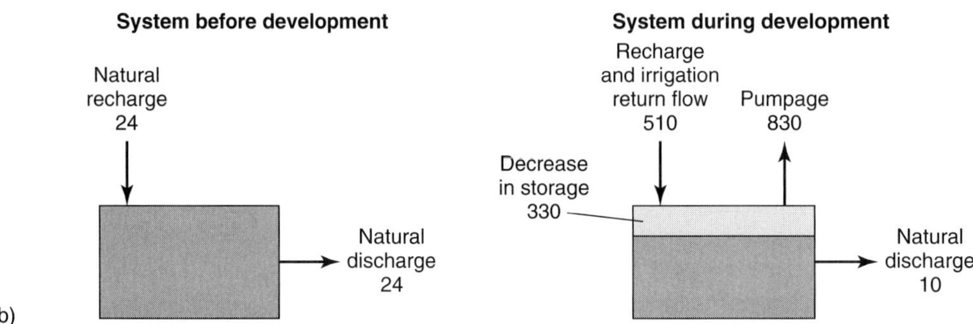

Figure 1.9 (a) Schematic geologic cross-section through the High Plains Aquifer, and (b) water budget for the High Plains Aquifer before and after development. Flow rates in (b) are in million cubic feet per day (one million cubic feet = 28,317 cubic meters).
(From Alley et al., 1999, p. 28.)

springs decreasing to 42% of pre-development levels. As of the 1990s, water levels in the High Plains Aquifer had continued to generally decline, however, a substantial amount of water remains in storage through much of the aquifer. Careful management of this resource should ensure a continued supply of water through the first decades of the twenty-first century (Alley et al., 1999).

The **tragedy of the commons** is the tendency to deplete and ultimately destroy a resource that has a common ownership. The concept was first developed by Garrett Hardin in his 1968 article, *The Tragedy of the Commons.* Hardin used the example of a pasture that is shared by herdsmen. Each herdsman will graze as many cows as possible, as each added cow enriches him further. This system works so long as the load imposed by the cumulative burden of all herdsmen does not exceed the carrying capacity of the pasture. At that point, Hardin (1968) wrote, "the inherent logic of the commons remorselessly generates tragedy." The logic of the commons is that each individual is

Thomas Chrowder Chamberlin: Multiple Working Hypotheses

Thomas Chrowder Chamberlin (1843–1928) grew up on a wheat farm near Beloit, Wisconsin. Chamberlin's father was a Methodist minister who had limited formal education, but was an incisive thinker and reader. He was known locally as a man of exceptionally clear speech and systematic ideas. Thomas and his brothers were thus raised in an atmosphere of serious and sharp intellectual debate. During this time, the young Thomas became acquainted with the wonders of nature first-hand on the prairie as he observed the migrations of the birds, the northern lights, and fossils in a nearby limestone quarry.

Chamberlin graduated from Beloit College in 1866 and began a lifetime largely devoted to teaching. His first appointment was as principal of the Delvan High School. Feeling a need for a broader background in natural science, in 1868 Chamberlin did a year of graduate study at the University of Michigan. This was followed by a teaching appointment at the State Normal School in Whitewater, Wisconsin. In the years that followed, Chamberlin held a series of academic and professional appointments, including President of the University of Wisconsin, the Wisconsin Geological Survey, and the United States Geological Survey. Chamberlin is best remembered today for his founding of the Department of Geology at the University of Chicago where he was Chairman from 1892 through 1919.

Thomas Chrowder Chamberlin (1843–1928)

The first great geological problem that Chamberlin worked on was the origin of the glacial deposits in the northcentral United States. It was Chamberlin's study of the glacial drift deposits that fostered in him an appreciation for the complexity of geological problems, and the necessity of adopting a rigorous and mature scientific method. In 1890, these philosophies were expressed by Chamberlin in his classic paper, *The Method of Multiple Working Hypotheses,* a work so enduring that it was reprinted 105 years later in the scientific journal that Chamberlin founded, *The Journal of Geology.*

In 1885, Chamberlin authored the first paper ever printed by the U.S. Geological Survey on the subject of hydrogeology, *The Requisite and Qualifying Conditions of Artesian Wells.* Chamberlin's clear exposition of the fundamental principles of artesian flow is one of the great classics of the geological literature. Chamberlin was perhaps the first to unequivocally state the maxim of the dynamic school of hydrogeology: "no stratum is entirely impervious."

It was Chamberlin's study of glacial deposits that led him to a consideration of deeper and more fundamental geologic problems. Pondering the ultimate cause of the Pleistocene glaciations, Chamberlin postulated that diastrophism (orogeny) of the solid crust of the Earth exerted a geochemical control on the concentration of carbon dioxide in the atmosphere that in turn affected mean planetary temperature through the greenhouse effect. Chamberlin was also the first to speculate on the climatic consequences of changes in oceanic circulation.

Thomas Chrowder Chamberlin: Multiple Working Hypotheses (continued)

In time, Chamberlin turned his attention to the ultimate geologic problem: the origin of the Earth. As Chamberlin studied the evolution of Earth's climate through geologic time, he became convinced that the Earth could not have formed by the condensation of a hot cloud of gas as had been maintained by the Laplacian nebular hypothesis for more than a century. Working with his colleague at the University of Chicago, F. R. Moulton, Chamberlin developed the planetismal theory, which held that the Earth had originated by the aggregation of small, solid bodies termed planetismals.

One of the great geological debates of the nineteenth century concerned the age of the Earth. The study of the succession of geologic strata and the evolution of life had convinced geologists that the Earth was at the very least hundreds of millions of years old. But, another answer was provided by the physicist Lord Kelvin, who had calculated how long it would have taken an initially molten Earth to cool to its present state. In 1897, Kelvin announced definitively that the Earth was between twenty and forty million years old. Kelvin's assumption of no internal heat sources would later be shown to be in error when radioactivity was discovered, however, at the time that Kelvin's remarks were published in the journal *Science* in 1899, atomic energy was unknown. It was Chamberlin who boldly challenged Kelvin's reasoning on philosophical grounds. With remarkable foresight, Chamberlin invoked the possibility of "unrecognized sources of heat." Chamberlin's son and biographer, R. T. Chamberlin, pointed out that his father anticipated by a quarter of a century the basic concepts of modern physics, and stepped in to prevent the geologic sciences from being led down a false path by erroneous reasoning.

University President, teacher, administrator, and researcher, Chamberlin's accomplishments were gigantic in their stature. Founder of the *Journal of Geology,* first recipient of the Penrose Medal from the Geological Society of America, Chamberlin strode like a colossus across the intellectual plain of his times. A man who was deeply religious in the truest sense, T. C. Chamberlin made intellectual honesty and devotion to duty his highest values. He had a deep faith in the evolution of mankind to greater possibilities and set an example for future generations to follow.

For Additional Reading

Back, W. 1996. T. C. Chamberlin, Early American Hydrogeologist. *Hydrogeology Journal* 4: 94–95.

Chamberlin, R. T. 1934. Biographical Memoir of Thomas Chrowder Chamberlin. *National Academy of Sciences Biographical Memoirs.* 15: 307–407.

Chamberlin, T. C. 1885. The Requisite and Qualifying Conditions of Artesian Wells. *Fifth Annual Report of the United States Geological Survey,* 131–173.

Chamberlin, T. C. and Raup, D. C. 1995. Historical Essay: The Method of Multiple Working Hypotheses, by T. C. Chamberlin. *Journal of Geology* 103: 349–354.

Willis, B. 1929. Memorial of Thomas Chrowder Chamberlin. *Geological Society of America Bulletin* 40: 23–33.

logically compelled to add further to the exploitation of the pasture, because altruistic sacrifice is not rewarded. If a herdsman were to withdraw cows from the common pasture, it would still be overwhelmed by other individuals less altruistic moving further cows onto the pasture. The only logical course for each herdsman to follow is to add still more cows to the pasture, and ultimately the common resource is ruined for all. The lesson of the tragedy of the commons is that a shared or common resource such as groundwater or surface water must be regulated by law or it may be destroyed. Alternatively, resources held in common may be sold to individuals. Individual ownership provides an economic motivation for the preservation of a resource.

1.6. Hydrogeology and the Philosophy of Science.

Hydrogeology is a science; it shares a common philosophical underpinning with other sciences. **Science** is a philosophy whose goal is the discovery of truth. **Philosophy** is the study of the fundamental nature of existence, of man, and of man's relationship to existence.

1.6.1. The Scientific Method.

The primary means that science uses to discover the truth are observation and experimentation. For untold millennia, there was no accepted means for discovering objective truth about the natural world. One man would observe a black cat walk across his path and then have an unlucky accident. He concluded that black cats are unlucky. His view had as much validity as anyone else's. Disputes were largely resolved by resort to authority. Before the advent of science, truth came from authoritarian structures in society. In our everyday lives, it is still largely the case that truth comes from authority. But, in science, if you can demonstrate through either observation or experimentation that authority is wrong, your different truth will be accepted (at least, ideally!).

The importance of observation that forms the foundation of the modern scientific method was laid by Francis Bacon (1561–1626). Bacon was an English philosopher, politician, and author. A man of enormous ambition, he rose to become Lord Chancellor of England and proclaimed "I have taken all knowledge to be my province." To understand the importance of Bacon's ideas, it is necessary to understand the world in which he lived. In the sixteenth century, virtually all scholars were bound by dogma. **Dogma** is a principle, belief, or statement of idea or opinion authoritatively considered to be absolute truth. In Bacon's day in Western Europe, the two primary sources of truth were held to be the teachings of the Catholic church and the writings of the Greek philosopher Aristotle (384–322 BC). The primacy of dogmatic belief at this time was most famously illustrated by the heresy trial of the Italian astronomer, Galileo Galilei (1564–1642). Galileo wrote a book in which he discussed how he had observed four moons circling the planet Jupiter. This was in direct contradiction to Aristotelian cosmology, which maintained that the heavens, unlike the corrupt earthly abode of man, were perfect and unchanging. Because Aristotle had not described the moons of Jupiter, they could not exist. When Galileo offered to prove the existence of Jupiter's moons by letting his critics observe them through his telescope, they merely laughed.

The essential thesis of Francis Bacon was that knowledge could only be built on observation of nature. Bacon was the first to explicitly advocate inductive reasoning. Inductive reasoning is the inference of general relationships from an accumulation of specific facts. He expounded upon his method in his book, *Novum Organum* (1620), literally translated as *The New Instrument.* Bacon was not a practicing scientist, his direct contributions to recorded knowledge were insignificant, however, his introduction of the scientific method was enormously influential.

As emphasized by Francis Bacon, the essence of science is the observation of nature, however, only certain types of observations qualify. For example, suppose that I observe a flying saucer land in my backyard. The door opens, and a tall man with a large head steps out and speaks with me. He says his name is Mentor, and that he is from the planet Xenon. The purpose of his visit here is to warn humans against self-destruction. Could I write up the above observation and submit it to a scientific journal for publication? I could, but I doubt if it would be accepted.

Scientific observations must be repeatable. The only types of observations that are accepted are those that can be repeated by other observers at different places and time. Repeatability tends to make science self-correcting. The strict requirement of repeatability also tends to make science a conservative philosophy. Observations of unusual phenomena that may be difficult to reproduce are not readily accepted; this makes it more difficult to overthrow existing theories and introduce new ideas. Some people think that science is too conservative in this respect. The American eccentric, Charles Hoy Fort (1874–1932) gathered together many of the observations that science had "damned" in his infamous *Book of the Damned* (1919) that starts out:

> A PROCESSION of the damned. By the damned, I mean the excluded. We shall have a procession of data that Science has excluded. Battalions of the accursed, captained by pallid data that I have exhumed, will march. You'll read them—or they'll march. Some of them livid and some of them fiery and some of them rotten.

Scientific observations are used to infer general relationships from specific instances. **Induction** is the process of inferring general relationships (physical laws) from specific instances. In chapter two, we will see how Darcy observed water flowing through a sand-packed pipe and inductively inferred the law of groundwater flow that bears his name. **Deduction** is the process of reasoning in which a conclusion follows necessarily from the stated premises; it is inference by reasoning from the general to the specific.

A **hypothesis** is an explanation for a scientific observation. Ideally, a scientist should use multiple working hypotheses. The method of multiple working hypotheses was first expounded by the American geologist Thomas Chamberlin (1843–1928) in a short paper that appeared in the journal *Science* in 1890. A revised and somewhat shortened version was published in the *Journal of Geology* in 1897 (Chamberlin, 1897), and was reprinted in the same journal 98 years later in 1995 (Raup, 1995). In his essay, Chamberlin (1897) decried the human instinct to focus upon a favored theory:

> The moment one has offered an original explanation for a phenomenon which seems satisfactory, that moment affection for his intellectual child springs into existence, and as the explanation grows into a definite theory his parental affections cluster about his offspring and it grows more dear to him.
>
> Instinctively, there is a special searching-out of phenomena that support it, for the mind is led by its desires. . . . When these biasing tendencies set in, the mind rapidly degenerates into the partiality of paternalism.
>
> The theory then rapidly rises to a position of control in the processes of the mind and observation, induction and interpretation are guided by it. From an unduly favored child it readily grows to be a master and leads its author whithersoever it will.
>
> When this last stage has been reached, unless the theory perchance to be the true one, all hope of the best results is gone.

As a remedy, Chamberlin advocated the **method of multiple working hypotheses.** The method of multiple working hypotheses was defined by Chamberlin as an

> . . . effort to bring up into view every rational explanation of the phenomenon in hand and to develop every tenable hypothesis relative to its nature, cause or origin, and to give all of these as impartially as possible a working form and a due place in the investigation.

Chamberlin noted that the study of geology dealt largely with complex phenomena, that are often not the subject of a single simple cause. The method of multiple working hypotheses was recommended in particular for encouraging thoroughness. Chamberlin wrote that "an adequate explanation often involves the coordination of several causes."

How do we discriminate between hypotheses? How do we determine which is the true hypothesis? Is it even possible to prove that a hypothesis is true? The answers to these questions were provided by the philosopher of science Karl Popper (1902–1994). According to **Popper's doctrine of falsifiability,** it is impossible to prove that a hypothesis is true (Popper, 1959). Even if all data are totally consistent with a given hypothesis, in the universe of all possibilities, it is always conceivable that there is a better hypothesis that has not yet been considered, or that new data will eventually appear that will disprove the hypothesis. Therefore, science advances by disproving hypotheses, and the most valuable type of scientific evidence is that which tends to falsify hypotheses.

As a simple example consider the following hypothesis: all swans are white. You cannot prove this hypothesis by observing white swans. Even if you study swans for twenty years and see a thousand white swans, this does not prove that a single black swan does not exist somewhere. In order to arrive at a unique conclusion, one must therefore disprove all alternative hypotheses, not prove a favored hypothesis. Science operates by disproving hypotheses. The principle was put forth most succinctly by that most famous fictional detective, Sherlock Holmes, who said:

> When you have eliminated the impossible, whatever remains, however improbable, must be the truth.

A corollary to Popper's Doctrine of Falsifiability is that if a hypothesis is incapable of falsification, it is not a scientific hypothesis. Consider the contentious debate of creationism vs. evolution. There are at least three possible hypotheses:

1. The Earth is young (few thousands of years).
2. The Earth is young, but was created by God to look old.
3. The Earth is old (billions of years).

Both hypotheses 1 and 3 are capable of being tested and potentially disproven. Therefore, they are both scientific hypotheses. But, hypothesis number 2 cannot be disproven, even theoretically. An omnipotent Being cannot be outsmarted. No matter what evidence is uncovered, the evidence could have been created to give whatever impression the Creator wanted. Therefore, hypothesis number 2 is not a scientific hypothesis.

Suppose that we end up with two alternative hypotheses both of which are supported by equal amounts of evidence? Which one is favored? In this circumstance, we apply the principle of parsimony or simplicity, otherwise known as **Occam's razor. Parsimony** is an economy or simplicity of assumptions in logical formulation. Occam's razor is the philosophical proposition that given two competing hypotheses, the favored one is the simple. In other words, simple explanations are favored over complex ones. Put more elegantly, "it is vain to do with more what can be done with less." The principle of parsimony was first formulated by the English philosopher, William of Occam (1285–1349).

In practice, it is usually impossible to absolutely disprove alternative hypotheses. Scientists usually tend to favor whichever hypothesis seems to be better supported by the data. As Chamberlin understood very well, most scientists have a tendency to embrace certain hypotheses and sometimes filter out data that do not agree with their preconceived ideas. The history of science shows that it often takes time for one hypothesis or paradigm to be replaced by another; the overturn of an established idea requires the accumulation of a large number of exceptions. Three years after Karl Popper published *The Logic of Scientific Discovery* (Popper, 1959), Thomas Kuhn published an enormously influential book titled *The Structure of Scientific Revolutions* (Kuhn, 1962). Kuhn maintained that the old view of science as a gradual progression of increasing knowledge was flawed. Instead, Kuhn postulated that science progresses in periodic revolutions where paradigms are overthrown. A **paradigm** is a world-view or model of how nature works. Kuhn maintained that typical scientists are not creative and original thinkers. Rather, they are conservative individuals who accept prevailing paradigms and apply their efforts to solving the problems that accepted theories dictate. Paradigms are overthrown only when the weight of the accumulated evidence becomes sufficient to do so. Kuhn developed his ideas primarily by studying the example of the Copernican revolution, where the Earth-centered universe was overthrown by the Sun-centered. In the decade immediately following the publication of Kuhn's book, the science of geology underwent a rapid paradigm shift with the introduction and acceptance of plate tectonics.

The ideas of Thomas Kuhn are somewhat contradictory to those of Karl Popper. Popper said that science advances by falsification. Kuhn acknowledged this, but explained that falsification generally does not occur as an everyday happenstance. Paradigm shifts require more than falsifiability; they require the availability of a new paradigm. The difference was expressed rather elegantly by Silver (1998, p. 105):

> Popper says that when the raft is uninhabitable we jump into the sea, while Kuhn says we jump only when another raft is available.

1.6.2. Benefits of Science and Technology.

An advantage science has over other philosophical systems for discovering truth is that science works; it is a proven method for discovering objective truths. Although science does not have as its domain ultimate metaphysical questions concerning the purpose of life, it has contributed immeasurably to the material progress of the human species. By material progress we mean increased wealth, health, and leisure time. The gains made by western civilization since the dawn of the scientific age in the sixteenth century have been astounding. All aspects of material human welfare have improved and continue to improve. The material gains humanity has made through science and technology (see Simon, 1995) include:

1. Increased life expectancy. Life expectancy at birth in the ancient world was around 22 years, mostly due to high infant mortality rates. By the sixteenth century in western Europe, life expectancy at birth had risen to around 35 years. As of the year 2000, average life expectancy at birth in countries with technological societies is around 75 years. Less developed countries also underwent dramatic increases in life expectancy as their societies became more technological. In China, the life expectancy in 1930 was only around 24 years. By 1990, it had risen to 61 years.
   Life expectancy increases can be attributed to a number of factors. One is reduced exposure to pathogens through the development of public water-supply systems. The discoverer of the law of groundwater motion, Henry Darcy, was an engineer who planned and supervised the construction of a public water-supply system for the city of Dijon, France. Another factor is the introduction of antibiotics, vaccinations, and other innovations made by medical science. Better nutrition through the increased availability of food may also contribute to longer life spans. The horrors of mass plagues and famines have been largely left behind us.
2. Increased productivity. One of the best measures of productivity is the proportion of the population engaged in agriculture. In 1800, 60% of the American population was engaged in producing raw food. By 1990, the percentage had fallen to 3%, implying a productivity gain of twenty-fold for each worker. The length of the work week for the average American worker in 1850 was 70 hours. By 1990, this had decreased to 40 hours.
3. Increased standard of living. At the beginning of the twenty-first century, it is difficult for Americans to imagine how their grandparents and great-grandparents lived at the beginning of the twentieth century. In the year 1900, few Americans had running water, bathtubs, hot water, or flush toilets. Central heating and electricity were rare. By 1980, 97% of American homes had indoor flush toilets, hot and cold piped water, and a shower or bathtub. Electricity and central heating were present in more than 99% of American homes. In 1990, even Americans defined as "poor" by the U.S. Census had a substantially higher standard of living than much of the world's population.
4. Increased educational opportunities. Over the periods of time for which data are available, the amount of education that youths receive has increased rapidly in both the United States and the rest of the world. The percentage of children enrolled in elementary and secondary schools in the United States increased from about 48% in 1850 to 90% in 1990. High school enrollments increased more dramatically, rising from about 5% of the eligible population in 1890, to 90% in 1990. In 1900, less than 3% of 23-year-olds had received bachelor or professional degrees. By 1990, this had risen 10-fold to more than 30%.

Although technological societies are undoubtedly more prosperous, there is no clear consensus on the difficult question of whether or not they are superior. John McCarthy, a computer scientist and professor at Stanford University, has suggested an objective criterion to answer this question. McCarthy suggests that we characterize as superior the societies that people choose to migrate to. If that criterion is adopted, then it is possible to render a value judgment. Experience has shown that people usually migrate to technological societies that are advanced in their development and application of scientific knowledge. It is not difficult to see why. Given the choice, most people prefer wealth over poverty, health over disease, and life over death.

REVIEW QUESTIONS

1. Define the following terms in the context of hydrogeology:
   a. hydrogeology
   b. hydrology
   c. potamology
   d. limnology
   e. cryology
   f. geohydrology
   g. groundwater
   h. porous media
   i. matrix
   j. permeability
   k. diagenesis
   l. aqueduct
   m. theory of infiltration
   n. Bernoulli's Law
   o. unsaturated zone
   p. vadose zone
   q. hydrologic cycle
   r. eustatic
   s. Phanerozoic
   t. greenhouse effect
   u. insolation
   v. Watt
   w. joule
   x. instream use
   y. offstream use
   z. Mining groundwater
   aa. High Plains Aquifer
   bb. tragedy of the commons
   cc. science
   dd. philosophy
   ee. dogma
   ff. induction
   gg. deduction
   hh. hypothesis
   ii. method of multiple working hypotheses
   jj. Popper's doctrine of falsifiability
   kk. paradigm
   ll. Occam's razor
   mm. parsimony

2. Why were ancient civilizations interested in hydrology?

3. What was the chief hydrological mystery in ancient times?

4. What is the "theory of infiltration" and who were its first advocates?

5. What was the chief hydrological advance made in the seventeenth century?

6. What was the purpose of most nineteenth-century hydrogeological studies?

7. What two developments marked the maturation of hydrogeology in the twentieth century?

8. What is the "dynamic paradigm" and what are its tenets?

9. Compare the continental and oceanic crust in terms of their average thickness and composition. Be specific, and use numbers where possible.

10. How much water is there on and near the surface of the Earth? What are the primary "reservoirs" in which this water is held? What percentage of the total is held in each reservoir? What is the residence time for water in each reservoir?

11. If the Earth were a smooth sphere with surface water evenly distributed over its surface, what would the water depth be?

12. List three ways in which the hydrologic cycle affects climate.
13. What is the natural greenhouse effect? How large is it? What percentage of the natural greenhouse effect is due to water vapor?
14. What two energy sources drive the hydrologic cycle? Give their average values and compare them.
15. How much water does an average human being require daily to maintain normal metabolic functions?
16. As of 1995, what was the average per capita daily water use in the United States?
17. How did water use in the United States change between 1900 and 1975? Between 1975 and 1995? In each case, explain whether it increased or decreased, and for what reasons.
18. Surface and groundwater use can be divided into what two categories?
19. What are the four categories of offstream use? What two categories represent the greatest use?
20. What are the primary means by which science discovers truth?
21. Scientific observations must be ________ .
22. Is it possible to prove that a scientific hypothesis is true? Explain.
23. Scientific hypotheses must be ________ .
24. List some specific ways in which science and technology have resulted in material progress.

SUGGESTED READING

Alley, W. M., Reilly, T. E., and Franke, O. L. 1999. Sustainability of ground-water resources. *U.S. Geological Survey Circular 1186.*

Biswas, A-K. 1970. A short history of hydrology in *The Progress of Hydrology, Proceedings of the First International Seminar for Hydrology Professors,* 2: 914–936.

Bredehoeft, J. D., Norton, D. L., Engelder, T., Nur, A. M., Oliver, J. E., Taylor, H. P., Jr., Titley, S. R., Vrolijk, P. J., Walther, J. V., and Wickham, S. M. 1990. Overview and Recommendations in *The Role of Fluids in Crustal Processes.* National Academy Press, Washington, D.C., 170 pp.

Cathles, L. M. III. 1990. Scales and effects of fluid flow in the upper crust. *Science* 248: 323–329.

Meinzer, O. E. 1942. Hydrology in Meinzer, O. E., ed. *Physics of the Earth—IX, Hydrology.* Dover Publications, New York, pp. 1–31.

CHAPTER 2

DARCY'S LAW AND HYDRAULIC HEAD

2.1. DARCY'S LAW.

In 1856, a French hydraulic engineer named Henry Darcy (1803–1858) published a report on the water supply of the city of Dijon, France. In that report, Darcy described the results of an experiment designed to study the flow of water through a porous medium (Freeze, 1994). Darcy's experiment resulted in the formulation of a mathematical law that describes fluid motion in porous media. **Darcy's Law** is the fundamental constitutive relationship that we use to understand the movement of fluids in the Earth's crust. Darcy's Law states that the rate of fluid flow through a porous medium is proportional to the potential energy gradient within that fluid. The constant of proportionality is the **hydraulic conductivity;** the hydraulic conductivity is a property of both the porous medium and the fluid moving through the porous medium.

Consider a pipe filled with sand as an idealized porous medium (Figure 2.1). The pipe has cross-sectional area A (m^2), and may be vertical, or tilted at some angle as shown. Water is introduced into the top of the pipe at a rate sufficient to maintain a steady flow of fluid through the porous medium; thus, the volumetric rate at which water enters the pipe at the top (Q, m^3-s^{-1}) is equal to the rate at which water leaves the pipe at the bottom.

Two tubes project from the side of the pipe; the distance that separates the tubes along the length of the pipe is Δs (m). Let h (m) be the height above an arbitrary datum such as sea or ground level. The height h to which water rises in the tubes is the water's **head** or **potentiometric surface.** The term potentiometric surface refers to a surface along which potential energy is constant. As we shall show later, if fluid density is constant, head is proportional to the potential energy per unit mass of fluid. The water in the top pipe rises to a level h_1 (m); the water in the bottom pipe rises to a level h_2 (m). The difference in the water level elevations is Δh (m).

The **Darcy velocity** (q, m-s^{-1}) or **specific discharge** through the pipe is defined as the volumetric flow rate (Q) divided by the area perpendicular to flow (A),

$$q\ (\text{m-s}^{-1}) = \frac{Q\ (\text{m}^3\text{-s}^{-1})}{A\ (\text{m}^2)} \qquad \textbf{(2.1)}$$

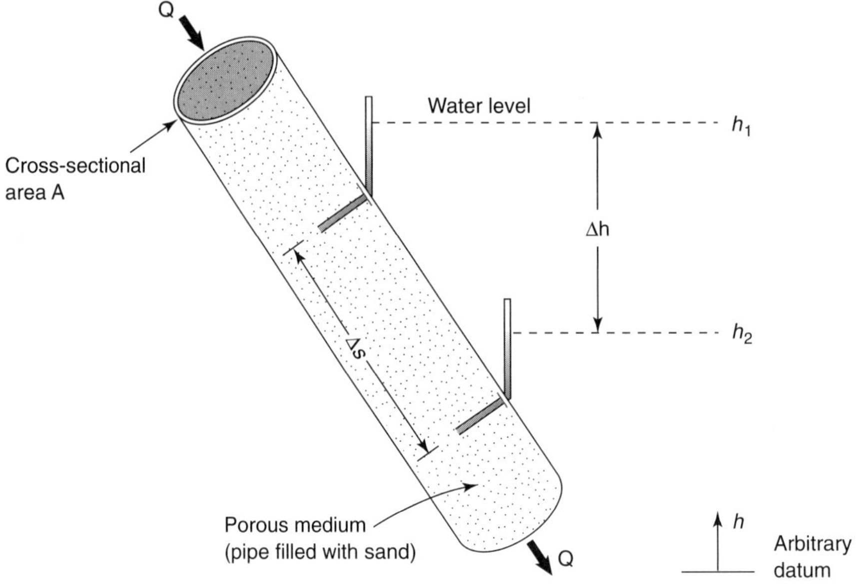

Figure 2.1 Experimental apparatus used to derive Darcy's Law.

Darcy found that

$$\frac{Q}{A} = q = -K\frac{\Delta h}{\Delta s} \tag{2.2}$$

where K (m-s^{-1}) is a constant known as the hydraulic conductivity. Darcy's Law can also be written in terms of derivatives,

$$q = -K\frac{dh}{ds} \tag{2.3}$$

where s (m) is an arbitrary direction along the length of the pipe. Later we shall substitute x, y, or z as appropriate. More formally, the Darcy velocity can be written as a vector

$$\vec{q} = -K\,\vec{\nabla}\,h \tag{2.4}$$

where the arrow ($\rightarrow$) denotes a vector quantity, and $\vec{\nabla}$ (m^{-1}) is the gradient operator, also a vector. Both the Darcy velocity and the gradient operator are vectors, while head is a scalar. A **scalar** is a quantity such as mass, length, temperature, or head, that does not depend upon direction. A scalar is completely specified by one number. In contrast, a **vector** quantity is one that depends upon direction, and can only be specified completely by quantitative representations of both magnitude and direction. The Darcy velocity is a vector as it has both magnitude and direction. The gradient of a scalar (such as head) is the spatial derivative of that scalar,

$$\vec{\nabla}h = \frac{dh}{dx}\,\hat{x} + \frac{dh}{dy}\,\hat{y} + \frac{dh}{dz}\,\hat{z} \tag{2.5}$$

where $\hat{x}$ is a unit vector in the x-direction, $\hat{y}$ is a unit vector in the y-direction, and $\hat{z}$ is a unit vector in the z-direction. Thus $\vec{\nabla}h$ is a vector with both magnitude and direction.

Darcy's Law is a statement that the rate at which fluid moves through a porous medium is proportional to the head (h) or potential energy gradient in that medium. The constant of proportionality is the hydraulic conductivity (K), and it is a property of both the porous medium and the fluid flowing through it.

Darcy's Law is the fundamental law that describes the motion of subsurface fluids. It is an empirical relationship; that is, it is derived from observation, not from theory. Note that we have said nothing of fluid pressure. Fluid movement through a porous medium does not necessarily occur in response to fluid pressure gradients. In fact, it is entirely possible for fluid to flow from regions of low fluid-pressure to regions of high fluid-pressure. Fluid pressure is an important component of fluid head, but it is not the only component. Later, we shall develop and express the relationship between fluid pressure and head in an explicit form.

Why does Darcy's Law incorporate a minus sign? Consider a head gradient that is positive in the positive x-direction (Figure 2.2). Darcy's Law then tells us that flow is in the minus x-direction; fluid flows from regions of high head to low head. In other words, fluid moves downhill from regions of high potential energy to regions of low potential energy. The presence of the negative sign in Darcy's Law is thus equivalent to the observation that water runs downhill.

The forms of Darcy's Law given in equations 2.2, 2.3, and 2.4 are appropriate only for fluids of constant density. Let the gravity vector be in the z-direction (vertical), and the x- and y-directions (horizontal) be perpendicular to the z-direction. A more generalized description of fluid flow in a porous medium that includes a buoyancy force term is (Bear, 1972)

$$\vec{q} = -\overline{\overline{K}}\left[\vec{\nabla} h + \frac{\rho - \rho_0}{\rho_0}\vec{\nabla} z\right] \tag{2.6}$$

In equation 2.6, $\vec{q}$ (m-s^{-1}) is the Darcy velocity, $\overline{\overline{K}}$ (m-s^{-1}) is the hydraulic conductivity tensor, $\vec{\nabla} h$ (dimensionless) is the gradient of the head field, ρ (kg-m^{-3}) is fluid density, ρ_0 (kg-m^{-3}) is a reference density, and $\vec{\nabla} z$ (dimensionless) is the gradient with respect to z. For our purposes, we may consider a tensor to be a physical quantity whose value is spatially variant. Thus hydraulic conductivity may change depending upon both the location and direction of flow.

Dropping the formality of vector notation, equation 2.6 can be expanded into three equations

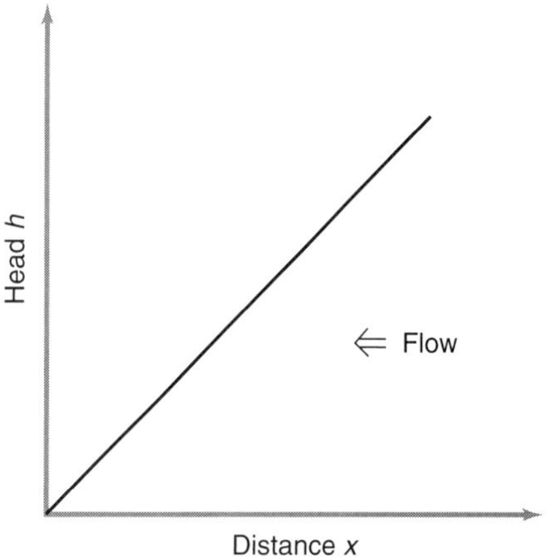

Figure 2.2 Darcy flow is in the $-x$ direction if $dh/dx > 0$.

$$q_x = -K_x\left[\frac{dh}{dx} + \frac{\rho - \rho_0}{\rho_0}\frac{dz}{dx}\right] \tag{2.7}$$

$$q_y = -K_y\left[\frac{dh}{dy} + \frac{\rho - \rho_0}{\rho_0}\frac{dz}{dy}\right] \tag{2.8}$$

$$q_z = -K_z\left[\frac{dh}{dz} + \frac{\rho - \rho_0}{\rho_0}\frac{dz}{dz}\right] \tag{2.9}$$

Note that $dz/dx = dz/dy = 0$, and equations 2.7 and 2.8 reduce to equation 2.3. In other words, there is no buoyancy force in directions perpendicular to the gravity vector, even for fluids whose density is not constant (this is not to imply that buoyancy forces in the z-direction cannot induce pressure and head gradients in the x- and y-directions and thus lead to convective circulation). However, $dz/dz = 1$ in equation 2.9, and there is a buoyancy force in the z-direction proportional to the normalized density, $(\rho - \rho_0)/\rho_0$.

In this text, we shall usually use the simpler form of Darcy's Law given by equation 2.4, and neglect buoyancy forces. Although density gradients are non-zero in virtually all geologic situations, the magnitude of the head gradient (∇h) that drives fluid flow is often much greater than the buoyancy forces associated with density gradients. In other words, we may obtain meaningful answers to many geologic problems by assuming buoyancy forces are not important. But, there are some geologic situations where the failure to consider buoyancy forces may lead to incorrect conclusions. In geothermal areas and the flanks of salt domes, large changes in temperature and salinity over short distances may lead to appreciable fluid-density gradients and associated buoyancy forces. In these areas it may be impossible to obtain an understanding of fluid flow unless buoyancy forces are taken into consideration. It is thus permissible to neglect buoyancy forces so long as the limitations of this simplification are kept in mind.

We may expand equation 2.4 to obtain

$$q_x = -K_x \frac{dh}{dx} \tag{2.10}$$

$$q_y = -K_y \frac{dh}{dy} \tag{2.11}$$

$$q_z = -K_z \frac{dh}{dz} \tag{2.12}$$

This, however, is not quite correct under all circumstances. It is also possible for the hydraulic conductivity in the *x*-direction (K_x) to affect fluid flow in the *y*-direction, etc. This seems counter-intuitive. Consider, however, the following situation (Figure 2.3). Let the flow be in some arbitrary direction *s* intermediate between the *x*- and *y*-directions. The Darcy velocities in both the *x*- and *y*-directions depend upon the Darcy velocity in the *s*-direction. In turn, the Darcy velocity in the *s*-direction (q_s) depends on *dh/ds,* which depends on both *dh/dx* and *dh/dy.* Thus q_x depends on *dh/dy,* and q_y depends on *dh/dx.* If we write equation 2.4 in terms of the *x*- and *y*-directions, we must have

$$q_x = -K_{xx} \frac{dh}{dx} - K_{xy} \frac{dh}{dy} \tag{2.13}$$

and

$$q_y = -K_{yx} \frac{dh}{dx} - K_{yy} \frac{dh}{dy} \tag{2.14}$$

where K_{xx} is the hydraulic conductivity that controls fluid movement in the *x*-direction in response to head gradients in the *x*-direction, and K_{xy} is the hydraulic conductivity that controls fluid movement in the *x*-direction in response to head gradients in the *y*-direction, etc. In three dimensions, the proper form of equation 2.4 is

$$q_x = -K_{xx} \frac{dh}{dx} - K_{xy} \frac{dh}{dy} - K_{xz} \frac{dh}{dz} \tag{2.15}$$

$$q_y = -K_{yx} \frac{dh}{dx} - K_{yy} \frac{dh}{dy} - K_{yz} \frac{dh}{dz} \tag{2.16}$$

$$q_z = -K_{zx} \frac{dh}{dx} - K_{zy} \frac{dh}{dy} - K_{zz} \frac{dh}{dz} \tag{2.17}$$

where hydraulic conductivity (K) is, in its most general form, a tensor.

Under what conditions can we use the simpler form of Darcy's Law given by equation 2.4?

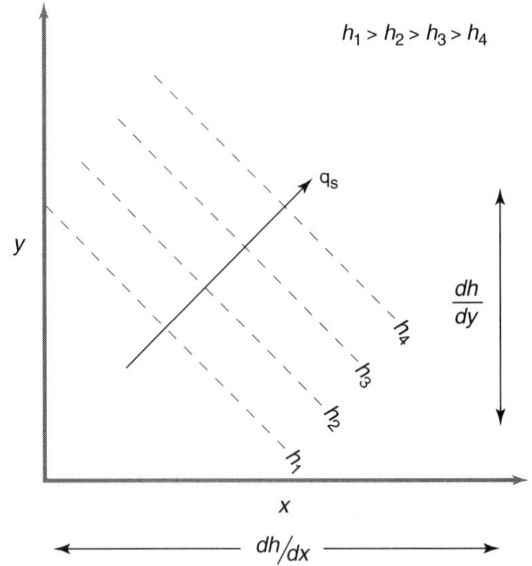

Figure 2.3 Darcy flow in an arbitrary *s*-direction depends on permeability in both the *x*- and *y*-directions. In turn, flow in the *x*-direction is determined by flow in the *s*-direction; thus, flow in the *x*-direction depends in part on permeability in the *y*-direction.

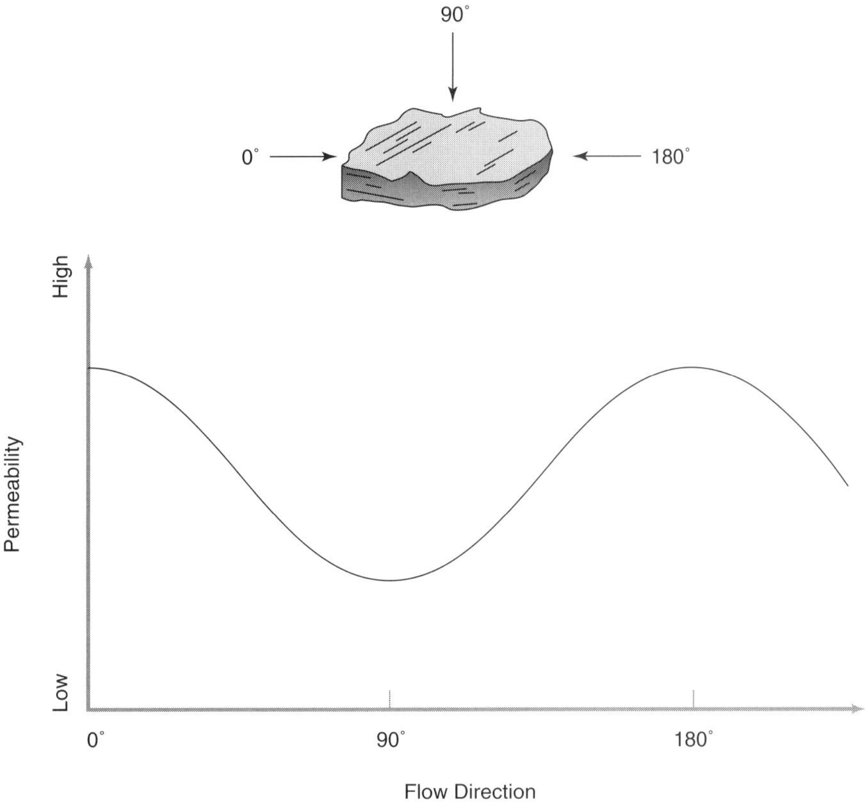

Figure 2.4 Permeability as a function of flow direction. An anisotropic material is one whose permeability varies with the direction of flow as shown. If the permeability does not vary with flow direction, the material is said to be isotropic. Most sedimentary rocks are anisotropic, having higher permeabilities in directions parallel to bedding planes.

Consider the following thought experiment. Suppose we were able to measure the hydraulic conductivity of an arbitrary rock sample as a function of flow in different directions. If we made continuous measurements through a full 360 degrees and plotted the hydraulic conductivity as a function of the flow angle, it is conceivable that we might find that hydraulic conductivity did not vary as function of the direction of flow. In that case, we would characterize the rock as possessing an **isotropic** hydraulic conductivity. An isotropic material is one whose properties do not depend upon direction. More commonly, however, we would find the rock to be **anisotropic.** An anisotropic material is one whose properties depend upon direction. For most rocks, we would find that the hydraulic conductivity was a maximum and a minimum in two directions, with 180 degree symmetry (Figure 2.4). These directions are the **principal directions of anisotropy.** For sedimentary rocks, the common case is that the principal directions of anisotropy are parallel and perpendicular to the bedding plane, with the higher conductivity parallel to the bedding plane.

In this text, we shall usually assume that the coordinate axes are the principal directions of anisotropy. Under this condition

$$K_{ij} = 0 \text{ for } i \neq j \tag{2.18}$$

and equations 2.15, 2.16, and 2.17 reduce to equation 2.4.

Henry Darcy: Discoverer of Darcy's Law

Henry Darcy (1803–1858) is remembered today as the discoverer of Darcy's Law, the fundamental constitutive relationship that we use to understand the movement of fluids in the Earth's crust. Darcy was born in Dijon, France, in 1803. He was educated at the Dijon Polytechnique, and graduated with honors in 1826 as a civil engineer. After graduation, Darcy received an appointment to Le Corps des Ponts et Chausseés, a government engineering agency. That Darcy was an exceptionally capable engineer can be inferred from his promotion to Chief Engineer in 1840 at the relatively young age of 37.

During his lifetime, Darcy was most noted for his design and construction of a water-supply system for his native town of Dijon. In the first half of the nineteenth century, most large cities had no central water-supply or sewage systems. Water was supplied by a few inadequate wells or drawn from a contaminated stream. There were no water-supply or drain pipes, and water was delivered daily to residences on carts. Darcy drew upon a nearby spring for a supply of fresh and pure water and designed a network of supply and drainage pipes. The entire system was constructed in 18 months. Although a relatively small and provincial town, thanks to Darcy, Dijon enjoyed the benefits of a

Henry Darcy (1803–1858). Portrait reproduced courtesy municipal library, Dijon, France. F. Perrodin, photographer.

2.2. Linear vs. Darcy Velocity.

The Darcy velocity (q, m-s^{-1}) should not be confused with the **linear velocity** (v, m-s^{-1}). Both the Darcy and linear velocities have units of length/time, but the linear velocity is the actual speed with which individual water molecules travel through a porous medium, while the Darcy velocity is the volume of water that passes through a certain cross-sectional area per unit time. In a pipe filled only with water (in other words, a porous medium with a porosity of 100%), $q = v$. But, in a porous medium with a porosity (ϕ, dimensionless) less than 100%, the Darcy velocity is always lower than the linear velocity, because less than 100% of that medium is available for fluid to flow through. Linear and Darcy velocities are related by the following equation

$$v = \frac{q}{\phi} \tag{2.19}$$

where v is linear velocity, q is Darcy velocity, and ϕ is fractional porosity ($0 < \phi < 1$). Consider, for example, a pipe filled completely with water (no sand, $\phi = 1$) with an area perpendicular to flow of 1 m^2. If water is moving through this pipe with a volumetric discharge rate of 1 m^3-s^{-1}, both the linear and Darcy velocities are 1 m-s^{-1}. Now take the same pipe and fill it with sand such that it has a modern water-supply system 20 years earlier than Paris. Darcy's efforts were so appreciated by his fellow townsmen, they renamed the main square in the town "Place Darcy" following his death in 1858. Notably, Darcy not only refused payment for his work in the construction of this water system, but even declined renumeration of his personal expenses. Philip (1995) noted that under the laws of the time, Darcy rightfully had a claim to the 1995 equivalent of 1.5 million US dollars. The only compensation Darcy would accept was free water for his family and household during his lifetime.

In the 1850s, Darcy fell into ill health. He resigned from his professional duties and devoted his time to hydraulic experiments. The first experiments that led to the discovery of Darcy's Law were conducted in October of 1855. Darcy's results were published in 1856 in a 647-page report titled *Les Fontaines Publiques de la Ville de Dijon.* Most of the report deals with the construction of Dijon's water-supply system. Darcy's Law is found in an appendix in a subsection titled *Determination of the Laws of the Flow of Water through Sand.* The obscure presentation of Darcy's Law, and the failure of its discoverer to expound fully on its implications, led Heath (1994) to speculate that Darcy may have been interested solely in water filtration and not realized the applications of the law to groundwater flow. But, this was disputed by Davis and Davis (1994), who pointed out that Dupuit (1857) noted the general applicability of Darcy's Law to groundwater flow in the year before Darcy's death.

For Additional Reading

Davis, S. N., and Davis, A. G. 1994. Filtration Experiments of Henry Darcy. *Groundwater* 32: 862–863.

Dupuit, A. J. E. J. 1857. Mémoire sur le mouvement de l'eau à travers les terrains perméables. *Comptes Rendus Hebdomadaires des Séances de L'Académie des Sciences* (Paris) 45: 92–96.

Freeze, R. A. 1994. Henry Darcy and the Fountains of Dijon. *Groundwater* 32: 23–30.

Heath, R. C. 1994. Comments on Henry Darcy Paper. *Groundwater* 32: 507.

Philip, J. R. 1995. Desperately seeking Darcy in Dijon. *Journal of the Soil Science Society of America* 59: 319–324.

porosity of $\phi = 0.1$. According to equation 2.19, if the flow is maintained such that the Darcy velocity remains unchanged, the linear velocity is now ten times higher than the Darcy velocity. Why? For the total volume of water per unit time that passes through the pipe to remain the same, the water molecules must move 10 times faster than before because now only 10% of the porous medium is open to flow.

2.3. Fluid Pressure.

In this section we derive a general expression for fluid pressure (P) as a function of depth that will find future application.

Consider an arbitrary column of cross-sectional area A (m^2) through a porous medium with porosity ϕ (dimensionless) that starts at the surface and extends to depth z (m) (Figure 2.5). Pressure (P) has units of Pascals (Pa $=$ kg-m^{-1}-s^{-2}) and is defined as force per unit area

$$P = \frac{\text{Force}}{\text{Area}} \tag{2.20}$$

$$\text{Force} = \text{Mass} \times \text{Acceleration} = mg \tag{2.21}$$

where g is the acceleration due to gravity (9.8 m-s^{-2}) and m is the total fluid mass (kg) from the surface to depth z.

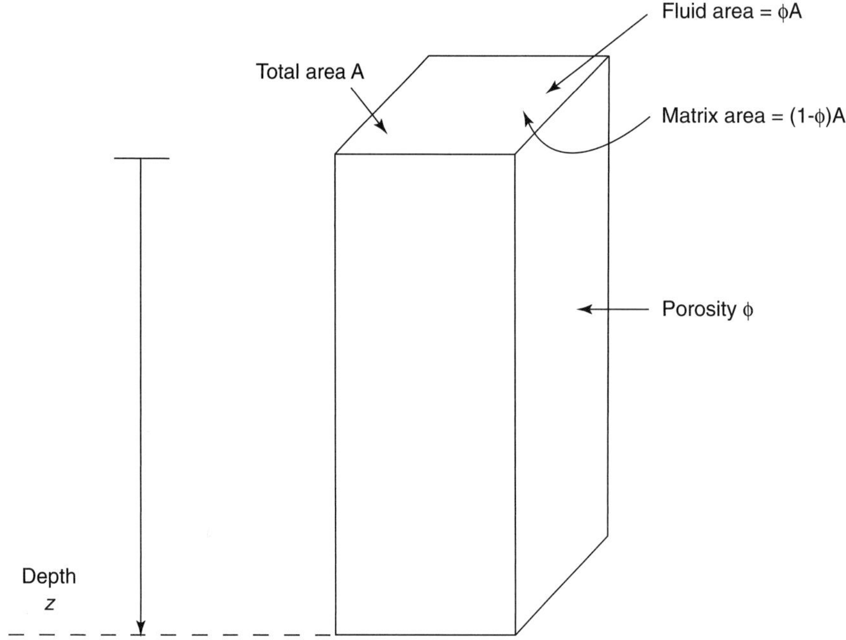

Figure 2.5 Column for calculating fluid pressure.

$$\text{Fluid Mass} = m = \text{density} \times \text{volume} \\ = \rho A z \phi \qquad \textbf{(2.22)}$$

where ρ is fluid density (kg-m^{-3}) and the relevant area is the fluid area $A\phi$, not the porous medium area A. Thus the fluid pressure (P) is

$$P = \frac{\rho A z \phi g}{A\phi} = \rho g z \qquad \textbf{(2.23)}$$

So long as the fluid does not bear the weight of the matrix grains, the fluid pressure at any depth depends only on the weight of the overlying fluid. If this is true, the fluid pressure is said to be **hydrostatic.** Note that the maintenance of hydrostatic fluid pressure requires good hydraulic communication between fluids in adjacent pore spaces. For example, if we were to insert a piece of glass horizontally into a porous medium, the fluid immediately below the glass would bear the weight not just of the overlying fluid, but also the weight of the matrix grains. Occurrences of fluid pressures above hydrostatic in geologic circumstances are usually related to low hydraulic conductivity.

The standard metric unit of pressure is the Pascal (Pa). Multiplying the units inherent in equation 2.23, we find

$$\text{kg-m}^{-3} \times \text{m-s}^{-2} \times \text{m} = \text{kg-m}^{-1}\text{-s}^{-2} \\ = \text{Pa} \qquad \textbf{(2.24)}$$

A Pascal is defined as a Newton per meter squared. The **Newton** (N) is the metric unit of force. From Newton's second law of motion we recall that force is equal to mass times acceleration. Thus

$$\text{kg} \times \text{m-s}^{-2} = \text{kg-m-s}^{-2} = \text{N} \qquad \textbf{(2.25)}$$

and if we divide equation 2.25 by area (m^2) we obtain equation 2.24.

Fluid pressures in the Earth are commonly expressed as megaPascals (MPa), 1 MPa = 10^6 Pa. For a fluid density (ρ) of 1000 kg-m^{-3}, acceleration due to gravity (g) of 9.8 m-s^{-2}, and depth (z) of one kilometer, the fluid pressure (from equation 2.23)

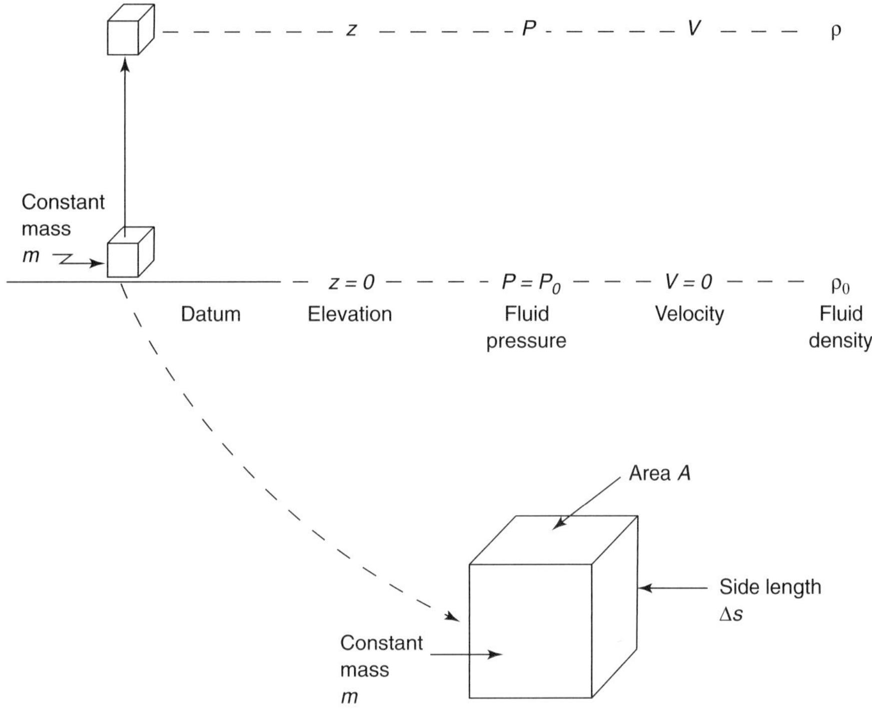

Figure 2.6 Movement of a cube of fluid with constant mass m from a standard state to an arbitrary point. The cube has sides of length Δs and faces of area A.

is 9.8×10^6 Pa, or about 10 MPa. Fluid pressure is also commonly expressed in bars. One **bar** is 10^5 Pa. One bar also equals 14.5 lb-in^{-2} or 0.98692 atm, thus one bar is approximately equal to atmospheric pressure at the Earth's surface.

The fluid pressure gradient with respect to depth in the Earth is the rate of change of fluid pressure with respect to depth,

$$\frac{dP}{dz} = \frac{d(\rho g z)}{dz} = \rho g \qquad \textbf{(2.26)}$$

which is about 10 MPa per kilometer under hydrostatic conditions.

2.4. HYDRAULIC HEAD.

In this section we formally define hydraulic head and show that it is proportional to the fluid potential energy per unit mass.

Suppose that we want to move a unit mass of fluid from some standard state to an arbitrary point (Figure 2.6). At the standard state, the elevation is zero, the fluid pressure is P_0, the fluid velocity is zero, and the fluid density is ρ_0. At our arbitrary point, the elevation will be z (m), the fluid pressure P (Pa), the fluid velocity v (m-s^{-1}), and the fluid density ρ (kg-m^{-3}).

Holding the unit mass constant, we calculate the energy needed to move a unit mass of fluid from the standard state to another arbitrary point by assuming conservation of energy. To move the unit mass of fluid, we consider three types of work or energy. First, there is the work needed to impart a potential energy to the fluid by raising it against gravity to an elevation z. Secondly, there is the work needed to impart a kinetic energy to the fluid by accelerating it from rest to a velocity v. Third, work is required to move the fluid against a pressure gradient, that, in the most general case, is not necessarily zero.

Energy is equivalent to work, which is the product of force times distance. Let m be mass (kg) and g (m-s^{-2}) the acceleration due to gravity. The force due to gravity is mg, thus the potential energy W_1 (J) is

$$W_1 = mgz \tag{2.27}$$

The kinetic energy W_2 (J) is

$$W_2 = \frac{mv^2}{2} \tag{2.28}$$

To calculate the work involved in moving a unit mass of fluid against a pressure gradient, suppose that the unit mass of fluid is a cube situated so that the top face is perpendicular to the pressure gradient as shown in Figure 2.6. Let the faces of the cube have area A (m^2) and the sides length Δs (m). Let the fluid pressure on the top face be P (Pa). The fluid pressure on the bottom face will then be $P + (dP/dz)\Delta s$. The net pressure acting on the cube is thus $(dP/dz)\Delta s$, and the net force = pressure × area = $A(dP/dz)\Delta s$. Since energy = work = force × distance, the differential work dW to move the cube a differential distance dz is

$$dW = A\frac{dP}{dz}\Delta s(dz) \tag{2.29}$$

The dz's cancel and we obtain

$$dW = A(dP)\Delta s \tag{2.30}$$

The product $A\Delta s$ is the volume of the cube, which is equal to the mass divided by the density, m/ρ. Thus the differential work dW required to move against a differential pressure dP is

$$dW = dP\frac{m}{\rho} \tag{2.31}$$

The total work required to work against the pressure gradient is obtained by integrating equation 2.31. Recall that we hold mass constant as we move our unit mass of fluid from the standard state, thus we can move m outside of the integral

$$W = m\int \frac{dP}{\rho} \tag{2.32}$$

and the total work W_3 (J) against the pressure gradient is

$$W_3 = m\int_{P_0}^{P} \frac{dP}{\rho} \tag{2.33}$$

Let the potential energy per unit mass be Φ (J-kg^{-1}). Then

$$\Phi m = W_1 + W_2 + W_3 \tag{2.34}$$

$$\Phi m = mgz + \frac{mv^2}{2} + m\int_{P_0}^{P} \frac{dP}{\rho} \tag{2.35}$$

$$\Phi = gz + \frac{v^2}{2} + \int_{P_0}^{P} \frac{dP}{\rho} \tag{2.36}$$

Equation 2.36 is the Bernoulli equation, or Bernoulli's Law. The **Bernoulli equation** states that the total mechanical energy of a fluid consists of the sum of the energies associated with the gravitational potential energy, the kinetic energy of motion, and the energy associated with fluid pressure, and that the sum of these energies is a constant. The equation was first derived and published by Daniel Bernoulli (1700–1782), a Swiss mathematician who made contributions to medicine, biology, physics, astronomy, and oceanography. Bernoulli invented the science of hydrodynamics with the publication of his book *Hydrodynamica* in 1738.

Now, we have to assume that the fluid density ρ is constant or we can't integrate equation 2.36 in a simple manner. This implies that we cannot define Φ for fluids of variable density. When we later show that Φ is proportional to head, this will similarly imply that head is proportional to the potential energy per unit mass of fluid only for fluids of constant density.

At this point, we shall also neglect the $v^2/2$ term, by arguing that for groundwater it is very small compared to the gz term. The gist of our argument is that groundwater velocities and kinetic energies are relatively low. Consider, for example, a unit mass of groundwater that accelerates from zero to a linear velocity v = 1 m-yr^{-1} and also moves a vertical distance z = 1 m. Converting units,

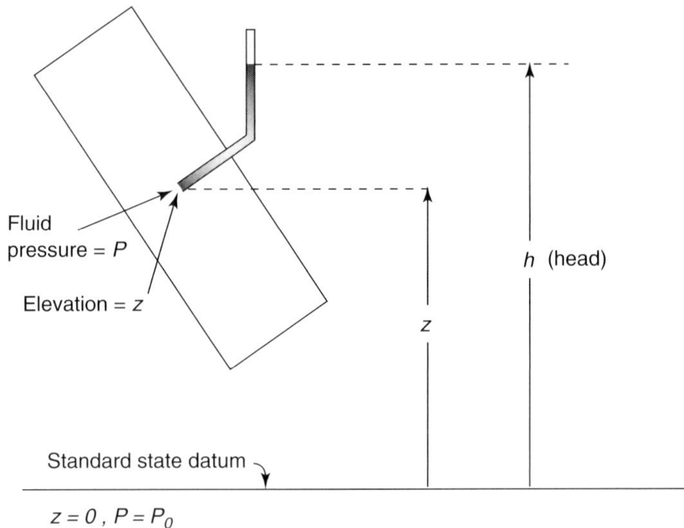

Figure 2.7 Relationship between head, elevation, and fluid pressure at an arbitrary point.

$$v = 1 \text{ m-yr}^{-1} \times \frac{1 \text{ yr}}{3.16 \times 10^{7} \text{ s}} = 3.17 \times 10^{-8} \text{ m-s}^{-1} \quad \textbf{(2.37)}$$

and

$$\frac{v^2}{2} = 5.0 \times 10^{-16} \text{ m}^2\text{-s}^{-2} \quad \textbf{(2.38)}$$

While the gz term is

$$gz = 9.8 \text{ m-s}^{-2} \times 1 \text{ m} = 9.8 \text{ m}^2\text{-s}^{-2} \quad \textbf{(2.39)}$$

The potential energy term is thus 16 orders of magnitude greater than the kinetic energy term.

Simplifying equation 2.36 we thus finally arrive at

$$\Phi = gz + \frac{P - P_0}{\rho} \quad \textbf{(2.40)}$$

Equation 2.40 is the potential energy content per unit mass of fluid. Let us now relate Φ to the hydraulic head h in Darcy's Law.

Consider a pipe filled with sand, wherein the standard state is defined with respect to an arbitrary datum as shown in Figure 2.7. Fluid pressure at the datum is P_0 (P_0 could be, for example, atmospheric pressure), and elevation is zero. Fluid rises in a tube stuck into the side of the pipe to an elevation h (m), where h is head by definition.

Fluid pressure at the base of the tube (P, Pa) is the sum of the weight of the fluid in the tube (equation 2.23) plus the pressure P_0 (Pa) at the base of the tube,

$$P = \rho g (h - z) + P_0 \quad \textbf{(2.41)}$$

Because there is no flow in the tube, hydraulic head (h, m) is constant throughout the tube (although, properly speaking, the tube is not a porous medium).

Substituting equation 2.41 into equation 2.40 we obtain

$$\Phi = gz + \frac{\rho g (h - z) + P_0 - P_0}{\rho} \quad \textbf{(2.42)}$$

$$\Phi = gz + \frac{\rho g h}{\rho} - \frac{\rho g z}{\rho} \quad \textbf{(2.43)}$$

$$\Phi = gh \quad \textbf{(2.44)}$$

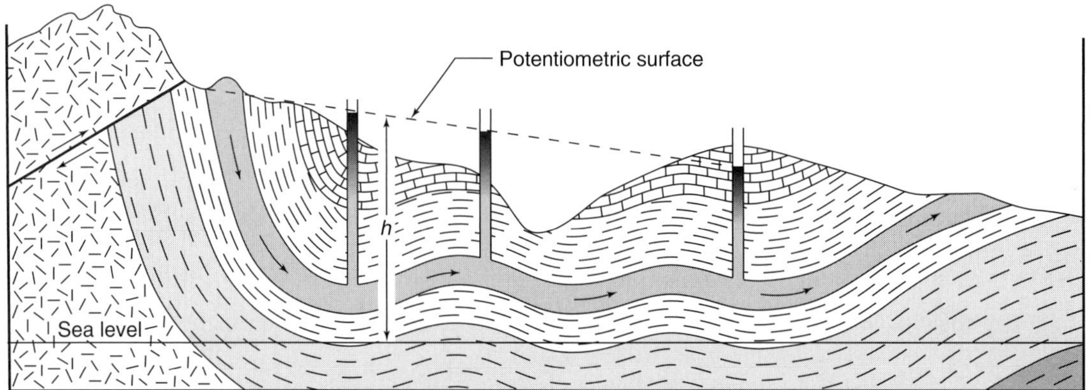

Figure 2.8 The water level in the wells is a direct measure of the head level in the aquifer.
(From Hubbert, 1953, p. 1973.)

Thus we have the simple and elegant result that the fluid potential energy per unit mass is the hydraulic head times the acceleration due to gravity—if the fluid density is constant and the kinetic energy negligible.

If we take the standard pressure P_0 to be zero, the fluid pressure is said to be **gauge pressure,** because a pressure gauge will commonly register atmospheric pressure as zero. If we let $P_0 = 0$ and substitute equation 2.44 into equation 2.40, we obtain a convenient expression for head

$$h = z + \frac{P}{\rho g} \quad \textbf{(2.45)}$$

Equation 2.45 shows that head (h) can be divided into two components: an elevation head (z), and a pressure head ($P/\rho g$).

The hydraulic head of groundwater can be measured by drilling a borehole to the depth of interest and measuring water level in the well (Figure 2.8). If the well is cased with some impermeable material such as metal or fiberglass and open only at the bottom, then the water level in the well is a measure of fluid head at the bottom of the well. Because head is measured with respect to an arbitrary datum (e.g., commonly ground or sea level), it is an arbitrary measure of the potential energy in a fluid. This is not an impediment to understanding and predicting the movement of subsurface fluids, because fluid motion occurs in response to head gradients—the absolute values are irrelevant. The level to which groundwater rises in a well is also known as the potentiometric surface, in reference to the fact that the water level demarcates a surface of constant potential energy. Potentiometric surface and head are one and the same thing.

As a further illustration, consider two glasses of water. One glass sits on a table, the other on the ground next to it. Which fluid has the higher head? The water in the glass on the table, or the water in the glass on the ground? Since both exist at the same pressure, the water on the table top has the higher head because it is at a higher elevation.

Now, consider this question. Is head constant throughout the fluid that occupies each glass? How can it be, if the elevation changes within the glass? Consider two points within one water glass (Figure 2.9) and apply equation 2.45. The lower point has elevation z_1 (m) and fluid pressure P_1 (Pa), the upper point has elevation z_2 (m) and fluid pressure P_2 (Pa). Thus

$$h_1 = z_1 + \frac{P_1}{\rho g} \quad \textbf{(2.46)}$$

$$h_2 = z_2 + \frac{P_2}{\rho g} \quad \textbf{(2.47)}$$

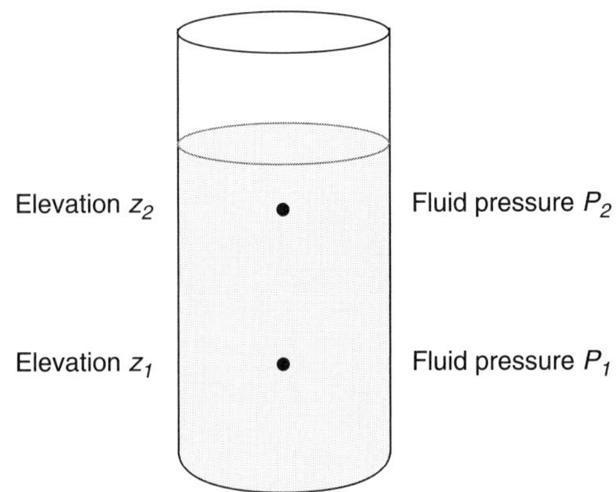

Figure 2.9 Two arbitrary points in a glass of water. What is the difference in head between the two points?

and

$$h_2 - h_1 = (z_2 - z_1) + \frac{P_2 - P_1}{\rho g} \quad \textbf{(2.48)}$$

Now, the fluid pressure at the bottom point (P_2) is the fluid pressure at the top point plus the additional fluid pressure due to the weight of the fluid column between the two points. Applying equation 2.23,

$$P_1 = P_2 + \rho g(z_2 - z_1) \quad \textbf{(2.49)}$$

Substituting equation 2.49 into equation 2.48 we find that

$$h_2 - h_1 = (z_2 - z_1) + \frac{P_2 - P_2 - \rho g(z_2 - z_1)}{\rho g} \quad \textbf{(2.50)}$$

$$h_2 - h_1 = (z_2 - z_1) - (z_2 - z_1) \quad \textbf{(2.51)}$$

$$h_2 - h_1 = 0 \quad \textbf{(2.52)}$$

Head is constant inside each glass, because elevation offsets are exactly countered by changes in fluid pressure. Note that this finding is consistent with our expectation that fluid movement occurs in response to head gradients, not pressure gradients. Even though the fluid in the bottom of the water glass is at a higher pressure than the fluid in the bottom, it does not spontaneously flow to the top of the glass. According to equation 2.45, head gradients are equivalent to pressure gradients only if elevation (z) is constant.

The requirement that fluid movement occur in response to head gradients instead of pressure gradients is equivalent to requiring conservation of energy. For example, suppose that we could drop a pipe into the oceans so that its open end was at the bottom where fluid pressure was very high. If fluid movement occurred solely in response to pressure gradients, fluid would run up the pipe to the surface. During the course of the movement up the pipe, the moving fluid could be used to turn a turbine and generate electricity. At the top of the pipe, the water could simply be allowed to run back into the oceans again, and we would have a limitless supply of energy. This does not happen; energy is conserved.

2.5. HYDRAULIC CONDUCTIVITY AND PERMEABILITY.

A little thought shows that the constant of proportionality in Darcy's Law, the hydraulic conductivity (K, m-s^{-1}), must depend on both properties of the porous medium itself as well as the fluids percolating through it. For example, if we were to substitute a viscous fluid such as molasses or a heavy oil for water, we should expect the Darcy velocity to be diminished even though no change was made in the nature of the porous medium itself. Similarly, were we to substitute a coarse-grained gravel for a fine-grained sand, we would expect fluid velocity to increase even if no change were made in the nature or composition of the fluid. Thus the single parameter K must depend on a number of other parameters, one or more of which characterize the porous medium, and one or more of which characterize the fluid.

The individual parameters and properties that are encompassed in hydraulic conductivity can be found by conducting a series of experiments wherein porous media are idealized by aggregates of uniform glass beads each with diameter d

Water Witching and Dowsing

Water witching is a form of divination that uses a forked stick, rod, or pendulum to locate underground water. Divination is the practice of foretelling the future by various natural or psychological techniques; it is found in all civilizations, ancient and modern, primitive and sophisticated, in all parts of the world. In the United States the practice of locating underground water through divination is most commonly called water witching, the term "dowsing" is more common in England.

Although various instruments can be used, the classic method is to employ a forked stick. The stick is commonly cut from the peach, willow, hazel, or witch hazel trees. One fork of the stick is held in each hand with the palms pointing upward. The bottom or butt of the Y is pointed upward at an angle of about 45 degrees. A typical description of the procedure was provided by Vogt and Hyman (1959, p. 2):

> Jeff Green seems like a man in a trance. His head is bent forward, and his eyes are focused upon the junction of the two forks of the peach limb that he holds in his hands. He clutches one fork of the branch in each hand in such a way that the junction points almost straight up in the air. For the past half-hour he has been pacing back and forth over Frank Brown's pasture. Suddenly, the peach limb quivers, and, as Jeff moves forward a few paces, it twists in his hands and points downward with such violence that the bark peels off. Jeff looks up and smiles at Frank Brown. "Dig here," he says, "and you'll find the water you need."

Cecil Bratton of Keota, Oklahoma, demonstrates his water-witching technique.
(Copyright 2000 The Oklahoma Publishing Company.)

The origin of water witching is lost in antiquity. Water witches are fond of referring to Moses as the first water witch, based on the biblical verse, Numbers 20:11:

> And Moses lifted up his hand, and with his rod he smote the rock twice: and the water came out abundantly, and the congregation drank, and their beasts also.

Notably, references to water witching are absent from Greek and Roman manuscripts. The first description of the practice appeared in the famous book on ores and mining, *De Re Metallica,* published in 1556 by Georgius Agricola (1494–1555). Agricola described how miners would search for mineral veins using a technique virtually identical to that of the modern water witch. The birthplace of the modern divining rod was in the mining districts of Germany, probably in the Harz mountains. German miners were imported into England during the reign of Queen Elizabeth (1558–1603), and brought the practice of the divining rod with them. By the end of the seventeenth century, divination with the rod had spread throughout Europe.

The widespread use of the divining rod for the location of underground water was first popularized by the French Baroness de Beausoleil. The Baroness and her husband were primarily employed in developing mines for the French government. In her book, *La restitution de Pluton,* published in 1640, the Baroness recommended the use of the divining rod for locating springs. The Baron and his wife were later imprisoned on charges of sorcery and died around 1645.

European exploration and colonization spread the practice of water witching throughout the world. In the United States, there are few publications relating to water witching prior to the year 1800. The first American academic paper on the subject was published in the *American Journal of Science* in 1821 by Reverend Ralph Emerson ("On the Divining Rod, with Reference to the Use of It in Exploring for Springs of Water"). Emerson reported on the use of divining rods for locating underground water in the states of New York and New Hampshire, and concluded that he was "totally skeptical of their efficacy, till convinced by my own senses." That same year (1826) the *American Journal of Science* published an article titled simply "The Divining Rod." Unlike Emerson, the anonymous author concluded that the "pretensions of diviners are worthless."

In 1917, the U.S. Geological Survey published *The Divining Rod: A History of Water Witching* by Arthur J. Ellis as *Water-Supply Paper 416.* In an introduction to the paper, O. E. Meinzer related that the U.S. Geological Survey received a large number of inquiries each year on the subject of water witching, as well as persistent demands that it be made a subject of investigation. Meinzer goes on to explain that *Water-Supply Paper 416* was written "merely to furnish a reply to the numerous inquiries that are continually being received from all parts of the country." The pamphlet contains an exhaustive bibliography listing 559 papers and books on the subject that were published from 1532 through 1916.

In 1959, Harvard anthropologist Evon Z. Vogt and University of Oregon psychologist Ray Hyman, published a study of water witching in the United States *Water Witching U.S.A.* (Vogt and Hyman, 1959). As part of this study, they sent questionnaires to a representative sampling of county agricultural extension agents throughout the United States. They found that 56% of the respondents expressed outright disbelief in the validity of water witching. But, 20% admitted to a belief in the efficacy of the practice, and another 24% indicated they were open-minded on the issue. Vogt and Hyman (1959) estimated that there were 25,000 water witches plying their trade in the United States. In 1998, the magazine *Popular Mechanics* reported that the American Society of Dowsers contained about 4,200 members (Wilson, 1998).

From the beginning, water witching has been a subject of great controversy. In *De Re Metallica* (1556), Agricola wrote:

> There are many great contentions between miners concerning the forked twig, for some say that it is of the greatest use in discovering veins, and others deny it.

In general, water witching is not now, nor has it ever been, accepted by the mainstream of science. Vogt and Hyman (1959) described it as an "outcast" opposed by geologists, water engineers, government officials, and other scientists for hundreds of years. Oscar E. Meinzer (1876–1948), the "father of groundwater geology" in the United States, did not mince words in his assessment of the practice. In U.S. Geological Survey *Water-Supply Paper 416* (Ellis, 1917, p. 5), Meinzer wrote:

> It is doubtful whether so much investigation and discussion have been bestowed on any subject with such absolute lack of positive results. It is difficult to see how for practical purposes the entire matter could be more thoroughly discredited, and it should be obvious to everyone that further tests by the United States Geological Survey of this so-called "witching" for water, oil, or other minerals would be a misuse of public funds.

In 1977, the U.S. Geological Survey published a fifteen-page pamphlet titled *Water Dowsing,* which indicated that inquiries concerning the subject had

Water Witching and Dowsing *(continued)*

continued, undiminished by Meinzer's rhetoric. Although toned down from Meinzer's bluntness, the conclusion 60 years later was the same: further testing of water witching is a waste of time and money.

Water witches and dowsers generally have the false and persistent notion that underground water exists in veins that may vary in magnitude from the diameter of a pencil to virtual underground rivers. Although this may be the case for areas underlain by crystalline bedrock or in Karst terrains, most groundwater is found in the interstitial pores of sediments and rocks. Vogt and Hyman (1959, p. 32) concluded that conceptions of most water witches are derived from the perpetuation of an ancient rural folklore.

Successful case histories are often offered as proof that water witching works, however, in many areas it is difficult to drill and not find water. Dowsers may also be responding, consciously or unconsciously, to other cues. The landscape itself often provides clues to the presence of underground water. The water table is apt to be closer to the surface in valleys, compared to hills. In arid regions, the presence of water-loving plants may indicate a shallow water table. The presence of springs, seeps, swamps, or lakes, is a sure sign of groundwater at the surface, although no guarantee that it exists in either sufficient quality or quantity.

In 1971, R. A. Foulkes reported on the results of a controlled series of experiments organized by the British Army and Ministry of Defense. Dowsers were asked to locate a series of buried objects simulating the presence of mines. Tests were also conducted to determine if dowsers could locate buried pipes through which water was running. The results of the tests were subjected to statistical analysis and reported in the journal *Nature* (Foulkes, 1971). Foulkes (1971) reported that the results of all trials was "frankly disappointing," concluding "there is no real evidence of any dowsing ability which could produce results better than chance or guessing."

The subject refuses, however, to die. In 1979, a report in the *New Scientist* claimed that Russian scientists had successfully deployed and tested dowsing techniques to locate metal ore deposits (Williamson, 1979). The same year that Foulkes had obtained negative results in England, two scientists from the Water Research Laboratory at Utah State University had positive results from a different experiment (*Utah Water Research Laboratory Progress Report 78-1,* p. 57, 1971). Researchers Duane Chadwick and Larry Jensen asked 150 novice dowsers chosen from staff and students at Utah State University to walk one-at-a-time along a test path chosen for the absence of visual features that could provide subconscious cues. Each dowser was given 30 wooden blocks, and asked to drop a block at locations where a "dowsing reaction" was obtained. It was found that there was a significant clustering of the locations in which blocks had been dropped by the test subjects. Dowsers tended to obtain more frequent reactions along path segments where changes in the gradient of the Earth's magnetic field were more pronounced.

In 1995, the *Journal of Scientific Exploration* published the results of a ten-year study that had been financed by the German government (Betz, 1995). The research was designed to test if there were cheap and reliable ways of finding drinking-water supplies in Third World countries. Over a ten-year period, researchers analyzed the successes and failures of

(Hubbert, 1969). The influence of the **viscosity** and density of various fluids may also be tested by varying these parameters while holding bead diameter constant (viscosity is the property of a fluid that describes its resistance to flow). The results of these experiments are that the Darcy velocity is found to be proportional to the square of the bead diameter (d^2, m^2). The Darcy velocity is also found to be proportional to the product of fluid density (ρ, kg-m^{-3}) and the acceleration due to gravity (g, m-s^{-2}), but inversely proportional to the fluid viscosity (μ, kg-m^{-1}s^{-1}). According to Darcy's original discovery, the Darcy velocity must also be proportional to the head gradient (∇h, dimensionless). These results may be used to express Darcy's Law as dowsers in locating water in arid regions of Sri Lanka, Zaire, Kenya, Namibia, and Yemen. They found that the dowser's success far exceeded chance probabilities. In Sri Lanka, 691 wells drilled at locations recommended by dowsers were 96 percent successful. The chances of finding water in Sri Lanka by random drilling are 30 to 50 percent. In an attempt to duplicate the work of Foulkes (1971), the German team tested to see if dowsers could locate buried pipes containing running water. Their results were the same as Foulkes (1971); although the dowsers could locate water-well sites, they could not find the location of buried pipes. German physicist Hans-Dieter Betz theorized that dowsers respond to subtle electromagnetic gradients that may result when water flowing through bedrock fractures changes its electrical properties.

Is water witching pure bunk, or do humans have the ability to respond to subtle natural cues in ways in which we do not yet fully understand? It seems likely that the answer shall remain controversial into the indefinite future. Agricola (1556) perhaps was wise when he said:

> Since this matter remains in dispute and causes much dissension amongst miners, I consider it ought to be examined on its own merits.

Vogt and Hyman (1959) have pointed out that the scientific questions aside, water witching is fascinating as a cultural phenomenon. Water witching is not a remnant of our primitive origins, nor a magical practice borrowed from non-literate societies. It was wholly invented in sixteenth-century Europe, and has come down to us virtually unchanged from its original form. Why does the practice not only survive, but flourish in a technological society that has never approved of it? When we understand the answer to this question, perhaps we shall obtain more insight into human nature.

FOR ADDITIONAL READING

Agricola, G. 1556. *De Re Metallica.* Translated from the first Latin Edition by H. C. Hoover and L. H. Hoover. New York: Dover Publications, 1950.

Betz, H-D. 1995. Unconventional water detection: field test of the dowsing technique in dry zones. *Journal of Scientific Exploration,* 9 (1, 2) 1–159.

Ellis, A. J. 1917. The divining rod, a history of water witching. *U.S. Geological Survey Water-Supply Paper 416.*

Foulkes, R. A. 1971. Dowsing experiments. *Nature,* 229: 163–68.

Mills, A. 1994. The dowsing rod: a most contentious instrument. *Geology Today,* 4 (10) 157–59.

U.S. Geological Survey. 1977. *Water Dowsing.* 15 pp.

Vogt, E. Z., and Hyman, R. 1959. *Water Witching U.S.A.* University of Chicago Press, 260 pp.

Williamson, T. 1979. Dowsing achieves new credence. *New Scientist,* 81 (1141): 371–73.

Williamson, T. 1987. A sense of direction for dowsers? *New Scientist,* 113 (1552): 40–43.

Wilson, J. 1998. Dowsing data defy the skeptics. *Popular Mechanics,* 175 (12): 46.

$$q \propto \frac{d^2 \rho g}{\mu} \nabla h \qquad \textbf{(2.53)}$$

where the symbol "$\propto$" means "proportional to." The only remaining factor that we have neglected is the influence of how the solid grains within a porous medium are arranged. Even if the grain size is identical, the arrangement of the grains as well as their shape will influence the resistance to flow. Let us postulate an arbitrary constant C (dimensionless) that encompasses the remaining factors that originate in the porous medium itself. We may then write Darcy's Law as

$$q = \frac{C d^2 \rho g}{\mu} \nabla h \qquad \textbf{(2.54)}$$

We have now succeeded in breaking down hydraulic conductivity into factors that are inherent in the porous medium (Cd^2), inherent to the fluid (ρ and μ), and external (g). The product Cd^2 that characterizes the capacity of a porous medium to transmit flow is the permeability ($k = Cd^2$). Rewriting Darcy's Law once again we finally end up with

$$q = \frac{k\rho g}{\mu} \nabla h \quad \textbf{(2.55)}$$

or

$$K = \frac{k\rho g}{\mu} \quad \textbf{(2.56)}$$

Permeability (k, m^2) is the capacity of a material to transmit a fluid in the presence of a potential energy gradient. Permeability is intrinsic to the medium itself, it does not depend on the nature of the fluid moving through the medium. Permeability has units of area (m^2); another common unit of permeability is the Darcy. 1 Darcy = 10^{-12} m^2.

2.6. Limits of Darcy's Law.

Darcy's Law has both an upper and lower limit of applicability; it does not hold at very high fluid velocities, and there is some question about whether or not it is an accurate description of fluid flow for very low head gradients, especially in materials of low permeability. The upper limit is given by the transition from laminar to turbulent flow. **Laminar flow** occurs when slowly moving fluids are dominated by viscous forces, it is characterized by smooth motion in laminae or layers. **Turbulent flow** is fast-moving flow in which inertial forces dominate; it is characterized by random, three-dimensional motions of fluid molecules superimposed on the overall direction of flow.

The nature of flow (laminar vs. turbulent) is quantified by the Reynolds Number (R_e). The **Reynolds Number** is a dimensionless ratio of the inertial to the viscous forces in a fluid. For porous media, the Reynolds Number is (Bear, 1972, p. 125)

$$R_e = \frac{q\rho d}{\mu} \quad \textbf{(2.57)}$$

where q (m-s^{-1}) is the Darcy velocity or specific discharge, ρ (kg-m^{-3}) is the fluid density, μ (kg-m^{-1}-s^{-1}) is the fluid viscosity, and d (m) is some characteristic length parameter of the porous medium. The characteristic length d is commonly taken as the mean matrix grain diameter. It has also been suggested that d be taken as the square root of permeability ($k^{1/2}$) (Bear, 1972; p. 125; Ward, 1964).

Experimental evidence indicates that Darcy's Law is valid as long as R_e does not exceed a critical value. Depending on the circumstances involved, the magnitude of the critical value is usually between 1 and 10. The critical value of the Reynolds Number is rarely exceeded in the Earth's crust. Possible exceptions would include flow in highly permeable rocks, such as Karst and cavernous volcanics, or flow near a pumping well.

Problem: (a) What is the maximum flow rate at which Darcy's Law applies for flow through a sandstone with a mean grain diameter of 0.5 millimeter? (b) How does this compare with groundwater velocities in the Earth's crust?

(a) We must solve equation 2.57 for the Darcy velocity, q

$$q = \frac{\mu R_e}{d\rho} \quad \textbf{(2.58)}$$

The mean grain diameter, d, is 0.5 millimeter = 5×10^{-4} m. We should specify $R_e = 1$, not $R_e = 10$. Although the problem asks for the maximum flow rate at which Darcy's Law is valid, we do not know *a priori* if the transition from laminar to turbulent flow in this particular sandstone begins at $R_e = 1$ or $R_e = 10$. To assure that we shall have laminar flow, it is thus prudent to calculate q for $R_e = 1$. The density of water is 1000 kg-m^{-3}, and water viscosity can be calculated from equation 4.23 as a function of temperature. Assuming a near-surface temperature of 20°C, $\mu = 1.0 \times 10^{-3}$ kg-m^{-1}-s^{-1}. We thus obtain

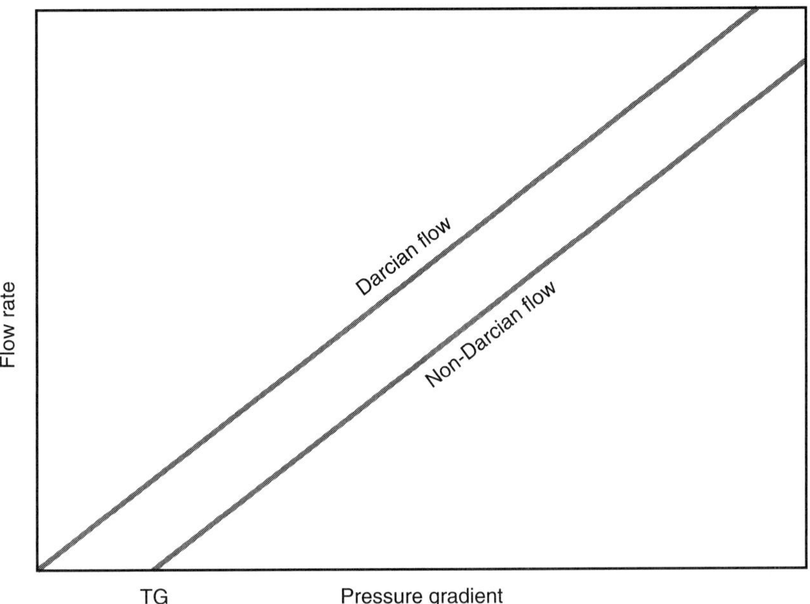

Figure 2.10 Schematic diagram of flow rate as a function of pressure (head) gradient through a Darcian and non-Darcian material. "TG" refers to a "threshold gradient" below which flow will not occur.
(From Byerlee, 1990, p. 2110.)

$$q = \frac{1.0 \times 10^{-3}\ \text{kg-m}^{-1}\text{-s}^{-1} \times 1}{5 \times 10^{-4}\ \text{m} \times 1000\ \text{kg-m}^{-3}} \quad \textbf{(2.59)}$$

$$q = 2.0 \times 10^{-3}\ \text{m-s}^{-1} \quad \textbf{(2.60)}$$

For groundwater, it is more intuitive to express the Darcy velocity in m-yr^{-1},

$$\begin{aligned} q &= 2.0 \times 10^{-3}\ \text{m-s}^{-1} \times 3.16 \times 10^{7} \\ &\quad \text{s} - \text{yr}^{-1} \\ &= 6.3 \times 10^{4}\ \text{m-yr}^{-1} \end{aligned} \quad \textbf{(2.61)}$$

(b) For deep flow (> 100 m depth), 1 m-yr^{-1} is a relatively high flow velocity (e.g., see Figure 1.4), much lower than the upper limit calculated above. Thus Darcy's Law is almost always valid in its application to fluid flow in the Earth's crust.

Darcy's Law may also not hold at very low permeabilities or very low head gradients; the exact nature of these thresholds is poorly understood at the present time. At low head gradients, water's rheological behavior may be non-Newtonian. **Rheology** is the study of how materials deform. A **Newtonian fluid** is one for which shear stress is proportional to shear rate. A shear stress is a stress that acts tangential to a surface and tends to deform a material. It is possible that there may exist a threshold head gradient below which Darcy Flow does not occur. Consider, for example, toothpaste in a tube. If the tube is squeezed, toothpaste will flow out. The harder the tube is squeezed, the faster the paste will flow. But, if the tube is simply laid down on a countertop, the paste will not flow under the influence of gravity. Similarly, if the tube is lightly touched, no flow will occur. A minimum amount of force must be exerted on the tube for the paste to flow.

Neuzil (1986) reviewed experimental studies of non-Darcian behavior in low-permeability media and concluded that reported deviations from

Darcy's Law were numerous. Mechanisms suggested to explain non-Darcian behavior primarily consisted of hypothesized changes in water properties near solid surfaces, however, Neuzil (1986, p. 1172) concluded that "the case for observed non-Darcian behavior in experimental data is weak," because many measurements have been made at or near the limit of resolution, and thus are subject to contamination from subtle experimental errors. Nevertheless, Neuzil (1986, p. 1172) cautioned that "it would be premature to dismiss the possibility of unanticipated flow behavior in low-permeability media...without experimental observations, the applicability of Darcy's Law at small gradients can only be inferred." Byerlee (1990) more definitively noted that experiments on water flow through dense clays had found the existence of a threshold pressure (head) gradient below which flow was zero (Figure 2.10). Thus, non-Darcian behavior apparently is well-documented for the case of flow at low head gradients through dense clays.

Review Questions

1. Define the following terms in the context of hydrogeology:
   a. Darcy's Law
   b. head
   c. potentiometric surface
   d. Darcy velocity
   e. specific discharge
   f. hydraulic conductivity
   g. vector
   h. scalar
   i. isotropic
   j. anisotropic
   k. principal direction of anisotropy
   l. linear velocity
   m. hydrostatic pressure
   n. Newton (the unit)
   o. bar
   p. gauge pressure
   q. viscosity
   r. permeability
   s. laminar flow
   t. turbulent flow
   u. Reynold's Number
   v. rheology
   w. Newtonian Fluid
2. Write down Darcy's Law. Give the full form that incorporates buoyancy forces, and then simplify to the proper form for fluids of constant density. Define each term and give its physical units.
3. (a) The Darcy velocity in a porous medium with a porosity of 12% is 0.5 m-yr^{-1}. What is the linear velocity? (b) The linear velocity in a sandstone whose porosity is 20% is 10 m-yr^{-1}. What is the Darcy velocity?
4. Write down the mathematical relationship between hydraulic head and fluid potential energy per unit mass. Define each term and give its physical units. Under what condition does this relationship hold?
5. Write down an equation that defines hydraulic head in terms of elevation and pressure heads. Define each term and give its physical units. How would your equation change if you expressed head in terms of depth instead of elevation?
6. Show that head is constant within a glass of water by applying equation 2.45 to any two points in the glass.
7. Write down the relationship between hydraulic conductivity and permeability. Define each term and give its physical units.
8. A laboratory test at room temperature and pressure finds that a rock has a hydraulic conductivity of 1.0×10^{-11} m-s^{-1}. What is its permeability? Is this rock likely to be a good aquifer? Is it likely to be a good reservoir for oil and/or gas? Hint: equation 4.23 may be useful.

Notations Used in Chapter Two

Symbol	Quantity Represented	Physical Units
A	cross-sectional area	m^2
C	constant	dimensionless
d	grain or bead diameter, characteristic length	m
Φ	energy per unit mass	$J\text{-}kg^{-1} = m^2\text{-}s^{-2}$
ϕ	porosity	dimensionless
g	acceleration due to gravity	$m\text{-}s^{-2}$
h	head	m
K	hydraulic conductivity	$m\text{-}s^{-1}$
k	permeability	Darcy = $10^{-12}\ m^2$
μ	fluid (dynamic) viscosity	Pa-s = $kg\text{-}m^{-1}\text{-}s^{-1}$
m	mass	kg
P, P_0	pressure	Pascal (Pa) = $kg\text{-}m^{-1}\text{-}s^{-2}$
Q	discharge	$m^3\text{-}s^{-1}$
q	Darcy velocity or specific discharge	$m\text{-}s^{-1}$
ρ, ρ_0	fluid density	$kg\text{-}m^{-3}$
R_e	Reynolds Number	dimensionless
$s, \Delta s$	distance	m
v	linear velocity	$m\text{-}s^{-1}$
W_1, W_2, W_3	work or energy	Joule (J) = $kg\text{-}m^2\text{-}s^{-2}$
x, y, z	distance	m
∇	gradient operator	m^{-1}

9. Under what conditions is Darcy's Law likely to break down?

10. Three formations, each 100 m thick overlie each other. If a steady-state vertical flow field is set up across the set of formations with $h_1 = 1000$ m at the top, and $h_4 = 690$ m at the bottom, find h_2 and h_3 at the two internal boundaries. Also find the Darcy velocity. The hydraulic conductivity of the top formation is $K_1 = 1\ m\text{-}s^{-1}$, the middle formation $K_2 = 10\ m\text{-}s^{-1}$, the bottom formation $K_3 = 0.5\ m\text{-}s^{-1}$.

11. Darcy's law is $q = -K\nabla h$. Explain why $q = -\nabla(Kh)$ is a form of Darcy's law that is valid only for a homogeneous, isotropic medium. Hint: don't work too hard!

SUGGESTED READING

Hubbert, M. K. 1940. The theory of ground-water motion. *Journal of Geology,* 48: 785–944.

Hubbert, M. K. 1987. Darcy's Law: Its Physical Theory and Application to Entrapment of Oil and Gas in Landa, E. R. and Ince, S., eds. *History of Geophysics, Volume 3, The History of Hydrology,* p. 1–26. Washington, D.C.: American Geophysical Union.

CHAPTER 3

PROPERTIES OF POROUS MEDIA

3.1. POROSITY.

Porosity (ϕ) is the fraction or percent void space in a rock or sediment. In the unsaturated zone, this void space is commonly filled with both fluid and air; below the water table, the void space is completely filled with fluid. Rocks and sediments are thus composed of two components: a solid matrix and a fluid that fills the void spaces between the matrix grains.

The porosity of a sediment depends upon the shape and arrangement of individual particles, the degree to which the constituent particles have been sorted, and mechanical compaction. In rocks, porosity is affected by the diagenetic processes of cementation and dissolution. Appreciable porosity can also be developed by fracturing; this is relatively more important in crystalline rocks that otherwise have low porosities. The porosity of a sediment is not determined by grain size; both silts and gravels may have either low or high porosities, however, sorting is an important determinant. Well-sorted materials containing uniform grain sizes tend to have higher porosities. Porosity is reduced in poorly sorted sediments, because small particles can fill the void spaces between the larger grains. If we were to fill a room with either bowling balls or BBs (well-sorted spheres), the total void space between the spheres in each case would be the same. But, if we poured BBs between the bowling balls (producing a poorly sorted mixture of sphere sizes), void spaces would be filled and the porosity dramatically reduced.

Slichter (1899) studied the theoretical arrangement of uniform spheres and found that the most compact arrangement resulted in a porosity of 26.0%, while the least efficient manner of packing spheres yielded an average porosity of 47.6%. It should therefore come as no surprise to find that the porosity of well-sorted sediments commonly lies between these two extremes. The porosities of most unconsolidated sediments such as sands, gravels, and clays, is less than 40%. Meinzer (1923) suggested that porosities greater than 20% be considered "large," those between 5 and 20% "medium," and porosities less than 5% to be "small." Fetter (2001, p. 75) gave the following porosity ranges for unconsolidated sediments: well-sorted sand or gravel, 25 to 50%; mixed sand and gravel, 20 to 35%; glacial till, 10 to 20%; silt, 35 to 50%, and

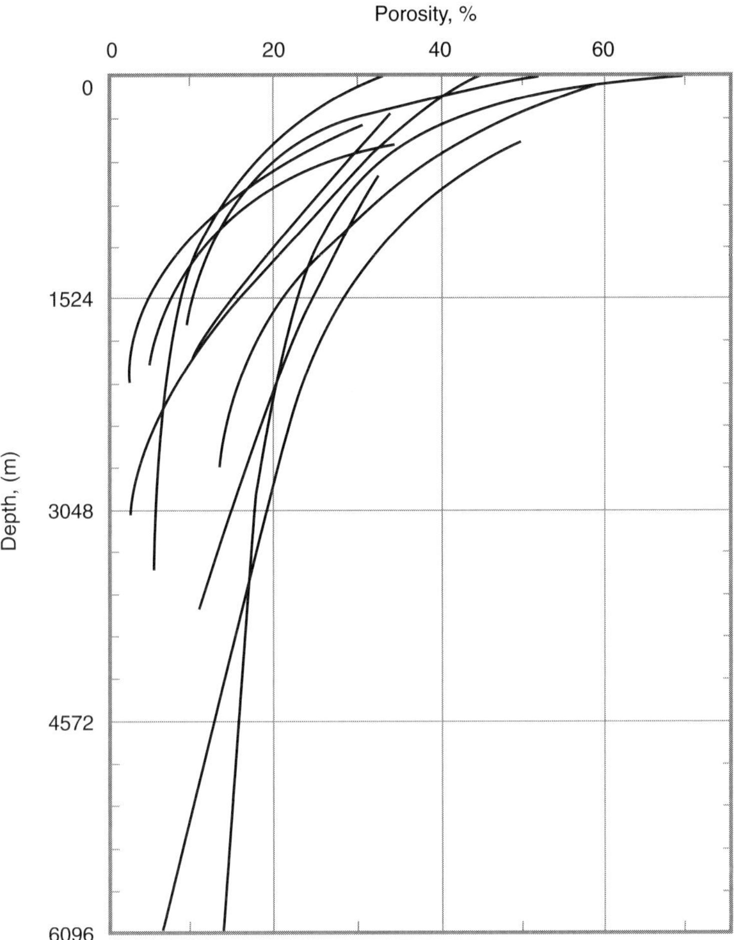

Figure 3.1 Shale porosity trends in different parts of the world.
(From Magara, 1980, p. 176.)

clay, 33 to 60%. The porosity of most soils is in the neighborhood of 50%.

As sediments lithify into rocks, porosity *tends* to decrease with increasing depth due to both mechanical compaction and chemical processes such as dissolution with subsequent precipitation and cementation. Near the ground surface, porosities typically range from 30 to 60% for rocks that have never been deeply buried and exhumed. At depths typical of oil wells (about 3 km), porosities range from 2 to 3% (0.02 to 0.03) up to about 20% (0.20) (Figures 3.1 and 3.2). Meinzer (1923, p. 9) sagaciously noted that "There is so much variation in the porosity of even rocks of the same kind that specific data are of little value." But, some general trends can be discerned from the available data.

Sandstone porosity tends to decrease linearly with increasing depth,

$$\phi(z) = \phi_0 + bz \tag{3.1}$$

where ϕ (dimensionless) is porosity, z (m) is depth, b (m^{-1}) is a constant, and ϕ_0 (dimensionless) is porosity at zero depth (Figure 3.2).

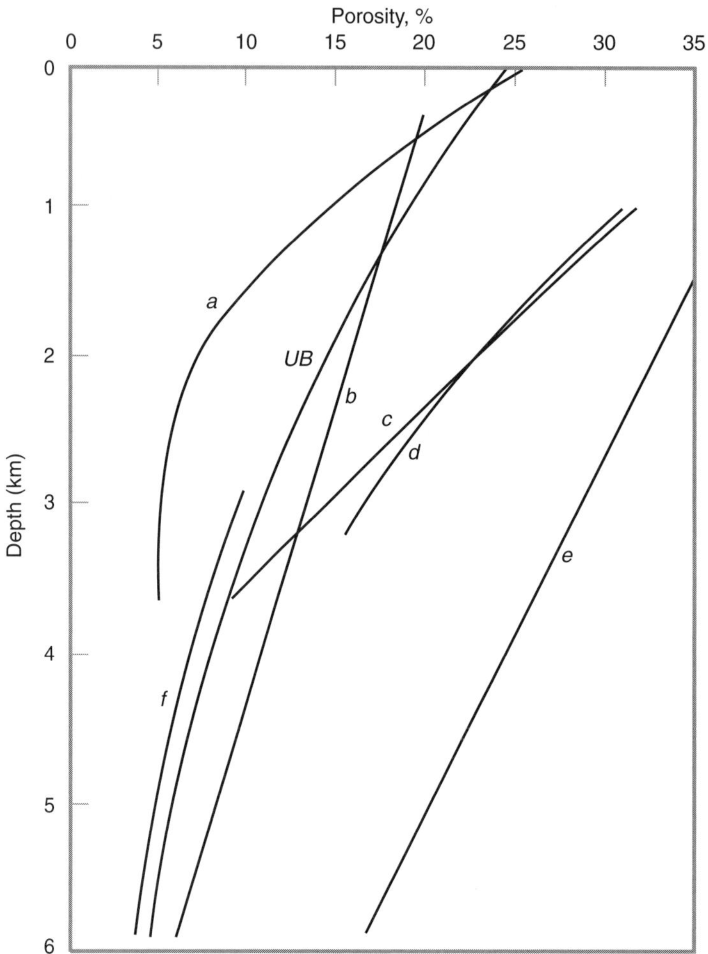

Figure 3.2 Porosity versus depth for (*a*) Jurassic-Cretaceous shale (Caucasus), (*b*) Pennsylvanian-Permian sandstone (Texas, Oklahoma), (*c*) Jurassic-Cretaceous sandstone (Caucasus), (*d*) Jurassic-Cretaceous quartz sandstone, (*e*) Quaternary sand (Louisiana), (*f*) Carboniferous silty sandstone. Curve marked *UB* is generalized porosity-depth curve used by Chapman et al. (1984) for modeling porosity in the Uinta Basin, Utah.
(From Chapman et al., 1984, p. 462.)

Shale porosity tends to decrease exponentially with increasing depth,

$$\phi(z) = \phi_0 e^{-bz} \qquad \textbf{(3.2)}$$

where z (m) is depth, ϕ_0 (dimensionless) is porosity at zero depth, e (dimensionless) is the base of the natural logarithms, and b (m^{-1}) is a constant (Figure 3.1). Equation 3.2 is known as **Athy's Law**, after Athy (1930) who first used equation 3.2 to describe the change of porosity with depth exhibited by Pennsylvanian and Permian Shales in northern Oklahoma.

Equation 3.1 implies that the rate at which sandstone porosity decreases with increasing depth is constant. In contrast, equation 3.2 implies that the rate at which shale porosity changes decreases with increasing depth. The different nature of these generalized curves is a reflection of the relative importance of mechanical versus chemical processes. In general, porosity reduction in

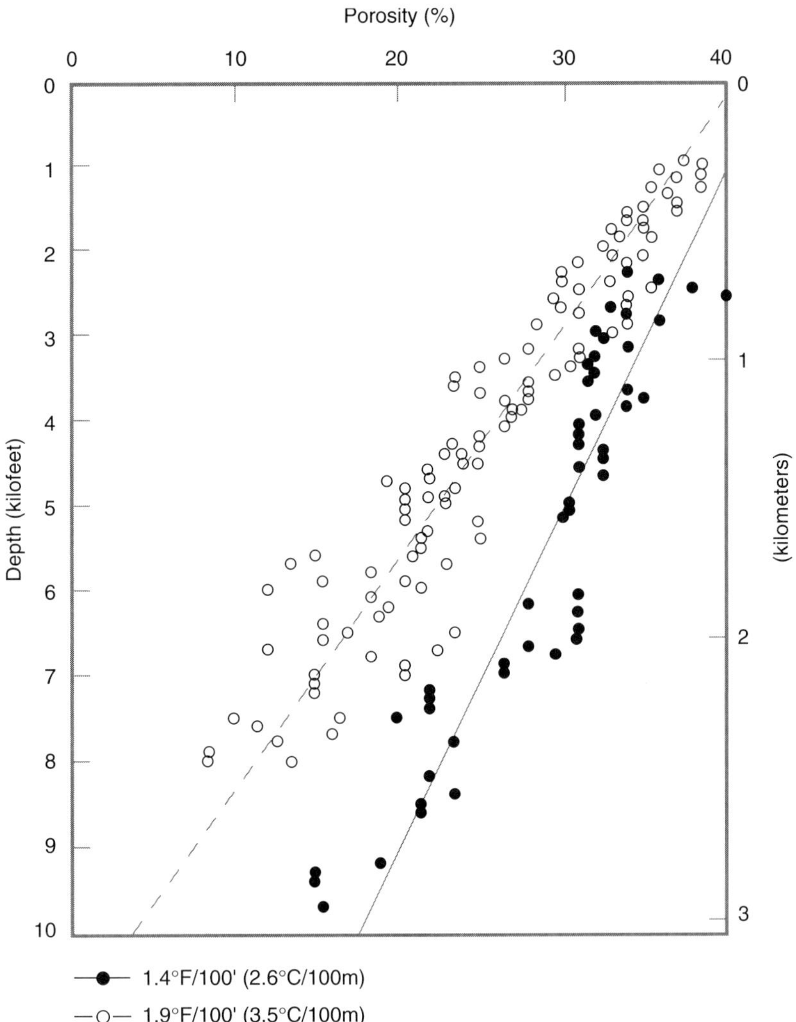

Figure 3.3 Sandstone porosity decreases more rapidly in the area with the higher geothermal gradient, suggesting that the primary mode of porosity reduction may be chemical instead of mechanical.
(From Magara, 1980.)

shale is dominated by mechanical compaction, especially near surface, however, chemical processes may be more important in sandstones. Most chemical reactions proceed more rapidly at higher temperature, and sandstones from areas with higher geothermal gradients tend to have lower porosities when compared to sandstones from cooler areas at the same depth (Figure 3.3).

As rocks or sediments at depth were once at the surface, the porosity of a rock or sediment evolves through time. The relationship between matrix and pore fluid is not passive. These two components interact both mechanically and chemically. Fluid may be squeezed out from collapsing pore spaces as matrix grains are subjected to increased stress levels, and there may be chemical reactions between solid and fluid components with resulting alterations in both solid and fluid composition.

The curves shown in Figures 3.1 and 3.2 are generalizations. Actual measurements of porosity

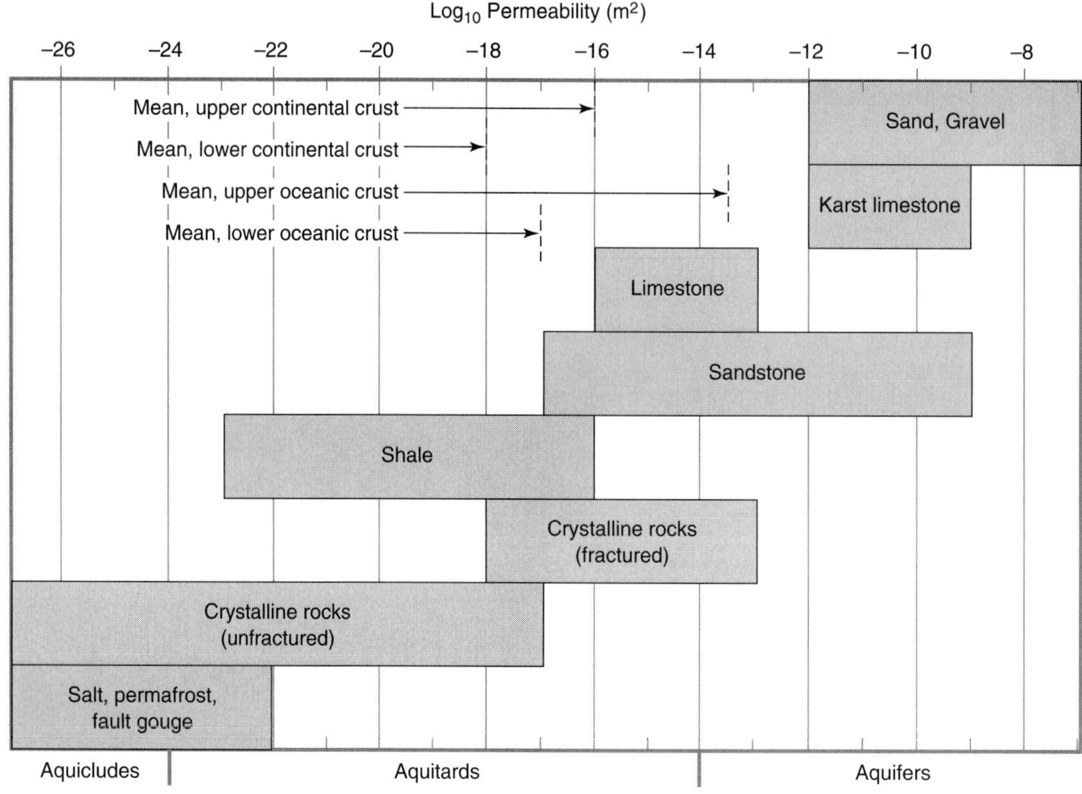

Figure 3.4 Average permeabilities of rocks and sediments.

at any depth will reveal a wide scatter (e.g., Figure 3.3). Porosity trends determined over a restricted depth range may not necessarily be safely extrapolated to other depths. For example, extrapolation of linear sandstone-porosity lines to great depths would produce impossible negative porosities.

3.2. Permeability.

Permeability is the capacity of a material to transmit fluid in the presence of a potential energy gradient (see discussion in chapter 2).

3.2.1. Definitions, Numbers, and Units.

Permeability (k) has units of length squared; in this text we shall usually use m^2. Another common unit for permeability is the Darcy, where 1 Darcy = 10^{-12} m^2. The common oil field unit of permeability is the milli-Darcy (mD), 1 mD = 10^{-15} m^2.

The permeability of rocks and sediments varies by 16 orders of magnitude (Figure 3.4). The permeability of well-sorted gravels is around 10^{-7} m^2, while the permeability of unfractured crystalline rock and shale, halite, and fault gouge, may be lower than 10^{-23} m^2 (Manning and Ingebritsen, 1999). As a generality, impermeable geologic materials are rare. Nearly all rocks have some capacity to transmit fluid.

An **aquifer** has traditionally been defined in the context of groundwater utilization as "a saturated permeable geologic unit that can transmit significant quantities of water under ordinary hydraulic gradients" (Freeze and Cherry, 1979, p. 47). Gen-

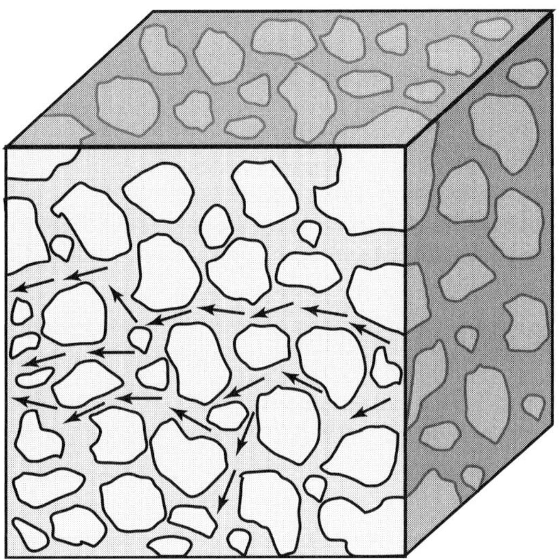

Figure 3.5 Fluid movement through a porous medium on a microscopic scale.

erally speaking, most aquifers have permeabilities greater than or equal to about 10^{-14} m^2. Aquifers can either be confined or unconfined. A **confined aquifer** is one that is bounded by less permeable layers (aquitards). For example, a sandstone layer underlain and overlain by shales may be a confined aquifer. An **unconfined aquifer** is one that is not completely bounded by less permeable strata. The typical unconfined aquifer consists of near-surface unconsolidated sediments such as sands and gravels.

Less permeable rock units are termed aquitards. An **aquitard** is a stratum that can transmit quantities of water that are very significant for a variety of geologic problems, but is inadequate for supplying economic quantities of water to wells. An **aquiclude** is a rock layer that completely excludes fluid flow through it. Geologic aquicludes are rare. Permafrost may constitute a true aquiclude, as any aqueous fluid attempting to move through permafrost will freeze. In general, any rock layer that is not an aquifer is an aquitard.

Like most generalities, the definitions above are useful conceptions that are fraught with imprecision and uncertainties. Later, as we develop the mathematical tools necessary to achieve a more quantitative understanding, we shall see that permeability, thickness, and time are inexorably linked, and it is impossible to separate these quantities. The question of whether or not a rock unit has a "high" or a "low" permeability can only be answered in the context of the application. Nevertheless, for most problems that are considered on a human time frame measured in years, we can consider "high" permeabilities to be those greater than 10^{-14} m^2, "moderate" permeabilities to be those in the range of 10^{-15} to 10^{-16} m^2, "low" permeabilities to be in the range of 10^{-17} to 10^{-20} m^2, and those lower than 10^{-20} m^2 to be very low.

3.2.2. Scale Dependence and the REV.

Permeability is a continuum-averaged physical property. In other words, the concept of permeability describes a rock property that has a physical significance only on a macroscopic scale.

Consider an enlarged view of a rock that shows the matrix grains and the fluid-filled spaces between them (Figure 3.5). As a fluid moves through the rock, the actual movement of the water molecules is through the spaces between the grains, not through the grains themselves. The permeability of the matrix grains is essentially zero, while the permeability of the channels between the grains is essentially infinite. If it were possible to measure permeability on a microscopic scale one would thus obtain either a very high value or a very low value, depending upon the location of the measurement. However, these extremes would average out as the scale of measurement was enlarged to include both matrix grains and pore channels. The point at which heterogeneities and resultant fluctuations in permeability average out is the **representative elementary volume** or **REV** (Figure 3.6). The REV is the smallest volume that can successfully represent the heterogeneous microscopic domain of a porous medium by a fictitious, homogeneous continuum.

The size of the REV depends upon the heterogeneity of the porous medium. The REV may be as small as a cube a few millimeters on a side,

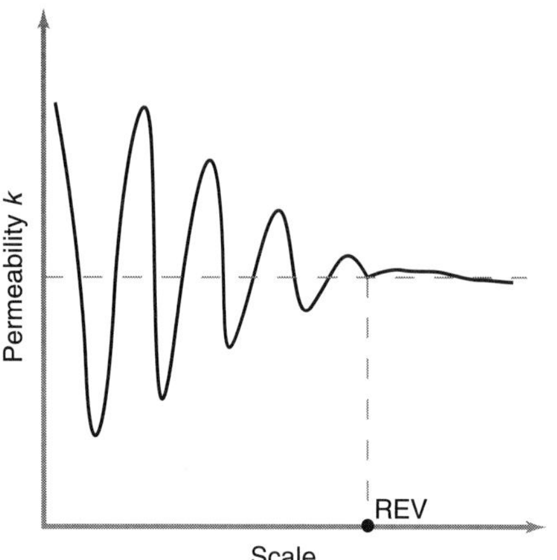

Figure 3.6 Permeability of a porous medium versus scale of measurement. The scale at which fluctuations in permeability average out is the representative elementary volume (REV).

or as large as a cube 100 to 1000 km on a side; heterogeneities exist at all geologic scales. Consider, for example, a fractured rock mass. The bulk permeability on the scale of the entire mass may be relatively high because of the effect of fractures. But, permeability measurements on unfractured core samples may yield numbers that are considerably lower. The problem can be exacerbated if fractured or broken cores are not measured but discarded as being unrepresentative. Failure to adequately characterize the REV may have catastrophic consequences. Hanor (1993) showed that the effective permeability of sediment layers underlying a toxic waste dump in Louisiana was up to 4 orders of magnitude higher than that inferred from core measurements, due to the presence of large-scale heterogeneities (see case study in Chapter 9). In general, as the scale

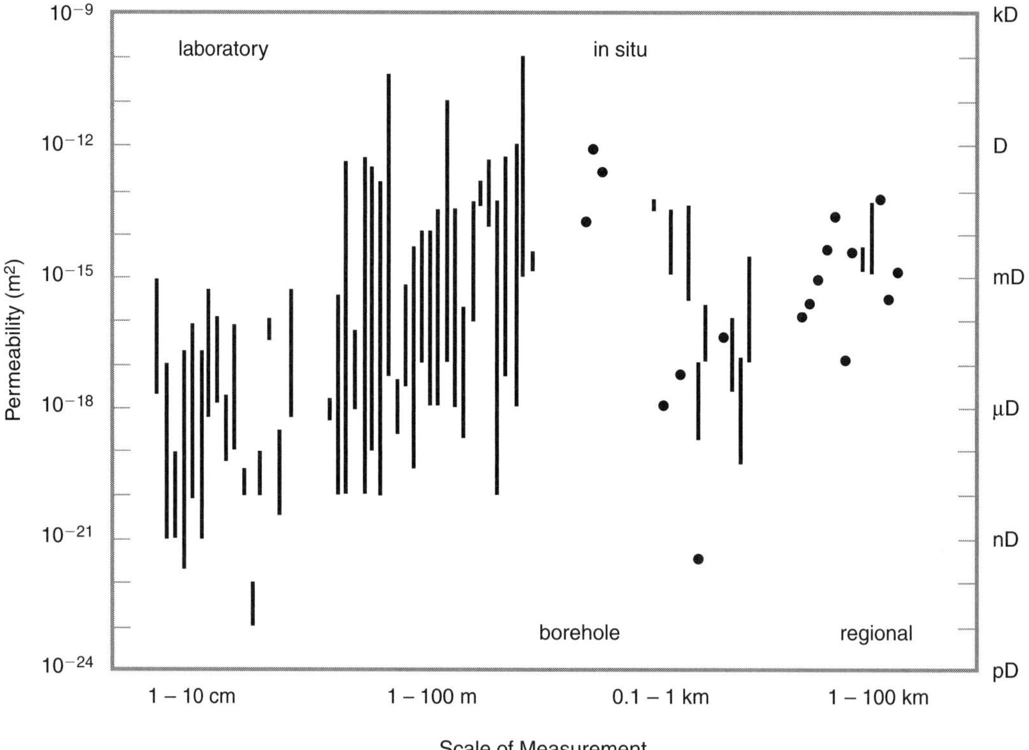

Figure 3.7 Permeability of crystalline rocks versus scale of measurement.
(From Clauser, 1992, p. 452.)

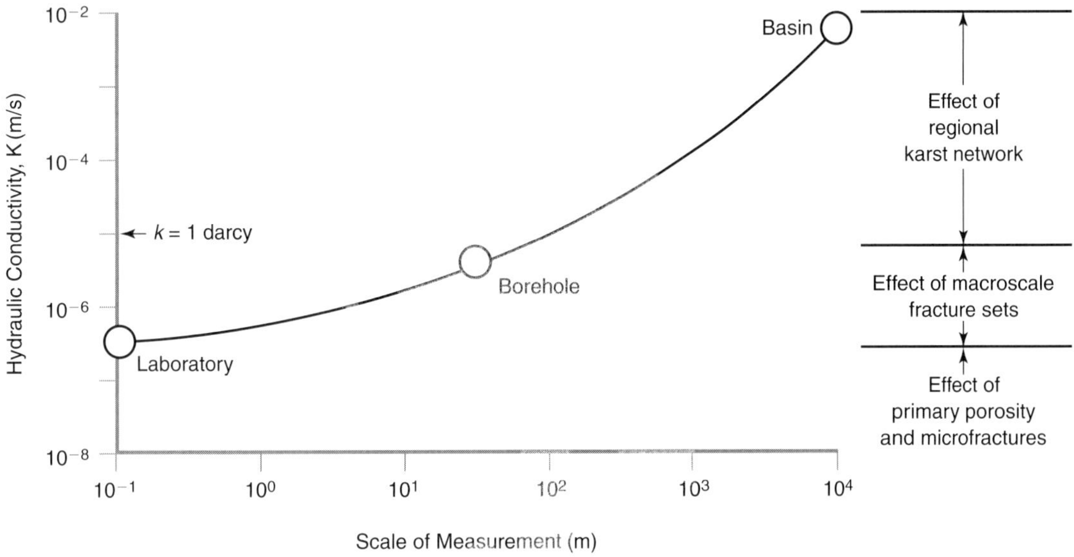

Figure 3.8 Hypothesized increase of hydraulic conductivity with scale for carbonate rocks in central Europe. (From Garven, 1986, p. 1017.)

of measurement increases, more heterogeneities are encountered and thus permeability tends to increase with scale (Figures 3.7 and 3.8). Most flow through crystalline rocks in the near-surface is through fractures. As a result, crystalline-rock permeability on a borehole scale (~1–100 m) tends to be about 3 orders of magnitude greater than permeability on a core scale (~1–10 cm) (Brace, 1980; Clauser, 1992).

The increase of permeability with scale is well-established for crystalline rocks (Brace, 1980; Clauser, 1992). Whether or not significant increases in permeability occur at larger scales in argillaceous media (clays and shales) is more controversial. Bredehoeft et al. (1983) found that the regional scale permeability of the Pierre Shale in South Dakota is 4 orders of magnitude higher than the value inferred from measurements on cores. Brace (1980) and Neuzil (1994), however, reviewed other data indicating no increase in regional scale permeabilities for argillaceous rocks in other settings.

Schulze-Makuch and Cherkauer (1997) and Schulze-Makuch et al. (1999) suggested that the increase of hydraulic conductivity with scale could be described by

$$K = cV^m \tag{3.3}$$

where K (m-s^{-1}) is hydraulic conductivity, c (m^{-2}-s^{-1}) is a characteristic of a geologic medium, which is determined by factors such as average pore size and interconnectivity, V (m^3) is volume of the porous medium under consideration, and m is a dimensionless exponent that varies from 0.5 to 1.0. Equation 3.3 applies up to the REV after which hydraulic conductivity (or permeability) is constant (Figure 3.9). The value of the exponent (m) depends upon the type of flow. For porous media in which the flow is through pores between solid grains, the exponent m tends to have values near 0.5. For porous media in which flow is dominantly through fractures, the exponent m is near 1.0. Porous media with significant amounts of both fracture and pore-flow tend to be characterized by intermediate values of m that range between 0.5 and 1.0.

3.2.3. Fractures and Faults.

Consider the idealized case of flow between parallel, smooth glass plates (Figure 3.10). Empirically, we find that

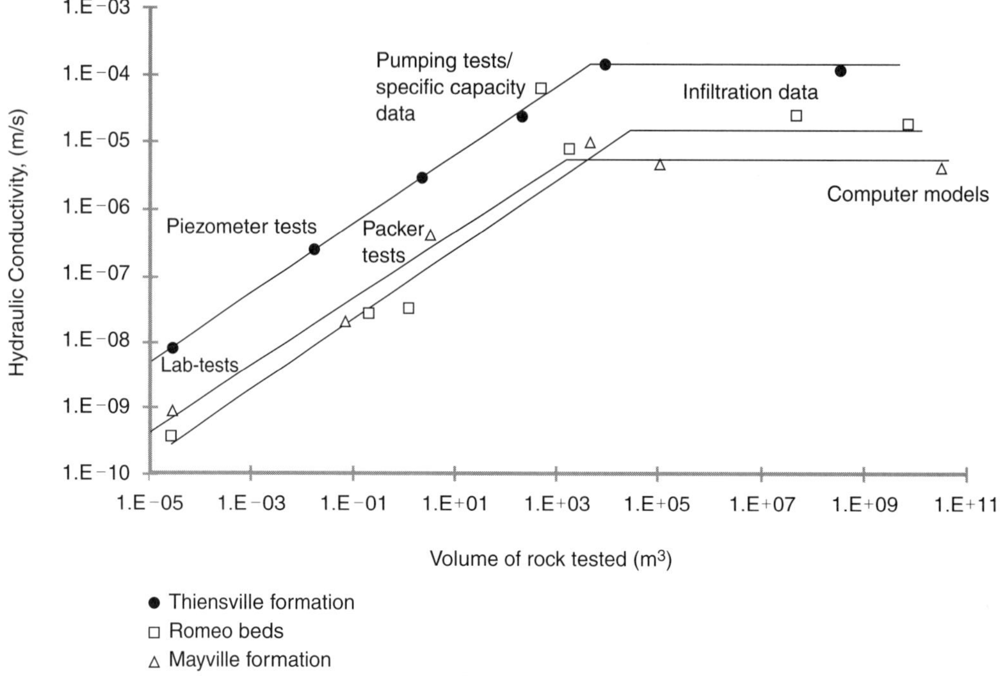

Figure 3.9 Hydraulic conductivity as a function of volume for three different geologic units. (From Schulze-Makuch, 1997, p. 3.)

$$k = \frac{N_f w^3}{12} \tag{3.4}$$

where k (m^2) is permeability, N_f (m^{-1}) is number of fractures per unit distance, and w (m) is fracture width (Brace, 1980). In other words, permeability is proportional to the cube of the fracture width. Thus the presence of fractures can dramatically increase permeability.

There is a tendency in the geological literature to alternatively invoke faults as both impermeable barriers to flow or conduits for flow, as suits the needs of any particular situation. In general, older, inactive faults, are lower in permeability as they tend to be filled by clay-rich fault gouge and hydrothermal cements. **Fault gouge** is soft, fine-grained material that fills a fault, vein, or fracture, and presumably originates from the grinding motion that occurs during fault movement. The term "gouge" originated with miners, who found they could remove this soft material from mineralized veins by gouging it out with a pick ax. Younger faults in tectonically active terrains subject to repeated earthquakes and fault movement tend to have higher permeabilities. The presence of thick layers of silica and other minerals deposited from hydrothermal fluids testifies to the role that many faults play as a conduit for fluids. Laboratory and field observations, however, indicate that fracture permeability created through fresh rupturing may be relatively short-lived. In some circumstances, the dissolution and reprecipitation of minerals by moving fluids reduces fracture width and permeability on time scales of hours to months (Sibson, 1994, p. 71). Thus the high permeability needed for significant fluid flow may be ephemeral unless continually reinvigorated through ongoing seismicity. In some cases, however, flow is known to occur through fracture systems in terrains that have been tectonically inactive for extended periods of time. Much of the spring flow in the Appalachian mountains is through faults and fractures, yet the Appalachian orogeny occurred hundreds of millions

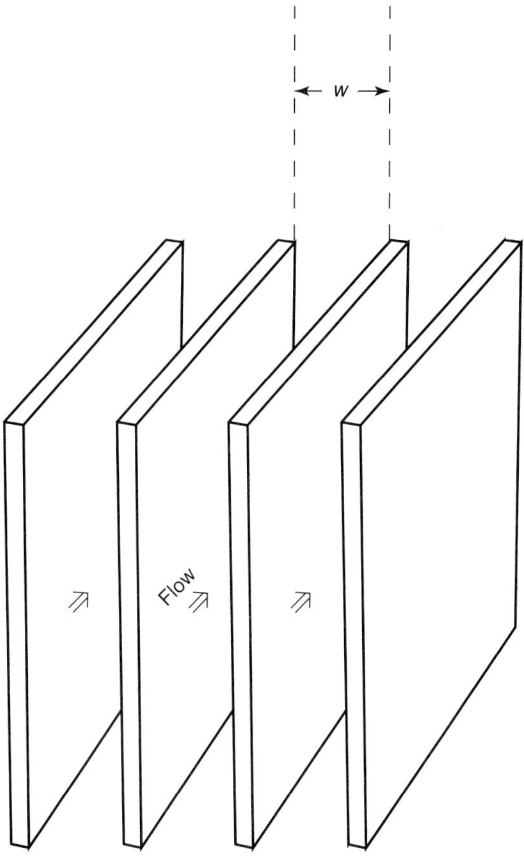

Figure 3.10 Flow between parallel glass plates separated by distance *w*.

of years ago. The geochemistry of each instance is unique. In some cases, fluid flow through a fault will result in precipitation; in other cases, dissolution may occur.

3.2.4. Anisotropy.

Consider the general case where permeability varies with direction. If we were to take a layered sedimentary rock and measure its permeability as a function of orientation, we would likely find a minimum and maximum at certain orientations. These orientations define the principal directions of anisotropy. For most sedimentary rocks, the bedding plane and its perpendicular are the principal directions of anisotropy.

Let x and y be horizontal directions parallel to bedding, and z be the vertical dimension perpendicular to bedding. Then, for most sedimentary rocks, $k_x \approx k_y \neq k_z$. If anisotropy Ψ is defined as the dimensionless ratio of the horizontal permeability to the vertical

$$\Psi = \frac{k_x}{k_z} \tag{3.5}$$

then usually we should find that $\psi \geq 1$. On a core scale, ψ is typically in the range of 1–2, however, on the scale of a sedimentary basin (100–1000 km), ψ may be as high as 100–1000 due to sedimentary layering. Sedimentary layering leads to high anisotropies because the average horizontal permeability for a layered sequence of rocks is computed differently than the vertical.

3.2.5. Averaging Methods.

There are three types of averages or means: **arithmetic, geometric,** and **harmonic.** For any group of N numbers x_i, $i = 1$ to N, the arithmetic average is defined

$$\overline{x_a} = \sum_{i=1}^{N} x_i w_i \tag{3.6}$$

where w_i is the dimensionless fraction of the *ith* of N components such that

$$\sum_{i=1}^{N} w_i = 1 \tag{3.7}$$

The w_i are weighting factors normalized such that their sum is one. The geometric mean is defined as

$$\overline{x_g} = \prod_{i=1}^{N} x_i^{w_i} \tag{3.8}$$

where again w_i is the fraction of the *ith* of N components such that equation 3.7 holds. The harmonic mean is defined as

$$\overline{x_h} = \frac{\sum_{i=1}^{N} w_i}{\sum_{i=1}^{N} \frac{w_i}{x_i}} \tag{3.9}$$

The weighting factors w_i in equation 3.9 do not necessarily have to be normalized to sum to one as they do in equations 3.6 and 3.8.

For any set of numbers that is averaged, it is always true that

$$\overline{x_a} \geq \overline{x_g} \geq \overline{x_h} \tag{3.10}$$

The arithmetic average is essentially controlled by the highest number in the averaged sequence, while the harmonic average is strongly affected by the lowest number.

Problem: Compute the arithmetic, geometric, and harmonic average of the following sequence of four numbers: 1, 10, 100, 1000.

Let each number be weighted equally, thus $w_1 = w_2 = w_3 = w_4 = 0.25$. Then

$$\overline{x_a} = \frac{1}{4} + \frac{10}{4} + \frac{100}{4} + \frac{1000}{4} \tag{3.11}$$

$$\overline{x_a} = 0.25 + 2.5 + 25 + 250 \tag{3.12}$$

$$\overline{x_a} = 277.75 \tag{3.13}$$

and

$$\overline{x_g} = (1^{0.25})\,(10^{0.25})\,(100^{0.25})\,(1000^{0.25}) \tag{3.14}$$

$$\overline{x_g} = 1 \times 1.778 \times 3.162 \times 5.623 \tag{3.15}$$

$$\overline{x_g} = 31.615 \tag{3.16}$$

and

$$\overline{x_h} = \frac{1}{\dfrac{0.25}{1} + \dfrac{0.25}{10} + \dfrac{0.25}{100} + \dfrac{0.25}{1000}} \tag{3.17}$$

$$\overline{x_h} = \frac{1}{.25 + .025 + .0025 + 0.00025} \tag{3.18}$$

$$\overline{x_h} = 3.6 \tag{3.19}$$

Which average should be used to compute the effective permeability of a group of rocks each with different permeabilities? Consider a layered sequence of isotropic rocks ($k_x = k_y = k_z = k$) with each layer of equal thickness. Let the three layers consist of a sandstone ($k = 10^{-12}$ m^2), a shale ($k = 10^{-16}$ m^2), and a limestone ($k = 10^{-15}$ m^2).

For flow parallel to bedding (through the layers), the effective bulk permeability k_x is the arithmetic average, 3.3×10^{-13} m^2. For flow perpendicular to bedding (across the layers), the effective bulk permeability k_z is given by the harmonic average, 2.7×10^{-16} m^2. It can be shown that the arithmetic and harmonic averages are theoretically correct for flow parallel and perpendicular to bedding by applying Darcy's Law, and this is left as an exercise. It is intuitive, though, that the arithmetic average is correct for flow parallel to bedding. Flow parallel to the layers may pass entirely through the high-permeability sandstone and bypass the low-permeability shale. The effective bulk permeability of the layered sequence is thus controlled by the highest permeability layer, just as the arithmetic average is controlled by the largest single number. In contrast, flow perpendicular to bedding must pass through all the layers, no layer can be bypassed. Thus the effective bulk permeability is controlled by the lowest permeability layer in the sequence, just as the harmonic average is dominated by the lowest number.

Note that although each of the three layers by itself is isotropic, the group of three taken together is highly anisotropic. The anisotropy, $\Psi = k_x/k_z = 3.3 \times 10^{-13} / 2.7 \times 10^{-16} = 1{,}222$. Thus sedimentary layering may lead to very high anisotropies. Rocks in sedimentary basins tend to be organized into layered sequences, albeit with lateral changes in lithology and permeability related to facies changes. On a large scale ($>$ 10 km or so) horizontal permeabilities that govern regional flow through layers of sedimentary rock are apt to be as much as 100 to 1000 times greater than vertical permeabilities.

The geometric average is theoretically correct for a random mixture of permeabilities (de Marsily, 1986, p. 82). If each of the three permeabilities used in the above example could be distributed at random through the rock layers considered, the average permeability for any direction of flow would be given by the geometric average of the three.

3.2.6. Measurement Techniques.

Permeability measurements may be divided into three categories on the basis of the REV sampled. Laboratory measurements may be made on cores on a scale of $10^{-2} - 10^{-1}$ m, well tests measure permeability on a scale of $10^{1} - 10^{3}$ m, and there are indirect methods that can be used to infer permeabilities on scales of $10^{4} - 10^{6}$ m.

For permeabilities higher than about 10^{-17} m^2, a permeameter (see box) may be used to measure core permeabilities in the laboratory (de Marsily, 1986, p. 75–76). Rocks with permeabilities lower than $\sim 10^{-18}$ m^2 will not attain steady-state flow on a reasonable human time scale, so transient methods must be used. In transient methods, a pressure pulse of some type is induced near the rock surface and the decay of fluid or gas pressure with time monitored. The observed pressure decay is then compared with some mathematical model of pressure dissipation that predicts pressure decay as a function of permeability. It is extremely difficult (and expensive) to measure permeabilities lower than about 10^{-20} m^2. The practical limit of resolution is about 10^{-22} m^2, although some laboratories have claimed resolution of permeabilities as low as 10^{-24} m^2.

Well tests measure in situ permeabilities directly by measuring the change of fluid pressure in a wellbore in response to fluid injection or withdrawal. In a **slug test**, fluid pressure within a well bore is instantaneously elevated (or lowered) by rapidly inserting (or withdrawing) a solid piece of pipe or slug into a well. The slug displaces fluid in the wellbore, causing the total fluid level and fluid pressure in the well to rise (Cooper et al., 1967). Alternatively, the fluid pressure can be lowered by rapidly withdrawing fluid from the well. The observed change of fluid pressure is then compared with some mathematical model that characterizes the change of fluid pressure with time in terms of permeability. The **drill-stem test** is a procedure commonly used in the petroleum industry to estimate formation pressure and permeability (Bredehoeft, 1965). A drill-string is a section of pipe lowered into a well; the lower end of the drill-string contains several tools such as valves and pressure gauges. A drill-stem test is conducted in the following manner. After the drill-string has been lowered to the desired depth in the well, a section of the surrounding strata are hydraulically isolated by activating packers at the top and bottom of the interval of interest. A common type of packer is an inflatable rubber collar. A valve on the drill-string is then opened. Fluid pressure drops, and fluids are free to enter the drill-string pipe and flow at the surface. The valve is then closed, and pressure in the well builds up asymptotically to the equilibrium fluid pressure. The rate at which pressure increases is a function of permeability. Permeability and equilibrium fluid pressure are estimated through a mathematical analysis. A typical procedure is to plot pressure as a logarithmic function of time. This usually produces a line whose intercept yields the equilibrium pressure, and whose slope is proportional to the permeability (Bredehoeft, 1965).

More recently, the use of drill-stem tests to determine in situ fluid pressures has been replaced by use of the **repeat formation tester (RFT)**. The RFT is a downhole tool that allows fluid-pressure measurements to be made repeatedly over short intervals. This capability allows the fluid-pressure gradient over short distances to be calculated, which in turn facilitates the location of the oil-water contact (as oil is generally less dense than water, the fluid-pressure gradient in oil-saturated rocks is lower).

Well tests are almost always preferable to core measurements, as they are more likely to accurately represent the REV of interest. As noted above, well tests in fractured media commonly yield average permeabilities 2 to 3 orders of magnitude higher than measurements on cores from the same formation (Figures 3.7, 3.8, and 3.9).

Permeability on scales greater than about 1000 m cannot be measured directly from well tests, because the time taken for the pressure transients to travel such distances would be inordinate. Permeability on scales of $10^{4} - 10^{6}$ m must be inferred by indirect means. These include modeling of heat flow (Deming, 1993), modeling of hydraulic head (Bredehoeft et al., 1983), radioactive isotopes (Phillips et al., 1986; Fabryka-Martin

Permeameters

A **permeameter** is a laboratory device used to measure permeability by applying Darcy's Law. There are two types of permeameters: constant-head and falling-head (see figures). In both cases, a sample is secured inside a cylinder bounded by disks whose hydraulic resistance is negligible.

The constant-head permeameter maintains a constant head difference Δh (m) across the sample by using overflow vents that allow water levels to remain constant. Note that head is approximately constant throughout the supply tube that feeds water through the bottom of the sample. Therefore, head at the bottom of the sample is given by the water level at the top of the supply tube. For a sample of thickness Δz (m), Darcy's Law is

$$q = -K\frac{\Delta h}{\Delta z}$$

where K (m-s^{-1}) is the hydraulic conductivity. The Darcy velocity (q, m-s^{-1}) is the volumetric flow rate (Q, m^3-s^{-1}) divided by the area perpendicular to flow. Therefore, if the Darcy velocity is multiplied by the cross-sectional area of the sample (A_s, m^2), we obtain the volumetric flow rate

$$Q = qA_s = -KA_s\frac{\Delta h}{\Delta z}$$

As a convenience, we can drop the negative sign because we know the direction of flow. If the permeameter is operated for a time of length t (s), a volume of water V (m^3) accumulates,

$$V = Qt = KtA_s\frac{\Delta h}{\Delta z}$$

or

$$K = \frac{V\Delta z}{tA_s\Delta h}$$

Having found the hydraulic conductivity, the permeability can be found from equation 2.56.

During operation of the falling-head permeameter, the head difference across the sample decreases (see figure). Head at the top of the sample is given by the water level above the sample. Head at the bottom of

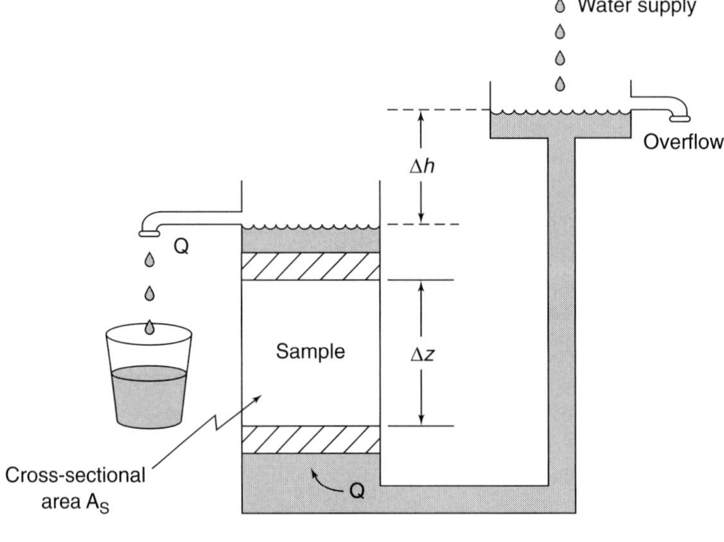

Constant-Head Permeameter

the sample is given by the water level in the supply tube. If we establish a datum $h = 0$ at an elevation equal to the water level at the top of the sample (see figure), then the water level h (m) in the supply tube is also equal to the head drop across the sample.

Initially, the water level in the supply tube is h_0 (m) at time zero. The permeameter then starts to operate, and water starts to move through the sample. After a time t (s) has elapsed, the water level in the supply tube has dropped to a level h_t (m).

The flow rate through the sample is not constant, but changing. Nevertheless, Darcy's Law applies.

$$q = -K\frac{h}{\Delta z}$$

where h (m) is now both the water level in the supply tube and the head change across the sample. The volumetric flow rate (Q, m^3-s^{-1}) is equal to the Darcy velocity (q, m-s^{-1}) multiplied by the cross-sectional area of the sample (A_s, m^2),

$$Q = qA_s = -KA_s\frac{h}{\Delta z}$$

Because mass is conserved, the volumetric flow rate through the sample (Q, m^3-s^{-1}) must be equal to the rate at which water is lost from the supply tube. The rate (m^3-s^{-1}) at which water leaves the supply tube is equal to the cross-sectional area of the tube (A_f, m^2) multiplied by the rate at which the water level in the tube is dropping (dh/dt, m-s^{-1}),

$$A_f\frac{dh}{dt} = -KA_s\frac{h}{\Delta z}$$

or

$$\frac{dh}{h} = -\frac{A_s}{A_f}\frac{K}{\Delta z}dt$$

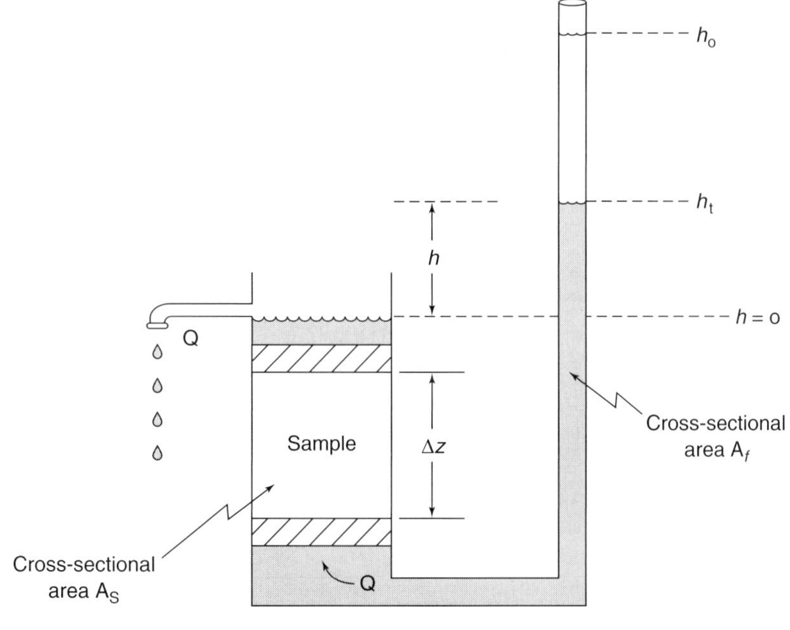

Falling-Head Permeameter

Permeameters (continued)

Integrating both sides of the above equation, we note that from time 0 to t, the water level h drops from h_0 to h_t,

$$\int_{h_0}^{h_t} \frac{dh}{h} = -\frac{A_s}{A_f}\frac{K}{\Delta z}\int_0^t dt$$

The integral on the left side is well-known to be the natural logarithm; the integral of dt on the right side is simply t.

$$\log_e(h)\Big|_{h_0}^{h_t} = -\frac{A_s}{A_f}\frac{K}{\Delta z}t$$

Recall that log (a) − log (b) = log (a/b), thus

$$\log_e\left[\frac{h_t}{h_0}\right] = -\frac{A_s}{A_f}\frac{K}{\Delta z}t$$

Solving for the hydraulic conductivity K,

$$K = -\frac{A_f}{A_s}\frac{\Delta z}{t}\log_e\left[\frac{h_t}{h_0}\right]$$

The permeability is found from equation 2.56, which relates permeability to hydraulic conductivity.

Constant-head permeameters are generally used for high-permeability materials, as the operation of the permeameter assumes a constant flow of water across the sample. Materials of lower permeability will not equilibrate so quickly; in these cases, a falling-head permeameter may be more useful.

As is the case for any scientific measurement, careful technique is required. There must be no leakage between the sample and the walls of the permeameter. If unconsolidated materials are removed from their natural state and packed into a permeameter, the measured permeability may not accurately reflect in situ values.

et al., 1987; Phillips et al., 1989), and inducing earthquakes by fluid injection (Ohtake, 1974).

An example of permeability inferred from radioactive isotopes is the dating of groundwater from the Great Artesian Basin in Australia (Figure 1.4). Chlorine-36 is a radioactive isotope that is produced in the atmosphere and enters rainwater. Infiltration of rain at the surface is the means by which chlorine-36 enters groundwater. At the time of infiltration, the ratio of chlorine-36 to total chlorine is a maximum. The longer groundwater remains in the Earth, the greater the amount of chlorine-36 decay, and the lower the ratio of chlorine-36 to total chlorine. Thus, the ratio of chlorine-36 to total chlorine can be used to estimate the age of groundwater, especially in regional flow systems maintained and driven by recharge at high elevations. In the case of the Great Artesian Basin in Australia, recharge occurs in the Great Dividing Range near the east coast. Groundwater flows westward, downhill, in a Jurassic-age confined aquifer composed primarily of sandstone. Radioactive dating of groundwater in the Great Artesian Basin indicates that its age increases westward, with increasing distance from the recharge area in the east (Bentley et al., 1986). Dividing the distance traveled by the age yields the linear velocity. The Darcy velocity can then be estimated by multiplying the linear velocity by the average fractional porosity (see equation 2.19). The head gradient can be found by comparing the potentiometric surface at different locations across the basin. Knowing both the Darcy velocity and head gradient, the hydraulic conductivity can be estimated from Darcy's Law. The permeability is then found from the hydraulic conductivity (equation 2.56), as the density and viscosity of groundwater are well known.

3.2.7. Effect of Depth and Pressure.

The permeability of both crystalline and sedimentary rocks tends to decrease with increasing depth and pressure. Permeability in sedimentary rocks decreases with increasing depth, because porosity decreases and fractures close. Crystalline rocks develop nearly all of their near-surface permeability from fractures. The closure of these fractures under the increasing pressures and effective stresses that are encountered at greater depths presumably accounts for a significant portion of the observed and

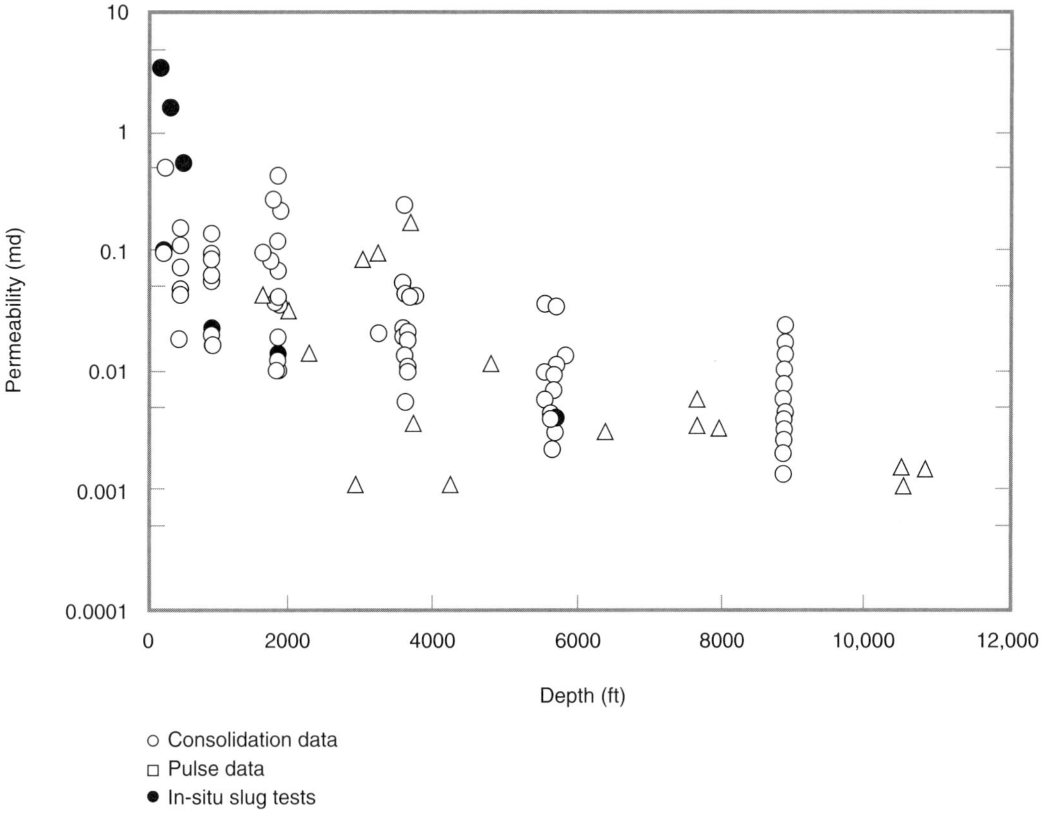

Figure 3.11 Permeability versus depth for the Pierre Shale in South Dakota.
(From Bredehoeft et al., 1992; after Bredehoeft et al., 1983; and Neuzil, 1986.)

inferred decrease of permeability with increasing depth.

Bredehoeft et al. (1992) found that the permeability of sedimentary rocks from the Big Horn Basin in Wyoming tended to decrease about 2 orders of magnitude from the surface to 3 km depth. Similarly, permeability measurements on the Pierre Shale in South Dakota show a decrease in permeability of three orders of magnitude from the surface to 3 km depth (Figure 3.11).

Brace et al. (1968) found that the permeability of unfractured Westerly Granite decreased by about 1 to 2 orders of magnitude when subjected to confining pressure roughly equivalent to an overburden thickness of 10 km. But, permeability data for fractured crystalline rocks in the near surface show a decrease in permeability of 3 to 4 orders of magnitude as depth increases from 50 m to 500 m (Brace, 1980, p. 244). Morrow and Lockner (1994) showed that both surface weathering and stress-relief fracturing in crystalline rocks have a strong influence on the pressure sensitivity of permeability. The permeability of crystalline rocks obtained from deep boreholes is often lower and more sensitive to pressure than the permeability of samples of surface rocks that have been exposed to weathering. Morrow and Lockner (1994) suggested that near-surface rocks contain weathering products in their fractures, which make the fractures more difficult to close under pressure. Intact cores of crystalline rocks from deep wells may exhibit permeability decreases of 4 to 5 orders of magnitude under confining pressures equivalent to depths of ~10 km. In contrast, crystalline rocks derived from near-surface samples (e.g., quarries) commonly exhibit permeability decreases of 1 to 3

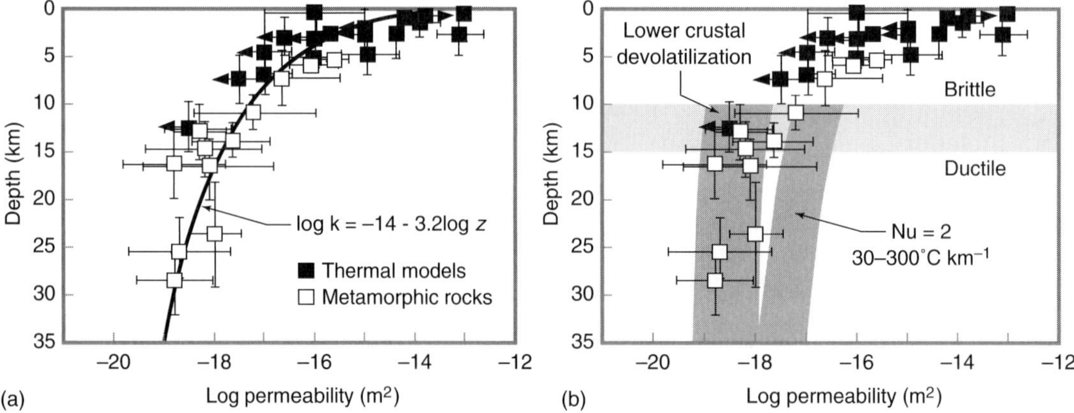

Figure 3.12 (*a*) Average permeability of the continental crust inferred from geothermal models and metamorphic systems. (*b*) Same data as in part (*a*), but with limiting curves added for lower crust.
(From Manning and Ingebritsen, 1999, p. 143.)

orders of magnitude when subjected to equivalent confining pressures.

Average permeabilities for the upper 5 km of the continental crust inferred from models of heat and groundwater flow are in the range of 10^{-14} m^2 to 10^{-17} m^2. The mean appears to be in the neighborhood of 10^{-16} m^2 (Townend and Zoback, 2000). The permeability at greater depths within the continental crust is largely constrained by analysis of metamorphic rocks and systems. From 5 to about 12 km, the average permeability of the continental crust ranges from 10^{-16} m^2 to 10^{-18} m^2. The deep continental crust (10 or 15 km down to 35 km) has an average permeability of around 10^{-18} m^2, and exhibits little depth dependence (Figure 3.12). The permeability of the lower crust is constrained to be lower than about 10^{-17} m^2, as larger permeabilities would lead to convective overturns and cooling.

In general, the upper 10 km of the continental crust is brittle, fractured, and relatively permeable. Below depths of about 10 km, the continental crust tends to be more ductile, less fractured, and less permeable. In the upper crust, processes such as regional groundwater flow and ore formation may be controlled by permeability, however, in the lower crust geologic processes control permeability. For example, metamorphic devolatization reactions produce fluid and create permeability through hydrofracturing.

Manning and Ingebritsen (1999) suggested that the depth dependence of permeability in the continental crust could be represented by a single function (Figure 3.12a),

$$\log_{10}(k) = 14 - 3.2\log_{10}(z) \qquad \textbf{(3.20)}$$

where k is permeability in m^2 and z is depth in kilometers. A closer inspection of the data, however, (Figure 3.12b) shows that there is little depth dependence below 10 km. What data and inferences are available at the present time suggest that the average permeability in the lower continental crust is nearly constant in the neighborhood of 10^{-18} m^2.

3.3. Porosity-Permeability Relationships.

Porosity tends to be linearly correlated with the logarithm of permeability (Archie, 1950)

$$\phi = a_1 + a_2\log(k) \qquad \textbf{(3.21)}$$

where ϕ is porosity (dimensionless) k is permeability (m^2), and a_1 and a_2 are dimensionless constants. Although well-defined $\phi - \log(k)$ relationships have been found for some rocks (Figures 3.13 and 3.14), it is not possible to define a universal correlation. In some circumstances, the correlation between porosity and $\log(k)$ is poor. In addition to

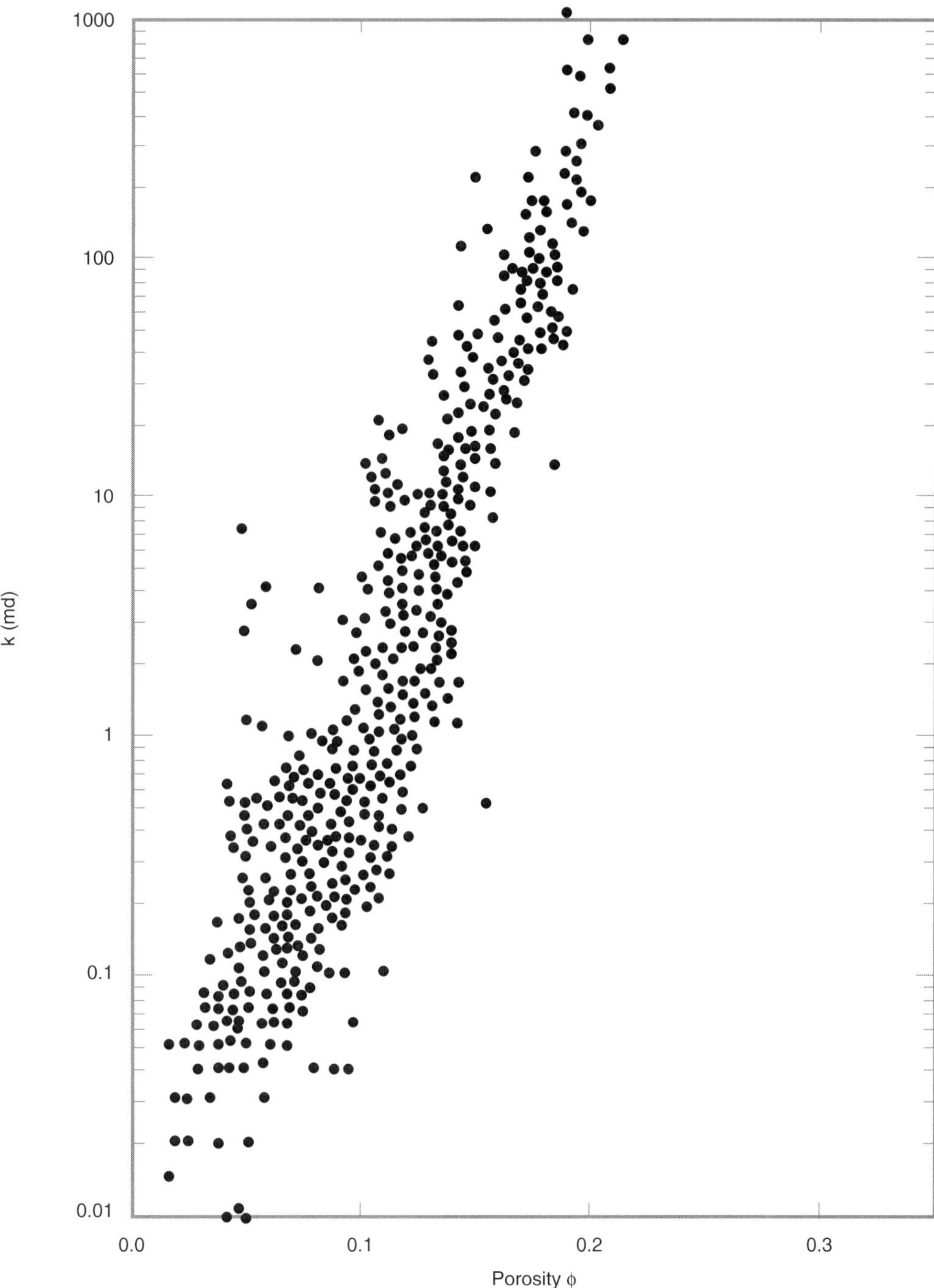

Figure 3.13 Log_{10} permeability vs. porosity for an Upper Carboniferous sandstone.
(From Nelson, 1994, p. 42.)

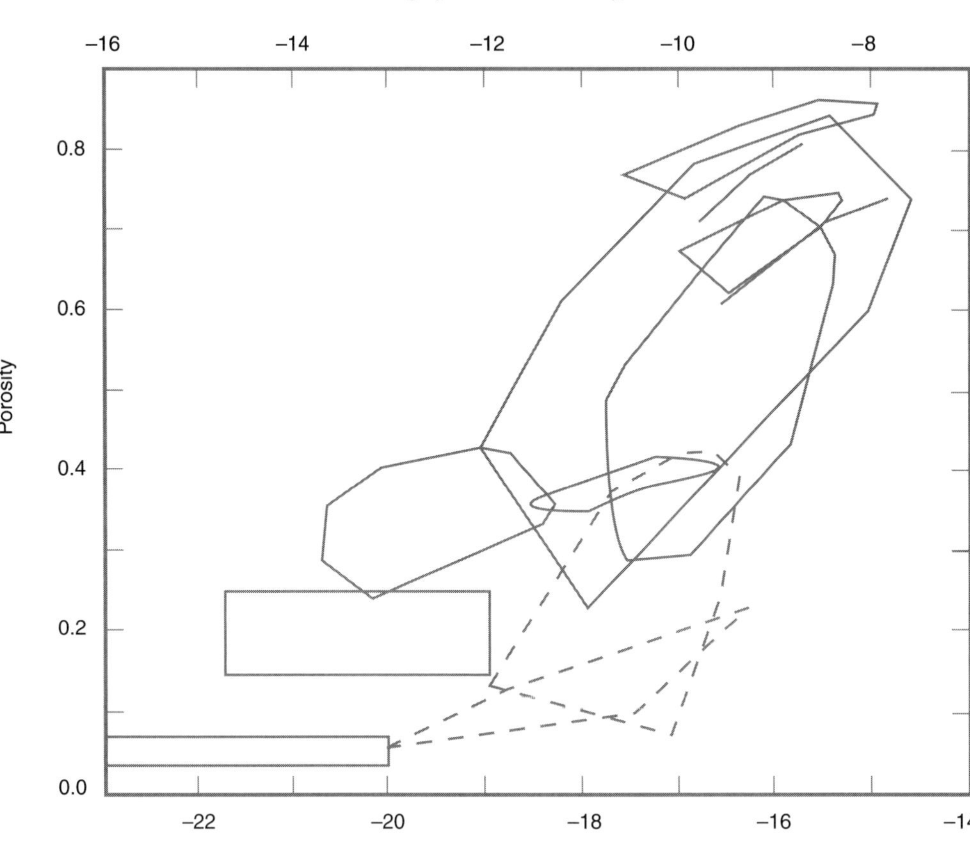

Figure 3.14 Log_{10} core-scale permeability derived from laboratory measurements vs. porosity for natural argillaceous media.
(From Neuzil, 1994, p. 146.)

porosity, permeability also depends on sorting, grain size, and lithology (e.g., sand vs. clay). Some unconsolidated clays have higher porosities than sands, but the sands have greater permeabilities. As grain size increases, porosity may decrease while permeability increases (Figure 3.15). Better sorting increases both porosity and permeability (Figure 3.15), but the increase may not be in accordance with equation 3.21 (see review by Nelson, 1994).

It is sometimes desirable to have a generalized model that specifies permeability in terms of porosity. Bethke (1985) used

$$\log_{10}(k_x) = -13 + 2\phi \qquad \textbf{(3.22)}$$

for sands, and

$$\log_{10}(k_x) = -19 + 8\phi \qquad \textbf{(3.23)}$$

for shales, where ϕ is fractional porosity ($0 < \phi < 1$), and k_x is permeability parallel to bedding in m^2.

About 65% of all sedimentary rocks are shale (Blatt, 1970, p. 257), and the porosity of shales tends to decrease exponentially with increasing depth (equation 3.2). As log permeability tends to be correlated with porosity, depth should correlate with log (log-permeability) in sedimentary rocks. Belitz and Bredehoeft (1988) studied permeability and porosity data from the Denver Basin in the central United States, and found that a log

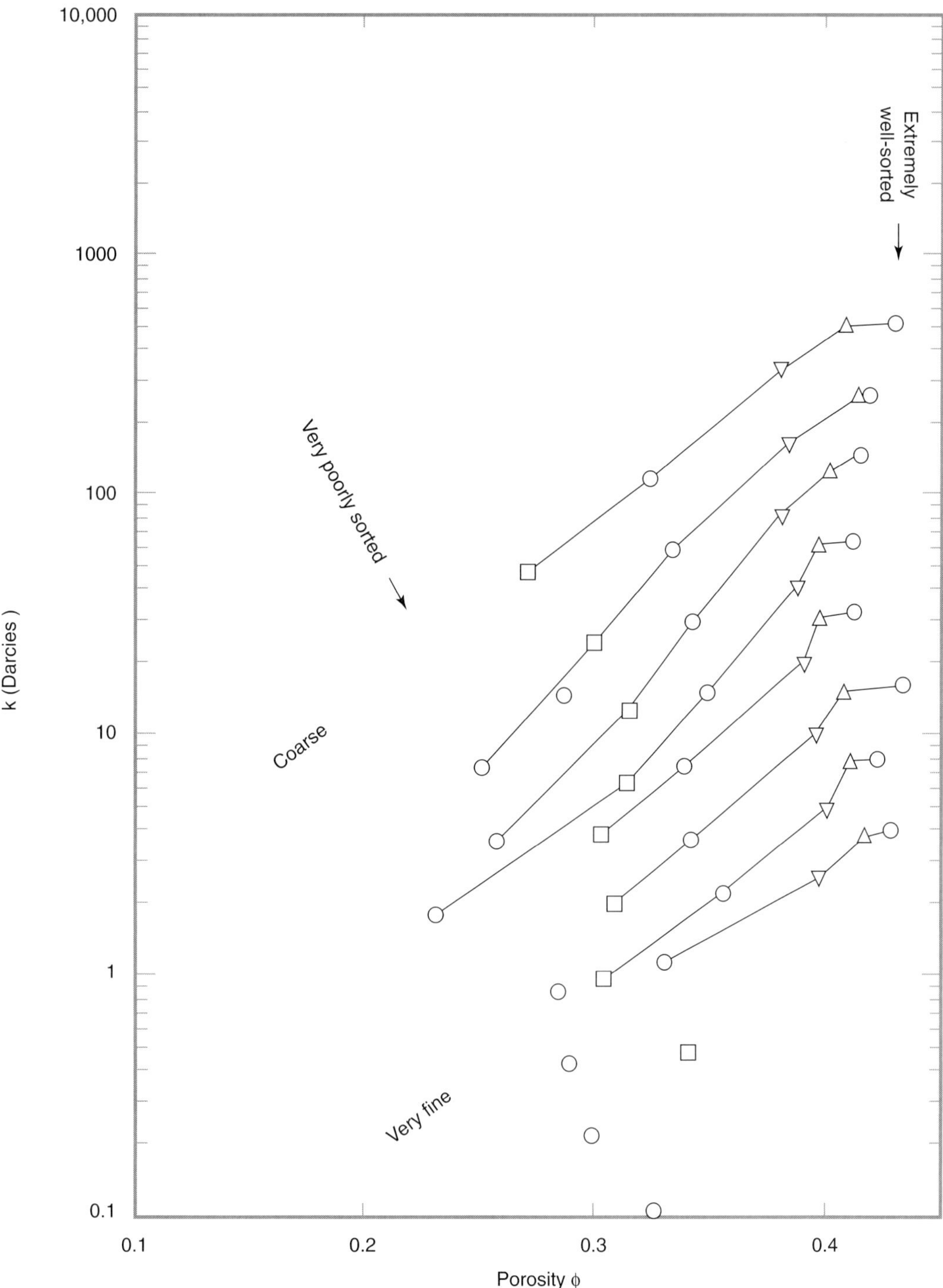

Figure 3.15 Log_{10} permeability vs. porosity for unconsolidated artificial sand packs.
(From Nelson, 1994, p. 40.)

Karl Terzaghi: Creator of the Science of Soil Mechanics

Karl Terzaghi (1883–1963) is known as the creator of the science of soil mechanics. He singlehandedly invented an interdisciplinary science that not only revolutionized foundation and earthwork engineering, but also led to an increased understanding of geologic processes such as compaction and consolidation.

Terzaghi was born into an Austrian family with a tradition of military service, however, a slight defect in vision prevented him from serving in the Austrian military and he instead chose to study engineering at the Technical University in Graz. During his undergraduate years Terzaghi was almost expelled for what Casagrande (1964) described as "excessive indulgence in academic freedom." Terzaghi's widow (Terzaghi, 1964) characterized the incident as an "imaginative prank." Terzaghi was saved from expulsion by the intercession of Professor Wittenbauer who reminded the faculty that the three students previously expelled from the University had become eminently successful. One of these expelled students was Nicolas Tesla.

Karl Terzaghi (1883–1963)

After graduation, Terzaghi served a compulsory year in the Austrian army and then returned to his alma mater and studied geology for a year. Following his geological studies, Terzaghi spent three years with an Austrian civil engineering and contracting firm and worked on a variety of projects. It was during these years that Terzaghi became aware of the general ignorance that prevailed in the field of foundation engineering and he decided to collect and correlate all available knowledge on the subject with the hope of discovering systematic principles that could be used to predict the behavior of soils under stress.

In 1912, Terzaghi received the degree of Doctor of Technical Science from Graz Technical University. A few weeks later, he departed for the United States. Terzaghi was under the impression that earthwork engineering was the most advanced in the United States and he was determined to learn all he could. In the U.S., Terzaghi visited one engineering site after another, supplementing his meager funds by accepting any type of work available. These temporary occupations included not only engineer, but boring foreman, time-keeper, and driller. In this way, Terzaghi became acquainted with the behavior of a variety of different soils and clays firsthand.

In 1916, Terzaghi was appointed as Professor of Foundation Engineering at the Imperial School of Engineering at Constantinople in Turkey. To supplement the field experience he had acquired in the United States, Terzaghi began to systematically read and digest all German, French, and English literature on the subjects of earthwork and foundation engineering. At the end of World War I, Terzaghi and other faculty from the defeated nations were summarily dismissed. Terzaghi found an appointment at the American Robert College in Istanbul. It was during this period of time that Terzaghi became convinced that the theoretical principles needed to understand the behavior of soils and their response to loading by foundations did not exist. He therefore launched an ambitious program to discover the principles himself through careful and thorough experimentation and observation.

Terzaghi assembled a soils laboratory by scrounging and salvaging parts from the college dump. He spent his days teaching, and his nights in experimentation. During this time he wrote his mentor Professor Wittenbauer that he had spent "six weeks of 12-hour days in which I experienced uninterrrupted failures." However, Terzaghi's luck soon changed and in the short time of a year he had completed experiments that created the basis of modern engineering geology. Terzaghi speculated to Wittenbauer that the vicissitudes to which he had been subjected had increased his creative powers, "I believe that the cares, misfortunes, and insecurity of my situation have multiplied my powers many times over." Five years later in 1925, the results of Terzaghi's work were published in the monumental book *Erdbaumechanik auf bodenphysikalischer Grundlage*.

In 1925, Terzaghi received an appointment at MIT in Boston. During this period of time Terzaghi's ideas were becoming well known, but were not yet widely accepted by the civil engineering community. Casagrande (1964) remarked that Terzaghi related to him that the ultimate acceptance of soil mechanics as a science would depend upon a new generation of engineers. Terzaghi was not known to mince words. Casagrande (1964) remarked that some of Terzaghi's written replies to criticisms were so "salty" that they were refused for publication. Other writings appeared only after substantive editing.

Terzaghi returned to his homeland in 1930 after receiving an appointment to a chair in civil engineering at the Technische Hochschule in Vienna, however, he returned permanently to the United States in 1938 after Hitler occupied Austria. From 1938 through his retirement in 1956, Terzaghi was associated with Harvard University. During his later years, Terzaghi divided his time between teaching, research, and consulting. Terzaghi accepted only the most important consulting offers, choosing those that offered him the opportunity to learn something new.

Terzaghi believed that engineering geology must depend upon geologic observation supplemented by physical tests carried out in a laboratory and judicious application of mathematical reasoning. Both theory and practice were indispensable.

Terzaghi's protege and oldest friend, Arthur Casagrande, described Terzaghi as possessing "bold vision, a brilliant analytical mind, and insatiable curiosity." Terzaghi was also possessed of a great power of concentration that enabled him to "write steadily by the hour" without interruption. In requiem, Casagrande (1964) suggested we let Terzaghi speak for himself:

> When I compare the importance of the various driving forces that have controlled my inner development, a passionate dislike of any lack of clarity in my thinking overshadows all the others. The satisfaction I have derived from a clear, orderly perception of relationships has always been so great that I valued material success only as a means of preserving my independence and my freedom to act in accordance with my own inner needs. The erratic and frequent changes which have characterized my life are rooted in this attitude. Before I recognized my mission an oppressive sense of dissatisfaction drove me from one extreme to another, and I never hesitated to abandon a field of activity if another environment promised broader stimulation and greater opportunities for growth.

FOR ADDITIONAL READING

Casagrande, A. 1964. Karl Terzaghi, 1883–1963. *Geotechnique,* 14: 1–12.

Goodman, R. E. 1999. Karl Terzaghi: The Engineer as Artist. *American Society of Civil Engineers.* 352 pages.

Legget, R. F. 1964. Karl Terzaghi, 1883–1963. *Canadian Geotechnical Journal,* 1: 122–124.

Terzaghi, R. D. 1964. Karl Terzaghi, 1883–1963. *Journal of the Boston Society of Civil Engineers,* 51: 289–293.

(log-permeability) vs. depth function fit data better than either a log-permeability vs. depth function or a log-permeability vs. log-depth function. It is not clear to what extent this relationship may be confirmed in other regions.

3.4. Compressibility.

There are three physical mechanisms by which fluid can be stored in a porous medium: (1) fluid compressibility, (2) compression of the porous medium, and (3) compression of solid matrix grains. The most important of these three mechanisms is usually compression of the porous medium itself. A porous medium may be compressed and fluids expelled through porosity reduction related to the normal compaction that occurs during burial. If a fluid is pumped into a rock at high pressure, however, it may actually push the matrix grains apart, thus increasing the storage capacity of the rock. The second most important mechanism for fluid storage in a porous medium is usually fluid compressibility. Although it is common to refer to fluids as incompressible, in fact they are slightly compressible. The third mechanism, compression of the solid matrix grains, is usually so small compared to the first two that it may be safely neglected in hydrogeological studies.

In the above discussion, the word "compression" is used generically to refer to a volume change that may be a compression or an expansion. It is useful to visualize the concept of fluid storage in rocks by considering an analogy to a sponge. A sponge is a porous medium that stores fluid and releases or absorbs fluid due to changes in applied stress. A rock behaves in a similar manner, it simply is much less compressible than a sponge.

3.4.1. Fluid Compressibility.

Fluid compressibility (β, Pa^{-1}) is defined as the negative ratio of the relative fluid-volume change that takes place in response to a change in fluid pressure,

$$\beta = \frac{\frac{-\Delta V_w}{V_w}}{\Delta P} \tag{3.24}$$

where V_w is fluid volume (m^3), and ΔV_w is the fluid volume (m^3) released by a porous medium in response to the change in fluid pressure ΔP (Pa). For pure water at standard temperature (25°C) and pressure (1 atm), $\beta = 4.4 \times 10^{-10}$ Pa^{-1}. The negative sign in equation 3.24 indicates that fluid volume will *decrease* in response to an *increase* in fluid pressure.

Equation 3.24 can be rearranged to yield

$$\Delta V_w = -\beta V_w \Delta P \tag{3.25}$$

Let the total volume of a porous medium be V_T(m^3). Then the volume of fluid V_w (m^3) in a saturated porous medium is

$$V_w = \phi V_T \tag{3.26}$$

or

$$\Delta V_w = -\beta \phi V_T \Delta P \tag{3.27}$$

If fluid pressure *decreases* ($\Delta P < 0$), the fluid in the pores of a rock will expand, ΔV_w will be positive, and the rock will *release* fluid. If the fluid pressure *increases* ($\Delta P > 0$), the fluid in the pores of a rock will contract, ΔV_w will be negative, and the rock will *take up* fluid.

3.4.2. Effective Stress.

To define the compressibility of a porous medium, we need to introduce the concept of effective stress. **Stress** is force per unit area; it is the same as pressure. **Effective stress** is the stress borne by the matrix grains in a rock. The essential idea behind effective stress is that the stress borne by the solid component of a rock is not simply that due to the weight of the overburden, but is reduced by a buoyancy force.

Consider a cube of material submerged in a pool of water (Figure 3.16). Its net weight will be the product of its mass and acceleration due to gravity minus a buoyancy force related to fluid pressure.

The force (N = kg-m-s^{-2}) due to gravity is

$$\begin{aligned} F_{\text{grav}} &= \text{mass} \times \text{acceleration} \\ &= \rho_{\text{cube}} \times A(z_2 - z_1) \times g \end{aligned} \tag{3.28}$$

where ρ_{cube} (kg-m^{-3}) is the density of the cube, A (m^2) is the area of a cube face, $z_2 - z_1$ (m) is the

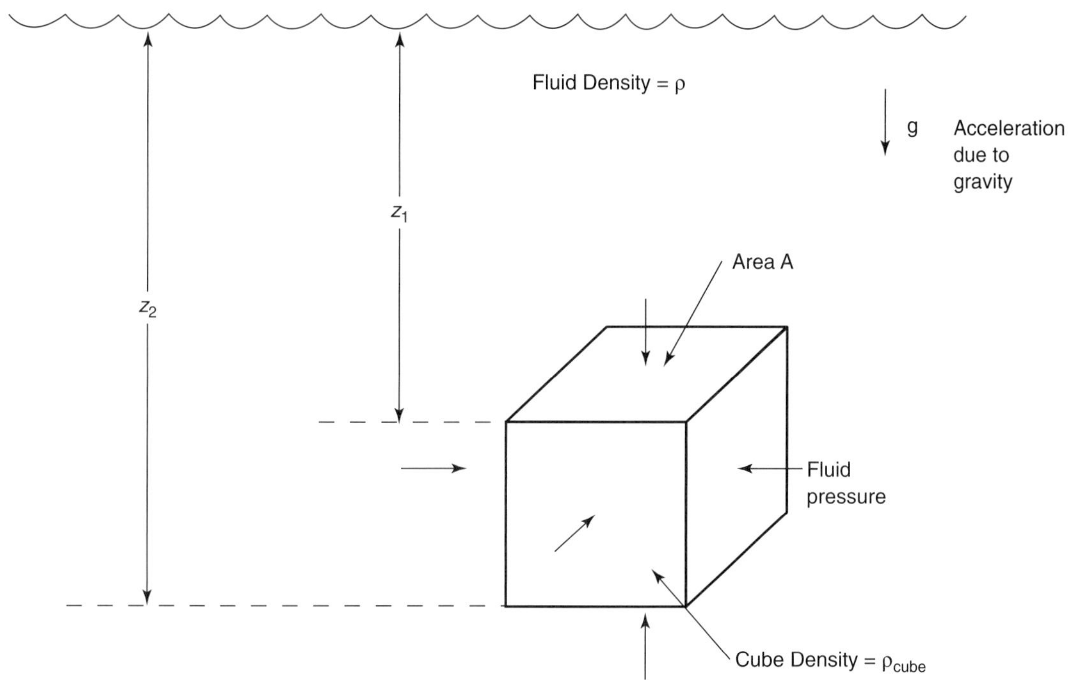

Figure 3.16 Forces acting on a cube suspended in water.

height of the cube, and g (m-s^{-2}) is the acceleration due to gravity.

As pressure is defined to be force per unit area, the forces on the cube faces due to fluid pressure are equal to the products of fluid pressure and area. The fluid pressure at any depth is simply given by equation 2.23 as $P = \rho g z$, where ρ is fluid density (kg-m^{-3}), g is the acceleration due to gravity (m-s^{-2}), and z (m) is water depth. Note that we need only consider the forces on the top and bottom of the cube. As long as we define our cube so that its top surface is parallel to the surface of the water, the fluid pressures on the sides will cancel. The total force (N = kg-m-s^{-2}) due to fluid pressure on the top face is

$$F_{top} = \rho g z_1 A \tag{3.29}$$

where z_1 (m) is the depth of the top of the cube and A (m^2) is the area of a cube face. The total force (N = kg-m-s^{-2}) due to fluid pressure on the bottom face of the cube is

$$F_{bottom} = -\rho g z_2 A \tag{3.30}$$

where z_2 (m) is the depth of the bottom of the cube, and we have defined a downwards directed force to be positive, and an upwards directed force to be negative.

The total force on the cube is thus

$$\begin{aligned} F_{total} &= F_{grav} + F_{top} + F_{bottom} \\ &= -\rho g A(z_2 - z_1) + \rho_{cube} g A(z_2 - z_1) \end{aligned} \tag{3.31}$$

The first term on the right side of equation 3.31 is a buoyancy force directed upwards. The second term is the downwards force due to gravity. Grouping terms,

$$F_{total} = gA(z_2 - z_1)(\rho_{cube} - \rho) \tag{3.32}$$

or

$$F_{total} = gV(\rho_{cube} - \rho) \tag{3.33}$$

where V (m^3) is the volume of the cube. Equation 3.33 gives the weight of the cube while suspended in water. Note that the weight of the cube if it were suspended in air would be $gV\rho_{cube}$. Upon suspension in the swimming pool, the weight of the cube is reduced by a bouyancy force equivalent to the weight of the water displaced ($gV\rho$). If ρ_{cube} (kg-m^{-3}) is greater than the fluid density, ρ(kg-m^{-3}), the weight of the cube will be reduced, but it will nevertheless sink to the bottom of the pool. If ρ_{cube} is less than the fluid density, ρ, the cube will float.

We have just derived what is known as Archimedes' principle. **Archimedes principle** is that a solid denser than a fluid will, when submerged in that fluid, have its weight reduced by the weight of the fluid it displaces. Now, take the pool of water we used to show the existence of buoyancy forces and fill it with sand so that we have a porous medium. The sand grains are suspended in a hydrostatic fluid-pressure regime, just as the rocks in the Earth's upper crust usually are. Even though our pool is now filled with sand, the buoyancy force is the same, because the fluid pressure at any depth is the same as before (see derivation of equation 2.23). In other words, the fluid pressure at any depth in a porous medium is simply that due to the overlying weight of the fluid, it is not (usually) affected by the weight of the matrix grains. The matrix grains bear the weight of only the overlying matrix grains, not the fluid. The stress, or force per unit area, due to the weight of the matrix grains, however, is reduced by a buoyancy force. The net stress borne by the matrix grains is termed the effective stress.

Across any plane through a porous medium (e.g., the Earth's crust), there is a total stress due to the weight of the overlying matrix grains and fluid. Consider a column of height z (m) and area A (m^2). The total weight of the matrix is the mass (kg) times the acceleration due to gravity (m-s^{-2}). The mass of the matrix (kg) is the product of matrix volume (m^3) and matrix density (ρ_m, kg-m^{-3}). Thus

$$\text{matrix weight} = \rho_m gz(1 - \phi)A \qquad \textbf{(3.34)}$$

where the area occupied by the matrix grains is not A, but $(1 - \phi)A$. Similarly,

$$\text{fluid weight} = \rho gz\phi A \qquad \textbf{(3.35)}$$

because the fluid occupies an area ϕA. Adding the matrix and fluid weights, the total weight is

$$\text{total weight} = \rho_m gz(1 - \phi)A + \rho gz\phi A \qquad \textbf{(3.36)}$$

The total stress (σ, kg-m^{-1}-s^{-2}) is the total weight (kg-m-s^{-2}) divided by the total area (A, m^2),

$$\sigma = \rho_m gz(1 - \phi) + \rho gz\phi \qquad \textbf{(3.37)}$$

where ρ_m (kg-m^{-3}) is the matrix density, ρ (kg-m^{-3}) is the fluid density, z (m) is depth, g (m-s^{-2}) is the acceleration due to gravity, and ϕ (dimensionless) is fractional porosity.

Effective stress (σ_e, kg-m^{-1}-s^{-2}) is the stress due to the weight of the overlying matrix grains minus the stress reduction due to a buoyancy force exerted by the fluid. The stress due to the weight of the matrix grains is given by the first term on the right side of equation 3.37. According to Archimedes Principle, the buoyancy force on any column of material is equal to the weight of a column of water with equivalent volume. If the column has volume V (m^3), the buoyancy force is $gV\rho$ (kg-m-s^{-2}) (equation 3.33). For a column of a saturated porous medium, the matrix volume is $(1 - \phi)Az$, thus the total buoyancy force is $\rho g(1 - \phi)Az$. When the buoyancy force is distributed over an area A, the net reduction in the stress due to the weight of the matrix grains is $\rho gz(1 - \phi)$. Thus the effective stress σ_e is

$$\sigma_e = \rho_m gz(1 - \phi) - \rho gz(1 - \phi) \qquad \textbf{(3.38)}$$

The first term on the right side of equation 3.38 is the stress or pressure due to the weight of the overlying matrix grains at depth z. The second term on the right side of equation 3.38 is the reduction in pressure or stress that occurs because of a buoyancy force equivalent to the weight of the water displaced by the matrix grains.

Let $P = \rho gz$ (Pa = kg-m^{-1}-s^{-2}) be fluid pressure (equation 2.23). Then,

$$\sigma_e + P = \rho_m gz(1\text{-}\phi) - \rho gz + \rho gz\phi + \rho gz \qquad \textbf{(3.39)}$$

$$\sigma_e + P = \rho_m gz(1\text{-}\phi) + \rho gz\phi \qquad \textbf{(3.40)}$$

Note that the right hand side of equation 3.40 is the definition of total stress given in equation 3.37. Thus

$$\sigma_e + P = \sigma \qquad \textbf{(3.41)}$$

where σ is the total stress (Pa = kg-m^{-1}-s^{-2}), P is the fluid pressure, and σ_e is the effective stress. Thus, the total stress at any depth is the sum of the fluid pressure and the effective stress.

The total stress (σ) is simply the force per unit area due to the total overburden weight (matrix plus fluid). Over a human time scale (although not perhaps a geologic), there will usually be no appreciable change in the total overburden weight (assuming no significant systematic change in the water table elevation as might occur through development of a groundwater resource). If the total stress σ is constant with respect to time (t),

$$\frac{d\sigma}{dt} = 0 = \frac{d\sigma_e}{dt} + \frac{dP}{dt} \tag{3.42}$$

and

$$\frac{d\sigma_e}{dt} = \frac{-dP}{dt} \tag{3.43}$$

Any change in the effective stress is equal to, but opposite in sign to, a change in fluid pressure. As fluid pressure *increases*, the effective stress *decreases*, and a porous medium will expand just as a sponge will if the pressure on it is lessened. If fluid pressure *decreases*, the effective stress *increases*, and a porous medium will contract, just as a sponge will compact if the stress on it is increased.

Now, recall equation 2.45 that defines hydraulic head in terms of elevation z (m) and fluid pressure P (Pa = kg-m^{-1}-s^{-2}),

$$h = z + \frac{P}{\rho g} \tag{3.44}$$

or,

$$P = (h - z)\rho g \tag{3.45}$$

If elevation is constant with respect to time ($dz/dt = 0$),

$$\frac{dP}{dt} = \rho g \frac{dh}{dt} \tag{3.46}$$

Substituting equation 3.46 into equation 3.43, changes in effective stress can be related to changes in head

$$\frac{d\sigma_e}{dt} = -\rho g \frac{dh}{dt} \tag{3.47}$$

In other words, changes in head may lead to changes in effective stress and expansion or contraction of a porous medium. If head is *increased* ($dh/dt > 0$), effective stress *decreases*, and a porous medium will expand. If head is *decreased* ($dh/dt < 0$), effective stress *increases*, and a porous medium will contract.

3.4.3. Porous Medium Compressibility.

We are now ready to define α, **porous medium compressibility** (Pa^{-1} = m-s^2-kg^{-1}) as the ratio of the normalized volume change of a porous medium to the change in effective stress

$$\alpha = \frac{\frac{-\Delta V_T}{V_T}}{\Delta \sigma_e} \tag{3.48}$$

where V_T (m^3) is the total volume of the porous medium (matrix plus fluid), ΔV_T (m^3) is the change in porous medium volume, and $\Delta\sigma_e$ (Pa = kg-m^{-1}-s^{-2}) is the change in effective stress.

The total change in porous medium volume is

$$\Delta V_T = -\alpha V_T \Delta\sigma_e \tag{3.49}$$

Changes in effective stress ($\Delta\sigma_e$) can be caused by changes in fluid pressure (ΔP). From equation 3.43, $\Delta\sigma_e = -\Delta P$. Substituting into equation 3.49,

$$\Delta V_T = \alpha V_T \Delta P \tag{3.50}$$

If matrix grains are assumed to be incompressible, the change in porous medium volume (V_T) must physically take place by a change in arrangement of the matrix grains. In other words, the porosity (ϕ) changes. If the porosity changes, there must be a change in the fluid volume (V_w) separate and distinct from the change in fluid volume that occurs as a result of the fluid itself contracting or expanding. The change in fluid volume (ΔV_w) must be equal in magnitude to ΔV_T and have the same sign. In other words, if the porous medium expands through an increase in porosity, the volume of water held by that medium must also increase. Thus,

$$\Delta V_w = \alpha V_T \Delta P \tag{3.51}$$

where ΔV_w is the volume of water *taken up* by a porous medium through expansion of that medium in response to a change in fluid pressure.

Archimedes: Founder of the Science of Hydrostatics

Archimedes (287–213 BC) was the most famous mathematician and inventor of ancient Greece. He was also the founder of the science of hydrostatics, which he described in his book, *On Floating Bodies*. Archimedes was spurred to discover the effect of buoyancy forces when he received a charge from King Hieron II. Hieron wanted to know how much silver and how much gold was present in a crown that had been made for him. Archimedes thought about the problem, but could not find a solution. One day, he lowered his body into a bath and noticed that the displaced water spilled over the sides. He obtained the instantaneous insight that because silver had a lower density than gold, the volume of a crown made of gold and silver could be measured and would be greater than a crown of the same weight made of pure gold. Archimedes became so excited that he leapt from his bath and ran through the streets naked shouting "Eureka!" (I have found it!). It was Archimedes who said in describing the physics of the lever, "Give me a lever long enough and a place to stand, and I will move the Earth." Archimedes also invented a device to lift water that became known as the Archimedean Screw. The Romans used Archimedean Screws to remove the groundwater that seeped into their mines. Archimedes was so given to intellectual preoccupation that at times he would have to be fed and bathed by force. At other times, he would make geometrical drawings in the ashes of fires or upon his own body.

Contemporary accounts describe Archimedes as not only a great mathematician, but also an ingenious prac-

The Death of Archimedes.
(From Mazzuchelli, 1737.)

tical inventor. Unfortunately, Archimedes considered engineering to be vulgar and his practical accomplishments to be unworthy of recording; nearly all of his books are devoted to pure mathematics. Archimedes lived in the city-state of Sicily, Italy, which was then occupied by the Greeks. In 213 BC, Sicily was attacked by the Roman General, Marcus Claudius Marcellus. However, the Romans were foiled almost single-handedly by Archimedes. King Herion II had placed Archimedes in charge of constructing the island's defenses. Polybius (c. 200–118 BC) said of the Romans:

> They failed to reckon with the talents of Archimedes or to foresee that in some cases the genius of one man is far more effective than superiority in numbers. This lesson they now learned by experience.

Archimedes devised catapults of various capacities and ranges that continuously harassed and damaged the enemy. The effect on the Romans was demoralizing. Archimedes had hidden most of his machines behind the city walls. Because the Romans could not see the source of the stones bombarding them, they half began to think they were battling gods instead of men. When the Romans finally reached the city walls, they found Archimedes had prepared yet another defense. The walls were pierced with several holes, about the width of a palm at a man's height. Behind each hole was an archer with a small catapult that discharged iron darts. The most fantastic of Archimede's defenses were polished metallic mirrors, which he used to reflect and focus sunlight that set the Roman ships afire.

Repulsed, the Romans resorted to a long-term siege. A year or two later, the Romans managed to successfully invade and occupy one of the fortified towers that looked down upon the city. Conquest was now assured. The Roman General Marcellus wept, for he knew that he would be unable to prevent his soldiers from plundering and sacking the beautiful city in front of him. Marcellus ordered his soldiers to not harm Archimedes. Meanwhile, Archimedes was so intent upon working out a problem in mathematics that he never noticed the Roman invasion. He was ordered to submit by a Roman solider who found him in his study, drawing geometric designs in the sand. The soldier did not recognize whom he was confronting. Archimedes submitted to capture, but warned the soldier not to disturb his sand drawings. Insulted, the soldier responded by running Archimedes through with his sword. The Roman General, Marcellus, was quite upset when he learned of Archimedes death and had a tomb built for him.

More than 2,000 years after Archimedes' death, people are still finding new applications for his mathematical work. In 1990, the journal *Science* reported that geochemist Joseph V. Smith at the University of Chicago used the geometrical concepts of Archimedes to deduce the structure of a mineral named boggsite. This set off a tremendous effort in the petrochemical industry to synthesize boggsite, which is a valuable catalyst in the manufacture of gasoline. Smith was quoted as saying that "Archimedes has helped make millions of dollars for the petrochemical industry." (Alper, 1990).

If we define ΔV_w to be the volume of water *released* by a porous medium, the sign changes

$$\Delta V_w = -\alpha V_T \Delta P \quad \textbf{(3.52)}$$

In general, unconsolidated materials tend to be more compressible than rocks. Amongst unconsolidated materials, clay tends to be more compressible than sand or gravel, while amongst rocks, shales tend to be more compressible than sandstones, carbonates, or igneous and metamorphic rocks (Freeze and Cherry, 1979, p. 55; Ge and Garven, 1992, p. 9124) (see Table 3.1).

Most compressibility estimates come from laboratory experiments where a rock or sediment sample is subjected to an applied pressure and its deformation measured. One of the problems with measuring rock compressibilities in this manner is that the time scale over which the rock stressed is different from that experienced in nature. Rocks subject to sustained stresses over millions of years conceivably

TABLE 3.1. Compressibility of Geologic Materials.

Material	Compressibility $\alpha(m^2 - N^{-1})$
Unconsolidated Sediments	
Clay	$10^{-6} - 10^{-8}$
Sand	$10^{-7} - 10^{-9}$
Gravel	$10^{-8} - 10^{-10}$
Rocks	
Shales	$10^{-9} - 10^{-10}$
Sandstones, Siltstones	$10^{-10} - 10^{-11}$
Limestone, Dolomite	$10^{-10} - 10^{-11}$
Igneous and Metamorphic	10^{-11}
Water (B)	4.4×10^{-10}

may undergo compression significantly greater than that observed in the laboratory over a period of a few hours. Compressibilities estimated from laboratory experiments should thus be regarded as *minimum* estimates.

In situ compressibilities that incorporate the influence of geologic time can be estimated from the change of porosity with depth, as the decrease of porosity with depth results in part from the compression of the rock due to increases in effective stress.

Assume an exponential porosity-depth relationship,

$$\phi(z) = \phi_0 e^{-z/b} \tag{3.53}$$

where ϕ (dimensionless) is porosity, z (m) is depth, ϕ_0 (dimensionless) is surface porosity, and b (m) is a constant. Rewriting the definition of porous medium compressibility (equation 3.48), in terms of derivatives,

$$\alpha = \frac{\dfrac{-dV_T}{V_T}}{d\sigma_e} \tag{3.54}$$

where V_T (m^3) is the volume of a porous medium. We will estimate α by the following procedure: (1) without loss of generality, let $V_T = 1 = (1 - \phi) + \phi$; (2) find dV_T/dz; (3) find $d\sigma_e/dz$; and (4) substitute into equation 3.54.

If we let the total volume $V_T = 1 = (1 - \phi) + \phi$, then

$$\frac{dV_T}{dz} = \frac{d(1-\phi)}{dz} + \frac{d\phi}{dz} \tag{3.55}$$

the first term on the right side of equation 3.55 is zero, because we assume the matrix grains to be incompressible. To evaluate the second term on the right side of equation 3.55, we take the derivative of equation 3.53,

$$\frac{d\phi}{dz} = -\frac{\phi_0}{b} e^{-z/b} \tag{3.56}$$

Substituting into equation 3.55,

$$\frac{dV_T}{dz} = -\frac{\phi_0}{b} e^{-z/b} \tag{3.57}$$

To find the change of effective stress with depth ($d\sigma_e/dz$), we first rearrange equation 3.38,

$$\sigma_e = (1 - \phi)gz\,(\rho_m - \rho) \tag{3.58}$$

$$\sigma_e = (z - \phi z)\,(\rho_m - \rho)g \tag{3.59}$$

$$\frac{d\sigma_e}{dz} = \left[1 - \phi - z\frac{d\phi}{dz}\right](\rho_m - \rho)g \tag{3.60}$$

But $\phi(z) = \phi_0 e^{-z/b}$ (equation 3.53), and $d\phi/dz = -(\phi_0/b)e^{-z/b}$. Substituting into equation 3.60,

$$\frac{d\sigma_e}{dz} = \left[1 - \phi_0 e^{-z/b} + \frac{z\phi_0}{b} e^{-z/b}\right](\rho_m - \rho)g \tag{3.61}$$

$$\frac{d\sigma_e}{dz} = \left\{1 + \phi_0 e^{-z/b}\left[\frac{z}{b} - 1\right]\right\}(\rho_m - \rho)g \tag{3.62}$$

If we multiply equation 3.57 by -1 and divided by equation 3.62, we obtain

$$\frac{\dfrac{-dV_T}{dz}}{\dfrac{d\sigma_e}{dz}} = \frac{-dV_T}{d\sigma_e} \tag{3.63}$$

Note that equation 3.63 now is equivalent to the definition of porous medium compressibility given in equation 3.54, because we have assumed $V_T = 1$ without any loss of generality. Substituting the right sides of equations 3.57 and 3.62 into equation 3.63 thus produces

$$\alpha = \frac{-dV_T}{d\sigma_e} = \frac{\frac{\phi_0}{b} e^{-z/b}}{\left\{1 + \phi_0 e^{-z/b}\left[\frac{z}{b} - 1\right]\right\}(\rho_m - \rho)g} \quad \textbf{(3.64)}$$

Recall that estimates of α derived from laboratory values range from 10^{-9} to 10^{-11} Pa^{-1}. To evaluate α as a function of depth (z) using equation 3.64, we need to choose "typical" numbers for b, ϕ_0, ρ_m, ρ, and g. From inspection of porosity-depth curves (e.g., Figures 3.1 and 3.2) we pick $\phi_0 = 0.25$, $b = 3000$ m. Using $\rho_m = 2650$ kg-m^{-3}, $\rho = 1000$ kg-m^{-3}, and $g = 9.8$ m-s^{-2}, we find that in the upper few kilometers, the derived compressibility values are as much as a factor of ten larger than those measured in the laboratory (Table 3.2, Figure 3.17). These estimates should be considered maximum estimates, just as the estimates that come from laboratory measurements should be considered minimum estimates. Although the above calculation incorporates the influence of geologic time, it may also be confused by porosity reduction from chemical processes. Our implicit assumption in calculating porous medium compressibility is that the decrease of porosity in the subsurface is solely mechanical in response to increases in effective stress. As Neuzil (1986, p. 1176) pointed out, however, porosity reduction may also occur through chemical means, such as mineral precipitation in pore spaces. Estimates of porous medium compressibility derived from porosity-depth relationships should thus be regarded as *maximum* estimates.

3.4.4. Specific Storage.

Having defined fluid and porous medium compressibility (B and α, respectively), we may now combine these two physical mechanisms to derive one physical parameter that totally describes the ability of a porous medium to store and release fluids (assuming that compression of matrix grains may be neglected).

TABLE 3.2 Compressibility Estimated from Porosity-Depth Data

Depth (km)	Porous Medium Compressibility (α) (Pa^{-1})
0	6.9×10^{-8}
1	4.2×10^{-8}
2	2.8×10^{-8}
3	1.9×10^{-8}
4	1.3×10^{-8}
5	9.4×10^{-9}
6	6.7×10^{-9}
7	4.8×10^{-9}
8	3.5×10^{-9}
9	2.5×10^{-9}
10	1.8×10^{-9}

Adding equations 3.27 and 3.52, the total fluid *released* from a porous medium of volume V_T due to a change in fluid pressure (ΔP) is

$$\Delta V_w = -\alpha V_T \Delta P - \phi B V_T \Delta P \quad \textbf{(3.65)}$$

$$\frac{\Delta V_w}{V_T}\frac{1}{-\Delta P} = \alpha + \phi B \quad \textbf{(3.66)}$$

Recall the relationship between head, elevation, and fluid pressure (equation 2.45),

$$h = z + \frac{P}{\rho g} \quad \textbf{(3.67)}$$

If the elevation of a porous medium is constant, then $\Delta z = 0$, and

$$\Delta P = \rho g \Delta h \quad \textbf{(3.68)}$$

Substituting equation 3.68 into equation 3.66,

$$\frac{\Delta V_w}{V_T}\frac{1}{-\Delta h} = \rho g(\alpha + \phi B) \quad \textbf{(3.69)}$$

The right side of equation 3.69, which includes both the fluid and porous medium compressibilities, defines a quantity known as the specific storage (S_s, m^{-1})

$$S_s = \rho g(\alpha + \phi B) \quad \textbf{(3.70)}$$

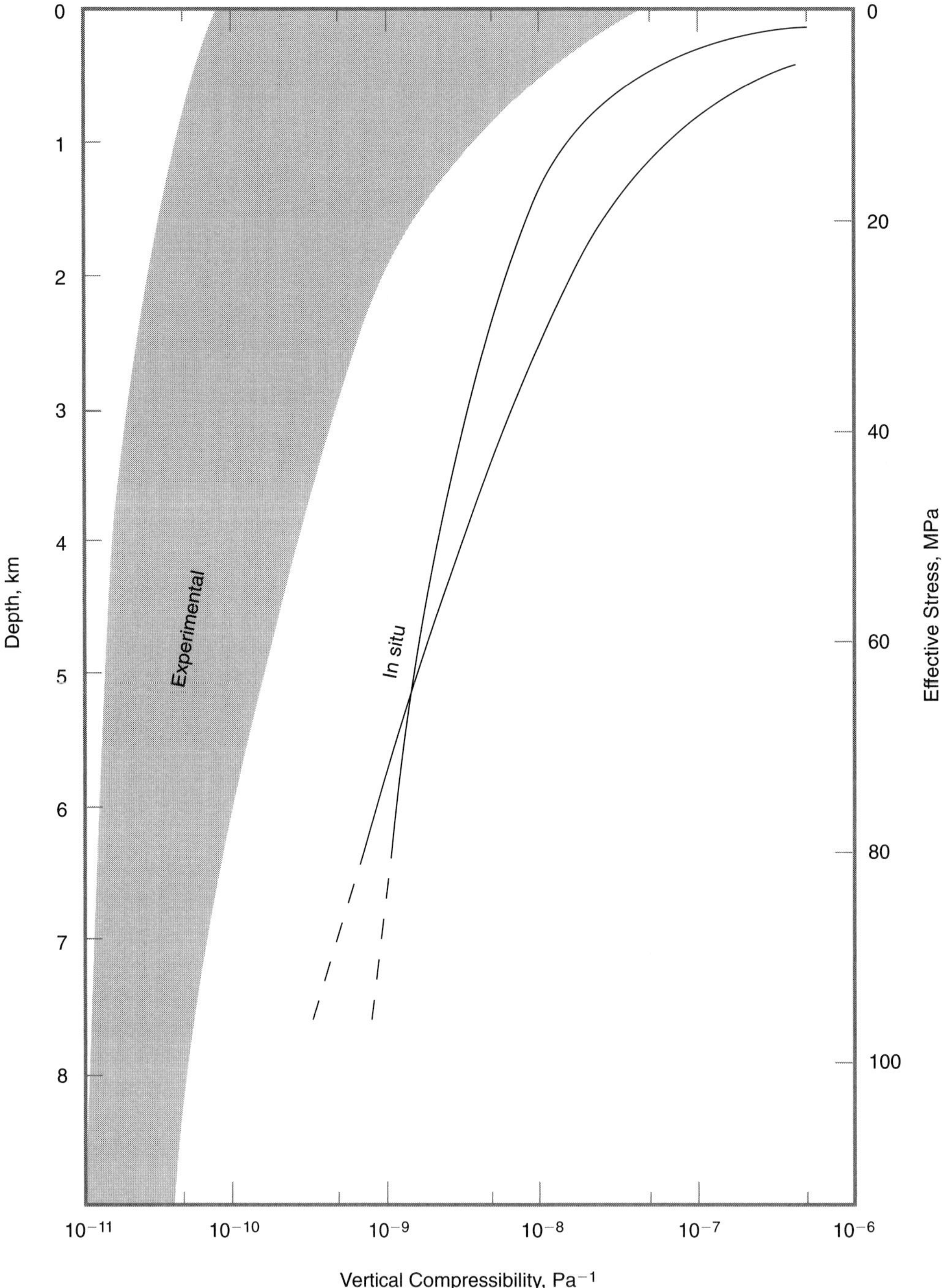

Figure 3.17 Plot compares values of vertical porous-medium compressibility obtained from laboratory experiments (left) with in situ values estimated from porosity-depth profiles (right).
(From Neuzil, 1986, p. 1176.)

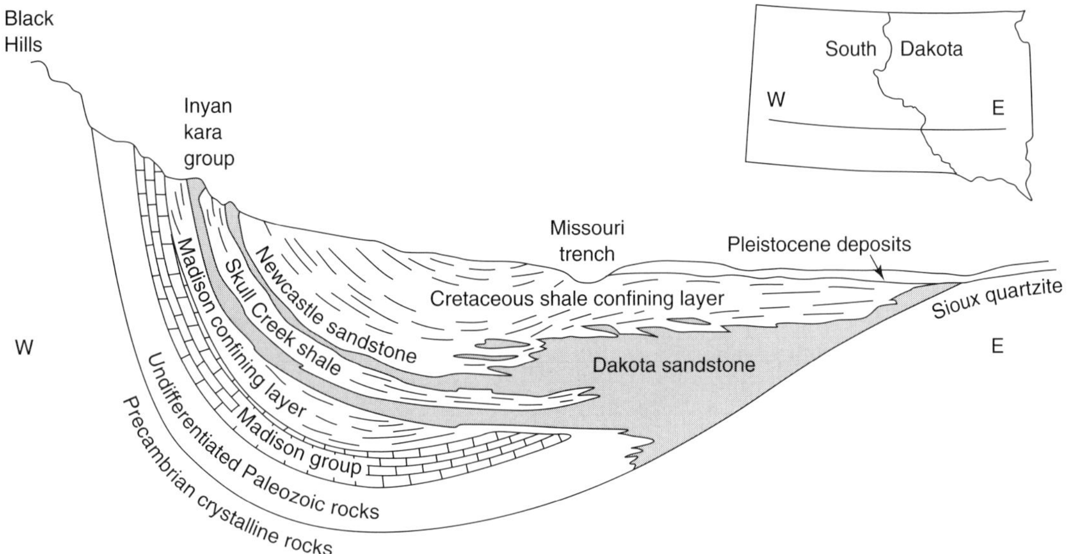

Figure 3.18 West-east geologic cross section through South Dakota showing the Dakota Sandstone, an important aquifer.
(From Bredehoeft et al., 1983, p. 5.)

The **specific storage** is the normalized volume of fluid ($\Delta V_w/V_T$) that is released from a porous medium in response to a change in head, it is a measure of an aquifer's ability to release or absorb groundwater.

Porous medium compressibility (α) is in the range of 10^{-8} to 10^{-11} Pa^{-1}, and water compressibility (B) is about 4.4×10^{-10} Pa^{-1}. If we assume a typical porosity of $\phi = 0.10$, then, at best, the α and ϕB terms that determine specific storage are equally important, while the α term may easily be three orders of magnitude greater than the ϕB term.

A minimum value of S_s for rocks can be estimated by setting $\alpha = 0$. If we assume a low porosity of $\phi = 0.05$, then $S_s = 2.2 \times 10^{-7}$ m^{-1}. A maximum value of S_s for rocks can be estimated by setting $\alpha = 10^{-8}$ Pa^{-1} (we may ignore the ϕB term), whereupon we find $S_s = 10^{-4}$ m^{-1}. Thus, for rocks (as opposed to unconsolidated sediments),

$$2 \times 10^{-7} \leq S_s \leq 10^{-4}\ \mathrm{m}^{-1} \qquad \textbf{(3.71)}$$

Thus if we know nothing about a rock, we nevertheless may be able to estimate its specific storage within three orders of magnitude. This is not terribly informative, but nevertheless the uncertainty is much less than for rock permeability, which may easily range over 12 orders of magnitude or more.

Specific storage should not be confused with specific yield. The **specific yield** of a porous medium is the volume of water that will drain from a saturated porous medium under the influence of gravity, divided by the total volume of the porous medium. The specific yield is always lower than the porosity because some water will be retained by capillary forces (see chapter 6).

3.5. THE DAKOTA SANDSTONE.

Prior to 1928, it was thought that all fluid storage in the subsurface took place as a result of fluid compression. In other words, it was thought that the α term in equation 3.70 was zero. Meinzer (1928) showed that this could not be the case by studying the Dakota aquifer in South Dakota (Figures 3.18 and 3.19).

The Dakota aquifer is a sandstone unit that outcrops in the Black Hills in western South Dakota

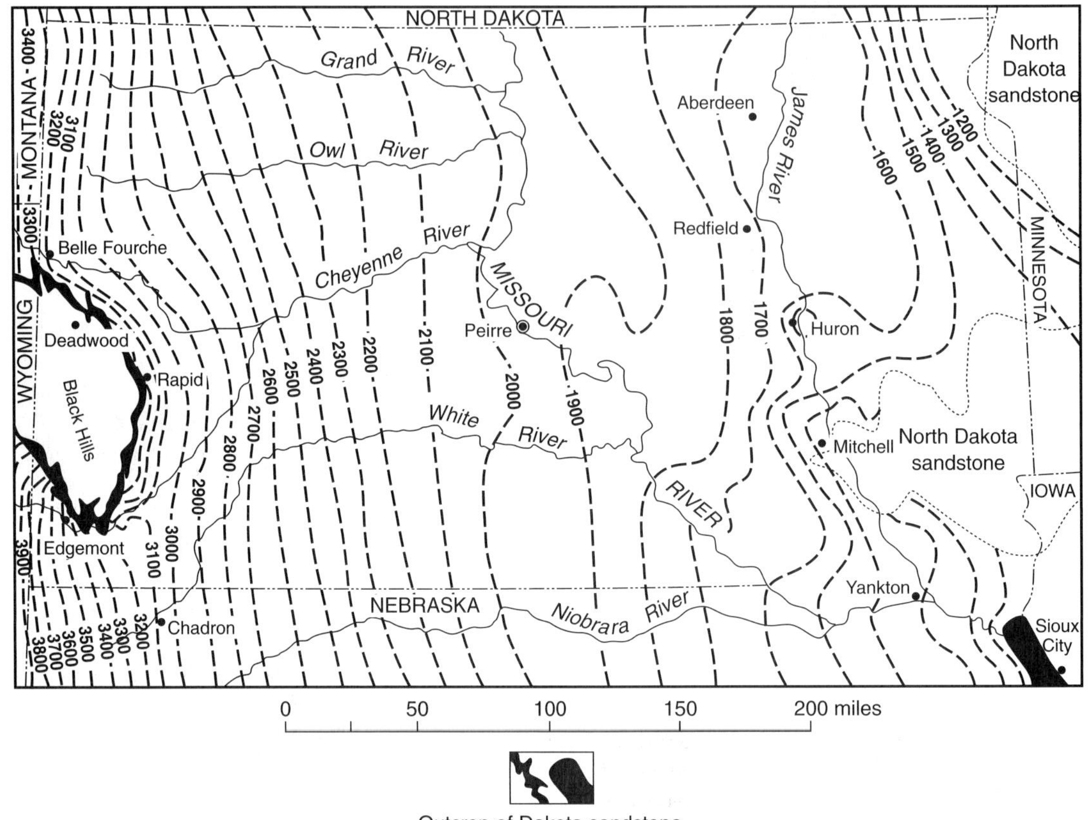

Figure 3.19 Head or potentiometric surface (in feet above sea level) of the Dakota Sandstone as mapped by Darton in 1909. Darton's map is believed to be a reasonable approximation of the potentiometric surface prior to development.
(From Darton, 1909.)

and dips eastward. Starting about the year 1880, the Dakota was heavily utilized as a groundwater resource (Figure 3.20). In 1910, the groundwater extraction rate from the Dakota peaked at about 250,000 gallons (946,250 liters) per minute. By 1920, the withdrawal rate had dropped to about 50,000 gallons (189,250 liters) per minute. The average withdrawal rate over 45 years was about 100,000 (378,500 liters) gallons per minute. The total volume of water withdrawn over 45 years was

$$\Delta V_w = 100{,}000 \text{ gals-min}^{-1} \times 3.785 \text{ liters-gal}^{-1} \times 10^{-3} \text{ m}^3\text{-liter}^{-1} \times 5.26 \times 10^5 \text{ min-yr}^{-1} \times 45 \text{ yr} = 9 \times 10^9 \text{m}^3 \quad (3.72)$$

Rearranging equation 3.69,

$$\Delta V_w = -\Delta h S_s V_T \quad (3.73)$$

In equation 3.73, the water volume (ΔV_w) is known from equation 3.72, and V_T, the total volume of the Dakota Sandstone, can be estimated as the product of the average formation thickness (100 m) and areal extent (750 km × 400 km) = 3×10^{13} m³. The head change, Δh, is known from observation to have been -100 m. Thus, the only unknown in equation 3.73 is the specific storage (S_s).

Solving equation 3.73 for S_s we find

$$S_s = \frac{\Delta V_w}{-\Delta h V_T} \quad (3.74)$$

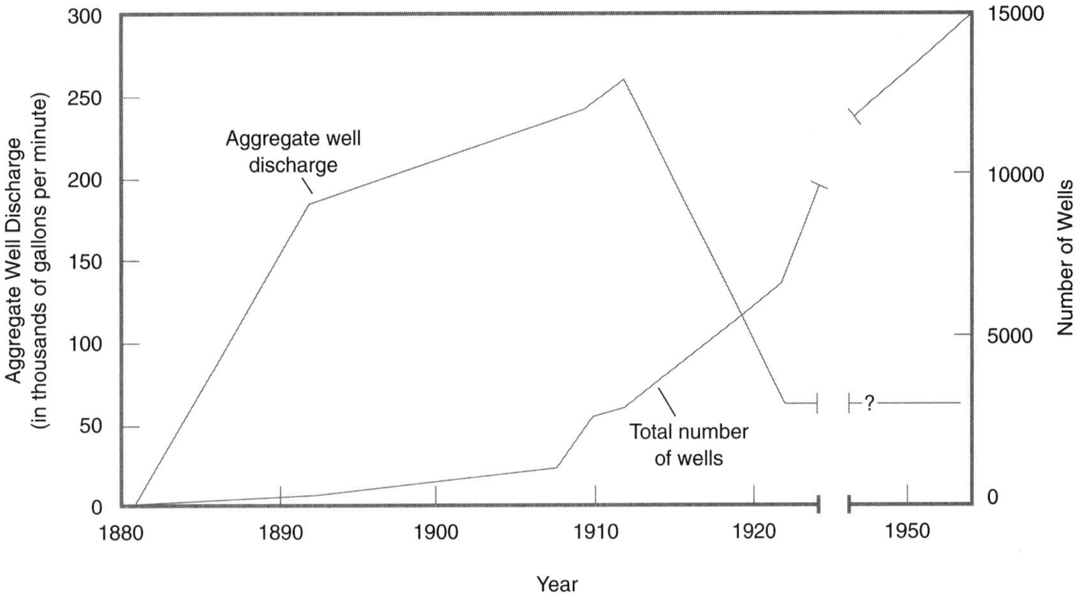

Figure 3.20 Estimated withdrawal rate of groundwater from the Dakota Sandstone in South Dakota. (From Bredehoeft et al., 1983, p. 21.)

$$S_s = \frac{9 \times 10^9 \text{ m}^3}{100 \times 3 \times 10^{13} \text{ m}^4} \quad \textbf{(3.75)}$$

$$S_s = 3 \times 10^{-6} \text{ m}^{-1} \quad \textbf{(3.76)}$$

Now compare this value with the specific storage that would result from fluid compression alone. Setting $\alpha = 0$ in equation 3.70,

$$S_s = \rho g \phi B \quad \textbf{(3.77)}$$

$$S_s = 1000 \text{ kg-m}^{-3} \times 9.8 \text{ m-s}^{-2} \times 0.10 \times 4.4 \times 10^{-10} \text{ kg-m}^{-1}\text{-s}^{-2} \quad \textbf{(3.78)}$$

$$S_s = 4 \times 10^{-7} \text{ m}^{-1} \quad \textbf{(3.79)}$$

where we have assumed a porosity $\phi = 0.10$.

The value calculated by assuming water compressibility as the only mechanism for fluid storage is too low by approximately an order of magnitude. Even though there are some uncertainties in the numbers we used for porosity, aquifer volume, and groundwater volume, the uncertainties are unlikely to be large enough to change our answer by a factor of ten. Thus the α term must exist. Meinzer's (1928) work is an elegant example of utilizing a large-scale natural experiment as an opportunity to study the properties of the Earth.

3.6. STORATIVITY AND TRANSMISSIVITY.

Consider a confined aquifer or tabular porous medium with thickness Δz (m). Its storativity (S, dimensionless) is defined to be the product of thickness and specific storage (S_s)

$$S = \Delta z S_s \quad \textbf{(3.80)}$$

Storativity is also defined as the volume of water released from (or taken into) storage per unit head drop (or increase) per unit surface area of a confined aquifer. Consider a confined aquifer of volume $\Delta x \Delta y \Delta z$ (m^3) and porosity ϕ (dimensionless). Let the aquifer thickness be Δz (m), and recall the definition of specific storage,

$$S_s = \frac{-\Delta V_w}{V_T} \frac{1}{\Delta h} \quad \textbf{(3.81)}$$

Oscar Edward Meinzer: Father of Groundwater Hydrology

O. E. Meinzer (1876–1948) was chief geologist of the Ground Water Division of the U. S. Geological Survey from 1912 through 1946, and is commonly referred to as the "father of modern groundwater hydrology" in the United States.

Meinzer was born on a farm near Davis, Illinois, one of six children. Sayre (1949) wrote that exposure to fossil-packed limestone outcrops and erratic granite boulders on the family farm stimulated an interest in geology. Whatever the case, Meinzer enrolled in Beloit College in Wisconsin and graduated magna cum laude in 1901. Meinzer was first appointed as a junior geologist at the U. S. Geological Survey (USGS) in 1907. In 1912, Meinzer was made acting Chief of the Ground Water Division of the USGS, several more senior geologists having quit or transferred due to a serious reduction in funding that had occurred earlier. Meinzer became the "right man in the right place at the right time" to determine the course and development of groundwater hydrology in the United States for the first half of the twentieth century (Hackett, 1964, p. 2).

During the time Meinzer came into charge of the USGS Ground Water Division, the United States was entering a phase of rapid growth. Many aquifers were being rapidly exhausted, and there was a need to map and evaluate the groundwater resources of the country. Hydrologists also needed to obtain a quantitative understanding of the physical principles that would lead to the most efficient utilization of groundwater resources. Maxey (1979) listed the following guidelines that Meinzer applied in his supervision of the Ground Water Division of the USGS: (1) Quantitative resource evaluation studies were needed throughout the country; (2) There was a need to systematically organize and collate groundwater knowledge; (3)

Oscar Edward Meinzer (1876–1948)

which can be rewritten as

$$S_s = \frac{-\Delta V_w}{\Delta x \Delta y \Delta z} \frac{1}{\Delta h} \tag{3.82}$$

where ΔV_w (m^3) is the fluid volume released due to the change in head Δh (m), and the aquifer volume V_T (m^3) is equal to the product of thickness (Δz, m) and area ($\Delta x \Delta y$, m^2). Thus

$$S = \Delta z S_s = \frac{-\Delta V_w}{\Delta x \Delta y} \frac{1}{\Delta h} \tag{3.83}$$

and the right side of equation 3.83 is the "volume of water released from (or taken into) storage per unit drop of head per unit surface area of aquifer" as defined in Figure 3.21.

Transmissivity (T, m^2-s^{-1}) is defined as the product of saturated aquifer thickness (Δz, m) and hydraulic conductivity (K, m-s^{-1})

Groundwater geology was an inherently multidisciplinary science that would benefit from the integrated efforts of geologists, physicists, chemists, and engineers; and (4) The USGS should emphasize public service, and data and information should be promptly and accurately released.

Meinzer understood that groundwater involved both geology and mathematical theory. For that reason, the group he assembled consisted of both geologists and engineers. Meinzer tended to look to particular people for theory, while in his mind others were best suited for fieldwork or other tasks. When C. V. Theis first developed his important theory, which predicted the change of head in a pumping well, Meinzer did not accept that the theory was correct. Meinzer had pegged Theis as a geologist, not a theoretician. Meinzer only came around when C. E. Jacob later showed that Theis' theory was correct. One may infer that Meinzer was authoritarian in his management of the Ground Water division from the statement of a latter-day scientist at the USGS, "*As I understood from the old timers, he was clearly the boss*."

Despite having formidable administrative duties at the USGS, Meinzer wrote an astonishing 110 technical papers during his career. Two of his most significant contributions were USGS *Water Supply Papers 489* and *494*. *Water Supply Paper 489* was titled "Occurrence of ground water in the United States, with a discussion of principles," and served as primary reference and handbook on groundwater geology for a number of years thereafter. Meinzer received his Ph.D in 1922 when this monograph was accepted by the University of Chicago as a dissertation. *Water Supply Paper 494,* "Outline of ground-water hydrology, with definitions," standardized much of the terminology used in the science. Most of Meinzer's definitions are still in use today.

Meinzer was described by Sayre (1949) as having "wide experience," "catholic tastes in reading," "phenomenal memory," "a keen, orderly mind," and an "unusual gift of concentration." During his lifetime, he received numerous awards and accolades, including the American Geophysical Union's Bowie Medal (1943), which is given "for outstanding contributions to fundamental geophysics and for unselfish cooperation in research." In 1965, the Geological Society of America established the O. E. Meinzer Award, which is given annually to the author of a paper that advances the science of hydrogeology.

FOR ADDITIONAL READING

Hackett, O. M. 1964. The father of modern ground water hydrology. *Ground Water,* 2: 2–5.

Maxey, G. B. 1979. The Meinzer era of U.S. hydrogeology, 1910–1940. *Journal of Hydrology,* 43: 1–6.

Meyer, G. 1995. Oscar E. Meinzer—Father of Modern Groundwater Hydrology in the United States. *Hydrogeology Journal,* 3: 76–78.

Sayre, A. N. 1949. Memorial to Oscar Edward Meinzer. *Proceedings Volume of the Geological Society of America, Annual Report for 1948,* p. 197–206.

$$T = \Delta z K \tag{3.84}$$

Transmissivity is also defined as "the volumetric flux or flow rate per unit head gradient per unit aquifer width." To reconcile this definition with equation 3.84, consider flow through an aquifer of saturated thickness Δz in the x-direction, where the area perpendicular to flow is $\Delta z \Delta y$ (Figure 3.22).

From Darcy's Law (equation 2.4),

$$K = \frac{q}{\nabla h} \tag{3.85}$$

where we have dropped the minus sign. Substituting into equation 3.84,

$$T = \frac{\Delta z q}{\nabla h} \tag{3.86}$$

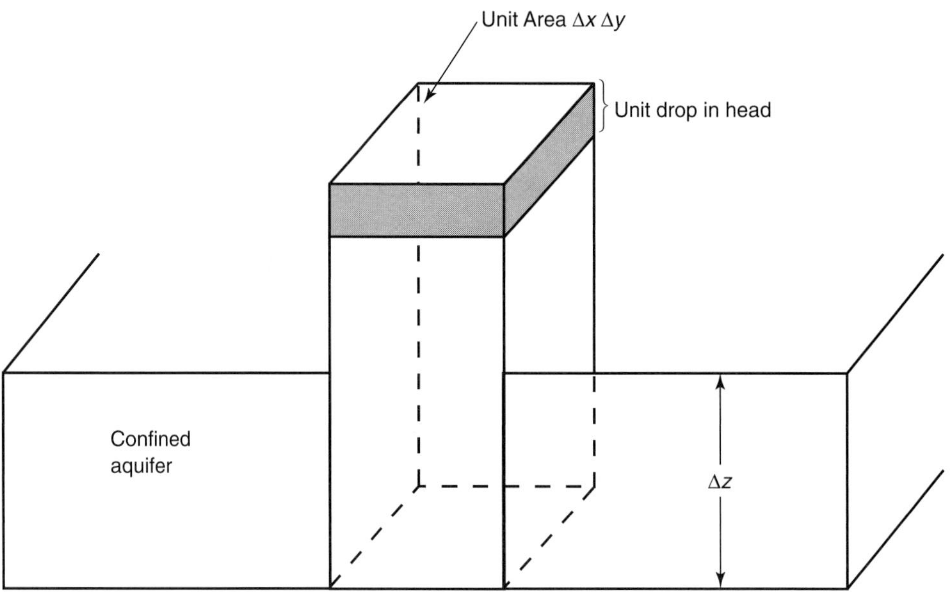

Figure 3.21 Storativity (S) is the volume of water released from (or taken into) storage per unit head drop (or increase) per unit surface area of aquifer.

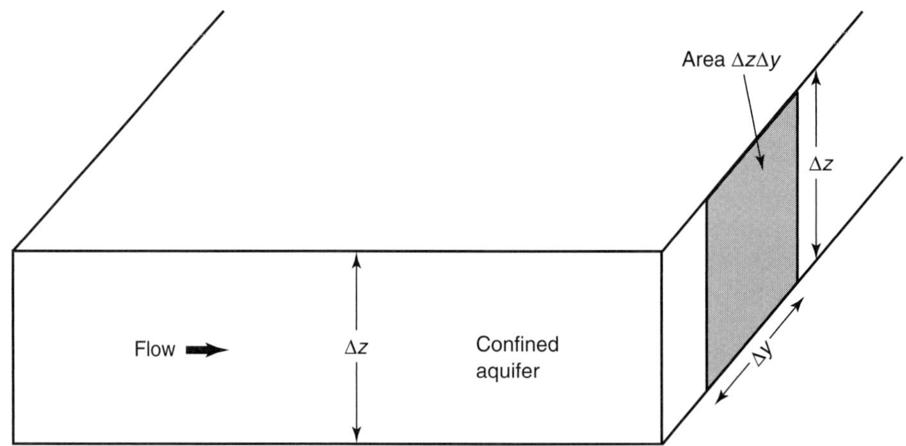

Figure 3.22 Transmissivity (T) is the volumetric flux or flow rate per unit head gradient per unit aquifer width.

The Darcy velocity (q, m-s^{-1}) is the volumetric flux (Q, m^3-s^{-1}) divided by the area perpendicular to flow (m^2). Thus

$$T = \frac{\Delta z Q}{\nabla h \Delta z \Delta y} \tag{3.87}$$

or

$$T = \frac{Q}{\nabla h \Delta y} \tag{3.88}$$

where Δy (m) is the unit aquifer width. Thus transmissivity is shown to be "volumetric flow per unit head gradient per unit aquifer width."

REVIEW QUESTIONS

1. Define the following terms in the context of hydrogeology:
   a. porosity
   b. Athy's Law
   c. aquifer
   d. confined aquifer
   e. unconfined aquifer
   f. aquitard
   g. aquiclude
   h. representative elementary volume (REV)
   i. fault gouge
   j. arithmetic average
   k. geometric average
   l. harmonic average
   m. permeameter
   n. slug test
   o. stress
   p. effective stress
   q. Archimedes' principle
   r. porous medium compressibility
   s. specific storage
   t. specific yield
   u. storativity
   v. transmissivity
2. What are two mechanisms for porosity reduction?
3. Describe how the porosity of (a) shales and (b) sandstones tends to change with depth. Invoke appropriate mathematical formulae and use a figure. Be sure that your figure is labeled with the correct numbers and units.
4. List permeability ranges for the following common rock types. Give both numerical values and physical units: gravel and sand, karst limestone, limestone (not karst), sandstone, shale, crystalline rocks (fractured), crystalline rocks (unfractured), salt and permafrost.
5. Explain the difference between confined and unconfined aquifers.
6. Explain the concept of "representative elementary volume." Use a figure. What is the relevance of this concept to environmental geology?
7. Discuss how the permeability of crystalline and argillaceous rocks changes with scale.
8. What is the relationship between permeability and fractures? Cite a quantitative formula.
9. Calculate the arithmetic, geometric, and harmonic mean of the following group of numbers: 1, 5, 50, 100.
10. Consider a layered sequence of rocks. Within each layer, permeability is constant and isotropic. The top layer is a 100-m-thick sandstone with a permeability of 10^{-14} m^2. The middle layer is a 200-m-thick limestone with a permeability of 10^{-15} m^2. The bottom layer is a 150-m-thick shale with a permeability of 10^{-17} m^2. What is the effective bulk permeability for all three layers for (a) flow parallel to the layering, and (b) flow perpendicular to the layering? What is the anisotropy of the layered group? *Note:* you need to take into account that the layers have different thicknesses.
11. Consider a layered sequence of three rocks. The layers have permeabilities k_1, k_2, and k_3; their thicknesses are ΔZ_1, ΔZ_2, and ΔZ_3. Show that (i.e., derive) the average permeability of the group of three beds is given by the harmonic average of the permeabilities for flow perpendicular to layers. Hint: proceed by applying Darcy's Law across the three layers and assume steady-state flow.
12. List three categories of permeability tests. You are hired to do a hydrogeologic characterization of a potential toxic waste dump. What type of permeability measurements are you going to order?
13. Why does the permeability of sedimentary rocks tend to decrease with increasing depth? Why does the permeability of

crystalline rocks tend to decrease with increasing depth?

14. What is the relationship between porosity and permeability? What other factors affect permeability besides porosity?

15. Using Figure 3.14, derive a relationship of the form of equation 3.23. If you know only the porosity of a shale, how accurately can you predict its permeability? What is the maximum likely error?

16. What percentage of sedimentary rocks are composed of shale?

17. How does the permeability of sedimentary rocks change with depth? Cite a quantitative relationship.

18. Define (a) isothermal compressibility of water and give a value for groundwater (with correct units) at 25°C; (b) aquifer compressibility and give a range of values (with correct units) for different lithologies; (c) specific storage and give a range of values (with correct units) for different lithologies.

19. What are three physical mechanisms by which rocks can store water? What is the relative importance of each of these mechanisms, and how is it affected by lithology?

20. Write down an expression that relates effective stress σ_e to total stress σ and fluid pressure P. What is the source of total stress, and is it likely to change? How are changes in effective stress related to changes in hydraulic head?

21. Show that the buoyancy force on an object suspended in a fluid is equal to the weight of displaced fluid (i.e., derive the expression $F_b = \rho g V$, where F_b is the buoyancy force, ρ is the density of the displaced fluid and V is the object volume). Use a figure. Show your work and state all assumptions.

Suggested Reading

Bredehoeft, J. D., Neuzil, C. E., and Milly, P. C. D., 1983. Regional Flow in the Dakota Aquifer: a study of the role of confining layers. *U.S. Geological Survey Water-Supply Paper 2237.*

Neuzil, C. E. 1994. How permeable are clays and shales? *Water Resources Research* 30: 145–150.

Notation Used in Chapter Three

Symbol	Quantity Represented	Physical Units
α	compressibility of a porous medium	$Pa^{-1} = m\text{-}s^2\text{-}kg^{-1}$
A, A_s, A_f	area	m^2
a_1, a_2	constants	dimensionless
B	fluid compressibility	$Pa^{-1} = m\text{-}s^2\text{-}kg^{-1}$
b	constant in porosity-depth equation	m^{-1}
c	constant	$m^{-2}\text{-}s^{-1}$
e	base of the natural logarithms	dimensionless
ϕ, ϕ_0	porosity, porosity at zero depth	dimensionless
F_{grav}	force due to gravitational acceleration	Newton (N) = $kg\text{-}m\text{-}s^{-2}$
$F_{\text{top}}, F_{\text{bottom}}$	force	Newton (N) = $kg\text{-}m\text{-}s^{-2}$
g	acceleration due to gravity	$m\text{-}s^{-2}$
$h, \Delta h$	head, change in head	m
K	hydraulic conductivity	$m\text{-}s^{-1}$
k	permeability	Darcy = $10^{-12}\ m^2$
m	constant	dimensionless
N_f	number of fractures per unit distance	m^{-1}
$P, \Delta P$	fluid pressure, change in fluid pressure	Pascal (Pa) = $kg\text{-}m^{-1}\text{-}s^{-2}$
Q	discharge	$m^3\text{-}s^{-1}$
q	Darcy velocity or specific discharge	$m\text{-}s^{-1}$
ρ	fluid density	$kg\text{-}m^{-3}$
ρ_{cube}	density of a cube suspended in a fluid	$kg\text{-}m^{-3}$
ρ_m	matrix density	$kg\text{-}m^{-3}$
S	storativity	dimensionless
σ	total stress	Pascal (Pa) = $kg\text{-}m^{-1}\text{-}s^{-2}$
$\sigma_e, \Delta\sigma_e$	effective stress, change in effective stress	Pascal (Pa) = $kg\text{-}m^{-1}\text{-}s^{-2}$
S_s	specific storage	m^{-1}
T	transmissivity	$m^2\text{-}s^{-1}$
t	time	s
$V, V_w, \Delta V_w$	volume, fluid volume, change in fluid volume	m^3
$V_T, \Delta V_T$	total (matrix plus fluid) volume of a porous medium, change in total volume	m^3
w	fracture width	m
w_i	ith weighting factor used in averaging	dimensionless
x, y	directions parallel to bedding	m
x_i	ith component of quantity being averaged	depends on quantity being averaged
Ψ	anisotropy	dimensionless
$z, \Delta z$	depth, thickness	m
$\overline{x_a}$	arithmetic mean	depends on quantity being averaged
$\overline{x_g}$	geometric mean	depends on quantity being averaged
$\overline{x_h}$	harmonic mean	depends on quantity being averaged

CHAPTER 4

PROPERTIES OF GEOLOGIC FLUIDS

4.1. OCCURRENCE AND COMPOSITION.

Geologic fluids are present in virtually all crustal rocks (Bodnar and Costain, 1991). About 20% of the volume of most sedimentary basins consists of geologic fluids. The composition of geologic fluids has the potential to provide important information on the geochemical, hydrological, thermal, and tectonic evolution of the Earth's crust (Hanor, 1994a, p. 151).

Most geologic fluids are aqueous brines, a mixture of water and NaCl. In water, NaCl dissociates into positively charged Na^+ ions and negatively charged Cl^- ions. An **ion** is a molecule with a net electrical charge. Positively charged ions are termed **cations** and negatively charged ions are **anions**. Besides Na, other important cations found in subsurface brines include Mg, K, Ca, and Sr.

Not all geologic fluids are water based. Examples of economically important geologic fluids that are not aqueous are oil and gas. Magma or liquid rock is another geologic fluid, however, 99% or more of all geologic fluids found in the Earth's crust are water-based, and our discussion is thus focused almost entirely on groundwater.

4.2. TERMINOLOGY AND UNITS.

The salinity of aqueous geologic fluids ranges from *fresh* to *brine* (Table 4.1). In this text, the term **salinity** is used to denote the total concentration of dissolved solids (**TDS**), not just the concentration of NaCl. The salinity of geologic fluids is controlled by solubility and availability of **solute**. The symbol ‰ denotes *parts per thousand,* or total mass of solute per 1000 mass units of **solution.** Seawater is about 35‰ TDS.

Example: If 100 g of salt is mixed with 900 g of water we have 1000 g of solution with a total dissolved solids (TDS) of 100‰. The solution is 10% salt by weight, or has a **weight percent** of 10.

TABLE 4.1 Salinity Terminology*

Description	TDS (‰)
Fresh	< 1
Brackish	1–10
Saline	10–100
Brine	> 100 (up to ~400)

*Seawater is 35 ‰

TABLE 4.2 Comparison of Salinity Units

Weight Percent Solute	‰	NaCl Molality
1	10	0.17
5	50	0.90
10	100	1.9
20	200	4.3
30	300	7.3
40	400	11.4

The **molality** (m) of a solution is the moles of solute per kilogram of *solvent*. Molality should not be confused with molarity (M). **Molarity** is the moles of solute per liter of *solution*. A **mole** is the amount of a substance that contains 6.02×10^{23} molecules (**Avogadro's number**). The **molecular weight** of a substance is a number equal to what a mole of a particular substance would weigh in grams. The molecular weight of any substance is the sum of the **atomic weights** of a molecule's constituent atoms and may thus be found from the **periodic table**. A comparison of salinity units is given in Table 4.2.

Example: From the periodic table we find that the atomic weight of Na is 22.99; the atomic weight of Cl is 35.45. Thus, the atomic weight of NaCl is 58.44, meaning that 58.44 g of NaCl constitutes one mole of NaCl. Mixing 58.44 g of NaCl with enough water to produce a liter of solution results in a solution with a molarity of one. If we mix 58.44 g of NaCl with 941.56 g of water, we would have a solution with a molality of 1/0.94156 = 1.06. This salt solution would be 58.44‰ NaCl or have a weight percent of 5.844.

Problem:

a. Find a general expression for conversion of weight percent solute to molality. (b) Apply the expression derived in part (a) to find the molality of a NaCl solution that is 10% solute by weight.

Molality is defined as moles of solute per kilogram of solvent.

1. Find the moles of solute in one kilogram of *solution.*

   The number of moles of solute can be found by dividing the mass of the solute (in grams) by the molecular weight (which is moles per gram). In a kilogram of solution, the mass of solute (in grams) is 1000 (g) × (W/100) = 10W, where W is the weight percent solute. The number of moles of solute is then the total mass of solute in grams divided by its molecular weight (MW). Thus, the moles of solute per kilogram of solution is 10W/MW.

2. Find the moles of solute in one kilogram of *solvent.*

   The mass of the *solvent* (in grams) is the mass of the solution minus the mass of the solute = 1000 − 10W. Thus, for every 10W/MW moles of solute we have (1000 − 10W)/1000 kilograms of solvent. The molality then is

$$\text{molality} = \frac{\dfrac{10W}{MW}}{\dfrac{(1000 - 10W)}{1000}} \tag{4.1}$$

$$\text{molality} = \frac{1000W}{MW(100 - W)} \tag{4.2}$$

where W is the weight percent solute in a solution and MW is the molecular weight of the solute.

b. The atomic weight of Na is 22.99, the atomic weight of Cl is 35.45. The molecular weight of NaCl is thus 22.99 + 35.45 = 58.44. If a solution is 10% NaCl by weight (W = 10), then substituting into equation (4.2) we find

$$\text{molality} = \frac{(1000)(10)}{(58.44)(100 - 10)} \tag{4.3}$$

$$\text{molality} = 1.90 \tag{4.4}$$

4.3. Composition and Origin of Solutes.

The discussion that follows is largely focused on solutes in fluids from sedimentary basins as opposed to basement or crystalline rocks. Although the bulk of the continental and oceanic crusts are composed of crystalline rock, our knowledge of the fluids in these rocks is limited. Most samples of crustal fluids have been obtained as a byproduct of petroleum exploration. Nearly all oil and gas are found in sedimentary basins, and the total number of boreholes drilled (and fluid samples collected) in such settings far exceeds similar studies in crystalline or basement rocks. Some notable exceptions are a study of fluids from basement rocks in the Canadian Shield (Frape and Fritz, 1987), and analysis of the brines found in the 9-km-deep borehole drilled by the German Continental Deep Drilling Project (Lodemann et al., 1997).

The salinity of aqueous fluids in sedimentary basins ranges from fresh, close to the surface, to over 400‰ in evaporite-rich basins such as the Michigan Basin (Figure 4.1). An **evaporite** is a sediment or a rock deposited by evaporation of a solution such as seawater. The highest known salinity (643 g-liter^{-1}) is for a $CaCl_2$-brine from the Saline Formation in the Michigan Basin (Case, 1945; Hanor, 1994a, p. 154). Although some exceptions are known, in general salinity tends to increase with depth (Hanor, 1979). Dickey (1969) noted that the increase of total salinity with depth in many oil fields tends to be linear (Figure 4.2).

Chloride makes up over 95% by mass of the anions in most sedimentary basin fluids with salinities greater than 10‰. Brackish and freshwater fluids more commonly have bicarbonate, sulfate, or acetate as their dominant anion (Hanor, 1994a, p. 155). There is a **cationic shift** in high-salinity waters. Sodium is the dominant cation by mass in fluids of moderate to low salinities (Figure 4.3), however, as salinity increases, the relative proportion of Na decreases while the proportions of K, Mg, and Ca increase. The most notable increase is in Ca, which typically becomes the dominant cation by mass in fluids whose salinities exceed 300‰.

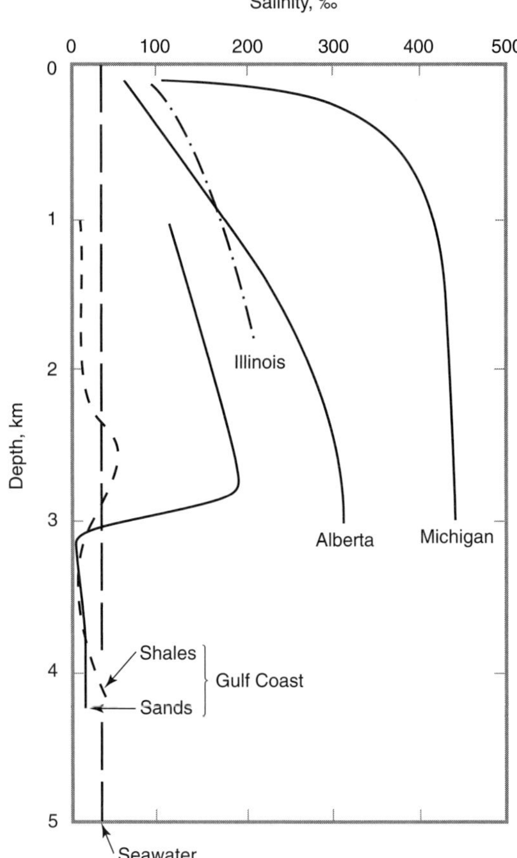

Figure 4.1 Maximum salinities versus depth for the Illinois, Alberta, Michigan, and Gulf Coast sedimentary basins. Note reversal of usual trend of increasing salinity with increasing depth for sands from the Gulf Coast basin.
(From Hanor, 1987, p. 82.)

At the present time, we do not have a definitive understanding of what mechanisms control the composition of solutes in the aqueous pore-fluids found in sedimentary basins, nor do we necessarily understand what geologic factors may affect the relative importance of different processes under different circumstances. The following mechanisms have all been proposed to account for the origin and composition of solutes: (1) subaerial evaporation of seawater or continental waters, (2) subsurface dissolution of evaporites, (3) membrane filtration, and (4) fluid-matrix interactions.

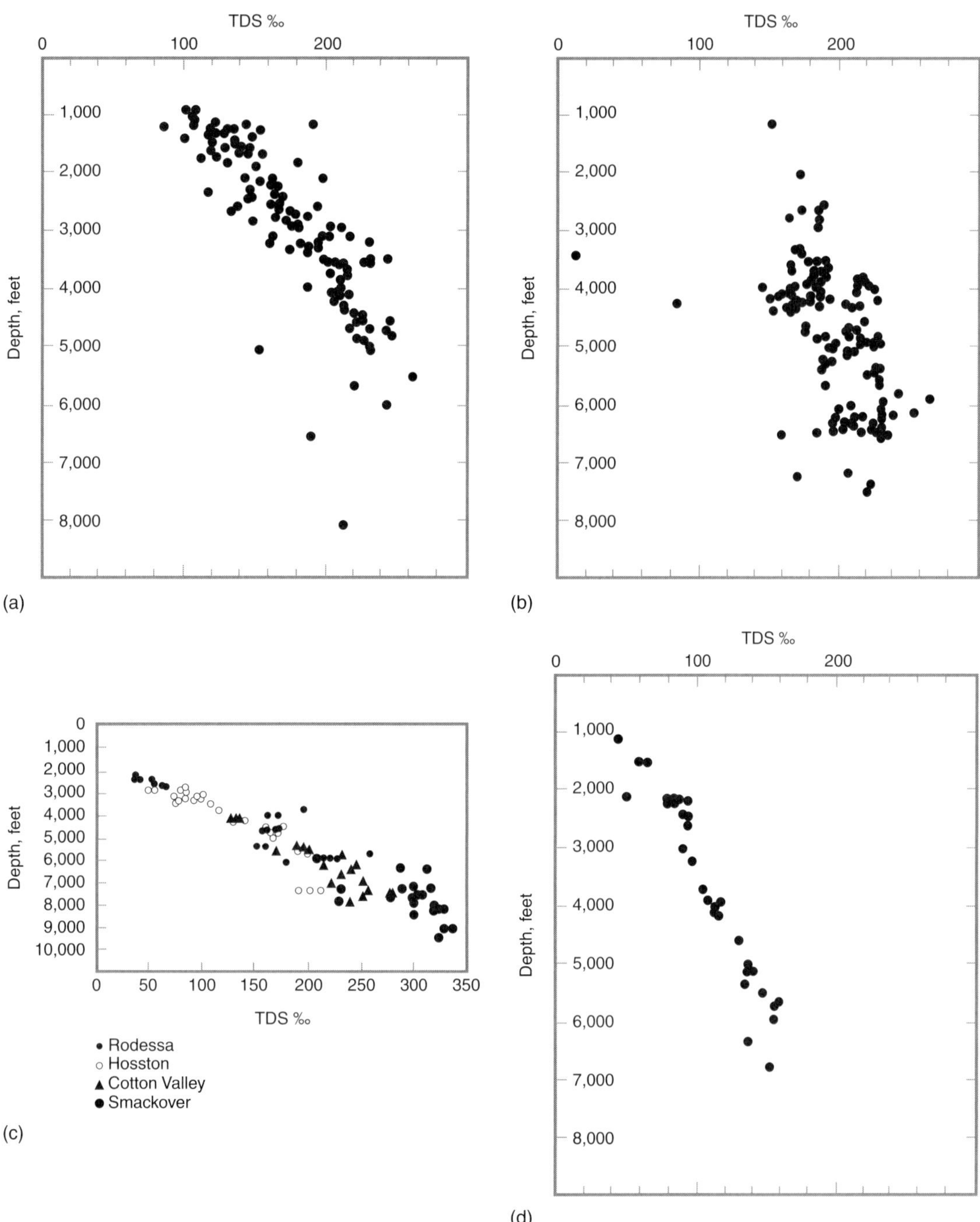

Figure 4.2 Solute concentration versus depth for oil field brines from (a) oil producing sands of the Pennsylvanian Cherokee Group in Oklahoma, (b) Ordovician Wilcox and Simpson sands of Oklahoma, (c) randomly selected waters from different horizons ranging from Jurassic to Cretaceous in southern Arkansas and northern Louisiana, (d) Eocene Wilcox sands of central Louisiana.

(From Dickey, 1969, p. 363–365.)

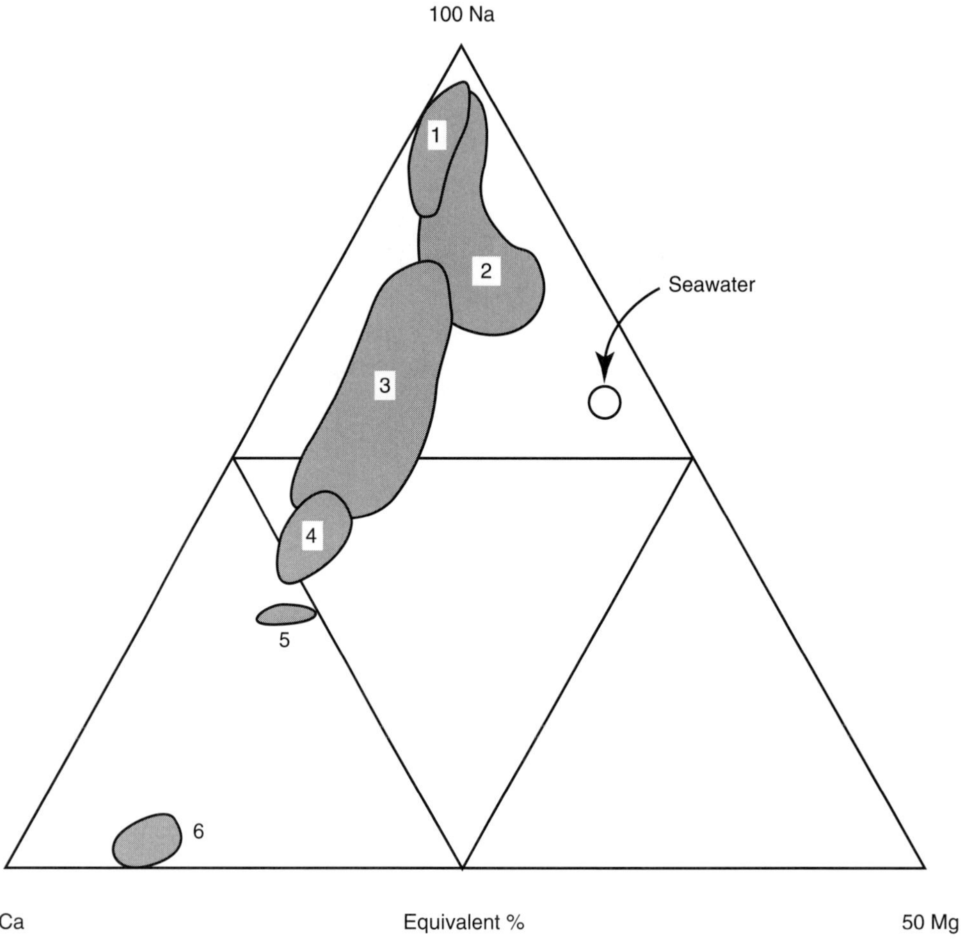

Figure 4.3 Variations in proportions of dissolved Na, Ca, and Mg in brines from (1) Texas, (2) California, (3) Kansas and Oklahoma, (4, 5) Appalachia, and (6) Arkansas
(From Hanor, 1987, p. 82; after DeSitter, 1947.)

Evaporation of seawater by itself cannot account for the major anion and cation composition of subsurface brines. One reason is that subaerially evaporated seawater has sulphate as a major solute over all stages of evaporation, yet sulphate is a minor constituent of most subsurface brines. Hanor (1994a, p. 156) concluded that "While it is probable that some subsurface brines have had evaporated marine waters as their ultimate precursors, it is obvious that other processes have been at work to account for their major solute composition." Some brines derived from continental waters (e.g., lakes, rivers, etc.) may resemble subsurface brines in overall chemical composition and salinity, but there can be significant differences, especially in anionic composition and the relative abundance of dissolved Mg.

The subsurface dissolution of evaporites has been documented to contribute to the solute composition of some aqueous brines in sedimentary basins. In the Gulf Coast Basin of the southeast US (Figure 4.4), the spatial variations in pore water salinity around some salt domes provide clear evidence for the dissolution of salt diapirs as the source

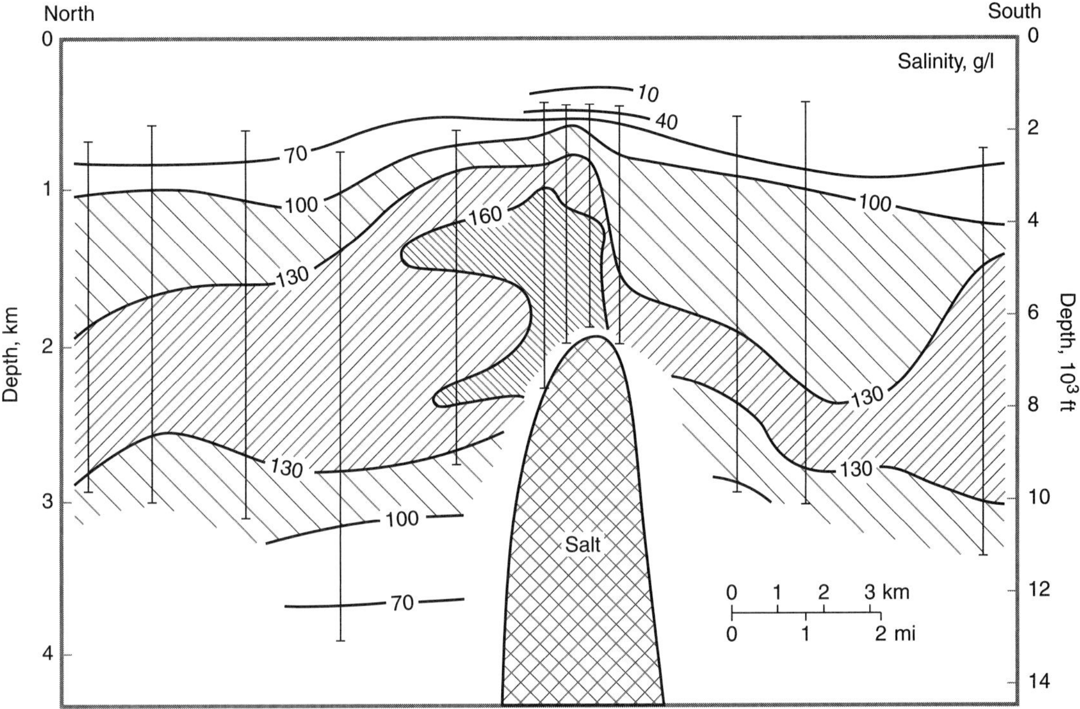

Figure 4.4 Variation in pore water salinity in the vicinity of the Welsh salt dome, Louisiana Gulf Coast. (From Hanor, 1994b, p. 41; after Hanor 1988.)

of the brines (Bennett and Hanor, 1987; Hanor, 1994a). Field mapping of salinity variations in the Gulf Coast Basin suggests that salts dissolved from halite domes have moved many kilometers vertically and tens of kilometers laterally throughout the surrounding sedimentary sequence (Bray and Hanor, 1990; Hanor and Sassen, 1990).

Membrane filtration has been proposed as a mechanism that could explain the existence of brines in evaporite-poor sedimentary basins. **Membrane filtration** refers to the filtering out of cations and anions by electrically polar clay molecules. Neutral water molecules pass through clay layers without difficulty, but charged anions and cations are electrostatically repulsed by the uneven distribution of charges on the surface of clay minerals. In theory, pore fluids on the influent side of a shale membrane will become progressively more saline due to membrane filtration.

Although membrane filtration has been demonstrated in the laboratory, there is *no known example* of subsurface brines having been produced by this mechanism. Hanor (1994a) pointed out that the Gulf Coast Basin of the southeast US is exemplary of the type of setting where membrane filtration should be producing brines. Upward flow through overpressured shale sections into overlying hydrostatically pressured sand layers should have produced higher salinities below the clay-rich shales. But, the opposite is observed: salinity actually decreases below the top of the shale sequence. Hanor (1994a, p. 159) speculated that the upward flow of fluids from the Gulf Coast overpressured zone takes place by way of fractures, bypassing the filtration that would be expected for flow through the porosity of the clay-rich shales.

Although both subaerial evaporation of marine and continental waters and subsurface dissolution of evaporites have the potential for producing brines with the range of salinities and dissolved chlorine concentrations of most subsurface brines

Nelson Horatio Darton: Master of Field Geology

Nelson Horatio Darton (1865–1948) was born on December 17, 1865, in Brooklyn, New York. At the age of 14, Darton quit public school to become an apprentice in his uncle's pharmacy. The young Darton showed an enormous aptitude for chemistry, supplementing the practical experience obtained in his uncle's pharmacy with supplemental readings. By the age of 15, Darton had acquired a reputation as a practical chemist and had developed his own consulting business. During his teenage years, Darton contributed several scientific papers to the chemical literature and was asked by the editor of *Scientific American* to take charge of the chemical and mineralogical part of the "notes and queries" column. Just before his sixteenth birthday, Darton was elected a member of the American Chemical Society.

Darton became interested in geology, and in 1882 (at the age of 17) published three papers on the occurrence of minerals in the vicinity of New York City. Darton spent more and more time on geologic collecting trips and field studies, and eventually his work caught the attention of G. K. Gilbert of the U.S. Geological Survey. Gilbert offered Darton a position with the USGS that he started in 1886 and kept for 50 years. At the USGS, Darton's exceptional aptitude for field geology was recognized, and he was assigned a series of geologic mapping duties. During his investigation of the geology of the Atlantic Coastal Plain, Darton became interested in the geology of artesian wells and underground water. In 1896, Darton published USGS Bulletin 138, *Artesian Well Prospects in the Atlantic Coastal Plain Region*, and became recognized as the Survey's leading expert on artesian wells.

Nelson Horatio Darton (1865–1948)

Darton's accomplishments, especially in the field of geologic mapping, seem almost superhuman in retrospect. In the field seasons of 1893 and 1894, Darton mapped 9,350 square miles in New York State at the rate of 100 to 150 square miles per day, much of this being covered by horse and buggy, or on foot. In 1905, Darton published what he considered to be his finest work, USGS Professional Paper 32, *Preliminary Report on the Geology and Underground Water Resources of the Central Great Plains*. This report consisted of 433 pages and contained 72 plates as well as many photographs and geologic maps. Although Darton did have the assistance of a team of coworkers, it is nevertheless remarkable that this monograph was one of eleven publications he authored that year. In the citation for the Geological Society of America Penrose Medal that Darton received in 1940, Douglas Johnson (Johnson, 1940, p. 82)

in sedimentary basins, these processes *cannot* explain their major cation composition. Thermodynamic buffering by silicate-carbonate mineral assemblages (i.e., matrix-fluid chemical reactions) has a first-order control on subsurface fluid compositions. The chemical potential of chloride, or the aqueous concentration of anionic charge, is a master variable in driving fluid-matrix interchanges and controlling fluid composition. Chloride composition, in turn, is controlled largely by testified not only to the enormous quantity of Darton's work, but also to its quality:

> I testify to our astonishment that one man could do so much and do it so uniformly well. Tonight we pay tribute to the fact that in the wealth of Darton's productive work there is so much that has withstood the acid test of time, so little of major importance that has required correction or revision.

Darton reputedly took a sturdy pride in his work. The story is told (King, 1949, p. 161) that one day Darton came upon a fellow geologist who was examining an outcrop. The other man said, half-humorously, "*I have made my guess as to what this formation this is, what is your guess*?" Darton replied, "*I never guess, I find the facts and I know*." Co-workers at the USGS would later relate that they considered Darton to have a big ego, without much of a sense of humor. A story illustrates Darton's view of himself. In the early days, the method of operation of the USGS was for many people to spend a season in the field and then return to Washington, DC in the winter. The Geologic Society of Washington (DC) was a prestigious group with many of the USGS scientists as members; as of the year 2000, the GSW still meet at the Cosmos Club in Washington. At one meeting, the Chairman was commenting on members in the audience, and said "the father of groundwater is here tonight" (or some such words) presumably intending to introduce Meinzer, at which point Darton stood up. In the opinion of some, Darton was entirely justified in his presumption.

Although Darton's interests and contributions to the science of geology were spread across many subdisciplines, he is especially remembered for his hydrogeological studies of the Dakota Aquifer in South Dakota. In 1900, Darton and his assistants began a study of the geology and underground waters of the central Great Plains. They collected data from thousands of wells, documenting the geologic controls on groundwater occurrence and flow. It was Darton who emphasized that "*underground water is a geologic problem that requires a study of the formations carrying it*" (Darton, 1941, p. 86).

During his career Darton received a long list of honors and awards. An abbreviated list includes the Daly Gold Medal of the American Geographical Society (1930), the Penrose Medal of the Geological Society of America (1940), and the Legion of Honor of the American Institute of Mining and Metallurgical Engineers (1944).

At age 70, Darton characterized his imminent forced retirement as "facing a terrible period of inactivity." He was granted a year's extension by US President Franklin D. Roosevelt, and arrangements were made for him to continue to use USGS facilities for another 12 years as an emeritus scientist. On February 28, 1948, Darton quietly passed in his sleep. During his lifetime, he published more than 220 contributions to the geologic sciences. Darton was one of the giant figures of the heroic age of American geology; he was the acknowledged master of field technique and the greatest reconnaissance geologist of all time.

FOR ADDITIONAL READING

Johnson, D., and Darton, N. H. 1941. Presentation of the Penrose Medal to Nelson Horatio Darton, Medalist, Address by Douglas Johnson and Response by Doctor Darton. *Proceedings of the Geological Society of America for 1940,* 81–88.

King, P. B. 1949. Memorial to Nelson Horatio Darton. *Proceedings Volume of the Geological Society of America,* April: 145–170.

Monroe, W. H. 1949. Memorial, Nelson Horatio Darton (1865–1948). *AAPG Bulletin,* 33: 116–124.

physical processes of fluid flow. The extensive fluid-matrix interactions that control the cation composition of subsurface brines in sedimentary basins may make it difficult to infer the ultimate origin of pore fluids from their present-day composition alone (Hanor, 1994a, p. 171–72). Studies of subsurface brines from basement rocks in the Canadian Shield led Frape and Fritz (1987) to a similar conclusion: solute composition is controlled by fluid-matrix interactions.

4.4. Isotopes.

Isotopic analysis is a valuable tool in the study of how fluids originated and evolved. An **isotope** is a variation of an **element** that has the same number of **protons** in the **nucleus**, but a different number of **neutrons**. The higher the atomic weight of an isotope, the heavier it is. For example, the heavy oxygen isotope is O^{18}, the light oxygen isotope is O^{16}. In this notation, O denotes oxygen, and the superscripted number is the total number of protons and neutrons in the oxygen atom. Thus O^{18} has two more neutrons in each atom compared to O^{16}.

Isotopes have the same gross chemical properties but different weights. There are subtle differences in the behavior of different isotopes of the same element, and these differences may be exploited to learn about natural systems.

Isotopic ratios are measured using **a mass spectrometer.** A mass spectrometer is "an instrument designed to separate charged atoms and molecules on the basis of their masses based on their motions in electrical and/or magnetic fields" (Faure, 1977, p. 65). In a mass spectrometer, a sample is first ionized by subjecting it to a stream of high-speed electrons. The collision of these electrons with a sample molecule strips that molecule of an electron, and the sample molecule acquires a positive charge. The positively charged ion is then accelerated through an electric field while simultaneously being subjected to a magnetic field. The magnitude of the deflection of the ion in the magnetic field gives a measure of the ratio of its charge to its mass. Thus isotopes with identical charges but slightly different masses can be separated and their relative proportions in a sample measured.

Isotopes are either radioactive or stable. Radioactive isotopes decay through nuclear reactions; their primary hydrogeologic use is in dating groundwater. Stable isotopes are not subject to radioactive decay and are largely used to understand the source and evolution of groundwater.

4.4.1. Stable Isotopes.

Stable isotopes are those that are not radioactive. The most common stable isotopes studied in connection with geologic fluids are the isotopes of oxygen and hydrogen. Oxygen has three naturally-occurring isotopes. These are O^{18} (0.2%), O^{17} (0.04%), and O^{16} (99.76%). Hydrogen has two stable isotopes, H^1 (99.99%) and H^2 (0.01%). H^2 is also known as deuterium and is denoted by the capital letter D.

The process that makes it possible to use stable isotopes in groundwater studies is fractionation. **Fractionation** is the separation of a mixture into parts that possess different properties. The oxygen and hydrogen isotopes found in water can undergo fractionation due to chemical and biological changes, or physical changes such as evaporation, condensation, freezing, and melting. Studies of the resulting isotopic ratios can then be used to infer the nature of the process that caused the fractionation or the source of the original water that underwent fractionation.

Fractionation is expressed as a ratio relative to a standard composition,

$$\delta_{heavy} = \frac{(\text{heavy/light})_{sample} - (\text{heavy/light})_{standard}}{(\text{heavy/light})_{standard}} \times 1000‰ \quad \textbf{(4.5)}$$

The isotopic ratio may alternatively be defined with respect to the light isotope as $\delta_{light.}$ The standard commonly used in studies of oxygen and hydrogen isotopes is **standard mean ocean water (SMOW)** (Craig, 1961). The present-day composition of ocean water is "exceedingly uniform" (Taylor, 1974, p. 850), and this is why it can be used as an isotopic standard. Exceptions are the areas subject to high rates of evaporation (e.g., the Red Sea) or dilution from freshwater (e.g., near the mouth of a large river).

Evaporation and condensation are both important fractionation processes for the hydrogen and oxygen isotopes found in water. The lighter isotopes of both hydrogen and oxygen tend to evaporate more easily than the heavier isotopes. Thus atmospheric water has a lighter isotopic composition than seawater, however, the opposite is true for condensation. Heavier isotopes tend to condense more easily. Most atmospheric water comes from evaporation of seawater, and this evaporation is disproportionately concentrated at low latitudes where

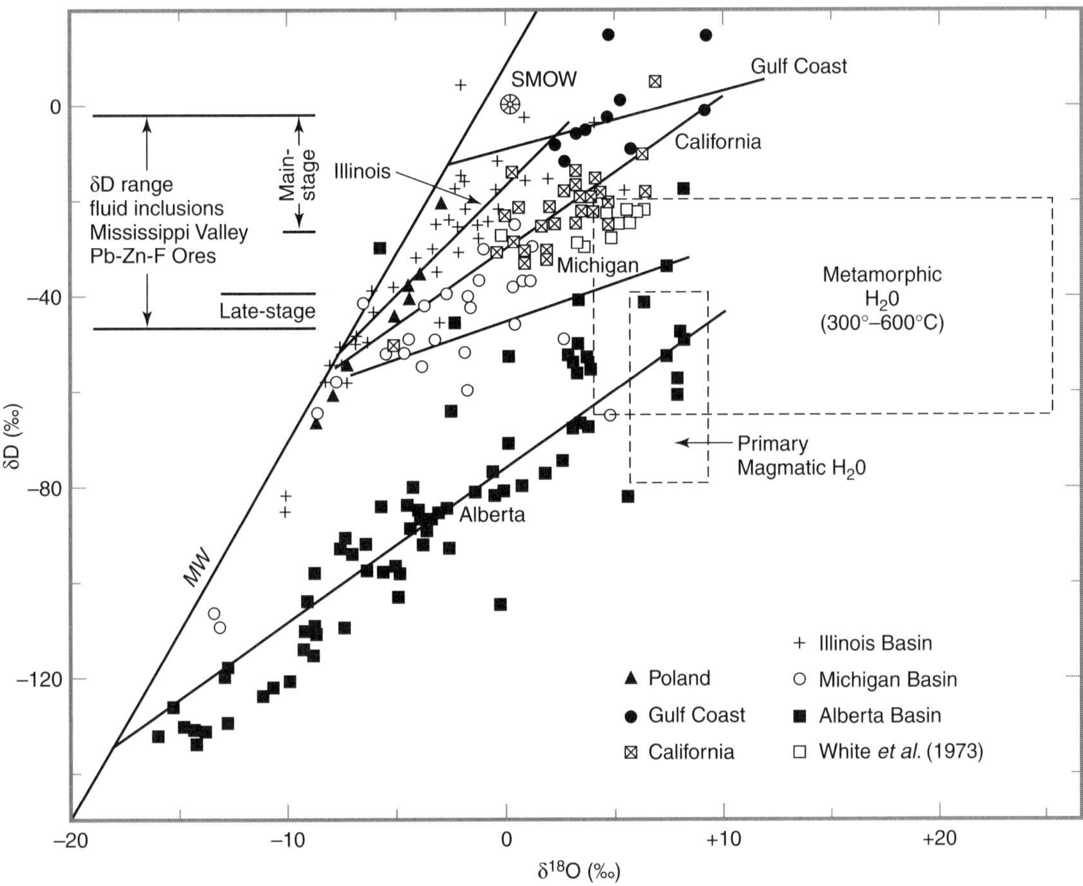

Figure 4.5 Values of δD and $\delta^{18}O$ for standard mean ocean water (SMOW), meteoric water (MW line), magmatic water, metamorphic water, and brines from several sedimentary basins. Also shown is the range of isotopic compositions, which characterize water found in fluid inclusions associated with Mississippi Valley-type lead-zinc deposits.

(From Taylor, 1974, p. 852.)

temperatures are higher. As atmospheric water vapor travels from the equator to the poles, it tends to progressively lose heavier oxygen and hydrogen isotopes through condensation. Thus atmospheric water and precipitation are increasingly depleted in the heavier isotopes with higher latitudes. This trend is known as the **meteoric water line** (Figure 4.5). Meteoric water is that which occurs in, or originates from, the atmosphere.

Oxygen isotopes can also be used to estimate the paleotemperature of seawater. The precipitation of $CaCO_3$ from seawater into the shells of marine organisms is affected by temperature, and thus the ratio O^{18}/O^{16} in carbonate skeletons is diagnostic of temperature at the time the shell was formed. But, interpretation of such ratios is not straightforward, because the mean isotopic composition of seawater has changed over time. During the last million years or so, Earth has been subjected to cyclic climatic variations known as ice ages. Ice ages tend to last about 100,000 years, and are separated by relatively brief interglacial periods about 10,000 years long. During an ice age, the mean planetary surface temperature typically falls about 5°C and continental ice sheets accumulate at high and middle latitudes. The accumulation of vast amounts of isotopically depleted ice in the form of great ice sheets enriches remaining seawater in the heavier isotopes. Thus the isotopic composition of ocean water was not constant in the geologic past. Paleotemperatures can also be inferred by direct analysis of the isotopic composition of ancient ice layers that have accumulated at the poles.

One way in which fractionation occurs in groundwater in the solid Earth is through the exchange of oxygen isotopes. Oxygen is the most abundant element in the Earth's crust, constituting 46% of the crust by weight (Press and Siever, 1986, p. 13). In contrast, hydrogen makes up only 0.14% of the crust by weight (Press and Siever, 1986, p. 595). As meteoric water enters the crust through infiltration, it tends to exchange the lighter O^{16} isotope for the heavier O^{18} isotope (Figure 4.5).

4.4.2. Radioactive Isotopes.

Radioactive isotopes are those that undergo nuclear decay and change from one element into another; they are used in geologic dating. Examples of radioactive isotopes that are employed in dating rocks include Uranium, Thorium, Rubidium-Strontium, and Potassium-Argon. Radioactive isotopes used to date groundwater include Carbon-14 (C^{14}), Tritium (H^3), Chlorine-36 (Cl^{36}), Helium-3 (He^3), and Helium-4 (He^4).

The rate at which radioactive isotopes decay is proportional to the number of atoms present,

$$\frac{dN}{dt} = -\lambda N \qquad \textbf{(4.6)}$$

where N (dimensionless) is the number of atoms remaining at any time, t (s) is time, and λ (s^{-1}) is the **decay constant**. Each radioactive isotope has a unique decay constant (λ). The negative sign in equation 4.6 indicates that the rate at which N is changing is negative. That is, the total number of atoms (N) is decreasing with increasing time. Equation 4.6 can be re-arranged and integrated,

$$\frac{dN}{N} = -\lambda dt \qquad \textbf{(4.7)}$$

$$\int \frac{dN}{N} = -\int \lambda dt \qquad \textbf{(4.8)}$$

$$\log_e (N) = -\lambda t + c \qquad \textbf{(4.9)}$$

where c is a dimensionless constant of integration. If at $t = 0$, there are N_o atoms present, then

$$c = \log_e (N_o) \qquad \textbf{(4.10)}$$

or

$$\log_e (N) = -\lambda t + \log_e (N_o) \qquad \textbf{(4.11)}$$

Exponentiating both sides and keeping in mind that $e^{(a+b)} = e^a e^b$,

$$N = N_o e^{-\lambda t} \qquad \textbf{(4.12)}$$

The **half-life** of a radioactive isotope is the time it takes for half of an initial amount of atoms of one element or isotope (the **parent isotope**) to change to atoms of another element or isotope (the **daughter isotope**). Let the half-life be denoted by $\tau_{1/2}$ (s). When one half-life has passed ($t = \tau_{1/2}$), $N = N_o/2$, thus

$$\frac{N_o}{2} = N_o e^{-\lambda \tau_{1/2}} \qquad \textbf{(4.13)}$$

Suppose that we have a radioactive parent isotope that decays into a stable daughter isotope. At time $t = 0$, we have N_o parent atoms. At time $t > 0$, we have (from equation 4.12) $N_o e^{-\lambda t}$ parent atoms left. The number of daughter atoms (D, dimensionless) generated must be equal to the number of decayed parent atoms. Thus

$$D = N_o - N_o e^{-\lambda t} = N_o(1 - e^{-\lambda t}) \qquad \textbf{(4.14)}$$

It is more convenient to express D in terms of N, because we cannot measure N_o, we can only measure D and N. From equation 4.12,

$$N_o = \frac{N}{e^{-\lambda t}} \qquad \textbf{(4.15)}$$

Substituting equation 4.15 into equation 4.14,

$$D = N(e^{\lambda t} - 1) \qquad \textbf{(4.16)}$$

because $1/e^{-\lambda t} = e^{\lambda t}$. Equation 4.16 can be solved for time (t),

$$t = \frac{1}{\lambda} \log_e \left[1 + \frac{D}{N}\right] \qquad \textbf{(4.17)}$$

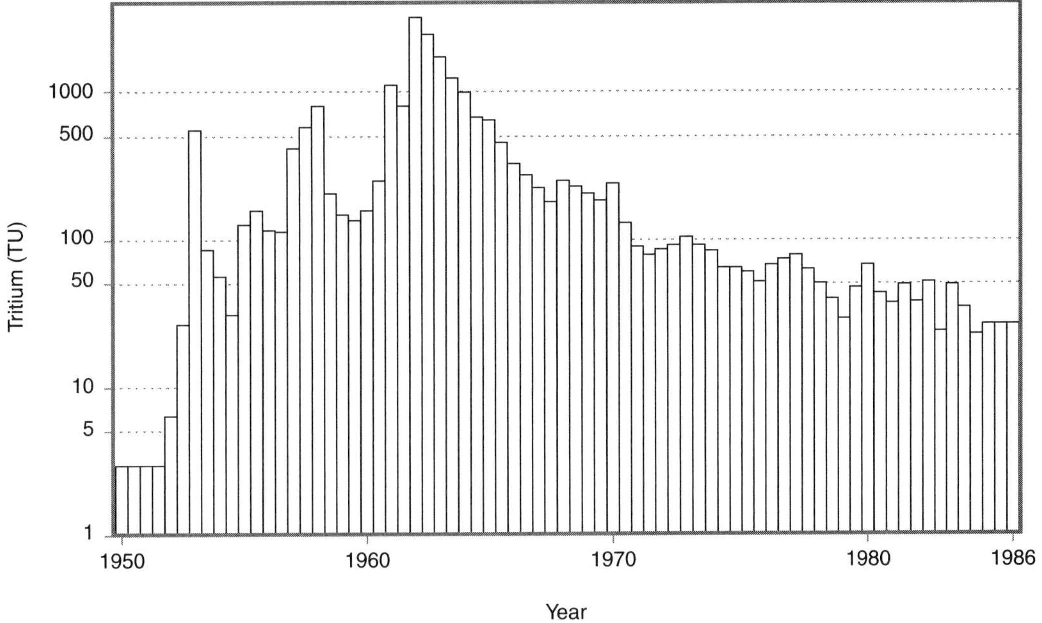

Figure 4.6 Tritium concentration in precipitation near Ottawa, Canada. Note logarithmic scale.
(From Robertson and Cherry, 1989, p. 1103.)

Thus, if the decay constant (λ) is known, and the ratio of parent and daughter atoms can be measured, the time since decay started (t) can be estimated. If the number of daughter atoms was zero at time $t = 0$, the estimated time (t) represents the age of a rock or groundwater sample.

Carbon-14 (C^{14}) is produced in the atmosphere by the interaction of cosmic-ray produced neutrons and nitrogen (N^{14}). The concentration of C^{14} in the atmosphere is in a dynamic equilibrium, with new C^{14} being produced at the rate necessary to replace that which decays. Radioactive C^{14} is incorporated into carbon dioxide, which is readily dissolved in surface waters. Once surface water has infiltrated into the subsurface, no further addition of radioactive C^{14} occurs. If no contamination occurs, C^{14} activity should therefore be a direct measure of the time elapsed since a water molecule has entered the subsurface through recharge. C^{14} has a half-life of 5,730 years and is generally useful for dating groundwater less than 50,000 years old. The accuracy of C^{14} dating may be problematical due to several sources of contamination. Groundwater exchanges carbon atoms with the solid Earth. Phillips et al. (1989) list possible sources of contamination that include dissolution of soil and aquifer carbonate, as well as exchange of groundwater calcium for matrix sodium.

Tritium (H^3) is a short-lived isotope of hydrogen; it has a half-life of 12.3 years. Tritium is produced naturally in the upper atmosphere by the reactions of protons and neutrons with nitrogen and oxygen. After oxidation, tritium is incorporated into atmospheric water and the hydrologic cycle. Tritium can also be produced by the detonation of nuclear bombs in the atmosphere. During the most active phase of nuclear above-ground testing, from 1952 through 1963, the concentration of tritium in the atmosphere rose by three orders of magnitude (Figure 4.6). The rapid rise in tritium concentration that occurred during this period of time is a useful tracer in shallow groundwater studies.

Tritium levels are expressed in units of tritium units. One **tritium unit (TU)** is equal to a natural or pre-bomb level of one tritium atom per 10^{18} hydrogen atoms. The utility of tritium studies is increased by simultaneous measurement of its decay product, Helium-3 (He^3). A typical application of tritium analyses is to estimate infiltration rates by modeling the migration of the tritium spike caused by nuclear testing. The interpretation of tritium measurements can be condensed into three rules of thumb (Mazor, 1991):

1. Groundwater with tritium levels below 0.5 TU predates 1952.
2. Groundwater with a tritium concentration above 10 TU postdates 1952.
3. Groundwater with a tritium concentration between 0.5 and 10 TU is a mixture of pre- and post-1952 water.

Chlorine-36 (Cl^{36}) is a radioactive isotope of chlorine with a half-life of 301,000 years. Cl^{36} is produced in the atmosphere by the interaction of cosmic rays with Argon-40. Argon is the third most abundant gas in the atmosphere; it constitutes about 1% of the atmosphere's volume. Cl^{36} can also be produced in the near-surface environment by interaction of cosmic rays with minerals in surface rocks. In the subsurface, the radioactive decay of Uranium and Thorium produces some Cl^{36}. In comparison to atmospheric production, surface and subsurface production of Cl^{36} are relatively small, and the Cl^{36} that is produced tends to be mineralogically bound and immobile. It is therefore usually safe to assume that most of the Cl^{36} found in groundwater originates from meteoric recharge. Two exceptions are geologic settings where evaporite dissolution is occurring or geothermal waters are circulating (Phillips et al., 1986).

Cl^{36} has two advantages in groundwater dating. First, it has (compared to tritium and carbon-14) a relatively long half-life. Secondly, chlorine usually undergoes few interactions with solid phases. Thus, it is much less subject to contamination. Cl^{36} dating was not possible prior to 1979 when advances in mass spectrometry techniques first made it possible to reliably measure Cl^{36} concentrations at the low levels at which it typically occurs ($Cl^{36}/Cl < 10^{-14}$). One of the first studies that demonstrated the utility of Cl^{36} in dating groundwater was an analysis of fluids from the Great Artesian Basin of Australia by Bentley et al. (1986) (see Figure 1.4).

Helium has two isotopes: the lighter and rarer He^3, and the heavier and more common He^4. Terrestrial helium isotopes are from one of two sources: (1) primordial helium trapped inside the Earth at the time of its accretion, or (2) helium produced radiogenically in the solid Earth by the radioactive decay of Uranium and Thorium. Although some crustal He^3 is produced radiogenically by the breakdown of Lithium-6 (Li^6), most of the He^3 found in the crust is primordial. Primordial helium is believed to have an isotopic ratio He^3/He^4 of the order of 10^{-4}, the value that characterizes gas-rich meteorites (O'Nions and Oxburgh, 1983). Helium that outgasses from the mantle today has a ratio (He^3/He^4) near 10^{-5}. The dilution from the primordial He^3/He^4 ratio (10^{-4}) is believed to be due to ongoing radioactive generation of He^4. In "normal" crust and mantle rocks, the production ratio (He^3/He^4) is 1.0×10^{-8} (O'Nions and Oxburgh, 1983, p. 429). The relative abundances of He^3 and He^4 are normally expressed by comparing them to the atmospheric ratio R_A for which (He^3/He^4) = 1.4×10^{-6}. Thus the isotopic ratios due to radioactive production of He^4 in the crust would be $R/R_A = 1.0 \times 10^{-8}/ 1.4 \times 10^{-6} = 0.01$. In contrast, areas of the crust where mass and heat are being transferred from the mantle show enrichment in He^3 (Oxburgh et al., 1986). Elevated levels of primordial He^3 in groundwater are a clear indication of mantle outgassing. Mid-ocean spreading ridges are characterized by isotopic helium ratios (R/R_A) of about 10. Other areas that show evidence of mass/heat transfer from the mantle include continental rifts, hot spots, and geothermal springs (Figure 4.7) (Ballentine et al., 1991).

In theory, the accumulation of He^4 can be used to date groundwater (Torgersen, 1980; Torgersen and Clarke, 1985). The theory depends on the more-or-less constant generation of He^4 by the decay of thorium and uranium that are "ubiquitous"

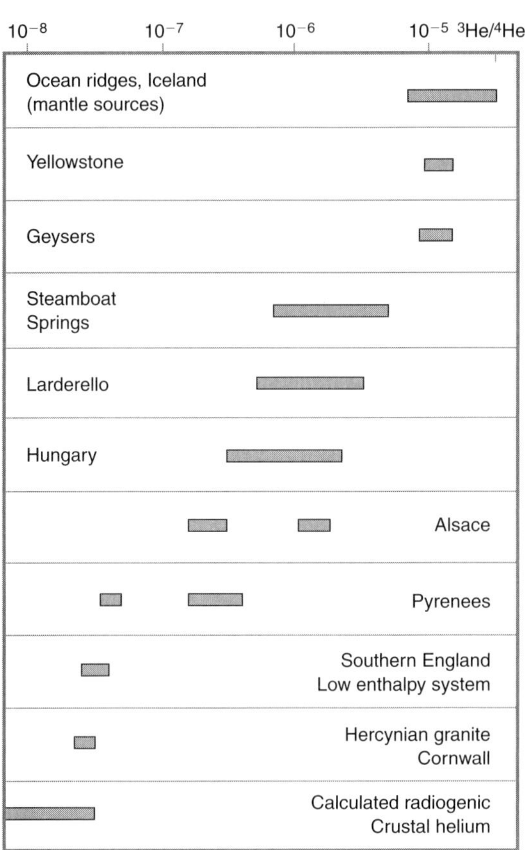

Figure 4.7 Characteristic distribution of helium isotopes in geothermal waters. Larger ratios of R/R_A indicate more mantle involvement.
(From Deak et al., 1988, p. 294.)

in nearly all crustal rocks (Torgersen and Clarke, 1985, p. 1211). In practice, uncertainties are introduced by the rate of release of He^4 from the solid to the liquid phase and the volume ratio of the solid phase (that releases He^4) to the liquid phase (that accumulates He^4). Torgersen and Clarke (1985, p. 1211) noted that the agreement between He^4 ages and ages calculated by C^{14} dating were often quite poor. He^4 dating usually yields an older age, often by factors up to 60 times. These discrepancies may reflect the fact that some aquifers can entrain not only locally produced He^4, but the entire crustal production. Torgersen and Clarke (1985) studied He^4 concentration in the Jurassic Aquifer of the Great Artesian Basin in Australia. They concluded that for relatively young groundwater (age less than 50,000 years) the concentration of He^4 was controlled by in situ radioactive decay of Uranium and Thorium. But, in the main part of the Great Artesian Basin they found that the rate of He^4 accumulation was 74 times the rate of production by radioactive decay in the aquifer. They concluded that the massive addition of He^4 to the Jurassic Aquifer was consistent with the production of nearly the entire production of crustal helium. The facilitation of He^4 release throughout the crust implies relatively high permeability in the lower crust.

4.5. ORIGIN OF PORE FLUIDS.

Water found in subsurface aqueous brines has traditionally been designated as meteoric, connate, or juvenile. **Meteoric water** is that which occurs in, or is derived from, the atmosphere. **Connate water** is water that was trapped in pore spaces at the time a rock was first formed; connate refers to pore fluid physically present at the time of sediment deposition. Hanor (1994b, p. 34) noted that connate means "born with," and thus should not be applied to pore water that has physically remained in its original pore spaces over geologic time but undergone chemical changes through fluid-matrix interactions. **Juvenile water** is water derived from the interior of the Earth that has not previously been present in the atmosphere or surface of the Earth.

Our present understanding suggests that the above three definitions are outdated and should be replaced by a new classification scheme. For example, the word "connate" strictly interpreted implies that a fluid has not evolved or interacted with a rock matrix. Today we understand that such a situation is rare to nonexistent. Furthermore, the word connate tells us nothing regarding the origin of the water. Rather, it refers to a condition of stagnation that usually does not exist in nature.

A more useful classification scheme is to divide water into oceanic, meteoric, and evolved. **Oceanic water** is water that is found in the Earth's oceans, or water that is found elsewhere but has not significantly changed its composition from the time it left the oceans. **Evolved water** is water that originated as either meteoric or oceanic water, but has subsequently changed its composition through chemical and/or physical processes. In this classification scheme, the term juvenile is discarded. Juvenile water constitutes a negligible portion of geologic fluids in the crust, even in hydrothermal or deep geothermal systems (Kharaka, 1986, p. 174). The term once applied to water vapor emitted from volcanoes, however, with our present understanding of plate tectonics and subduction, it appears likely that most volcanic emissions may contain substantial amounts of recycled water that originated in the oceans or atmosphere and subsequently evolved.

The new classification scheme avoids much of the confusion inherent in the old terminology. Consider, for example, the pore water in a newly deposited layer of terrigenous sediment. By definition, this is connate water, but was derived immediately from meteoric water. If the pore fluid had been derived from oceanic water, it would also have been termed connate. The word "connate" does not help us understand the origin of the fluid.

Up to 1966, it was thought that the pore fluids in sedimentary rocks were largely connate. In that year, our understanding started to change when Clayton et al. (1966) studied brines from the Illinois, Michigan, and Alberta sedimentary basins. Plots of the isotopic composition of water from these basins (Figure 4.5) extrapolated back to the local meteoric water line in each case. The conclusion of Clayton et al. (1966) was that the connate pore fluids in each of these basins had been flushed out by meteoric water. Meteoric water in sedimentary basins commonly evolves isotopically by adding the heavier isotopes of both oxygen and hydrogen. A correlation between the amount of total dissolved solids and the degree of isotopic enrichment in O^{18} suggested to Clayton et al. (1966) that the isotopic evolution of meteoric water in sedimentary basins occurs mostly through fluid-matrix interaction. Most of the O^{18} enrichment is due to exchange with O^{18}-rich calcite. Calcite is ubiquitous in sedimentary basins and is enriched in O^{18} at the time of its formation due to equilibrium partitioning. During burial at higher temperatures, calcite is dissolved, exchanges oxygen isotopes with the surrounding fluid, and precipitates again. During this process, calcite becomes isotopically lighter and pore fluid becomes isotopically heavier (Knauth and Beeunas, 1986). In the case of the hydrogen isotopes, subsurface water can become either enriched or depleted in deuterium through exchanges with hydrous minerals such as clays. The nature (heavier or lighter) and extent of the isotopic evolution depends upon the isotopic composition of the hydrous matrix minerals and the original meteoric water. Knauth and Beeunas (1986, p. 425) stated that there should be little hydrogen-isotope exchange between meteoric water and hydrous minerals at temperatures below 100°C.

Not all pore fluids originate as meteoric water. Kharaka (1986) showed that some pore fluids from the Gulf Coast Basin in the southeast United States originated as ocean water. Extrapolation of the oxygen and hydrogen isotopic ratios of these fluids leads back to SMOW, not the meteoric water line (Figure 4.8). The deuterium depletion seen in pore fluids from the Gulf Coast Basin was attributed by Kharaka (1986) to isotopic exchanges with clay minerals possessing a very light original δD value. The Gulf Coast basin is undergoing active subsidence, and the younger sediments have not yet had the opportunity to be uplifted and flushed with meteoritic water.

Kharaka (1986) characterized formation waters in sedimentary basins as generally "highly mobile." In support of this assertion, Kharaka (1986) cited studies of fluids from the Western Canadian Sedimentary Basin, the Dniepor-Donets Basin and Ukrainian Shield, the Upper Silesian Coal Basin in Poland, and the Sacramento Valley in California. There is evidence at all of these sites that fluids of different ages and origins have mixed. Thus, in general, isotopic analyses suggest that the fluid regime in sedimentary basins usually involves the active circulation of meteoric water.

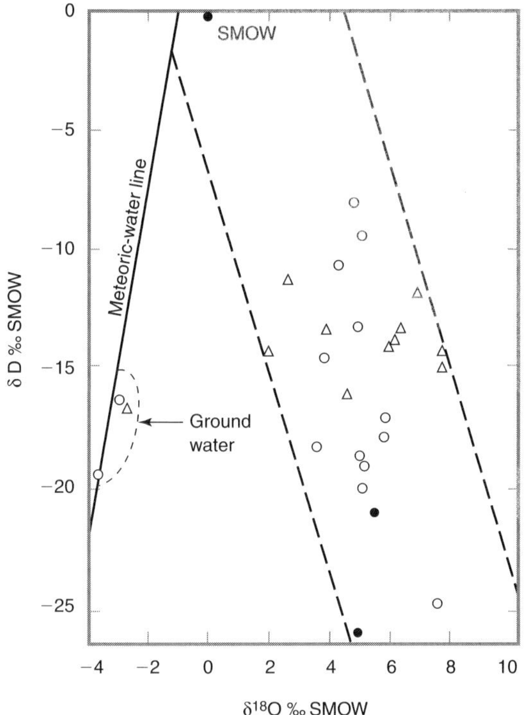

Figure 4.8 Isotopic composition of formation waters from northern Gulf of Mexico Basin. Trend of decreased δD values with increasing $\delta^{18}O$ values goes through SMOW and away from the meteoric water line. (From Kharaka, 1986, p. 176.)

In contrast to conclusions drawn from stable isotope analyses, some recent studies of Br-Cl-Na systematics (Figure 4.9) suggest that at least some brines in sedimentary basin originated from seawater evaporation, and that the original ocean water has not been completely replaced by meteoric water, although it may have been diluted. **Br-Cl-Na systematics** is the study of similarities and differences amongst the composition and concentration of these ions (Bromine, Chlorine, and Sodium) in geologic fluids. The main utility of Br-Cl-Na systematics is that the method can be used to discriminate between brines formed from evaporation of seawater, and those formed by dissolution of evaporites. As seawater begins to evaporate, the ratios Cl/Br and Na/Br remain constant (Figure 4.9). When the brine concentration reaches about 162 g/liter, halite (NaCl) begins to precipitate and the ratios Cl/Br and Na/Br begin to decrease as both Na and Cl are removed by halite precipitation while Br remains dissolved. Thus brines formed by evaporation of seawater tend to have high concentrations of bromine. In contrast, brines formed through the dissolution of halite by freshwater tend to have low concentrations of bromine. Walter et al. (1990) and Stueber and Walter (1994) studied pore fluids from the Illinois Basin (Figure 4.10) and concluded on the basis of Br-Cl-Na systematics that remnant marine fluids had contributed substantially to overall pore fluid composition, especially in older rocks. These data indicate that stable isotope plots, which extrapolate back to the meteoric water line are evidence of dilution by meteoric waters, not complete replacement.

4.6. Physical Properties of Water.

4.6.1. Phase Diagram for Water.

Water phase is determined by pressure and temperature (Figure 4.11). The point at which all three phases may coexist is the **triple point**. Under the high fluid pressures that may be characteristic of the subsurface, water may be liquid at temperatures well exceeding its boiling point (~100°C) at standard pressure. Below the **critical point** (Figure 4.11), liquid water and water vapor exist as distinct and separate phases. Above the critical point, a single supercritical fluid exists at all temperature and pressure conditions, and all physical, chemical and transport properties of water vary in a continuous manner with changing pressure or temperature—there are no discontinuities in these properties. The properties of water gradually become more liquid-like or more gas-like as temperature and/or pressure changes in the supercritical fluid (Bodnar and Costain, 1991, p. 983).

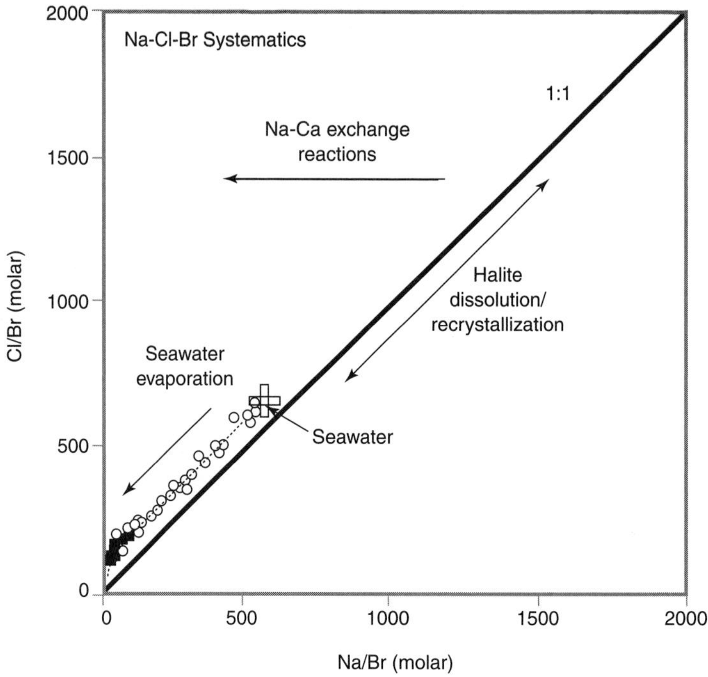

Figure 4.9 Na-Cl-Br (Sodium-Chlorine-Bromine) systematics.
(From Kesler et al., 1995, p. 641.)

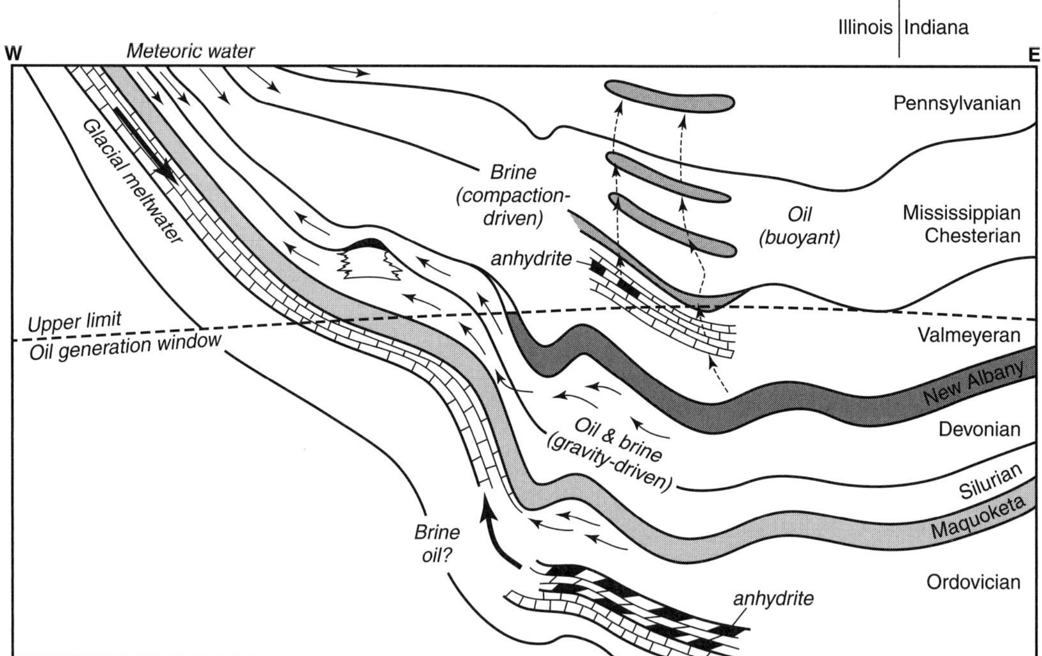

Figure 4.10 Schematic cross-section through the Illinois Basin showing distribution of meteoric water and brine derived from seawater evaporation. Also shown are driving forces that include sediment compaction, topographic gradients, and buoyancy forces.
(From Stueber and Walter, 1994, p. 1438.)

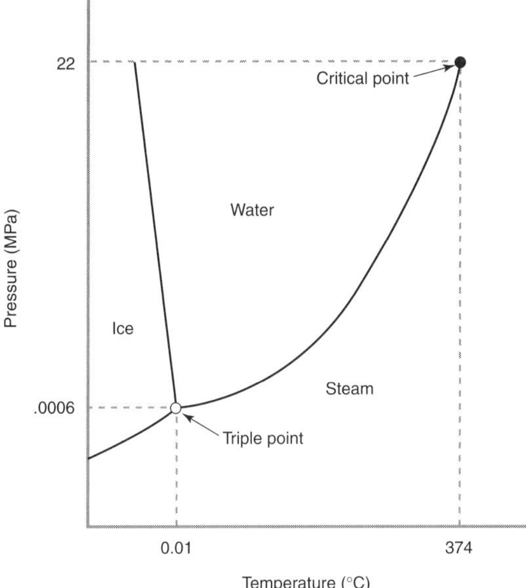

Figure 4.11 Phase diagram for water.

Near the critical point, the physical and thermodynamic properties of water reach maxima or minima and change rapidly with small changes in either temperature or pressure. As a consequence, small changes in pressure or temperature near the critical point may have large influences on a fluid's capacity for heat and mass transport. Buoyancy forces and fluid heat capacity become large, and the viscosity of the fluid is reduced. An increase in heat capacity and decrease in viscosity near the critical point imply that a fluid's capacity for heat and mass transport is at a maximum at or near the critical point.

Phase diagrams for the aqueous brines found in the Earth's crust may differ significantly from that for pure water (Figure 4.11). In many crustal environments, the physical properties and phases of aqueous brines in the crust are adequately represented by those of the H_2O-CO_2-NaCl system (Bodnar and Costain, 1991, p. 983). Addition of NaCl to H_2O causes the critical point to shift to higher temperatures and pressures, while addition of CO_2 causes the critical point to migrate to slightly lower temperatures but much higher pressures. The shift of the critical point due to the presence of dissolved solutes such as NaCl or dissolved gasses like CO_2 may have important consequences for understanding geologic processes such as ore formation that rely upon heat and mass transport by moving fluids. For example, the region of optimum heat and mass transport for a 20% NaCl solution by weight is in the neighborhood of ~560–620°C and 700–1100 bars, compared to ~400°C and 300 bars for pure water (Bodnar and Costain, 1991, p. 985).

4.6.2. Specific Heat Capacity, Heats of Fusion and Vaporization.

The **specific heat capacity** of a substance is the amount of energy or heat necessary to raise the temperature of a specified mass of that substance by a specified amount. The specific heat capacity (C, J-kg^{-1}-°K^{-1}) of pure water at 0°C and 1 atm pressure is

$$C = 4184 \text{ (J-kg}^{-1}\text{-°K}^{-1}) \qquad \textbf{(4.18)}$$

Specific heat capacity may also be expressed in units of calories per gram per degree Celsius (cal-g^{-1}-°C^{-1}). To convert from cal-g^{-1}-°C^{-1} to J-kg^{-1}-°K^{-1}, multiply by

$$1\,\frac{\text{cal}}{\text{g-°C}} \times 4.184\,\frac{\text{J}}{\text{cal}} \times 1000\,\frac{\text{g}}{\text{kg}} \times 1\,\frac{\text{°K}}{\text{°C}} = 4184\,\frac{\text{J}}{\text{kg-°K}} \qquad \textbf{(4.19)}$$

The **heat of fusion** is the amount of heat or energy necessary to change a specified mass of a substance from a solid to a liquid while maintaining a constant temperature and pressure. The heat of fusion of pure water is about 80 cal-g^{-1}. To convert to J-kg^{-1}, multiply by

$$80\,\frac{\text{cal}}{\text{g}} \times 4.184\,\frac{\text{J}}{\text{cal}} \times 1000\,\frac{\text{g}}{\text{kg}} = 3.35 \times 10^5\,\frac{\text{J}}{\text{kg}} \qquad \textbf{(4.20)}$$

The **heat of vaporization** is the amount of heat or energy necessary to change a specified mass of a substance from a liquid to a vapor while maintaining a constant temperature and pressure. The heat of vaporization of pure water is about 540 cal-g^{-1}. To convert to J-kg^{-1},

$$540 \frac{\text{cal}}{\text{g}} \times 4.184 \frac{\text{J}}{\text{cal}} \times 1000 \frac{\text{g}}{\text{kg}} = 2.26 \times 10^6 \frac{\text{J}}{\text{kg}} \quad \textbf{(4.21)}$$

4.6.3. Density.

The density of pure water at STP is about 1000 kg-m^{-3}. To convert kg-m^{-3} to g-cm^{-3}, multiply by

$$1000 \frac{\text{kg}}{\text{m}^3} \times 1000 \frac{\text{g}}{\text{kg}} \times \frac{1\text{m}^3}{10^6\text{cm}^3} = 1 \frac{\text{g}}{\text{cm}^3} \quad \textbf{(4.22)}$$

Water density is affected by temperature, pressure, and salinity. Water density tends to *decrease* with *increasing* temperature, but *increase* with *increasing* pressure and salinity. In the Earth's crust, salinity and temperature tend to be more important in controlling fluid density than pressure (Phillips et al., 1981).

Temperature in the Earth's crust is controlled by the geothermal gradient and the ground surface temperature. The **geothermal gradient** is the rate at which temperature increases with depth. The geothermal gradient is largely determined by the thermal conductivity of crustal rocks and heat flow from the Earth's interior. Surface ground temperature is controlled by climate, and ranges from about 25°C near the equator to as low as −22°C at polar extremes. The mean annual air temperature for Earth is about 15°C, however, ground temperatures tend to be 2–3°C warmer than air temperatures, so the average ground surface temperature on Earth is about 17 to 18°C. The average geothermal gradient ranges from about 10°C-km^{-1} to as high as 60°C-km^{-1}. A nominal estimate of the average geothermal gradient on the continents is 25°C-km^{-1} (see Kappelmeyer and Haenel, 1974; Jessop, 1990).

As long as we are not in an unusual geologic situation, fluid pressure in the upper continental crust is determined simply by the weight of the overlying fluid as given by equation 2.23. Consider a fluid with a constant salinity of 5 molal that is about 22–23% TDS by weight. How does its density change with increasing depth in an area characterized by a typical ground surface temperature of 17°C and average geothermal gradient of 25°C-km^{-1}? Using data given by Phillips et al. (1981), water density can be calculated as a function of pressure and temperature (Table 4.3). The tendency of *increasing* temperature to *decrease* fluid density is usually greater than the tendency of *increasing* pressure to *increase* fluid density. In other words, temperature is usually a more important determinant of fluid density in the Earth than pressure.

TABLE 4.3 Effect of Pressure and Temperature on Fluid Density

Depth (km)	Fluid Pressure (MPa)	Temperature (°C)	Fluid Density (kg-m^{-3})
1	10	42	1146
2	20	67	1125
3	30	92	1113
4	40	117	—
5	50	142	1092

The above calculations show that a constant-salinity fluid would usually decrease in density with increasing depth and temperature in the Earth's crust. This situation could be gravitationally unstable; a buoyancy force would tend to promote the overturn of the less dense fluid. Whether or not the fluid in a given area overturns would depend on the geothermal gradient and the permeability of the rocks in that area, however, the situation illustrated above usually does not occur because salinity usually increases with depth. The increase of salinity with depth is sufficient to counteract the decrease of fluid density associated with temperature increases, and fluid density usually increases with increasing depth. The fluids found in the upper continental crust are thus usually stable, and should not spontaneously overturn. There are some exceptional circumstances where this generalization is violated. One example might be near an igneous intrusion where the thermal gradients and buoyancy forces due to thermal expansion would be unusually high.

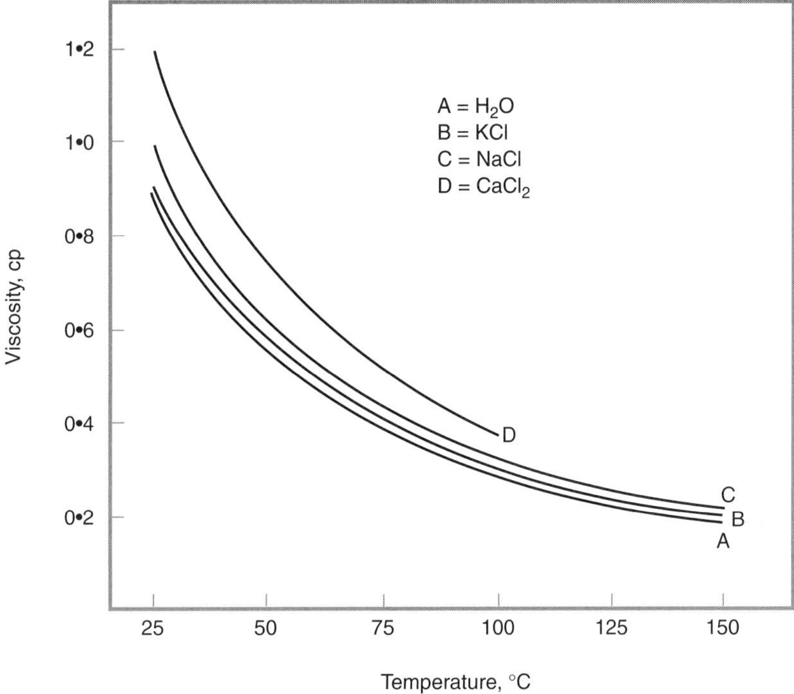

Figure 4.12 Viscosity of water and 1 molal solutions of NaCl, KCl, and CaCl, as a function of temperature. (From Phillips et al., 1981, p. 24.)

4.6.4. Viscosity.

Viscosity is the property of a fluid that offers resistance to flow. More formally, viscosity (μ, kg-m^{-1}-s^{-1}) is defined as the constant of proportionality between the shear stress applied to a fluid and its deformation rate. In this text, the term viscosity means absolute or **dynamic viscosity**. **Kinematic viscosity** is the dynamic viscosity (μ) divided by fluid density (ρ).

The viscosity of pure water is strongly dependent on temperature (Figure 4.12). Empirically, we find that (Touloukian et al., 1975)

$$\mu = 2.4 \times 10^{-5} \times 10^{[248/(T+133)]} \quad \textbf{(4.23)}$$

where μ is viscosity in kg-m^{-1}-s^{-1} and T is temperature in °C.

Note that water viscosity decreases by a factor of ten from 0 to 150°C (Table 4.4). For an average geothermal gradient of 25°C-km^{-1}, 150°C corresponds to depths of about 5–6 km. The decrease of fluid viscosity with increasing temperature (and thus increasing depth) implies that, all other factors being equal, deep flow is promoted. Of course, all other factors are never equal, and the decrease of permeability with depth usually overshadows the decrease of fluid viscosity. But, a viscosity decrease of an order of magnitude or more is large enough to potentially have important implications for fluid flow in the upper crust. Salinity also has an effect on viscosity, but the effect of temperature is usually much greater (Figure 4.12). An alternative viscosity unit is the **poise**. One poise = 0.1 kg-m^{-1}-s^{-1}.

4.6.5. Surface Tension.

The water molecule (H_2O) is dipolar. While the overall electrical charge is zero, one end of the water molecule has a net negative charge, and the

Table 4.4 Effect of Temperature on Water Viscosity

Temperature (°C)	Water Viscosity ($kg\text{-}m^{-1}\text{-}s^{-1}$)
0	1.8×10^{-3}
50	5.4×10^{-4}
100	2.8×10^{-4}
150	1.8×10^{-4}

other end has a net positive charge. As a result, water molecules tend to stick together, with the positive end of one attracted to the negative end of another. **Surface tension** is the strength of the attractive force between liquid molecules. The surface tension of water causes it to form a film on its surface strong enough to support a steel needle. Surface tension also causes water to form drops, instead of spreading out evenly and smoothly. Amongst common liquids, only mercury has a greater surface tension. In addition to sticking to each other, water molecules also tend to adhere to certain substances, such as glass. The combination of surface tension and adherence is responsible for the important phenomenon of capillarity that is discussed in chapter 6.

The surface tension of water can be decreased through the addition of a class of chemicals known as surfactants. Detergent and soap are common surfactants. The decreased surface tension that is produced by adding surfactants increases the cleaning ability of water by increasing its ability to spread, wet surfaces, and enter small spaces. The human body produces a natural surfactant in the lungs that aids respiration. Surface tension also decreases with increasing temperature. Thus hot water is a better cleanser than cold. The manufacture of lead shot takes advantage of the phenomenon of surface tension by pouring molten lead through a screen mesh at the top of a tower. By the time the lead drops have fallen to the bottom of the tower, surface tension has pulled them into a spherical shape.

Review Questions

1. Define the following terms in the context of hydrogeology:
   a. ion
   b. cation
   c. anion
   d. salinity
   e. solute
   f. solution
   g. TDS
   h. weight percent
   i. ‰
   j. fresh
   k. brackish
   l. saline
   m. brine
   n. molality
   o. molarity
   p. mole
   q. Avogadro's number
   r. molecular weight
   s. atomic weight
   t. periodic table
   u. evaporite
   v. cationic shift
   w. membrane filtration
   x. isotope
   y. element
   z. proton
   aa. nucleus
   bb. neutron
   cc. mass spectrometer
   dd. stable isotope
   ee. fractionation
   ff. SMOW
   gg. meteoric water line
   hh. meteoric
   ii. radioactive isotope
   jj. decay constant

kk. half-life
ll. parent isotope
mm. daughter isotope
nn. TU
oo. connate
pp. juvenile
qq. oceanic water
rr. evolved water
ss. Br-Cl-Na systematics
tt. triple point
uu. critical point
vv. specific heat capacity
ww. joule
xx. Kelvin
yy. heat of fusion
zz. heat of vaporization
aaa. geothermal gradient
bbb. dynamic viscosity
ccc. kinematic viscosity
ddd. poise
eee. surface tension

2. What percent of the volume of sedimentary basins consists of fluids?

3. What are the most important cations in the aqueous brines found in the Earth's crust? What is the most important anion?

4. What three factors control the composition of solutes in the aqueous brines found in the Earth's crust?

5. What is the level of TDS in seawater?

6. What is the molecular weight of H_2O? of $CaCl_2$? of $CaCO_3$?

7. A solution is 5% NaCl by weight. What is its molality?

8. A solution is 10% $CaSO_4$ by weight. What is its molality?

9. An NaCl solution has a molality of 3.0. What is its weight percent?

10. How does salinity change with increasing depth in sedimentary basins?

11. How many known examples are there of subsurface brines being produced by membrane filtration?

12. What process explains the major cation composition of aqueous brines in sedimentary basins?

13. What is the "master variable" that drives fluid-matrix interchanges and controls fluid composition?

14. What stable isotopes are most commonly analyzed in groundwater studies?

15. Why is meteoric water increasingly depleted in heavy isotopes with increasing latitude?

16. How has the isotopic composition of the oceans changed over the last million years?

17. A sample of groundwater from a well has a C^{14} concentration that is 1/8 that of local meteoric water. How much time has passed since this water entered the subsurface? What are likely sources of error in your estimate?

18. A sample of groundwater from a well has a Cl^{36} concentration that is 1/4 that of local meteoric water. How much time has passed since this water entered the subsurface?

19. What areas show enrichment in the He^3 isotope? What is the significance of He^3 enrichment?

20. A classification scheme divides terrestrial water into what three categories?

21. Why is the water found in sedimentary brines commonly enriched in O^{18}? Where does the O^{18} come from?

22. Do most of the aqueous brines found in sedimentary basins originate as meteoric or oceanic water? Summarize the evidence and arguments for each possibility.

23. Where is a fluid's capacity for heat and mass transport at a maximum?

24. What is the specific heat capacity of pure water at STP?

25. How much energy does it take to vaporize 100 g of ice at 0°C? (assume STP).

26. What is the average geothermal gradient on the continents? What is the range? What is the average ground surface temperature on Earth?

27. What three factors affect fluid density in the crust? Which two factors are the most important?

28. Calculate the viscosity of water at 75, 100, and 125°C.

Suggested Reading

Hanor, J. S. 1987. History of thought on the origin of subsurface sedimentary brines. In Landa E. R., and Ince, S., *The History of Hydrology: History of Geophysics Serial,* 3: 81–91. Washington, D. C.: AGU.

Hanor, J. S. 1994a. Origin of saline fluids in sedimentary basins. In *Geofluids: Origin, Migration and Evolution of Fluids in Sedimentary Basins,* ed. J. Parnell, Geological Society Special Publication No. 78, 151–174.

Hem, J. D. 1970. Study and interpretation of the chemical characteristics of natural water. *USGS Water-Supply Paper 1473* (2d ed.).

Notation Used in Chapter Four

Symbol	Quantity Represented	Physical Units
C	specific heat capacity	$J\text{-}kg^{-1}\text{-}°K^{-1}$
c	constant of integration	dimensionless
D	number of daughter atoms	dimensionless
$\delta_{heavy}, \delta_{light}$	isotopic fractionation ratio	dimensionless
λ	radioactive decay constant	s^{-1}
M	molarity	$mole\text{-}m^{-3}$
m	molality	$mole\text{-}kg^{-1}$
μ	water viscosity	$Pa\text{-}s = kg\text{-}m^{-1}\text{-}s^{-1}$
MW	molecular weight	$kg\text{-}mole^{-1}$
N, N_o	number of atoms, number of atoms present at time zero	dimensionless
‰	parts per thousand	dimensionless
T	temperature in degrees Celsius	°C
t	time	s
$\tau_{1/2}$	half-life	s
TDS	total dissolved solids	dimensionless
W	weight percent solute	dimensionless

CHAPTER 5

TRANSIENT FLOW

5.1. THE DIFFUSION EQUATION.

Darcy's Law describes the steady-state flow between two points at which head is fixed, but it cannot predict how head will change with time in an evolving flow system. The mathematical formulation that describes the change of head with time is known as the **diffusion equation**. The diffusion equation is a mathematical statement of the conservation of mass. We will derive the diffusion equation for flow in one dimension (x); the results can be readily extended to multiple dimensions by adding additional terms to the derivation.

Consider a cube or control volume in a porous medium of porosity ϕ (dimensionless) (Figure 5.1). The cube has sides of length Δx, Δy, and Δz (m). Fluid is flowing through the cube from left to right in the $+x$-direction. The fluid has density ρ (kg-m^{-3}) and Darcy velocity q_x (m-s^{-1}).

If fluid mass inside the cube is to be conserved, then the rate at which the fluid mass inside the cube changes must be equal to the difference between the inflow and outflow rates,

$$\underset{(1)}{\frac{\text{change in mass}}{\text{unit time}}} \underset{=}{=} \underset{(2)}{\text{mass inflow rate}} \underset{-}{-} \underset{(3)}{\text{mass outflow rate}} \tag{5.1}$$

It is convenient to number each term (1, 2, 3) of equation 5.1 and consider them in turn. The final step in the derivation is to assemble the expanded terms. Starting with term (1), the fluid mass inside the cube is the product of density and the *fluid* volume, $\rho\phi\Delta x\Delta y\Delta z$ (kg). Thus,

$$\frac{\text{change in mass}}{\text{unit time}} = \frac{d(\rho\phi\Delta x\Delta y\Delta z)}{dt} \tag{5.2}$$

where t (s) is time. If the cube volume is held constant with respect to time, then any change in fluid mass must take place by a change in fluid density (ρ, kg-m^{-3}) or porosity (ϕ, dimensionless),

$$\frac{\text{change in mass}}{\text{unit time}} = \frac{d(\rho\phi)}{dt}\Delta x\Delta y\Delta z \tag{5.3}$$

Expanding term (2) of equation 5.1,

$$\text{mass inflow rate} = \rho q_x \Delta y\Delta z \tag{5.4}$$

We may check the physical units to verify that this is the correct expression

$$\text{kg-m}^{-3} \times \text{m}^3\text{-m}^{-2}\text{-s}^{-1} \times \text{m}^2 = \text{kg-s}^{-1} \tag{5.5}$$

The units (kg-s^{-1}) are the same as in equation 5.3, as they must be. A porosity term is not needed in equation 5.4, because porosity is implicit in definition of the Darcy velocity.

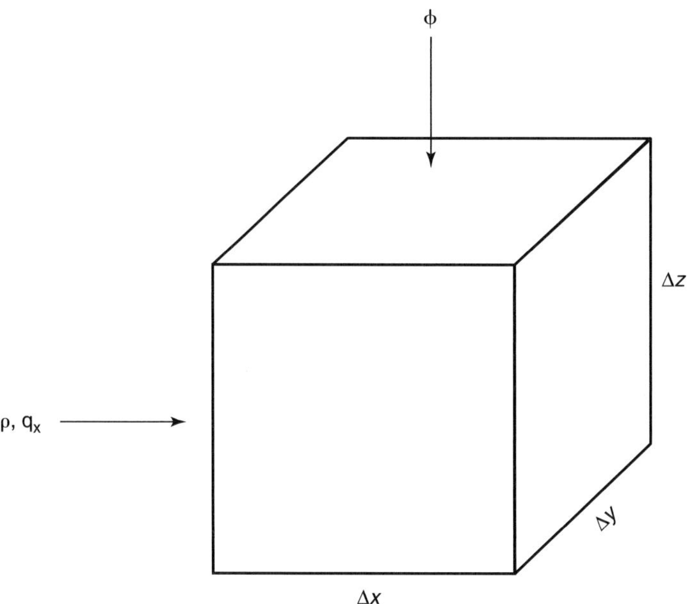

Figure 5.1 Control volume for derivation of diffusion equation.

The mass outflow rate, term (3) of equation 5.1, is equal to the mass inflow rate plus any change in the flow rate that takes place over the length of the cube, Δx (m)

$$\text{mass outflow rate} = \text{mass inflow rate} + \frac{d(\text{mass inflow rate})}{dx}\Delta x \quad \textbf{(5.6)}$$

Equation 5.6 is the key to the entire derivation. We have approximated the total change that takes place over the length of the cube Δx by multiplying the rate of change with respect to length (x) by the total length Δx. Because the rate of change with respect to x is not constant over the cube, our approximation is imprecise, however, the approximation becomes more exact as the cube length Δx becomes smaller. As we eventually replace Δx by the differential length dx, the approximation becomes exact.

Continuing to expand equation 5.6,

$$\text{mass outflow rate} = \rho q_x \Delta y \Delta z + \frac{d(\rho q_x \Delta y \Delta z)}{dx}\Delta x \quad \textbf{(5.7)}$$

$$\text{mass outflow rate} = \rho q_x \Delta y \Delta z + \frac{d(\rho q_x)}{dx}\Delta x \Delta y \Delta z \quad \textbf{(5.8)}$$

where $\Delta y \Delta z$ can be taken outside of the derivative because it does not depend on x. We are now ready to assemble terms (1), (2) , and (3), and substitute into equation 5.1,

$$\frac{d(\rho\phi)}{dt}\Delta x \Delta y \Delta z = \rho q_x \Delta y \Delta z - \rho q_x \Delta y \Delta z - \frac{d(\rho q_x)}{dx}\Delta x \Delta y \Delta z \quad \textbf{(5.9)}$$

The volume term $\Delta x \Delta y \Delta z$ (m^3) cancels, as it should: if we derive a generalized description of fluid flow it should not depend upon the specific size of the control volume considered. We are left with

$$\frac{d(\rho\phi)}{dt} = \frac{-d(\rho q_x)}{dx} \quad \textbf{(5.10)}$$

Now, let us assume the fluid density ρ (kg-m^{-3}) is constant with respect to distance x (m), but *not* necessarily constant with respect to time t (s),

$$\frac{d(\rho\phi)}{dt} = \frac{-\rho d(q_x)}{dx} \quad \textbf{(5.11)}$$

Substituting Darcy's Law, $q_x = -K(dh/dx)$, into equation 5.11,

$$\frac{d(\rho\phi)}{dt} = -\rho \frac{d\left[-K\dfrac{dh}{dx}\right]}{dx} \quad \textbf{(5.12)}$$

Assume the hydraulic conductivity K (m-s^{-1}) is constant with respect to x. Then

$$\frac{d(\rho\phi)}{dt} = \rho K \frac{d^2h}{dx^2} \quad \textbf{(5.13)}$$

or,

$$\frac{1}{\rho}\frac{d(\rho\phi)}{dt} = K\nabla^2 h \quad \textbf{(5.14)}$$

where ∇^2 (m^{-2}) is the **Laplacian**, the second spatial derivative. The Laplacian is a scalar, whereas the first spatial derivative, the gradient ($\vec{\nabla}$, m^{-1}), is a vector.

Conceptually, the left side of equation 5.14 represents the time rate of change of fluid mass inside the control volume. According to equation 5.14, the fluid mass may change either due to changes in porosity or changes in fluid density. If the fluid mass inside the control volume is constant with respect to time, $d(\rho\phi)/dt = 0$ and we have **Laplace's equation**,

$$\nabla^2 h = 0 \quad \textbf{(5.15)}$$

More generally, the fluid mass inside the cube may change with respect to time. The left side of equation 5.14 can be expanded,

$$\frac{1}{\rho}\frac{d(\rho\phi)}{dt} = \frac{d\rho}{dt}\frac{\phi}{\rho} + \frac{d\phi}{dt} \quad \textbf{(5.16)}$$

The first term on the right side of equation 5.16, $(d\rho/dt)$, represents the compressibility of the fluid. The second term on the right side of equation 5.16, $(d\phi/dt)$, represents the compressibility of the solid matrix grains and the compressibility of the porous medium itself. Assuming that matrix grains are incompressible, we ought to be able to relate the right side of equation 5.16 to fluid compressibility (B, m-s^2-kg^{-1}), porous medium compressibility (α, m-s^2-kg^{-1}), and ultimately to specific storage (S_s, m^{-1}).

Recall the definition of fluid compressibility (B) from equation 3.24,

$$B = \frac{\dfrac{-\Delta V_w}{V_w}}{\Delta P} \quad \textbf{(5.17)}$$

where V_w (m^3) is fluid volume, and P (kg-m^{-1}-s^{-2}) is fluid pressure. A change in fluid volume can be related to a change in fluid density by introducing the definition of fluid density,

$$V_w = \frac{M_w}{\rho} \quad \textbf{(5.18)}$$

where M_w (kg) is fluid mass and ρ (kg-m^{-3}) is fluid density. If fluid mass is constant, then a change in fluid volume (dV_w, m^3) is equal to

$$dV_w = M_w \frac{-d\rho}{\rho^2} \quad \textbf{(5.19)}$$

because $d(\rho^{-1}) = -\rho^{-2}d\rho$. Substituting $M_w = \rho V_w$ into equation 5.19, we obtain

$$dV_w = -\rho V_w \frac{d\rho}{\rho^2} \quad \textbf{(5.20)}$$

or

$$\frac{dV_w}{V_w} = \frac{-d\rho}{\rho} \quad \textbf{(5.21)}$$

Substituting equation 5.21 into equation 5.17 and switching from differentials to derivatives,

$$B = \frac{\dfrac{d\rho}{\rho}}{\Delta P} \quad \textbf{(5.22)}$$

From equation 2.45, if elevation z (m) is constant, $\Delta P = \rho g \Delta h$, or $dP = \rho g dh$, where g is the acceleration due to gravity (m-s^{-2}), and h is head (m). Substituting into equation 5.22,

$$B = \frac{\dfrac{d\rho}{\rho}}{\rho g dh} \quad \textbf{(5.23)}$$

or

$$\frac{d\rho}{\rho} = \rho g dh B \qquad \textbf{(5.24)}$$

or

$$\frac{d\rho}{dt} = \rho^2 g B \frac{dh}{dt} \qquad \textbf{(5.25)}$$

where t is time. Multiplying both sides of equation 5.25 by (ϕ/ρ) we finally obtain

$$\frac{d\rho}{dt}\frac{\phi}{\rho} = \frac{dh}{dt}\phi\rho g B \qquad \textbf{(5.26)}$$

Note that the left side of equation 5.26 is identical to the first term on the right side of equation 5.16. Let us now relate the second term on the right side of equation 5.16 to porous medium compressibility (α).

From equation 3.51, a change in fluid volume (ΔV_w, m^3) due to porous medium compressibility is related to changes in fluid pressure (ΔP, kg-m^{-1}-s^{-2}) by

$$\Delta V_w = \alpha V_T \Delta P \qquad \textbf{(5.27)}$$

where V_T is the volume of a saturated porous medium. From equation 2.45, if elevation is constant, $\Delta P = \rho g \Delta h$,

$$\Delta V_w = \alpha V_T \rho g \Delta h \qquad \textbf{(5.28)}$$

For a saturated porous medium, $\Delta V_w = \Delta(\phi V_T)$,

$$\frac{\Delta(\phi V_T)}{V_T} = \alpha\rho g \Delta h \qquad \textbf{(5.29)}$$

If the total volume of the porous medium is constant, $\Delta(\phi V_T) = V_T(\Delta\phi)$, thus

$$\Delta\phi = \alpha\rho g \Delta h \qquad \textbf{(5.30)}$$

Switching from differentials to derivatives,

$$\frac{d\phi}{dt} = \alpha\rho g \frac{dh}{dt} \qquad \textbf{(5.31)}$$

Note that the left side of equation 5.31 is identical to the second term on the right side of equation 5.16. Substituting equations 5.26 and 5.31 into equation 5.16 and then into equation 5.14, we arrive at

$$\alpha\rho g \frac{dh}{dt} + \phi\rho g B \frac{dh}{dt} = K\nabla^2 h \qquad \textbf{(5.32)}$$

Gathering terms,

$$\frac{dh}{dt}\rho g(\alpha + \phi B) = K\nabla^2 h \qquad \textbf{(5.33)}$$

From equation 3.70, $S_s = \rho g(\alpha + \phi B)$

$$\frac{dh}{dt} S_s = K\nabla^2 h \qquad \textbf{(5.34)}$$

or

$$\frac{dh}{dt} = \frac{K}{S_s}\nabla^2 h \qquad \textbf{(5.35)}$$

The ratio K/S_s is known as the **hydraulic diffusivity**. The diffusion equation (5.35) says that the rate at which head is changing (dh/dt) is proportional to the curvature of the head field ($\nabla^2 h$). The constant of proportionality, the hydraulic diffusivity, is a physical property of the fluid and the porous medium.

At this point it is worth digressing slightly to explain some discrepancies that occurred during the derivation of equation 5.35, the diffusion equation. To obtain equation 5.19, we assumed that fluid mass (M_w) was constant, even though it is implicit that the fluid mass inside the control volume changes with respect to time. Similarly, in the derivation of equation 5.30, we assumed that the total volume of the porous medium (V_T) was constant, even though the porous medium is compressible. Strictly speaking, these simplifications violate our assumptions, however, their virtue is they enable us to obtain the correct result in an instructive manner without the mathematics becoming too complex. A formally correct definition of equation 5.35 requires starting with two conservation statements, one for the fluid, the other for the solid grains (Cooper, 1966; Domenico and Schwarz, 1990, p. 113).

What is $\nabla^2 h$? The geometrical interpretation of the second spatial derivative is that it is the *curvature*. Consider the case where h is constant with respect to distance x (Figure 5.2a). If h is constant, both dh/dx and $\nabla^2 h$ are zero. Thus, the diffusion equation tells us that $dh/dt = 0$, or head will not

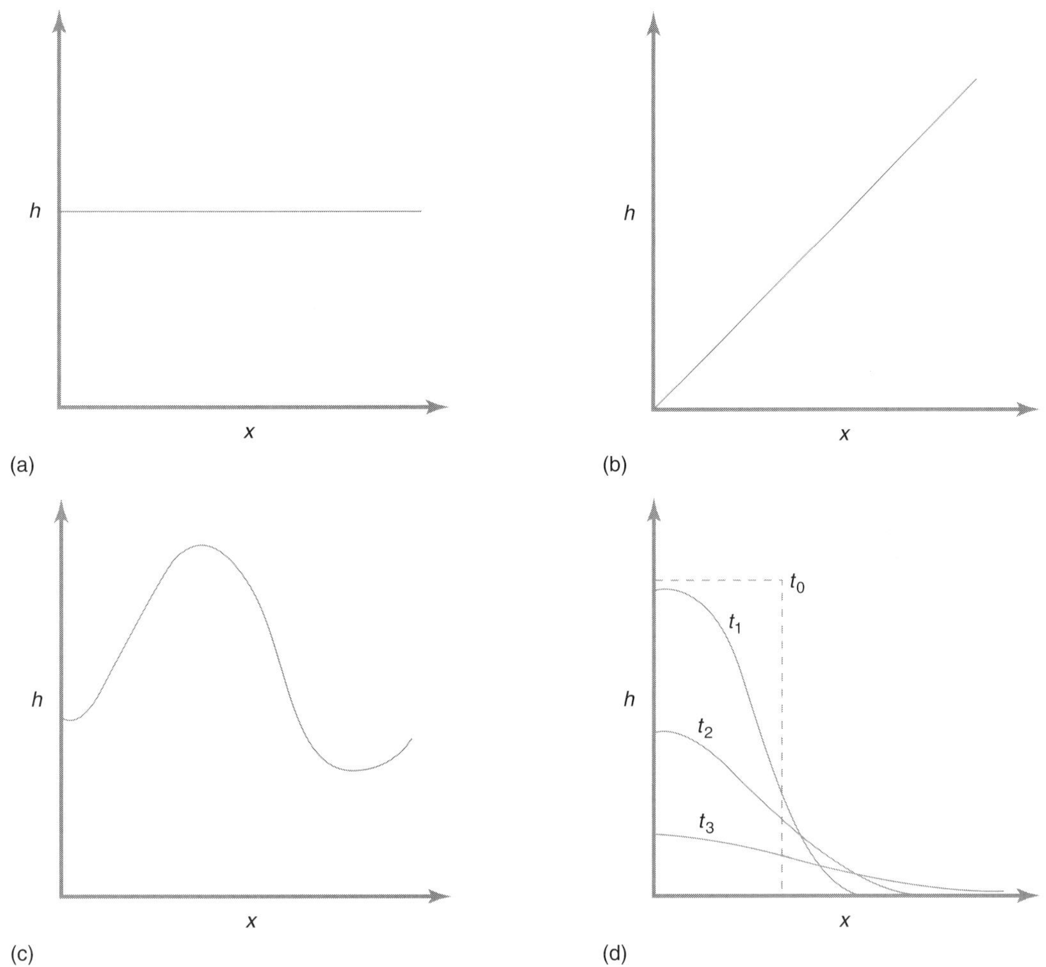

Figure 5.2 Geometric interpretations of the diffusion equation. Plots show head (h) as a function of distance (x) in a homogeneous domain. In part (a), head is constant, the head gradient is zero, and there is no flow. In part (b), head is not constant, the head gradient is constant, and flow is constant and non-zero. In part (c), neither head nor the head gradient is constant, and flow is generally non-zero and changes with time. Part (d) shows schematically the evolution of head with time supposing an initial condition as shown for time t_o. The head field evolves as described by the diffusion equation for times $t_2 > t_1 > t_0$, etc.

change with time. Darcy's Law also tells us that there is no flow as $dh/dx = 0$. If dh/dx is constant but not zero, $\nabla^2 h = 0$, and $dh/dt = 0$ (Figure 5.2b). In this case, Darcy's Law predicts a non-zero flow, but the flow is constant with respect to time.

Suppose that dh/dx is not constant with respect to x (Figure 5.2c). In this case, $\nabla^2 h$ does not equal zero and the head field will evolve and change through time. In contrast to the straight lines shown in Figures 5.2a and 5.2b, the head field in Figure 5.2c is *curved*.

Finally, consider the situation shown in Figure 5.2d, wherein the initial condition is a rectangular-shaped head distribution. At the corner of the rectangle, the curvature $\nabla^2 h$ approaches infinity, while elsewhere the curvature is zero. If the initial condition is relaxed, the head distribution will evolve through time as shown in Figure 5.2d. The rate at which head changes, dh/dt, will be highest at the points of greatest curvature. The head distribution will *diffuse* away from regions where it is high to those regions where it is low. The nature of the final head distribution will depend upon the boundary conditions, which in this example are unspecified.

The rate at which fluid transients diffuse in a porous medium is controlled by the hydraulic diffusivity, which is the ratio of the hydraulic conductivity to the specific storage. If the hydraulic conductivity (K, m-s^{-1}) is high, transients diffuse away quickly. In contrast, if the specific storage (S_s, m^{-1}) is high, transients diffuse away more slowly.

Consider a pipe filled with sand as a simple example of a one-dimensional porous medium. There are fluid pressure gauges at each end of the pipe. Assume that the pipe is horizontal, so that we may speak of changes in fluid pressure as being equivalent to changes in head (recall equation 2.45, $h = z + (P/\rho g)$, so if elevation, z, is constant, dh is proportional to dP).

If you start pumping at one end, how much time will pass before the pressure gauge at the other end changes? In other words, if we disturb the fluid pressure at one point, how quickly (or slowly) will the pressure transient diffuse (or travel) through the pipe? According to the diffusion equation (equation 5.35), the greater the hydraulic conductivity, K, the faster the pressure transient will diffuse through the medium. On the other hand, if the hydraulic conductivity is zero, the pressure transient cannot diffuse through the medium and dh/dt (or, in this case, dP/dt) is zero. Suppose the medium has no storage capacity ($S_s = 0$). According to the diffusion equation, the pressure transient will diffuse through the medium at infinite speed—it has to, there is no provision for storing energy by compressing the fluid or expanding the porous medium by pushing the sand grains apart. On the other hand, if the storage medium has an infinite capacity to store fluid ($S_s = \infty$), the fluid transient never reaches the other end of the pipe.

To utilize the diffusion equation to study fluid flow, it is necessary to solve it. In general, solutions to any differential equation may be (1) analytical, (2) numerical, or (3) graphical. Analytical solutions are those that can be written down in a simple, exact form. To obtain an analytical solution, one is forced to reduce the geologic complexity of a problem to a simplified form that has a simple geometry, uniform properties, and idealized boundary and initial conditions (Konikow and Bredehoeft, 1992). Despite these limitations, the ability to write down an exact solution to the problem may provide valuable insight into the physics. Many analytical solutions to groundwater problems are borrowed from heat conduction theory. Fourier's Law of Heat Conduction is analogous to Darcy's Law, and the diffusion of temperature by heat flow is similar to the diffusion of head by groundwater flow (see chapter 11). The classic reference for analytical solutions to heat conduction problems is the book *Conduction of Heat in Solids* (1959) by H. S. Carslaw and J. C. Jaeger.

The disadvantage to analytical solutions is that the simplifications introduced into how the problem is posed may remove some or most of the essential physics. The advantage to numerical solutions is that they can be found for domains of arbitrary geometric and geologic complexity. Although simple numerical solutions can be exact, numerical methods and computer models are usually applied to obtain approximate answers to complex problems. Whether or not enough data exist to adequately constrain the geological complexity of a numerical model is another question. Graphical solutions are approximate and intuitive. In the following sections we briefly discuss examples of each of these types of solutions.

5.2. Laplace's Equation—Solutions by Integration.

Recall Laplace's equation (equation 5.15)

$$\nabla^2 h = 0 \qquad \textbf{(5.36)}$$

We arrived at this form of Laplace's equation by assuming that hydraulic conductivity was spatially invariant and thus simplified equation 5.12. If we

Charles Edward Jacob: Groundwater Theoretician

by John D. Bredehoeft

C.E. Jacob (1914–1970), known to his friends as Jake, was the preeminent groundwater theoretician of the 1940s and 1950s.

The groundwater profession in the 1920s recognized that the Thiem steady-state theory was inadequate to describe the data from observation wells. It was clear to the hydrogeologists of the day, especially the members of the Groundwater Branch of the U.S. Geological Survey, that there had to be some form of elastic aquifer response and a non-equilibrium theory. But, the form of the new theory was unclear. It was not until C. V. Theis suggested the analogy to the theory of heat flow in a 1935 paper that the new theory began to take shape. Theis' 1935 paper was met with skepticism, especially by his boss the Chief of the Groundwater Branch, Oscar Meinzer. It was Jacob who derived the groundwater flow equation (the so-called Theis equation) from first principles.

Jacob received his B.S. degree in Civil Engineering from the University of Utah in 1935, graduating with honors. He went on to Columbia University where he obtained an M.S. degree in Civil Engineering (hydraulics) in 1936. From there he joined the Groundwater Branch of the USGS on Long Island, New York. His first paper, published in 1939 in the *Transactions of the American Geophysical Union*, dealt with observations in an artesian well on Long Island of passing railroad trains. In this paper Jacob demonstrated his understanding of the elasticity of aquifer systems.

Jacob's studies in aquifer elasticity led him to publish his derivation of the groundwater flow equation in the *Transactions of the American Geophysical Union* in 1940. In this paper he derived the flow equation from first principles. The heart of the derivation is the storage coefficient; Jacob showed that it involved both the compressibility of the rock framework as well as that of the water. He went on to show that the general groundwater flow equation is analogous to the equation for heat flow; this vindicated Theis' earlier suggestion. It convinced the groundwater community of the validity of this approach.

It is the general groundwater flow equation, suggested by Theis and derived by Jacob, upon which the modern theory of groundwater is built. Once the general groundwater flow equation was derived it was quickly applied to numerous problems of flow to wells. The 1940s and early 1950s were the period of well hydraulics in groundwater hydrology. A plethora of solutions to problems groundwater flow to wells were derived. Jacob was the leader in this effort.

Charles Edward Jacob (1914–1970)

From Long Island, Jacob went to Washington, DC with the Groundwater Branch of the U.S. Geological Survey. Jacob was a devoted Mormon; in 1947 he returned to the University of Utah as Head of the Department of Geophysics. In 1953 he joined the faculty of Brigham Young University.

Jacob's prized student was Mahdi Hantush, a native of Iraq; Hantush earned his Ph.D. with Jacob at the University of Utah. Perhaps the most important conceptual extension of the groundwater theory in

Charles Edward Jacob: Groundwater Theoretician *(continued)*

the 1940s was the recognition that the confining layers associated with artesian aquifers were not impermeable. This was not a new idea, Darton and Chamberlain had discussed flow through the confining layers in the 1890s and early 1900s. But, including the flow through confining layers in the general equation that described groundwater flow was new. Theis in his 1935 derivation had assumed that the confining layers were impermeable, this simplified his analysis. Jacob and Hantush derived a solution for non-steady flow to a well in a "leaky" aquifer in 1955. Hantush later revised the leaky aquifer solution to include the effects of storage in the confining layer.

Jacob's bibliography is not long–18 published papers. One of the most lucid statements of groundwater theory is a chapter he wrote entitled *Flow of Groundwater* that was published in the book *Engineering Hydraulics* (1950). In this single chapter Jacob developed the theory of groundwater in great elegance. This chapter is still relevant and worth reading by any serious hydrogeologist; it demonstrated Jacob's full understanding of the discipline.

Mahdi Hantush established a groundwater program at New Mexico Institute of Mining and Hydrology in Socorro, New Mexico. Hantush had a distinguished program that graduated a number of hydrogeologists who are leaders in the profession today. In 1965 Hantush returned to Baghdad, Iraq, to become a Dean at the National University; this was something of a family obligation. Jacob replaced Hantush at New Mexico Tech; he continued the tradition begun by Hantush. Jacob was instrumental in uniting the Departments of Geology, Geophysics, and Hydrology into an expanded Department of Geosciences. There is still a strong graduate program in hydrogeology at New Mexico Tech.

In 1961, Jacob formed the consulting firm C.E. Jacob and Associates. The firm was successful with clients from throughout the world. He died in 1970, at the age of 56 from a heart attack. His premature death ended a distinguished career. Jake was to his many friends an energetic, personable, and friendly man. Professionally he was dynamic, imaginative and productive. His colleagues, most of whom are now dead, spoke highly of Jake—to a man they held him in highest esteem.

Jacob's derivation of the storage coefficient, the key to deriving the groundwater flow equation, has stood the test of time. In 1966, Roger de Wiest published a paper in *The Journal of Geophysical Research* ("On the Storage Coefficient and the Equations of Groundwater Flow") suggesting that Jacob had overlooked an important term in the storage coefficient. The suggestion was refuted by Hilton Cooper, in his paper "The Equation of Groundwater Flow in Fixed and Deforming Coordinates," that was published the same year. Cooper showed that de Wiest had misinterpreted Darcy's Law as applied to a deforming media; Jacob's storage coefficient was correct—Cooper received the Meinzer Award of the Geological Society of America for this paper.

For Additional Reading

Cooper, H. H., Jr. 1966. The equation of groundwater flow in fixed and deforming coordinates. *J. Geophysical Res* 71: 4785–4790.

De Wiest, R. J. M. 1966. On the storage coefficient and the equations of groundwater flow: *J. Geophysical Res* 71: 1117–1122.

Jacob, C. E. 1950. On the flow of water in an elastic artesian aquifer. *Transactions, American Geophysical Union* 574–86.

Jacob, C. E. 1950. Flow of ground water. In, *Engineering Hydraulics*, ed. H. Rouse New York: John Wiley & Sons, 321–86.

Titus, F. B. 1973. Memorial to Charles Edward Jacob, 1914–1970. *Geological Society of America Memorials* 2: 76–78.

Spiegel, Z. 1971. The Contributions of C. E. Jacob to Scientific Hydrology and Engineering Works. In *Proceedings of the National Ground Water Quality Symposium,* August 25–27, 1971. Denver, Colorado, 7–9.

do not make this simplifying assumption, Laplace's equation may take on the form

$$\frac{d\left[-K_X\frac{dh}{dx}\right]}{dx}=0 \tag{5.37}$$

or, in multiple spatial dimensions,

$$\frac{d\left[-K_x\frac{dh}{dx}\right]}{dx}+\frac{d\left[-K_y\frac{dh}{dy}\right]}{dy}+\frac{d\left[-K_z\frac{dh}{dz}\right]}{dz}=0 \tag{5.38}$$

In general, there are two methods for solving differential equations: integration and guessing. The simplified form of Laplace's equation given in equation 5.36 can be solved by integration. Working in one dimension (x),

$$\frac{d^2h}{dx^2}=0 \tag{5.39}$$

$$\frac{dh}{dx}=c_1 \tag{5.40}$$

$$h=c_1x+c_2 \tag{5.41}$$

where c_1 (dimensionless) and c_2 (m) are constants of integration. Equation 5.41 has the geometric form of a line (Figure 5.3). The slope of the line (c_1) and the intercept (c_2) depend upon the boundary conditions. Because there are two unknowns (c_1 and c_2), two boundary conditions are required to arrive at a unique solution. The boundary conditions may either be the value of h at two distinct points (points on the line), or one point on the line and the value of dh/dx (the slope of the line). Note that for any given set of boundary conditions, Laplace's equation has only one solution: there is only one line that passes through any two given points, and there is only one line with a specified slope that passes through a given point.

The above solution presupposes that hydraulic conductivity is homogeneous and does not vary spatially. If hydraulic conductivity is piecewise constant over specified domains, it is possible to obtain one-dimensional solutions by solving equation 5.36 in each domain and requiring that the solutions match at the boundaries.

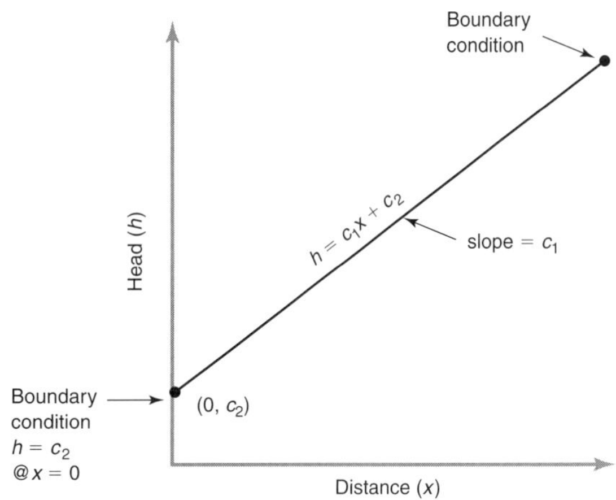

Figure 5.3 One-dimensional solution to Laplace's equation in a homogeneous medium is a line.

Problem: Solve Laplace's equation ($\nabla^2h = 0$) in one dimension (x). Boundary conditions are $h(0)=1$ m and $h(10) = 13$ m. For $5 \geq x \geq 0$, hydraulic conductivity is $K_1 = 6$ m-s^{-1}; for $10\geq x \geq 5$, hydraulic conductivity $K_2 = 3$ m-s^{-1}. (a) What is value of head at $x = 5$? (b) What is the Darcy velocity? (c) Graph the solution.

a. It is useful to first sketch the problem, laying out each domain and the demarcation between them (Figure 5.4a). The solution in each domain will be linear. Because Laplace's equation applies, we have steady-state flow. Thus, the Darcy velocity in one domain must be equal to the Darcy velocity in the other.

$$q_1=-K_1\,(\nabla h)_1=q_2=-K_2(\nabla h)_2 \tag{5.42}$$

Let h at $x = 0$ be denoted h_0, etc., and substitute into equation 5.42,

$$K_1\frac{h_5-h_0}{5}=K_2\frac{h_{10}-h_5}{5} \tag{5.43}$$

Solving for h_5,

$$K_1h_5-K_1h_0=K_2h_{10}-K_2h_5 \tag{5.44}$$

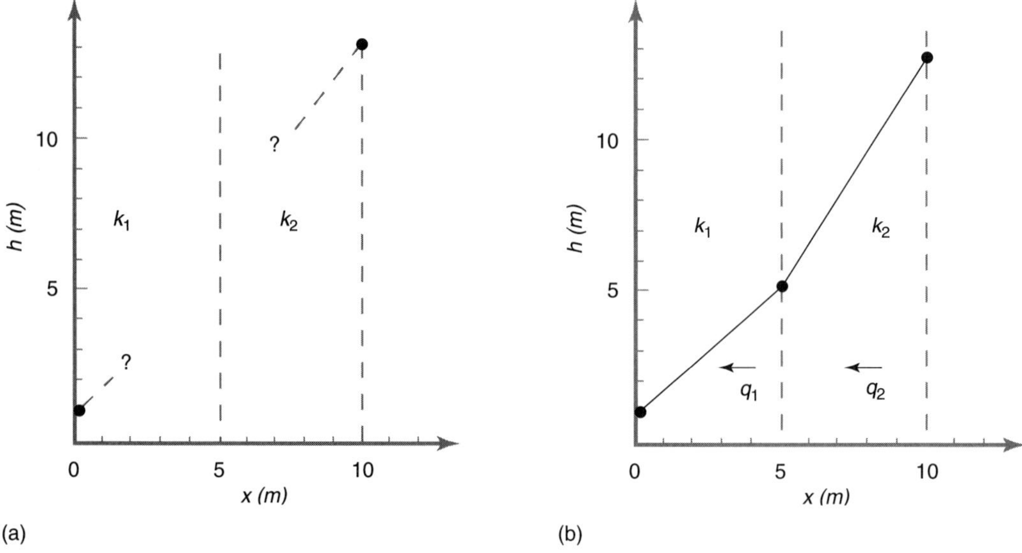

Figure 5.4 One-dimensional solution to Laplace's equation in a heterogeneous medium is piecewise linear.

$$h_5(K_1+K_2) = K_2h_{10} + K_1h_0 \tag{5.45}$$

$$h_5 = \frac{K_2h_{10} + K_1h_0}{K_1 + K_2} \tag{5.46}$$

$$h_5 = \frac{3(\text{m-s}^{-1}) \times 13(\text{m}) + 6(\text{m-s}^{-1}) \times 1(\text{m})}{6(\text{m-s}^{-1}) + 3(\text{m-s}^{-1})} \tag{5.47}$$

$$h_5 = 5(\text{m}) \tag{5.48}$$

b. The Darcy velocity is

$$q_1 = -K_1(\nabla h)_1 = -K_1\frac{h_5 - h_0}{5} \tag{5.49}$$

$$q_1 = -6(\text{m-s}^{-1}) \times \frac{5(\text{m}) - 1(\text{m})}{5(\text{m})} \tag{5.50}$$

$$q_1 = \frac{-24}{5}(\text{m-s}^{-1}) \tag{5.51}$$

where the minus sign indicates flow is in the $-x$ direction. We can check the answer to part (a) by calculating q_2 and seeing if $q_1 = q_2$.

$$q_2 = -K_2(\nabla h)_2 = -K_2\frac{h_{10} - h_5}{5} \tag{5.52}$$

$$q_2 = -3(\text{m-s}^{-1}) \times \frac{13(\text{m}) - 5(\text{m})}{5(\text{m})} \tag{5.53}$$

$$q_2 = \frac{-24}{5}(\text{m-s}^{-1}) \tag{5.54}$$

and we find that $q_1 = q_2$ as required.

c. The solution may now be graphed (Figure 5.4b).

5.3. Laplace's Equation — Solutions by Flow Nets.

If head is assumed to be a potential field (true if fluid density is constant), head contours are lines of equal potential or **equipotential** lines. Lines representing the direction of fluid flow are **flow lines** or **streamlines**. Flow lines are perpendicular to equipotential lines. The area between two streamlines is a **flow tube**. A **flow net** is a plot of equipotential and streamlines; it is a graphical solution to Laplace's equation. Flow nets are usually constructed in two dimensions, although there is no theoretical reason that three-dimensional flow nets cannot be made.

Casagrande (1937, p. 135) attributed the origin of flow-net solutions to Philipp Forcheimer (1852–1933). Forcheimer was an Austrian hydrologist who is remembered as the author of the monumental work *Hydraulik* (1930), first published in 1914. Rouse and Ince (1957) described Forcheimer's *Hydraulik* as "the outstanding compilation of and commentary upon hydraulic data of all time."

The rules for constructing flow nets are as follows:

1. Flow lines must be perpendicular to equipotential lines.
2. Equipotential lines must be perpendicular to impermeable boundaries.
3. Equipotential lines are parallel to constant-head boundaries.
4. Groundwater divides are planes of symmetry and constitute no-flow boundaries.
5. Points on the water table that have equal elevation must be connected by an equipotential line.
6. Equipotential lines should be drawn so that the difference in head between any two lines is the same.
7. Flow lines should be drawn so that the total discharge (m^3-s^{-1}) per unit length (m) between any two flow lines is the same.

If rules (6) and (7) are followed, then the ratio of the sides of any quadrilateral bounded by two flow lines and two equipotentials is constant throughout the flow net. Both Casagrande (1937, p. 137) and Cedergren (1989, p. 127) offer a number of practical suggestions for constructing flow nets.

1. Before starting, examine the boundary conditions carefully for fixed equipotential and no-flow boundaries. Also look for planes of symmetry. Some flow lines and equipotentials are predetermined by boundary conditions for every flow net. Mark the direction of flow lines and/or equipotentials at these locations.
2. Do not clutter up the drawing with too many lines. Use only enough to include essential features. Four or five flow tubes are generally sufficient for the first attempt.
3. Pay attention to the appearance of the entire flow net. Do not focus on details until the flow net has become refined through iteration.
4. Make smooth transitions between the smooth and curved sections of flow lines and equipotentials.
5. Do not expect to draw a perfect flow net upon the first attempt. Use pencil, and erase and refine your flow net through iteration.

Example: Flow through a homogeneous, isotropic substrate underneath a dam (Figure 5.5). The dam is assumed to be impermeable. The substrate, which may be either sediment or sedimentary rock, is assumed to be of sufficiently high permeability that the underlying rock may be treated as an impermeable boundary. The bottom of the water body on the left side of the dam is at constant pressure and elevation, thus it is an equipotential. Pressure and elevation are also fixed on the right side of the dam at ground level, thus it also is an equipotential.

Example: Topographically driven groundwater flow through a homogeneous, isotropic earth (Figure 5.6). The boundary condition at the upper surface is fixed head, however, head is not constant along the upper surface, but varies according to the elevation of the water table. The elevation at which each equipotential intersects the ground surface gives the value of head along that equipotential. The midpoint of the hill defines a plane of symmetry that is a no-flow boundary. As

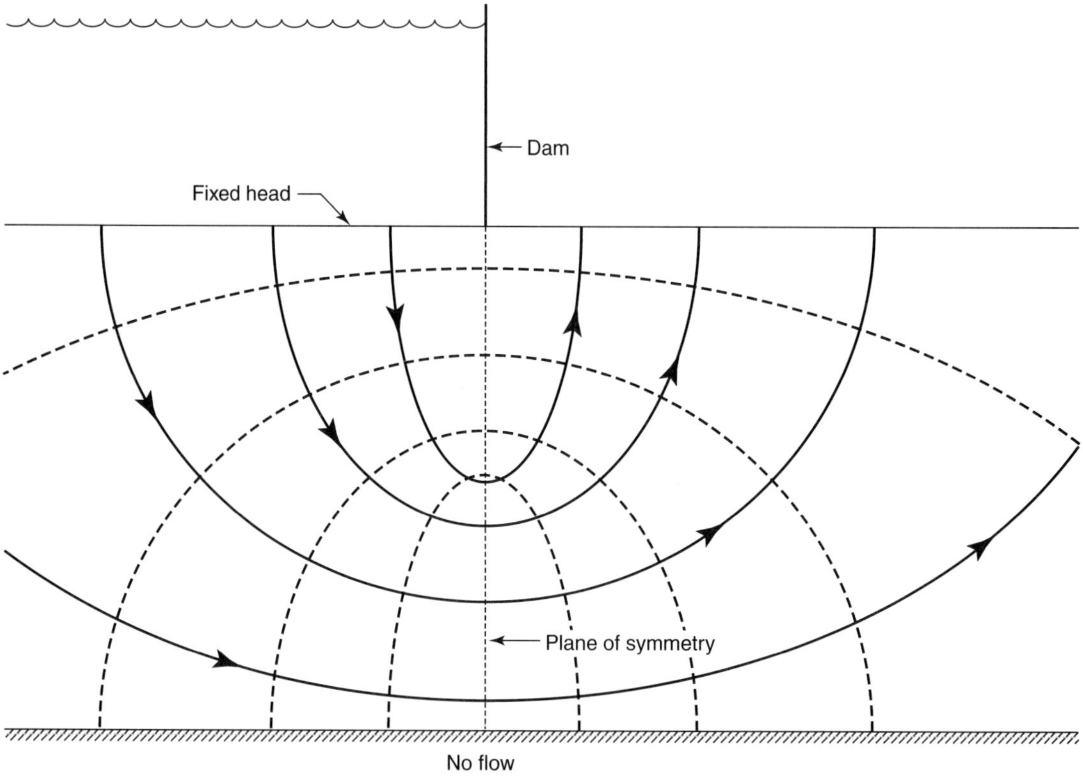

Dam
Fixed head
Plane of symmetry
No flow

Figure 5.5 Flow net for flow underneath a dam.
(After Casagrande, 1937, p. 135.)

Water table

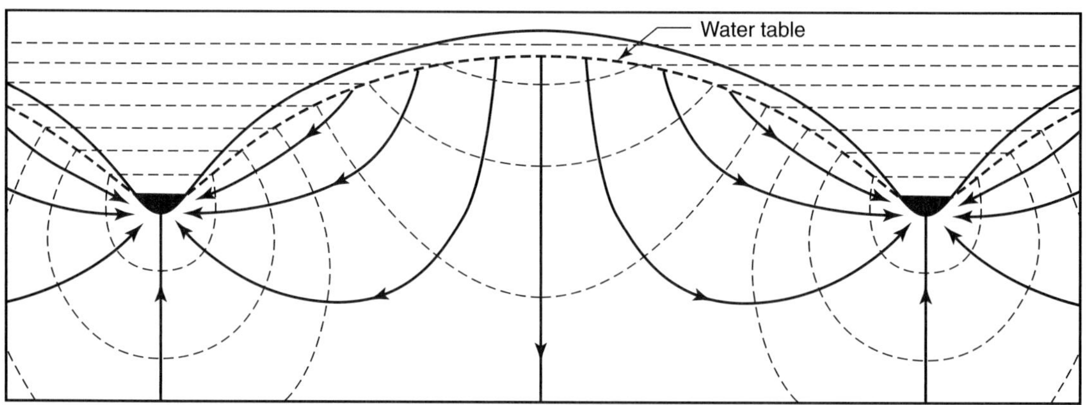

Figure 5.6 Flow net for topographically driven flow.
(From Hubbert, 1940, p. 930.)

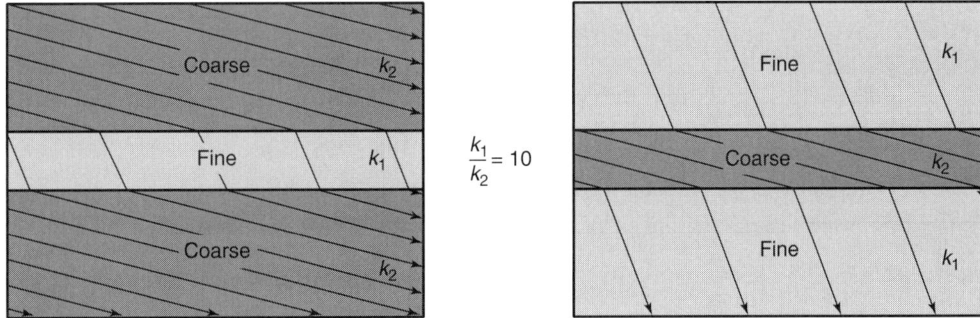

Figure 5.7 Refraction of flow lines across porous media with a permeability contrast of 10:1.
(From Hubbert, 1940, p. 846.)

water flows downhill, streamlines must indicate flow from upper elevations to lower elevations. Thus the equipotentials must be curved as shown.

5.4. REFRACTION OF FLOW LINES.

In a heterogeneous medium where hydraulic conductivity is not constant, flow lines are **refracted** or bent (Figure 5.7).

Consider flow across a boundary between two media (Figure 5.8a). The top medium has a hydraulic conductivity K_1(m-s^{-1}) $<$ K_2(m-s^{-1}), where K_2 is the hydraulic conductivity in the bottom layer. As drawn (Figure 5.8a), the flow lines define flow tubes. In the top medium, the flow tube has width a(m); in the bottom medium, the flow tube has width c(m). Each tube has the same depth in the third dimension, perpendicular to the two dimensions of the figure as drawn. If the flow is steady-state the rate at which fluid moves through the top tube must be equal to the flow rate through the bottom tube,

$$Q_1 = Q_2 \tag{5.55}$$

where Q_1 is the volumetric flow rate (m^3-s^{-1}) in the top layer, and Q_2(m^3-s^{-1}) is the volumetric flow rate in the bottom layer. Because the volumetric flow rate is equal to the Darcy velocity multiplied by the area perpendicular to flow,

$$q_1 a = q_2 c \tag{5.56}$$

According to Darcy's Law,

$$q_1 = K_1 \frac{\Delta h}{\Delta s_1} \tag{5.57}$$

where Δs_1(m) is the length shown in Figure 5.8(b), and Δh(m) is the head drop which occurs along the length Δs_1. Equation 5.57 can be written as

$$q_1 a = K_1(a) \frac{\Delta h}{\Delta s_1} \tag{5.58}$$

Similarly,

$$q_2 c = K_2(c) \frac{\Delta h}{\Delta s_2} \tag{5.59}$$

where Δs_2(m) is the length shown in Figure 5.8(c), and the head drop Δh(m) is the change in head that occurs along the length Δs_2. Note that Δh is the same for each flow tube. Substituting equations 5.58 and 5.59 into equation 5.56,

$$K_1(a) \frac{\Delta h}{\Delta s_1} = K_2(c) \frac{\Delta h}{\Delta s_2} \tag{5.60}$$

where a(m) is the distance between the flow lines in the top layer, c(m) is the distance between the flow lines in the bottom layer, $\Delta h = h_1 - h_2$(m), Δs_1(m) is the physical distance between the head contours h_1(m) and h_2(m) in the top layer, and Δs_2(m) is the distance between the head contours h_1 and h_2 in the bottom layer (Figure 5.8).

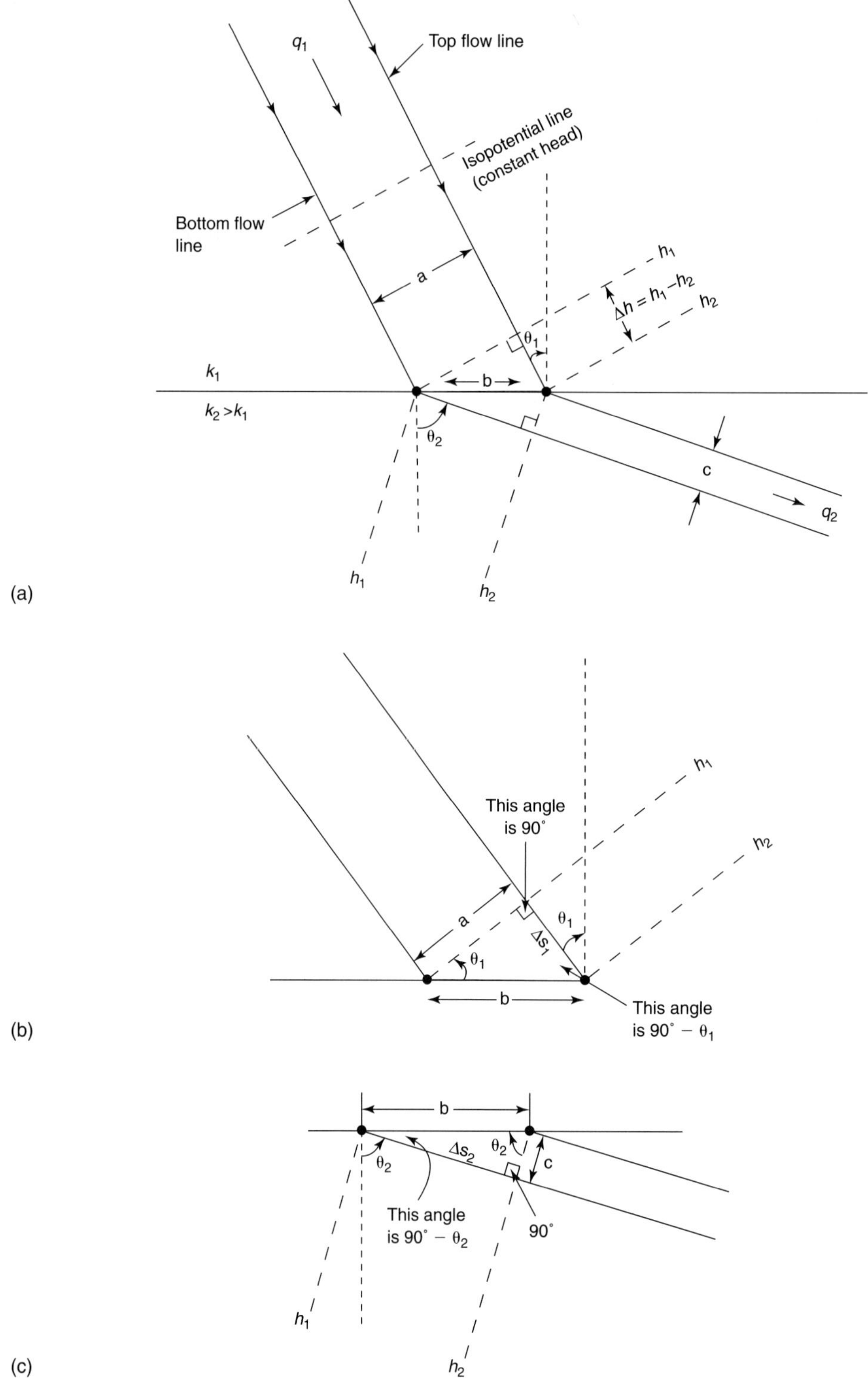

Figure 5.8 a, b, c Figures for calculating refraction of flow lines.

The angle Θ_1 (dimensionless) is the angle between the *top* flow line and a line drawn perpendicular to the boundary between the top and bottom layers (Figure 5.8b). The angle between the top flow line and the boundary must therefore be $90^\circ - \Theta_1$, and the angle between the boundary and the equipotential h_1 must be Θ_1, as the sum of the angles in any triangle is 180°. Thus (refer to Figure 5.8(b))

$$a = b\cos\Theta_1 \quad \textbf{(5.61)}$$

and

$$\Delta s_1 = b\sin\Theta_1 \quad \textbf{(5.62)}$$

The angle Θ_2 (dimensionless) is the angle between the *bottom* flow line and a line drawn perpendicular to the boundary between the top and bottom layers (Figure 5.8c). The angle between the bottom flow line and the boundary must therefore be $90^\circ - \Theta_2$, and the angle between the boundary and the equipotential h_2 must be Θ_2 as the sum of the angles in a triangle is 180°. Thus (refer to Figure 5.8(c))

$$c = b\cos\Theta_2 \quad \textbf{(5.63)}$$

and

$$\Delta s_2 = b\sin\Theta_2 \quad \textbf{(5.64)}$$

Substituting equations 5.61, 5.62, 5.63, and 5.64 into equation 5.60 yields

$$K_1(b\cos\Theta_1)\frac{\Delta h}{b\sin\Theta_1} = K_2(b\cos\Theta_2)\frac{\Delta h}{b\sin\Theta_2} \quad \textbf{(5.65)}$$

or

$$K_1\frac{\cos\Theta_1}{\sin\Theta_1} = K_2\frac{\cos\Theta_2}{\sin\Theta_2} \quad \textbf{(5.66)}$$

or

$$\frac{K_1}{K_2} = \frac{\tan\Theta_1}{\tan\Theta_2} \quad \textbf{(5.67)}$$

As permeability (k, m^2) is proportional to hydraulic conductivity (equation 2.56), we can also write

$$\frac{k_1}{k_2} = \frac{\tan\Theta_1}{\tan\Theta_2} \quad \textbf{(5.68)}$$

So if $K_2 > K_1$ (Figure 5.8), $\tan\Theta_2 > \tan\Theta_1$, $\Theta_2 > \Theta_1$, and flow is refracted *away* from the normal. Conversely, if $K_2 < K_1$, flow is refracted *towards* the normal. A practical consequence of equation 5.67 is that flow in aquifers tends to be horizontal, while flow in confining layers or aquitards tends to be vertical.

Problem: Flow moves from an overlying shale ($k_1 = 10^{-17}$ m^2) into an underlying sandstone ($k_2 = 10^{-14}$ m^2). The direction of flow in the shale is defined by the angle Θ_1, where Θ_1 is the angle between a flow line and a line perpendicular to bedding. Let Θ_2 be the angle of flow in the underlying sandstone aquifer. (a) if Θ_1 is 10°, what is Θ_2? (b) if Θ_1 is 1°, what is Θ_2?

a. Rewriting equation 5.68,

$$\tan\Theta_2 = \frac{k_2}{k_1}(\tan\Theta_1) \quad \textbf{(5.69)}$$

or

$$\Theta_2 = \tan^{-1}\left[\frac{k_2}{k_1}(\tan\Theta_1)\right] \quad \textbf{(5.70)}$$

$$\Theta_2 = \tan^{-1}\left[\frac{10^{-14}}{10^{-17}}\tan(10^\circ)\right] \quad \textbf{(5.71)}$$

$$\Theta_2 = \tan^{-1}[(1000)(0.176326)] \quad \textbf{(5.72)}$$

$$\Theta_2 = 89.7^\circ \quad \textbf{(5.73)}$$

b. and if $\Theta_1 = 1^\circ$,

$$\Theta_2 = \tan^{-1}[(1000)(0.017455)] \quad \textbf{(5.74)}$$

$$\Theta_2 = 86.7^\circ \quad \textbf{(5.75)}$$

A practical application of flow-line refraction is understanding topographically driven flow in a foreland basin consisting of a basal aquifer overlain by a confining layer or aquitard (Figure 7.15). The refraction of the flow lines acts to concentrate flow in the basal aquifer. Flow in the basal aquifer is dominantly horizontal, while flow in the confining layer is primarily vertical. The path of least resistance from one end of the basin to the other is through the basal aquifer. The shortest flow path

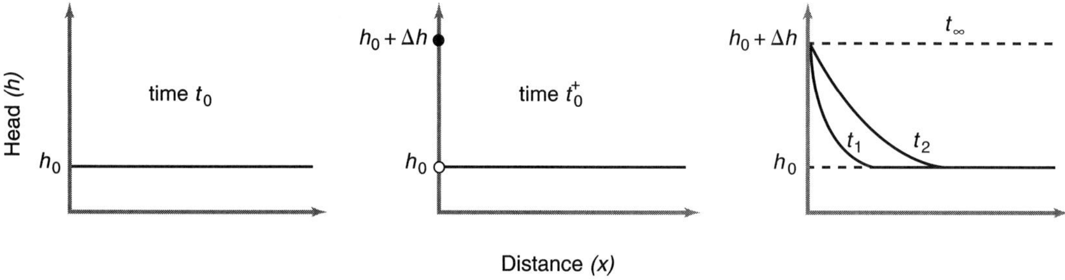

Figure 5.9 Propagation of an instantaneous head change of magnitude Δh at one boundary of a one-dimensional porous medium.

from an energy standpoint is for the fluid to move directly down through the aquitard to the underlying aquifer. The fluid then moves laterally through the aquifer as far as possible before moving vertically through the aquitard to eventual discharge at the surface.

5.5. Characteristic Length-Time.

It is often the case that we lack the data needed to meaningfully constrain detailed computer models. A geologic situation may also be too complex to be readily amenable to an analytical solution, most of which rely upon a simplified representation of the geometry. In these circumstances it is still possible to quickly obtain an approximate answer to a flow problem through a **scale analysis**. A scale analysis utilizes the relationship between characteristic time and length, and allows us to not only obtain approximate answers, but also provides insight into the physics of transient fluid flow into a porous medium.

Consider a pipe filled with sand as an idealized one-dimensional porous medium. The hydraulic diffusivity is constant, and the sides of the pipe do not interact with the porous medium or the fluid. The initial condition is $h = h_0$(m) everywhere at time t_0 (s). At an increment of time infinitesimally later, t_0^+, head at $x = 0$ is instantaneously incremented by an amount Δh(m). At succeeding times t_1, t_2, etc., the increase in head propagates along the pipe (Figure 5.9).

What is the characteristic time t (s) it will take for a transient hydraulic disturbance to propagate a characteristic distance y (m)? The rate at which the change in head propagates through the porous medium is determined by the diffusion equation (equation 5.35)

$$\frac{dh}{dt} = D\nabla^2 h \qquad \textbf{(5.76)}$$

where h (m) is head, t (s) is time, and D (m^2-s^{-1}) is the hydraulic diffusivity. Recall that the hydraulic diffusivity (D) is the hydraulic conductivity (K, m-s^{-1}) divided by the specific storage (S_s, m^{-1}).

The solution to the problem described above is (Turcotte and Schubert, 1982, p. 159)

$$h(x, t) = \Delta h \operatorname{erfc}\left[\frac{x}{\sqrt{4Dt}}\right] + h_0 \qquad \textbf{(5.77)}$$

where x (m) is distance, t (s) is time, Δh (m) is the head increment at $x = 0$, erfc $= 1 -$ erf, and erf is the error function. The error function is

$$\operatorname{erf}(u) = \frac{2}{\sqrt{\Pi}} \int_0^u e^{-s^2} ds \qquad \textbf{(5.78)}$$

The erf(0) = 0, erf(1) = 0.84, erf(2) = 0.9953 (Figure 5.10).

How long does it take the disturbance at $x = 0$ to propagate a distance y? The answer is given by equation 5.77,

$$h(x = y, t) = h_0 + \Delta h \operatorname{erfc}\left[\frac{y}{\sqrt{4Dt}}\right] \qquad \textbf{(5.79)}$$

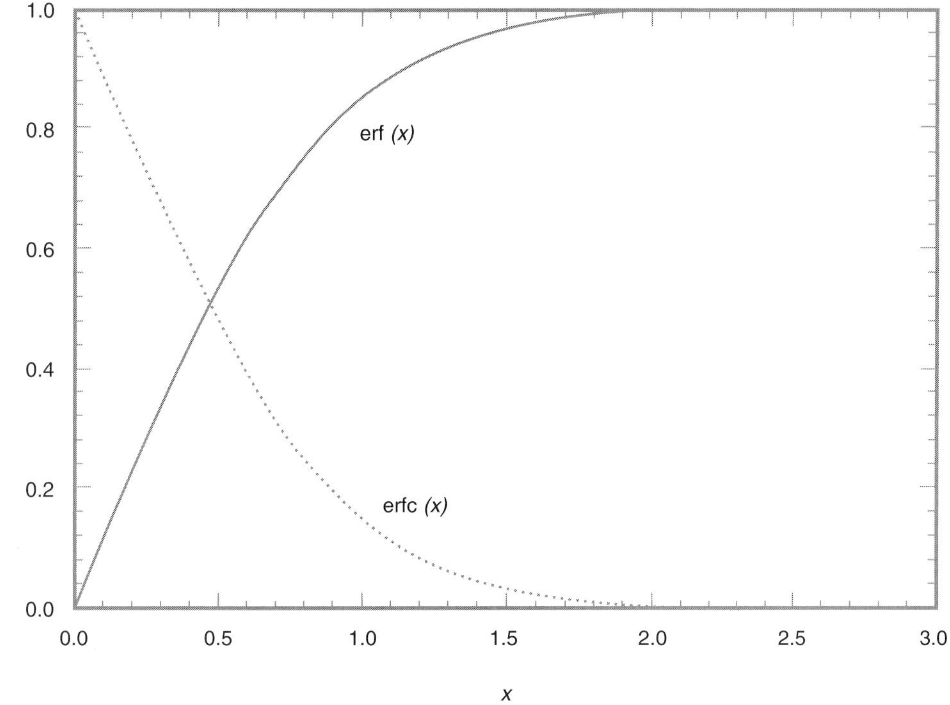

Figure 5.10 The error function (erf x) and complementary error function (erfc x).

But, there is some ambiguity in what the word "propagate" means. Let us define "propagate" as the point in space-time (x, t) where the change in head equals 16% of the eventual change (Figure 5.11). In other words, the characteristic length-time will be defined essentially as the point in space and time at which the disturbance first has a *significant* appearance. It is convenient to use 16% (instead, of, say, 15%) for reasons that will become apparent below. If y is chosen such that $h(x = y, t) = h_0 + 0.16\Delta h$, then

$$h_0 + 0.16\Delta h = h_0 + \Delta h \operatorname{erfc}\left[\frac{y}{\sqrt{4Dt}}\right] \quad \textbf{(5.80)}$$

$$0.16 = \operatorname{erfc}\left[\frac{y}{\sqrt{4Dt}}\right] \quad \textbf{(5.81)}$$

Recall that $\operatorname{erfc}(x) = 1 - \operatorname{erf}(x)$. Thus

$$0.16 = 1 - \operatorname{erf}\left[\frac{y}{\sqrt{4Dt}}\right] \quad \textbf{(5.82)}$$

$$\operatorname{erf}\left[\frac{y}{\sqrt{4Dt}}\right] = 0.84 \quad \textbf{(5.83)}$$

But erf(1) =0.84. Thus

$$\frac{y}{\sqrt{4Dt}} = 1 \quad \textbf{(5.84)}$$

or

$$y = \sqrt{4Dt} \quad \textbf{(5.85)}$$

Equation 5.85 describes the characteristic distance y (m) a hydraulic disturbance propagates through a medium of hydraulic diffusivity D (m^2-s^{-1}) in a characteristic time t (s). Note that t depends not upon y, but upon y^2.

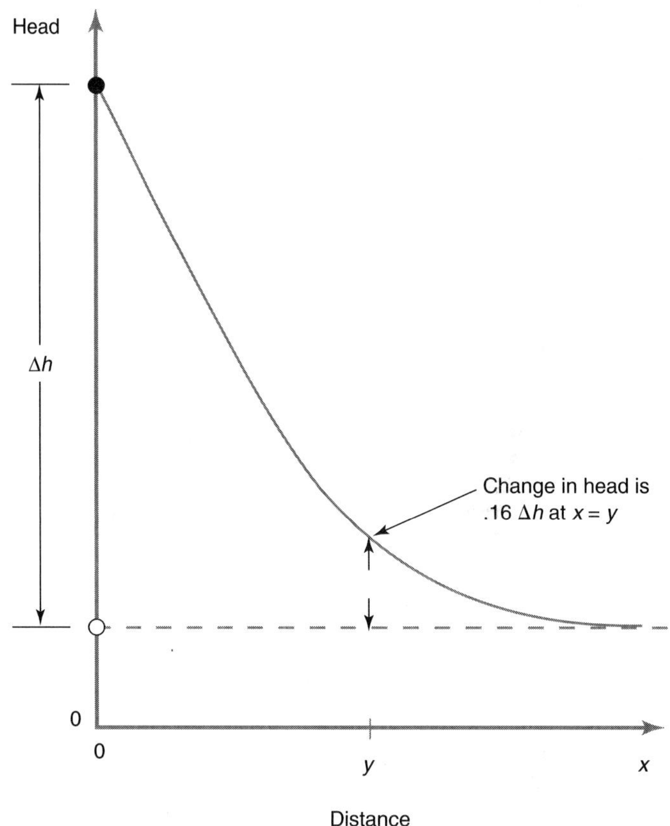

Figure 5.11 At a characteristic time t, the head change at the characteristic distance $x = y$ is considered to be significant if it is 16% of the head change Δh at $x = 0$.

Problem: Your neighbor drills a well 1000 m from your well and starts pumping water out of the same aquifer that you rely upon. How long will it be before the disturbance affects your well?

Rearranging equation 5.85, we solve for t (s),

$$t = \frac{y^2}{4D} \tag{5.86}$$

The hydraulic diffusivity, D (m^2-s^{-1}), is the hydraulic conductivity (K, m-s^{-1}) divided by the specific storage (S_s, m^{-1}). The hydraulic conductivity is related to permeability (k, m^2) by equation 2.56.

$$D = \frac{K}{S_s} = \frac{k\rho g}{\mu S_s} \tag{5.87}$$

where ρ is fluid density (kg-m^{-3}), g is the acceleration due to gravity (m-s^{-2})and μ is fluid viscosity (kg-m^{-1}-s^{-1}). From equation 3.71, the specific storage of most rocks tends to be in the range

$$2 \times 10^{-7} \leq S_s \leq 10^{-4}\text{m}^{-1} \tag{5.88}$$

The dynamic viscosity of water is temperature dependent (equation 4.23). At 20°C, $\mu = 1.0 \times 10^{-3}$ (kg-m^{-1}-s^{-1}). Assume a permeability of $k = 10^{-14}$ m^2, and a specific storage of $S_s = 10^{-6}$ m^{-1}. For these numbers, the hydraulic diffusivity is

$$D = \frac{10^{-14}(\text{m}^2) \times 1000(\text{kg-m}^{-3}) \times 9.8(\text{m-s}^{-2})}{1.8 \times 10^{-3}(\text{kg-m}^{-1}\text{s}^{-1}) \times 10^{-6}(\text{m}^{-1})} \tag{5.89}$$

$$D = 5.44 \times 10^{-2} \text{m}^2\text{-s}^{-1} \tag{5.90}$$

Substituting into equation 5.86 to find time,

$$t = \frac{(1000)^2(\text{m}^2)}{4 \times 5.44 \times 10^{-2}(\text{m}^2\text{-s}^{-1})} \tag{5.91}$$

$$t = 4.6 \times 10^6(\text{s}) \times \frac{1(\text{hr})}{3600(\text{s})} \times \frac{1(\text{day})}{24(\text{hr})} = 53 \text{ days} \tag{5.92}$$

Depending on the permeability and specific storage of the aquifer, it thus seems likely that the effects of the neighbor's pumping will start appearing at your well within a few months. There is no way to discern from the above analysis whether or not the effects will be significant. The scale analysis only tells us the characteristic time it takes the disturbance to propagate—it does not tell us the magnitude of the disturbance.

Note that in order to find t we had to make some assumptions about the permeability and specific storage of the aquifer. Even if we lack these data, the value of a scale analysis is that it defines the relationship between time and hydraulic diffusivity in a quantitative manner. In other words, we now understand for what values of hydraulic diffusivity the pumping disturbance will occur within a relevant timespan, even if we lack a specific knowledge of what the hydraulic diffusivity is. Having the insight to know what data are important to solve the problem, it is now possible to conduct the proper investigations to obtain these data.

The above estimate must be considered a minimum estimate of the time necessary before the disturbance propagates to the neighboring well. Even if we have characterized accurately the hydraulic properties of the groundwater and aquifer, one of the assumptions inherent in the derivation of equation 5.85 has been violated. We assumed a one-dimensional geometry in our derivation, whereas in the case of the aquifer the geometry is at least two-dimensional, and perhaps three-dimensional. In the well example above, the disturbance is likely to take longer to propagate to the neighboring well than we would expect from our one-dimensional calculation. This is because the head disturbance associated with pumping of the new well may diffuse away into two or three dimensions, diminishing its effect.

In general, equation 5.85 illustrates that for transient flow problems it is impossible to separate time and distance from hydraulic diffusivity. The three are inexorably linked through the process of diffusion. An important result is that for transient problems, the characteristic time for a hydraulic disturbance depends upon the square of the distance. It takes four times as long for a transient to propagate twice as far. Thus, if we seek to confine a hydraulic transient of some type, we are better off with twice the thickness of a given confining material rather than the same thickness of a different material with half the hydraulic diffusivity. In contrast, for steady-state flow problems, the resistance to flow is proportional to the hydraulic resistance, which is distance divided by the hydraulic conductivity.

5.6. Finite Differences—A Numerical Method.

Not all problems are amenable to exact, or analytical solutions, however, the diffusion equation can always be solved with a numerical method. Numerical methods are usually employed when hydraulic properties are sufficiently heterogeneous and/or the geometry complex enough that the problem is not amenable to obtaining an exact solution, and a more precise answer is desired than may be obtained from a scale analysis.

As a simple example we consider the solution of Laplace's Equation in one dimension (equation 5.15) by finite differences. The essential idea of the **finite difference method** is to approximate derivatives by small, or finite, differences. In other words, we approximate dh/dx by $\Delta h/\Delta x$. The approximation becomes better as Δx becomes smaller. Thus the method is convergent; that is, we can approximate the true answer as closely as practical limitations will allow.

The first step in numerical methods is to take the geometry of interest and break it up into a discrete

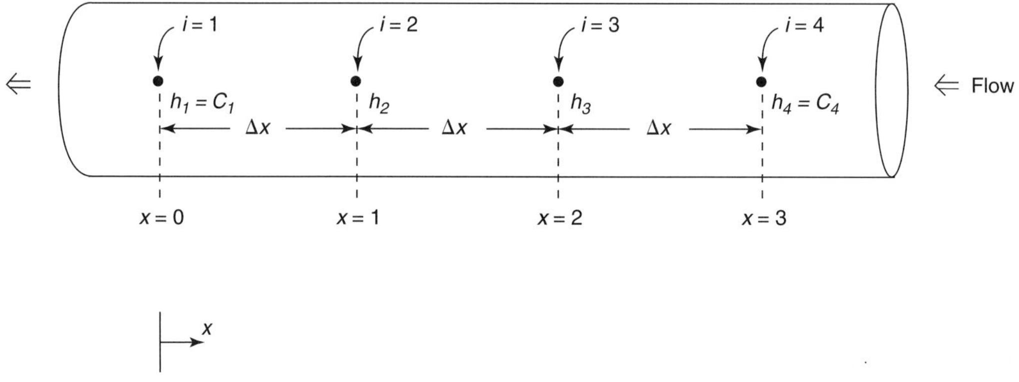

Figure 5.12 Discretization for one-dimensional finite-difference problem.

series of nodes. A **node** is a location or point in space at which the solution to a differential equation is to be found. For our simple example, we will consider one-dimensional flow through a pipe filled with sand (Figure 5.12). We will calculate the value of head and its derivatives at each node.

Let there be only four nodes. The subscript i will be used to distinguish each node. At $i = 1$, $h = h_1$, etc. The first node will be located at $x = 0$, the second at $x = 1$, the third at $x = 2$, and the fourth at $x = 3$, such that the distance between each node $\Delta x = 1$.

The value of head between the ends of the pipe depends on the value of head at the ends. In other words, the boundary conditions. We cannot find a solution unless we specify two boundary conditions. Let head be fixed at each end such that $h_1 = C_1$ and $h_4 = C_4$, where C_1 and C_4 are constants. We now proceed to find h_2 and h_3 by writing the finite difference approximation to Laplace's equation at every node, as Laplace's equation must hold everywhere.

Laplace's equation in one dimension (x) is

$$\frac{d^2h}{dx^2} = 0 \quad \textbf{(5.93)}$$

or

$$\frac{d\left[\frac{dh}{dx}\right]}{dx} = 0 \quad \textbf{(5.94)}$$

We now replace dh/dx by $\Delta h/\Delta x$, where Δh is the difference in head between two nodes, and Δx is the distance between two nodes. At the *ith* node, equation 5.94 becomes

$$\frac{(dh/dx)_{i+1/2} - (dh/dx)_{i-1/2}}{\Delta x} = 0 \quad \textbf{(5.95)}$$

where $(dh/dx)_{i+1/2}$ is the value of (dh/dx) between the *ith* and $i + 1$ nodes, and $(dh/dx)_{i-1/2}$ is the value of (dh/dx) between the *ith* and $i - 1$ nodes. Replacing the remaining derivatives in equation 5.95 by finite-difference approximations,

$$(dh/dx)_{i+1/2} \approx \frac{h_{i+1} - h_i}{\Delta x} \quad \textbf{(5.96)}$$

$$(dh/dx)_{i-1/2} \approx \frac{h_i - h_{i-1}}{\Delta x} \quad \textbf{(5.97)}$$

Substituting equations 5.96 and 5.97 into equation 5.95,

$$\frac{\frac{h_{i+1} - h_i}{\Delta x} - \frac{h_i - h_{i-1}}{\Delta x}}{\Delta x} = 0 \quad \textbf{(5.98)}$$

$$\frac{h_{i+1} - 2h_i + h_{i-1}}{\Delta x^2} = 0 \quad \textbf{(5.99)}$$

Equation 5.99 is a finite difference approximation to Laplace's equation. It must be true at every *ith* node. Therefore, we may write equation 5.99 for each of four nodes, $i = 1, 2, 3$, and 4. At the first and fourth nodes, head is fixed so that

$$h_1 = C_1 \tag{5.100}$$

and

$$h_4 = C_4 \tag{5.101}$$

At the second node, $i = 2$

$$h_3 - 2h_2 + C_1 = 0 \tag{5.102}$$

where $\Delta x = 1 = \Delta x^2$. At the third node,

$$C_4 - 2h_3 + h_2 = 0 \tag{5.103}$$

We now have 2 equations (5.102 and 5.103) in two unknowns (h_2, h_3). Rewriting the above set of four equations,

$$\begin{array}{llllll} \text{at } i=1 & h_1 & & & & = C_1 \\ \text{at } i=2 & h_1 & -2h_2 & +h_3 & & = 0 \\ \text{at } i=3 & & h_2 & -2h_3 & +h_4 & = 0 \\ \text{at } i=4 & & & & h_4 & = C_4 \end{array}$$

The four equations could also be written

$$\begin{array}{lllll} (1)\,h_1 & +(0)h_2 & +(0)\,h_3 & +(0)\,h_4 & = C_1 \\ (1)\,h_1 & +(-2)\,h_2 & +(1)\,h_3 & +(0)\,h_4 & = C_2 = 0 \\ (0)\,h_1 & +(1)\,h_2 & +(-2)\,h_3 & +(1)\,h_4 & = C_3 = 0 \\ (0)\,h_1 & +(0)\,h_2 & +(0)\,h_3 & +(1)\,h_4 & = C_4 \end{array}$$

or, using matrix notation,

$$\begin{bmatrix} h_1 \\ h_2 \\ h_3 \\ h_4 \end{bmatrix} \begin{bmatrix} 1 & 0 & 0 & 0 \\ 1 & -2 & 1 & 0 \\ 0 & 1 & -2 & 1 \\ 0 & 0 & 0 & 1 \end{bmatrix} = \begin{bmatrix} C_1 \\ C_2 \\ C_3 \\ C_4 \end{bmatrix}$$

or, more simply,

$$\bar{\boldsymbol{h}}\,\bar{\bar{\boldsymbol{G}}} = \bar{\boldsymbol{C}} \tag{5.104}$$

where, in the terminology of linear algebra, $\bar{\boldsymbol{h}}$ and $\bar{\boldsymbol{C}}$ are vectors and $\bar{\bar{\boldsymbol{G}}}$ is a matrix. We can easily handle problems that involve hundreds or thousands of nodes by using matrix inversion algorithms that solve for the vector $\bar{\boldsymbol{h}}$.

For our simple, four-node example, however, we can solve for the value of head at the interior nodes more directly. Returning to our simple set of two equations (5.102 and 5.103) in two unknowns,

$$h_1 - 2h_2 + h_3 = 0 \tag{5.105}$$

$$h_2 - 2h_3 + h_4 = 0 \tag{5.106}$$

where $h_1 = C_1$ and $h_4 = C_4$ are the specified boundary conditions. For demonstration purposes, let us assign values to the constants C_1 and C_4 so that we may compute h_2 and h_3. Let $C_1 = 1$ and $C_4 = 4$. Then, substituting into equations 5.105 and 5.106 we obtain

$$-2h_2 + h_3 = -1 \tag{5.107}$$

$$h_2 - 2h_3 = -4 \tag{5.108}$$

We may now find h_3 by multiplying equation 5.108 by 2 and adding it to equation 5.107. Doing so, we find

$$-3h_3 = -9 \tag{5.109}$$

and $h_3 = 3$. We can now substitute for h_3 back into either equation 5.107 or 5.108 and find that $h_2 = 2$.

5.7. PHILOSOPHY OF MODELS.

In the context of hydrogeology, a **model** is a representation of a geologic system or process. In hydrogeology we usually deal with computer models that seek to predict quantities such as the direction and magnitude of subsurface groundwater flow, the concentration and movement of contaminants, and how these variables will change with time. A typical use of a computer model is to predict in which direction a contaminant plume will migrate and how fast it will move. Models are a valuable tool, however, they have limitations.

There are three types of errors that may arise in computer modeling: conceptual errors, errors arising from lack of constraints or data, and numerical errors. The first two categories encompass what are usually the most significant types of errors. A conceptual error is a failure to accurately represent

Qanats: Ancient Persian Waterworks

A **qanat** (also kanat) is an underground tunnel that transports groundwater from a well to the ground surface. Most qanats are found in Iran, which receives nearly all of its rainfall during three winter months. Surface runoff is rapid, and streams are dry during the rest of the year. Qanats furnish a means of tapping groundwater for irrigation; in 1933, the American engineer Millard Butler (1933) estimated that the water supplied by qanats supported more than half the population of Iran.

Irrigation has sustained life in the arid lands of the Middle East for thousands of years. Near what is present-day Iraq, the city of Babylonia was founded around 4000 BC. One of the most famous of the Babylonian Kings was Hammurabi (1792–1750 BC), who wrote the most ancient of all known laws, the Code of Hammurabi. On the stone tablet that contains Hammurabi's code, the King is described in part as [he who] "made great the name of Babylon" and "brought plenteous water to its inhabitants." Some of the laws of Hammurabi dealt with irrigation issues:

> If any one open his ditches to water his crop, but is careless, and the water flood the field of his neighbor, then he shall pay his neighbor corn for his loss.

The Babylonian civilization developed the first system of writing, the earliest known laws, the city-state, the potter's wheel, the sailboat, plow, and a variety of literary, musical, and architectural forms that influenced all of Western civilization.

Qanats in Iran tap aquifers found in the upper reaches of vast alluvial fan deposits that fill the valleys of the high plateaus. Larger than the alluvial fans found in North America, these deposits are hundreds of miles (1 mile = 1.6 kilometers) in breadth. Their length from mountain slope to valley was described by Tolman (1937) as "several scores of miles" (a score is twenty). In the upper reaches of the fans, the deposits are largely permeable gravels. Grain size decreases with increasing distance into the valleys; the silt deposits at the toe of the fans can be a highly productive soil if irrigation water is brought in by a qanat.

Surface shaft of a qanat in the Northern Tafilalt Oasis, Morocco, as it appeared in 1992.
Photograph courtesy of Dr. Dale Lightfoot, Oklahoma State University.

Qanats consist of one semi-horizontal tunnel about a meter in diameter that gently slopes downhill, punctuated by a series of vertical wells (see cross-section). The wells in the upper sections of the alluvial fan extend 1 or 2 meters below the water table and collect water. The groundwater that seeps into these wells then travels downslope through the semi-horizontal tunnel that eventually emerges at the ground surface. The water collected by a qanat is usually used for irrigation.

Qanats are usually constructed by a wealthy individual or consortium of landowners. The ancient legal principle that applies is that if a man constructs a qanat, he then owns all the land irrigated by that qanat. A qanat is considered to be a quality investment. Unless the water table sinks, a high return is assured with

Qanats: Ancient Persian Waterworks

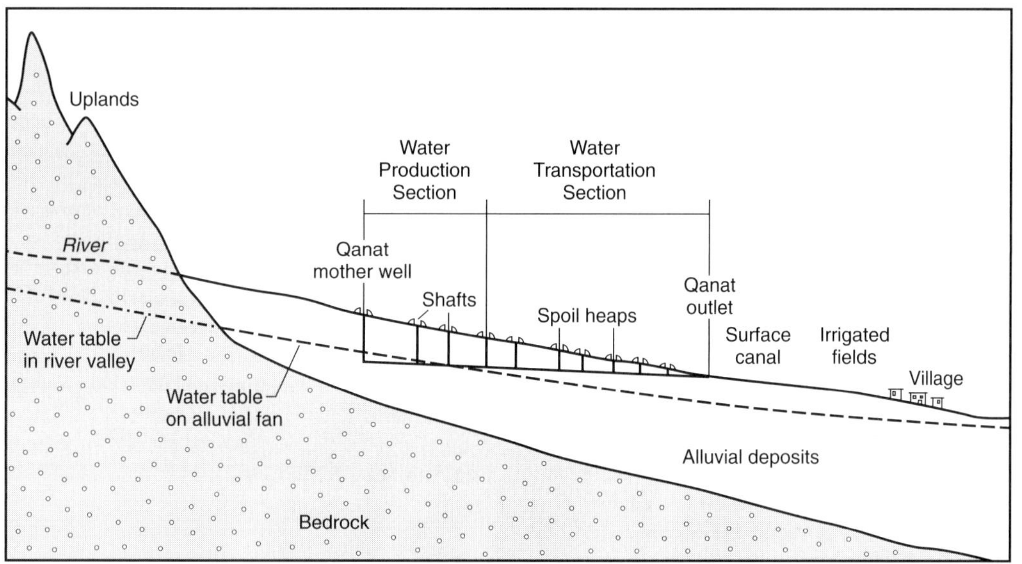

Cross-section through a typical qanat.
(From Beaumont, 1973, p. 24.)

minimal maintenance. The first step in the construction of a qanat is to bring in the local water-finding experts. They proceed partly by intuition, and partly by intelligence. The lie of the land, the proximity of springs, the occurrence of vegetation— all of these things are noted and an educated guess is made as to where to dig.

Once the digging spot has been chosen, the surveyors are brought in. They dig a well until water is reached. If the water table is not encountered before a depth of 91 meters (300 feet) is reached, the well is abandoned and a new group of diviners consulted. If the well succeeds in finding water, it is continued downward for about 1 to 2 meters below the water table. This first well is known as the mother well, and its depth is measured and recorded by cutting a piece of string as long as the well is deep.

A semi-horizontal tunnel must now be excavated that extends to the area to be irrigated. The downhill slope of the tunnel must be high enough that water flows downhill, but not so great that the high velocity of the flowing water will erode the tunnel. The digging is done by specialized workers known as muqannis. They lack formal education; their engineering knowledge has been acquired by experiences handed done over countless generations. The semi-horizontal tunnel that transports water is punctuated by a series of vertical shafts that are usually spaced at intervals of about 30 meters. The primary purpose of these shafts is to provide access and ventilation for the muqannis; the wells in the upper reaches of the qanat also collect water, just as the mother well does.

The depth to which each vertical shaft must be dug is found using a crude but effective system of leveling. After the mother well has been dug, a pole is placed on the ground at an appropriate distance for the construction of the second shaft. A string is stretched between the ground level of the mother well, and a point on the downslope pole. When a drop of water on the string has no tendency to run either up or downhill, the string is considered to be level. The distance between the ground surface and the point at which the string is tied to the downslope pole gives the elevation drop from the mother well. The elevation drop is then recorded

Qanats: Ancient Persian Waterworks *(continued)*

by knotting off the string whose length matches the depth of the mother well. The remaining length is the depth to which the downslope shaft must be dug. A few centimeters are added to account for the downhill grade in the qanat tunnel, and to account for the string length lost by the tying of the knot.

The digging is done with hand tools and the sediments are removed in buckets hoisted by means of a portable windlass. There is no ventilation in the tunnels, and illumination is provided by sunlight reflected by mirrors. To keep the excavation proceeding in a straight line, the muqannis place two lamps several meters behind them. If, on looking back, they see that the two flames are superimposed, the tunnel is straight. This system is not entirely successful; sharp kinks are usually found in the tunnels just before they reach a well. Tolman (1937, p. 14) related that "*Accidents are frequent, and the loss of life is great, but there are always others to carry on.*"

In 1973, there were estimated to be 10,000 to 40,000 qanats in Iran (Beaumont, 1973). The depths of some exceptional mother wells exceeds 250 meters; however, most are less than 50 meters deep. The discharge of most qanats is less than 100 m^3 of water per hour. Some qanats have lengths of up to 30 kilometers, but most are shorter than 5 kilometers. The origin of qanats is lost in history; they are thought to have been in existence at least as early as 800 BC, and possibly considerably earlier.

Following World War II, some landowners began drilling deep wells and extracting water by means of electric or diesel pumps. In a number of areas, this new pumping caused the water table to fall dramatically, and several qanats dried up. It seems likely that water use in Iran in the future will depend more on drilled wells than the ancient system of qanats. Properly managed, pumped wells can utilize groundwater more efficiently than qanats, because the qanats have high rates of flow during the winter months when no crops are grown and the water is wasted (Beaumont, 1973).

For Additional Reading

Beaumont, P. 1973. A traditional method of ground-water utilization in the Middle East: *Ground Water* 11: (5) 23–30.

Butler, M. A. 1933. Irrigation in Persia by Kanats. *Civil Engineering* 3: (2) 69–73.

Smith, A. 1953, *Blind white fish in Persia.* London: George Allen and Unwin Ltd.

Tolman, C. F, 1937. *Ground water.* New York: McGraw-Hill.

the physical processes one is trying to model. One type of conceptual error is a failure to represent or include some of the physical processes involved. Another type of conceptual error is to incorrectly characterize the physical processes that are included. A model that contains a conceptual error may generate precise answers, but the answers will have little to do with physical reality. Consider a common and stubborn type of groundwater contaminants, a class of organic chemicals known as non-aqueous phase liquids (NAPLs). NAPLs are very difficult to clean up; they are not readily flushed from porous media because capillary forces hold them in place at residual saturations. Any computer model of NAPL behavior that does not include capillary forces contains a significant conceptual error and will not be capable of making accurate predictions.

The second type of error is that which results from limited data or model constraints. Geologic environments are complex, and the modeler typically never has all of the data necessary to represent the complexity of the underground environment with 100% accuracy. The best that can usually be hoped for is to have an approximate representation that will not lead to gross errors. Numerical errors arise in the implementation of the discretization process. Virtually all computer models have numerical errors, but the errors are usually small. Numerical errors may arise from coarse discretizations. If that is the case, the error can be reduced by adopting a finer grid. In the modeling of contaminant transport, a special type of error, numerical diffusion, arises. Numerical diffusion is diffusion that arises spontaneously as a numerical error; it is not physical. Numerical diffusion is a stubborn and difficult problem to overcome, and several different schemes have been proposed to reduce the magnitude of this type of error.

The first step in the adoption of a computer model is to apply the model to a simplified geometry with homogeneous properties for which an analytical solution is known. The computer solution is compared to the analytical solution. The creator of the model then makes a subjective assessment as to whether or not the numerical errors in the model are small enough. Typically, the numbers provided by the computer model should be within, say, 1% of the correct answer. Another procedure that is commonly done is to check and see if the model conserves mass and energy. These procedures raise the obvious question: how does the modeler know that the model will provide an accurate solution when it is applied to a complex domain for which no analytical solution is known? The answer is quite simple: he doesn't. There is no way to be 100% sure that the model is not producing significant numerical errors in some circumstances. This is one reason a scientist needs to develop a sound physical intuition as to what appears reasonable and what does not.

The next step in model development is calibration. In calibration, the model is run with historical data to see if it can accurately predict the past behavior of a system. For example, historical data regarding the movement of a contaminant plume through an aquifer may be available from a network of monitoring wells. If the model predicts that the plume will migrate faster than what was observed, the modeler may reduce the value chosen for the average permeability of the aquifer. This will produce a concomitant reduction in the velocity of groundwater and entrained contaminants.

It is important to distinguish between model calibration and model validation (Konikow and Bredehoeft, 1992). There is no guarantee that a calibrated model that reproduces past behavior will be able to provide accurate predictions of future behavior. Model validation, sometimes known as model verification, is an unfortunate term that arises from philosophical ignorance. No matter how many times a model is shown to be an accurate reproducer of past behavior there is never a basis for concluding that the model will be accurate under different circumstances. A truly accurate model requires infinite knowledge and infinite data. At the present state of scientific development both are lacking. It seems likely that this will continue to be the case for some time to come.

A calibrated model is not necessarily unique. That is, there may be more than one model capable of matching the past behavior of a geologic system, however, these models may make very different predictions of future behavior. A corollary is that although a model accurately reproduces what data are available, it may not be capable of making accurate predictions of future behavior. Konikow and Bredehoeft (1992) discuss the example of the Dakota Aquifer in South Dakota. The Dakota Aquifer is a sandstone unit confined by a Cretaceous shale. A forty-hour pumping test was done on the Dakota Aquifer near the city of Wall, South Dakota. The Theis model, which assumes no leakage from the impermeable layers both above and below an aquifer, matches the data from the pumping test perfectly (for all intents and purposes). But, so does a Hantush model, which assumes that confining layers leak. Which model is accurate? As the Cretaceous confining layer is known to leak, the Hantush model is more accurate, however, the forty-hour pumping test would have to be conducted for more than 1000 years before the superiority of the Hantush model became apparent. The ability of the Theis model to represent the behavior of the Dakota Aquifer over a short period of time does not necessarily imply its ability to predict future behavior accurately.

Properly understood, a computer model is a tool deployed to obtain an understanding of a physical system, not a reproduction of that system. A few years ago, some professors at a distinguished university in the United States proposed a prototype course in geophysics where students would apply geophysical computer models without any knowledge of the underlying mathematics or physics. The argument was made that learning geophysics was analogous to playing baseball. Physics is involved in baseball at every step, yet players learn to play the game perfectly without any knowledge whatsoever of physics. The baseball player learns by doing—why can't the students of geophysics and hydrogeology do the same? The answer is that the analogy is false. The baseball player deals with 100% reality. The computer models applied in geophysics and hydrogeology are substantially short of this mark. Not understanding the basic physics of the problems they are analyzing, students will not understand the limitations of the model they are applying nor the assumptions inherent in the model's formulation. This inevitably results in misapplications and inferences that turn out to be wrong. One cannot learn physics by playing computer games, nor can a human mind be replaced by a computer algorithm.

Review Questions

1. Define the following terms in the context of hydrogeology:
   a. diffusion equation
   b. Laplacian
   c. Laplace's equation
   d. hydraulic diffusivity
   e. equipotential
   f. flow line
   g. streamline
   h. flow tube
   i. flow net
   j. refraction
   k. scale analysis
   l. finite-difference method
   m. node
   n. model
2. Derive $\frac{dh}{dt} = \frac{K}{S_s} \frac{d^2h}{dx^2}$ the diffusion equation in one dimension (x) by assuming conservation of mass and Darcy's law. State simplifying assumptions as they are introduced into your derivation.
3. List seven rules for constructing flow nets.
4. Draw flow nets for the following situations: (a) flow underneath a dam, (b) topographically driven flow in a foreland basin, (c) underground flow in an undulating terrain consisting of alternating hills and valleys.
5. A homogeneous, isotropic porous medium of hydraulic conductivity K_1 overlies a homogeneous, isotropic porous medium of hydraulic conductivity K_2. There is steady-state flow through and across the layers. Flowlines in the top layer are at an angle Θ_1 to the vertical; flowlines in the bottom layer are at an angle Θ_2 to the vertical. Show that (i.e., derive) $K_1/K_2 = \text{Tan}\ \Theta_1/\text{Tan}\ \Theta_2$. Use a figure.
6. Explain why flow in confining layers tends to be vertical and flow in aquifers tends to be horizontal. Invoke a mathematical formula for the refraction of flow lines in your explanation.
7. (a) A shaley confining layer overlies a sandstone aquifer. Flow vectors in the confining layer are oriented 10° from the vertical. If $k_{shale} = 10^{-17}$ m^2 and $k_{ss} = 10^{-14}$ m^2, what is the orientation of flow vectors in the sandstone aquifer? (b) A limestone is overlain by a material of unknown composition and hydraulic properties. The permeability of the limestone $k_{ls} = 10^{-14}$ m^2. Flow vectors in the limestone are oriented 0.1° from the vertical. The orientation of flow vectors in the overlying layer is 89.9° from the vertical. What is the permeability of the overlying layer? What is its probable composition?

8. You have a farm in S. Dakota. Your water supply comes from the Dakota aquifer. A neighbor drills a new well and starts pumping. What will happen to your well, and when will it happen? What sort of measurements or data do you need to make an approximate estimate of the time at which your well will be affected, and what is your estimate? (note: you need to find numbers and validate them instead of just calculating an arbitrary number based on arbitrary assumptions).
9. Discuss the relative advantages and disadvantages of analytical and numerical solutions.
10. Solve $\nabla^2 h = 0$ using finite-differences. Use 5 nodes. Fix head at the first and fifth node as $h_1 = 0$, $h_5 = 100$. Solve for h_2, h_3, and h_4. Assume homogeneous hydraulic properties.
11. Graph the solution to $\nabla^2 h = 0$ in one dimension (x). Assume all x-values in meters. Boundary conditions are $h(0) = 1$ m and $h(10) = 10$ m. For $5 \geq x \geq 0$, hydraulic conductivity is $K_1 = 1$ m-s^{-1}; for $10 \geq x \geq 5$, hydraulic conductivity $K_2 = 2$ m-s^{-1}. What is the Darcy velocity? What is value of head at $x = 5$?
12. Graph the solution to $\nabla^2 h = 0$ in one dimension (x). Assume all x-values in meters. Boundary conditions are $h(0) = 0$ m and dh/dx at $(x=10) = 1$. For $5 \geq x \geq 0$, hydraulic conductivity is $K_1 = 10$ m-s^{-1}; for $10 \geq x \geq 5$, hydraulic conductivity $K_2 = 5$ m-s^{-1}. What is the Darcy velocity? What is value of head at $x = 5$?
13. What are three types of errors inherent in computer models?
14. Explain the significance of non-uniqueness in the application of computer models in hydrogeology.

SUGGESTED READING

Konikow, L. F., and Bredehoeft, J. D. 1992. Ground-water models cannot be validated. *Advances in Water Resources* 15: 75–83.

Notation Used in Chapter Five

Symbol	Quantity Represented	Physical Units
α	compressibility of a porous medium	$Pa^{-1} = m\text{-}s^2\text{-}kg^{-1}$
a, c	flow tube widths	m
B	fluid compressibility	$Pa^{-1} = m\text{-}s^2\text{-}kg^{-1}$
C_1, C_2, C_3, C_4,	constant head values	m
c_1	constant of integration	dimensionless
c_2	constant of integration	m
$\bar{\boldsymbol{C}}$	vector of constant head values	m
D	hydraulic diffusivity	$m^2\text{-}s^{-1}$
Δs	distance between equipotentials in a flow tube	m
Δx, Δy, Δz	length of control volume sides	m
erf	error function	dimensionless
erfc	complementary error function	dimensionless
ϕ	porosity	dimensionless
g	acceleration due to gravity	$m\text{-}s^{-2}$
h, Δh, h_o	head, change in head, head at time zero	m
K	hydraulic conductivity	$m\text{-}s^{-1}$
$\bar{\bar{\boldsymbol{G}}}$	matrix of constants	dimensionless
$\bar{\boldsymbol{h}}$	vector of head values	m
k	permeability	Darcy $-$ 10^{-12} m^2
μ	fluid (dynamic) viscosity	$Pa\text{-}s = kg\text{-}m^{-1}s^{-1}$
M_w	fluid mass	kg
P, ΔP	fluid pressure, change in fluid pressure	Pascal (Pa) $= kg\text{-}m^{-1}\text{-}s^{-2}$
Q	volumetric flow rate	$m^3\text{-}s^{-1}$
q, q_x	Darcy velocity, Darcy velocity in the x-direction	$m\text{-}s^{-1}$
Θ_1, Θ_2	angles of flow refraction	dimensionless
ρ	fluid density	$kg\text{-}m^{-3}$
S_s	specific storage	m^{-1}
t, t_0	time, time zero	s
u, s	dummy variables used in definition of error function	dimensionless
V_T, ΔV_T	total (matrix plus fluid) volume of a porous medium, change in total volume	m^3
V_w, ΔV_w	fluid volume, change in fluid volume	m^3
x	distance	m
y	distance	m
∇	gradient operator	m^{-1}
∇^2	Laplacian, the second spatial derivative	m^{-2}

CHAPTER 6

NEAR SURFACE FLOW

6.1. THE UNSATURATED ZONE.

The **unsaturated zone** (Figure 6.1) is the region between the ground surface and the water table where soil and/or rock pores are partially filled with air and partially filled with fluid. The unsaturated zone is also known as the **vadose zone.** Near the base of the unsaturated zone, there is some ambiguity as to where the unsaturated zone ends and the water table begins. The base of the unsaturated zone contains a **capillary fringe,** which is a fully saturated layer containing water drawn up into the unsaturated zone by capillary forces (see section 6.1.1). In this text, we follow others (e.g., Meinzer, 1923) by considering the capillary fringe to be part of the unsaturated zone, even though it is completely saturated by fluid. The capillary forces in the unsaturated zone result in fluid pressures that are lower than atmospheric (see section 6.1.1). In terms of gauge pressure, where atmospheric pressure is considered to be the zero datum, pressures in the unsaturated zone are "negative." Thus the water table begins where the earth is fully saturated and the fluid pressure is at atmospheric.

6.1.1. Capillary Forces.

One of the chief distinctions between the unsaturated and saturated zones is that capillary forces are relatively more important in the unsaturated zone.

Consider the following experiment (Figure 6.2). If we place a small glass tube in a pan of water, the water in the tube will move up the sides of the tube until it reaches a height y (m) above the level of the water in the pan. The top of the water column in the tube is a meniscus. A **meniscus** is a curved upper surface of a liquid in a container. We observe that the meniscus is curved, and the angle γ at which the meniscus intersects the glass tube is the **wetting angle.**

The result of our experiment is quite interesting, because the water is apparently moving uphill without any energy source, contrary to our expectation that energy is conserved. One might speculate that the situation could be exploited by drilling a hole in the side of the tube just beneath the water meniscus. If the water ran out the hole and back into the pan below, we would have created a perpetual motion machine. Furthermore, we could hook up a turbine (admittedly, a small one) to the stream of water flowing back down to the pan and generate electricity, thus violating the principle of conservation of energy.

An examination of the fluid pressure levels is also intriguing. Consider pressures P_1 and P_2 (kg-m^{-1}-s^{-2}) at the points 1 and 2 as shown (Figure

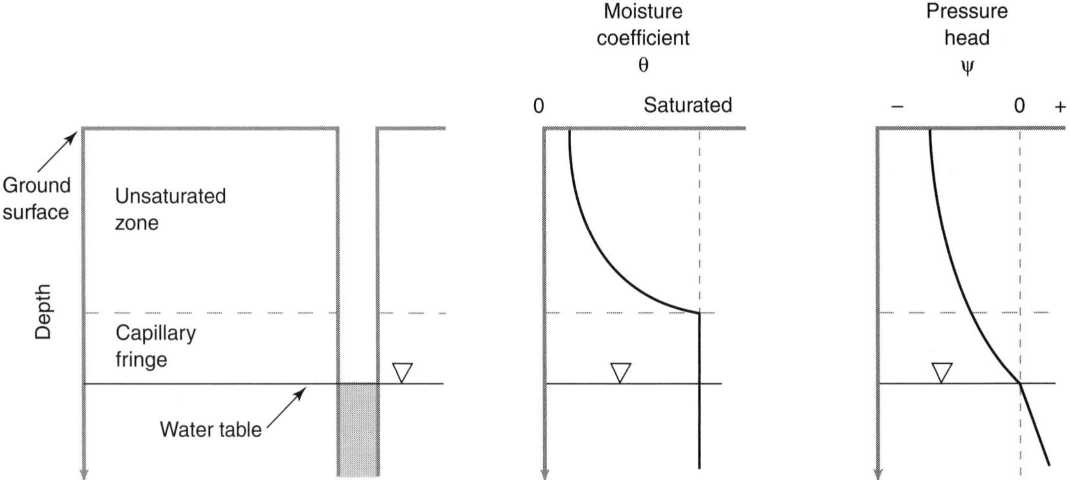

Figure 6.1 The distribution of moisture (θ) and the capillary pressure head (Ψ) in the unsaturated zone.
(After Freeze and Cherry, 1979, p. 40.)

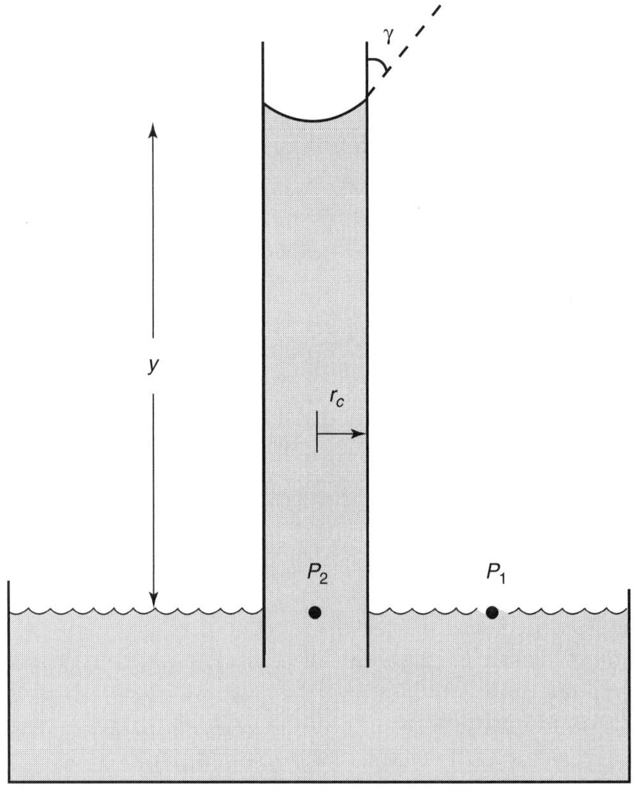

Figure 6.2 The rise of water in a capillary tube.
(After Furbish, 1997, p. 65.)

6.2). The fluid pressure P_1 is the atmospheric pressure. The pressure at point P_2 must therefore be the sum of the fluid pressure at point 1 (P_1) and the fluid pressure due to the weight of the water in the small-diameter tube, $P_2 = P_1 + \rho g y$, where the water has density ρ (kg-m^{-3}), g is the acceleration due to gravity (m-s^{-2}), and the fluid column in the tube has height y (m). Thus we must have $P_2 > P_1$, and because the fluid at these two points is at the same elevation, fluid must move from point 2 to point 1, however, it does not move. Once the fluid moves up the tube to a certain height it remains at that height, motionless. Therefore, there must exist a capillary pressure in the tube that cancels the additional pressure due to the height of the water column. The capillary pressure must be lower than atmospheric pressure, or the water would not move up the tube. If we work in terms of gauge pressure where atmospheric pressure is considered to be zero, the capillary pressure is negative.

The negative capillary pressure arises from an attractive force between the water molecules and the glass molecules that make up the walls of the tube. This force is directed upwards and is equal in magnitude to the weight of the fluid column in the tube. The force must act only along the circle that defines the air-water-glass contact. If the force acted uniformly along the cylindrical water-glass contact throughout the entire length of the tube, water would never cease to rise in the tube. Let us denote the attractive force between the water and glass molecules by F_{cap} (kg-s^{-2}), where F_{cap} is defined to be the force per unit length along the contact circle. F_{cap} is a **capillary force,** and the small tube is a **capillary tube.** The upward component of the capillary force is $F_{cap}\cos\gamma$ (Figure 6.2). Let the capillary tube have radius r_c (m). The contact circle along which the capillary force acts has length $2\pi r_c$ (m). As F_{cap} is defined to be the force per unit length, the total upward force acting on the water column must be $2\pi r_c F_{cap}\cos\gamma$ (kg-m-s^{-2}). The total upward force acting on the water column must also be equal to the weight of the water in the capillary tube. The weight of the water in the capillary tube is equal to the mass of the water multiplied by the acceleration due to gravity (g, m-s^{-2}). The mass of the water in the capillary tube is equal to the volume of the water ($\pi r_c^2 y$, m^3) multiplied by water density (ρ, kg-m^{-3}). Setting the two forces equal,

$$2\pi r_c F_{cap}\cos\gamma = \pi r_c^2 y\rho g \qquad \textbf{(6.1)}$$

The height (y, m) to which water will rise in a capillary tube is a measure of the relative importance of the capillary force,

$$y = \frac{2F_{cap}\cos\gamma}{r_c\rho g} \qquad \textbf{(6.2)}$$

We have found empirically (through observation) that there exists a capillary force that tends to impel water to move into and remain inside small tubes. The relative magnitude of the capillary force as expressed by equation 6.2 is inversely proportional to the radius of the capillary tube. In other words, capillary forces tend to be more important when flow through small tubes is considered. The magnitude of the capillary force depends upon the nature of the fluid and solid with which it is contact. Some fluid-solid combinations exhibit wetting behavior, where the fluid molecules are more attracted to the solid molecules than they are to adjacent water molecules. Other fluid-solid combinations exhibit nonwetting behavior, where fluid molecules are attracted more strongly to each other than to the solid.

Finally, note that the initial movement of a fluid up a capillary tube does not violate the conservation of energy. The movement is simply an expression of the potential electrical energy that exists in the water and glass tube.

The significance of capillary forces is that rock and soil pores tend to act like capillary tubes. If a soil or rock is wet, a positive pressure or head must be applied to extract the water. If the applied pressure is not larger than the capillary pressure, the fluid will remain in the pores.

6.1.2. Flow in the Unsaturated Zone.

The relative saturation of the unsaturated zone is quantified as the volumetric water or volumetric moisture content (θ) of the unsaturated zone. The **volumetric moisture content** (θ) of a porous medium is the dimensionless ratio of the water or fluid volume to the total volume of the porous medium. In a fully saturated porous medium, the

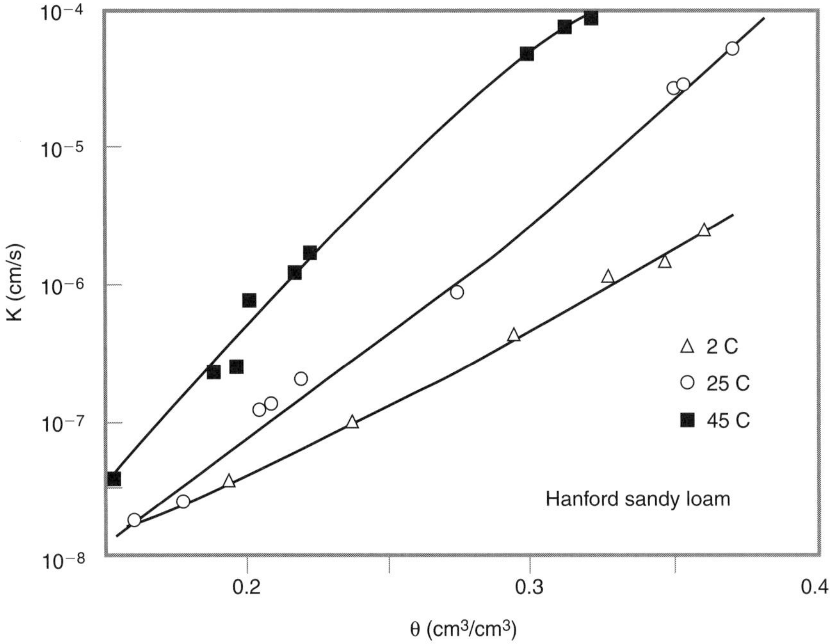

Figure 6.3 The dependence of hydraulic conductivity (K) on the volumetric moisture coefficient (θ) in the unsaturated zone.
(From Nielson et al., 1986, p. 935.)

volumetric moisture coefficient (θ) is equal to the porosity (ϕ). If a porous medium is not fully saturated, $\theta < \phi$.

Recall that head consists of an elevation head (z) and a pressure head ($P/\rho g$). In the saturated zone, pressure heads are normally positive. That is, fluid pressures in the saturated zone are usually greater than atmospheric. The negative pressure produced by capillary forces in the unsaturated zone produces a **capillary-pressure head,** $\Psi = P/\rho g$ (m), where P (kg-m^{-1}-s^{-2}) is now capillary pressure. The hydraulic head h (m) that drives fluid flow in the unsaturated zone is thus

$$h = z + \Psi \tag{6.3}$$

Strictly speaking, equation 6.3 also applies in the saturated zone where capillary forces are usually neglected because they are relatively less important than in the unsaturated zone.

The capillary-pressure head (Ψ) in the unsaturated zone is measured with a device called a **tensiometer.** A tensiometer consists of a piece of pipe whose end is covered with a porous and permeable cup. The porous end of the tensiometer is inserted into the ground and the pipe is filled with water and capped. The capillary-pressure head will draw water from the pipe into the soil lowering the pressure of the water remaining in the pipe below atmospheric. The drop in water pressure is monitored with a gauge. Dividing the pressure drop (ΔP) by the product of fluid density and the acceleration due to gravity yields the capillary-pressure head ($\Psi = \Delta P/\rho g$). The total head can be found by adding the capillary-pressure head (Ψ) to the elevation (z) at the point of measurement.

Fluid movement in the unsaturated zone obeys Darcy's Law. However, the situation is more complex than in the saturated zone because both the capillary-pressure head (Ψ) and the hydraulic conductivity (K) are functions of the volumetric moisture content (θ). In general, the hydraulic conductivity tends to increase as the moisture content increases (Figure 6.3). There are at least three reasons for this (Philip, 1969, p. 218):

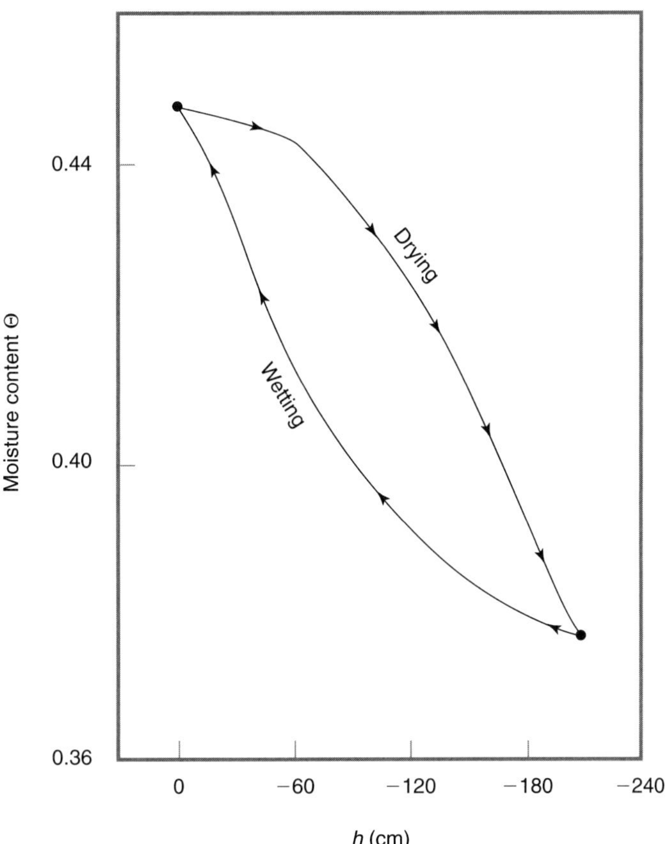

Figure 6.4 Hysteresis in the moisture content (θ) of an unsaturated porous medium as a function of head (h). (After Nielsen et al., 1986, p. 925.)

1. The total cross-sectional area available for flow decreases as θ decreases;
2. As θ decreases, the larger pores empty first. The relative contribution to the bulk hydraulic conductivity (K) depends upon the square of the pore throat radius;
3. As θ decreases, continuity is lost as pore fluids become isolated from the flow network.

The capillary pressure head (Ψ) is also a function of the volumetric moisture content (Figure 6.4). In general, the greater the degree of saturation, the smaller the absolute value of the capillary pressure head. A porous medium typically contains pores of various sizes. The capillary pressure head is more negative in the smaller pores, just as a capillary tube of a small diameter will support a water column of a greater height. When a porous medium begins to fill with water (the "wetting" curve, Figure 6.4), the smaller pores that have the larger (absolute values) of capillary pressure head fill first and the (absolute value of) average capillary pressure head decreases.

Not only do hydraulic conductivity and the capillary pressure head depend upon the volumetric moisture coefficient, they also exhibit hysteresis (Figure 6.4). A material that exhibits **hysteresis** is one whose behavior is not reversible. One reason porous media exhibit hysteresis during wetting and drying is that pores are not uniform in their size. Consider the behavior of a pore with an enlarged section sandwiched between two sections with

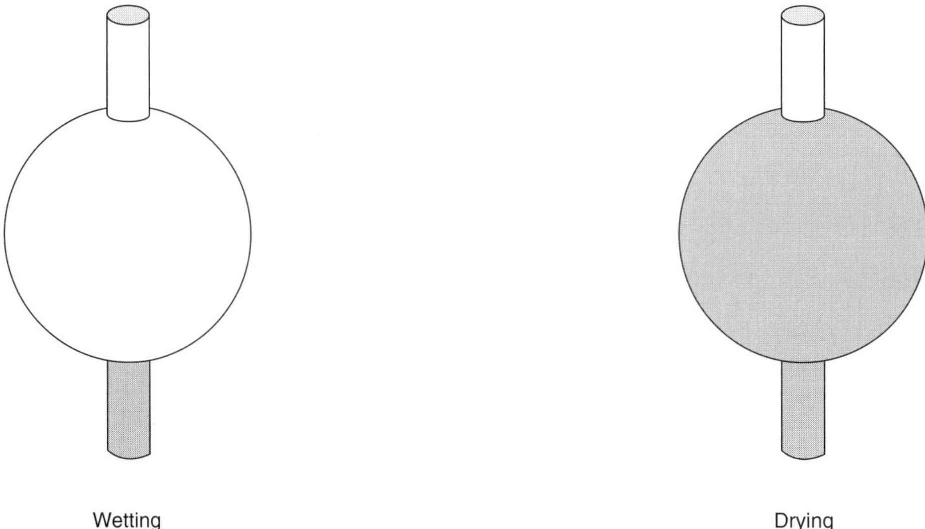

Figure 6.5 A pore channel with an enlarged section between two restricted areas exhibits hysteresis during wetting and drying. The enlarged section is more difficult to fill during wetting, but retains water during drying because of the capillary pressure head in the smaller-diameter tube (see discussion in text).

smaller diameters (Figure 6.5). Suppose the pore is filled (wetted) or emptied (dried) from below as shown in Figure 6.5. Let the capillary-pressure head in the smaller pore channels be −0.5 m, and the capillary-pressure head in the larger channel be −0.25 m. If the ambient fluid pressure at the filling point is −0.4 m, the smaller-diameter pore channel will fill, but the larger will not. If the ambient fluid pressure at the intake point increases to −0.25 m or larger, the larger-diameter pore channel will fill. But, if the process is reversed the larger-diameter pore channel will not empty until the ambient pressure at the outlet reaches the capillary-pressure head of the smaller-diameter pores, −0.4 m. The small-diameter pore channels that surround the larger-diameter channel will retain fluid until their capillary-pressure heads are exceeded. Thus porous media tend to retain more fluid during drying than wetting, even if the ambient pressure head is equivalent (Figure 6.4).

Vertical fluid flow in the unsaturated zone is described by **Richards Equation,** so called because it was originally derived by Richards (1931). Consider vertical (in the z-direction) fluid movement through a cube of a porous medium whose sides have lengths Δx, Δy, and Δz (m) (Figure 6.6). Conservation of fluid volume inside the cube requires that the rate at which the fluid volume inside the cube changes must be equal to the difference between the volumetric inflow and outflow rates,

$$\begin{array}{ccccc} (1) & = & (2) & - & (3) \\ \dfrac{\text{change in volume}}{\text{unit time}} & = & \begin{array}{c}\text{volumetric}\\\text{inflow rate}\end{array} & - & \begin{array}{c}\text{volumetric}\\\text{outflow rate}\end{array} \end{array} \tag{6.4}$$

It is convenient to number each term of equation 6.4 (1), (2), and (3), and consider each in turn.

Term (1): the total volume of fluid inside the cube is the product of the volumetric moisture coefficient (θ, dimensionless) and the total volume ($\Delta x \Delta y \Delta z$, m^3). The change in fluid volume per unit time is the derivative with respect to time (t),

$$\frac{\text{change in volume}}{\text{unit time}} = \frac{d[\theta \Delta x \Delta y \Delta z]}{dt} \tag{6.5}$$

Term (2): the volumetric inflow rate (m^3-s^{-1}) is the Darcy velocity in the z-direction (q_z, m-s^{-1})

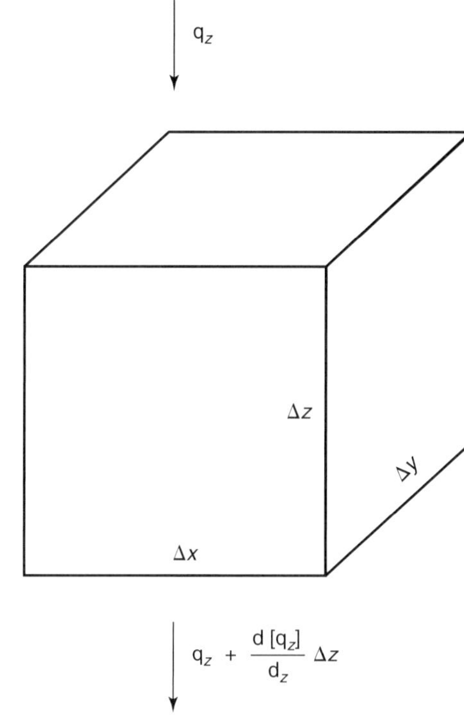

Figure 6.6 Control volume for derivation of Richard's equation.

multiplied by the area perpendicular to flow ($\Delta x \Delta y$, m^2).

$$\text{volumetric inflow rate} = q_z \Delta x \Delta y \quad \textbf{(6.6)}$$

Term (3): the volumetric outflow rate is equal to the volumetric inflow rate, plus whatever change in flow rate takes place over the length of the cube (Δz). Thus

$$\begin{matrix}\text{volumetric}\\ \text{outflow rate}\end{matrix} = q_z \Delta x \Delta y + \frac{d[q_z \Delta x \Delta y]}{dz} \Delta z \quad \textbf{(6.7)}$$

where the change in flow rate that takes place over the length of the cube (Δz) has been approximated by the rate of change with respect to z multiplied by the total length Δz. As Δz becomes infinitesimally small, this approximation becomes exact.

Because the control volume is a cube, the cross-sectional area ($\Delta x \Delta y$) does not change with respect to the z-coordinate and can be taken outside of the derivative sign,

$$\begin{matrix}\text{volumetric}\\ \text{outflow rate}\end{matrix} = q_z \Delta x \Delta y + \frac{d[q_z]}{dz} \Delta x \Delta y \Delta z \quad \textbf{(6.8)}$$

Assembling terms (1), (2), and (3) by substituting equations 6.5, 6.6, and 6.8 into equation 6.4,

$$\frac{d[\theta \Delta x \Delta y \Delta z]}{dt} = -\frac{d[q_z]}{dz} \Delta x \Delta y \Delta z \quad \textbf{(6.9)}$$

If the cube is not deforming with respect to time, the volume ($\Delta x \Delta y \Delta z$) can be taken outside of the time derivative on the left side of equation 6.9 and cancels with the volume term on the right side. We are left with

$$\frac{d\theta}{dt} = -\frac{d[q_z]}{dz} \quad \textbf{(6.10)}$$

Darcy's Law also applies in the unsaturated zone,

$$q_z = -K(\theta)\nabla h = -K(\theta)\frac{d[z + \Psi]}{dz} \quad \textbf{(6.11)}$$

where q_z (m-s^{-1}) is the Darcy velocity in the z-direction, head (h) in the unsaturated zone is equal to the sum of the elevation head (z) and the capillary pressure head (Ψ), and we have written the hydraulic conductivity (K, m-s^{-1}) as $K(\theta)$ to remind us that in the unsaturated zone the hydraulic conductivity depends upon the volumetric moisture coefficient.

$$q_z = -K(\theta)\nabla h = -K(\theta)\left[\frac{d\Psi}{dz} + 1\right] \quad \textbf{(6.12)}$$

Substituting equation 6.12 into equation 6.10,

$$\frac{d\theta}{dt} = \frac{d\left[K(\theta)\left(\frac{d\Psi}{dz} + 1\right)\right]}{dz} \quad \textbf{(6.13)}$$

Equation 6.13 is known as Richards equation (Richards, 1931). Richards equation can be solved to find the distribution and change of moisture within the unsaturated zone. Note, however, that the problem is necessarily more complex than in the saturated zone because the hydraulic conductivity is not constant but depends on θ. The mathematical

Wells Throughout History

A well is a pit or shaft that is usually constructed for the purpose of extracting underground fluids. Wells are used not only to obtain potable water, but also oil, gas, and other substances. In Louisiana and Texas, sulfur is obtained from wells by drilling salt domes. The solid mineral found in the cap rocks of salt domes is dissolved by pumping hot water down the wells. A mixture of sulfur and water returns to the surface and the sulfur is separated from the water. More recently, wells have been used not to extract fluids from the Earth, but to inject them. Injection into deep wells is one method of disposing of toxic wastes. Wells can also be installed for the purpose of monitoring the migration of contaminated groundwater.

The first well may have originated when people scraped away soil or sand at the bottom of a dry riverbed. Holes that collected water during wet seasons could have been deepened as they shrank during droughts. The Bushmen of the Kalahari desert extract water from beneath dry riverbeds with straws made of hollow reeds. The end of the reed is stuffed with grass to filter out sand particles. The reed is pushed into the dry sand to a depth of two to three feet (0.6 to 0.9 meters) until the shallow water table is reached; water can then be sucked out of the ground (Chapelle, 1997). In ancient times, natural springs were improved by shallow excavation; the pit dug around the spring was then lined with stones or timbers. The lining collected a supply of water that could be dipped out with a ladle, and also kept the water relatively free from suspended sediment. Dug wells cannot be regarded as a necessary or inevitable development. For primitive societies of hunter-gatherers, the spring and lined spring were quite sufficient to supply all of their water needs. The well was a revolutionary innovation brought about by the pressure of increasing population. The development of well-digging was simultaneously a cause, index, and consequence of increased human control over nature (Clark, 1944).

Shallow wells dug by hand are probably the oldest method of well construction. Most of the early well-digging took place in the arid and semi-arid regions of the Middle East where the technology to obtain groundwater was essential to survival. Beaumont (1973) estimated that the peoples of the Middle East have used groundwater for at least 10,000 years. The first permanent wells were constructed in Mesopotamia around 6000 BC. The oldest existing well is a brick-lined well at Chanhu-daro in Pakistan, which was constructed around 3000 BC (Hardcastle, 1987). With the advent of the iron age around 1200 BC, the availability of superior iron tools may have made it easier to dig wells.

The Roman philosopher Pliny advised how to locate an optimal location for well-digging: lie face down on the ground at sunrise and look for the places that the most water vapor rises.
(From Rusconi, 1592, p. 118.)

There are many references to wells in the Bible. O. E. Meinzer (1934) stated that "the twenty-sixth chapter of Genesis . . . reads like a water-supply paper." The ability to construct wells allowed Abraham, the first Hebrew Patriarch, to colonize the land of Canaan around 2000 BC. Unlike the fertile valleys of the region, the land Abraham migrated to was composed of limestone hills, poorly suited for agriculture, however, the land was suitable for grazing sheep and goats if a reliable source of water could be found. At places where groundwater seeped out of the ground, Abraham and his family constructed wells by digging conical holes in the ground. The holes were lined with rocks to prevent their collapse. In this manner, patches of marshy ground were turned into pools of clear water that livestock could drink from (Chapelle, 1997). The grandson of Abraham, Jacob, constructed a well 3500 years ago that is still in existence in the city of Nablus in Palestine. It was at this very well that Jesus met a Samaritan woman and told her (John, 4):

> But whosoever drinketh of the water that I shall give him shall never thirst; but the water that I shall give him shall be in him a well of water springing up into everlasting life.

Today, Jacob's well is the site of a Greek Orthodox Church that has been built around the well. Unlike other antiquities of questionable pedigree, there seems to be no dispute that this well is the authentic well dug by Jacob. The well is 3 meters wide and about 32 meters deep. In earlier times, the well was said to have been deeper.

A dug well must have a minimum diameter of about one meter to accommodate the workers. Many such wells have diameters of one to three meters, with some as large as 12 meters in diameter. The great majority of such wells are less than 100 feet (30 meters) deep, but some are known to reach depths as great as 150 meters. Traditionally, small towns had one dug well that functioned not only as the town's water supply, but was also a social center. The sides of hand-dug wells in unconsolidated materials may cave in unless reinforced by timbers or masonry. An advantage that a large-diameter dug well has over its smaller-diameter drilled counterpart is a large storage capacity. Prior to the year 1900, such wells commonly yielded water at rates greater than it could be extracted.

The ancients dug wells not only for water, but also for oil. Sennacherib (705/704–681 BC) king of Assyria, had a 37-foot (11 meters) deep well dug to recover oil. The well was lined with blocks of alabaster. Oil and asphalt were recovered from the well; the oil was burned for heat and light, asphalt was used to waterproof boats (Brantly, 1961).

The practice of throwing a coin in a "wishing well" or fountain for good luck is very old; it originates in ancient beliefs that springs and water wells were inhabited by spirits who would reward such an offering by bestowing good luck in return for the gift. As Clark (1944, p. 5) wrote,

> Bubbling from the ground, ever renewed and ever pure, it is hardly surprising that springs should have impressed early man by their magical potency.

The Romans in particular believed that springs and wells were inhabited by natural spirits and had the habit of leaving offerings at these places. In the year 1852, some Jesuit priests in Italy decided to repair and refurbish an ancient well in their care. The well was drained and excavations began. At the bottom of the well, they found a layer of coins dating to the 4th century AD. Lower down, was a layer containing gold and silver coins from the reign of Augustus Caesar (63 BC–AD 14), first Emperor of Rome. As the digging proceeded, workers encountered coins of even greater age. Finally, they came upon arrowheads and stone knives at the bottom.

Groundwater, springs, and wells have been readily associated with the supernatural throughout history. Most groundwater is hidden from view, and its unexplained appearance at the ground in the form of seeps and springs may be mysterious (Chapelle, 1997). In the early Middle Ages, the worship of wells and springs became rampant. Some wells became known as holy wells and associated with various saints. Other wells were known as "cursing wells" and associated with devils. In AD 960, the Canon Law of the Catholic Church forbade the "worship of fountains" (Hardcastle, 1987).

Obtaining a pure and abundant supply of water has been one of the most important duties of city governments so long as men have dwelt in cities. The Romans found that wells were inadequate to supply densely populated urban centers, because they would soon become contaminated with sewage. For this reason, these master plumbers constructed a great system of aqueducts to supply Rome with water, however, the Romans were prodigious well-diggers in other parts of their empire. Beginning sometime between the second century BC and the first century AD, they dug the first wells in central and western Europe. Roman wells were lined with either stone or timber and were tens of meters deep. At the Saalburg, a Roman fort in Germany, 98 wells were found by archeological excavation, the deepest of which reached a depth of 24 meters. An excavated Roman well in England was found to be 57 meters deep (Clark, 1944). The Roman

Wells Throughout History *(continued)*

philosopher Pliny gave advice on how to locate an optimal location for well-digging: lie face down on the ground at sunrise and look for the places that the most water vapor rises (see figure). These instructions were followed until the eighteenth or nineteenth century (Hardcastle, 1987).

In addition to digging, there are some other methods that can be used to construct shallow wells. One technique is called augering. An **auger** is a tool for boring a hole in the Earth. Augers come in all sizes; some are hand-operated, others are driven by machinery. Augering works best in unconsolidated materials such as clays that are not easily susceptible to caving and collapse. Another technique for constructing shallow wells is to simply drive a piece of pipe into the ground. Commonly, the end of the pipe is pointed and has a hole that allows water to enter. Just inside the end of the pipe is a screen of some type to filter out sand and sediment. The pipe can be driven by hydraulic machinery or a sledge hammer. Most driven wells are completed in unconsolidated formations. The Biscayne Aquifer in southeastern Florida is an interesting example of a consolidated formation, which is so soft that wells can be constructed simply by hammering a piece of steel or iron pipe into it. The end of the pipe is left open, and acts like a coring device. When the pipe is driven to its full depth, it is extracted and emptied. The pipe is then reinserted into the hole. Parker et al. (1955) estimated that a well 20 to 40 feet (6 to 12 meters) deep can be constructed in a day in this manner.

The optimal technique for drilling a modern water well depends entirely on the local geology. The best person to choose an appropriate technique is usually not a hydrogeologist, but an experienced driller whose knowledge of local conditions has been proven by trial-and-error. The primary objective of the design and construction is to extract water at the highest rate possible, while excluding suspended sand and silt. A typical well has the following elements (see figure). The well is cased from the surface to the aquifer to prevent collapse. The space between the inside and outside casings is filled with a grout, commonly cement, to prevent surface water from entering and contaminating the well. For the same reason, the well casing extrudes a few inches (1 inch = 2.54 centimeters) above the ground surface. A screen and a gravel pack prevent sediment from entering the well, ensuring a supply of clean water. In some consolidated aquifers, a gravel pack and screen may not be necessary.

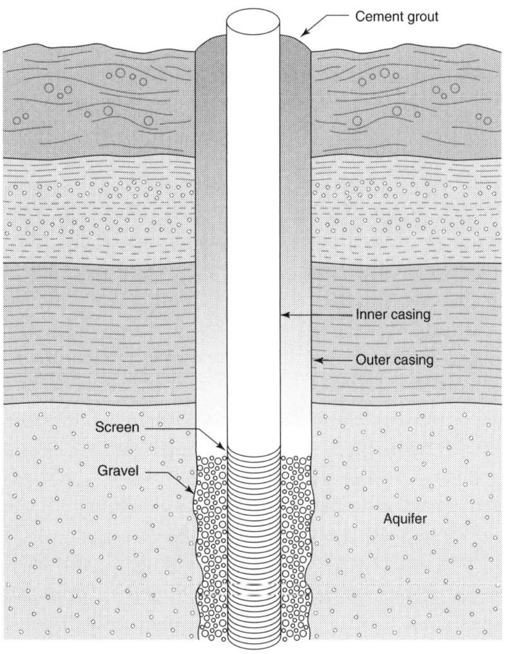

Elements of a typical modern water well.
(After Harlan et al., 1989.)

Nearly all wells deeper than about 30 meters must be completed by drilling. There are two primary methods of drilling wells: cable-tool drilling, also known as percussion drilling, and rotary drilling. **Cable-tool drilling** is a method of drilling in which a tool suspended from a cable is alternately lowered and raised. The lowest tool on the end of the cable is a bit; the bit is usually made of sharpened steel or iron and is massive enough to crush all types of rocks. The drilling action comes from the impact of the bit breaking the rock in the bottom of the well. The drill string is also rotated, and the successive hammering breaks up the rock into small fragments. If water is not naturally present in the bottom of the well, it is added so that the mixture of broken rock and water forms a slurry. When the slurry becomes thick enough so as to reduce the impact of the drill bit, the slurry is removed by bailing or pumping. Cable-tool drilling in unconsolidated formations progresses mainly by mixing and loosening material in the bottom of the well; in consolidated formations, the cable tool bit works primarily by crushing rock at the bottom of the well. In unconsolidated strata, casing must be set immediately behind the cable tool to prevent the drilled hole from collapsing. **Casing** is a tube or pipe used to line wells. No casing is usually necessary in consolidated formations.

Cable-tool drilling is a simple and relatively inexpensive technique that historically has proven to be quite suitable for drilling to moderate depths (few hundred meters). The bit has to be brought to the surface periodically to clean out the hole, but the bit can be sharpened on site. Unlike rotary drilling, there is no mud or drilling fluid in the hole, so there is no risk of leakage contaminating groundwater. The lack of drilling mud on the other hand makes cable-tool wells susceptible to blowouts if overpressured strata are encountered. McLaurin (1896) described a nineteenth-century oil geyser that occurred on the island of Okestra, which juts into the Caspian Sea. In 1886, the Nobel Brother's No. 50 well hit oil. The pressure of the oil was so high that it sent a stream of petroleum 400 feet (122 meters) into the air for seventeen months. The total volume of oil disgorged was three million barrels (477 million liters).

The origin of cable-tool drilling is lost in antiquity; however most indications are that it started in ancient China. Most of the ancient Chinese wells were drilled not for the purpose of obtaining potable water, but to obtain brine that was evaporated to make salt. Chinese cable-tool technology may be thousands of years old, and can be documented to have been in existence since at least the second century AD. But, there is no evidence that the Chinese were able to drill wells deeper than 100 meters prior to the eleventh century. The literature is replete with offhand references to wells "thousands of feet" deep being completed "thousands of years" ago. Most of these claims appear to have arisen when various authors mistakenly mixed the depths of nineteenth-century Chinese wells with the recognition that the technique itself dates back thousands of years.

The center of Chinese drilling was in the province of Sichuan. Sichuan is far enough inland that drilling for salt was more economical than importing salt made by evaporating seawater. The impetus for developing well technology was explained by the eleventh-century Chinese scholar Su Shi, "People never miss an opportunity to make a profit" (Vogel, 1993). By the year 1132, there were more than 4900 brine wells in Sichuan. The typical well had the diameter of a "drinking bowl" and could be as deep as 120 meters. The Chinese were ingenious in the way they used bamboo for almost everything except the iron drill bit. Bamboo tubes were used as casing to prevent near-surface freshwater from infiltrating into the boreholes and diluting the brines. Brine was extracted from a completed well by using a hollow bamboo tube with a leather valve at the bottom. Bamboo was also used to remove excess dirt and water from boreholes. Bamboo stalks have several sections, each of which is hollow and naturally sealed at the top and bottom. By opening a hole in the top of each bamboo section, the Chinese turned a length of bamboo stalk into a series of small buckets. It has been suggested that at least part of the reason the early Chinese were more successful than European drillers is that they suspended their cable tools with strips of bamboo instead of rope. Bamboo has more than 10 times the strength of ordinary hemp rope (Vogel, 1993).

The Chinese maintained their lead in well technology until the middle of the nineteenth century. By the late sixteenth century, wells as deep as 300 meters could be drilled. This increased to 500 meters by the eighteenth century. The zenith of Chinese cable-tool drilling was the Xinhai well, which reached a depth of 1000 meters in 1835.

Cable-tool well drilling was also in use in medieval Europe, although this is disputed by Hardcastle (1987). Nearly contemporaneous with Chinese achievements in the eleventh century, the world's first artesian wells were drilled in AD 1126 in northern France. The wells were several hundred feet (1 foot = 0.30 meters) deep, and were cased 11.3 feet (3.4 meters) above ground level so as to provide a driving force for a mill. The term "artesian" to describe a water well that spontaneously flows at the ground surface is derived from Artois, the name of the region in

Wells Throughout History *(continued)*

"Kicking Down a Well," a crude and inexpensive method of cable-tool drilling widely used in nineteenth-century America. (From McLaurin, 1896, p. 70.)

which these wells are located (Davis and DeWiest, 1966).

From 1800 to 1859, there was rapid improvement in drilling techniques in both Europe and the United States. In 1818, the French government funded the drilling of a well to a depth of 1780 feet (543 meters) to supply the city of Paris with artesian water. The well took 10 years to complete. Later, the Passy well of Paris, completed in 1857, reached a depth of 1923 feet (586 meters). Its diameter was 28 inches (0.7 meters) and it flowed 5.6 million gallons (21 million liters) of water daily that, under artesian pressure, reached a height of 54 feet (16 meters) above the ground surface (Bowman, 1911).

Like China, much of the early development of well drilling in the United States was for the purpose of producing salt from brine. The most famous of the early brine wells was the well drilled by the Ruffner brothers near Charleston, West Virginia. The brothers were in the business of recovering salt from salt springs and conceived the idea of drilling so as to obtain larger quantities of a more concentrated brine. Their first step was to take a hollowed-out log of a sycamore gum tree 4 feet (1.2 meters) in diameter and place it on top of the area where the well was to be dug. One man worked inside the hollow log with a pick and shovel, excavating the ground beneath the log. As he dug, the soil he removed was lifted up in buckets by two men working from a platform erected above him. As the hole was deepened, the log sank and formed a casing that kept the well from collapsing. At a depth of 13 feet (4 meters), bedrock was encountered and cable-tool drilling began. The drilling was slow and tedious, but when they had penetrated the rock to a depth of 17 feet (5 meters) below the base of the gum log, a fissure was struck that yielded a strong flow of brine. The Ruffner brothers were encouraged, and carried the hole down to a total depth of 58 feet (18 meters) below the ground surface where they were rewarded by an ample flow of concentrated brine. Following the success of the Ruffner brothers, the drilling of salt wells became common in West Virginia and Ohio. By 1823, there was a newspaper report of a brine well 750 feet (229 meters) deep (Carlston, 1943).

The primitive cable-tool technique used by the Ruffner brothers was known as "kicking down a well" and was used for the better part of the nineteenth century not only to drill for brine and water, but also to construct shallow oil wells. Unconsolidated sediments were first dug out to the bedrock, and the well cased by a wooden tube. To "kick down a well" an elastic pole (spring-pole) of ash or hickory wood 12 to 20 feet (4 to 6 meters) long was suspended over a fulcrum (see figure). One end of the spring-pole had either stirrups or a tilting platform attached to it. Two or three men worked the stirrups or platform to produce a jerking motion that drew the pole down, with a resultant upward motion being supplied by the elasticity of the spring-pole. At a distance of about two to three feet (0.6 to 0.9 meters) from the workmen, the cable-tool was suspended from the spring-pole. The strokes were rapid, and the slurry at the bottom of the well was removed with a device known as a "sand pump," a hollow tube with an inward opening valve at its bottom. "Kicking down" a well afforded a means of drilling for those with heavy muscles but light purses. In general, it was totally inadequate for deep drilling (McLaurin, 1896, p. 70).

Levi Disbrow was an observer of the early salt-well drilling in West Virginia who became the first water-well driller in America. He drilled his first well in May of 1824 for a distillery in New Brunswick, New Jersey. The well was 175 feet (53 meters) deep, and flowed 1.5 gallons (5.7 liters) of water per minute at an artesian head 3 feet (0.9 meters) above the ground surface. Levi Disbrow's technique was more sophisticated than the primitive method employed by the Ruffner brothers. The United States issued three patents for his cable-tool technology, one in 1825, two in 1830.

In 1859, in Titusville, Pennsylvania, Colonel Drake drilled the first commercial oil well in the United States. The actual drilling was not done by Colonel Drake, but by a blacksmith named William Smith, better known by his nickname of "Uncle Billy Smith." Uncle Billy was knowledgeable both in the working of iron and the drilling of wells. For this reason, Colonel Drake hired him and his nephew to make the tools, assemble the drilling rig, and drill the well. The well was completed at a depth of 69 feet (21 meters) on August 27, 1859. At this point, no oil was observed in the well and the operators became discouraged. The well was shut down for several days, however, when the operators later returned they found that the well had partly filled with oil. The Drake well was considered to be a success, and the first oil boom began. Over the next six years, the population of Titusville swelled from 150 inhabitants to more than 5,000.

The Drake well was not by any means the first oil well in the world. Shallow wells for oil had been dug by hand for thousands of years (Brantly, 1961). Nevertheless, the Drake well was the right well, in the right place, at the right time. It started the American oil business. The importance of the Drake well for wells in general was that the beginning of the oil business provided a powerful economic stimulus for the development of drilling technology. The basic technique remained the same, but the mechanics were perfected and mechanical power was used. The first oil well drilled by steam power was completed in 1860; later, internal combustion engines were used. By 1911, oil wells were being drilled to depths of 5000 feet (1524 meters) (Bowman, 1911).

Rotary drilling is a method of well drilling where a cutting bit is attached to the lower end of a string of hollow drill pipe. The entire string is rotated; it is the rotation of the drill bit that provides the cutting action. The drill bit is somewhat larger in diameter than the drill string, so its action evacuates an annulus between the drill string and the ground. As the drill string rotates, a liquid (mud) is flushed down its interior and reaches the bottom of the hole. The circulating mud picks up drill cuttings in suspension and carries them back to the surface, traveling on the return journey in the space between the outside of the drill string and

Wells Throughout History *(continued)*

the surrounding bedrock. On the surface, the mud is screened to remove rock cuttings and re-circulated.

Rotary drilling is the single method most commonly used today to drill wells. Rotary drilling has several advantages. It is a comparatively rapid, low-cost drilling method. The filling of the well with drilling mud maintains hydrostatic pressures, preventing blowouts. Great depths can be reached. In 1971, the Bertha Rogers well in Oklahoma reached a world record depth of 31,441 feet (9,583 meters). This record has reportedly been exceeded by a Russian well on the Kola Peninsula that reached a depth of 12,226 meters. More recently, in 1995, a scientific research well in Germany reached a total depth of 9,101 meters. One of the disadvantages of the rotary method is that the circulating mud develops a coating on the surface of the well annulus called a filter cake. If a water well is being drilled, it may be difficult to remove the filter cake that otherwise impedes the flow of groundwater into the well.

The history of rotary drilling extends back to the ancient Egyptians, who used this method to quarry stone for the pyramids. Rotary drills were used in Louisiana in 1823 to drill water wells, but the drill cuttings were removed by bailing. The first systematic engineering of rotary methods dates back to 1833 when a French engineer named Fauvelle noticed that when a cable-tool bit struck an artesian aquifer the pressure of the water carried the drill cuttings in suspension up and around the drill string. Fauvelle went on to develop his own rotary drilling system and successfully drilled his first well in 1845. In 1844, a patent was issued to an Englishman named Robert Beart for a rotary drilling system (Lehr et al., 1988).

Although there were rotary drilling systems introduced in the first half of the nineteenth century, the system did not gain widespread acceptance until it was used to bring in the famous Spindletop oil well. Spindletop was the name of a low-lying hill located about 4 miles (6.4 kilometers) south of Beaumont, Texas. The hill was the site of occasional seepages of natural gas, and a local entrepreneur named Pattillo Higgins became convinced that oil could be found there. Higgins drilled his first well in 1893, and a second in 1895. Both wells failed. His drilling contractor tried a third hole on their own; that also failed. The standard cable-tool drilling rigs of the day could not make progress in the thick sequences of unconsolidated sediments found on the Texas Gulf Coast. After reaching depths of 300 to 400 feet (91 to 122 meters), the holes collapsed due to what the drillers described as "running quicksands." Unsuccessful, Higgins advertised in an engineering trade journal, hoping to find someone who could invest the capital necessary to complete a well at Spindletop. His advertisement was answered by Captain Anthony Lucas, a naturalized American citizen and engineer who had explored for sulfur on the salt domes of South Louisiana. Captain Lucas was convinced that the gentle hills found throughout the Gulf Coast were the surface expressions of salt domes, great bodies of salt that had flowed and intruded upwards, tilting adjacent rock strata. Lucas' main interest was in the discovery of sulfur; oil was a secondary concern. What interest he had in oil was not helped along by local geologists. In 1898, William Kennedy of the Texas Geological Survey published a newspaper article where he warned that looking for oil in the Beaumont area was a waste of money and time. His opinion was not unique. The US Geological Survey studied the Spindletop area and also issued a negative report. When Spindletop later produced the greatest oil well in American history, oilmen came to the conclusion that geologists and the application of geology were worthless in the search for oil. This conception persisted for decades.

Captain Lucas began drilling at Spindletop in July of 1899, however, his casing collapsed at 575 feet (175 meters) and the well was lost. Two days earlier, there had been a significant show of oil in the hole. Lucas now knew there was oil at Spindletop, but his money was gone. He traveled to Pittsburgh and presented his ideas to the greatest American oil explorationists, the team of John Galey and James Guffey. Galey and Guffey had drilled wildcat wells from Washington, DC, to the wilds of the Mojave desert, and they were looking for a new opportunity.

Galey and Guffey decided to drill at Spindletop, and they hired the best rotary drillers in the business, the Hamill brothers of Corsicana, Texas. At that time, the rotary method had been widely used in oil drilling for only about five years. After its success at Spindletop, it became increasingly popular, displacing cable-tool techniques almost entirely by 1930. The Hamill brothers found drilling at Spindletop to be extremely difficult. No one had ever successfully drilled a deep well in this type of geology. The water that was flushed through the rotary drill pipe was failing to bring up the loose sands. Curt Hamill recognized that he had to increase the viscosity and density of the circulating fluid—but how? Hamill invented the world's first drilling mud by driving some cattle through a shallow pond. The cattle stirred up the water enough to generate a thick mud. When the muddy water was poured down the Spindletop well bore, the hole was stabilized and the sands flushed out. Drilling had started on October 27, 1900. After encountering and overcoming enormous difficulties, a total depth of 1020 feet (311 meters) was reached on January 8, 1901. The bit then became stuck and could not be freed. The rotary chain broke. It looked as if the well was lost. At 10:30 AM on the morning of January 10, 1901, the drillers tried to re-enter the hole. The drill string had reached a depth of 700 feet (213 meters) when mud began gushing from the well. The mud was followed by six tons (5,443 kilograms) of pipe. The pipe shot into the air; where it came down it stuck in the ground like giant spikes. The eruption stopped, and all was quiet for a moment. Then mud started to flow again, followed by gas. Finally, a thick stream of greenish crude oil gushed out. The oil stream erupted into the sky as a fantastic geyser, reaching a height of 100 feet (30 meters). It was the greatest oil well ever seen in the United States. The flow continued for nine days at a rate of 84,000 barrels (13.3 million liters) per day, until workers succeeded in capping the well. Prior to Spindletop, a substantial oil well was considered to be one that yielded 50 barrels (7,900 liters) per day. Oil production from the Spindletop well increased total world oil production by 20%, and United States oil production by 50%. This single well by itself produced as much oil as 37,000 typical oil wells in the Eastern United States, and more than twice as much oil as the entire state of Pennsylvania, which heretofore had been the leading producer of oil in the United States.

Spindletop changed everything. It proved the value of rotary drilling methods, but the consequences were far more widespread and significant. The cheap energy provided by abundant oil changed the United States from a rural, agrarian society to an urban, industrialized nation. In 1900, 39% of the population lived on farms. By 1997, only 2% of the population was engaged in farming. Perhaps the greatest facilitator of social change in the twentieth century of the United States was the automobile. In 1900, there were only about 8,000 cars in the entire United States. It was thought that the gradual decline of the Pennsylvania oil fields would ensure that there would never be enough gasoline available to make the automobile a common possession. Horseless carriages, like yachts, were a toy for the rich to enjoy. When the Spindletop gusher came in, the price of oil quickly dropped from $2 dollars to 3 cents a barrel, recovering to 83 cents a barrel two and a half years later. With an abundant supply of cheap fuel, by 1910 the number of automobiles in the United States had risen to 450,000. The new automobiles needed paved roads, and asphalt, a petroleum byproduct, provided an inexpensive means of paving. Things would never be the same again (Gillespie, 1995a, b).

FOR ADDITIONAL READING

Brantly, J. E. 1961. "Percussion-drilling system." *History of Petroleum Engineering,* p. 133–269. New York: American Petroleum Institute.

Carlston, C. W. 1943. Notes on the early history of water-well drilling in the United States. *Economic Geology* 38 (2), 119–36.

Chapelle, F. H. 1997. *The hidden sea: groundwater, springs, and wells.* Tucson, Arizona: Geoscience Press, Inc., 238 pp.

Clark, G. 1944. Water in antiquity. *Antiquity* 18 (69), 1–15.

Gillespie, R. H. 1995a. Rise of the Texas oil industry. Part 1. *Exploration at Spindletop:* The Leading Edge, 14 (1), 22–24.

Gillespie, R. H. 1995b. Rise of the Texas oil industry: Part 1. *Spindletop changes the world:* The Leading Edge, 14 (2), 113–115.

Hardcastle, B. J. 1987. Wells ancient and modern—an historical review. *Quarterly Journal of Engineering Geology,* 20: 231–38.

Vogel, H. U. 1993. The great well of China. *Scientific American,* 268 (6) 116–20.

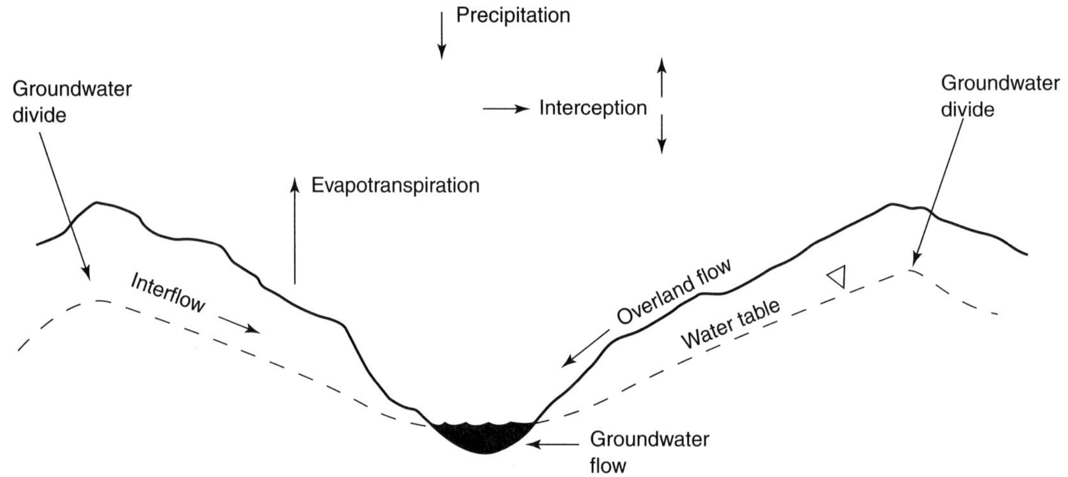

Figure 6.7 Processes by which water moves in a drainage basin.

description of fluid movement in the unsaturated zone and corresponding solutions are discussed in some detail by Philip (1969).

6.1.3. Capillary Barriers.

One of the fascinating aspects of the unsaturated zone is that coarse materials with otherwise high conductivities may act as barriers to flow in some circumstances. A coarse-grained sediment such as gravel contains large pore spaces in which the absolute value of the capillary-pressure head is small compared to a fine-grained material such as a clay. The fine-grained clay is thus preferentially saturated compared to the coarse-grained gravel. As hydraulic conductivity in the unsaturated zone increases with the degree of saturation (Figure 6.3), an unsaturated gravel whose large pores are filled with air may have a lower hydraulic conductivity than a saturated and fine-grained clay unit.

6.2. INFILTRATION.

The smallest unit that is usually used to analyze near-surface water flow is the drainage basin, also referred to as a **watershed** or **catchment** (Figure 6.7). A **drainage basin** is an area that has a common drain, usually a stream. Drainage basins are separated by topographic divides that usually (but not necessarily) coincide with groundwater divides. A **groundwater divide** is a ridge on the water table that separates areas of lesser head values. Precipitation that falls upon opposite sides of a topographic divide moves in opposite directions, both over land and underground. With the exception of the draining stream, drainage basins are semi-closed hydrologic systems. Occasionally, there may exist a drainage basin that lacks a draining stream; an example is the area surrounding the Great Salt Lake in Utah. Drainage basins of larger scales encompass and contain drainage basins of smaller scales.

The movement of water in the near-surface environment of a drainage basin can be divided into a number of different flow mechanisms and paths (Figure 6.7). Water enters a drainage basin through **precipitation.** Water leaves a drainage basin by **evapotranspiration, overland flow, interflow,** and **groundwater flow. Evapotranspiration** is the sum of **transpiration** and **evaporation. Transpiration** is the movement of water from the soil to the atmosphere by plants. **Evaporation** is the transformation of liquid water near and at the ground surface to water vapor in the atmosphere. **Overland flow** is the direct movement of water over the land surface to the nearest stream or body of water. **Interception** refers to precipitation caught by vegetation before it may reach the ground surface. The surface at which

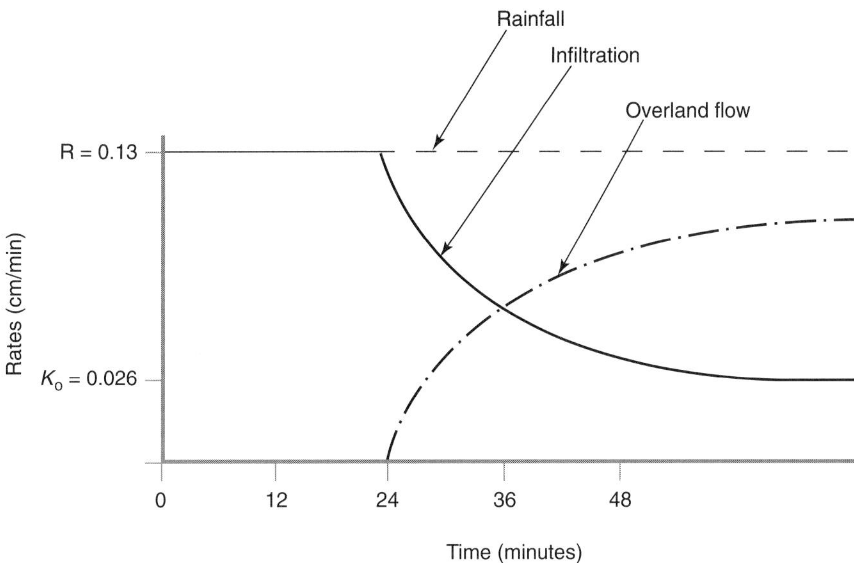

Figure 6.8 Rate of infiltration as a function of time elapsed since the beginning of rainfall. Overland flow begins when the infiltration rate falls below the rainfall rate. "R" is the rainfall rate (0.13 centimeters per minute), "K_o" is the saturated hydraulic conductivity of the soil (0.026 centimeters per minute.)
(From Freeze, 1974, p. 639.)

the solid Earth is saturated with fluid at atmospheric pressure is the **water table. Infiltration** is the entry of water into the ground surface and the subsequent flow of water away from the ground surface. Subsurface flow that takes place in the unsaturated or vadose zone between the ground surface and water table is termed **interflow.** The movement of groundwater below the water table is **groundwater flow.**

When it rains, the rate at which infiltration takes place decreases with time as the soil pores fill and the absolute value of the capillary pressure head decreases (Horton, 1933). Any given soil has a maximum rate at which infiltration can take place; this rate is the **infiltration capacity** (Figure 6.8). The infiltration capacity of a soil tends to approach a constant value as time increases. This limiting value is the average value of the saturated hydraulic conductivity (the demonstration of this is left as an exercise). If the rainfall rate exceeds the infiltration capacity, water either accumulates (ponds) on the ground surface or runs downhill as overland flow.

It is possible to construct a simple model of the infiltration process if we assume that the soil moisture travels downward as a "slug." This is often called **plug** or **piston flow.** Piston flow is flow in which the moving material does not diffuse; the advancing front is sharply and clearly demarcated. If we treat infiltration as piston flow in which a completely saturated area marches downward, it is possible to derive a simple expression for the position of the advancing front as a function of time. This equation was first derived by Green and Ampt (1911), and is known as the Green-Ampt equation.

Let the capillary pressure head in the unsaturated zone at the advancing moisture front be $-\Psi_f$ (m), the depth of the moisture front be Z_f (m), and the saturated hydraulic conductivity be K_s (m-s^{-1}) (Figure 6.9). The capillary pressure head at the surface is zero, thus the capillary pressure gradient across the saturated "slug" of moisture is $-\Psi_f/Z_f$. Substitution into equation 6.12 (Darcy's Law) results in

$$i = -K_s\left[\frac{-\Psi_f}{Z_f} + 1\right] \tag{6.14}$$

where i (m-s^{-1}) is the infiltration rate. Note that the infiltration rate (i) is a specific discharge or

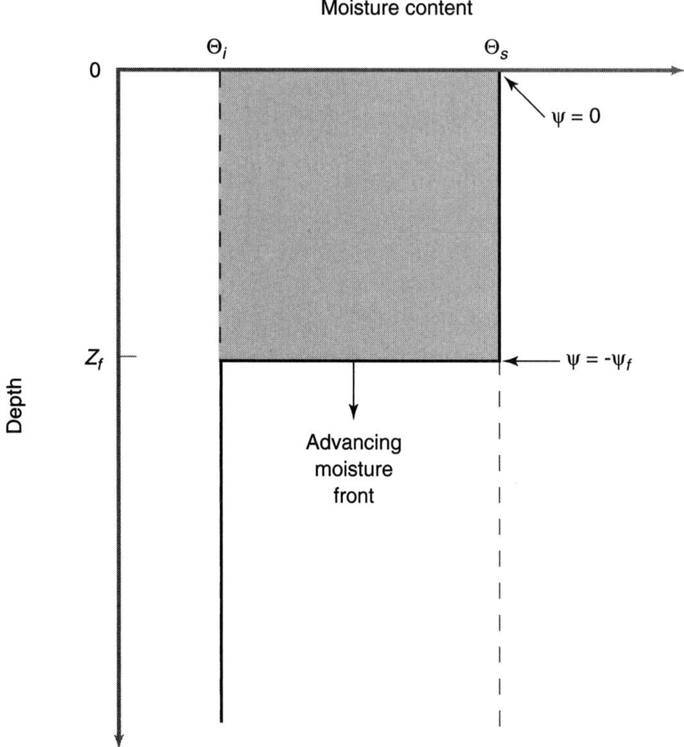

Figure 6.9 Infiltration can be approximately modeled by an advancing moisture front moving without diffusion. This is known as slug or piston flow.

Darcy velocity, not a linear velocity. Rewriting "1" as Z_f/Z_f, we obtain

$$i = -K_s\left[\frac{-\Psi_f + Z_f}{Z_f}\right] \quad \textbf{(6.15)}$$

Denote the cumulative amount of water per unit area that has infiltrated by the capital letter I (m). I is equal to the difference between the saturated moisture coefficient (θ_s) and the initial moisture content (θ_i) multiplied by the saturated depth (Z_f). The cumulative water volume per unit area thus is

$$I = Z_f\,[\theta_s - \theta_i] = Z_f\,[\Delta\theta] \quad \textbf{(6.16)}$$

The infiltration rate (i) must be the time derivative of the total cumulative water added through infiltration,

$$i = \frac{-dI}{dt} = -\Delta\theta\,\frac{dZ_f}{dt} \quad \textbf{(6.17)}$$

where the negative sign indicates downward flow. Substituting equation 6.17 into equation 6.15,

$$\Delta\theta\,\frac{dZ_f}{dt} = K_s\left[\frac{-\Psi_f + Z_f}{Z_f}\right] \quad \textbf{(6.18)}$$

Equation 6.18 can be integrated to find the time t it takes for the infiltration front to advance to a position Z_f (Hornberger et al., 1998, p. 187),

$$\frac{K_s t}{\Delta\theta} = Z_f + \Psi_f \log_e\left[1 + \frac{Z_f}{-\Psi_f}\right] \quad \textbf{(6.19)}$$

Note that the time t it takes for the saturated moisture front to advance a distance Z_f is proportional to the amount of moisture added to the soil ($\Delta\theta$), but inversely proportional to the saturated hydraulic conductivity (K_s).

If greater accuracy is desired, the process of infiltration can be mathematically modeled by solving Richards equation numerically. Figure 6.10

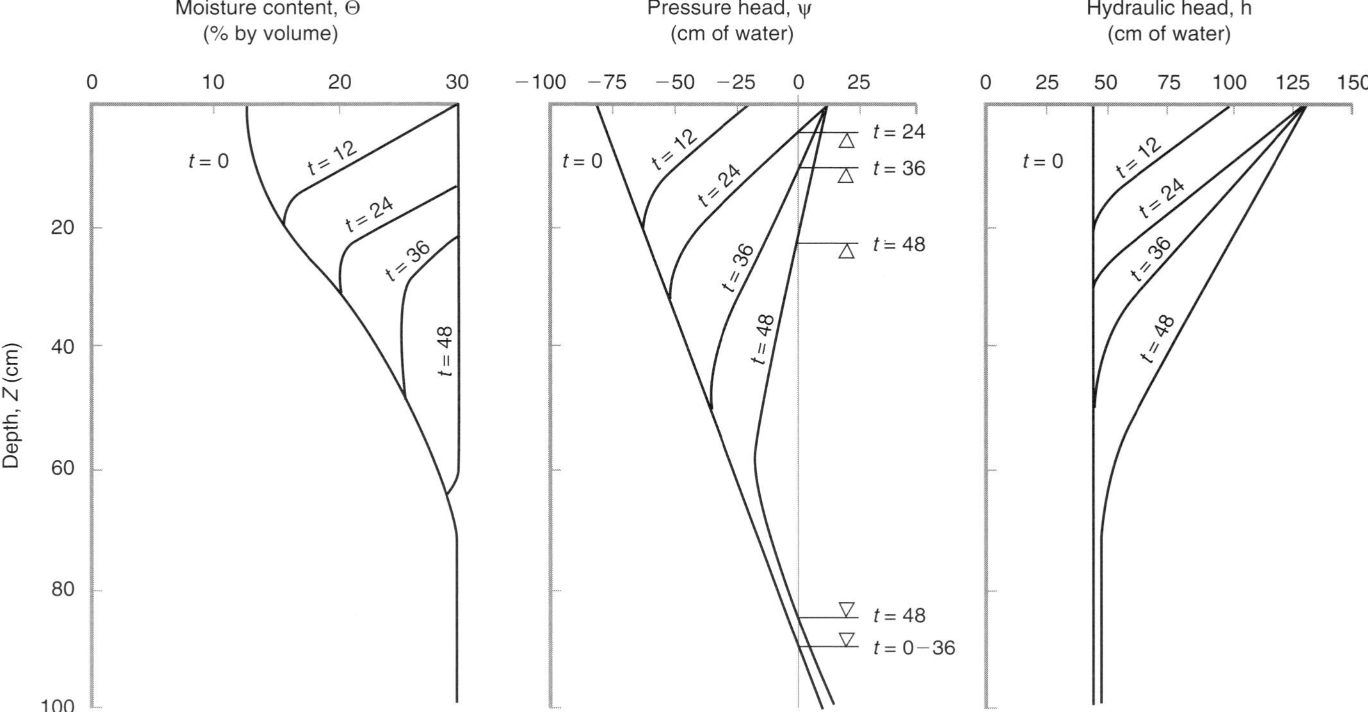

Figure 6.10 Moisture content (Θ), capillary pressure head (Ψ), and hydraulic head (*h*), in the unsaturated zone calculated by a numerical model as functions of time (*t*) elapsed since infiltration began.

(From Freeze, 1974, p. 639.)

shows the 48-hour evolution of moisture content (θ), pressure head (Ψ), and head (h) as modeled for a hypothetical infiltration event where the rainfall rate is a constant 0.13 cm per minute. The relationship between rainfall rate, infiltration, and overland flow is shown in Figure 6.8. At first, the infiltration capacity of the soil is high and there is no overland flow, however, the infiltration capacity decreases with time. Water that does not infiltrate the ground surface goes into overland flow (or ponds at the surface).

6.3. Origin of Streamflow.

In ancient times, it was thought that streams and springs were fed by some mysterious mechanism in the Earth's interior that pulled up water from the seas into the continents. Only in the seventeenth century was it realized that streams and springs are part of a hydrologic cycle fed by rainfall. As Freeze (1974, p. 627) elegantly stated, "indifferent rains provide sufficient water to feed mighty rivers."

6.3.1. Stream Hydrographs.

The primary means by which contributions to streamflow are assessed is through hydrographs. A **hydrograph** is a record of the volumetric discharge (volume per unit time) at a particular point in a stream over time (Figure 6.11). Most streams show a rapid rise and subsequent fall in flow rate during and immediately following a precipitation event. Streams may also continue flowing for several months in the absence of any precipitation. This observation suggests that a hydrograph may be divided conceptually into two components: quickflow and baseflow. **Quickflow** is the component of stream flow that takes place coincident with, and shortly after, a precipitation event; it rapidly rises and then falls. **Baseflow** is the component of stream flow that is much longer in duration and declines more slowly than quickflow; baseflow is supplied to a stream by groundwater flow.

It is difficult to separate a hydrograph into quickflow and baseflow components. The separation can be done empirically by attributing the rapid rise in streamflow during a precipitation event, and assuming that quickflow ceases within a few days following the end of the precipitation. A more precise and modern method is to use chemical tracers. One surprising result that has emerged from the use of such tracers is that much quickflow consists of "old" water that was driven out of storage by a precipitation event. This is contrary to our intuition, which suggests that the surge of water delivered to a stream following a storm consists of the same water that fell as precipitation during the storm. But, it does appear that much quickflow consists of "old" water, at least this appears to be the case for forested watersheds in temperate regions (Hornberger et al., 1998, p. 204).

There are four different means by which rainfall reaches streams: direct precipitation on a stream's surface, overland flow, subsurface storm flow, and groundwater flow (Figure 6.12). Direct precipitation on the surface of a stream is usually insignificant. The relative importance and contribution of the other three processes to quickflow and baseflow depend upon climate, geology, topography, soil characteristics, vegetation, and land use (Freeze, 1974; Hornberger et al., 1998).

6.3.2. Overland flow.

Overland flow is the dominant mechanism by which quickflow occurs. Overland flow (defined above) occurs when the topographic gradient promotes runoff instead of ponding, and one of two conditions exists: either the precipitation rate exceeds the infiltration rate, or infiltration has progressed to the point that the water table rises to the ground surface. If overland flow occurs because rainfall is faster than infiltration, the resulting overland flow is termed **Hortonian overland flow** after Horton (1933) who first described the process.

Overland flow occurs in urban areas with large expanses of directly connected, impervious surfaces (Hornberger et al., 1998). Overland flow may also be common throughout large sections of deserts and semi-arid areas that lack a vegetative cover. The lack of organic matter in the soil

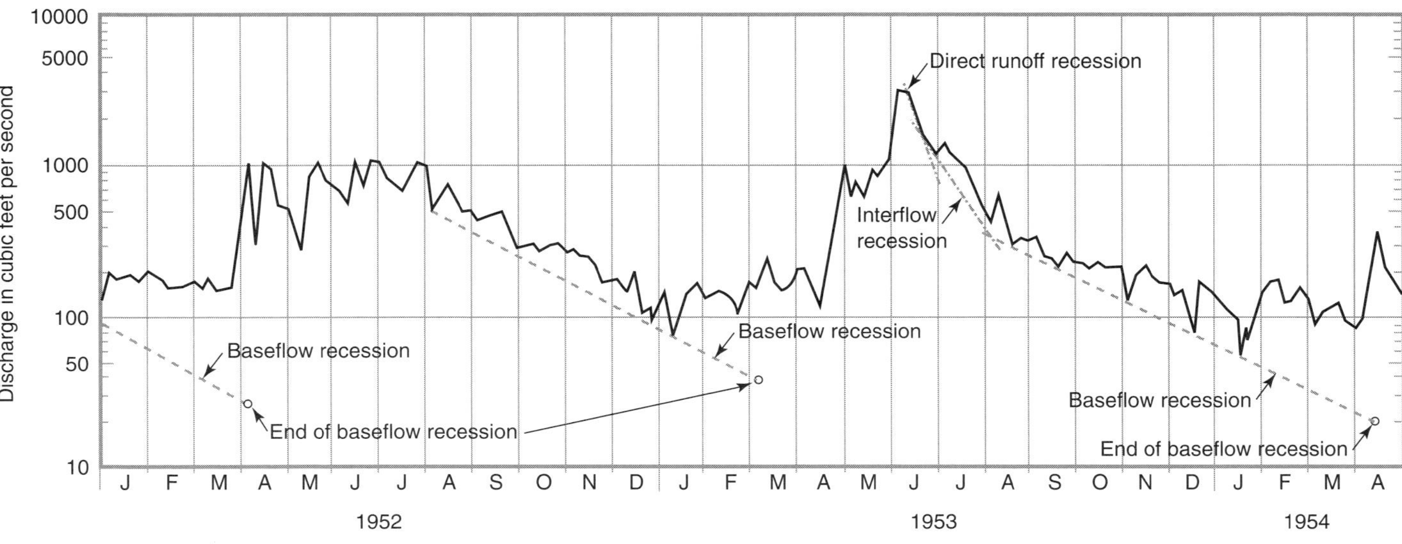

Figure 6.11 Streamflow hydrograph of the Elbow River in Alberta, Canada.
(From Meyboom, 1961, p. 1208.)

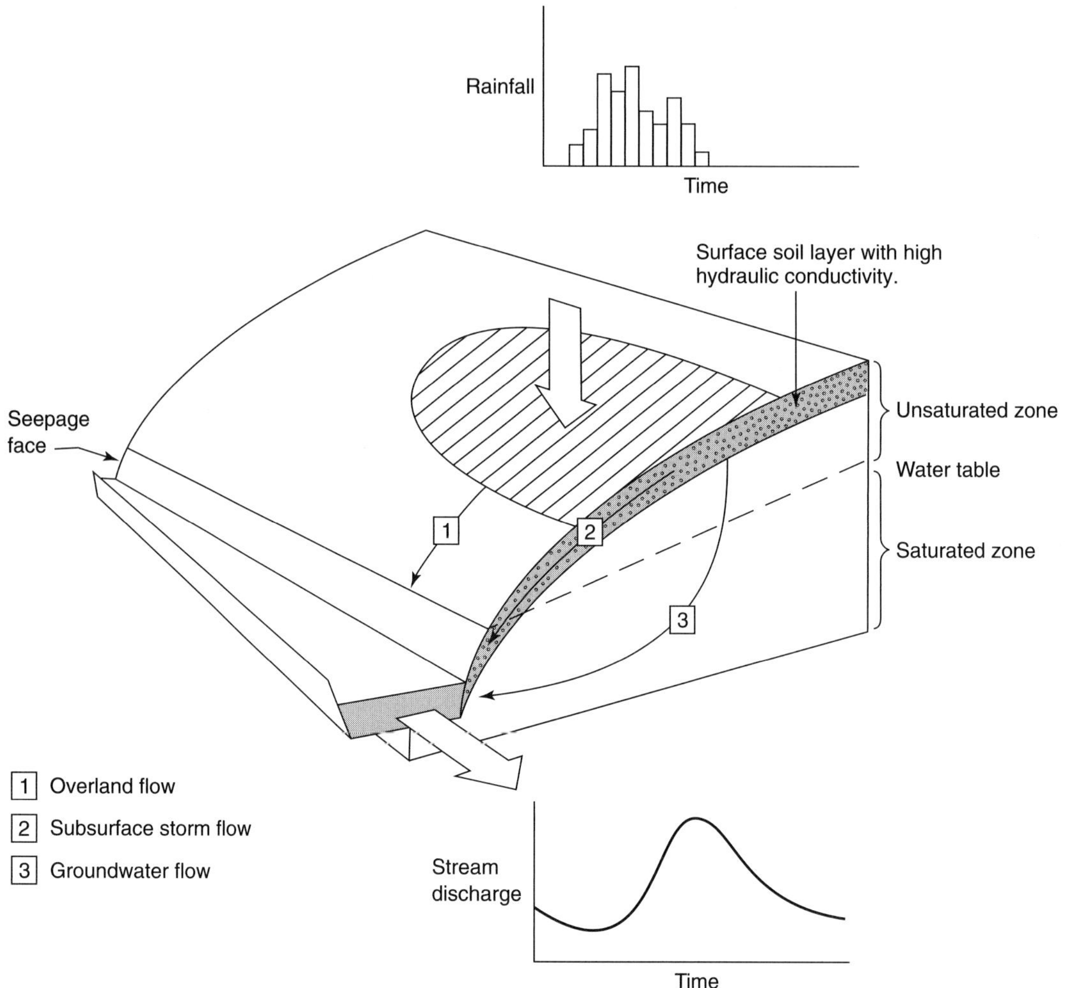

Figure 6.12 Three most important processes that contribute to streamflow: overland flow, subsurface storm flow, and groundwater flow.
(From Freeze, 1974, p. 630.)

prevents the development of porous channels through which water may move easily, and infiltration is inhibited. Low infiltration rates and high rates of overland flow make these areas prone to flash floods due to the rapidity with which large amounts of water can be delivered to streams.

In contrast to urban environments and deserts, overland flow in forested areas may be confined to a small portion of the total area. Forest soils tend to have a high infiltration capacity. Live and decaying vegetation on the surface protects the soil surface from compaction caused by raindrop impact, while roots keep the soil permeable and porous. Freeze (1974, p. 631) concluded that overland flow was "a (spatially) rare occurrence in humid, vegetated basins." Although it may still be the most significant component of quickflow, overland flow in forests comes from an area that is typically about 1 to 3% of the total watershed area (Freeze, 1974). Although Hortonian overland flow in forested areas may be uncommon, overland flow can also be generated from expanding source areas. The **expanding source area concept** is the idea that during a precipitation event the water table in low-lying areas next to distributaries may rise to the surface and overland flow may occur. Following the end of the precipitation event, the water table drops and the "source areas" shrink in size.

6.3.3. Subsurface Storm Flow.

Subsurface storm flow is lateral flow that occurs along the top of unsaturated zone when that area temporarily becomes saturated when precipitation is high (Figure 6.12). As infiltration rate typically decreases with increasing time, the top of the unsaturated zone may become completely saturated in a relatively short period of time. The primary requirement for subsurface storm flow is the presence of a shallow soil horizon of high permeability at the surface. There is considerable evidence that such a horizon commonly occurs, however, Freeze (1974) concluded that subsurface storm flow makes a significant contribution to stream flow only under exceptional circumstances.

6.3.4. Groundwater Flow.

Groundwater flow is flow through the saturated Earth below the water table. In general, groundwater flow is the slowest and steadiest component of stream flow. Groundwater flow is seldom the cause of major runoff during storms. Its primary role is sustaining streams between precipitation events (Freeze, 1974). Groundwater flow is the source of baseflow to streams.

6.4. BASEFLOW RECESSIONS.

During the absence of precipitation, streamflow is composed entirely of baseflow. The exponential decline of baseflow that occurs over time is a **baseflow recession** (Figure 6.11).

After the termination of quickflow, baseflow decreases exponentially with increasing time. This is the beginning of a baseflow recession. Depending on the size of the watershed and the permeability of the rocks and sediments within it, baseflow may continue for several months. It is baseflow that allows streams to continue to flow for extended periods of time in the absence of precipitation. A stream that gains water from baseflow is termed gaining or **effluent.** A stream that loses water to the subsurface (and thus recharges groundwater) is termed losing or **influent.**

Let Q (m^3-s^{-1}) be the stream discharge rate, or volume of water that passes through an effluent stream per unit time. If overland flow has ceased, all the water that leaves the watershed via the stream must be equal to the water entering the stream by baseflow. Thus by recording the stream discharge on a stream hydrograph we are indirectly measuring the baseflow. Both empirically and theoretically (Hall, 1968), we find that

$$Q = Q_o e^{-at} \quad \textbf{(6.20)}$$

where Q (m^3-s^{-1}) is the instantaneous stream discharge (volume per unit time), t (s) is the time since the baseflow recession started, e (dimensionless) is the base of the natural logarithms, and Q_o (m^3-s^{-1}) and a (s^{-1}) are constants. If quickflow has ceased, then the stream discharge is equal to the baseflow.

The nature of the recession curve (Figure 6.11) depends on the permeability of the ground materials in a watershed. In regions characterized by low-permeability crystalline rocks, the recession curve will be steep, reflecting the fact that lower permeability effectively allows only areas close to a stream to drain within the recession time. Regions characterized by higher-permeability bedrock (e.g., **karst** limestone) have flatter recession curves. The presence of higher permeability bedrock allows a greater area to be drained, extending the duration of the baseflow (Domenico and Schwartz, 1990, p.16).

Problem: 10 days after the start of a baseflow recession, Q = 100 m^3-s^{-1}; 30 days after the start of a recession, Q = 20 m^3-s^{-1}.

(a) what are Q_0 and the recession constant a?

Recall equation 6.20, $Q = Q_0 e^{-at}$; the properties of logarithms dictate that $\log_e(xy) = \log_e(x) + \log_e(y)$, thus

$$\log_e Q = \log_e Q_0 - at \quad \textbf{(6.21)}$$

Let $Q_1 = 100$ m^3-s^{-1}, $t_1 = 10$ days, $Q_2 = 20$ m^3-s^{-1}, and $t_2 = 30$ days as given. Substitute into equation 6.21:

$$\log_e Q_1 = \log_e Q_0 - at_1 \quad \textbf{(6.22)}$$

$$\log_e Q_2 = \log_e Q_0 - at_2 \quad \textbf{(6.23)}$$

or

$$4.61 = \log_e Q_0 - 10a \quad \textbf{(6.24)}$$

$$3.00 = \log_e Q_0 - 30a \quad \textbf{(6.25)}$$

We now find the constant a by subtracting equation 6.25 from equation 6.24

$$1.61 = 20a \quad \textbf{(6.26)}$$

$$a = 0.0805 \text{ (days}^{-1}) \quad \textbf{(6.27)}$$

Note that the constant a must have units of days^{-1} so that the product at is dimensionless. We may now find Q_0 by substituting back into equation 6.24 (or 6.25 if we prefer).

$$4.61 = \log_e Q_0 - 0.805 \quad \textbf{(6.28)}$$

$$\log_e Q_0 = 5.415 \quad \textbf{(6.29)}$$

$$Q_0 = 225 \text{ (m}^3\text{-s}^{-1}) \quad \textbf{(6.30)}$$

Q_0 must have units of m^3-s^{-1}, as both Q_1 and Q_2 were given in m^3-s^{-1}. We can now check our result by substituting back into equation 6.20:

$$100 \text{ (m}^3\text{-s}^{-1}) = 225 \text{ (m}^3\text{-s}^{-1})\, e^{-(0.0805 \times 10)} \quad \textbf{(6.31)}$$

$$20 \text{ (m}^3\text{-s}^{-1}) = 225 \text{ (m}^3\text{-s}^{-1})\, e^{-(0.0805 \times 30)} \quad \textbf{(6.32)}$$

(b) What is the instantaneous stream discharge Q at $t = 15$ days?

Recall equation 6.20, $Q = Q_0 e^{-at}$, and substitute:

$$Q = 225\, e^{-(0.0805 \times 15)} \text{ (m}^3\text{-s}^{-1}) \quad \textbf{(6.33)}$$

$$Q = 67.3 \text{ (m}^3\text{-s}^{-1}) \quad \textbf{(6.34)}$$

(c) What is the *cumulative* volumetric discharge from the beginning of the baseflow recession through a period of 6 months? Note that the problem is not asking for Q (m^3-s^{-1}), the instantaneous discharge. Rather, we are asked to find the *total* volume of water integrated over time. Let the total volume be designated Q^* (m^3).

To find the total discharge Q^* in m^3 we must multiply the discharge rate Q (m^3-s^{-1}) by time. Q is not constant, but changes as time t changes. Therefore, we must integrate from time zero to time t.

$$Q^* = \int_0^t Q(t)\, dt = \int_0^t Q_0 e^{-at} dt = Q_0 \int_0^t e^{-at} dt \quad \textbf{(6.35)}$$

Equation 6.35 is easily integrated by recalling that $\int e^u\, du = e^u$. Let $u \equiv -at$, then $du/dt = -a$, or $du = -adt$. Thus

$$Q^* = \frac{-Q_0}{a} \int_0^t -ae^{-at} dt \quad \textbf{(6.36)}$$

where we have placed equation 6.35 in the requisite form for integration by multiplying it by $(-a/-a)$. Integrating,

$$Q^* = \frac{-Q_0}{a} [e^{-at}]_0^t \quad \textbf{(6.37)}$$

$$Q^* = \frac{Q_0}{a} [1 - e^{-at}] \quad \textbf{(6.38)}$$

Equation 6.38 gives the total volumetric discharge Q^* after a time t. Note that in equation 6.38 Q_0 and the constant a must have the same time units. Formerly, we expressed Q_0 in m^3-s^{-1} and a in days^{-1}. Similarly, a and t in the expression e^{-at} must have the same time units. In part (a) we found that $Q_0 = 225$ m^3-s^{-1} and $a = 0.0805$ days^{-1}. Converting units,

$$a = 0.0805 \frac{1}{\text{day}} \times \frac{1 \text{ day}}{86400 \text{ s}} = 9.317 \times 10^{-7} \text{s}^{-1}$$

$$t = 6 \text{ months} \times \frac{30.4 \text{ days}}{\text{month}} = 182.4 \text{ days}$$

Thus,

$$Q^* = \frac{225}{9.317 \times 10^{-7}} [1 - e^{-(0.0805)(182.4)}] \quad \textbf{(6.39)}$$

Note that the term in brackets is essentially one as the exponential term is close to zero. In other words, the total volumetric discharge after 6 months is not significantly different than it would be if the baseflow recession continued indefinitely. We finally arrive at

$$Q^* = 2.41 \times 10^8 \text{ m}^3 \quad \textbf{(6.40)}$$

How much water is 10^8 m^3? Is the answer physically reasonable? If we take the cube root of 2.41×10^8, we obtain 623. In other words, the total water available for discharge to the stream by baseflow is a cube of water 623 meters on a side. The answer seems reasonable enough, depending, of course, on the size of the drainage basin under consideration. If we had obtained a cube of water 6.23 m on a side, or a cube of water 623 km on a side, we would have been alerted to check our calculations for a numerical error.

(d) How long does it take for 90% of the available groundwater to reach the stream?

We first find quantitatively what is meant by *available groundwater*. The total available groundwater is Q^* for $t = \infty$. It follows from equations 6.35 and 6.38 that

$$Q^*(t = \infty) = {}_0\!\int^{\infty} Q(t)\, dt = \frac{Q_0}{a}[1 - e^{-\infty}] = \frac{Q_0}{a} \quad \textbf{(6.41)}$$

We now have to find the time t it takes for Q^* to equal 90% of Q_0/a. In other words, we have to find t such that:

$${}_0\!\int^{t} Q_0 e^{-at} dt = 0.9 \times \frac{Q_0}{a} \quad \textbf{(6.42)}$$

Integrating as we did for equation 6.35, we obtain

$$\frac{Q_0}{a}[1 - e^{-at}] = 0.9 \times \frac{Q_0}{a} \quad \textbf{(6.43)}$$

Q_0 now drops out. Note that an interesting corollary is that the answer does not depend on Q_0, only on the constant a.

$$1 - e^{-at} = 0.9 \quad \textbf{(6.44)}$$

$$e^{-at} = 0.1 \quad \textbf{(6.45)}$$

$$-at = \log_e(0.1) = -2.30 \quad \textbf{(6.46)}$$

$$t = \frac{2.30}{a} = \frac{2.30}{0.0805 \text{ days}^{-1}} = 29 \text{ days} \quad \textbf{(6.47)}$$

6.5. DUPUIT THEORY AND FLOW IN UNCONFINED AQUIFERS.

Dupuit theory is applied to understand flow in an unconfined aquifer with homogeneous hydraulic conductivity (Figure 6.13). The theory was first derived by the French engineer, Arsene Jumes Emile Juvenal Dupuit (1804–1866) and published in 1863. Like Henry Darcy, Dupuit was an engineer who made considerable contributions to both hydraulics and hydrogeology (Rouse and Ince, 1957).

Consider two points in an unconfined aquifer (Figure 6.13). At $x = 0$ (m), $h = h_1$ (m), and at $x = L$ (m), $h = h_2$ (m). The elevation of the water table, h (m), is the head or potentiometric surface, while simultaneously being the saturated thickness of the unconfined aquifer above an arbitrary datum. It is possible to find relatively simple expressions for the Darcy velocity (q_x, m-s^{-1}) and head (h) as function of distance (x) if we invoke the **Dupuit** (doo-pwee) **Assumptions.** They are

1. The slope of the water table is dh/dx.
2. If dh/dx is "small," equipotentials are vertical and flowlines are horizontal. In other words, there is no flow in the vertical (z) direction.

Dupuit theory results in replacing the water table by a parabola; it is usually applied to calculate shallow flow in a porous medium and is invalid in certain regions where the above assumptions are violated (see text after Eq. 6.75).

The Darcy velocity is the volume of water per unit area perpendicular to flow per unit time. Consider a section of an unconfined aquifer that has a width Δy (m) but whose height decreases "downstream" (Figure 6.14). Under conditions of steady flow, the volumetric rate of groundwater flow (m^3-s^{-1}) is constant as required by the conservation of mass. But, the Darcy velocity must increase downstream because the area perpendicular to flow decreases as the saturated thickness of the

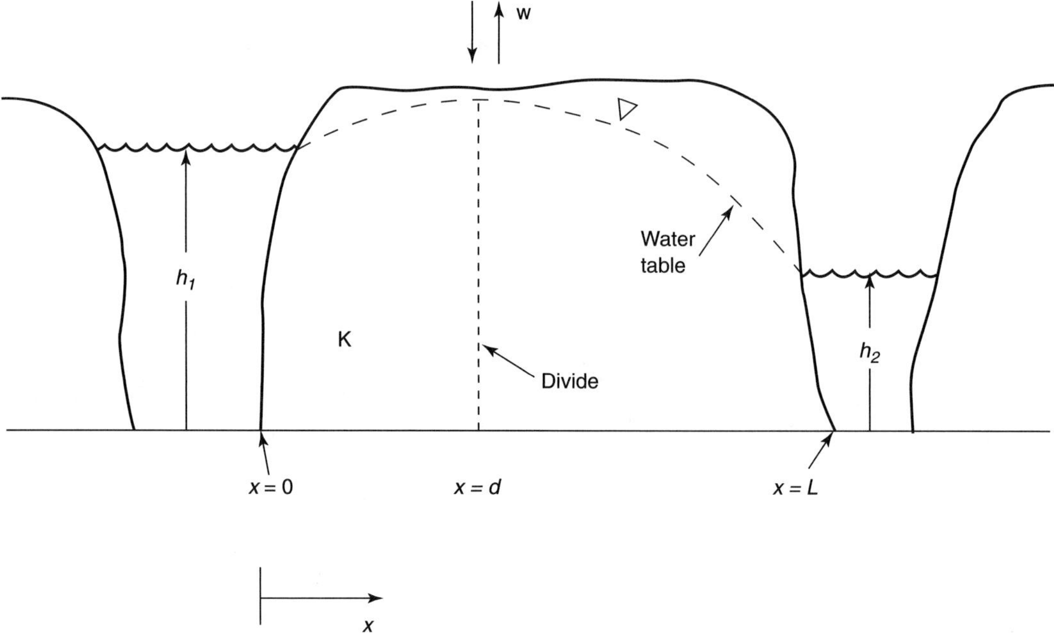

Figure 6.13 Flow in an unconfined porous medium.

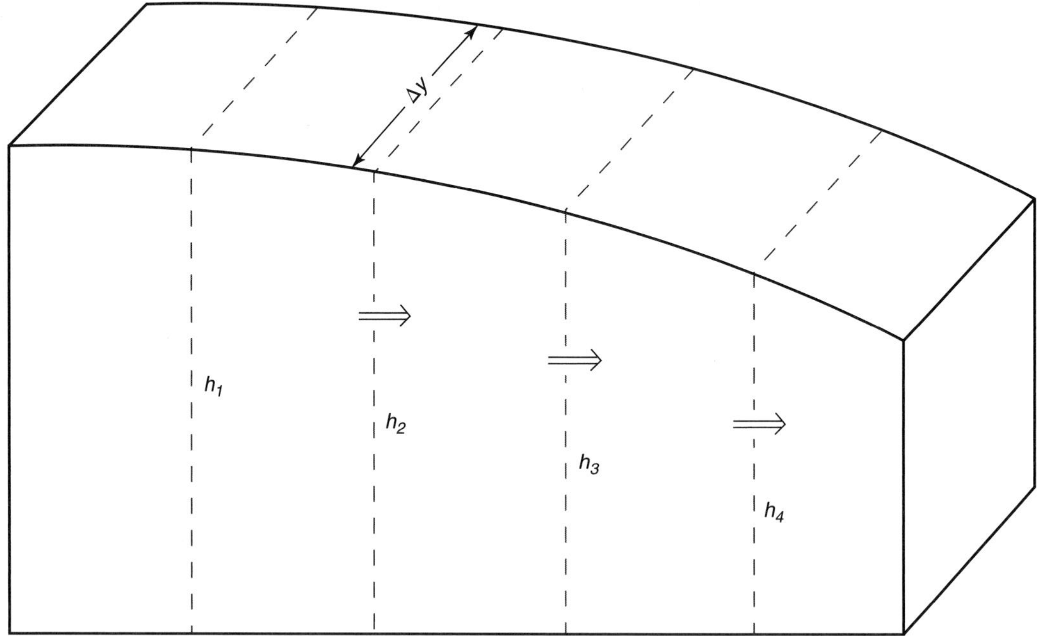

Figure 6.14 If flow per unit width (Δy) through an unconfined porous medium is constant, the Darcy velocity must increase downstream.

The Mississippi River: An Aquifer-Making Machine

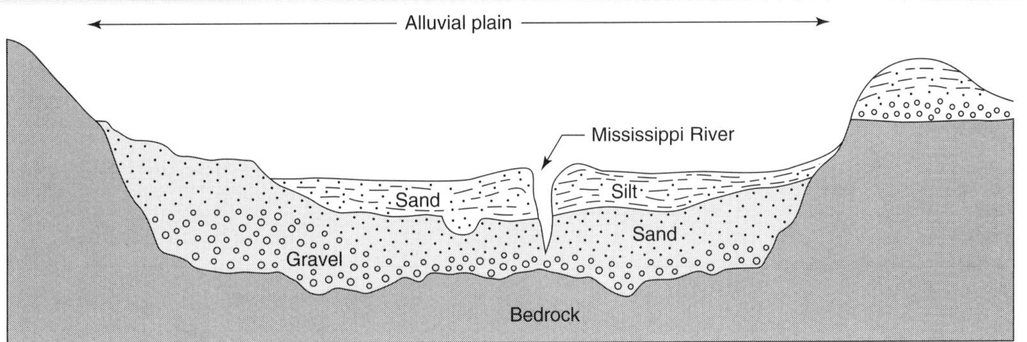

Schematic cross section through the Mississippi River alluvial aquifer.
(After Boswell et al., 1968.)

The Mississippi is the largest river in North America. Including its major tributaries, it drains about one-eighth of the continent. The Mississippi River is inexorably woven into the history of the United States; its lure and lore have fascinated Americans for centuries. The Mississippi is a metaphor for life itself. Its murky depths are impenetrable by human vision, and its flow to the sea is inexorable and unrestrained.

South of the confluence of the Mississippi with the Ohio River, near Cairo, Illinois, the Mississippi has created a vast floodplain underlain by an alluvial aquifer (see cross-section). The average breadth of the floodplain is about 129 kilometers; its length from Cairo down to the Gulf of Mexico is about 885 kilometers (Fisk, 1944; Ackerman, 1996). Rivers like the Mississippi are never constant in their paths, but wander back and forth over their floodplains. In the nineteenth century, the Mississippi was a major route for commerce as steamboats moved their goods up and down the river. The constantly changing course of the river made the piloting of these steamboats a task for only the most skillful pilots. The American author, Mark Twain (1835–1910) was a riverboat pilot from 1856 through 1860. In his book, *Life on the Mississippi*, he described the challenge of piloting a steamboat (see figure) on the Mississippi:

> Piloting becomes another matter when you apply it to vast streams like the Mississippi and the Missouri, whose alluvial banks cave and change constantly, whose snags are always hunting up new quarters, whose sand-bars are never at rest, whose channels are forever dodging and shirking, and whose obstructions must be confronted in all nights and all weathers without the aid of single lighthouse or a single buoy; for there is neither light nor buoy to be found anywhere in all this three or four thousand miles of villainous river.

The primary mechanism by which the Mississippi and other meandering rivers accomplish their sideways wandering is through the creation of meanders. A meander is a curve in a river. If some slight, natural curve occurs in the course of a river's path, the water velocity on the outside of the curve tends to be faster than the water velocity on the inside part of that same curve. The water moving along the outside curve has to move faster because it has to traverse a farther distance in the same amount of time. The faster flow on the outside bank tends to erode that bank, while the slower flow on the inside curve allows sediment to accumulate. Over time, the outside bank is undercut by the river's flow and collapses. As the outward path is enlarged, the inside path is correspondingly narrowed by the formation of a sand bar so that the river maintains a constant width. This process continues until the meander becomes quite exaggerated and pronounced. Eventually, the river enters a flood stage and cuts across the sand bar formed during its lateral

The Mississippi River: An Aquifer-Making Machine *(continued)*

Steamboat *Baton Rouge.*

(From Twain, 1883; reproduced courtesy of the Western History Collections, University of Oklahoma Libraries.)

migration. The formation of a cutoff shortens the course of a river and leaves the abandoned segment as an oxbow lake. In considering how cutoffs had shortened the course of the Mississippi, Mark Twain madc a wry comment on the application (or misapplication, should we say) of uniformitarianism:

> In the space of 176 years the Lower Mississippi has shortened itself 242 miles. That is an average of a trifle over one mile and a third per year. Therefore, any calm person, who is not blind or idiotic, can see that in the Old Oolitic Silurian Period, just a million years ago next November, the Lower Mississippi River was upwards of 1,300,000 miles long, and stuck out over the Gulf of Mexico like a fishing rod. And by the same token any person can see that 742 years from now the Lower Mississippi will be only a mile and three-quarters long, and Cairo and New Orleans will have joined their streets together, and be plodding comfortably along under a single mayor and a mutual board of aldermen. There is something fascinating about science. One gets such wholesale returns of conjecture out of such a trifling investment of fact.

It is this process of shifting and sorting sediment that makes the Mississippi an efficient machine for producing aquifers (Chapelle, 1997). The finer-grained clays and silts that, when deposited, tend to produce confining layers, are carried off by the river's flow to the great Mississippi Delta in the Gulf of Mexico. The coarser-grained sands and gravels are cleaned, sorted, and piled up.

The Mississippi alluvial aquifer consists of Quaternary age (0–1.6 Ma) sediments overlying less permeable bedrock. The upper part of the aquifer consists of silt, clay, and fine-grained sand. It averages 9 meters in thickness, and Ackerman (1996) suggested that this layer acts more as a confining layer than an aquifer. The lower part of the alluvial aquifer averages 30 meters in thickness and consists mostly of gravel and sand. Its average permeability is about 7×10^{-11} m^2 (Ackerman, 1996). Most of the sands and gravels in the lower part of the alluvial aquifer were deposited at the termination of the last ice age, from about 18,000 to 5,000 years before present. The exact chronology is uncertain, but as the glaciers retreated massive amounts of outwash turned the Mississippi into a braided stream choked with

unconfined aquifer height decreases. The Darcy velocity (q_x) is

$$q_x = -K\frac{dh}{dx} \tag{6.48}$$

where K (m-s^{-1}) is the hydraulic conductivity. The Darcy velocity is also the volumetric flow per unit area per unit time, so we can write

$$\frac{\text{volume}}{h\Delta y\Delta t} = -K\frac{dh}{dx} \tag{6.49}$$

where $h\Delta y$ (m^2) is the area perpendicular to flow and Δt (s) is unit time. Now, note that the reason q_x changes with respect to distance (x) is that h changes—Δy is constant with respect to x. It is therefore useful to define a quantity q'_x (m^2-s^{-1}) that is constant with respect to x, large amounts of coarse-grained sediments (Autin, 1991). As the glacial meltwaters waned, there was a transition to finer-grained sediment and the Mississippi took on its present status of a muddy, meandering stream.

It is the interaction of the Mississippi with the alluvial aquifer that it formed which accounts for one of Twain's puzzling observations. Every spring, the Mississippi rises. In late summer, the river falls to low stage. As the amount of water carried by the Mississippi increases progressively downstream, one would naturally expect the amount of rise or fall to increase proportionately. But, that is not what happens. Twain remarked:

> The difference in rise and fall is also remarkable—not in the upper, but in the lower river. The rise is tolerably uniform down to Natchez (three hundred and sixty miles above the mouth)—about fifty feet. But at Bayou La Fourche the river rises only twenty-four feet; at New Orleans only fifteen, and just above the mouth only two and one-half.

As Chapelle (1997) has pointed out, the moderation of the Mississippi's rise and fall downstream can be attributed to interaction with the alluvial aquifer. The Mississippi carries an immense amount of water. South of Cairo, Illinois, the Mississippi is about 610 meters wide and 11 meters deep (Chapelle, 1997). The average width and depth of the alluvial aquifer are 129 kilometers and 30 meters, respectively (Ackerman, 1996). If we assume that the sand and gravel aquifer has an average porosity of 30% (Chapelle, 1997), the amount of water contained in the alluvial aquifer at any time is more than a hundred times larger than the amount of water in the river. It is this reservoir that buffers the rise and fall of the lower Mississippi. When the river is rising, the aquifer absorbs water; when the river is falling, it releases it. As awe-inspiring as the mighty Mississippi is, it is only the surface manifestation of a vast underground reservoir of water.

FOR ADDITIONAL READING

Ackerman, D. J. 1996. Hydrology of the Mississippi River Valley Alluvial Aquifer, South-Central United States. *U.S. Geological Survey Professional Paper 1416-D.*

Autin, W. J., Burns, S. F., Miller, B. J., Saucier, R. T., Snead, J. I. 1991. Quaternary geology of the Lower Mississippi Valley in Morrison, R. B. (ed) Quaternary nonglacial geology; Conterminous U.S. Boulder, Colorado, Geological Society of America, The Geology of North America, v. K-2, p. 547–581.

Chapelle, F. H. 1997. *The hidden sea: groundwater, springs, and wells.* Tucson, Arizona: Geoscience Press, Inc., 238 pp.

Fisk, H. N. 1944. *Geological investigation of the alluvial valley of the lower Mississippi River.* U.S. Dept. Army, Mississippi River Commission, 78 pp.

Twain, M. 1883. *Life on the Mississippi.* Boston: J. R. Osgood, 624 pp.

$$q_x' = q_x h = \frac{\text{volume}}{\Delta y \Delta t} = -Kh\frac{dh}{dx} \qquad \textbf{(6.50)}$$

or

$$q_x' dx = -Kh\, dh \qquad \textbf{(6.51)}$$

where q_x' (m^2-s^{-1}) is the volumetric flow rate *per unit width perpendicular to the flow direction.* As x increases from 0 to L, h decreases from h_1 to h_2 (Figure 6.14). Integrating both sides of equation 6.51,

$${}_0\!\int^L q_x'\, dx = -K_{h_1}\!\int^{h_2} h\, dh \qquad \textbf{(6.52)}$$

$$q_x' L = -K\left[\frac{h^2}{2}\right]_{h_1}^{h_2} = -K\left[\frac{h_2^2}{2} - \frac{h_1^2}{2}\right] \qquad \textbf{(6.53)}$$

$$q_x' = \frac{-K}{2L}(h_2^2 - h_1^2) \qquad \textbf{(6.54)}$$

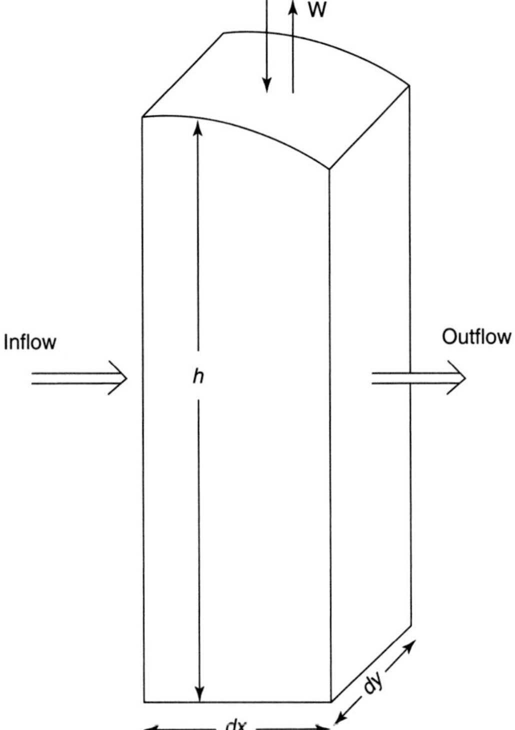

Figure 6.15 Control volume for deriving equations that describe flow in an unconfined porous medium with a source term.

Equation 6.54 gives the flow velocity q'_x in terms of the hydraulic conductivity and the head values at any two points. Note that the flow velocity q'_x is constant with respect to x.

Let us now find an expression for $h(x)$. Assume conservation of mass in a control volume (Figure 6.15) such that

$$\begin{matrix}\text{net outflow}\\ \text{rate}\end{matrix} - \begin{matrix}\text{net inflow}\\ \text{rate}\end{matrix} = W \quad \textbf{(6.55)}$$

where W (m-s^{-1}) is a source term representing infiltration ($W > 0$) or evapotranspiration ($W < 0$). The total inflow through the left side of the control volume (Figure 6.15) is found by multiplying equation 6.51 by the width perpendicular to the flow direction (dy),

$$\begin{matrix}\text{net inflow}\\ \text{rate}\end{matrix} = q'_x\,dy = -Kh\frac{dh}{dx}dy \quad \textbf{(6.56)}$$

The total flow rate out the right side is sum of the inflow rate on the left side plus any change in flow rate that takes place over the length (dx) of the control volume. The change in flow rate that takes place over the length of the control volume is the product of the rate of change of flow rate with respect to distance and the total distance. Thus

$$\begin{matrix}\text{net outflow}\\ \text{rate}\end{matrix} = q'_x\,dy + \left[\frac{d(q'_x\,dy)}{dx}\right]dx \quad \textbf{(6.57)}$$

Substituting in the definition of q'_x from equation 6.56,

$$\begin{matrix}\text{net outflow}\\ \text{rate}\end{matrix} = -Kh\frac{dh}{dx}dy + \frac{d\left[-Kh\dfrac{dh}{dx}\right]}{dx}dxdy \quad \textbf{(6.58)}$$

Thus,

$$\begin{matrix}\text{net outflow}\\ \text{rate}\end{matrix} - \begin{matrix}\text{net inflow}\\ \text{rate}\end{matrix} = \frac{d\left[-Kh\dfrac{dh}{dx}\right]}{dx}dxdy \quad \textbf{(6.59)}$$

The source term W has units of m-s^{-1}; it is the volume of fluid added or subtracted per unit area per unit time. Thus the total rate at which fluid volume is added or subtracted is the product $Wdxdy$ (m^3-s^{-1}). Substituting into equation 6.59, we obtain

$$\frac{d\left[-Kh\dfrac{dh}{dx}\right]}{dx}dxdy = Wdxdy \quad \textbf{(6.60)}$$

If we assume that the hydraulic conductivity K is homogeneous, we may take it outside of the derivative

$$-K\frac{d\left[h\dfrac{dh}{dx}\right]}{dx} = W \quad \textbf{(6.61)}$$

Now, note that the following relations are true:

$$\frac{d^2(h^2)}{dx^2} = \frac{d\left[\dfrac{d(h^2)}{dx}\right]}{dx} = \frac{d\left[2h\dfrac{dh}{dx}\right]}{dx} = 2\frac{d\left[h\dfrac{dh}{dx}\right]}{dx} \quad \textbf{(6.62)}$$

Substituting equation 6.62 into equation 6.61,

$$-K\left[\frac{d^2(h^2)}{dx^2}\right] = 2W \quad \textbf{(6.63)}$$

or

$$\frac{d^2(h^2)}{dx^2} = \frac{-2W}{K} \quad \textbf{(6.64)}$$

Equation 6.64 can be solved for h^2 by integration. Integrating once,

$$\frac{d(h^2)}{dx} = \frac{-2Wx}{K} + c_1 \quad \textbf{(6.65)}$$

Integrating again,

$$h^2 = \frac{-Wx^2}{K} + c_1x + c_2 \quad \textbf{(6.66)}$$

The constants of integration, c_1 (m) and c_2 (m^2), are found by applying boundary conditions (Figure 6.13). At $x = 0$, $h = h_1$, and at $h = L$, $h = h_2$. Substituting into equation 6.66,

$$h_1^2 = c_2 \quad \textbf{(6.67)}$$

$$h_2^2 = \frac{-WL^2}{K} + c_1L + h_1^2 \quad \textbf{(6.68)}$$

$$c_1 = \frac{(h_2^2 - h_1^2)}{L} + \frac{WL}{K} \quad \textbf{(6.69)}$$

Substituting equations 6.67 and 6.69 into equation 6.66,

$$h^2 = \frac{-Wx^2}{K} + x\left[\frac{(h_2^2 - h_1^2)}{L} + \frac{WL}{K}\right] + h_1^2 \quad \textbf{(6.70)}$$

Equation 6.70 gives the square of the water table elevation between any two points at which head is fixed and there is a non-zero source term, W (if we can find h^2, then we can find h simply by taking the square root).

We can also find $q'_x(x)$ for a non-zero source term by taking the derivative of equation 6.70,

$$2h\frac{dh}{dx} = \frac{-2Wx}{K} + \frac{h_2^2 - h_1^2}{L} + \frac{WL}{K} \quad \textbf{(6.71)}$$

Multiplying each side of equation 6.71 by $-K/2$, we obtain

$$-Kh\frac{dh}{dx} = Wx + K\left[\frac{h_1^2 - h_2^2}{2L}\right] - \frac{WL}{2} \quad \textbf{(6.72)}$$

The left side of equation 6.72 is the definition of q'_x (by equation 6.51), thus

$$q'_x = K\left[\frac{h_1^2 - h_2^2}{2L}\right] - W\left[\frac{L}{2} - x\right] \quad \textbf{(6.73)}$$

Suppose there is a water divide at $x = d$ (Figure 6.13). At the divide we must have $q'_x = 0$. The position of the divide can be found by setting $q'_x = 0$ in equation 6.73 and solving for $x = d$,

$$0 = K\left[\frac{h_1^2 - h_2^2}{2L}\right] - W\left[\frac{L}{2} - d\right] \quad \textbf{(6.74)}$$

or

$$d = \frac{L}{2} - \frac{K}{W}\left[\frac{h_1^2 - h_2^2}{2L}\right] \quad \textbf{(6.75)}$$

In summary,

1. Dupuit theory applies to unconfined porous media.
2. Dupuit theory assumes horizontal flow and vertical equipotentials. Therefore, it does not apply where there is a significant quantity of vertical flow.
3. Dupuit theory is invalid (a) near seepage faces, and (b) at a crest on the water table (a divide with recharge).
4. Dupuit theory is generally applicable where the horizontal lengths of interest are much greater than the thickness of the unconfined aquifer. This means shallow flow.

One application of Dupuit theory is calculating flow through a dam (Figure 6.16).

Problem: Find the volume of water that flows through a soil dam each day. Assume steady-state conditions. The dam is 225 m long, 30 m wide, and has a hydraulic conductivity of 8.0×10^{-7} m-s^{-1}. On the upstream side, the water is 25 m deep; on the downstream side, the water is 5 m deep.

The total volumetric flow rate (Q, m^3-s^{-1}) is the flow per unit width (q'_x, m^2-s^{-1}) times the width. The flow per unit width is given by equation 6.73,

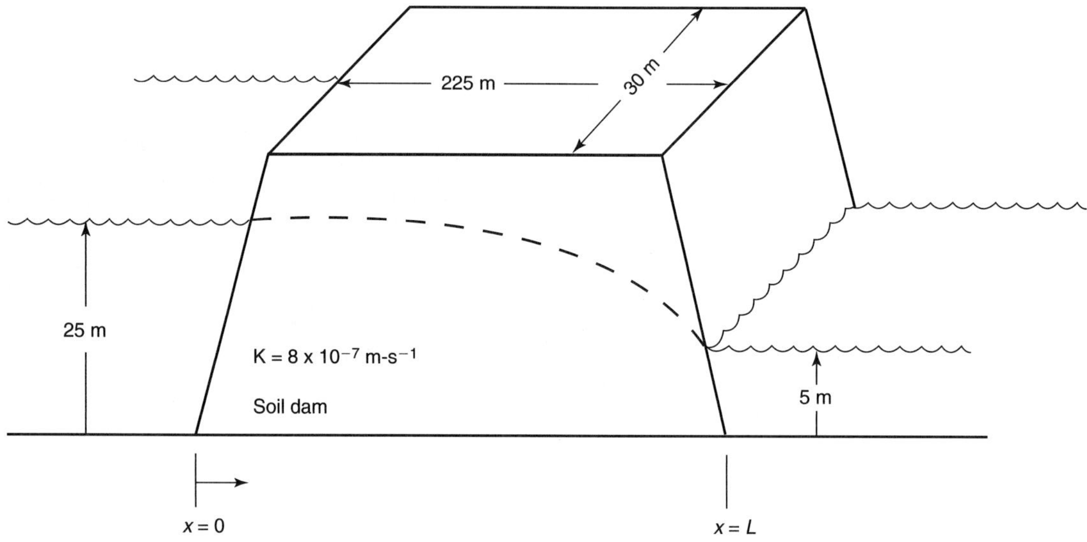

Figure 6.16 Application of Dupuit theory to calculate flow through a dam.

$$q'_x = K\left[\frac{h_1^2 - h_2^2}{2L}\right] - W\left[\frac{L}{2} - x\right] \quad \textbf{(6.76)}$$

In the absence of infiltration into the top of the dam, $W = 0$. Thus

$$Q = 30(\text{m}) \times 8.0 \times 10^{-7}(\text{m-s}^{-1}) \times \frac{(25)^2(\text{m}^2) - (5)^2(\text{m}^2)}{(2 \times 225)(\text{m})} \quad \textbf{(6.77)}$$

$$Q = 3.2 \times 10^{-5}\,(\text{m}^3\text{-s}^{-1}) \quad \textbf{(6.78)}$$

Is the answer physically reasonable? It is useful to convert the units to $\text{m}^3\text{-day}^{-1}$. Then

$$Q = 2.76\ \text{m}^3\text{-day}^{-1} \quad \textbf{(6.79)}$$

This seems like a reasonable leakage rate.

Equation 6.76 shows that hydraulic conductivity and length are equally important in determining leakage through the dam. The hydraulic conductivity of different materials may vary by several orders of magnitude; thus, the most efficient design will likely incorporate a dam of low-conductivity material as thin as strength considerations permit.

The above problem can also be formulated in terms of how wide a dam must be in order to keep leakage down to an acceptable level. In that case, equation 6.76 should be rearranged to solve for L. Note that there is no hydraulic difference between a thick dam of high-conductivity material and a thinner dam of low-conductivity material. In each case, the **hydraulic resistance** (L/K) is the same.

6.6. Flow to Wells.

It is important to understand groundwater flow to wells for several reasons. Most obvious is the necessity for estimating the rate at which groundwater may be delivered as an economic resource and the effect of withdrawal on water storage. Well tests may also be used to make in situ estimates of hydraulic conductivity and permeability. These tests are useful not only in the exploitation of near-surface aquifers for potable water, but also find use in the petroleum industry for estimating the permeability of petroleum reservoirs.

The mathematical analysis of pumping wells can be complex. To simplify flow modeling, a

Charles Vernon Theis: The Physical Conception of the Process

Charles Vernon Theis (1900–1987) was born on March 27, 1900, in Newport, Kentucky. As a young boy, Theis had a precocious aptitude for schoolwork and graduated from high school two years early at the age of 16. Following graduation, Theis worked for a year primarily as an office boy for the Rotary Club in Cincinnati, Ohio. Theis managed to save enough money to enter the civil engineering program at the University of Cincinnati in the fall of 1917. During his third year at Cincinnati, Theis was elected to the engineering honorary society, Tau Beta Pi. It was at this time that he first met the man who would help him write one of the most influential papers in the history of hydrogeology, Clarence Lubin. Lubin had been nominated for membership in Tau Beta Pi, and Theis at first voted against him on the basis of some hearsay from other members. At the next meeting, a more senior member of the fraternity spoke in defense of Lubin, and Theis discovered that some of the Tau Beta Pi membership had voted against Lubin because he was Jewish. Theis called for a new vote, and Lubin was once again voted down, however, Theis and Lubin became friends.

Charles Vernon Theis (1900–1987)

Theis graduated in 1922 with a degree in civil engineering. He was offered a graduate assistantship, which he accepted, in the geology department at the University of Cincinnati. Theis thought some education in geology might help him in his engineering work. His interest in geology soon caused him to change the direction of his career. Theis' most influential professor in graduate school was Nevin Fenneman (1865–1945). Fenneman emphasized the importance of what he called "getting the physical conception of the process." Years later, it was this advice that enabled Theis to correctly formulate and solve the problem of determining the hydraulic conductivity of an aquifer.

In 1929, Theis received the first Ph.D. in geology awarded by the University of Cincinnati. On July 1, 1930, he was offered a position with the Division of Ground Water of the U. S. Geological Survey. Theis stayed with the U. S. Geological Survey for the duration of his career.

During the early 1930s, one of Theis' first assignments for the USGS was to characterize the hydraulic conductivity of the High Plains aquifer in New Mexico. The existing history of water-level data indicated that the aquifer was in transient decline, however, at that time the only method available for determining aquifer characteristics was the steady-state Thiem analysis.

Over a period of several years, Theis finally obtained a correct "physical conception of the process." He could describe the problem adequately as an initial-value problem subject to certain boundary conditions. Theis lacked the specific mathematical knowledge to solve the problem, but recognized that the solution might already exist in heat conduction theory. The flow of groundwater was analogous to the flow of heat by conduction that had been studied theoretically for more than a hundred years. Theis subsequently wrote to his college friend, Clarence Lubin, asking him if an analogous solution existed in heat conduction theory. Lubin, relying to some extent

Charles Vernon Theis: The Physical Conception of the Process *(continued)*

upon the classic book by H. S. Carslaw (1921), *Introduction to the Mathematical Theory of the Conduction of Heat in Solids*, solved the problem and mailed the solution back to Theis. Theis wanted to include Lubin as a co-author, but Lubin declined, stating the "insignificance" of the mathematical contribution. Theis published his solution in the *Transactions of the American Geophysical Union* in 1935, and followed up with expositions in *Economic Geology* in 1938 and *Civil Engineering* in 1940. The exposition of Theis' method resulted in its wide application, and the "Theis Equation" quickly became the standard method of aquifer analysis. White and Clebsch (1994, p. 51) assessed the importance of Theis' contribution by stating that "it would not be an exaggeration to state that the publication of this paper revolutionized the science of ground-water hydrology."

To publish his work, Theis had to persevere. O. E. Meinzer, who was head of the USGS Ground Water Division, either did not recognize the importance of Theis' work, or lacked confidence in Theis' ability to solve the problem. Theis was told that a fellow scientist at the USGS, Van Orstrand, had done some work on the non-equilibrium problem. Theis discussed this with Van Orstrand, and concluded that either Meinzer or someone else had taken the problem to Van Orstrand for solution after Theis had shown it to them. Van Orstrand, however, was never able to figure out the "physical conception" of the problem. Theis' paper, which was ultimately published in the journal *Economic Geology* in 1938, was returned by Meinzer with the note that it had made a "very unfavorable impression" with several men at USGS headquarters.

During his later years, Theis worked on a number of important hydrogeological problems. These include radioactive-waste disposal and the importance of geologic heterogeneities for contaminant transport problems. After retiring in 1970, Theis continued to come into his office at the USGS on a regular basis. Only during the last six months of his life, when he was ill with emphysema and lung cancer, did he fail to show up. Even at the last stage of his life, however, he continued to work by dictating notes on a tape recorder.

In 1984, Theis received the Robert E. Horton medal from the American Geophysical Union for "outstanding contributions to the geophysical aspects of hydrology." The American Institute of Hydrology established the C. V. Theis award, which is given annually for outstanding contributions in groundwater hydrology.

For Additional Reading

White, R. R. 1994. Memorial to Charles Vernon Theis, 1900–1987. *Geological Society of America Memorials* 25: 155–57.

White, R. R., and Clebsch, A. 1994. C. V. Theis, the man and his contributions to hydrogeology in "Selected contributions to ground-water hydrology by C. V. Theis, and a review of his life and work." *U. S. Geological Survey Water-Supply Paper 2415,* p. 45–56.

number of assumptions are usually made. Aquifers are assumed to be homogeneous and isotropic, and two-dimensional problems are analyzed in polar coordinates (Figure 6.17). Radial symmetry is also assumed, thus the only variable of interest is the distance r (m) from the borehole center. If radial symmetry is assumed, the diffusion equation (equation 5.35) rewritten in **polar coordinates** (Figure 6.17) becomes

$$\frac{dh}{dt} = \frac{K}{S_s}\left[\frac{d^2h}{dr^2} + \frac{1}{r}\frac{dh}{dr}\right] \tag{6.80}$$

Recalling that storativity $S = \Delta z S_s$ (dimensionless) and transmissivity $T = \Delta z K$ (m^2-s^{-1}), where Δz (m) is aquifer thickness, equation 6.80 can be rewritten

$$\frac{S}{T}\frac{dh}{dt} = \left[\frac{d^2h}{dr^2} + \frac{1}{r}\frac{dh}{dr}\right] \tag{6.81}$$

As groundwater is withdrawn from an aquifer, head in the aquifer decreases. The decrease of head in an aquifer due to water withdrawal is termed **drawdown**. If head in the aquifer is plotted as function of radial distance from a well, the resulting geometric form is called a **cone of depression** (Figure 6.18).

Consider a horizontal aquifer sandwiched between two confining layers (Figure 6.18). If the interface between the confining layers and aquifer are impermeable boundaries, all flow originates from water storage in the aquifer and there is no leakage of groundwater from the confining layers to the aquifer. For this situation, a plot of head (at the pumping well) versus time will be exponential in nature and a plot of head vs. log time will be largely linear (Figure 6.19). If the layers that confine the aquifer leak, then the initial response will be the same as for the non-leaky case, but at later times head will not decrease as quickly. The nature of the head drawdown will also be affected by the storage capacity of the confining layers (Figure 6.19). In nature, confining layers both leak and store groundwater. Thus in most situations drawdown will be less than predicted by a simple analysis that does not consider the influence of water movement from confining layers into an aquifer. The relative significance of confining-layer leakage increases with increasing time (Figure 6.19).

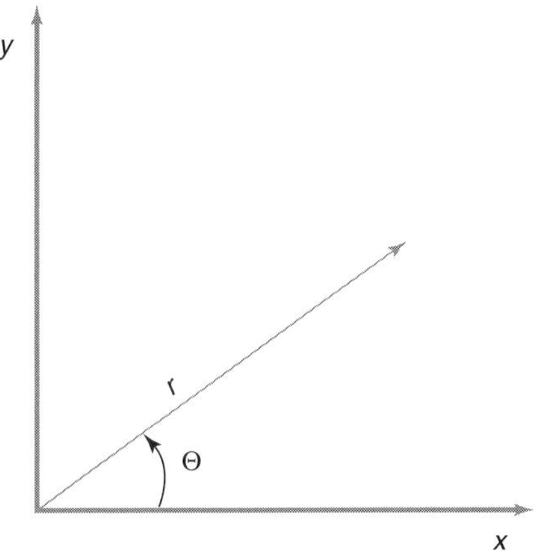

Figure 6.17 Comparison of cartesian (x, y) and polar (r, Θ) coordinates.

6.6.1. Steady-state flow to a confined aquifer, no leakage.

The simplest situation we can analyze is the steady-state withdrawal of water from a confined aquifer with no leakage. This situation was first analyzed by Thiem (1906).

Recall that the Darcy velocity is the volumetric rate of flow per unit area. For an aquifer of height

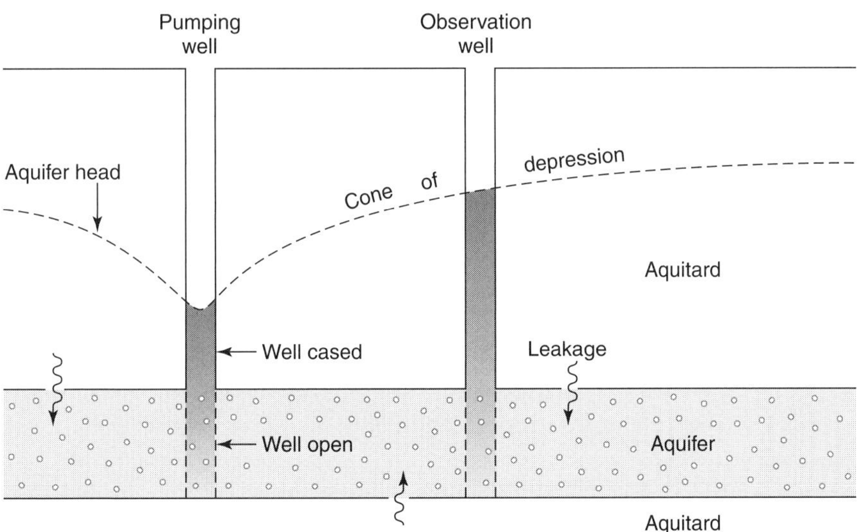

Figure 6.18 Head in a confined aquifer. Drawdown due to pumping creates a geometric form known as a *cone of depression.*

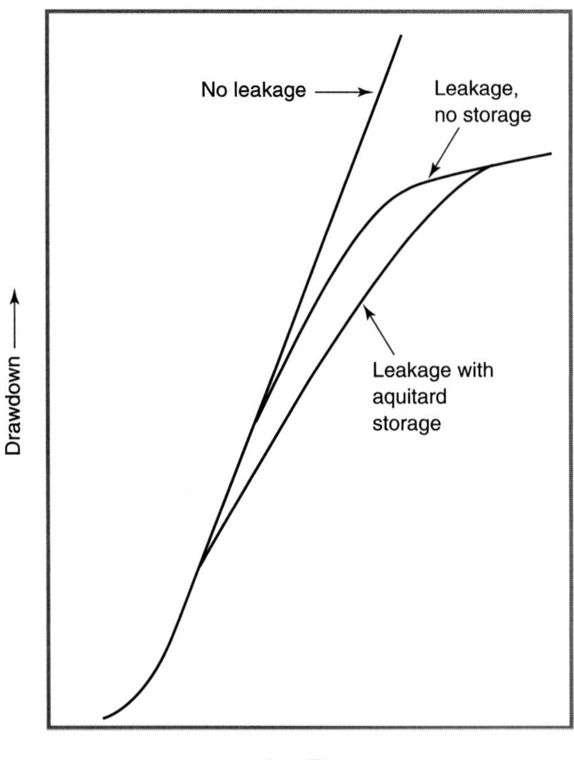

Figure 6.19 Drawdown of a confined aquifer versus log time. In general, there are three types of situations. If there is no leakage from the confining aquitards, the drawdown is a linear function of log (time). If there is appreciable leakage from surrounding aquitards that have no storage capacity, head in the aquifer does not decrease as quickly as it otherwise would. Finally, if there is leakage across the aquifer-aquitard interface from aquitards with finite storage capacity, drawdown is further inhibited. In general, the most realistic case includes both leakage and aquitard storage. In most situations we can therefore expect aquifer drawdown to be a linear function of log (time) for only a limited period of time.

(After Domenico and Schwarz, 1990, p. 143, and Hantush, 1960.)

Δz (m) penetrated by a well of radius r (m), the area of flow is $2\pi r\Delta z$ (m^2). Thus the volumetric rate of flow (Q, m^3-s^{-1}) is (by Darcy's Law)

$$Q = (2\pi r \Delta z)\, K \frac{dh}{dr} \qquad \textbf{(6.82)}$$

or

$$Q = (2\pi r T) \frac{dh}{dr} \qquad \textbf{(6.83)}$$

Equation 6.83 can be rearranged,

$$dh = \frac{Q}{2\pi T} \frac{dr}{r} \qquad \textbf{(6.84)}$$

If we have data from two observation wells, equation 6.84 can be solved by integration. The first observation well is at a distance r_1 from the pumping well (where $r = 0$ at the center of the well) and has head h_1. Similarly, the second well is at a distance r_2 and has head h_2. Integrating both sides of equation 6.84,

$$\int_{h_1}^{h_2} dh = \frac{Q}{2\pi T} \int_{r_1}^{r_2} \frac{dr}{r} \qquad \textbf{(6.85)}$$

Evaluating the integrals,

$$h_2 - h_1 = \frac{Q}{2\pi T} \log_e \left(\frac{r_2}{r_1}\right) \qquad \textbf{(6.86)}$$

Equation 6.86 can be rearranged to solve for *T*,

$$T = \frac{Q}{2\pi(h_2 - h_1)} \log_e\left(\frac{r_2}{r_1}\right) \quad \textbf{(6.87)}$$

If we know the aquifer thickness (Δz) we can thus estimate the aquifer hydraulic conductivity (K), as $T = \Delta z K$. Note that this procedure only applies to steady-state conditions. If head is changing with time at any of the three wells, equation 6.87 cannot be used to reliably estimate transmissivity or hydraulic conductivity.

Problem: Groundwater has been pumped from a confined aquifer steadily for a number of years at a rate of 500 gallons per day. The aquifer is 10 meters thick. At a distance of 50 meters from the pumping well, head in an observation well is 10 meters. At a distance of 100 meters from the pumping well, head is 12 meters. What is the hydraulic conductivity and permeability of the aquifer? Based on its permeability, is the aquifer likely composed of sound rock or an unconsolidated sediment?

Because the well has been pumped steadily for several years, we assume that the hydraulic regime is at steady state. Recall that transmissivity is the product of aquifer thickness (Δz) and hydraulic conductivity. Applying equation 6.87,

$$T = \Delta z K = \frac{Q}{2\pi(h_2 - h_1)} \log_e\left(\frac{r_2}{r_1}\right) \quad \textbf{(6.88)}$$

or

$$K = \frac{Q}{2\Delta z\pi(h_2 - h_1)} \log_e\left(\frac{r_2}{r_1}\right) \quad \textbf{(6.89)}$$

We need to convert Q to metric units.

$$500\frac{\text{gal}}{\text{day}} \times \frac{1 \text{ liter}}{3.78541 \text{ gal}} \times \frac{10^3\text{cm}^3}{\text{liter}} \times \frac{1 \text{ m}^3}{10^6\text{cm}^3} \times \frac{1 \text{ day}}{86400 \text{ s}} = 1.529 \times 10^{-6}\frac{\text{m}^3}{\text{s}} \quad \textbf{(6.90)}$$

Substituting into equation 6.89,

$$K = \frac{1.529 \times 10^{-6}\text{m}^3\text{-s}^{-1}}{2 \times 10(\text{m}) \times 3.1416 \times (12 - 10)\text{m}} \log_e\left(\frac{100}{50}\right) \quad \textbf{(6.91)}$$

$$K = 8.4 \times 10^{-9}\,(\text{m-s}^{-1}) \quad \textbf{(6.92)}$$

To convert from hydraulic conductivity to permeability, we write equation 2.56 in terms of permeability (k, m^2),

$$k = \frac{K\mu}{\rho g} \quad \textbf{(6.93)}$$

where μ (kg-m^{-1}s^{-1}) and ρ (kg-m^{-3}) are groundwater viscosity and density, respectively, and g (m-s^{-2}) is the acceleration due to gravity. Water viscosity as a function of temperature is given by equation 4.23. For a typical near-surface temperature of 15°C, water viscosity is 1.14×10^{-3} kg-m^{-1}-s^{-1}. Thus permeability is

$$k = \frac{8.4 \times 10^{-9}(\text{m-s}^{-1}) \times 1.14 \times 10^{-3}(\text{kg-m}^{-1}\text{-s}^{-1})}{1000(\text{kg-m}^{-3}) \times 9.8(\text{m-s}^{-2})} \quad \textbf{(6.94)}$$

$$k = 9.8 \times 10^{-16}\text{ m}^2 \quad \textbf{(6.95)}$$

According to Figure 3.4, this permeability is characteristic of a number of different types of both rocks and unconsolidated sediments. Thus it is impossible to discern the nature of the aquifer's composition from the pumping test alone.

6.6.2. Transient flow to a confined aquifer, no leakage.

We now consider the more general case of *transient* flow to a confined aquifer without leakage. We again assume a two-dimensional geometry with radial symmetry. The aquifer is assumed to be homogeneous and isotropic with regard to its hydraulic properties. The initial condition is that head is at a constant value h_0 everywhere. At times $t > 0$, the boundary conditions are

$$\text{as } r \to \infty,\ h \to h_0 \quad \textbf{(6.96)}$$

$$\text{as } r \to 0,\ r\frac{dh}{dr} \to \frac{Q}{2\pi T} \quad \textbf{(6.97)}$$

where the arrow symbol ($\to$) means "approaches." Equation 6.97 is equivalent to a constant volumetric flow rate from the well. It is convenient to introduce a dimensionless group designated by the letter u

$$u = \frac{r^2 S}{4Tt} \tag{6.98}$$

The solution to equation 6.80 for the above initial and boundary conditions is

$$h_0 - h = \frac{Q[-\mathrm{E}_i(-u)]}{4\pi T} \tag{6.99}$$

where $h_0 - h$ is the drawdown and $E_i(u)$ is the exponential integral,

$$-\mathrm{E}_i(-u) = {}_u\!\int^{\infty} \frac{\mathrm{e}^{-u}}{u}\, du \tag{6.100}$$

The exponential integral can be written as an infinite series,

$${}_u\!\int^{\infty} \frac{\mathrm{e}^{-u}}{u}\, du = [-0.5772 - \log_e u + u - \frac{u^2}{2\cdot 2!} + \frac{u^3}{3\cdot 3!} - \cdots\cdots] \tag{6.101}$$

As t becomes large, u becomes small. Note that for *small u,*

$$-\mathrm{E}_i(-u) \approx -\log_e(u) - 0.577 \tag{6.102}$$

Thus

$$h_0 - h \approx \frac{Q}{4\pi T}[-\log_e(u) - 0.577] \tag{6.103}$$

or

$$h_0 - h = \frac{Q}{4\pi T}\left[-\log_e\left(\frac{r^2 S}{4Tt}\right) - 0.577\right] \tag{6.104}$$

Noting that $\log(a/b) = \log(a) - \log(b)$,

$$h_0 - h = \frac{Q}{4\pi T}\left[-\log_e\left(\frac{r^2 S}{4T}\right) + \log_e(t) - 0.577\right] \tag{6.105}$$

or,

$$h_0 - h = \frac{Q}{4\pi T}\left[-\log_e\left(\frac{r^2 S}{4T}\right) - 0.577\right] + \frac{Q}{4\pi T}[\log_e(t)] \tag{6.106}$$

Thus for a given and constant distance r;

$$h_0 - h = \frac{Q}{4\pi T}[\log_e(t)] + \text{constant} \tag{6.107}$$

A plot of drawdown $(h - h_0)$ versus log time should have the form of a line with slope $Q/4\pi T$. If Q is known, the transmissivity (T) can be determined from the line slope. Because $T = \Delta z K$, if the aquifer thickness (Δz) is known, the hydraulic conductivity (K) can be found. Note that once T has been determined, substitution back into equation 6.104 for any value of r allows storativity S and the specific storage to be estimated.

Strictly speaking, it is meaningless to take the log of "time" in equation 6.107, because the number whose log is taken will be different depending upon the scale we use. But, the slope $Q/4\pi T$ remains the same, no matter what time units are chosen. Changing from one time unit to another is equivalent to multiplying by a constant (c). Note that $\log(ct) = \log(c) + \log(t)$. Thus changing time units changes the intercept of equation 6.107, but does not affect the slope of the line nor the determination of T or K.

Problem: Groundwater is pumped from a confined aquifer 20 meters thick at a rate of 100 m^3-day^{-1}. Drawdown is measured at an observation well 10 meters from the pumping well as a function of time. The data are given in Table 6.1. Plot drawdown versus $\log_e$(time) at the observation well and deduce the hydraulic conductivity and specific storage of the aquifer.

Plotting the data as a function of log(time), the drawdown is seen to be linear for most of observation period. Note that at early times the drawdown curve does not have the linear form predicted by equation 6.107. This is for two reasons. At early times, the aquifer does not respond as predicted because the mathematical model upon which equation 6.107 is based assumes water is withdrawn from a well of infinitesimal radius—a line sink, not a well with a finite radius. The approximation given by equation 6.107 also becomes better as time (t) becomes larger. If pumping were to continue for a longer period of time, however, the rate of drawdown would eventually decrease below that predicted by equation 6.107 as the influence of leakage from neighboring formations became significant (Figure 6.19). Finally, note that the data do not fall on a perfect line, even for the period of time during which the model is a valid approximation of reality. This is because the aquifer being pumped is not perfectly homogeneous nor isotropic.

TABLE 6.1 Drawdown of an idealized confined aquifer.

Time	Drawdown (meters)
1.0 (s)	0.01
2.5 (s)	0.10
5.0 (s)	0.26
25.0 (s)	0.84
1.0 (min)	1.2
2 (min)	1.5
5 (min)	1.9
10 (min)	2.3
1.0 (hr)	3.1
24.0 (hr)	4.5

The slope of the line plotted in Figure 6.20 can be determined either by an "eyeball" fit or more formally by a procedure such as least squares. The slope of the line is found to be 0.46 m (the vertical axis has the dimension of length, and the horizontal axis is dimensionless). Thus, from equation 6.107,

$$0.46\ (\mathrm{m}) = \frac{Q}{4\pi T} \tag{6.108}$$

Solving for $T = \Delta z K$,

$$T = \Delta z K = \frac{100}{4\pi(0.46)} \frac{\mathrm{m}^3}{\text{day-m}} \tag{6.109}$$

$$T = \Delta z K = 17.3 \frac{\mathrm{m}^2}{\mathrm{day}} \times \frac{1\ \mathrm{day}}{86400\ \mathrm{s}} \tag{6.110}$$

$$T = \Delta z K = 0.0002 \frac{\mathrm{m}^2}{\mathrm{s}} \tag{6.111}$$

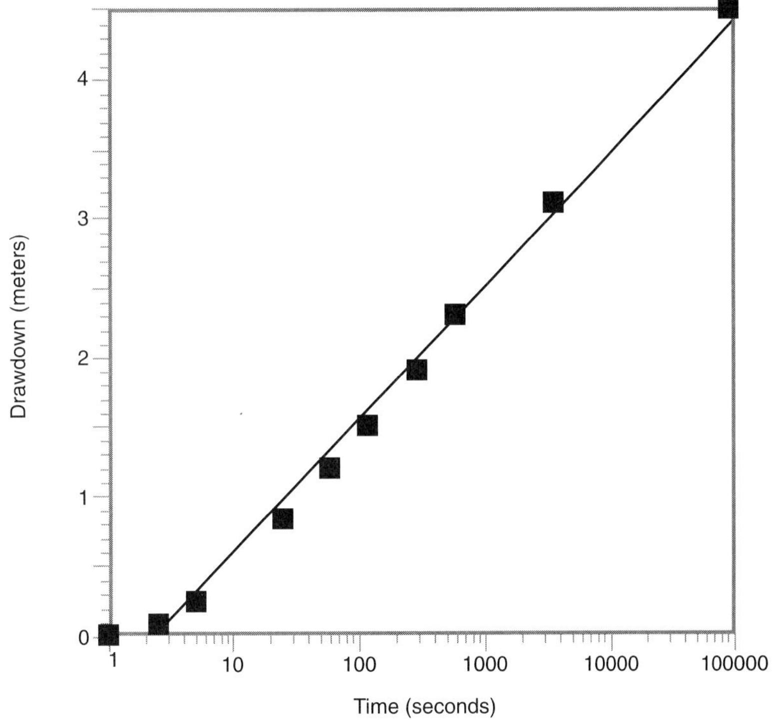

Figure 6.20 Drawdown in a confined aquifer.

$$K = \frac{0.0002}{20} \frac{\text{m}^2}{\text{s-m}} \quad \textbf{(6.112)}$$

$$K = 1.0 \times 10^{-5} \frac{\text{m}}{\text{s}} \quad \textbf{(6.113)}$$

To find storativity and the specific storage, we rearrange equation 6.104 and solve for u,

$$\frac{h_0 - h + \frac{Q}{4\pi T}(0.577)}{\frac{Q}{4\pi T}} = -\log_e(u) \quad \textbf{(6.114)}$$

Taking care to pick a time-drawdown point from the linear part of the well's response, we choose $h_0 - h = 2.3$ m at $t = 10$ min.

$$\frac{2.3\ (\text{m}) + 0.46\ (\text{m})(0.577)}{0.46\ \text{m}} = -\log_e(u) \quad \textbf{(6.115)}$$

$$5.58 = -\log_e(u) = \log_e \frac{1}{u} \quad \textbf{(6.116)}$$

$$264.3 = \frac{1}{u} \quad \textbf{(6.117)}$$

$$u = \frac{r^2 S}{4Tt} = 0.0038 \quad \textbf{(6.118)}$$

$$S = \frac{4Tt \times 0.0038}{r^2} \quad \textbf{(6.119)}$$

$$S = \frac{4 \times 0.0002\ (\text{m}^2\text{-s}^{-1}) \times 600(\text{s}) \times 0.0038}{100\ (\text{m}^2)} \quad \textbf{(6.120)}$$

$$S = 1.8 \times 10^{-5} \quad \textbf{(6.121)}$$

Recall the storativity is the product of specific storage and aquifer thickness ($S = \Delta z S_s$). Thus the specific storage is

$$S_s = \frac{1.8 \times 10^{-5}}{20}\ \text{m}^{-1} \quad \textbf{(6.122)}$$

$$S_s = 9.1 \times 10^{-7}\ \text{m}^{-1} \quad \textbf{(6.123)}$$

Review Questions

1. Define the following terms in the context of hydrogeology:

a. unsaturated zone
b. vadose zone
c. capillary fringe
d. meniscus
e. wetting angle
f. capillary force
g. capillary tube
h. volumetric moisture content
i. capillary-pressure head
j. tensiometer
k. hysteresis
l. Richards equation
m. drainage basin
n. watershed
o. catchment
p. groundwater divide
q. precipitation
r. evapotranspiration
s. overland flow
t. interflow
u. groundwater flow
v. transpiration
w. evaporation
x. interception
y. water table
z. infiltration
aa. interflow
bb. infiltration capacity
cc. plug or piston flow
dd. hydrograph
ee. quickflow
ff. baseflow
gg. Hortonian overland flow
hh. expanding source area concept
ii. subsurface storm flow
jj. baseflow recession

kk. effluent
ll. influent
mm. karst
nn. Dupuit assumptions
oo. hydraulic resistance
pp. polar coordinates
qq. drawdown
rr. cone of depression
ss. auger
tt. cable-tool drilling
uu. casing
vv. rotary drilling

2. Explain how capillary forces allow water to run "uphill." Does the existence of a capillary force violate the conservation of energy? Why not?

3. In what type of rocks and/or soils would you expect capillary forces to be most important?

4. How is the hydraulic conductivity changed when the moisture content of the unsaturated zone changes? Why does it increase or decrease?

5. Why do porous media (e.g., soils) exhibit hysteresis during wetting/drying cycles?

6. Explain how a gravel layer in the unsaturated zone could potentially have a lower hydraulic conductivity than a clay layer.

7. Apply the Green-Ampt equation to the data shown in Figure 6.10 and estimate the saturated hydraulic conductivity (K_s).

8. A soil has a saturated hydraulic conductivity of 10^{-6} m-s^{-1}, a porosity of 50%, an initial capillary pressure head (Ψ) of -0.5 m, and an initial moisture content (θ) of 0.10. How long will it take a steady rainfall to completely saturate the top meter of the soil? Assume piston flow and apply the Green-Ampt equation.

9. Where does streamflow come from? What are four different mechanisms by which water reaches streams?

10. Stream discharge is described by $Q = Q_0 e^{-at}$, where t (s) is time measured from the start of a baseflow recession, and Q_0 (m^3-s^{-1}) and a (s^{-1}) are constants. Five days after the start of a baseflow recession, stream discharge $Q = 150$ m^3-s^{-1}. Ten days after the start of the same recession, stream discharge $Q = 100$ m^3-s^{-1}. (a) What are the constants Q_0 and a? (b) What is the instantaneous discharge Q at $t = 20$ days? (c) What is the cumulative volumetric discharge Q^* at the end of 3 months in m^3? (d) How long does it take for 75% of the available groundwater to reach the stream?

11. What are the Dupuit assumptions? List them.

12. Where is Dupuit theory invalid?

13. An unconfined aquifer has a hydraulic conductivity of $K = 10^{-5}$ m-s^{-1}. There are two observation wells 100 meters apart. Both penetrate the aquifer to the bottom. In one well, water stands 7 meters above the bottom. In the other well, water is 6 meters above the bottom. (a) What is the discharge per 30 m wide strip of the aquifer in m^3-day? (b) What is the water-table elevation at a midpoint between the two wells?

14. You have been hired as a consultant to advise on the construction of an earthen dam. The water level on the reservoir side is to be maintained at 100 m. The water level on the downstream side is to be maintained at 10 m. The dam is to be 75 m wide. How thick does the dam have to be to limit the leakage through the dam to acceptable levels? (Hint: proceed in the following manner. Define what an acceptable leakage level is. Research the hydraulic properties of soils. Proceed to solve the problem. Make use of the library if necessary.) Do you need to build in a "safety factor"? What tests might you want to conduct?

15. What is the difference between a confined and unconfined aquifer?

16. Groundwater is pumped from a 10-meter-thick confined aquifer at a rate of 500 cubic

meters per day. Head is monitored at a nearby observation well, located 30 meters from the pumping well. What is the permeability and specific storage of the confined aquifer? Head data are given in the following table (after Domenico and Schwarz, 1990, p. 153)

Time (minutes)	Head (meters)
6	9.7
8	9.4
10	9.2
20	8.4
50	7.4
100	6.7
200	6.0
500	5.0
1000	4.2

Suggested Reading

Freeze, R. A. 1974. Streamflow generation. *Reviews of Geophysics,* 12: 627–47.

Notation Used in Chapter Six

Symbol	Quantity Represented	Physical Units
a	constant in baseflow recession equation	s^{-1}
c_1	constant of integration	m
c_2	constant of integration	m^2
d	position of groundwater divide (distance from $x = 0$)	m
$\Delta x, \Delta y, \Delta z$	length of control volume sides	m
Δy or dy	width of an unconfined aquifer in a direction perpendicular to the flow direction	m
Δz	aquifer thickness	m
F_{cap}	capillary force per unit length	Newton per meter ($N\text{-}m^{-1}$) = $kg\text{-}s^{-2}$
g	acceleration due to gravity	$m\text{-}s^{-2}$
γ	wetting angle	dimensionless
$h, \Delta h$	head, change in head	m
I	cumulative water infiltration	m
i	infiltration rate	$m\text{-}s^{-1}$
K, K_s	hydraulic conductivity, saturated hydraulic conductivity	$m\text{-}s^{-1}$
L	length of an unconfined aquifer	m
μ	fluid (dynamic) viscosity	Pa-s = $kg\text{-}m^{-1}s^{-1}$
$P, \Delta P$	fluid pressure, change in fluid pressure	Pascal (Pa) = $kg\text{-}m^{-1}\text{-}s^{-2}$
Q^*	cumulative volumetric discharge	m^3
$\theta, \Delta\theta$	volumetric moisture content, change in volumetric moisture coefficient	dimensionless
Q, Q_0	volumetric flow rate or stream discharge rate, stream discharge rate at time zero	$m^3\text{-}s^{-1}$
q, q_z	Darcy velocity, Darcy velocity in z-direction	$m\text{-}s^{-1}$
q'_x	volumetric flow rate per unit width perpendicular to the flow direction (x-direction)	$m^2\text{-}s^{-1}$
ρ	fluid density	$kg\text{-}m^{-3}$
r	radius of well bore	m
r_c	radius of capillary tube	m
S	storativity	dimensionless
T	transmissivity	$m^2\text{-}s^{-1}$
$t, \Delta t$	time, change in time	s
u	dimensionless group of terms used in derivation of Theis Equation	dimensionless
W	infiltration rate into unconfined aquifer	$m\text{-}s^{-1}$
Ψ	capillary pressure head	m
y	height to which water will rise in a capillary tube	m
z	elevation	m
Z_f	depth of moisture front	m

CHAPTER 7

Driving Forces and Mechanisms of Fluid Flow

7.1. Sediment Compaction.

As sediments accumulate in downwarped areas of the Earth's crust known as sedimentary basins pore fluids are expelled through the process of compaction. Historically, geologists have tended to invoke sediment compaction as a mechanism to explain phenomena such as ore deposits that seem to require fluid movement for their existence (e.g., Jackson and Beales, 1967).

7.1.1. Sedimentary Basins and Sedimentation Rates.

Any downwarped area of the Earth's crust where sediments accumulate and lithify to form sedimentary rocks over geologic time is a **sedimentary basin.** Sedimentary basins can be subdivided into categories that include foreland basins, intracratonic or platform basins, rift basins that may evolve into passive margin basins, and other diverse types that may include strike-slip (pull-apart) basins and forearc basins.

A **foreland basin** is a sedimentary trough that forms in an area adjacent to a fold and thrust belt, usually in an area characterized by a plate collision and convergent tectonics (Figure 7.1). As one plate advances over another, the overridden plate is forced down. Sediment erodes from the uplifted area on the overriding plate and accumulates in the downwarped area in front of its advance. The accumulation of sediment constitutes a load sufficient to force further subsidence that in turn leads to greater sediment accumulation. Examples of foreland basins include the Appalachian Basin in eastern North America, the Alberta or Western Canadian Sedimentary Basin, the Andean Basin in South America, and the Ganga and Indus Basins, formed south of the suture between the Indian and Euro-Asian Plates (Beaumont, 1981).

An **intracratonic basin** is a sedimentary basin found in the central, tectonically stable interior of a continent. Intracratonic basins are typically oval-shaped and extend laterally to a width of a few hundred kilometers (Sleep et al., 1980). These basins usually contain a few kilometers of sediments whose accumulation over long intervals of geologic time (of the order of 10^8 yrs) indicates continual subsidence for periods greater than 100 million years. Intracratonic basins are also known as platform basins. Sleep (1971) was the first to note that the tectonic subsidence of these basins was proportional to the square root of time, and suggested that their subsidence was thermally controlled, as is the ocean basins. But, thermal

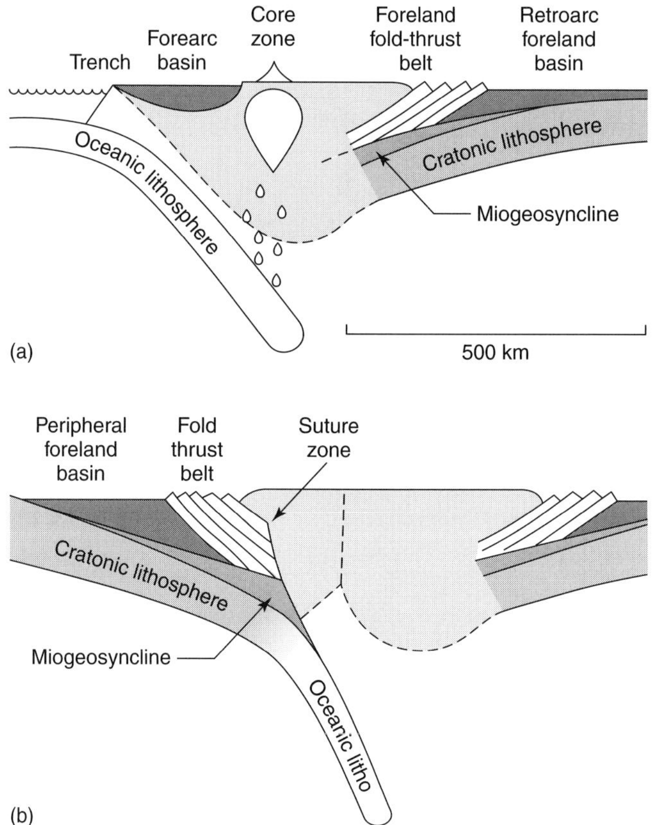

Figure 7.1 Plate tetonic settings in which foreland basins form.
(From Beaumont, 1981, p. 293.)

subsidence alone is generally insufficient to explain the origin of these basins. Klein (1991) concluded that the subsidence histories of several intracratonic basins in North America reflected an initial fault-controlled mechanical subsidence during rifting, followed by thermal subsidence and finally by subsidence related to the excess mass of a cooling igneous body at depth. Examples of intracratonic basins in North America include the Illinois, Michigan, and Williston Basins (Figure 7.2).

A **rift basin** is a sedimentary basin formed by the extension of the lithosphere. The **lithosphere** is the outer, rigid section of the Earth that deforms elastically on a geologic time scale; it is about 100 kilometers thick on average. An initial period of rapid, fault-controlled mechanical subsidence associated with lithospheric extension is followed by a longer period of thermally controlled subsidence as the extended lithosphere cools (McKenzie, 1978). If the rifting progresses far enough, the lithosphere may be stretched to the point where its thickness is essentially zero. At that point, sea floor spreading begins and a new ocean basin is formed. The ocean basins are the largest set of sedimentary basins on Earth, covering about 71% of its surface area. An ocean basin that abuts a continental margin in the absence of orogenesis is termed a **passive margin basin.**

A **strike-slip basin** is a sedimentary basin that forms where the Earth's crust separates due to an

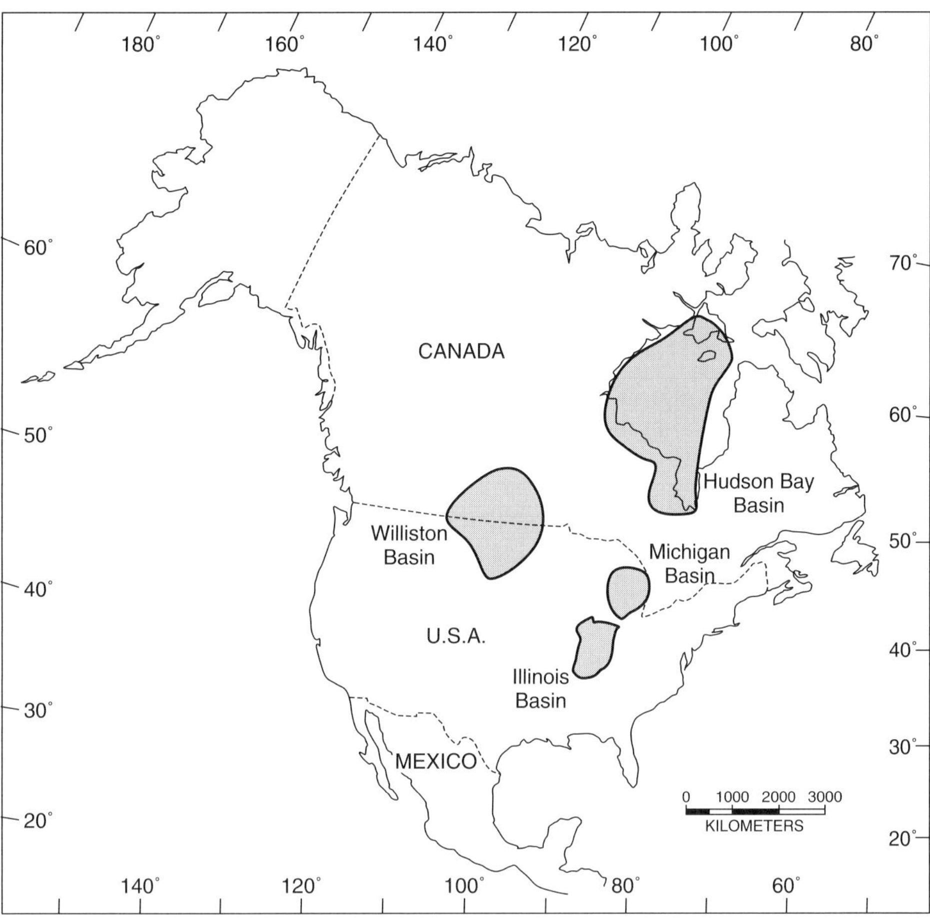

Figure 7.2 Location of major intracratonic basins in North America.
(From Klein, 1991, p. 4.)

offset in a strike-slip fault (Figure 7.3). The crust is literally pulled apart; thus, strike-slip basins are also known as pull-apart basins. A well-known example of such a basin is the Los Angeles Basin in Southern California, formed by an offset in the San Andreas Fault. A **forearc basin** is a sedimentary basin formed between a subduction-zone trench and a volcanic island arc. There are also other types of sedimentary basins, their categorization is potentially as diverse as the heterogeneity of the Earth's crust.

Sedimentation rates in sedimentary basins (Figure 7.4) can vary by as much as 3 orders of magnitude, from 1 m-Ma^{-1} to 1000 m-Ma^{-1} (an *Ma* is a mega-annum, a million years). Rates below or above these values can be important locally, but burial histories between these limits are most common. The sedimentation rate for a passive margin changes as it evolves from a rift basin (50–100 m-Ma^{-1}) to a passive margin basin (10–20 m-Ma^{-1}). The lowest sedimentation rates (~10 m-Ma^{-1}) are found in intracratonic basins such as the Michigan, Illinois, and Williston basins in North America (Sleep, 1971; Schwab, 1976; Sleep et al., 1980). Strike-slip and forearc basins are characterized by much higher rates (100–1000 m-Ma^{-1}). Foreland basins experience the most varied sedimentation rates, but generally occupy the

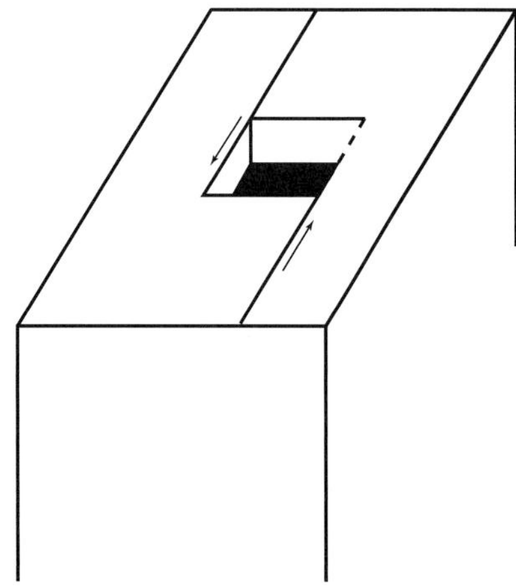

Figure 7.3 Conceptual illustration of how a strike-slip or pull-apart basin forms at an offset in a strike-slip fault.

TABLE 7.1 Average Sedimentation Rates.

Geologic Setting	Sedimentation Rate (m-Ma^{-1})
intracratonic basins	~10
rift basins	~50–100
foreland basins	~100
strike-slip/forearc	~100–1,000
river deltas	~1,000–10,000

middle ground. The highest sedimentation rates are found in areas of rapidly prograding river deltas (e.g., the Gulf Coast basin of the southeastern U.S.), where the sediment deposition rate can be as high as 1000–5000 m-Ma^{-1} (Sharp and Domenico, 1976; Bethke, 1986; Bredehoeft et al., 1988).

7.1.2. Fluid Velocities Scale with Sedimentation Rates.

Porosity-depth curves (e.g., Figures 3.1 and 3.2) show that porosity in a sedimentary basin typically decreases with increasing depth. Thus fluid must be expelled as pore space decreases with increasing time and depth of burial. What is not obvious are answers to questions such as these: Where does the fluid go? Upward? Downward? Sideways? How fast does the fluid move? Does it move fast enough to be an efficient mechanism for heat and mass transport?

Consider a column through a sedimentary basin, analogous to a wastebasket, that accumulates sediment through time. The basin has cross-sectional area A (m^2), which we may conveniently take to be 1 m^2. Sediments at the top of the column are relatively uncompacted, with a high porosity, while those at the base of the column are relatively compacted, with a low porosity. Let R (m-s^{-1}) be the sedimentation rate. The volume of sediment per unit time that enters the column is

$$\frac{\text{volume}}{\text{unit time}} = RA \qquad \textbf{(7.1)}$$

Let us be optimistic and suppose that fully half of the sediment is composed of pore fluid at the surface, and that all of this fluid is expelled during the process of compaction. Then the rate at which pore fluids are expelled is

$$\frac{\text{fluid volume expelled}}{\text{unit time}} = 0.5RA \qquad \textbf{(7.2)}$$

Recall that the definition of Darcy Velocity (q, m-s^{-1}) is fluid volume per unit area per unit time. If we assume that all of the fluid expelled leaves through the top of the column,

$$q = \frac{0.5RA}{A} = 0.5R \qquad \textbf{(7.3)}$$

In other words, the average Darcy velocity of fluid flow due to sediment compaction is proportional to the sedimentation rate and is of the same order of magnitude. Sedimentation rate varies with the geologic setting (Table 7.1).

The maximum depth of sedimentary basins is about 20,000 m; thus, it is worth noting that sedimentation rates as high as 1,000 to 10,000 m/Ma cannot be sustained for very long. The rates cited in Table 7.1 are averages. Actual sedimentation events may be discontinuous, and some basins

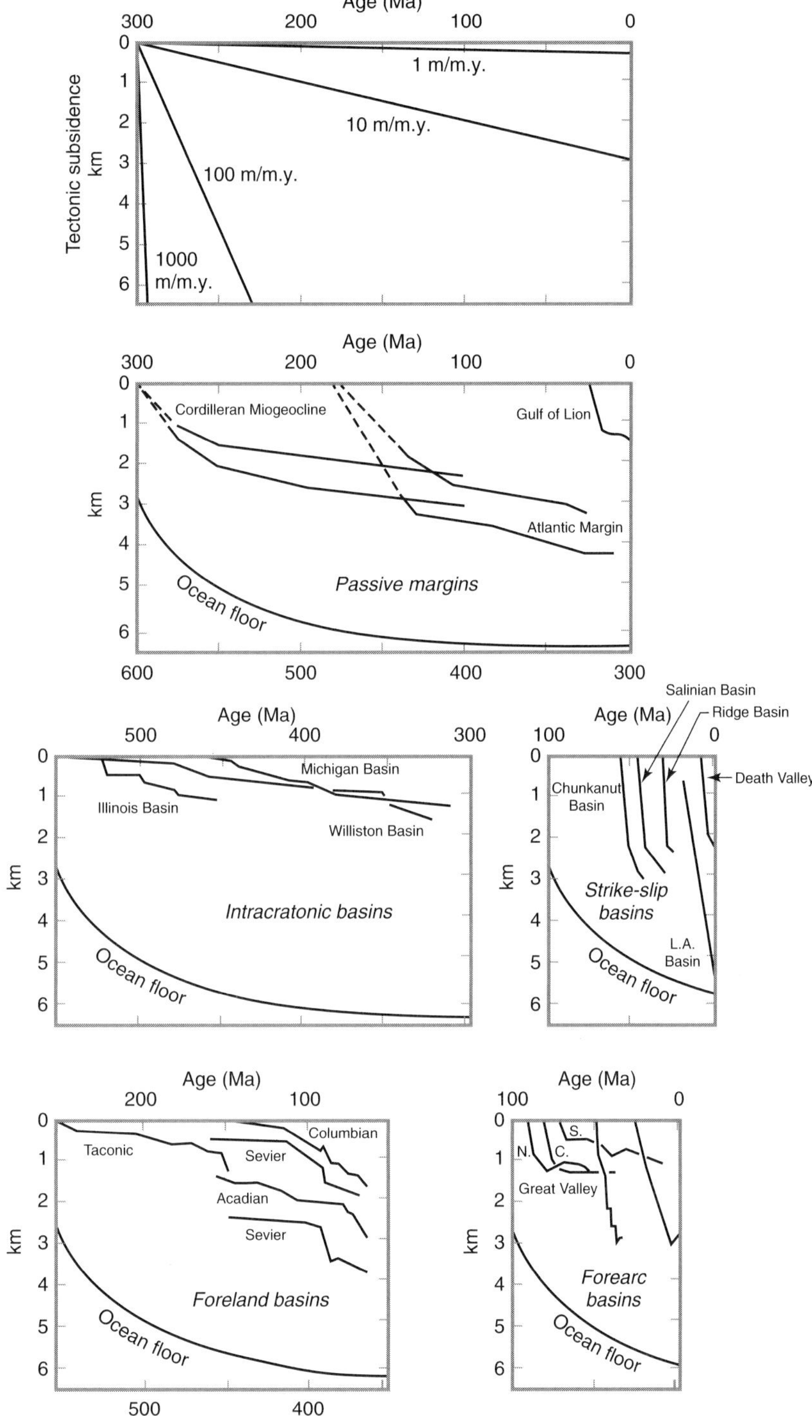

Figure 7.4 Tectonic subsidence rates for different types of sedimentary basins. Tectonic subsidence is the subsidence that would occur in the absence of sedimentation. Total subsidence is usually 2 to 3 times greater than tectonic subsidence.

(From Deming, 1994c, p. 166; after Angevine et al., 1990.)

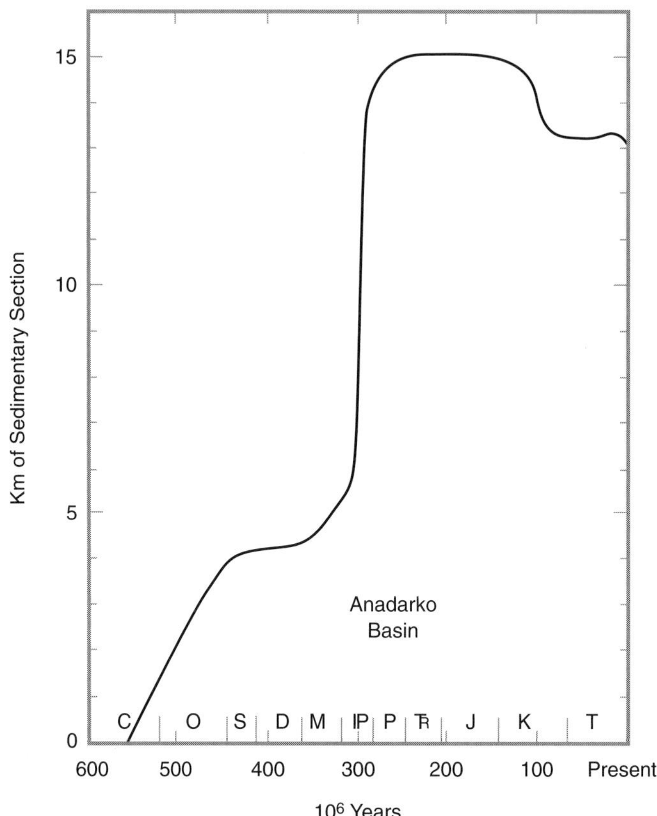

Figure 7.5 Sediment thickness/subsidence of the Anadarko Basin as a function of time. (From Gilbert, 1992, p. 203.)

may undergo transitions from one tectonic setting to another. For example, the **Anadarko Basin** in southwestern Oklahoma originated as a rift basin in Paleozoic time and averaged sedimentation rates of ~20 m-Ma^{-1} over 200 Ma (Figure 7.5). In Pennsylvanian time, there was a transition to compressional tectonics, and the Anadarko became a foreland basin with a relatively short period (~10–20 Ma) characterized by high sedimentation rates of 100–1000 m-Ma^{-1}. For the last 250 Ma, the Anadarko has been tectonically quiescent, with erosion dominating over sedimentation.

If we take 100 m-Ma^{-1} as an average to high sedimentation rate, the average Darcy velocity (by equation 7.3) of fluid expelled by sediment compaction is 0.00005 m-yr^{-1}. This is very low in comparison to flow velocities in some topographically driven flow systems. For example, isotopic dating shows that the average linear velocity of groundwater in the Great Artesian Basin in Australia is ~ 1 m-yr^{-1} (see Figure 1.4). For a porosity of 10%, this is equivalent to a Darcy Velocity of 0.1 m-yr^{-1} (see equation 2.19), 200 times faster than in our compaction-driven flow example.

As the above comparison illustrates, typical compaction-driven flow rates are 2 to 3 orders of magnitude lower than those found in many flow systems driven by topographic gradients. As a result, we are forced to conclude that compaction-driven flow is a relatively inefficient mechanism for heat and/or mass transport unless flow can be focused spatially or temporally.

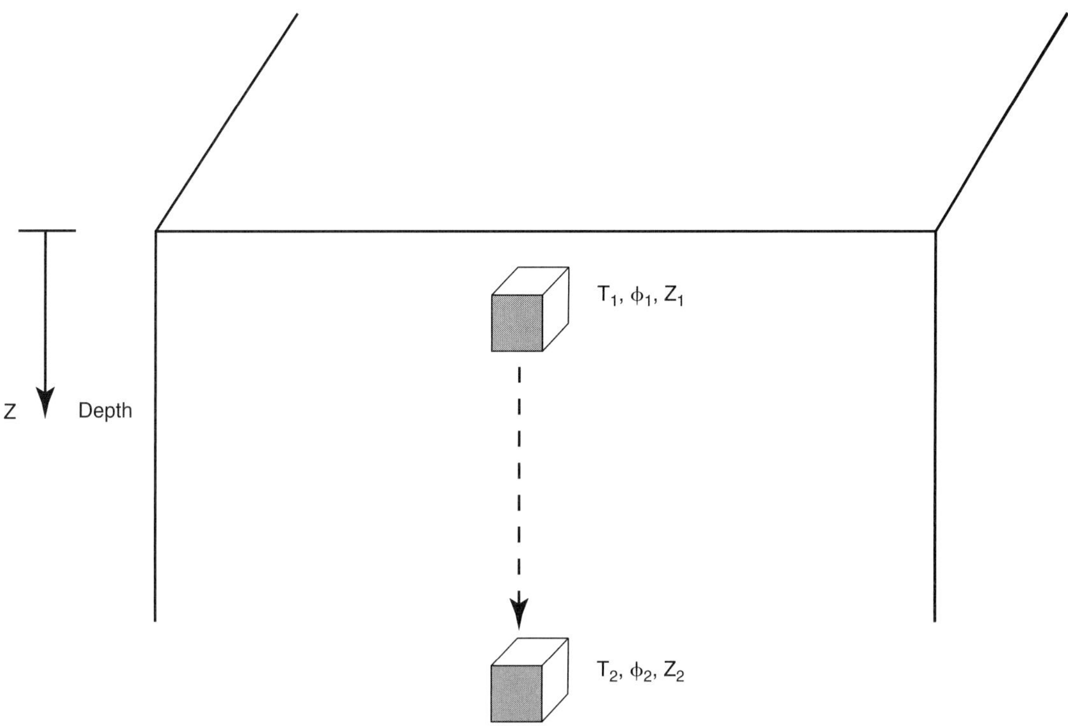

Figure 7.6 A control volume of rock or sediment in a sedimentary basin evolves from a depth Z_1, temperature T_1, and porosity ϕ_1, to a depth Z_2, etc.

7.1.3. Pore Collapse versus Aquathermal Pressuring.

As a control volume of sediment or rock becomes buried through geologic time and its depth increases, it compacts through pore collapse and the pore fluid also heats up (Figure 7.6). Both of these processes may act to expel pore fluid from the control volume and drive fluid flow. The increase in fluid pressure associated with the thermal expansion of the heated pore fluid is termed **aquathermal pressuring.**

We can determine the relative importance of these two processes through a simple calculation. We start with an equation of state that describes the change of fluid density that takes place with respect to changes in fluid pressure and ambient temperature (Domenico and Palciauskas, 1979, p. 956)

$$\frac{d\rho}{\rho} = BdP - \alpha_T dT \tag{7.4}$$

where ρ (kg-m^{-3}) is fluid density, B (m-s^2-kg^{-1}) is the isothermal fluid compressibility (as defined in equation 3.24), P (kg-m^{-1}-s^{-2}) is fluid pressure, α_T (K^{-1} or °C^{-1}) is the coefficient of thermal expansion of the fluid, and T (K or °C) is temperature. The fluid density, ρ, is the fluid mass (M, kg) divided by the fluid volume (V, m^3),

$$\rho = \frac{M}{V} \tag{7.5}$$

If we hold the fluid mass (M) of the control volume constant,

$$d\rho = Md\left[\frac{1}{V}\right] \tag{7.6}$$

$$d\rho = -M\frac{dV}{V^2} \tag{7.7}$$

Thus

$$\frac{d\rho}{\rho} = -M\frac{\frac{dV}{V^2}}{\frac{M}{V}} = \frac{-dV}{V} \tag{7.8}$$

Substituting equation 7.8. into equation 7.4,

$$\frac{dV}{V} = -BdP + \alpha_T dT \tag{7.9}$$

Now, to calculate the increase in fluid pressure associated with heating the pore fluid, suppose the pore fluid is heated but cannot expand because the pore space is constant. Constant fluid volume implies that $dV/V = 0$, thus

$$BdP = \alpha_T dT \tag{7.10}$$

or

$$dP = \alpha_T \frac{dT}{B} \tag{7.11}$$

Alternatively, to calculate the increase in fluid pressure that occurs from pore collapse alone, suppose that temperature is held constant, but that the pore space changes. Then $dT = 0$ in equation 7.9, and

$$\frac{dV}{V} = -BdP \tag{7.12}$$

or

$$dP = \frac{-\frac{dV}{V}}{B} \tag{7.13}$$

Equation 7.11 gives the change in fluid pressure due to changes in temperature, while equation 7.13 gives the change in fluid pressure due to changes in fluid volume related to compaction of a porous medium. By comparing the magnitude of these two equations we may estimate whether aquathermal pressuring or pore collapse is more important in driving fluid flow in a sedimentary basin.

The magnitude of the relative changes in pressure due to these two different mechanisms can be found by dividing equation 7.11 by equation 7.13,

$$\text{ratio} = \frac{\alpha_T dT}{\frac{dV}{V}} \tag{7.14}$$

where we have dropped the negative sign. Converting from derivatives to differentials, we may equivalently write

$$\text{ratio} = \frac{\alpha_T \Delta T}{\frac{\Delta V}{V}} \tag{7.15}$$

To evaluate the magnitude of this ratio, consider an average sedimentation rate of 50 m-Ma^{-1} for 1 Ma. Assume that the change of porosity of depth may be described by an exponential model, $\phi = \phi_0 e^{-z/b}$, and take $\phi_0 = 0.25$ and $b = 3000$ m. Let the coefficient of fluid thermal expansion (α_T) be 5×10^{-4} K^{-1}. Choose $z_1 = 1000$ m depth, $z_2 = 1050$ m depth, and assume an average geothermal gradient of 25°C-km^{-1}. Then $\phi_1 = 0.17913$, $\phi_2 = 0.17617$, and $\Delta\phi = \phi_1 - \phi_2 = 0.00296$. Define the average porosity as $\bar{\phi} = (\phi_1 + \phi_2)/2 = 0.17765$. We can then approximate $\Delta V/V$ by $\Delta\phi/\bar{\phi} = 0.00296/0.17765 = 0.01667$. The change in temperature from z_1 to z_2 is 1.25 °C. Thus

$$\text{ratio} = \frac{(5 \times 10^{-4})(1.25)}{0.01667} = \frac{1}{27} \tag{7.16}$$

or pore collapse is 27 times more important in creating a driving force for fluid flow than aquathermal pressuring. Clearly, our result will vary according to the geologic setting, however, in most cases pore collapse is a much stronger mechanism for driving fluid flow than aquathermal pressuring.

The perceptive reader may note that there is a potential source of error in the above calculation: we have considered fluid pressure instead of head, even though elevation is not constant. It is true that head decreases as depth increases from z_1 to z_2; however, this decrease in head is offset exactly by an increase in the hydrostatic fluid pressure.

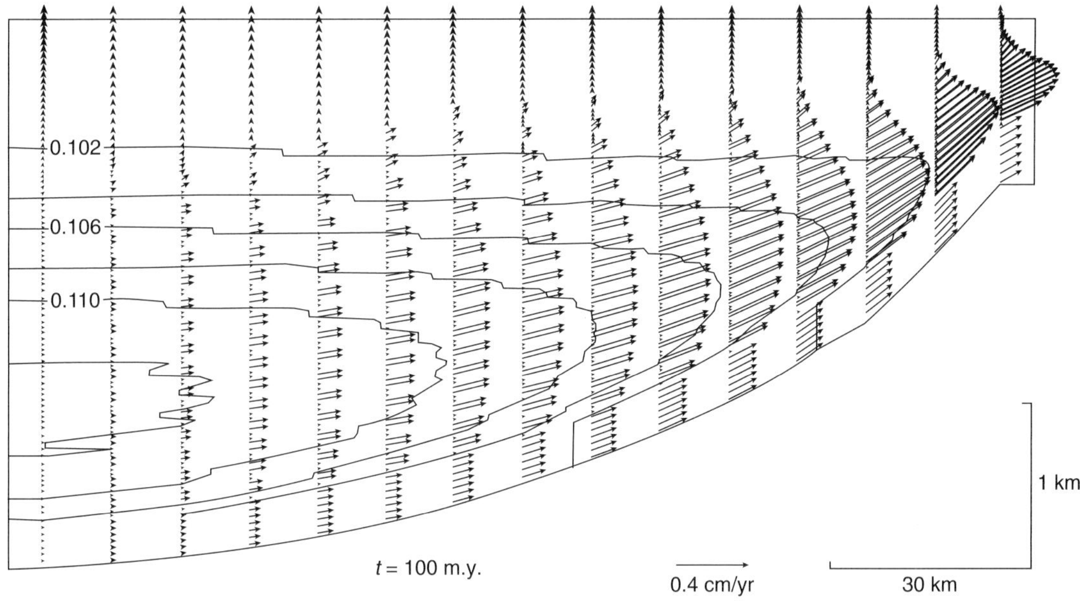

Figure 7.7 Linear flow velocities as calculated by a two-dimensional model of fluid flow in an intracratonic basin driven by sediment compaction. Heavy lines are fluid equipotentials in Mpa. Vertical exaggeration is 25:1. (From Bethke, 1988, p. 3502.)

Although we have neglected the head decrease that occurs from z_1 to z_2, we have also neglected the offsetting increase in the hydrostatic fluid pressure. Our calculations of fluid pressure generated from pore collapse and aquathermal pressuring have been calculations of the *excess* pore pressure above the ambient hydrostatic regime.

7.1.4. Compaction-Driven Flow in an Intracratonic Basin.

Bethke (1985) studied compaction-driven flow in an intracratonic basin with a two-dimensional finite-difference model. In general, Bethke (1985) found that:

1. Fluids expelled from shallow sediments move upwards, while deeper fluids move laterally to basin edges (Figure 7.7).
2. Very small excess pressures (exceeding hydrostatic) develop.
3. Fluid velocity scales with sedimentation rate. Maximum linear velocity is about 0.5 cm-yr^{-1}; average linear fluid velocity is about 0.2 cm-yr^{-1}.
4. Compaction-driven fluid flow in intracratonic basins is not an efficient mechanism for heat and mass transport (due to the low flow velocities), and thus is probably not responsible for the formation of hydrothermal ore deposits (e.g., Mississippi Valley-type lead-zinc deposits) found on basin margins. Other researchers have suggested that compaction-driven flow could remain a viable mechanism for ore formation by invoking spatial or temporal focusing of flow (Cathles and Smith, 1983; Cathles, 1993b).
5. Interlayered sands and shales lead to a very high basin-scale anisotropy, $k_x/k_z \sim 1000$. This implies that although the lateral dimension of intracratonic and

other sedimentary basins exceeds the vertical extent by more than 100; one-dimensional (vertical) analyses of fluid flow are thus perhaps invalid.

6. Expelled pore fluids tend to move vertically out of compacting shales (aquitards) into neighboring sandstone aquifers and then travel horizontally to basin edges (Figure 7.7).
7. Some pore fluids move downwards (inspect the isopotentials in Figure 7.7). Near the center of the basin, the path of least resistance may be for pore fluids to move downwards to a basal aquifer and then laterally to the basin edge.
8. In stark contrast to topographically driven flow, flow velocities are not dependent upon permeability so long as the actual compaction process itself is not inhibited or retarded. Pore collapse not only provides the driving force for fluid flow, it also provides the fluid source. Thus, *if* pore collapse occurs, fluid velocities are only dependent upon the rate of pore collapse and fluid expulsion. Low permeabilities will tend to inhibit pore-fluid expulsion, but this results in higher pore-fluid pressure and head gradients that offset the low permeability. *If* the permeability is so low, however, as to prevent or retard pore collapse, then overpressures develop and flow velocities are lower than they would otherwise be.
9. Because of the low flow velocities, Bethke's results argue against compaction-driven flow as a viable mechanism for secondary oil migration or the formation of ore deposits in and near intracratonic basins in North America (Figure 7.2).

7.2. Topographically Driven Flow.

The topography of the water table usually mimics the topography of the ground surface. Unlike the solid Earth, however, fluids readily flow downhill, from regions of high potential energy (head) to regions of low potential energy. If precipitation and infiltration in regions of high elevation are sufficient to recharge the water table, a continuous supply of groundwater is available to maintain flow. The time constant (equation 5.86) that describes the diffusion of transients in regional flow systems is usually orders of magnitude lower than the characteristic time scale over which geologic processes that create topographic gradients (e.g., mountain building) operate, and quasi steady-state flow systems result. It is scarcely too strong a statement that topographically driven fluid-flow occurs anywhere there is topographic relief; it is virtually pervasive throughout the upper continental crust. Bredehoeft et al. (1982, p. 302) put it most succinctly: "Land elevation and relief are clearly the driving forces that determine most regional ground-water flow . . . "

A characteristic situation is a foreland basin (Figure 7.8). In the mountain range and its foothills recharge takes place and fluid moves downward and out into the neighboring foreland basin in the direction of decreasing hydraulic head. The moving fluid seeks the path of least resistance, or highest hydraulic conductivity. Often, this may be the lowest stratigraphic unit in a sedimentary basin. Through a coincidence of stratigraphy, Cambrian rocks are often permeable sandstone aquifers. The moving fluid then follows a largely horizontal path through the basin, with eventual discharge probably taking place near the distal edge, depending upon the geometry and hydraulic conductivity structure of the basin. Total transport distances of several hundred to a thousand kilometers or more are conceivable. For example, isotopic and trace-element analyses of saline groundwater discharging from natural springs and artesian wells in central Missouri (Banner et al. 1989) show that these fluids most likely originated as meteoric recharge in the Rocky Mountains, ~1000 km to the west. The existence of regional, topographically driven groundwater flow-systems has also been documented for several sedimentary

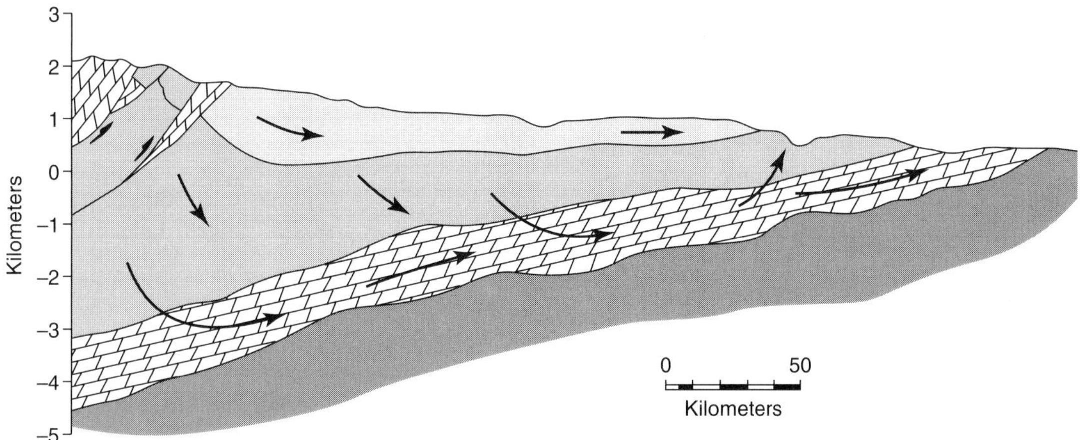

Figure 7.8 Conceptual model of topographically-driven flow in a foreland basin.
(From Garven and Freeze, 1984, p. 1091.)

basins in western North America by John Bredehoeft and his co-workers at the U.S. Geological Survey through measurements of hydraulic head and permeability. These include the Kennedy Basin in South Dakota (Bredehoeft et al. 1983), the Denver Basin in Colorado (Belitz and Bredehoeft 1988), and the Big Horn Basin in Wyoming (Bredehoeft et al. 1992). Other documented examples of regional scale flow driven by topographic gradients include the Dakota Sandstone in South Dakota (Figures 3.18 and 3.19) and the Great Artesian Basin in Australia (Figure 1.4).

If an aquifer is overlain by a layer of lower permeability, or confining layer, the aquifer may be artesian. An **artesian aquifer** is one in which the head of the groundwater is commonly found to be greater than the elevation of the land surface. A well that penetrates an artesian aquifer will therefore usually flow at the ground surface, depending on variations in the local topography. The requirement for artesian flow is that the head drop of the fluid in the aquifer from recharge to a well be less than the elevation drop from recharge to the same well. The classic example of an artesian aquifer is the Dakota Sandstone in South Dakota (see Figures 3.18 and 3.19) as studied by Darton (1909). Other sources define "artesian aquifer" to be any aquifer in which the head of the groundwater is sufficient to enable it to rise above the top of the aquifer. There are two problems with this definition. The first is that the definition is conceptually useless, because it is virtually identical to "confined aquifer." The second problem is that the definition is etymologically incorrect. The term "artesian" derives from the Artois region of France where wells drilled in 1126 AD spontaneously flowed at the surface. The word "artesian" thus implies that the potentiometric surface is greater than the elevation of the ground surface. These difficulties could be avoided if use of the word "artesian" were restricted to wells instead of aquifers. In cases such as the Dakota Sandstone, however, it is sensible to use term "artesian aquifer" as there is considerable historical and physical significance associated with artesian flow from this formation.

7.2.1. Flow in Homogeneous, Isotropic Domains.

One of the first quantitative studies of topographically driven flow was done by Toth (1962). Toth

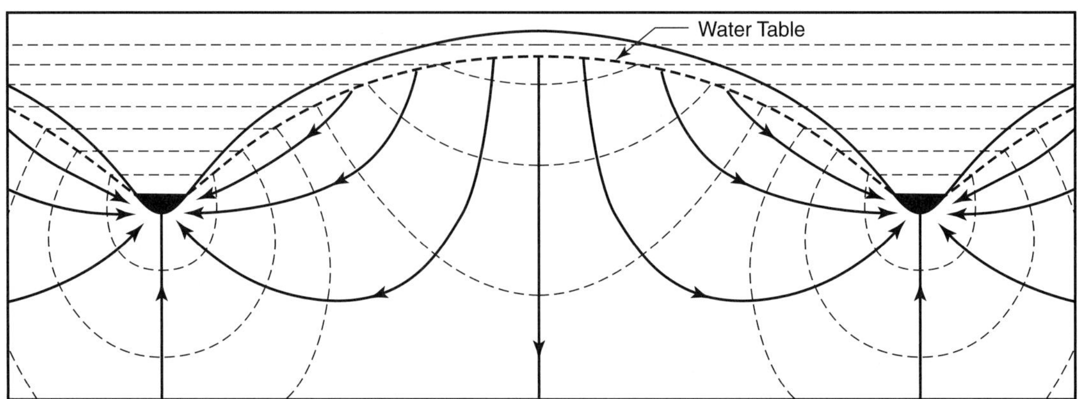

Figure 7.9 Topographically driven flow in a setting with alternating valleys and hills.
(From Toth, 1962; after Hubbert, 1940.)

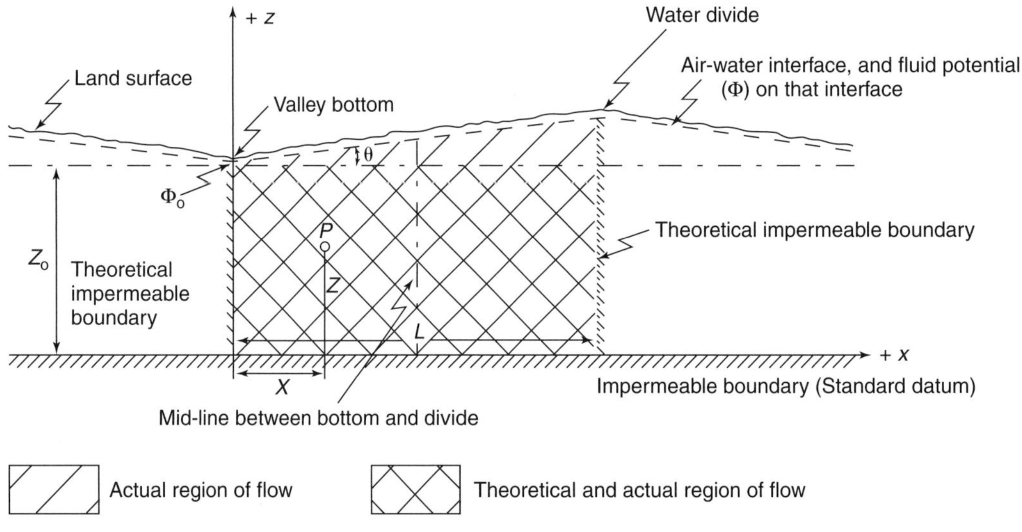

Figure 7.10 Toth model of topographically-driven flow in which alternating hills and valleys are approximated by trapezoids.
(From Toth, 1962, p. 4379.)

considered a situation of symmetrically alternating valleys and hills (Figure 7.9). In this situation, a vertical line drawn between both valleys and hills is a plane of symmetry and constitutes a no-flow boundary. Thus, the entire flow system can be analyzed by considering one symmetrical segment between a valley and hill.

Toth (1962) approximated alternating series of valleys and hills by an alternating set of trapezoidal shapes (Figure 7.10). He then further simplified by approximating a trapezoid by a rectangle and imposing a boundary condition at the top of the rectangle that mimicked the head change due to the topographic gradient (Figure 7.11). Replacing the

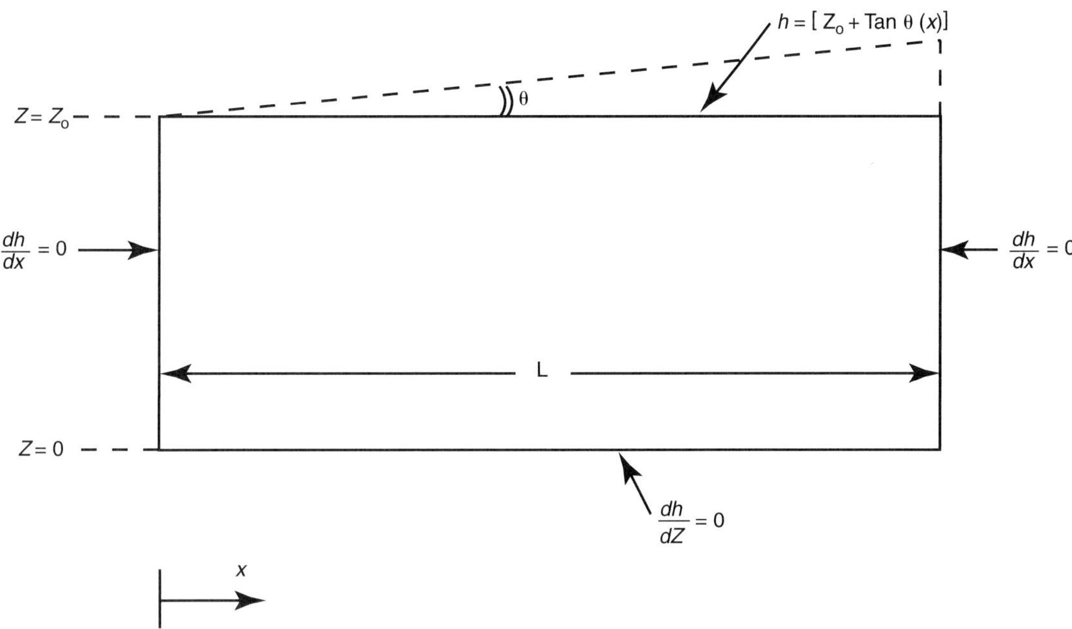

Figure 7.11 Boundary conditions and physical dimensions of Toth model of topographically driven flow.

trapezoidal geometry by a rectangle allowed Toth to derive an exact solution for head as a function of two-dimensional space.

Toth (1962) solved Laplace's equation in a two-dimensional, homogeneous domain,

$$\nabla^2 h = 0 \tag{7.17}$$

Boundary conditions for Toth's problem are as follows (Figure 7.11). The horizontal head gradient was fixed at zero on the lateral boundaries (no flow boundaries). The vertical head gradient was fixed on the bottom boundary (no flow boundary). Head was fixed at the top boundary to be equal to the value that would occur in a trapezoidal shape with the water table at the surface,

$$h = [z_0 + \tan\theta(x)] \tag{7.18}$$

where h (m) is hydraulic head, z_0 (m) is elevation of the top of the rectangle or top left side of the trapezoid, θ (dimensionless) is the angle between the horizontal and the trapezoid's top surface, and x (m) is distance measured from the left side of the trapezoid (see Figure 7.11).

Toth's solution for a domain with homogeneous and isotropic hydraulic conductivity is (Figure 7.12)

$$h = g[z_o + (\tan\theta)L/2] - 4g(\tan\theta)L/\Pi^2 \sum_{m=0}^{\infty} \frac{\cos[(2m+1)\Pi x/L]\cosh[(2m+1)\Pi z/L]}{(2m+1)^2\cosh[(2m+1)\Pi z_0/L]} \tag{7.19}$$

where g (m-s^{-2}) is the acceleration due to gravity, θ is the angle between the horizontal and the trapezoid's top surface, x (m) is distance measured from the left side of the trapezoid, L (m) is the total length of the trapezoid, and z_0 (m) is the height of the left side of the trapezoid.

In 1963, Toth extended his work on the quantitative analysis of topographic flow by publishing a solution for an undulating topography described by a sine function (Toth, 1963). Toth's solution (Figure 7.13) shows that when local topographic

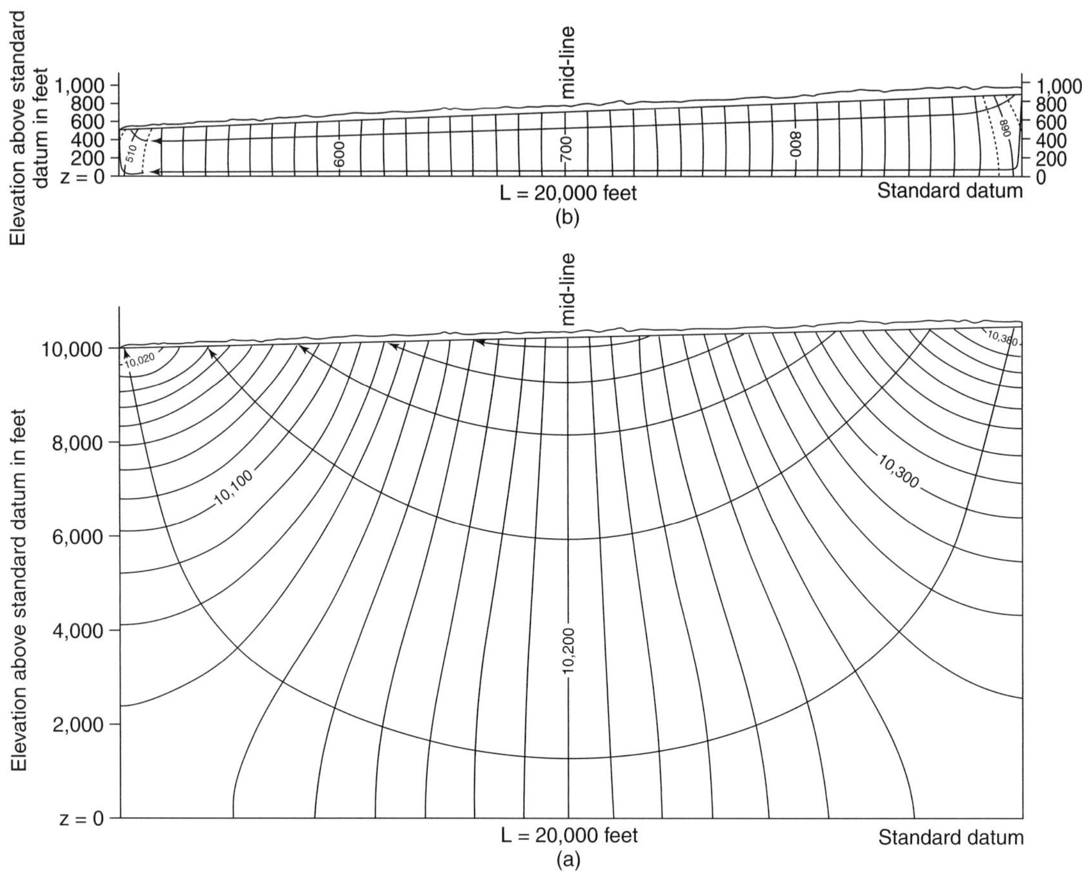

Figure 7.12 Head contours and streamlines in Toth model of topographically driven flow. (From Toth, 1962, p. 4380.)

variations are superimposed on a regional topographic gradient, local flow systems develop. Note that the local flow systems may have a direction of motion opposite to that of the regional system. The local systems tend to be shallow in nature, while regional scale flow dominates at depth (see Figure 7.13).

7.2.2. Flow in Inhomogeneous, Anisotropic Domains.

The advent of digital electronic computers enabled Freeze and Witherspoon (1967) to conduct the first study of topographically driven flow for inhomogeneous, anisotropic media. They solved Laplace's equation in two dimensions,

$$\frac{d\left[-K_x \frac{dh}{dx}\right]}{dx} + \frac{d\left[-K_z \frac{dh}{dz}\right]}{dz} = 0 \qquad \textbf{(7.20)}$$

where K_x (m-s^{-1}) is the hydraulic conductivity in the horizontal dimension, and K_z (m-s^{-1}) is the hydraulic conductivity in the vertical dimension. Note that the absolute values of conductivity do not affect the solutions to equation 7.20; it is only

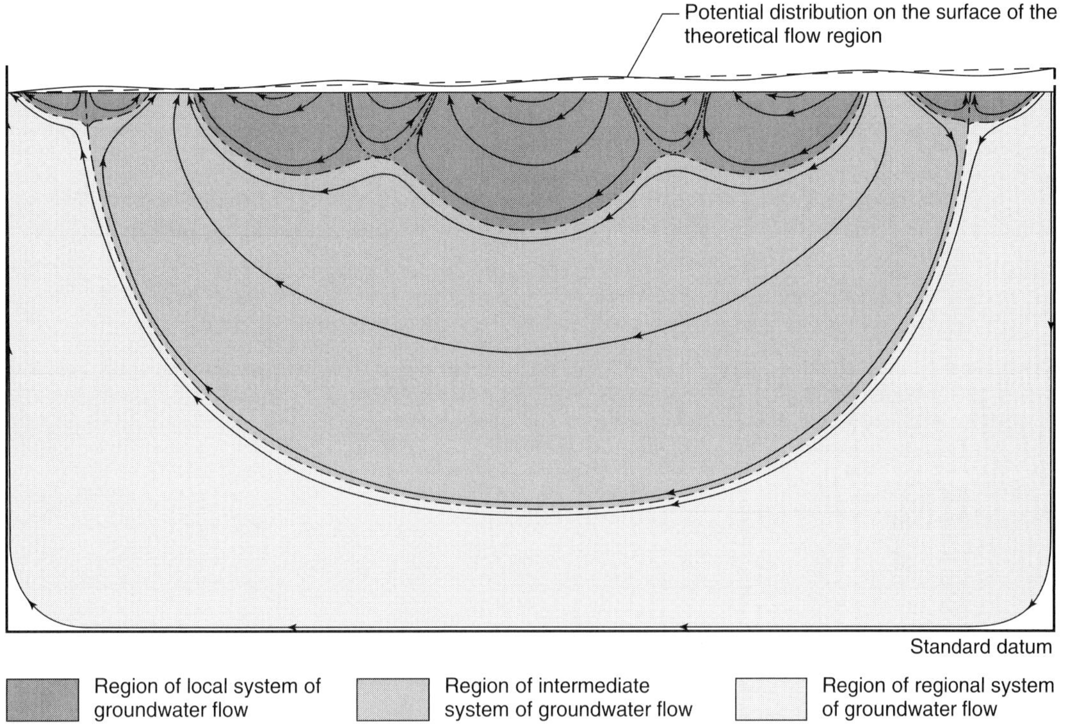

Figure 7.13 Head contours and streamlines in topographically-driven flow model with a sinusoidal topography. (From Toth, 1963, p. 4807.)

the ratio of the vertical and horizontal conductivities that is relevant. The solutions obtained by Freeze and Witherspoon (1967) were for a rectangular region with vertical, impermeable sides, and a horizontal, impermeable base. Head was fixed at the top of the model to be equal to the elevation of the ground surface (Figure 7.14).

1. Effect of Permeability Variation. The effect of a basal aquifer is to concentrate flow (Figure 7.15a, b, c). As the permeability contrast between the basal aquifer and overlying aquitard becomes more pronounced, so does the refraction of flow vectors. Flow in the overlying aquitard becomes essentially vertical, while flow in the underlying aquifer is essentially horizontal. As the permeability of the basal aquifer increases relative to the overlying aquitard, the hinge line moves upslope, creating a larger discharge area. The larger discharge area is necessary to accommodate the greater volume of water moving through the system. The **hinge line** is an imaginary line that demarcates the region of downward moving flow from the region of upward moving flow.

   If the higher permeability layer is on top, there is little difference between the flow directions that would result from layered and homogeneous permeability distributions (compare Figures 7.14a and 7.15e). The total flow rate through the higher permeability system is, of course, much greater. Note that

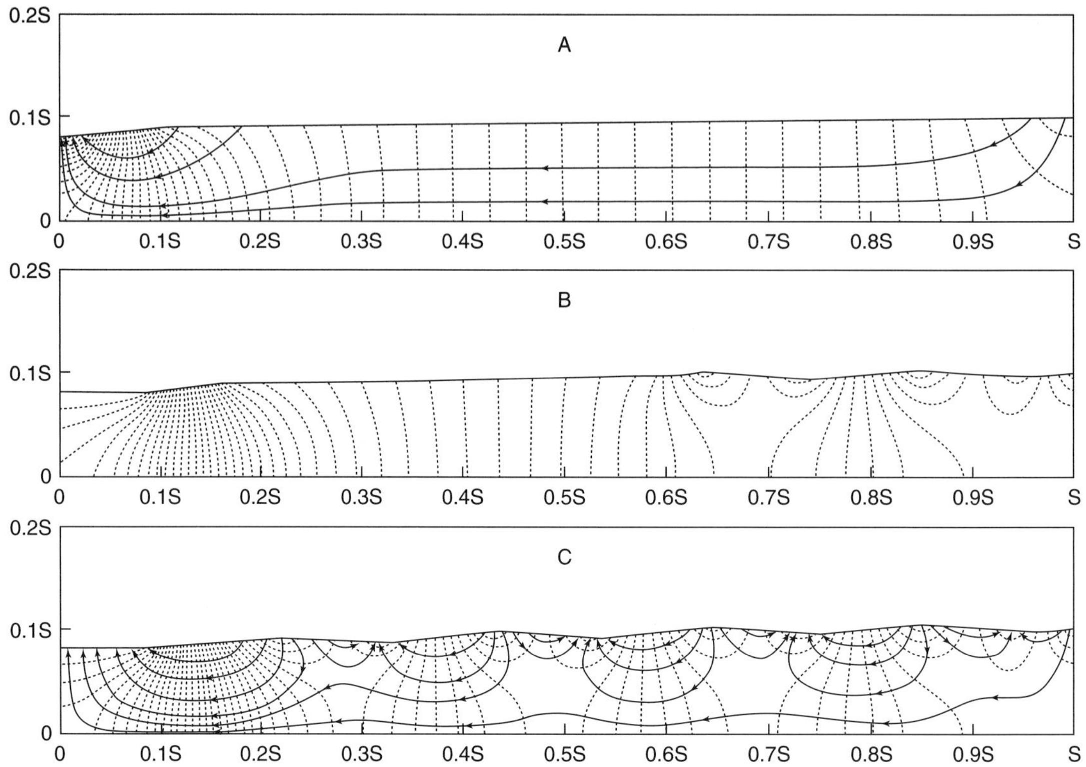

Figure 7.14 Effect of water table configuration and topography on regional groundwater flow through domains with homogeneous, isotropic hydraulic conductivity.
(From Freeze and Witherspoon, 1967, p. 625.)

the equipotentials in the aquifer (Figure 7.15e) are virtually vertical, indicating there is very little flow into the underlying aquitard. *It is thus usually permissible to treat the interface where a high permeability layer overlies a low permeability layer as a no-flow boundary.*

(2) Effect of Partial Layers and Lenses. The effect of high permeability "lenses" and partial layers is to concentrate flow (Figure 7.16). High permeability areas attract flow like a magnet. A buried high permeability layer in what would otherwise be a recharge area, can turn substantial portions of that recharge area into a discharge area (Figure 7.16a). If the buried high permeability region occurs downstream, in what otherwise might be a discharge region, the total recharge area is enlarged, and the zone of most intensive recharge overlies the buried aquifer (Figure 7.16b).

In summary, Freeze and Witherspoon's modeling studies of topographically-driven groundwater flow showed that

1. Groundwater discharge tends to be concentrated in major valleys.
2. Recharge areas are invariably larger than discharge areas.
3. In areas with hummocky terrain, numerous sub-basins are superimposed on the regional flow system.
4. Buried aquifers tend to concentrate flow toward the principal discharge area, limit the

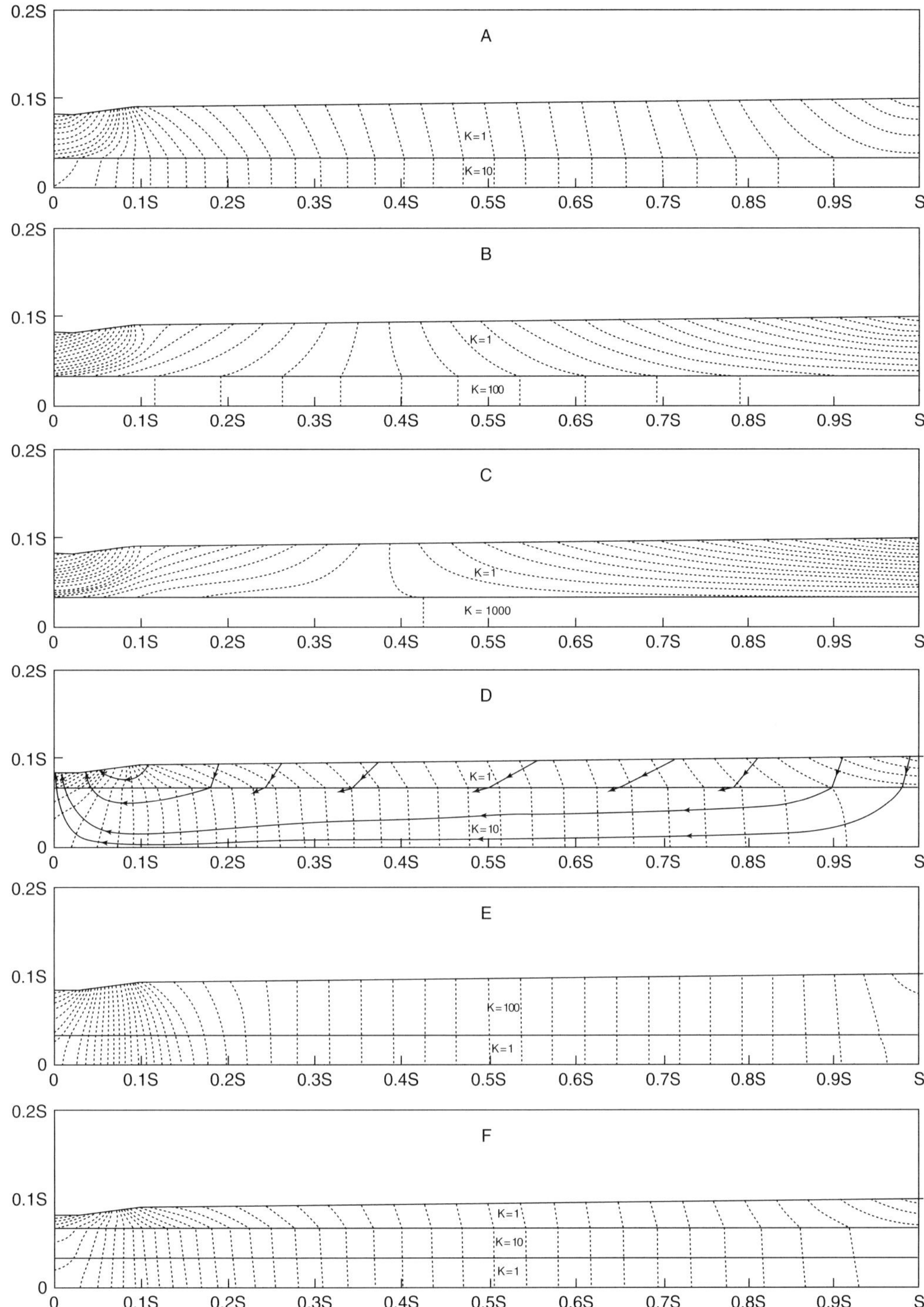

Figure 7.15 Regional groundwater flow through layered media with a simple water-table configuration, but contrasts in hydraulic conductivity between layers.

(From Freeze and Witherspoon, 1967, p. 627.)

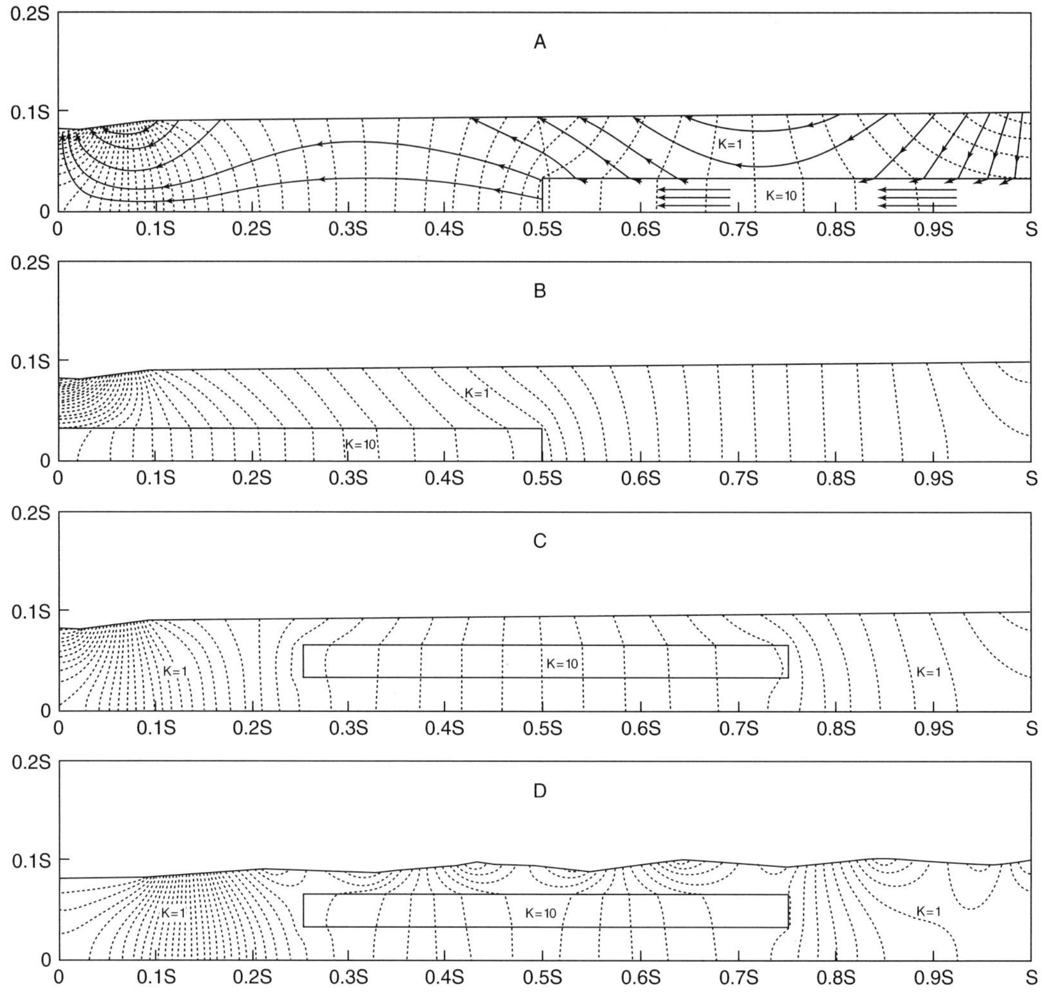

Figure 7.16 Regional groundwater flow through partial layers and lenses. The figure shows the effect on flow of a high permeability layer (K = 10) in a low-permeability (K = 1) medium.
(From Freeze and Witherspoon, 1967, p. 629.)

importance of sub-basins in producing small scale flow systems, and need not outcrop to produce artesian flow conditions.

5. Stratigraphic discontinuities can lead to distributions of recharge and discharge areas that are difficult to anticipate and that are largely independent of the water table configuration.

7.3. Buoyancy Forces.

Buoyancy forces arise from density gradients due to changes in salinity or temperature.

7.3.1. Salinity Gradients.

Consider two points in the crust. Each point is at a depth of 5 km ($z_1 = z_2 = 5$ km), and they are also

József Tóth: Father of Petroleum Hydrogeology

by

Benjamin J. Rostron

Department of Earth and Atmospheric Sciences, University of Alberta

József Tóth's (born in 1933) contributions to the field of hydrogeology over the past 40 years have changed its direction, expanded its scope, and modified the definition of the science. He is internationally known for his research on the theoretical and applied aspects of regional groundwater flow. It would be unlikely to find a textbook on hydrogeology in any country or in any language without a chapter on his theory of regional groundwater flow. Tóth's work, in over 70 technical publications and unpublished reports, stimulated extensive research internationally on soil salinization, wetland hydrology, slope stability, geothermal heat transport, transient flow on geologic time scales, radioactive waste isolation, regional groundwater development, and land-use planning and management. His personal research and work with students have contributed immeasurably to the birth and development of a new approach to exploration for hydrocarbons, namely the new discipline of *petroleum hydrogeology*.

József Tóth

József Tóth was born in the rural town of Bekes in eastern Hungary in 1933. He completed his undergraduate studies in exploration geophysics at the School of Mining and Geodesy of Sopron, Hungary. In the spring of 1956, weeks prior to being awarded his undergraduate degree, he and his wife Elizabeth fled Hungary to Austria and later settled in Holland. There he received an M.Sc. in geophysics from the State University of Utrecht, in 1960.

In 1960, Dr. Tóth was recruited by Dr. Bob Farvolden at the Alberta Research Council as a hydrogeologist to explore for groundwater in the province. Thus, in October of 1960 he immigrated to Canada and overnight became a hydrogeologist exploring for water supplies for many small towns across Alberta (including Red Deer, Olds, Trochu, Three Hills, and others). Admittedly, at the time his hydrogeological training was somewhat limited, based on his exploration geophysics background, supplemented by pouring over current textbooks of the day and rapid on-the-job training. He performed his first pumping test in a blizzard over Christmas huddled in a station wagon near Rocky Mountain House. His understanding and interest in hydrogeology quickly grew. Based on his early fieldwork in Alberta, Dr. Tóth developed the backbone of his theory of regional groundwater flow, for which he received his Ph.D. from the State University of Utrecht in 1965 under the supervision of Dr. Van Bemmelen. For this research, he was the first recipient of the Geological Society of America's *O.E. Meinzer Award* for "Distinguished Contribution to Hydrogeology" later in 1965. He spent twenty years with the Alberta Research Council as a hydrogeologist, rising to the Head of the Groundwater

József Tóth: Father of Petroleum Hydrogeology *(continued)*

Department and earning the distinction of Research Fellow. At the Research Council, he formulated, directed and administered the scientific and technical activities of the department, including a 10-year-long hydrogeological mapping program covering the majority of the Province of Alberta. The hydrogeological maps produced during that time remain in demand 20 years later.

Prof. Tóth's teaching contributions over the past 40 years are similarly unparalleled. He introduced and taught hydrogeology courses at the University of Alberta (1966) and the University of Calgary (1978), where he taught part time while still at the Research Council. Dr. Tóth joined the Geology Department at the University of Alberta as a full professor in 1980. He directly supervised 7 Ph.D. and 15 M.Sc. students until his "semi"-retirement in 1996. Joe Tóth continues to contribute as an Emeritus Professor at the University of Alberta and has recently established close ties with Eövös Lóránd Science University in Budapest, Hungary, where he was a Visiting Professor in 1996 and 1997, and is currently supervising graduate students there.

Over his distinguished career he has worked on hydrogeological problems all over the world, presented invited- and keynote-speeches at many conferences, organized numerous hydrogeological field trips (most recently a two-week hydrogeological excursion across most of Hungary), and given a multitude of presentations. He has received many scientific awards in addition to his Meinzer Award, most recently receiving the International Association of Hydrogeologists' *President's Award 1999*, which is presented annually to a senior hydrogeologist who ". . . has made outstanding contributions to the advancement of hydrogeology."

There are few hydrogeologists, if any, more dedicated to the discipline. His field trip participants have come to expect the un-written rule that "the field trip always expands to fit the allotted daylight" (remembering that summers in Alberta have 18+ hours of daylight!). Not a day passes when he does not work on, or at least think about, a hydrogeological problem of some sort. His enthusiasm for his work has been an inspiration to his students and colleagues.

József Tóth remains an active gymnast, avid photographer, outdoorsman, world traveler, and fan of certain fine "potable liquids" other than water. Always fond of dogs, many of his field photographs include a black Labrador for scale. Joe has often been quoted saying "What other job would pay me to do hydrogeology and travel all over the world doing it?"

at the same elevation. The two points are separated by a lateral distance of 10 km. Because the two points are at the same elevation, the head gradient ($\vec{\nabla} h$) is proportional to the pressure gradient ($\vec{\nabla} P$) (see equation 2.45). The fluid pressure at each point will be determined by the weight and thus the average density of the pore fluid above that point. Let the average fluid density (kg-m^{-3}) *above* point #1 be $\bar{\rho}_1$, and the average fluid density above point #2 be $\bar{\rho}_2$.

From equation 2.45, the head difference Δh between these two points is

$$\Delta h = \frac{P_1 - P_2}{\bar{\rho} g} \tag{7.21}$$

where $\bar{\rho} = (\bar{\rho}_1 + \bar{\rho}_2)/2$ (kg-m^{-3}) is an average fluid density, g is the acceleration due to gravity (m-s^{-2}), $P_1 = \bar{\rho}_1 g z_1$ is the fluid pressure at point 1 (kg-m^{-1}-s^{-2}), and $P_2 = \bar{\rho}_2 g z_2$ is the fluid pressure at point 2 (kg-m^{-1}-s^{-2}). Thus,

$$\Delta h = \frac{(\bar{\rho}_1 z_1 - \bar{\rho}_2 z_2)}{\bar{\rho}} \tag{7.22}$$

Differences in fluid density are caused by differences in salinity, temperature, and pressure. All of these factors are variable throughout the Earth's crust. Suppose that the difference between $\bar{\rho}_1$ and $\bar{\rho}_2$ is only 1%, such that $\bar{\rho}_1 = 1010$ kg-m^{-3}, and $\bar{\rho}_2 = 1000$ kg-m^{-3}. Then

$$\Delta h = \frac{10\ (\text{kg-m}^{-3}) \times 5000\ \text{m}}{1005\ (\text{kg-m}^{-3})} = 50\ (\text{m}) \quad \textbf{(7.23)}$$

The Darcy Velocity of the horizontal fluid flow between these two points is (from equation 2.55)

$$q = \frac{-k\bar{\rho}g}{\mu} \nabla h \quad \textbf{(7.24)}$$

where k (m^2) is permeability and μ is fluid viscosity ($\text{kg-m}^{-1}\text{-s}^{-1}$). If we assume a typical crustal permeability of 10^{-15} m^2, fluid viscosity of 10^{-3} $\text{kg-m}^{-1}\text{-s}^{-1}$ (pure water at 150 °C, equation 4.23), then

$$q = \frac{10^{-15}\text{m}^2 \times 1005\ (\text{kg-m}^{-3}) \times 9.8\ (\text{m-s}^{-2})}{10^{-3}\ (\text{kg-m}^{-1}\text{-s}^{-1})} \times \frac{50}{10000}\frac{\text{m}}{\text{m}} \quad \textbf{(7.25)}$$

$$q = 4.9 \times 10^{-11}\ (\text{m-s}^{-1}) \times 3.156 \times 10^{7}\ (\text{s-yr}^{-1}) = 0.0015\ (\text{m-yr}^{-1}) \quad \textbf{(7.26)}$$

Assuming a typical porosity of 10%, the linear velocity is ten times higher than the Darcy velocity (equation 2.19), thus

$$v = 0.015\ (\text{m-yr}^{-1}) \quad \textbf{(7.27)}$$

This may not seem like much, but the time it takes to completely exchange the fluid between point 1 and point 2 is now

$$\text{time} = \frac{\text{distance}}{\text{velocity}} = \frac{10000\ (\text{m})}{0.015\ (\text{m-yr}^{-1})} = 6.7 \times 10^{5}\text{yr} \quad \textbf{(7.28)}$$

In other words, there is a complete fluid exchange between points 1 and 2 every 670,000 years. Over an appreciable span of geologic time (10^8 yrs), there consequently can be hundreds of complete fluid exchanges in the crust as a consequence of relatively small density gradients associated with slight lateral variations in salinity and temperature.

The complexity of the situation becomes apparent when we consider that the moving fluid itself carries heat and changes salinity as it interacts with solid rock components. Thus, even in the absence of external driving forces for fluid flow such as topography or sediment compaction, the fluid regime of the continental crust may be quite dynamic when viewed from a chronological perspective that is geologic.

Bethke (1989 p. 132) also pointed out that a consideration of buoyancy forces (equation 2.6) implies that vertical and lateral flow must take place anytime a non-zero lateral fluid-density gradient exists. It is thus evident that density-driven flow is probably both pervasive and perpetual throughout the upper continental crust, although its magnitude may be small in comparison to topographically driven flow. The cumulative effect of low rates of mass transport over geologic time may, as Bethke (1989) suggested, go a long way towards explaining diagenetic alteration in basin sediments. But, fluid motion probably does not occur if a fluid-density gradient falls below a critical value; there must be a lower limit to the physical applicability of Darcy's Law. That is, there must be a point at which the inertial and frictional forces that are usually neglected are large enough to prevent any fluid movement. Exactly where this point occurs is unknown at the present time, and more work in this area is needed (see section 2.6).

7.3.2. Thermal Free Convection—Rayleigh Analysis.

Thermal free convection occurs as a result of a fluid-density inversion due to thermal expansion (review section 4.6.3).

Consider a horizontal porous medium situated between two impermeable boundaries. "Horizontal" here means the boundaries are *exactly* perpendicular to the gravity vector. If it is assumed that salinity is constant and the fluid incompressible, then free convection will occur whenever the Rayleigh number exceeds a critical value. The **Rayleigh number** is a dimensionless ratio that determines the onset of thermal free convection in an idealized situation. The Rayleigh number (R_a, dimensionless) for a porous medium situated between two impermeable, isothermal boundaries is (Turcotte and Schubert, 1982, p. 404)

$$R_a = \frac{\alpha_T g \rho^2 C k y^2 \gamma}{\mu \lambda} \quad \textbf{(7.29)}$$

where α_T (K^{-1} or $°C^{-1}$) is the coefficient of thermal expansion for a fluid, g (m-s^{-2}) is the acceleration due to gravity, ρ (kg-m^{-3}) is fluid density, C (J-kg^{-1}-°K^{-1}) is fluid specific heat capacity, k (m^2) is permeability, y (m) is height of the porous medium between boundaries, or cell height, γ (K-m^{-1}) is the thermal gradient, μ (kg-m^{-1}-s^{-1}) is the fluid dynamic viscosity, and λ (W-m^{-1}K^{-1}) is the thermal conductivity of the saturated porous medium.

The critical value of R_a at which convection begins in the porous medium described above is $4\Pi^2 \approx 40$ (Turcotte and Schubert, 1982, p. 405). If we apply a Rayleigh analysis to the continental crust, we may ask: for what value of permeability does convection begin? Rearranging equation 7.29,

$$k = \frac{40\mu\lambda}{\alpha g \rho^2 C y^2 \gamma} \quad \textbf{(7.30)}$$

Let $\alpha_T = 10^{-3}$ K^{-1}, $g = 9.8$ m-s^{-2}, $\rho = 1000$ kg-m^{-3}, $C = 4200$ J-kg^{-1}-K^{-1}, $y = 5000$ m, $\gamma = 0.025$ K-m^{-1}, $\mu = 5 \times 10^{-4}$ kg-m^{-1}-s^{-1}, and $\lambda = 2.5$ W-m^{-1}-K^{-1}. We then find that

$$k = 1.9 \times 10^{-15} \text{ m}^2 \quad \textbf{(7.31)}$$

Compilations of data by Brace (1984) and Clauser (1992) show that average permeabilities of crystalline rocks in the continental crust on the kilometer scale are of the order 10^{-15} m^2; thus convection is apparently feasible over a representative range of thicknesses and thermal gradients. This analysis does not consider, however, the increase of fluid density with depth that occurs as salinity increases. Data from sedimentary basins show that salinity increases with increasing depth, although the rate of increase tends to decrease as depth increases (see Chapter 4). Similarly, salinity data from relatively shallow boreholes (~1600 m, maximum depth) drilled in crystalline rocks of the Canadian Shield found that the total dissolved solids content of the crystalline crust increased exponentially with increasing depth to an average value ~150 milligrams per liter at 1500 m depth (Frape and Fritz, 1987). This salinity gradient, however, must decrease at greater depths, because in situ fluids would soon become saturated. For example, extrapolating the rate of increase observed from 0-1500 m to a depth of 2000 m yields a total dissolved solids ~400 milligrams per liter. In the presence of a flat salinity gradient it is conceivable that the geothermal gradient may be sufficiently high to lead to a density inversion and convective overturn if the average crustal permeability is high enough.

If convective overturns occur on a crustal scale, they must be temporally rare. Lachenbruch and Sass (1977) showed that the sudden onset of convection in a formerly quiescent layer dramatically cools the area through which fluid is circulating and also conductively mines heat from below the convection cell. The cooling continues until insufficient heat is available to drive the convective process, or the background flux of heat from the Earth is sufficient to maintain a quasi steady-state circulation system (as it does in geothermal areas). Once a convection cell shuts down, it may take tens of millions of years for the lithosphere to regain lost heat as the thermal recharge must occur conductively.

The inherently catastrophic nature of convective episodes has led some authors to speculate that free convection in the crust may be responsible for unexplained phenomena that apparently require large amounts or sudden releases of heat. Deming (1992) suggested that episodic releases of heat from free convection triggered by orogeny may have played a role in the formation of Mississippi Valley-type lead-zinc ores in the North American midcontinent. Nunn (1994) speculated that what appears to be episodic thermal subsidence recorded in the stratigraphy of intracratonic sedimentary basins (e.g., the Michigan Basin) could be due to catastrophic releases of heat from recurrent episodes of free convection.

7.3.3. Thermal Free Convection in Tilted Media.

Rayleigh-type analyses (see section 7.3.2 above) assume convection between two plates that are exactly parallel to each other and perpendicular to

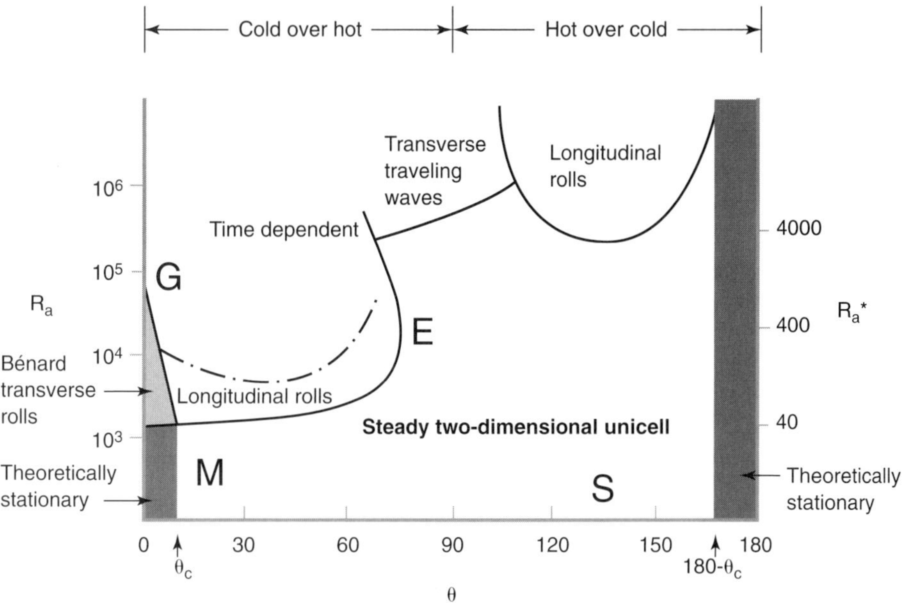

Figure 7.17 Convective flow regimes in a tilted porous medium as a function of tilt angle (Θ) and Rayleigh number (R_a).
(From Criss and Hofmeister, 1991, p. 200.)

the gravity vector, however, in many geologic situations these assumptions are violated (Criss and Hofmeister, 1991). In porous media that are tilted with respect to the gravity vector by more than a small critical angle (about ~5°), convection *always* occurs, regardless of the Rayleigh number, for any non-zero permeability (Figure 7.17). In media tilted by more than the critical angle, convection occurs even if the upper fluids are hotter than the bottom fluids. The magnitude of heat transported by free convection under these circumstances may be of the same order as that due to conduction. Criss and Hofmeister (1991) suggested that crustal intrusions and subduction zones are specific examples of geologic situations where convective circulation may be common although its presence would not be indicated by a Rayleigh-type analysis. In general, the ubiquitous presence of convection in tilted porous media implies that fluid circulation will be universally present in the tectonically active parts of the Earth's crust.

7.4. SEISMIC PUMPING.

The concept of seismic pumping was originally developed by Sibson et al. (1975) to describe how fluids could be drawn into a fractured rock mass as tectonic strains increased, and then released or literally pumped out as an earthquake occurred and the strain released. In this text we adopt a more relaxed definition of **seismic pumping,** defining it to be any fluid movement that occurs in response to a change in tectonic stress and/or strain.

7.4.1. Stress and Strain.

Let us first review the definitions of stress and strain. **Stress** is force per unit area; it is a pressure. **Strain** is a deformation; a change in length, area, or volume, that takes place in response to a change in stress. The essential idea behind seismic pumping is simple. The upper crust is a porous medium that contracts and expands in response to changes in stress. As the crust contracts or expands, it expels or absorbs fluids. Fluid-filled spaces in the Earth's crust may be large fractures, pores between rock grains, or microfractures. If a compressive stress increases, pores and fractures close and fluids are expelled. If an extensional stress increases, pores and fractures expand and fluids are taken up. The situation is analogous to a sponge. If a sponge is squeezed, fluids are expelled, however, if the external stress on a sponge is reduced, fluids are absorbed.

There is extensive evidence for the existence and efficacy of seismic pumping. A **vein** is essentially a fracture or fault filled with ore minerals. The texture of the minerals that fill veins commonly indicate hundreds of episodes of ore deposition by moving hot fluids. The inferred intermittent nature of fluid movement along with the obvious association with faulting strongly suggest seismic pumping is the likely mechanism responsible for fluid movement. Sibson (1994) described the gold-quartz Mother Lode vein in the western Sierra Nevada foothills in California as exemplary evidence for large amounts of highly focused and intermittent fluid flow. In this area, veins averaging 1 meter in thickness have been traced for kilometers along strike and up to one kilometer in depth. Given the low solubility of quartz in water, more than 10^9 m^3 of groundwater (a cube of water 1 kilometer on a side) would had to have flowed through a vein for each kilometer of length.

7.4.2. Normal and Reverse Faulting.

In general, crustal fluids are *expelled* following earthquakes on normal faults in extensional terrains, and *taken up* following earthquakes on reverse faults in compressional terrains (Wood, 1994) (Figure 7.18). A **fault** is a fracture in the Earth's crust along which motion occurs; the motion must be in a direction parallel to the fault plane. A **normal fault** is a fault in an area that is undergoing extension. Normal faults usually occur at relatively high angles to the ground surface (60°). During an earthquake that nucleates on a normal fault, the two sides move *apart*. For an earthquake to occur, stress and strain must gradually increase over a period of time until they are suddenly released during the temblor. As stress builds up in an area dominated by extensional tectonics, strain is accommodated by increases in the volume of fractures and pores. Correspondingly, the volume of pore fluid stored in the crust increases. An earthquake then releases the built-up stress and strain, pores and fractures close, and fluids are expelled. Fluid expulsion may quietly increase well levels or river discharges for a few days, or it may actually result in spectacular geysers of groundwater discharging from the Earth. Following the 1983 magnitude 7.3 earthquake at Borah Peak, Idaho, groundwater fountained from a series of fissures parallel to the surface trace of the main earthquake fault (Wood et al., 1985). Another major earthquake on a normal fault occurred in southeast Montana in 1959. Following the 1959 Lake Hebgen earthquake, discharges increased dramatically in three nearby rivers even though there was no precipitation event (Figure 7.19). The total volume of water displaced from the crust in the vicinity of the 1959 Hebgen Lake earthquake was about 0.5 km^3.

A **reverse fault** is a fault in an area that is undergoing compression and shortening. Reverse faults normally occur at relatively low angles to the ground surface (30°). During an earthquake that nucleates on a reverse fault, the two sides move closer together. As stress builds up in an area dominated by compressional tectonics, pores and fractures are closed and pore fluids are gradually expelled over a period of years. When the built-up stress is relieved by an earthquake, pores and fractures expand and fluid is taken up by the crust. Wood (1994) described the August 31, 1896, earthquake in Rikuu, Japan, as an example of increased fluid storage following an earthquake on a reverse fault. In the case of the 1896 Rikuu earthquake, the surface was ruptured over a distance of 36 km with a maximum uplift of 3.5 meters. No changes in river flows were reported following the earthquake (which does not mean they did not occur). But, hot springs supplying bath houses at Oshuku, Tsunagi, and Osawa dried up, and there was a significant reduction in flow at the thermal springs at Namari and Yuda. All of the hot springs

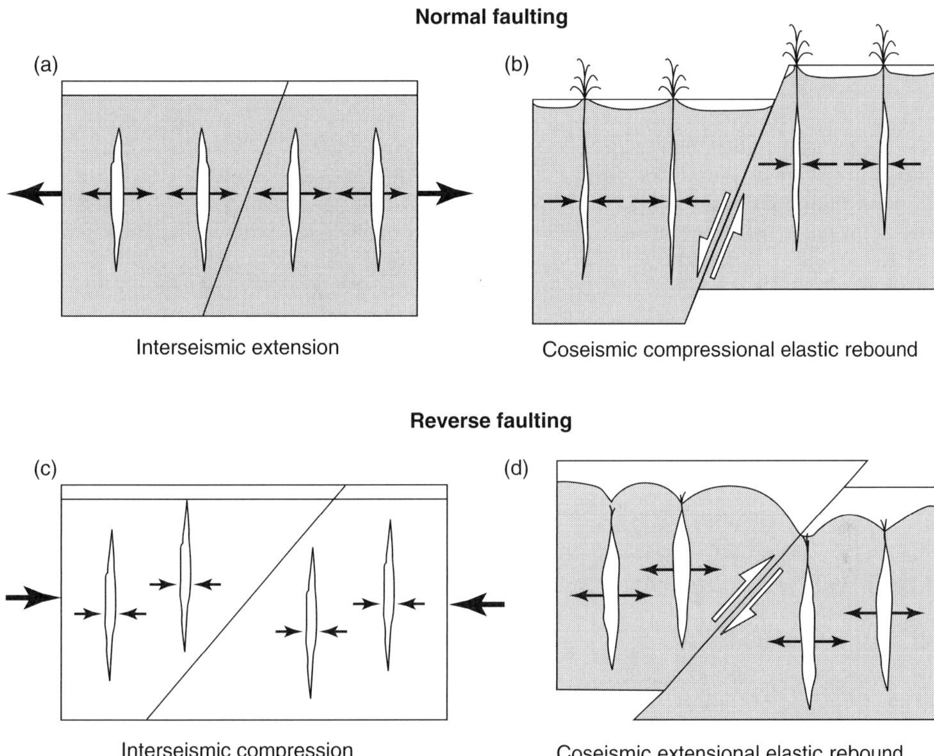

Figure 7.18 (a) In extensional terrains, porosity and fluid storage increase as strain builds up between earthquakes. (b) During and immediately following an earthquake on a normal fault, strain is released, porosity decreases, and fluids are expelled. (c) In compressional terrains, porosity and fluid storage decrease as strain builds up between earthquakes. (d) During and immediately following an earthquake on a reverse fault, strain is released, porosity increases, and fluids are absorbed.
(From Wood, 1994, p. 86.)

eventually resumed their normal flow rates, indicating that the flow had temporarily been diverted to fill up pore spaces and fractures opened after the release of compressive stress.

7.4.3. Dilatational Jogs.

Due to changes in tectonic stress fields and the mechanical properties of rocks, the surface traces of earthquake faults are usually not straight lines. Instead, fault traces are often offset at one or more **fault jogs** (Figure 7.20). Deformation and strain may be concentrated at these fault jogs, especially immediately following an earthquake. The strain built up over the entire length of a fault may be quickly transferred to a short jog segment. **Dilatation** is a volumetric expansion, and a **dilatational jog** is an area between two fault segments where extensional deformation may be concentrated during fault movement and an earthquake. The nearly instantaneous expansion that takes place here results in a dramatic drop in fluid pressure. The rush of fluids into a dilatational jog from surrounding areas may be so intense that rocks are fractured or brecciated. As fluids encounter the low pressures in the vicinity of a dilatational jog, boiling and ore deposition may result. Dilatational jogs act as suction pumps, and are often characterized by multiply

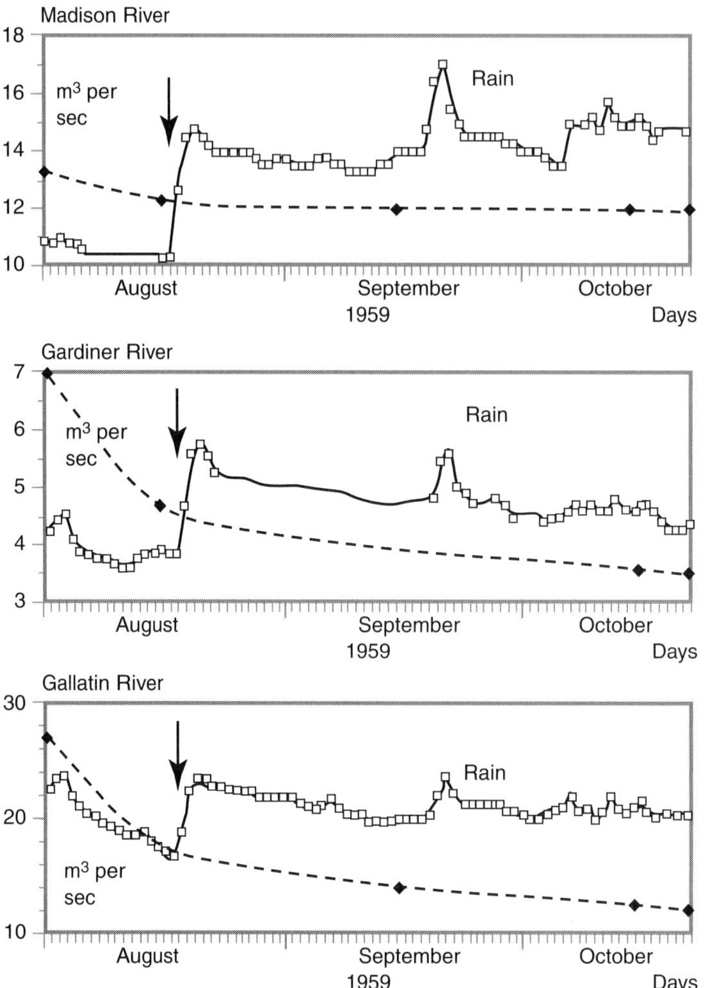

Figure 7.19 Stream discharge (thick line) as a function of time for three rivers in the vicinity of the 1959 Hebgen Lake earthquake in Montana. Stream flow increased immediately following the earthquake (arrow) even though there was no precipitation event. Dashed lines show average monthly discharge.
(From Wood, 1994, p. 87.)

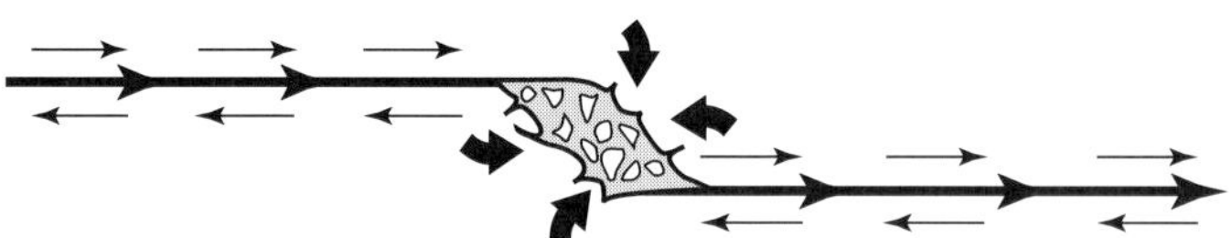

Figure 7.20 Arrows illustrate flow into a dilatational jog immediately following fault movement.
(After Sibson, 1987, p. 702.)

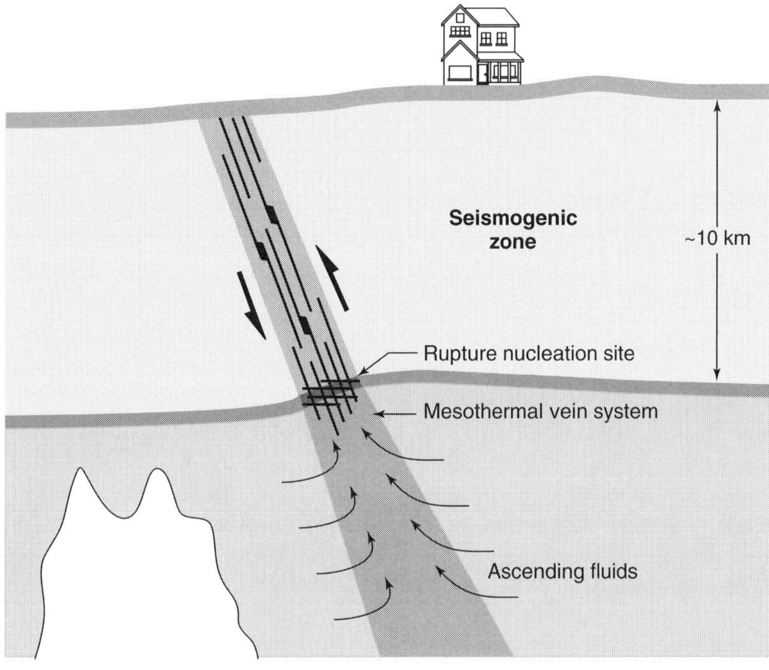

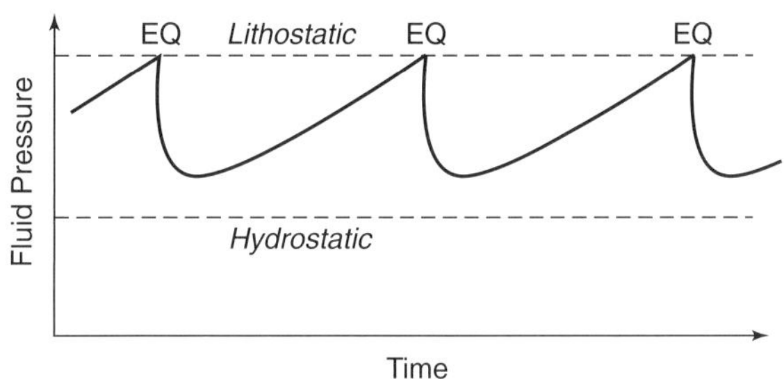

Figure 7.21 Schematic illustration (top) of fault-valve action wherein lithostatically-pressured fluids from the lower continental crust are released into the upper crust by rupturing related to a seismic event. Bottom figure shows cyclic nature of fluid pressure through time as repeated rupturing occurs.
(From Sibson, 1994, p. 80.)

re-cemented wallrock breccias created by repeated hydraulic implosions (Sibson, 1994).

7.4.4. Fault-valve Action.

Fault-valve action occurs when a fault or fracture ruptures the seal that maintains an area of fluid pressures above hydrostatic. When the seal is breached, fluids move out from the overpressured zone into the surrounding hydrostatic environment (Figure 7.21). One likely scenario for fault-valve action is the upper-lower crust interface. At depths of about 10 to 15 km the brittle-ductile transition occurs. The **brittle-ductile transition** is the zone where pressures and temperatures are sufficiently high that crustal rocks no longer deform by fracturing, instead they flow. As a result, there are no open fractures. Permeability is essentially zero, and fluids must bear the weight of overlying rocks. Thus the fluid-pressure gradient in the lower crust is lithostatic. The brittle upper crust is also known as the seismogenic zone, because faulting and earthquakes are largely confined to this region. In the lower crust, the stresses that lead to fracturing and earthquakes in the upper crust are resolved by plastic deformation.

Sibson (1994) suggested that the frequent presence of minor veins in fault zones developed during periods of tectonic activity is evidence that minor fault-valve activity is widespread throughout the Earth's crust. Fault-valve action may be responsible for many gold-quartz vein deposits that occur in steep reverse faults that may have once penetrated to the base of the upper crust.

7.5. Osmosis.

Osmosis is the tendency of water to diffuse through a semi-permeable membrane to make the concentration of water on one side of the membrane equal to that on the other side. A **semi-permeable membrane** is a material that neutral water molecules may move through, but charged ions such as Na^+ or Cl^- may not.

Consider a water tank divided into two halves by a semi-permeable membrane. One half of the tank contains a salt (NaCl) solution, the other contains freshwater. Water molecules will move through the membrane from the fresh side to the salty side in an attempt to equalize the salinity. As water moves from the fresh side to the salty side, the fluid pressure in the salty half will increase. The increased fluid pressure on the salty side will tend to drive water molecules back towards the fresh side of the tank. These two processes, osmosis and pressure-driven flow, will continue until a dynamic equilibrium is reached. Osmosis can lead to substantial pressure and head gradients. If the concentration of salt (NaCl) on the salty side is equal to that of seawater (about 34‰), the pressure differential between the salty and freshwater sides will be equal to the pressure generated by a column of water about 240 m in height (Duxbury and Duxbury, 1994, p. 156).

Reverse osmosis is the process of increasing the fluid pressure on the salty side of a semi-permeable membrane until water molecules move from the salty to the fresh side. Reverse osmosis is one way of desalinizing seawater, and is used in the country of Saudi Arabia. As of 1994, Saudi Arabia had 22 reverse-osmosis plants generating 230 million gallons of freshwater per day (Duxbury and Duxbury, 1994, p. 157).

In the Earth, compacted clays and shales can act as semi-permbeable membranes (Freeze and Cherry, 1979, p. 104; Neuzil, 2000). Historically, the significance of osmosis in geological environments is controversial, because the membrane properties of soils and rocks are poorly understood. But, a recent experiment reported by Neuzil (2000) suggests that osmosis may be a common means of generating anomalous pressures in the subsurface. Neuzil (2000) reported the results of a nine-year in situ study of osmotic pressure generation in the Cretaceous-age Pierre Shale in South Dakota. The shale at the study site had a relatively low permeability (10^{-20} to 10^{-21} m^2), and consisted of about 70–80% clay. Four boreholes were drilled into the Pierre Shale. Two boreholes were filled with a fluid whose salinity was close to the pore fluid in the surrounding rock, and two boreholes were filled with a fluid whose salinity was substantially higher. Experimental results showed the development of significant fluid-pressure differentials from osmosis. Boreholes penetrating the Pierre Shale that contained low-salinity water maintained more or less constant head and salinity, however, fluid levels rose in boreholes containing salty water, while salinities fell. These increases in fluid levels combined with decreases in salintiy were an indication that osmosis drove water from the surrounding Pierre Shale into two of the boreholes.

Extrapolation of Neuzil's (2000) results imply that osmosis is capable of generating fluid-pressure anomalies of the order of 20 MPa that could persist for tens of millions of years. The conditions necessary for the generation of anomalous pressures by osmosis may be common in the subsurface. Many sedimentary basins exhibit large contrasts in solute concentration. If low-permeability shales are hydraulically intact over large distances, it is possible that osmosis could explain the presence of anomalous fluid pressures that heretofore had been attributed to dynamic processes such as diagenesis, oil and gas generation, and tectonic forces (Neuzil, 2000).

REVIEW QUESTIONS

1. Define the following terms in the context of hydrogeology:
   - a. sedimentary basin
   - b. foreland basin
   - c. intracratonic basin
   - d. passive margin
   - e. rift basin
   - f. lithosphere
   - g. forearc basin
   - h. strike-slip basin
   - i. Anadarko Basin
   - j. aquathermal pressuring
   - k. artesian aquifer
   - l. hinge line
   - m. Rayleigh number
   - n. seismic pumping
   - o. stress
   - p. strain
   - q. vein
   - r. fault
   - s. normal fault
   - t. reverse fault
   - u. fault jog
   - v. dilatation
   - w. dilatational jog
   - x. fault-valve action
   - y. brittle-ductile transition
   - z. osmosis
   - aa. semipermeable membrane
   - bb. reverse osmosis

2. Explain the "wastebasket" model for compaction-driven flow (use a figure), and apply it to show that *average* flow rates due to pore collapse scale with sedimentation rates. What are average sedimentation rates for different types of sedimentary basins? Compare your estimates of average flow rates from sediment compaction with those of topographically driven flow and comment on the relative efficacy of these mechanisms for processes that involve heat and/or mass transport such as oil migration and the formation of hydrothermal ore deposits.

3. Starting with the equation of state for fluid compressibility (equation 7.4), derive an expression that gives the ratio of fluid pressure changes with time due to pore collapse and aquathermal pressuring. Compare the magnitude of pore collapse and aquathermal pressuring as driving forces for fluid flow in a "typical" sedimentary basin.

4. Read the paper by Bethke (1985) on compaction-driven flow in intracratonic basins and answer the following questions. Don't get hung up on the complexity of the mathematics; concentrate on the results and their importance for geologic processes such as oil migration and ore formation.
   - a. In what geologic processes is compaction driven flow thought to play a role?
   - b. What is a possible mechanism for concentration of brines in sedimentary basins, and how does it work?

c. Discuss the validity of one-dimensional models of compaction-driven flow in sedimentary basins. Under what circumstances are they good approximations to geologic reality?
d. Draw a schematic representation of Bethke's idealized intracratonic basin. Include a vertical and horizontal scale. Add head contours (you need not give specific values). Add representative Darcy velocity vectors and a scale. Explain why flow in part of the basin is downwards. In what part of the basin is fluid velocity the highest?
e. Discuss the effect of permeability on both compaction-driven and topographically driven fluid flow.
f. What is aquathermal pressuring? Is it important in generating excess pressures within slowly subsiding sedimentary basins?
g. Is compaction-driven flow a valid hypothesis for explaining oil migration in the Williston Basin? Why or why not?
h. How is it possible to get downwards flow from sediment compaction and pore collapse?
i. In what part of an intracratonic basin would the primary direction of fluid flow due to sediment compaction be vertical upwards?

5. Read the paper by Freeze and Witherspoon (1967) and answer the following questions.
a. What equation did Freeze and Witherspoon solve? Did they derive an analytical or numerical solution? Why did they use that particular solution method (analytical vs. numerical)?
b. What are the parameters that control regional-scale topographically driven flow? What is a "regional" scale?
c. Show how a buried lens of high permeability material can affect regional flow in an otherwise homogeneous medium. Draw a flow net and show vertical and horizontal scales.
d. You find a natural spring. Draw a cross-section with topography to show how you could be in either the discharge or recharge area of a *regional* flow system (draw a flow net on your cross-section).
e. Test whether or not Freeze and Witherspoon's computer solutions are correct by seeing if the refraction of flow lines in Figure 7.15c is consistent with theory.
f. In Figure 7.15c, where would artesian flow occur if a well penetrated the basal aquifer?

6. (a) Write down an expression for the Rayleigh number in saturated porous media. Define each term, and give its SI units. (b) What is the Rayleigh number (what does it tell us)? What is the critical value (give a number)?

7. Read the paper by Criss and Hofmeister (1991) and answer the following questions.
a. What relevant assumptions inherent in a Rayleigh number analysis are likely to be violated in geologic situations? Give two geologic examples.
b. According to Criss and Hofmeister (1991) free convection *always* occurs when what criterion is satisfied (give a number)?

8. Does an earthquake that occurs on a normal fault tend to release or take up fluids from crustal rocks? Is the situation different for an earthquake that occurs on a reverse fault?

9. Salt water is poured into an uncased well, which is then sealed at the surface. How will the salinity of the well water change over time? What factors determine any change that takes place?

SUGGESTED READING

Bethke, C. M. 1985. A numerical model of compaction-driven groundwater flow and heat transfer and its application to the paleohydrology of intracratonic basins. *Journal of Geophysical Research,* 90: 6817–6828.

Criss, R. E. and Hofmeister, A. M. 1991. Application of fluid dynamics principles in tilted permeable media to terrestrial hydrothermal systems. *Geophysical Research Letters,* 18: 199–202.

Freeze, R. A., and Witherspoon, P. A. 1967. Theoretical analysis of regional groundwater flow. 2. Effect of Water-table configuration and subsurface permeability variation. *Water Resources Research,* 3: 623–34.

Wood, R. M. 1994. Earthquakes, strain-cycling and the mobilization of fluids in *Geofluids: Origin, Migration and Evolution of Fluids in Sedimentary Basins,* ed. by J. Parnell, Geological Society Special Publication No. 78, p. 85–98.

Notation Used in Chapter Seven

Symbol	Quantity Represented	Physical Units
A	area	m^2
α_T	coefficient of thermal expansion	K^{-1} or $°C^{-1}$
B	fluid compressibility	$Pa^{-1} = m\text{-}s^2\text{-}kg^{-1}$
b	constant in porosity-depth equation	m^{-1}
C	specific heat capacity	$J\text{-}kg^{-1}\text{-}K^{-1}$
$\phi, \bar{\phi}$	porosity, mean porosity	dimensionless
g	acceleration due to gravity	$m\text{-}s^{-2}$
γ	geothermal gradient	$K\text{-}m^{-1}$
$h, \Delta h$	head, change in head	m
k, k_x and k_z	permeability, permeability in x- and z-directions	Darcy = 10^{-12} m^2
K_x, K_z	hydraulic conductivity in x- and z-directions	$m\text{-}s^{-1}$
L	length in Toth solution	m
λ	thermal conductivity	$W\text{-}m^{-1}$ K^{-1}
M	fluid mass	kg
μ	water viscosity	$Pa\text{-}s = kg\text{-}m^{-1}\text{-}s^{-1}$
‰	parts per thousand	dimensionless
P, ΔP or dP	fluid pressure, change in fluid pressure	Pascal (Pa) = $kg\text{-}m^{-1}\text{-}s^{-2}$
θ	angle that characterizes slope of water table in Toth solution	dimensionless
q	Darcy velocity or specific discharge	$m\text{-}s^{-1}$
R	sedimentation rate	$m\text{-}s^{-1}$
$\rho, \bar{\rho}$	fluid density, average fluid density	$kg\text{-}m^{-3}$
R_a	Rayleigh number	dimensionless
T, dT	temperature, change in temperature	K or °C
v	linear velocity	$m\text{-}s^{-1}$
V, ΔV or dV	fluid volume, change in fluid volume	m^3
x	distance	m
y	cell height	m
z	depth	m
z_0	elevation in Toth solution	m
∇	gradient operator	m^{-1}
∇^2	Laplacian, the second spatial derivative	m^{-2}

CHAPTER 8

ABNORMAL FLUID PRESSURES

8.1. OVERPRESSURES AND UNDERPRESSURES.

Abnormal fluid pressures are those that are above or below hydrostatic. **Hydrostatic** fluid pressures are those in which the fluid pressure at any depth is due to the weight of the overlying fluid (as defined by equation 2.23). Nominal hydrostatic fluid-pressure gradients are usually about 10.5 kPa-m^{-1} (0.465 psi-ft^{-1}). **Lithostatic** pressure is the pressure due to the weight of the entire overburden (fluid plus matrix). Fluid pressures generally cannot exceed lithostatic, as fluid pressures in excess of lithostatic cannot be contained by the total overburden weight. Because rocks have lateral strength, however, it is possible to find isolated occurrences of fluid pressures which are slightly in excess of lithostatic.

Fluid pressures below hydrostatic are termed **underpressures.** Fluid pressures in excess of hydrostatic are termed **overpressures** or **geopressures**. Overpressures in sedimentary basins tend to be more common than underpressures. Sedimentary basins with overpressures typically consist of hydrostatically pressured sediments extending from the surface to depths of 2 to 3 kilometers. The hydrostatic section is underlain by a transition interval, followed by a deep section of abnormally high fluid pressure and fluid-pressure gradients (Figure 8.1). There is considerable interest in understanding the origin and evolution of overpressures, as abnormally high pressures represent a hazard in drilling for petroleum. Under normal circumstances, high fluid pressures at depth are balanced by the weight of the drilling fluid in the wellbore. Drilling engineers usually tend to make the overall weight of the drilling fluid column slightly higher than is necessary so as to avoid a catastrophic blowout. If a zone of abnormally high fluid pressure is unexpectedly encountered, there is danger of a destructive blowout wherein formation fluids rush up the wellbore at great speeds.

8.2. STATIC VERSUS DYNAMIC HYPOTHESES.

As discussed by Bredehoeft et al (1994), there are two distinct schools of thought on the creation and maintenance of anomalous fluid pressures in the Earth. The **static school** (Bradley, 1975; Hunt, 1990; Bradley and Powley, 1994) believes that abnormal pressures, regardless of origin, are maintained by pressure seals (Hunt, 1990; Ortoleva,

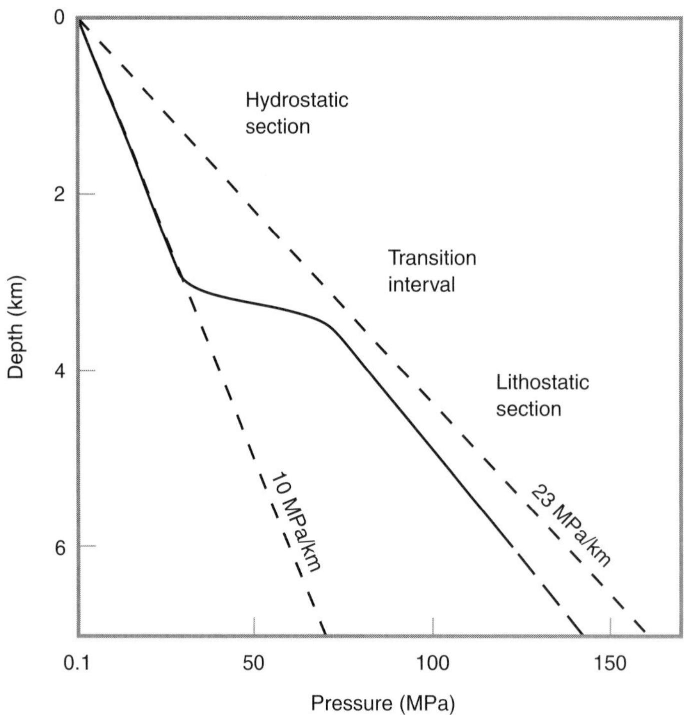

Figure 8.1 General trend of fluid pressure versus depth in basins with Gulf Coast-type geopressures. (From Bethke, 1986, p. 6536.)

1994a, b) (Figure 8.2). A **pressure seal** was defined by Hunt (1990, p. 2) as (emphasis added):

> . . . a zone of rocks capable of hydraulic sealing, that is, preventing the flow of oil, gas, and water. The term does *not* refer to capillary seals . . . the term refers to seals that prevent essentially *all* pore fluid movement over substantial intervals of geologic time.

Other authors have implicitly defined the term "pressure seal" more loosely by using it to refer to any low-permeability unit, however, as the above quote from Hunt (1990) shows, the original intention was to apply the term to units that essentially behave as if they have zero permeability over substantial intervals of geologic time (10^7-10^8 yrs).

Pressure seals are one aspect of a paradigm wherein it is thought that sedimentary basins have two superimposed hydrogeological systems: a shallow system characterized by hydrostatic pressures, and a deeper system consisting of a series of overpressured, hydraulically isolated pressure compartments (Hunt, 1990; Bradley and Powley, 1994) (Figures 8.3 and 8.4). A **pressure compartment** is a three-dimensional hydraulically isolated volume of the Earth's crust that has a fluid pressure different from the ambient surroundings. The role of the pressure seal is to maintain anomalous (above hydrostatic) pressures in the lower system by preventing the movement of fluid across compartment boundaries. Compartments may be breached by fracturing when fluid pressures exceed lithostatic, but the seal regains its integrity when the fluid pressure drops below lithostatic.

There are several difficulties with the pressure-seal pressure-compartment concept. One is the lack of a known geochemical/geologic mechanism to create pressure seals. Some authors have claimed that pressure seals exhibit distinct diage-

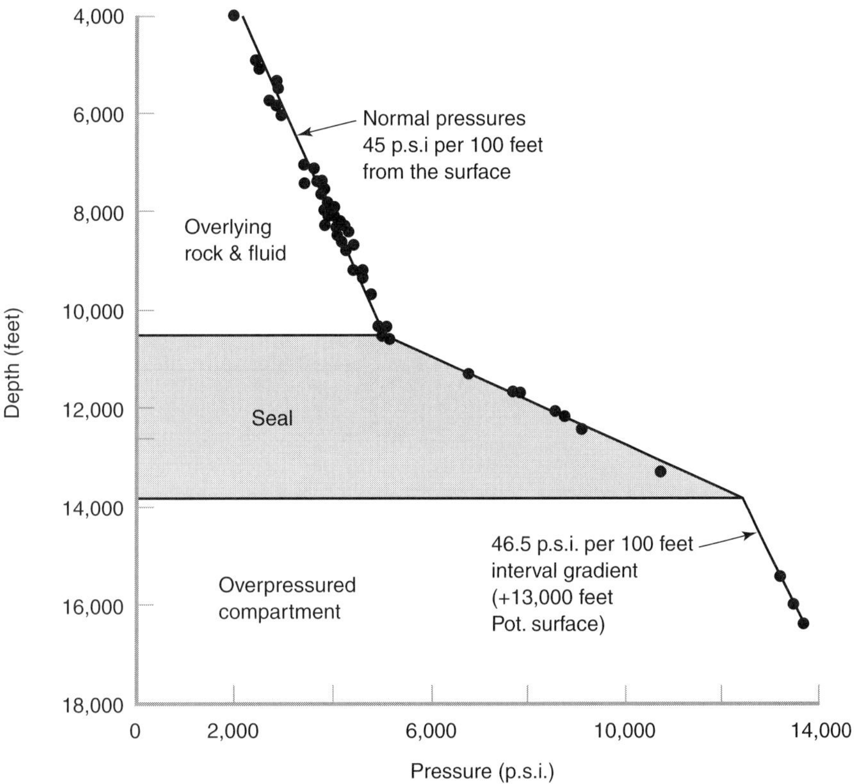

Figure 8.2 Fluid pressure as a function of depth, Cook Inlet fields, Alaska. Pressure seal is inferred to exist in region of high fluid pressure gradient.
(From Hunt, 1990, p. 6; after Powley, 1980.)

netic bands (Al-Shaieb et al., 1994), suggesting the possibility that diagenetic processes such as calcite precipitation in shale pore spaces may be an effective sealing mechanism. But, this idea remains largely untested. As Bradley and Powley (1994) noted, "*no* example of a directly measured, permeability-defined seal is known." The existence of pressure seals is inferred from differences in hydraulic potential measured across relatively permeable reservoir rocks. Even if top- and bottom-bounding pressure seals could come into existence through diagenetic processes, it is difficult to imagine what geologic features could function as lateral seals. Faults have been invoked as likely candidates, but faults tend to be conveniently invoked alternately as either seals or conduits for flow as suits the need of specific circumstances.

Unlike Hunt (1990) who explicitly specified that pressure seals are *not* capillary seals, Revil et al (1998) proposed that sealing could occur in sedimentary basins through the formation of capillary gas seals. A capillary force is an attractive force that exists between two different substances (see section 6.1.1). Rock pores and channels that exist between solid matrix grains will act as capillary tubes, drawing water into them and holding it there until forcefully displaced. Thus a fine-grained sedimentary rock saturated with water may act as a layer of zero permeability to prevent the entry of fluids such as oil or gas unless the capillary force can be overcome. If alternating layers of fine- and coarse-grained sediments are present, gas may accumulate in the coarse-grained sediments with capillary forces preventing it from entering into adjacent

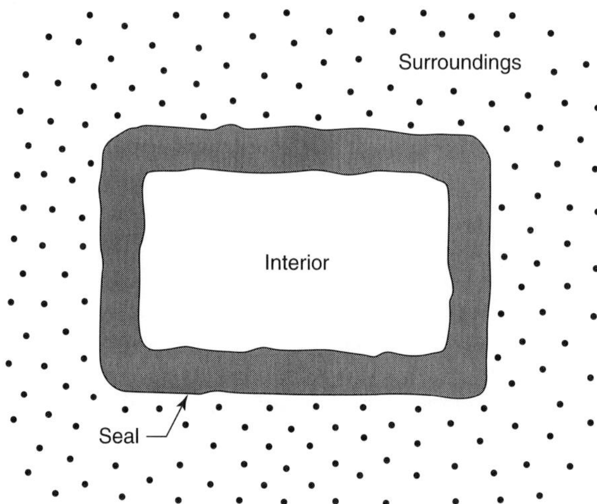

Figure 8.3 Conceptual model of generic pressure compartment. Interior of compartment is hydrostatically pressured, and is of relatively high-permeability. The interior is separated from its surroundings by a pressure seal.
(From Ortoleva, 1994b, p. 40.)

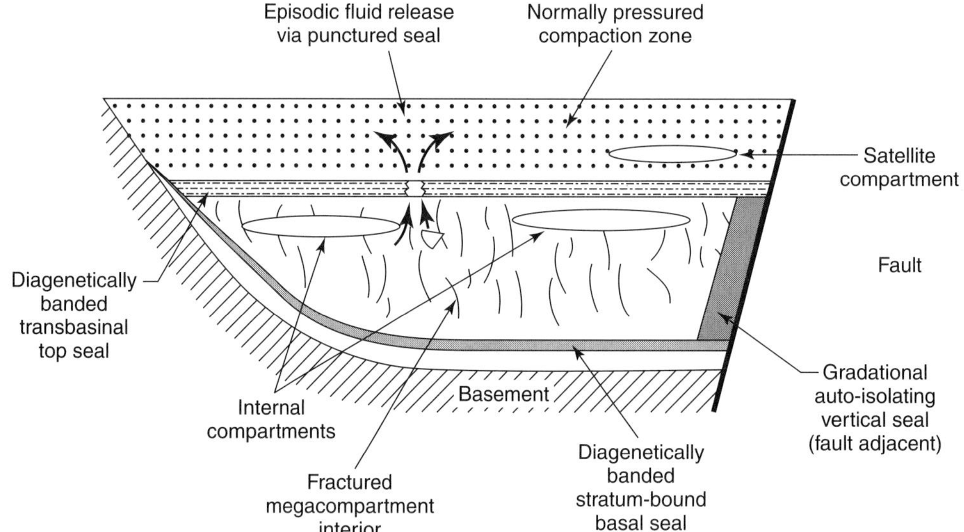

Figure 8.4 Pressure compartment/seal paradigm of sedimentary basin hydrogeology showing different levels of compartmentalization.
(From Ortoleva, 1994b, p. 44.)

water-saturated fine-grained rocks. The immobile gas thus constitutes a seal of zero permeability—unless the hydraulic gradient is large enough to overcome the capillary force.

A hypothesis that invokes sealing by gas capillary forces has several advantages. As gas generation also tends to produce abnormally high fluid pressures, the hypothesis is parsimonious in that it simultaneously provides a mechanism for both generating and maintaining anomalous fluid pressures. Capillary sealing also provides a mechanism for achieving zero permeabilities without

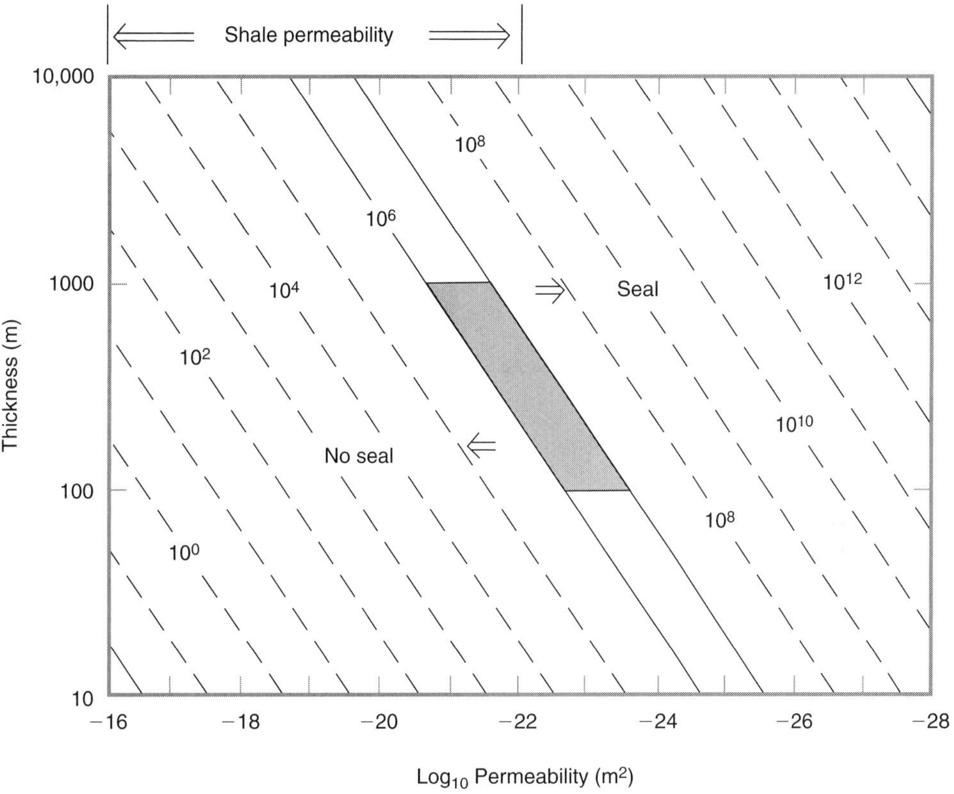

Figure 8.5 Maximum time (in years) over which a layer of given thickness and permeability may confine excess pressures. Shaded area indicates approximate permeability required to sustain a 100 to 1000-m-thick seal over geologic time (about 1 Ma).
(From Deming, 1994a, p. 1008.)

contradicting established dynamic paradigms, which maintain that aquicludes are rare to nonexistent. For example, Neuzil (1994) reviewed the permeability of shales and clays and found that most argillaceous media have permeabilities greater than 10^{-20} m^2, too high by three to four orders of magnitude to preserve abnormal fluid pressures for 100 Ma. Capillary sealing also explains how pressure compartments can be sealed in all three dimensions. Gas follows fluids along the path of least resistance until all escape routes are plugged. Finally, sealing by capillary forces can produce the type of compartmentalization apparently observed in basins such as the Anadarko Basin in southwest Oklahoma. The existence of levels of pressure compartmentalization is difficult to explain unless some type of physical or chemical sealing-mechanism is invoked.

Deming (1994a) quantified the conditions under which pressure seals may retain abnormal pressures by calculating the characteristic time a seal of a specified thickness and permeability may confine a pressure transient (review section 5.5). He found that to confine abnormal pressures for more than 10^8 yrs with a seal 100 to 1000 m thick would require seal permeabilities in the range of 10^{-23} to 10^{-25} m^2 (Figure 8.5). This range is near or below the lowest extreme of measured shale permeabilities (Figure 3.14). It is thus difficult to maintain abnormal fluid pressures over geologic time without either pressure seals or the influence of an active and ongoing geologic process to offset

John D. Bredehoeft: Fluids in Geologic Processes

by L. F. Konikow and P. A. Hsieh

U.S. Geological Survey, Menlo Park, California

John D. Bredehoeft (born in 1933) is a quantitative hydrogeologist who has consistently pushed outward the frontiers of the science. He is still active and has a long productive history of accomplishments, which have won him many awards and international recognition. He has a talent for identifying critical problems, simplifying each down to a tractable question, and then deriving and publishing a solution that has great implications and transfer value. Some of his work has been earthshaking. Literally!

John received his undergraduate education at Princeton University, where his major was geological engineering. His graduate studies were at the University of Illinois, where he received an M.S. and Ph.D. in Geology. Reflecting on his early career, John remarked (after receiving the Horton Medal from the American Geophysical Union in 1997),

> I was lucky to go to the University of Illinois, where my major professor and mentor was Burke Maxey. He instilled in those of us who were associated with him a demand for excellence. Upon receiving my Ph.D. in the early 1960s, I was lucky to go to work at the U.S. Geological Survey. I arrived at a time when I could apprentice with some of the best professionals engaged in the study of groundwater.

John's remarks clearly reflect the generosity of his outlook on doing scientific work. He shares credit and he shares ideas. And those who have been lucky enough to be mentored by John know that his ideas are usually gems. Many recipients of his generous support and encouragement of young scientists have carried on the tradition of excellence of work and generous sharing of ideas.

John's work often reflects a multidisciplinary approach to solving difficult problems. He has made important advances linking groundwater hydrology with geophysics, geochemistry, tectonics, petroleum engineering, economics, and numerical methods. John's interests and work have great breadth and depth, as they extend beyond the purely technical realm of science and engineering into the social realms of management of natural resources, the management and administration of research organizations, and even philosophy of science.

John D. Bredehoeft

A common thread running through much of John's work is his interest in the role of fluids in geologic processes. John's first publication, in 1963, was the first quantitative examination of membrane filtration in the subsurface. His 1968 papers on anomalous fluid pressure were the first cogent analyses of geologic processes as hydrologic driving forces and the first recognition of anomalous pressures as hydrodynamic transients. His 1967 paper on the response of aquifers to earth tides is extensively cited as the seminal paper on that topic. His analysis of thermal profiles for estimating groundwater flow rates is elegantly simple, yet has proven to be of tremendous utility. He was among the first to use hydraulic fracturing as a method for determining the state of stress in the subsurface. In 1970, many geologists barely recognized the existence, let alone the importance, of subsurface fluids. That is no longer true today, and geologists in great numbers are now looking at how groundwater controls or influences ore deposits, hydrocarbon reservoirs, tectonic processes, volcanic events, and almost every other sub-field of geological and geophysical sciences.

During the time that John was creating and applying new computer simulation models of groundwater processes, he was also "shaking things up" as a participant in the well-known Rangely, Colorado, experiments (where earthquakes were created and controlled by high-pressure fluid injection). He followed up on this by contributing to the Parkfield, California, earthquake studies where he was a proponent of using water wells as strain meters to monitor deformation of the earth near faults, partly in search of potential earthquake precursors.

In the realm of groundwater systems analysis, John has made several fundamental contributions to methods of well test analysis. He was instrumental in the development of the rigorous theory of slug tests, now one of the basic tools of the field hydrogeologist. He also extended the slug test technique to solve the difficult problem of field measurement of very low permeabilities.

Most practicing hydrogeologists today routinely apply computer simulation models to help them understand and solve the particular problem being addressed. They all owe a debt of gratitude to John Bredehoeft, who helped pioneer the development and application of digital simulation of groundwater systems when most hydrologists were still using analog models. His papers, particularly those co-authored with George Pinder, were widely recognized as standard references in groundwater model analysis. Many model developers built upon the basics that John laid out, and many of today's flow and transport modelers use programs based on his work. In the early 1970s, John was among the few who saw the significance and pervasiveness of groundwater contamination problems: this was a motivating factor for his pushing strongly for the development and application of solute-transport models. Less than 20 years later, dealing with groundwater contamination had become a multi-billion dollar a year industry.

John wanted society to benefit from tax-supported research. His interest in promoting a "practical payoff" of science is illustrated in the area of groundwater management, where he showed how economic theories can be applied in light of realistically variable hydrogeologic conditions to develop policies for water allocation or development of groundwater resources. John analyzed topics such as groundwater depletion, conjunctive use, and artificial recharge. In countering what he considered a common "groundwater myth," he demonstrated the fallacy of basing groundwater management rules (such as restrictions on pumpage) on computed water budgets (or recharge rates) for conditions prevailing prior to development.

John is not only a leading scientist, but a leader of scientists. John served for a number of years as Chief of the National Research Program of the Water Resources Division (WRD) of the USGS, which at that time employed close to 300 scientists and engineers. In this position, John substantially increased the relevance and visibility of this hydrologic research program. He later served for several years as the Regional Hydrologist for the operational program in the eight-state Western Region of WRD. What is perhaps the most amazing feat is that John remained a productive scientist and researcher during the years he served as a manager.

For Additional Reading

Freeze, R. A., Bredehoeft, J. D., and Pinder, G. F. 1976. Presentation of the O. E. Meinzer Award to John D. Bredehoeft and George F. Pinder, Citation and Responses. *GSA Bulletin,* 87 (8): 1212–1213.

Konikow, L. F., and Bredehoeft, J. D. 1998. Bredehoeft Receives Robert E. Horton Medal, Citation and Response. *EOS, Transactions, American Geophysical Union,* 79 (8): 101.

Wolff, R. G., and Bredehoeft, J. D. 1997. Penrose Medal presented to John D. Bredehoeft, Citation and Response. *GSA Today,* 8 (3): 13–14.

the natural tendency for pressure equalization. If pressure seals exist, it thus seems likely that some type of active geochemical or capillary mechanism is necessary. Otherwise, rock permeabilities are simply too high to allow high fluid pressures to exist in the Earth's crust over geologic time.

The **dynamic school** embraces the classical hydrogeologic tenet that "there are no totally impermeable geologic materials" (Bredehoeft et al., 1982, p. 297) (with, of course, rare exceptions for materials such as permafrost and salt). Pressure seals simply do not exist in this paradigm. Abnormal formation pressures may be caused by a transient disturbance related to some ongoing geologic process (e.g., rapid sedimentation), or an equilibrium process such as topographically driven fluid flow (Tóth, 962; Neuzil, 1995). The existence of abnormal pressures is seen as an indication that the rate of pressure generation (either positive or negative) is sufficiently high so as to maintain abnormal pressures in the presence of low-permeability rocks for substantial periods of geologic time.

8.3. Causes of Abnormal Fluid Pressures.

Hypothetical mechanisms for the creation of abnormal fluid pressures in the Earth's crust can be either steady-state or transient. Generally speaking, there is only one steady-state mechanism, but inumerable transient processes that can lead to abnormal fluid pressures.

8.3.1. A Steady-state Mechanism.

Topographically driven flow is a steady-state mechanism that can lead to both under- and overpressures. As long as the topography and permeability remain unchanged, the pattern of under- and overpressures due to topographically driven flow will not be subject to pressure equalization by flow. In fact, it is the flow itself that is responsible for the abnormal pressures.

A simple example of abnormal fluid pressures due to topographically driven flow is provided by considering flow between an alternating series of hills and valleys (Figure 8.6). The level to which water will rise in a cased well open at the bottom is determined by fluid head at the bottom of the well. This is equal to the elevation of the water table at the point the head contour at that depth intersects the water table. Note that in areas of descending flow, head contours are bent concave downwards. As a result, the head contours that intersect the bottom of the well will not rise to the top of the well, and the water level in the well will be depressed below the top of the well. In the discharge area at the bottom of the hill, the opposite situation prevails. Head contours are bent convex upwards, and water levels in wells located here will rise above the top of the wells and artesian flow will occur.

What is the fluid pressure (P, kg-m^{-1}-s^{-2}) at the bottom of each well? It is simply equal to the product of fluid density (ρ, kg-m^{-3}), acceleration due to gravity (g, m-s^{-2}), and height of the fluid column (z, m) in the well ($P = \rho g z$, equation 2.23). Thus, in areas of high elevation and descending flow where the fluid level falls below the water table, the fluid pressures at depth are below hydrostatic. These regions are therefore *underpressured*. Conversely, wells in the valleys where flow is ascending are *overpressured*. Thus fluid is moving from regions of underpressures to regions of overpressures! This is a striking example of the importance of using head to characterize flow regimes instead of pressure. Consider, for example, the disastrous consequences of placing a hazardous waste dump in an underpressured area with active flow.

8.3.2. Transient Mechanisms.

Following Neuzil (1995), we can rewrite the diffusion equation (equation 5.35) as

$$K\nabla^2 h = S_s \frac{dh}{dt} + \Gamma \qquad \textbf{(8.1)}$$

where K (m-s^{-1}) is hydraulic conductivity, h (m) is head, ∇^2 is the Laplacian (m^{-2}), S_s is the specific storage (m^{-1}), t (s) is time, and Γ (s^{-1}) is a geologic forcing term that represents the geologic agent responsible for abnormal pressure generation. In general, transient processes responsible for

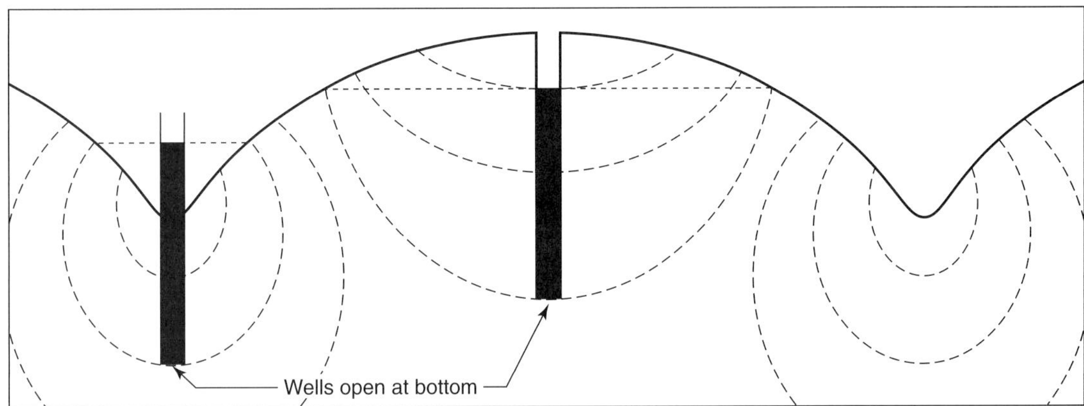

Figure 8.6 Cross-sectional plot of head (dashed lines) underneath a hill symmetrically flanked by valleys. Wells are present at the apex of the hill and the bottom of the valley as shown. The wells are open only at the bottom, so that the water level in each well is determined by head at the bottom of each well. Water rises in each well to the point at which head contours at the bottom of the wells intersect the water table at the surface. Fluid regime in recharge areas is underpressured; fluid regime in discharge areas is overpressured. Fluid thus flows from regions of underpressures to regions of overpressure.
(After Hubbert, 1940.)

abnormal pressure generation can be divided into three categories: (1) thermomechanical response of the fluid and matrix, (2) porosity changes due to stress changes and diagenesis/metamorphism, and (3) fluid sources and sinks (Neuzil, 1995, p. 748).

An example of abnormal pressures created by a thermal change in a fluid is aquathermal pressuring, that we considered earlier in section 7.1.3. Aquathermal pressuring is generally not as important as porosity changes, although the relative importance depends upon the local geothermal gradient and burial rate.

Underpressuring due to erosion is an example of abnormal pressure generation through the mechanical response of a porous medium to a change in stress (Toth and Millar, 1983). As erosion occurs, overburden weight decreases and the void spaces in a porous rock will tend to expand, just as they contracted in response to an increase in burial depth and effective stress. Generally speaking, a rock will not recover all of the porosity it had during its burial at the same depth. The failure to completely recover original porosity upon exhumation is an example of hysteresis. **Hysteresis** is the failure of a property that has been changed by an external agent to return to its original value when the cause of the change is removed. In this case, the external agent that causes a porous rock to compact is an increase in effective stress due to an increase in overburden weight from sedimentation. If and when the process reverses and the overburden is removed by erosion, porosity increases, but not all of the original porosity is recaptured.

Underpressuring can occur when the hydraulic diffusivity of a porous medium is so low that fluid flow cannot occur quickly enough to equalize head gradients created by pore space expansion. Pore space expansion is essentially synchronous with erosion, as stress is transmitted through rocks virtually instantaneously. The elastic response of the rock is also relatively fast, however, pressure equalization by fluid movement is a diffusive process whose rate is described by a characteristic time constant (equation 5.86) that is determined by the hydraulic diffusivity of a medium undergoing a change and its thickness.

Erosional unloading has been invoked to explain the existence of underpressures in Cretaceous shales of the Western Canadian Sedimentary Basin in Alberta, Canada (Parks and Toth, 1995).

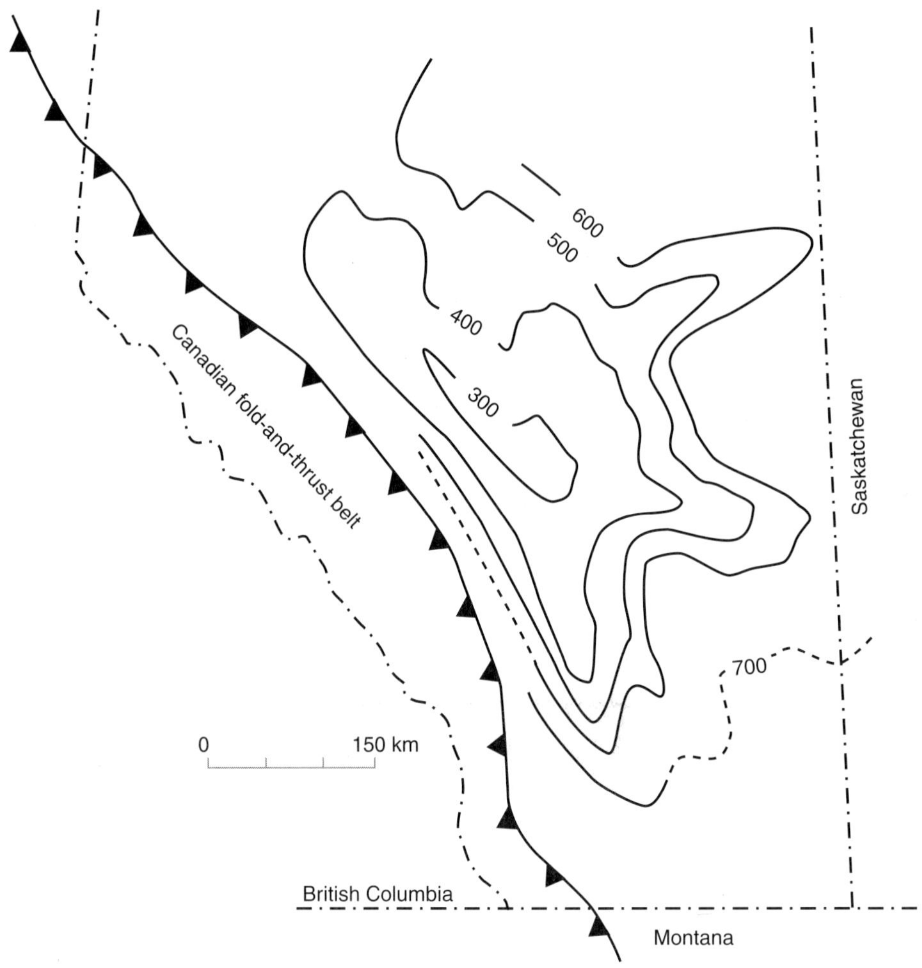

Figure 8.7 Fluid head (meters) in Lower Cretaceous rocks of the Western Canadian Sedimentary Basin. Closed minimum suggests the presence of a transient underpressure possibly related to pore expansion from erosion and unloading.
(From Corbet and Bethke, 1992, p. 7204, after Hitchon, 1969.)

The existence of head minima, which appear to be closed in three dimensions (Figure 8.7), suggests a disequilibrium condition that cannot be sustained over substantial periods of geologic time. Corbet and Bethke (1992) studied underpressures in this area and concluded that for the underpressures to be due to erosional unloading, shale permeabilities would have to be lower than 3×10^{-20} m^2. This is near the lower end of shale permeability, but not unusually low. In such situations it is difficult to show uniquely that a specific mechanism is responsible for an observed hydraulic phenomenon. The best that can usually be done is to delineate the circumstances under which a particular mechanism may operate, and then compare those findings with observational data.

Compaction disequilibrium is another example of a geologically important mechanism that is responsible for the existence of overpressures through porosity changes. Compaction disequilibrium occurs when the pore fluid sustains part of the matrix overburden weight, due to the failure

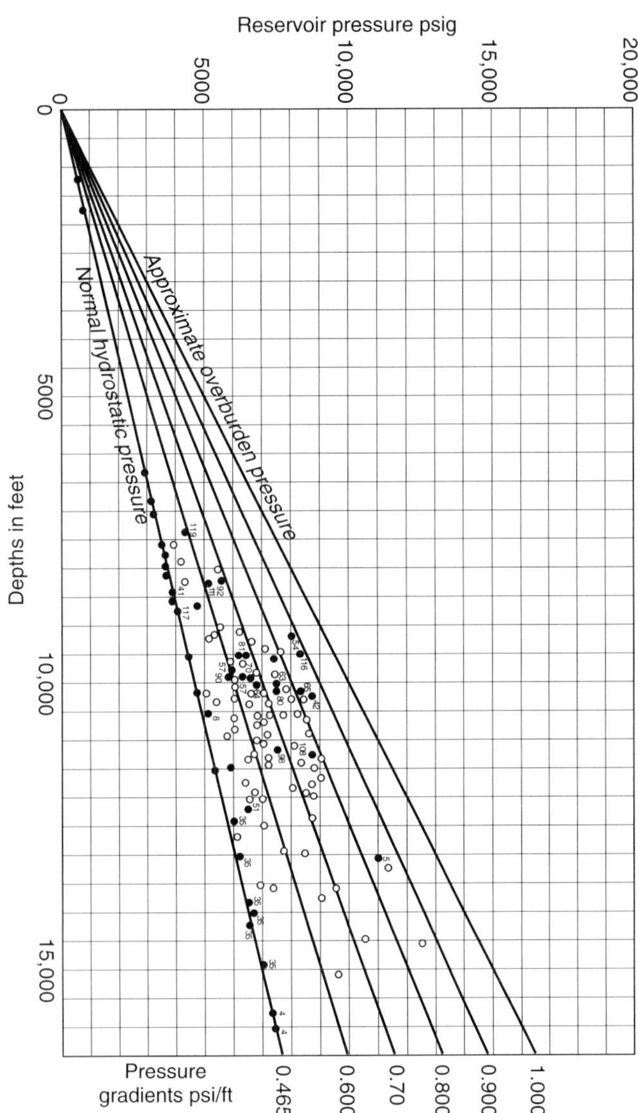

Figure 8.8 Reservoir fluid-pressure versus depth for Louisiana Gulf Coast wells. Solid circles correspond to measured pressures, open circles represent estimated fluid pressures.
(From Dickinson, 1953, 414.)

of sediments to reach equilibrium compaction conditions quickly enough. Porosity reduction is inhibited by the difficulty in expelling pore fluids from low-permeability shales and clay-rich sediments. The creation of overpressures by compaction disequilibrium requires high sedimentation rates and a predominance of low-permeability sediments or rocks.

Two examples of geopressures that are thought to be due to compaction disequilibrium are the Gulf Coast Basin of the southeast U.S. (Figure 8.8) and the South Caspian Basin of the former USSR (Bredehoeft and Hanshaw, 1968; Sharp and Domenico, 1976; Bethke, 1986; Bredehoeft et al., 1988; Mello et al., 1994). In the South Caspian Basin, sedimentation rates as high as ~1300 m-Ma^{-1} have resulted

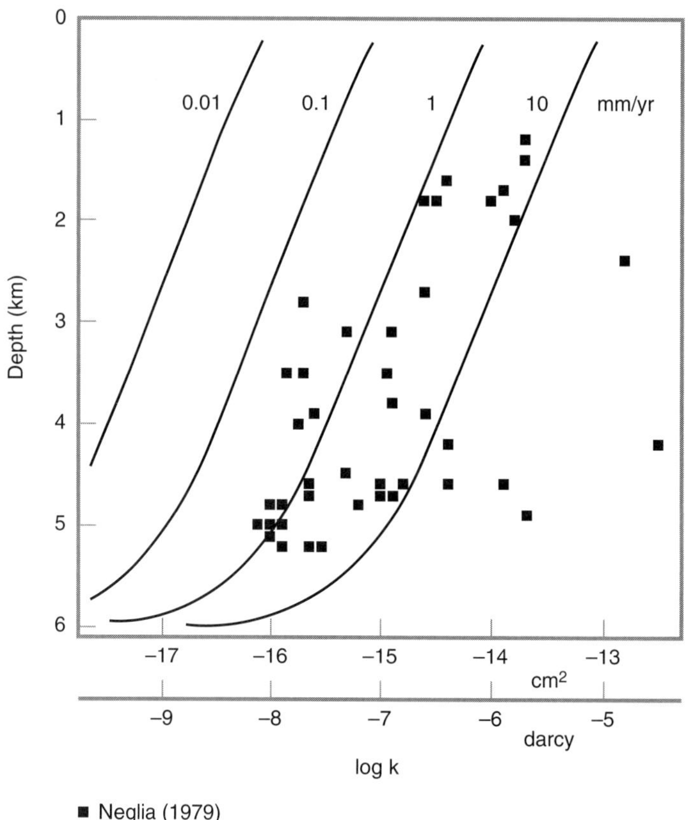

Figure 8.9 Permeability profiles required for overpressuring in a 6-km-deep sedimentary basin as a function of sedimentation rate in millimeters per year. Solid squares represent permeability measurements on sediments from the Gulf Coast basin as reported by Neglia (1979).
(From Bethke, 1986, p. 6538.)

in enormous thicknesses of sediments (25 km in deepest sections) and substantial overpressures (Bredehoeft et al., 1988). The Gulf Coast Basin in the southeast United States has been studied more intensively. Most rocks in the Gulf Coast Basin are shales overlain by deltaic systems containing alternating series of sands and shales grading vertically upward into massive sandstones (Mello, 1994, p. 2776). The sandstones are hydrostatically pressured; overpressures develop either in the mixed sands/shales, or in the underlying shale sequences. Overpressures result when pore fluids are not able to escape from low-permeability shales quickly enough to maintain a "normal" porosity-depth curve. The hydrological properties of the shales thus play a critical role in the maintenance and dissipation of overpressure in the Gulf Coast Basin, which contains more than 85% shale and shaley sediment (Bethke, 1986, p. 6539). For example, Bethke (1986, p. 6538) showed that anomalous formation pressures in the Gulf Coast basin could be maintained at sedimentation rates of 100-10,000 m-Ma^{-1} (0.1–10 millimeters-yr^{-1}) if average shale permeabilities were in the range of 10^{-18}-10^{-20} m^2. These values are consistent with laboratory measurements of shale permeabilities from the Gulf Coast (Neglia, 1979) as well as current sedimentation rates of 1 to 5 millimeters per year (Figure 8.9). Bethke (1986) also showed that deeper basins can develop overpressures in more permeable sediments than shallow basins (review section 5.5) (Figure 8.10). As overpressuring

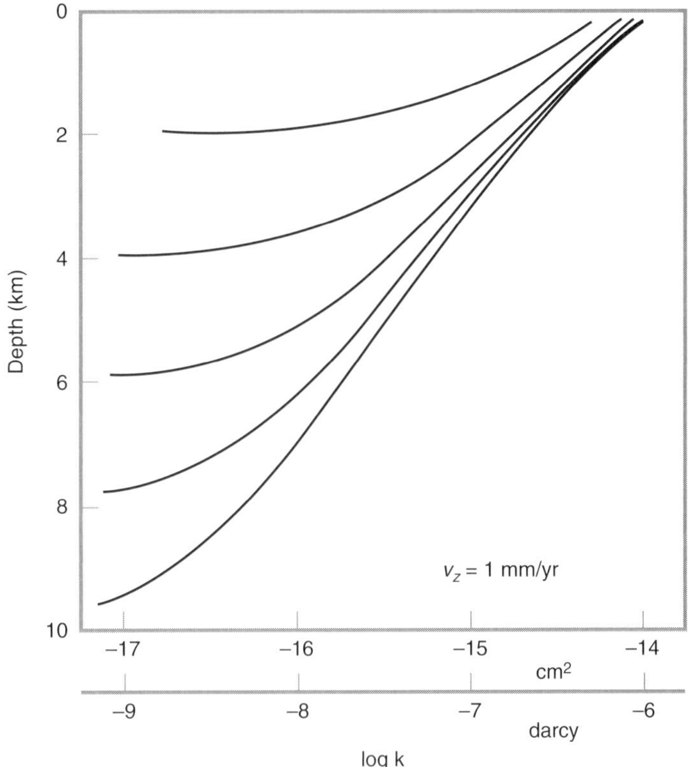

Figure 8.10 Permeability profiles required for overpressuring at a sedimentation rate of 1 millimeter per year as a function of total basin depth. Overpressuring can occur in deeper basins with higher average permeabilites.
(From Bethke, 1986, p. 6537.)

from compaction disequilibrium is a transient phenomenon, this follows directly from the relationship between characteristic time and length (equation 5.86).

Despite the low permeability needed for the development of overpressures from compaction disequilibrium, there is evidence that large volumes of fluid have escaped from the Gulf Coast overpressured zone. Crude oils in the thermally immature Tertiary reservoirs of the shallow hydropressured zone are generally believed to be derived from Cretaceous or Jurassic source rocks within the geopressured zone (Nunn and Sassen, 1986; Kennicutt et al., 1992; Whelan et al., 1994). The presence of significant volumes (~10 percent) of secondary porosity and diagenetic cements in some Gulf Coast sandstones from the geopressured zone has also been interpreted as evidence for large volumes of fluid circulation within the geopressured zone itself.

At the present time, it is poorly understood exactly how fluids escape from, or move through, the overpressured zone in the Gulf Coast. Relatively low salinities in the overpressured zone tend to suggest that membrane filtration is not taking place, and thus uniform flow through pore spaces seems unlikely (see Figure 4.1 and discussion in section 4.3). Fluid may move laterally until it encounters a fault or fracture that enables it to escape into the overlying hydrostatically pressured zone. Nunn (1996) reviewed evidence that geopressured sediments in the Gulf Coast basin are mechanically weak, and suggested that upward fluid movement could be caused by buoyancy-driven propagation of isolated fluid-filled fractures. Nunn's (1996) calculations showed that isolated, fluid-filled fractures with lengths of a few meters or more can propagate through geopressured sediments with velocities of 1000 m-yr^{-1}.

The conversion of the clay mineral **smectite** to **illite,** another clay mineral, releases water, and may contribute to the development of overpressures in some sedimentary basins. Smectite is a common mineral found in shales and contains abundant water between the layers of its crystal structure. Under conditions of high pressure and temperature, the water is expelled from smectite and it is converted to illite. The precise nature of the diagenetic reactions involved and their exact dependence upon pressure and temperature conditions is poorly known. Some water may be expelled at temperatures lower than 60 °C; however, temperatures as high as 200 °C may be required for complete expulsion. The net volume change that takes place is also uncertain. Estimates range from about 4 to 40%. The primary evidence that implicates the smectite-illite conversion in overpressuring is an observed change in the ratio of smectite to illite that occurs near the top of the overpressured zone. This coincidence has been observed, for example, in the Gulf Coast Basin of the southeast United States (Bethke, 1986), however, geopressures are already well developed in the Caspian Basin, even though the smectite-illite ratio remains unchanged down to depths of 6 km. This counterexample shows that the conversion by itself cannot be responsible for the generation of overpressures in rapidly subsiding basins similar to the Caspian or Gulf Coast Basin. It may be that the smectite-illite conversion is related to the development of overpressures by reducing permeability, instead of acting as a fluid source.

Petroleum generation is probably the most theoretically significant fluid source capable of creating abnormal fluid pressures. Neuzil (1995) estimated the magnitude of Γ (the forcing function, eq. 8.1) for different geologic mechanisms, and concluded that the mechanism with the largest probable magnitude was petroleum generation. Neuzil (1995, p. 758) estimated that the magnitude of Γ for petroleum generation can be as large as 10^{-14} s^{-1}, while other mechanisms evaluated by Neuzil result in Γ estimates typically on the order of 10^{-15} s^{-1}. Other researchers (e.g., Barker, 1990), however, have shown that gas generation is a much more effective mechanism for overpressuring than oil generation. Overpressure from oil generation is a consequence of higher-density kerogen being replaced by lower-density oil, which requires more volume for the same mass. Kerogen is the solid organic material that breaks down to form oil and gas at high temperatures. As natural gas has a much lower density than liquid oil, it follows that gas generation is a more efficacious mechanism for overpressure generation than oil generation (Figure 8.11). Barker (1990) estimated that 85 to

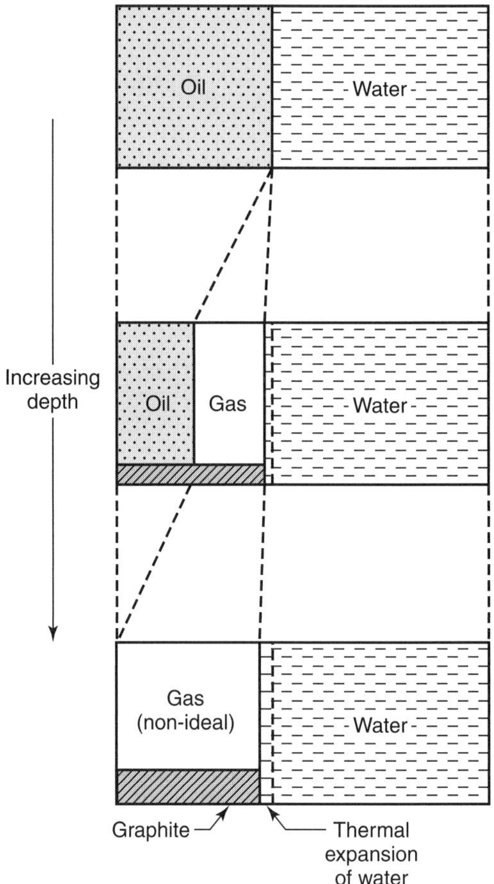

Figure 8.11 Conceptual model of liquid petroleum cracking to gas with increasing depth, time, and temperature. Cracking process also produces a graphite residue. Volume changes not shown to scale.
(From Barker, 1990, p. 1257.)

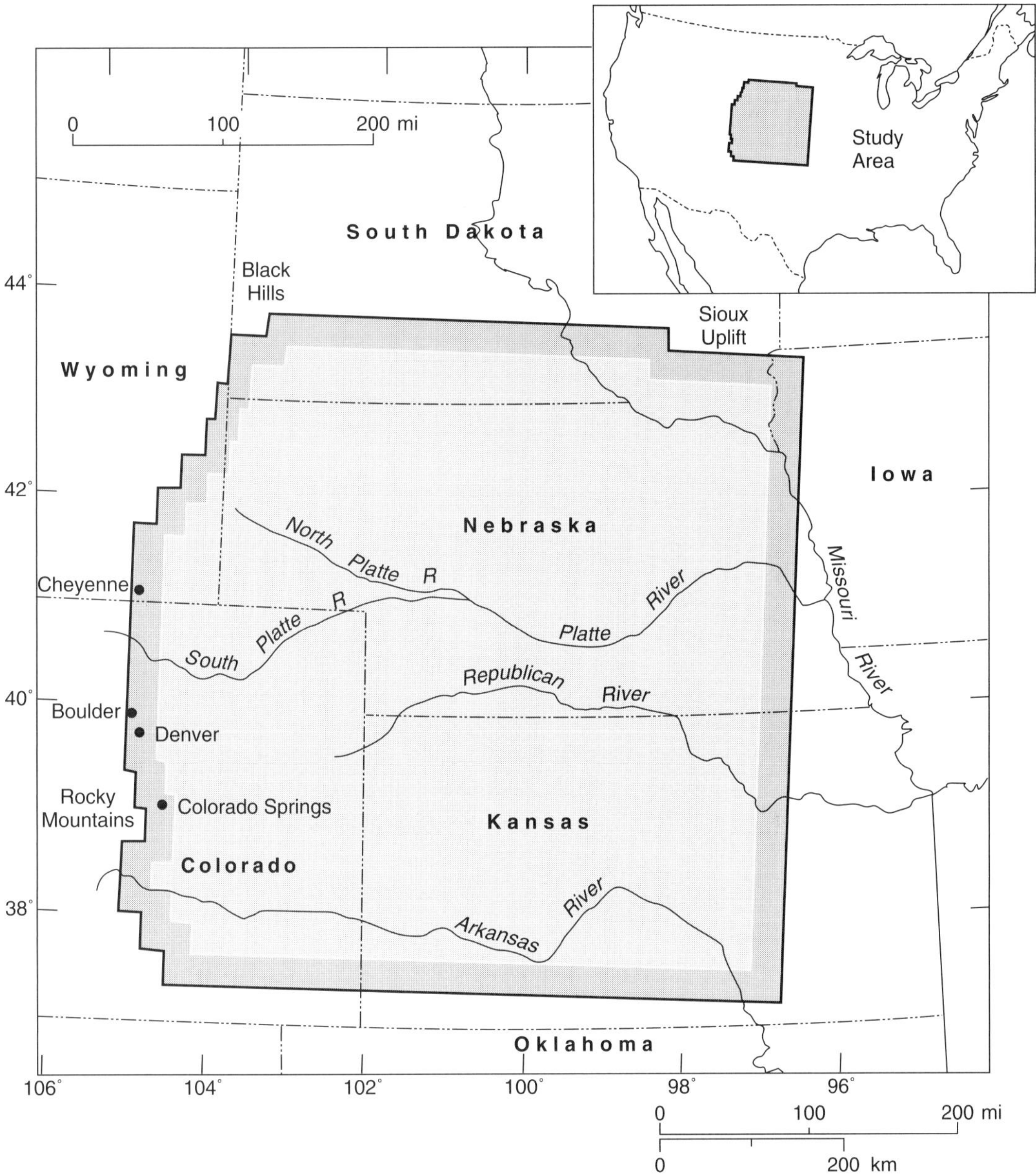

Figure 8.12 Location of the Denver Basin in the Central U.S.
(From Belitz and Bredehoeft, 1988, p. 1335.)

113 m^3 of gas is generated by each barrel (1 barrel = 42 gallons = 158.98 liters = 0.159 m^3) of oil that turns into gas at high temperatures. Lithostatic pressures can thus be reached after only 1% of the oil in a reservoir cracks into gas.

8.4. CASE STUDY: UNDERPRESSURES IN THE DENVER BASIN.

The Denver Basin in the central United States (Figures 8.12 and 8.13) is known to have extensive

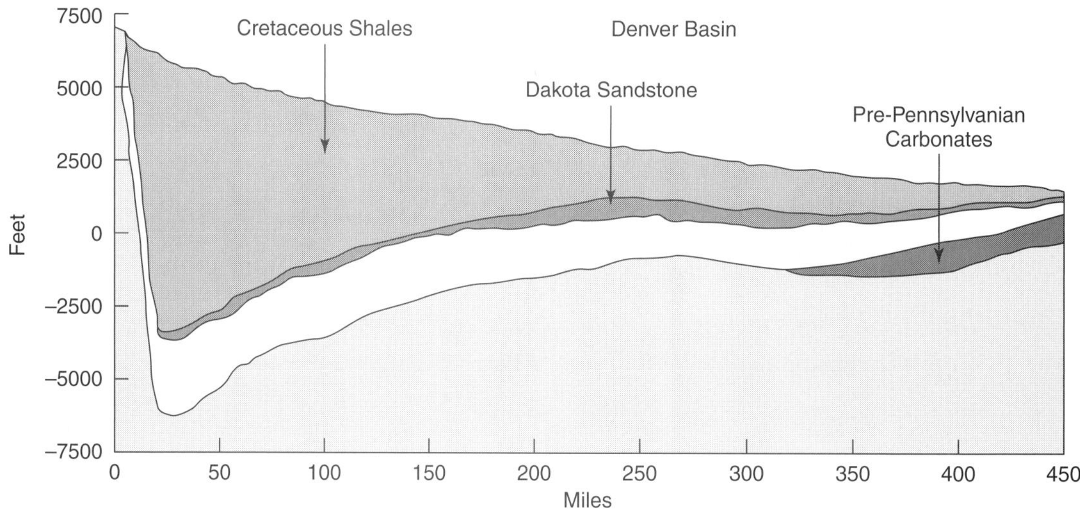

Figure 8.13 Generalized geologic cross-section through the Denver Basin.
(From Belitz and Bredehoeft, 1990.)

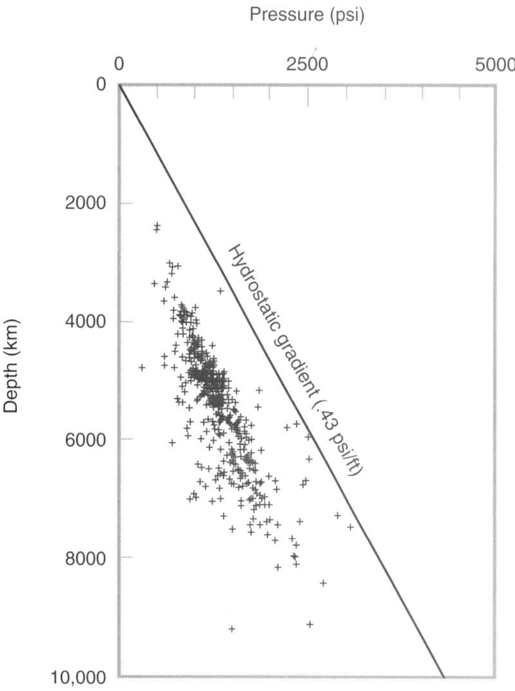

Figure 8.14 Fluid pressure versus depth for Denver Basin sandstones.
(From Belitz and Bredehoeft, 1988, p. 1335.)

areas of underpressures. The average fluid pressure gradient is about 57% of hydrostatic (Figure 8.14).

Belitz and Bredehoeft (1988) showed, using a simple "pipe model," that underpressures in the Denver Basin could be the result of topographically driven groundwater flow (Figures 8.15, 8.16). In a gross sense, the stratigraphy of the Denver Basin consists of an aquitard composed of Cretaceous shale overlying the Dakota Sandstone aquifer (Figure 8.15). We know, from refraction of head contours (review section 5.4) that flow in confining layers tends to move vertically. Thus, flow in the recharge region must be nearly vertical downwards through the Cretaceous shales into the Dakota Sandstone. In the Dakota Sandstone the direction of flow must be nearly horizontal, parallel to bedding. It is thus possible to analyze flow in the Denver Basin with a simple "pipe model" (Figure 8.16). The vertical pipe represents flow through the Cretaceous Shale aquitard; the nearly horizontal pipe represents flow through the Dakota Sandstone aquifer. At the top of the vertical pipe, head at the recharge site (h_r, m) is fixed by elevation z_r (m) (we assume that the water table is at the ground surface). The length of the vertical pipe is equal to the depth of the basin, D (m). The hydraulic conductivity of the vertical pipe

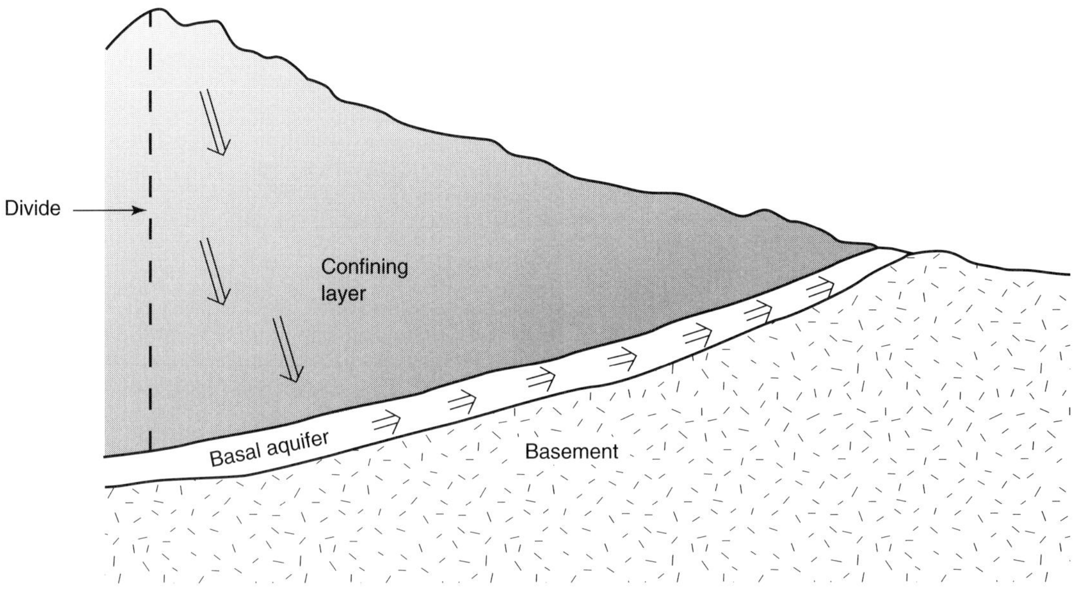

Figure 8.15 Topographically-driven flow in the Denver basin can be approximated by near-vertical flow through a Cretaceous Shale confining layer into a basal aquifer where flow is nearly horizontal.

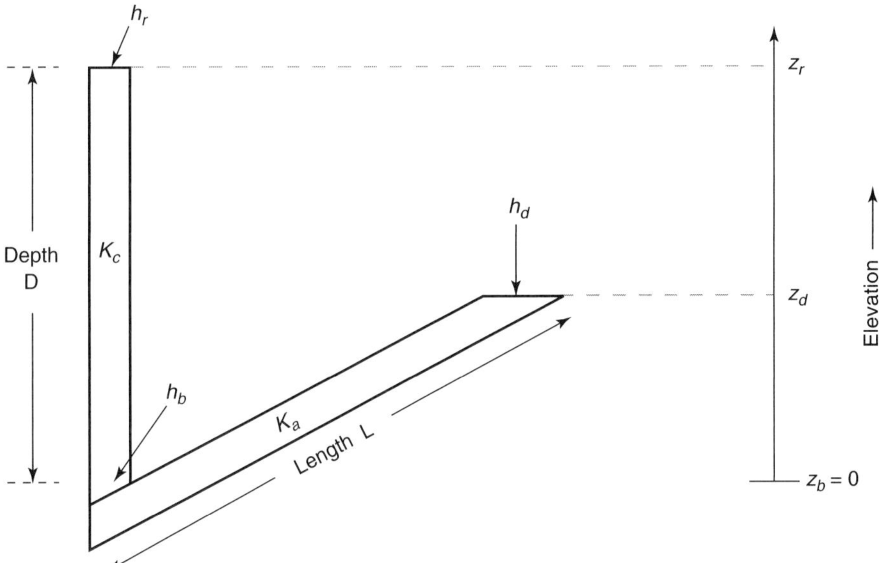

Figure 8.16 Pipe model for topographically-driven flow through the Denver Basin.
(After Belitz and Bredehoeft, 1988.)

representing the confining layer is K_c (m-s^{-1}). The horizontal pipe has length L (m), equal to the length of the basin. Head is fixed at the end of the horizontal pipe in the discharge region (h_d, m) by elevation (we again assume the water table is at the ground surface). The horizontal pipe representing the

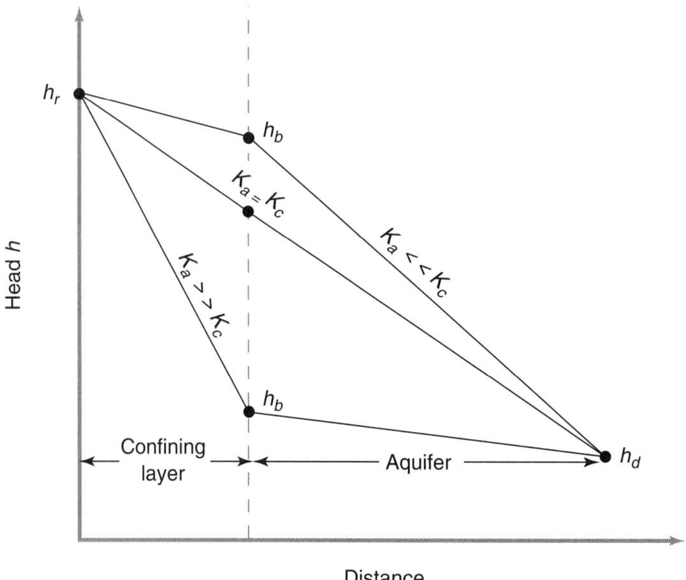

Figure 8.17 Schematic illustration of head gradients for pipe-model scenarios with contrasting hydraulic conductivities.

aquifer has hydraulic conductivity K_a (m-s^{-1}). Head at the boundary between the two pipes (h_b, m) is unknown. For simplicity, we will define an arbitrary datum for head and elevation by defining the elevation of the boundary between the vertical and horizontal pipes to be zero. Note that both h_r and h_d are greater than zero.

If we assume conservation of mass and steady-state flow, then the Darcy velocity in the confining layer (q_c, m-s^{-1}) must be equal to the Darcy velocity in the aquifer (q_a, m-s^{-1}). Applying Darcy's Law, we obtain

$$\frac{K_c(h_r - h_b)}{D} = \frac{K_a(h_b - h_d)}{L} \quad \textbf{(8.2)}$$

or, rearranging equation 8.2,

$$\frac{K_c}{K_a}\frac{L}{D} = \frac{(h_b - h_d)}{(h_r - h_b)} \quad \textbf{(8.3)}$$

Equation 8.3 implies that the value of head at the boundary between the confining layer and the basal aquifer (h_b) is determined by the geometry of the basin (the ratio L/D) and the ratio of the hydraulic conductivities (K_c/K_a) (Figure 8.17). If $K_a >> K_c$, the head gradient in the confining layer is much greater than in the aquifer. Consequently, h_b is relatively low. Conversely, if $K_c >> K_a$, the head gradient in the confining layer would be much lower than in the aquifer, and h_b would be relatively high. Of course, in the latter situation, the designations "confining layer" and "aquifer" would have to be reversed.

Recall that head (h, m) has an elevation and pressure component (equation 2.45)

$$h = z + \frac{P}{\rho g} \quad \textbf{(8.4)}$$

where z is elevation (m), P is fluid pressure (kg-m^{-1}-s^{-2}), ρ is fluid density (kg-m^{-3}), and g is the acceleration due to gravity (m-s^{-2}). Thus,

$$h_b = 0 + \frac{P_b}{\rho g} \quad \textbf{(8.5)}$$

$$h_d = z_d + 0 \quad \textbf{(8.6)}$$

$$h_r = D + 0 \quad \textbf{(8.7)}$$

where the subscript b indicates the boundary between the vertical and horizontal pipes, and we have specified fluid pressure (P) in terms of gauge pressure. Thus, at the recharge and discharge sites where fluid pressure is equal to atmospheric pressure, the fluid pressures P_r and P_d are zero. Substituting equations 8.5, 8.6, and 8.7 into equation 8.3 we obtain

$$\frac{K_c}{K_a}\frac{L}{D} = \frac{\frac{P_b}{\rho g} - z_d}{D - \frac{P_b}{\rho g}} \tag{8.8}$$

$$\left[D - \frac{P_b}{\rho g}\right]\frac{K_c}{K_a}\frac{L}{D} = \frac{P_b}{\rho g} - z_d \tag{8.9}$$

$$D\frac{K_c}{K_a}\frac{L}{D} - \frac{P_b}{\rho g}\frac{K_c}{K_a}\frac{L}{D} - \frac{P_b}{\rho g} = -z_d \tag{8.10}$$

$$\frac{P_b}{\rho g}\left[\frac{K_c}{K_a}\frac{L}{D} + 1\right] = \frac{K_c}{K_a}L + z_d \tag{8.11}$$

$$\frac{P_b}{\rho g} = \frac{\frac{K_c}{K_a}L + z_d}{\frac{K_c}{K_a}\frac{L}{D} + 1} \tag{8.12}$$

$$P_b = \rho g\left\{\frac{\frac{K_c}{K_a}L + z_d}{\frac{K_c}{K_a}\frac{L}{D} + 1}\right\} \tag{8.13}$$

$$P_b = \rho g D\left\{\frac{\frac{K_c}{K_a}L + z_d}{\frac{K_c}{K_a}L + D}\right\} \tag{8.14}$$

Equation 8.14 gives the fluid pressure at the boundary between the confining layers and aquifer in terms of the hydrostatic fluid pressure ($\rho g D$) and a term (in brackets) that depends upon the hydraulic conductivity of the confining layer and aquifer as well as the basin geometry (length and depth).

Note that the basin is underpressured even if $K_c = K_a$. In this case, the ratio of the pressure gradient in the basin to a hydrostatic gradient is

$$\frac{L + z_d}{L + D} \tag{8.15}$$

If we take $L = 644$ km (400 miles), $z_d = 1.676$ km (5500 ft), and $D = 3.048$ km (10,000 ft) for the Denver Basin (see Figure 8.13), the fluid pressure gradient is 99.8% of hydrostatic.

Problem: What must the ratio K_c/K_a be for topographically driven flow in the Denver Basin to result in the fluid pressures being 57% of hydrostatic?

Denote the dimensionless ratio of fluid pressure (P_b at depth D) to hydrostatic fluid pressure ($\rho g D$) as γ. From equation 8.14,

$$\frac{P_b}{\rho g D} = \gamma = \frac{\frac{K_c}{K_a}L + z_d}{\frac{K_c}{K_a}L + D} \tag{8.16}$$

Now solve for the ratio K_c/K_a in terms of γ. Rearranging equation 8.16,

$$\frac{K_c}{K_a}L + z_d = \gamma\left[\frac{K_c}{K_a}L + D\right] \tag{8.17}$$

Gathering terms,

$$\frac{K_c}{K_a}L(1-\gamma) = \gamma D - z_d \tag{8.18}$$

$$\frac{K_c}{K_a} = \frac{\gamma D - z_d}{L(1-\gamma)} \tag{8.19}$$

For $\gamma = 0.57$ as observed, $D = 3.048$ km, $z_d = 1.676$ km, and $L = 644$ km,

$$\frac{K_c}{K_a} = 2.2 \times 10^{-4} \tag{8.20}$$

Equation 8.20 implies that the aquifer is about 4,500 times more permeable than the overlying confining layer.

Review Questions

1. Define the following terms in the context of hydrogeology:
   a. hydrostatic
   b. lithostatic
   c. underpressures
   d. overpressures
   e. geopressures
   f. static school
   g. pressure seal
   h. pressure compartment
   i. capillary force
   j. dynamic school
   k. hysteresis
   l. compaction disequilibrium
   m. smectite
   n. illite
   o. kerogen
2. Which is more common in sedimentary basins—overpressures or underpressures?
3. If a sedimentary basin is overpressured, at what depths are overpressures usually first encountered?
4. What are the two schools of thought on the creation and preservation of abnormal pressures in the Earth's crust?
5. What are some of the difficulties with the pressure seal concept?
6. What are the advantages of hypothesizing gas capillary forces as pressure seals?
7. Use a scale analysis to estimate the minimum permeability necessary for a layer 50 m thick layer to confine abnormal fluid pressures for 10 Ma (review section 5.5).
8. Show how overpressures and underpressures may both be due to steady-state topographically driven flow. Use a figure. Explain how fluid can be flowing from areas where fluid pressure is below hydrostatic to areas where fluid pressure is above hydrostatic.
9. Is a good idea to locate a toxic waste dump in an underpressured area? Why or why not?
10. What are two geologic processes that can lead to underpressuring?
11. What process is believed to be responsible for the existence of overpressures in the Gulf Coast Basin of the southeast U.S.? What geologic factors have contributed to the development of geopressures in this area?
12. In what two ways could the conversion of smectite to illite contribute to overpressuring?
13. Explain how oil and/or gas generation can lead to overpressuring? Which (oil or gas generation) is more likely to lead to overpressuring? Why?
14. What is the *maximum* underpressuring that can occur in the Denver Basin?

Suggested Reading

Belitz, K., and Bredehoeft, J. D. 1988. Hydrodynamics of Denver Basin. Explanation of subnormal fluid pressures. *American Association of Petroleum Geologists Bulletin,* 72: 1334–1359.

Bethke, C. M. 1986. Inverse hydrologic analysis of the distribution and origin of Gulf Coast-type geopressured zones. *Journal of Geophysical Research,* 91: 6535–6545.

Osborne, M. J., and Swarbrick, R. E. 1997. Mechanisms for generating overpressures in sedimentary basins: a reevaluation. *AAPG Bulletin,* 81: 1023–1041.

Notation Used in Chapter Eight

Symbol	Quantity Represented	Physical Units
D	basin depth	m
Γ	forcing term for geologic overpressuring	s^{-1}
g	acceleration due to gravity	m-s^{-2}
γ	ratio of fluid pressure to hydrosatic	dimensionless
h	head	m
h_b	head at boundary between confining layer and aquifer	m
h_d	head in discharge region	m
h_r	head in recharge region	m
K	hydraulic conductivity	m-s^{-1}
K_a	hydraulic conductivity of basin aquifer	m-s^{-1}
K_c	hydraulic conductivity of confining layer	m-s^{-1}
L	basin length	m
P	fluid pressure	Pascal (Pa) = kg-m^{-1}-s^{-2}
P_b	fluid pressure at boundary between confining layer and aquifer	Pascal (Pa) = kg-m^{-1}-s^{-2}
P_d	fluid pressure of discharge region	Pascal (Pa) = kg-m^{-1}-s^{-2}
P_r	fluid pressure of recharge region	Pascal (Pa) = kg-m^{-1}-s^{-2}
ρ	fluid density	kg-m^{-3}
S_s	specific storage	m^{-1}
z	elevation	m
z_d	elevation of discharge region	m
z_r	elevation of recharge region	m
∇^2	Laplacian, the second spatial derivative	m^{-2}

CHAPTER 9

Environmental Hydrogeology

9.1. Land Subsidence.

Land subsidence is a gradual settling or sudden sinking of the Earth's surface due to subsurface movement of soil or rocks. Land subsidence is a problem throughout the world; in the United States a total area of more than 17,000 square miles (44,000 square kilometers) has been affected by subsidence, an area greater than that covered by the states of New Hampshire and Vermont (Galloway et al., 1999).

Although subsidence can and does occur on geologic time scales as a result of natural forces, in this chapter we confine our discussion to subsidence that happens on a human time scale due to human actions. There are three basic mechanisms by which human actions can cause land subsidence. These are (1) compaction of aquifer systems, (2) dewatering and oxidation of organic soils, and (3) dissolution and collapse of susceptible earth materials. Interestingly, land subsidence can be induced by both the extraction and addition of water depending on the geologic circumstances involved.

9.1.1. Aquifer Compaction.

The most serious and widespread type of land subsidence is that which occurs in response to the mining of groundwater and resultant compaction of aquifers. Land subsidence may also occur in response to withdrawal of oil from reservoir rocks. Fluid withdrawal causes head to decrease that in turn causes effective stress to increase. The response of a compressible porous medium to increases in effective stress is compaction. Recall equation 3.43, which states that rate of change of fluid pressure is equal to, but opposite in sign to, the rate of change in effective stress

$$\frac{d\sigma_e}{dt} = \frac{-dP}{dt} \tag{9.1}$$

where σ_e is effective stress (kg-m^{-1}-s^{-2}), t is time (s), and P is fluid pressure (kg-m^{-1}-s^{-2}). If water is extracted from an aquifer, or oil from a petroleum reservoir, the fluid pressure will drop. The total change in fluid pressure ΔP will be negative. Equation 9.1 implies that a negative change in fluid pressure will create a positive increase in effective stress, $\Delta\sigma_e$, or

$$-\Delta P = \Delta\sigma_e \tag{9.2}$$

If elevation (z) is constant, then $\Delta P = \rho g \Delta h$ and

$$-\rho g \Delta h = \Delta\sigma_e \tag{9.3}$$

where h is head (m), ρ is fluid density (kg-m^{-3}), and g is the acceleration due to gravity (m-s^{-2}).

Recall the definition of porous medium compressibility (equation 3.48)

$$\alpha = \frac{\frac{-\Delta V_T}{V_T}}{\Delta\sigma_e} \qquad \textbf{(9.4)}$$

where V_T is the volume of porous medium (m^3), and α is porous medium compressibility ($m\text{-}s^2\text{-}kg^{-1}$). If the reduction in volume occurs through vertical compaction of a porous medium with original thickness b (m), $(\Delta V_T/V_T) = (\Delta b/b)$ and the total change in thickness Δb is

$$\Delta b = -b\Delta\sigma_e\,\alpha \qquad \textbf{(9.5)}$$

Substituting equation 9.3 into equation 9.5,

$$\Delta b = \rho g \Delta h b \alpha \qquad \textbf{(9.6)}$$

Note that if head declines ($\Delta h < 0$) in equation 9.6, $\Delta b < 0$ implies that the porous medium contracts and land subsidence may occur. Conversely, if head increases, $\Delta b > 0$ implies that the porous medium expands.

How much land subsidence is possible? The compressibility (α) of unconsolidated clays is in the range of 10^{-6} to 10^{-8} Pa^{-1} (Freeze and Cherry, 1979, p. 55). For a nominal water density (ρ) of 1000 $kg\text{-}m^{-3}$ and acceleration due to gravity (g) of 9.8 $m\text{-}s^{-2}$, a decline in head of 1 meter will lead to a 0.1% compression for a nominal compressibility of 10^{-7} Pa^{-1}. If the decline in head is 100 m, and the total unit thickness is 100 m, total land subsidence will be 10 m.

Land subsidence in areas such as the San Joaquin Valley (see case study below) is facilitated by a geologic structure consisting of alternating layers of high permeability (but relatively incompressible) sand aquifers interbedded with low permeability (but compressible) clay aquitards. Withdrawal of water from sand and sandstone aquifers alone usually does not lead to land subsidence, because sand and its lithified counterpart are relatively incompressible. The compressibility of sand ranges from 10^{-7} to 10^{-9} Pa^{-1}, while sandstone tends to range from 10^{-9} to 10^{-11} Pa^{-1} (Freeze and Cherry, 1979, p. 55). Clay, on the other hand, tends to be more compressible, but it is difficult to extract groundwater from clay aquitards due to their relatively low hydraulic conductivity. A layered structure allows groundwater to be pumped from the sand aquifers. The clay aquitards in turn slowly drain into the sand aquifers, contracting as fluid escapes. The process was first modeled mathematically by Terzaghi (1925). Terzaghi's model assumes that head is drawndown instantaneously at the boundaries of a homogeneous aquitard by rapid pumping of surrounding sand aquifers. Head then decays through time as the transient disturbance at the aquitard boundaries propagates into the interior (Figure 9.1).

Although in theory land subsidence can be reversed by pumping groundwater back into the subsurface and increasing head to its original level, in practice this is rarely possible. Most clay layers tend to exhibit hysteresis, and will not expand to original dimensions after contracting. The collapse that occurs when head is lowered is usually irreversible. When groundwater withdrawals go beyond a certain point, the water that is extracted is pulled from permanent storage in clay-rich aquitards. This water can never be returned to the subsurface, as aquitard compaction is largely irreversible. Thus the groundwater that is withdrawn has been mined from the subsurface; an otherwise renewable resource is permanently lost through overuse.

One area in which partial success was achieved in reversing land subsidence is the Wilmington oil field in the lowland harbor area of Los Angeles and Long Beach, California. In 1962, decades of oil production had produced a maximum subsidence of 8 meters at the center of the subsiding area. As much of the affected area was originally only 1.5 to 3.0 meters above sea level, extensive remediation was necessary. Massive levees, retaining walls, and landfills were constructed, and structures were raised. Subsidence also ruptured pipelines, oil-well casings, utility lines, and damaged buildings. By 1962, the total cost of remediation exceeded \$100 million. In 1961, a full-scale water injection program began. Water from shallow wells was injected into petroleum reservoirs. The goals of the injection program were twofold:

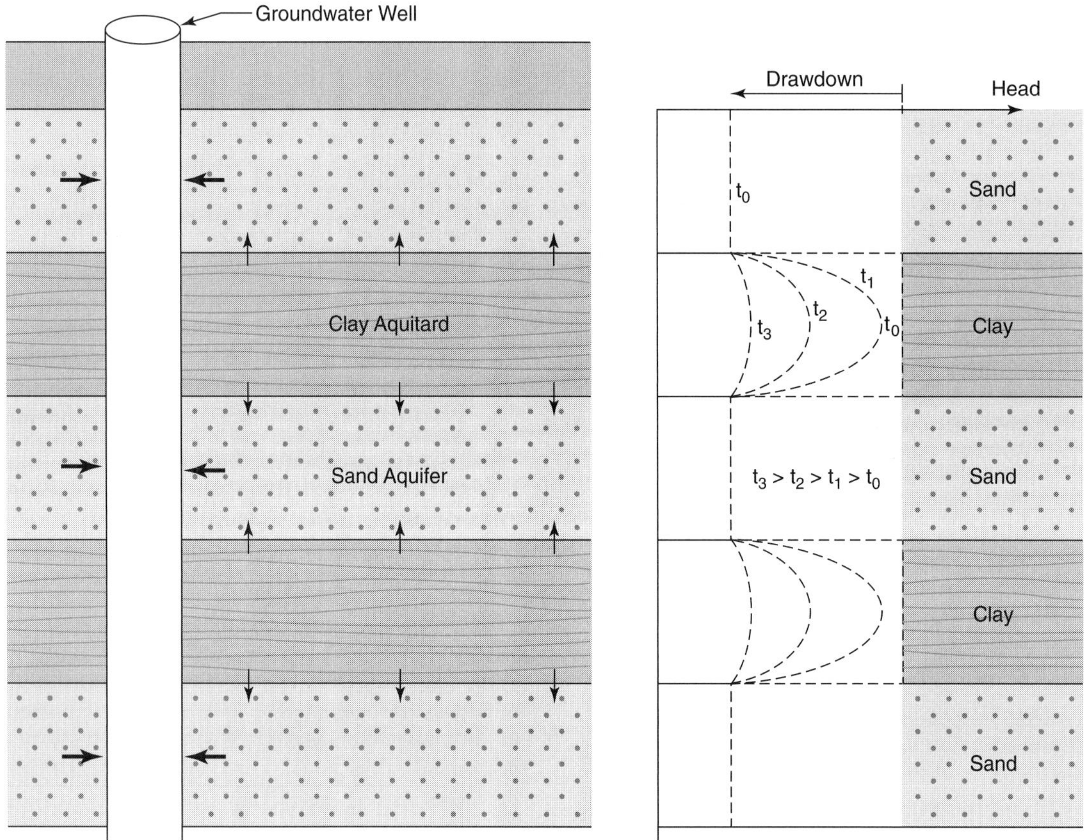

Figure 9.1 Head change through time in an alternating sand-clay sequence. At time t_0, head is lowered by pumping water from sand layers. Subsequently, at later times t_1, t_2, etc., water drains from clay aquitards into surrounding sand layers as head change slowly propagates into the aquitards.

first, to decrease or stop subsidence; second, to enhance oil recovery. The land surface stabilized shortly after injection began in 1961, and by 1969 had rebounded as much as 0.3 meters (Poland, 1972, p. 63–64).

Land subsidence can lead to severe environmental problems, or it may be relatively benign. Poland (1972, p. 52) listed 11 areas around the world where substantial subsidence up to 9 meters had occurred. These areas included Osaka and Tokyo, Japan, Mexico City, and several areas in the United States. Although the magnitude of land subsidence in Japan is not so large as the San Joaquin Valley in California, the effects and consequences have been more severe. Many of the large cities in Japan are built on low, flat alluvial plains underlain by unconsolidated Quaternary deposits. Heavy extraction of groundwater for industrial and residential supply has significantly lowered the artesian head leading to compaction and land subsidence. Poland (1972, p. 54) characterized land subsidence in eastern Tokyo as "probably the most serious environmental subsidence problem in the world." The subsidence began around 1920 and was accompanied by a drawdown in artesian head from original heights above sea level to 60 m below sea level by 1967. Land subsidence resulted in an area of 80 km^2 populated by more than two million people sinking below mean high-tide level. As of 1972, the lowest ground level was 2.3 m below mean sea level. To prevent inundation and flooding, a wall was built surrounding the entire area that had subsided below high-tide level and drainage pumps were installed. Despite these measures, much of Tokyo remains at risk of serious flooding due to typhoons or possible failure of a water-retaining structure caused by a major earthquake (Poland, 1972, p. 54–55).

9.1.2. Case Study: Subsidence in the San Joaquin Valley, California.

One of the most dramatic examples of land subsidence anywhere in the world is the San Joaquin Valley in California. Galloway et al. (1999) described land subsidence in the San Joaquin Valley as "the single largest human alteration of the Earth's surface topography." Much of what follows is taken from Galloway et al (1999).

The San Joaquin Valley comprises the southern two-thirds of the Central Valley of California. Bounded by the Sierra Nevada Mountains on the east, the Diablo and Temblor Ranges on the west, and the Tehachapi Mountains on the south, the valley is a trough created by the collision of the Pacific and North American tectonic plates. The Central Valley contains sediments deposited by streams draining the mountains, and sediments laid down in lakes that have formed on the valley floor and subsequently disappeared over geologic time. Total sediment thickness in the Central Valley in some places is thousands of feet (1 foot = 0.3 meters). More than half of the sediment thickness is composed of fine-grained clays, silts, and sands susceptible to compaction. The climate is arid to semiarid; total annual precipitation averages from 5 to 16 inches (13 to 41 centimeters).

There are two aquifer systems in the San Joaquin Valley that supply groundwater for human consumption. Shallow, unconfined, and partially confined aquifers exist near-surface throughout the valley, especially near the margins of the valley and the toes of younger alluvial fans. A deeper, confined aquifer system underlies a laterally extensive clay layer. The deeper aquifer system is composed of interbedded coarse and fine-grained sediment layers.

Most land subsidence in the Central Valley of California has directly resulted from pumping water for irrigation of crops. Although land subsidence is an undesirable side effect of irrigation, the irrigation of Central Valley soils has made the area one of the most productive agricultural regions in the world. As of 1997, the Central Valley of California produced 25% of the food in the United States from 1% of its farmland. Irrigated agriculture in the Central Valley began following the California gold rush of 1849. By 1900, much of the flow of the Kern River and the entire flow of the Kings River had been diverted through canals and ditches to irrigate land throughout the southern part of the valley. Ten years later, in 1910, all of the available surface water had been used for agriculture. Surface water supplies also tended to be undependable due to droughts. Because of these two factors, farmers turned to groundwater for irrigation of crops. At first, groundwater extractions were limited to shallow aquifers in the central valley where flowing wells were commonplace. As the near-surface, unconfined aquifers were emptied, farmers began installing deeper wells. Irrigation withdrawals (Figure 9.2) accelerated in the 1930s with the introduction of improved pump designs and rural electrification. By 1955, one-fourth of the total groundwater used for irrigation in the entire United States was extracted from the San Joaquin Valley. Water levels in wells that penetrated the deep aquifer system in the western and southern parts of the basin had declined by more than 100 feet (30 meters). In some areas, the decline in the potentiometric surface was 400 feet (122 meters). In 1960, the rate of decline was 10 feet (3 meters) per year.

Although land subsidence in the San Joaquin Valley probably first began in the 1920s with the initiation of significant groundwater withdrawals, it was first noted in 1935 by a consulting engineer, I. H. Althouse. The first documentation of the subsidence problem was done by Ingerson (1941), who published a map and topographic profiles

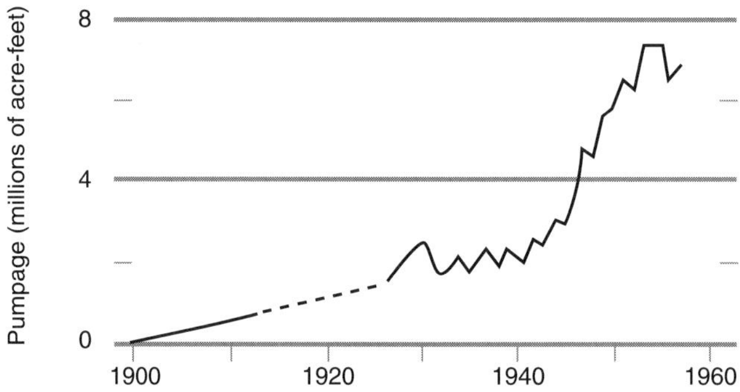

Figure 9.2 Extraction of groundwater for irrigation in the San Joaquin Valley. One million acre − feet = 1.2 billion cubic meters.
(From Galloway et al, 1999, p. 28.)

showing subsidence. Negative effects of land subsidence in the San Joaquin valley include (1) ruptured well casings, (2) damage to highways and water-transport structures from differential subsidence, and (3) unexpected flooding as differential subsidence caused streams and creeks to change their courses. However, over most of the valley, subsidence occurred so slowly and over such a broad area that its effects went largely unnoticed by most residents. Land subsidence in the San Joaquin Valley first became a public concern in 1951 when it posed a potential threat to the Delta-Mendota Canal whose construction had just been completed by the U.S. Bureau of Reclamation. Subsequently, the U.S. Geological Survey began a thorough investigation of the causes, rates, and extent of land subsidence in the San Joaquin Valley. As of 1970, about 10,875 km^2 of land had subsided more than 0.3 m. A maximum subsidence of 9.1 meters was reached in 1981 at a point 64 km west of Fresno, California (Figure 9.3).

By the late 1960s, the completion of several water-supply projects allowed the importation of increased amounts of surface water to the San Joaquin Valley, reducing the need to utilize groundwater for irrigation. Groundwater levels began to recover, and subsidence was slowed or arrested over a large part of the affected area. From 1967 through 1974, water levels in wells that penetrated the deep aquifer system rose by as much as 200 feet (61 meters). Since 1981, subsidence in the San Joaquin Valley has been managed through the importation of surface water through canals and the California Aqueduct. Surface-water imports have replaced groundwater for agriculture. By 1982, aquifer head had recovered to 1940–50 levels and land subsidence had either stopped or been considerably slowed.

While water levels were increasing, subsidence in some areas continued, albeit at a slowed rate. Near Cantua Creek, 2 feet (0.6 meters) of subsidence occurred from 1968 through 1974 while water levels simultaneously rose by more than 200 feet (61 meters). The delay observed in the reversal of subsidence trends can be attributed to the slow diffusion of pressure transients through clay aquitards.

Since 1974, land subsidence in the San Joaquin Valley has been greatly slowed or stopped, but will return in the future if droughts cause aquifer pumping to increase. During the droughts of 1976–77 and 1987–91, surface water supplies were diminished, and high rates of groundwater pumping resumed. During the 1976–77 drought, although the annual volume of water pumped was only one-third what it was during the peak withdrawal rates of the 1960s, water levels in deep wells fell by more than 150 feet (46 meters) and 0.5 feet (15 centimeters) of subsidence occurred. Similar drops in head and renewed subsidence

Figure 9.3 U.S. Geological Survey scientist Joseph Poland stands next to a telephone pole in the San Joaquin Valley marking the approximate location of maximum land subsidence in the United States. Signs on the pole mark elevation of land surface in 1925, 1955, and 1977.
(From Galloway et al, 1999, p. 23.)

occurred during the drought of 1987–91. The dramatic drop in head levels that occurred in response to relatively modest water extractions reflected the permanent loss of aquifer storage through aquifer compaction that had taken place in response to the overwithdrawals of water, which began in the 1920s.

9.1.3. Hydrocompaction.

Hydrocompaction is compaction of soil in response to wetting (Figure 9.4). It is a near-surface phenomenon entirely different from aquifer compaction in its causes and effects. Hydrocompaction in the San Joaquin Valley in California began to be noticed in the 1940s and 1950s when farmers began cultivating soils that had never before been farmed or irrigated. The standard technique of flood irrigation caused an irregular settling of the ground, producing an undulating surface with highs and lows offset by 3 to 5 feet (0.9 to 1.5 meters). In certain cases, localized settlings of 10 feet (3 meters) or more occurred. In contrast to the uniform and progressive regional subsidence due to aquifer compaction, the effects of hydrocompaction were irregular, localized, and could develop quickly. As a consequence, land subsidence due to hydrocompaction disrupted the infrastructure devoted to the distribution of irrigation water, and damaged pipelines, power lines, roadways, airfield runways, and buildings.

Studies of hydrocompaction in the San Joaquin Valley showed that it occurred only in alluvial-fan sediments above the highest level of the prehistoric water table. Affected soils also lay in areas where low annual rainfall had never permitted infiltration to penetrate below the near-surface zone subject to summer desiccation. Sediments deposited in these areas typically contained significant amounts of the clay mineral montmorillonite. When dry, montmorillonite acts as a strong bonding agent to hold dry sediment particles together and preserve structural strength. When soils containing these montmorillonite bonds were wet for the first time since deposition, they lost the strength necessary to support the weight of the overlying soil. Compaction followed structural failure, with volume of the affected soils being reduced by as much as 10%.

The hazards of hydrocompaction are mitigated somewhat by the fact that it is a one-time only occurrence. When dry soils are thoroughly wet, the compaction occurs once and is not cyclic in nature, however, depending upon the thickness of the susceptible sediments, it may take time for the wetting

Figure 9.4 Surface cracks and land subsidence due to hydrocompaction in the San Joaquin Valley. (From Galloway et al, 1999, p. 32.)

front to migrate to the bottom of the dry zone. Sediments suffering from hydrocompaction may also undergo further compaction if presented with a surface load such as a building or road.

9.1.4. Dewatering and Oxidation of Organic Soils.

Organic soils are those that are rich in carbon. When these soils are drained for agriculture or other purposes, microbial decomposition oxidizes organic carbon, producing carbon dioxide gas and water.

In the Everglades ecosystem in southern Florida, land subsidence has been caused by drainage and oxidation of peat soils. **Peat** is a light, spongy material formed in temperate humid environments by the accumulation and partial decomposition of vegetable remains under conditions of poor drainage. The Everglades formed in a 40 to 50 mile (64 to 80 km) wide limestone basin that stretches 130 miles (209 km) from Lake Okeechobee in the north to the southern tip of Florida (Figure 9.5). A 5 to 15 mile (8 to 24 km) wide bedrock ridge of higher elevation separates the Everglades basin from the eastern coastline. Under natural conditions, the entire Everglades Basin was subject to annual floods, with much of the land being perpetually inundated by freshwater. The flooding took place in vast, sheet-like flows that moved slowly over the land surface under the influence of low topographic gradients. The dominant plant species found in the plains south of Lake Okeechobee was sawgrass. Sawgrass is actually not a grass at all, but a member of the sedge family. It derives its name from leaves that are lined with upward pointing teeth. The growth and decay of sawgrass deposited organic matter on top of limestone bedrock at a rate of about 0.03

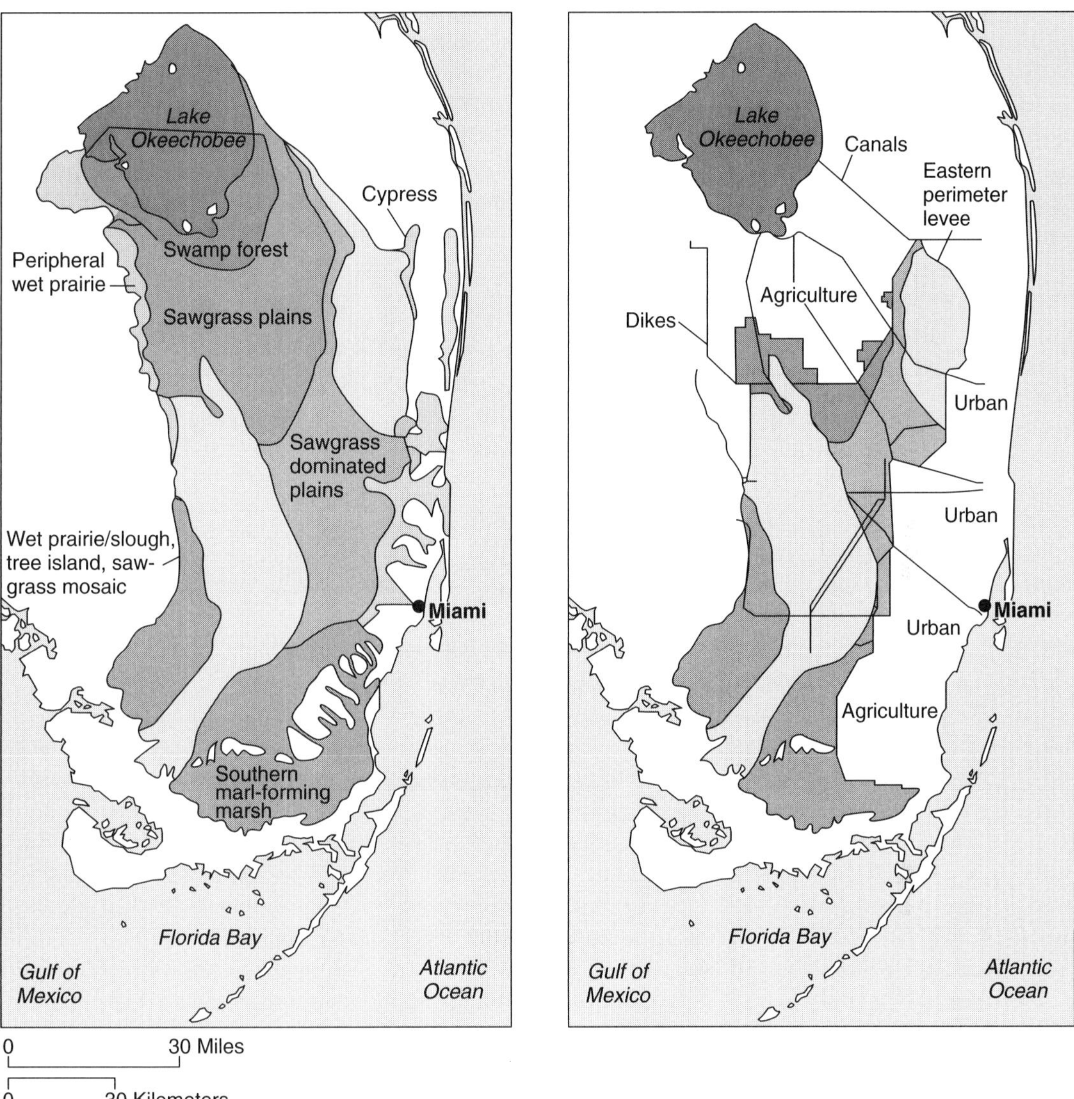

Figure 9.5 Distribution of vegetation and land use in the Everglades prior to development around 1900 (left) and as of 1990 (right).
(From Galloway et al, 1999, p. 100.)

inches (0.7 millimeters) per year. Over 5,000 years, a peat soil about 12.5 feet (3.8 meters) thick was laid down.

Agriculture in the Everglades began around 1913, with farming of the natural levee immediately south of Lake Okeechobee. Early attempts to farm the peat soil of the sawgrass plains were unsuccessful, due to flooding, winter freezes, and a deficiency of copper in the soil. It was discovered that the soil could be farmed by the addition of necessary trace minerals; however, the area was still subject to catastrophic floods. The great hurricane of 1928 caused 2,000 deaths and flooded the area for several months. To prevent further inundations, a permanent dike was built around the southern perimeter of Lake Okeechobee. The flow of water from Lake Okeechobee, which had fed the areas to the south for thousands of years, was stopped. Drainage of Lake Okeechobee was now accomplished artificially, with flow managed by a system of canals. An area of thick peat soil south of the lake was designated the "Everglades agricultural area" and opened to farming (Figure 9.5).

With the addition of trace minerals, the peat soil underneath the sawgrass proved to be extremely productive, however, drainage caused the land to subside at an average rate of about 1 inch (2.5 centimeters) per year. Total subsidence up to 1999 was estimated to be in the range of 3 to 9 feet (0.9 to 2.7 meters). In a wetlands area where water flow is driven by a normal topographic gradient of 2 inches per mile (3 centimeters per kilometer), this is a significant amount of subsidence.

The main crop grown today in the Everglades agricultural area is sugarcane. It is uncertain if agriculture in this area can be sustained. As long ago as 1951, it was predicted that by the year 2000, the peat soil would have shrunk from an original thickness of 12 feet (3.7 meters) to less than 1 foot (0.3 meters). In 1951, it was thought that such catastrophic shrinkage would result in the end of cultivation, as it was not thought possible to grow crops on less than 1 foot (0.3 meters) of soil. Although the estimates of soil shrinkage made in 1951 appear to have been correct, little land had been retired as of 1999. One reason is that farmers had learned to grow sugarcane in as little as 6 inches (15 centimeters) of soil by piling it up in windrows to allow successful germination. In recent years, the rate of subsidence also appears to have declined. As of 1999, the future of agriculture in the Everglades was uncertain. Unless unforeseen technological advances are made, it seems likely that agriculture in the area south of Lake Okeechobee cannot be sustained for more than decades.

Unfortunately, land subsidence in the Everglades agricultural region has made restoration of the predevelopment ecosystem impossible. Even if it were politically and economically feasible, it is technically impossible to reverse the land subsidence. Urban development has also made restoration of the natural flow system impossible. Managed water flow through a system of canals will be required indefinitely. The tragedy of the Everglades is that a unique ecosystem, which took thousands of years to develop, was apparently destroyed for the sake of a short-lived unsustainable agriculture.

9.1.5. Case Study: Sinkholes, West-Central Florida.

A **sinkhole** is a depression in the land surface caused by collapse of an underground cavity (Figure 9.6). The formation of sinkholes is usually sudden and can be catastrophic. In addition to the obvious danger to buildings, roads, and infrastructure, sinkholes can create open routes for contaminants to enter regional aquifers. One of the most susceptible areas to sinkhole formation is the Karst region of West-Central Florida. **Karst** is a type of topography formed in regions underlain by carbonate bedrock. Karst topography is characterized by the presence of sinkholes, caves, and other dissolution features. The name Karst comes from the Karst Region of Yugoslavia, which borders the Adriatic sea, where "barren, whitish limestone ridges and islands contrast vividly with the wine-dark waters of the sea" (Birkeland and Larson, 1978).

The Florida Peninsula of North America is part of a larger, mostly submerged carbonate platform capped by a layer of sand and clay deposits. Over the last million years, a series of ice ages have alternately raised and lowered mean ocean levels, exposing and covering the carbonate platform. Prior to the end of the last ice age, about 18,000 years ago, sea level was lower than present day by about 300 feet (91 meters). The low sea levels that occurred during the ice ages exposed the carbonate bedrock of Florida to weathering and dissolution. A drop in sea level also increased the head gradient between land and sea, promoting the formation of Karst through the increased and deeper circulation of freshwater in the carbonate bedrock that underlies Florida.

Figure 9.6 Sinkhole in Winter Park, Florida, 1981.
(From Galloway et al, 1999, p. 5.)

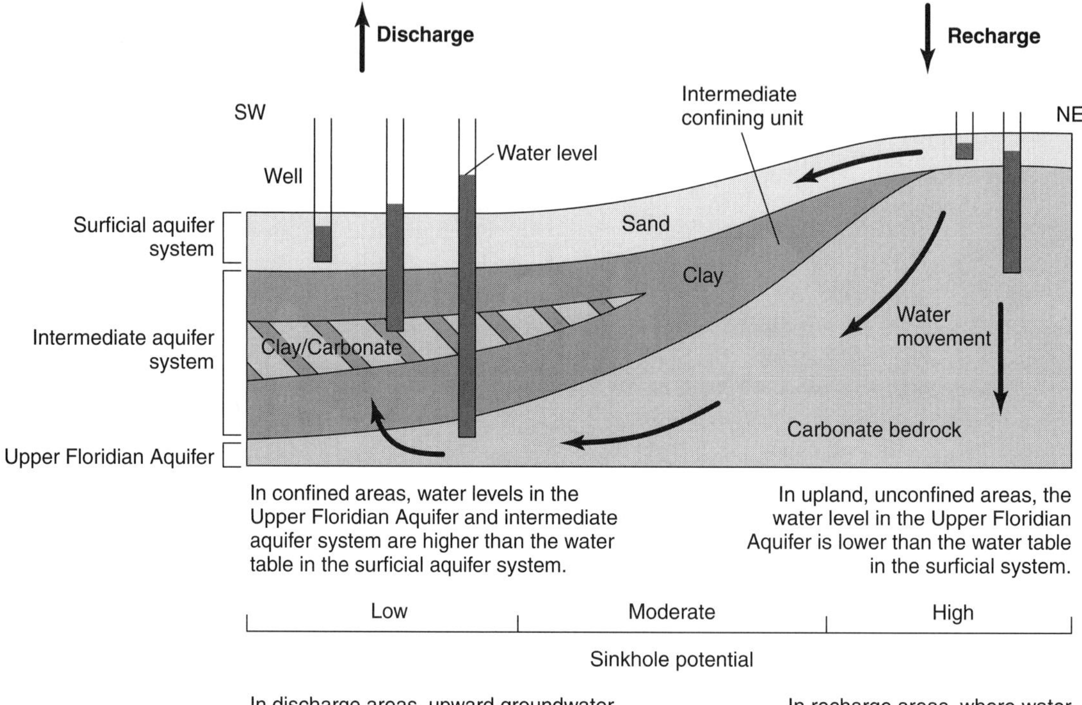

Figure 9.7 Hydrostratigraphy of west-central Florida, showing (1) surficial aquifer system, (2) intermediate layer, which can act as a confining layer or aquifer in different areas, and (3) Upper Floridian Aquifer.
(From Galloway et al, 1999, p. 130.)

The ultimate cause of all the sinkholes found in Florida is the dissolution of carbonate bedrock by weakly acidic groundwater. Most groundwater is not neutral or alkaline in its natural state, but slightly acidic. The dissolution of carbon dioxide in falling rain as well as soil water turns most groundwater into a weak solution of carbonic acid (H_2CO_3) with a pH of about 5.7. Like most rocks, carbonates contain joints, bedding planes, and other natural fractures. As groundwater infiltrates and percolates along these openings, it reacts chemically to dissolve either limestone or dolomite. As dissolution proceeds, what may have started out as a narrow fracture may be enlarged to a cavernous opening as further dissolution opens a route for increased flow. Groundwater flowing through carbonate rocks eventually becomes saturated with dissolved minerals, at which point no further dissolution occurs, however, carbonate salts of calcium and magnesium may precipitate from the groundwater, forming the stalactites and stalagmites commonly found in caves.

The water table in west-central Florida is close to the surface, and most groundwater recharge occurs by infiltration of rainwater. The hydrostratigraphy of west-central Florida consists of three units: a surficial layer of unconsolidated sediments, an intermediate clastic layer that acts as a confining layer in places, and the deep Floridian Aquifer (Figure 9.7). The top layer is composed of unconsolidated sand, shell, and clay, varies in thickness from 10 to 100 feet (3 to 30 meters), and is a viable aquifer in places. The intermediate layer is composed of a heterogeneous mixture of clastic sediments that thicken from north to south. To the north, the intermediate layer contains a larger percentage of clays and acts as a confining layer. In the south, the intermediate layer contains higher amounts of sands and interbedded carbonates, and is known as the "intermediate aquifer system." The presence, composition, and thickness of the intermediate layer determine the type and frequency of the sinkholes that form in west-central Florida. The bottom hydrostratigraphic unit is the Upper Floridian Aquifer; it is one of the most productive aquifer systems in the world. Containing dissolution cavities formed during previous glacial periods when active downward flow was promoted by low sea levels, the permeability of the Upper Floridian Aquifer is in the neighborhood of 10^{-10} to 10^{-11} m^2. In upland regions to the north, the Upper Floridian Aquifer is partially exposed at the surface and is recharged by infiltration. Artesian conditions may be found in the south where the intermediate layer forms a confining layer.

Although sinkholes form naturally as a result of ongoing dissolution, there are two ways in which human actions can cause the formation of sinkholes: lowering of head and loading. Both of these actions increase the effective stress borne by a carbonate skeleton that has been weakened by dissolution and the formation of underground cavities. If the bedrock cannot bear the increased effective stress, the bedrock structure collapses and a sinkhole forms at the surface.

Sinkholes are often formed by increases in effective stress due to lowering of fluid pressure by pumping. Some instructive case histories are given by Tihansky (1999). By the early 1930s, groundwater development along the west coast of Florida had caused upconing and saltwater intrusion. Municipalities began to look inland for new groundwater sources. The city of St. Petersburg opened a well field north of Tampa in Section 21, and began pumping in 1963. In April of 1964, the pumping rate was tripled. Within one month, 64 sinkholes formed within 1 mile (1.6 kilometers) of the field. Most of the sinkholes formed in the vicinity of the well designated number 21-10, that was pumping groundwater at twice the rate of other wells in the field. The spatial distribution of sinkholes around well 21-10 was not random, but followed linear trends reflecting pre-existing joints in the carbonate bedrock.

On February 25, 1998, hundreds of new sinkholes ranging in diameter from 1 to 150 feet (0.3 to 46 meters) formed within a 6-hour period when a new irrigation well started pumping (Figure 9.8). The sinkholes formed in an upland area near the Florida coast that straddled Pasco and Hernando counties. The area contained cavernous limestone bedrock overlain by 20 feet (6 meters) of sand. The well was drilled through 140 feet (43 meters) of limestone, penetrating a cavern from 148 to 160 feet (45 to 49 meters) depth. Drilling stopped and water production began. Two small sinkholes formed near the well shortly after pumping started. Sinkholes of various sizes started to appear throughout the area as groundwater extraction continued. Extensional cracks and crevices formed throughout the landscape. Trees fell over, and there was widespread collapse and slumping of the surficial sediment layer. The first two sinkholes that formed near the well continued to grow in size to become the largest examples of the hundreds of sinkholes that formed. They swallowed several 60-foot-tall (18 meters) pine trees, and eventually consumed more than 20 acres (80,937 square meters) of forest.

In April of 1988, several sinkholes formed in northwestern Pinellas County as an apparent result of increased surface loading from the spraying of effluent from a wastewater treatment plant. As wastewater was sprayed over 118 acres (477,529 square meters), the surficial sediments became saturated increasing the total stress and thus the effective stress on the underlying karst limestone bedrock. The volume of wastewater sprayed over one year was equivalent to 290 inches (737 centimeters) of rainfall. As surficial sediments became saturated, ponding of the sprayed effluent occurred. During the beginning of the rainy season, several sinkholes formed and the effluent ponds drained into the underlying aquifer system. Within

Figure 9.8 Sinkholes in west-central Florida formed as a result of the 6-hour operation of an irrigation well.
(From Galloway et al, 1999, p. 138.)

2 weeks, traces of effluent appeared at a spring 2,500 feet (762 meters) away, indicating that the formation of sinkholes had formed a pathway for contaminants to enter the regional aquifer. Subsequent tracer tests indicated that the groundwater velocity between the spring and effluent-treatment area was 160 feet (49 meters) per day.

9.2. SALTWATER INTRUSION.

Freshwater in the continental crust generally exists as a relatively thin layer underlain by brines and surrounded on coastal areas by salty ocean water. Problems result when an aquifer supplying freshwater becomes contaminated with saltwater. In 1969, the American Society of Civil Engineers found one or more types of saltwater intrusion in 42 of the 50 states of the United States (ASCE, 1969). A unique aspect to saltwater contamination is that its effects are usually hidden. In 1977, the United States Environmental Protection Agency (EPA, 1977) characterized the problem in the following way:

> Waste from municipal and industrial sources entering natural streams or reservoirs are responsible for the more visible types of pollution; their detection is rapid, their source can usually be identified, and their elimination will result in rapid natural improvement of water quality. In contrast, the clandestine movement of saltwater through a freshwater aquifer continues, defying early detection, concealing its origins, and creating long-term problems with expensive remedies.

In the case of seawater, the addition of as little as 2% seawater to freshwater can cause it to be undrinkable. In general, the amount of saltwater contamination that can be tolerated depends on the use to which the water is to be put. Water with a total dissolved solids (TDS) content of less than about 0.5 grams per liter is suitable for domestic purposes. Higher levels of TDS (3 to 12 grams per liter) can be tolerated in water used for irrigation or livestock consumption.

There are a number of different ways in which freshwater aquifers can become contaminated by saltwater; the three most important mechanisms are (1) changes in groundwater head that occur as a result of development, (2) accidental or inadvertent destruction of natural barriers that formerly separated fresh and saline waters, and (3) disposal of waste saline water. Most saltwater intrusions fall into one of the following three categories: (1) seawater intrusion into coastal aquifers, (2) intrusion into inland aquifers, and (3) surface contamination. The most important and prevalent type of contamination is the movement of seawater into coastal aquifers.

9.2.1. Seawater Intrusion into Coastal Aquifers.

As land elevations and water tables are higher for continents than for oceans, freshwater tends to flow through coastal aquifers that may discharge into the sea. If groundwater head is lowered for any reason, the flow of water may be reversed, with a tongue of saltwater infiltrating a freshwater aquifer in a landward direction (Figure 9.9). As salty ocean water is denser than freshwater, the advancing body of saltwater tends to move along the base of the aquifer. The interface between the fresh and saltwater tends to be parabolic in its shape.

It is generally recognized that the first analyses of seawater intrusion were made by a Dutch Captain of Engineers, W. Badon-Ghyben, in 1889. An identical analysis was arrived at independently by Alexander Herzberg, who studied the freshwater-saltwater balance on the island of Norderney off the north coast of Germany in 1901. Badon-Ghyben and Herzberg calculated the depth of the saltwater-freshwater interface in terms of the height of the water table on land above sea level by assuming hydrostatic conditions. The Badon-Ghyben-Herzberg relationship is based on a hydrostatic balance between salt and freshwater. Suppose, as a thought experiment, we were able to insert a U-shaped glass tube into an aquifer so that one side of the tube contained a freshwater column, and the other side contained a saltwater column. If the position of the saltwater-freshwater interface is not moving, the fluid pressure at the

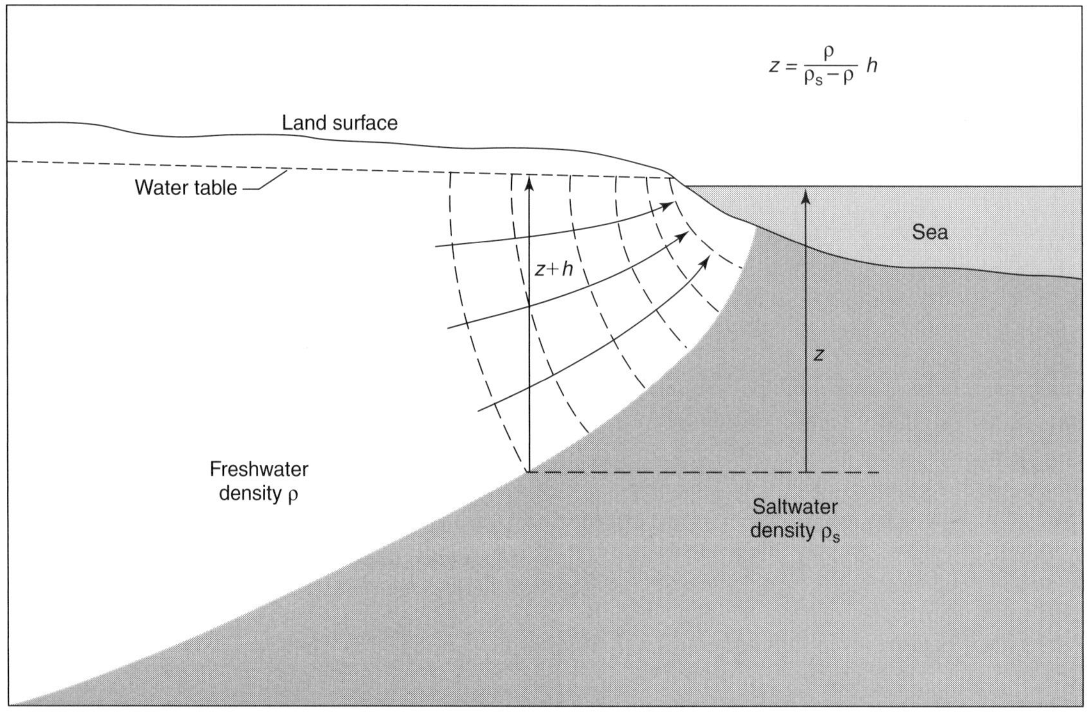

Figure 9.9 Badon-Ghyben-Herzberg approximation is based on a static balance between saltwater and freshwater and predicts that the depth to the saltwater-freshwater interface below sea level (z) will be 40 times the height of the terrestrial water table above sea level(h).
(From Cooper, 1964, p. C2.)

base of the freshwater column must be equal to the fluid pressure at the base of the saltwater column. Because saltwater is denser than freshwater, the column of freshwater must be higher by an amount h to balance the higher pressure created by the denser saltwater column. Applying equation 2.23, which gives fluid pressure as a function of fluid density, depth, and the acceleration due to gravity, the condition of hydrostatic equilibrium can be stated mathematically as

$$\rho g(z + h) = \rho_s g z \tag{9.7}$$

where ρ is freshwater density (kg-m^{-3}), ρ_s is saltwater density (kg-m^{-3}), g is the acceleration due to gravity (m-s^{-2}) z is depth (m), and h is elevation (m) of the terrestrial water table above sea level (Figure 9.9). Solving equation 9.7 for the depth of the freshwater-saltwater interface,

$$z = \frac{h\rho}{\rho_s - \rho} \tag{9.8}$$

For a nominal freshwater density of 1000 kg-m^{-3}, and a nominal seawater density of 1025 kg-m^{-3},

$$z = 40h \tag{9.9}$$

or the depth (below sea level) of the saltwater tongue intruding onto land is about 40 times the height of the terrestrial water table above sea level.

As an interesting historical note, Carlston (1963) pointed out that the Badon-Ghyben-Herzberg relationship had first been discovered and published in the *American Journal of Science* in 1828 by Dr. Joseph Du Commun, a teacher of French at the West Point Military Academy. Du Commun had attempted to answer the question of

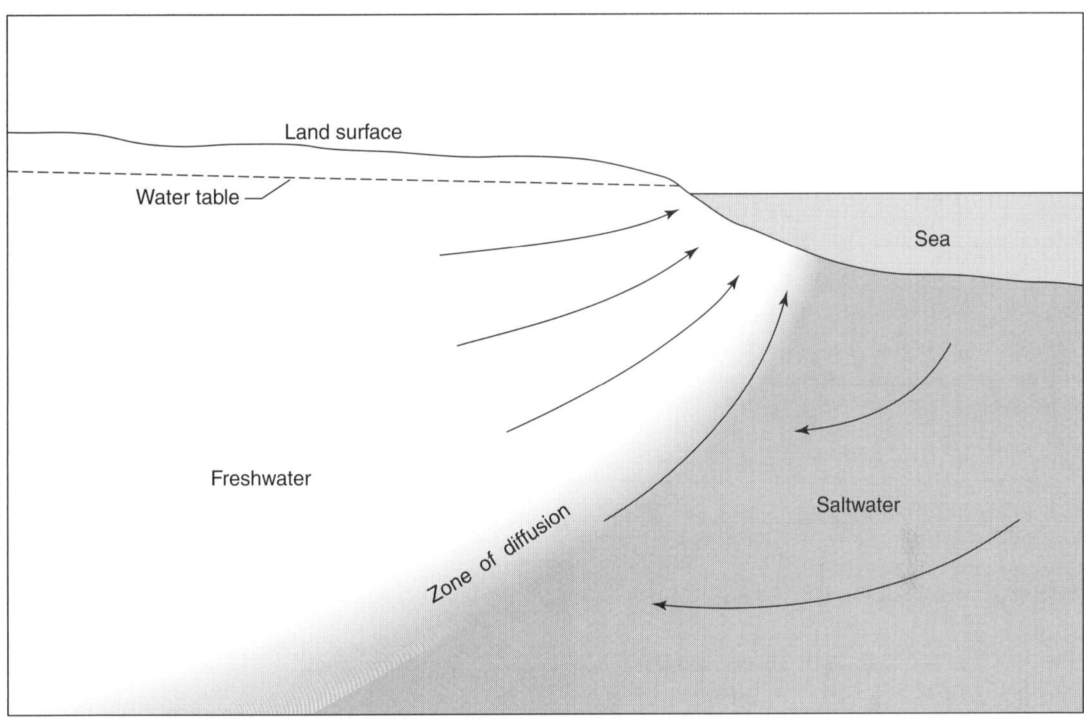

Figure 9.10 The saltwater-freshwater interface is not sharp, but is graduated due to the dispersion and diffusion of salt into freshwater.
(From Cooper, 1964, p. C3.)

why the water level in a distillery well near the Raritan River in New Jersey rose and fell in accordance with the rise and fall of ocean tides in the river. Du Commun explained that when saltwater entered the river mouth during high tide, the greater density of the saltwater caused fluid pressure to rise, increasing the freshwater head on land. Accordingly, water level in the distillery well rose in phase with the ocean tide.

Hubbert (1940) noted that the Badon-Ghyben-Herzberg relationship should be more properly termed the Badon-Ghyben-Herzberg "approximation," as the situation near coastal areas is almost never in hydrostatic equilibrium. Because the elevation of the water table on land is almost always higher than sea level, there is a flow from land to sea. In other words, the hydrostatic equilibrium assumed in a Badon-Ghyben-Herzberg analysis does not exist. The usual depiction of the freshwater-saltwater interface as a sharp and distinct front is also an approximation. Fresh and saltwater are miscible fluids. **Miscible** fluids are those that can be mixed in each other. An example of immiscible fluids would be oil and water. The primary reason that the saltwater-freshwater interface is not sharp is that dispersion tends to mix the fluids in the area in which they come into contact (Figure 9.10). **Dispersion** is mechanical mixing. In a coastal aquifer, dispersion is augmented by the daily variations in tide levels and seasonal variations in the height of the terrestrial water table. During times of high tide, head on the seaward side of the aquifer increases and saltwater may invade the land. Heavy precipitation events may cause the terrestrial water table to rise, and the increase in head on the landward side may cause the tongue of invading saltwater to be pushed back towards the sea. The environment is dynamic.

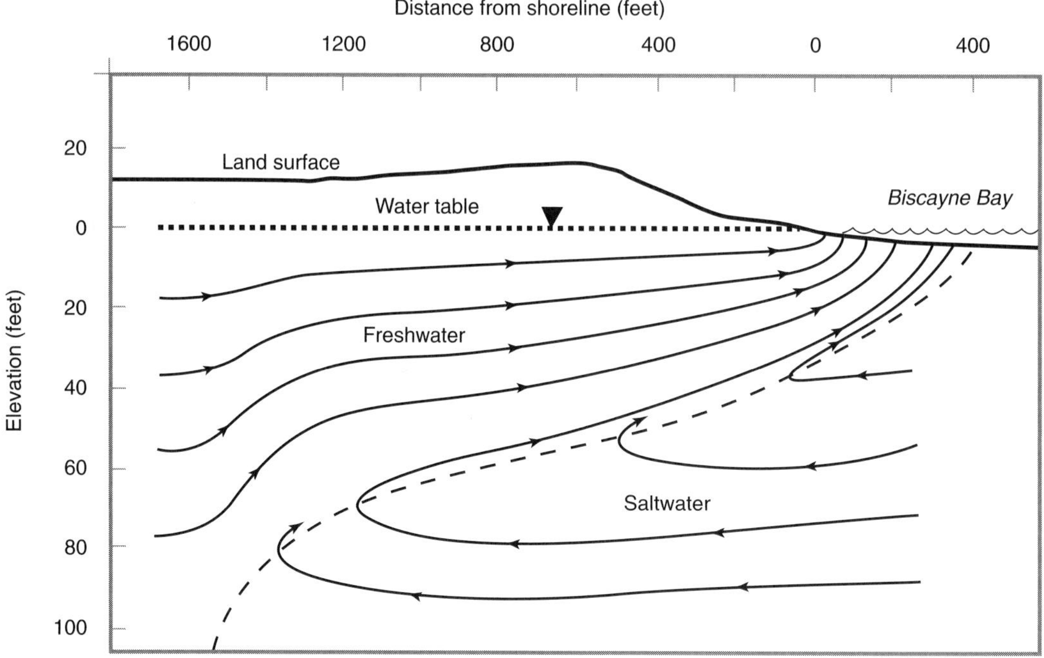

Figure 9.11 Streamlines show the movement of freshwater and seawater near a coast. If the freshwater-seawater interface is not moving, there must exist a line marking a point of net zero horizontal movement. Salt moves vertically across this line into the zone of freshwater and is carried back to the sea. Thus, even if the saltwater encroachment has been stabilized, there is still an active circulation of salt to and from the sea. 1 foot = 0.3 meters. (From Kohout, 1964, p. C31.)

If the overall position of the saltwater-freshwater interface is not changing, there must exist an imaginary line somewhere in the aquifer where the horizontal component of saltwater transport is zero (Figure 9.11). Even if the position of the saltwater-freshwater interface is stagnant, there is still an active transport of saltwater to the land and back to the sea. Dispersion moves salt vertically across the interface where it is caught up in the seaward flow of freshwater. Thus, the invading salt tongue exists in a state of dynamic equilibrium where salt is being continually transported landward and then back to the sea. Kohout (1964) estimated that in the case of the Biscayne Aquifer in southeastern Florida, about seven-eighths of the total discharge at the shoreline originated as freshwater, with one-eighth being a return of seawater.

Seawater intrusion usually results from decreases in terrestrial head, or the upstream movement of seawater through canals and streams; both causes are related to urban and industrial development. Development can cause head decreases in several ways. The most obvious cause is pumping of an aquifer to supply water to an increasing population. Continuous and heavy pumping of a deep, regional limestone aquifer near the Savannah area of Georgia and South Carolina has resulted in the development of a massive cone of depression. As of 1969, the American Society of Civil Engineers (ASCE, 1969) reported that the cone of depression was 50 miles (80.5 km) wide and had lowered the terrestrial water table to 120 feet (37 meters) below sea level. Seawater accordingly started to move toward shore, but as of 1969 was still located a considerable distance offshore. Less obvious are the decreases in head that may occur as incidental results of urbanization when vegetated areas are replaced by pavement. Increases in paved

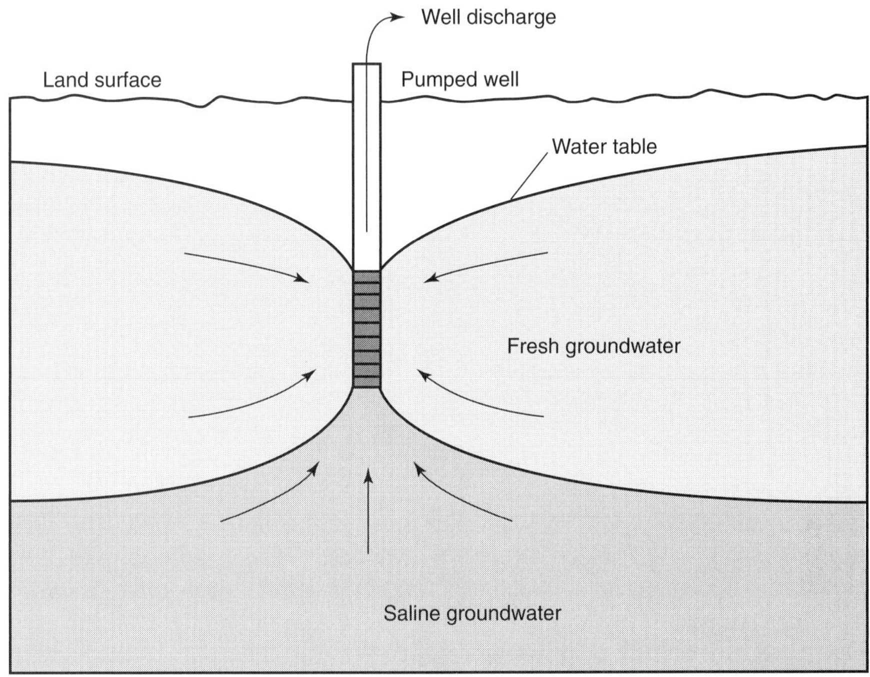

Figure 9.12 Upconing of saline groundwater from pumping.
(From Alley et al, 1999, p. 64.)

areas lead to increases in runoff and decreases in recharge that lower the water table. Sometimes, canals and other drainage channels are constructed for the purpose of draining swamps and wetlands. Drainage has the beneficial effect of making land available for the construction of houses and roads, but also lowers the water table.

9.2.2. Intrusion into Inland Aquifers.

Saltwater intrusion into inland aquifers generally results from either upconing or the breaching of a confining layer by a deep well. **Upconing** is the rise of a cone-shaped body of saltwater that may occur when head is lowered in the overlying freshwater strata (Figure 9.12). The depth at which salinity begins to appreciably increase in the continental crust depends upon a number of factors and is variable. As a rule of thumb, a depth of about 100 meters is probably a good average number. The American Society of Civil Engineers (ASCE, 1969) cited a depth of 400 feet (122 meters) for the depth to "mineralized" water in the midwestern states of the United States.

Lowering of the water table leading to upconing may occur not only through pumping but also through dewatering operations that may be conducted in connection with quarrying, mining, excavating, or road building. Deepening of a gaining stream will also cause the water table in the vicinity of the stream to drop. In order for upconing to occur, the interval separating the underlying saltwater from the near-surface freshwater must have relatively high permeability in comparison to the layer containing the freshwater.

In addition to upconing, saltwater may find its way into the near-surface freshwater zone if confining layers are breached. Typically, this occurs when abandoned oil well casings become broken or corroded. State laws and regulations now require that abandoned wells be cemented shut, however, this was not the case for many older wells that were simply capped or plugged near the top upon abandonment. When the casings of these wells break or corrode, a pathway is created for brines at depths of thousands of meters to travel to the freshwater zone near the surface. Each instance is unique, but the American Society of Civil Engineers (ASCE, 1969) described the following situation as "typical." In 1955, three water wells were drilled near Terre Haute, Indiana. The wells were completed in a shallow sand and gravel aquifer that normally yielded potable water with a chloride concentration of around 16 parts per million (ppm). But, samples of water from these wells had chloride concentrations of about 550 ppm, thirty-four times higher than normal. The source of the contamination was an abandoned oil well located about 2000 feet (610 meters) from the water wells. The abandoned oil well was 1700 feet (518 meters) deep; the water at that depth had a salinity of 8,500 ppm chloride. The head of the water in the oil well was 28 feet (8.5 meters) above head levels in the nearby water wells, causing saltwater to flow towards the freshwater wells. To remediate the situation, the abandoned oil well was properly plugged and the water wells were pumped for several months until the saltwater had been flushed from the aquifer.

Eddy (1965) related the case of the Saginaw Valley in Michigan as a classic case of aquifer contamination resulting from saltwater leaking upward from improperly plugged deep wells. As of 1965, it was almost impossible to obtain potable groundwater from aquifers in this area that otherwise should have yielded freshwater of good quality. Saginaw is located at the head of navigation on the Saginaw River, which empties into Lake Huron, and was an historic sawmill town from about 1840 through 1880. Waste wood products from the lumber industry supplied low-cost fuel to produce salt by boiling brine. Wells were drilled solely for the purpose of obtaining very salty brines to be used to manufacture salt. As the lumber industry declined in the late nineteenth century, the salt industry closed down. Drilling for brines was followed by a period of exploration drilling for coal. Except for areas in which coal was found, test wells were abandoned without plugging. In 1925, oil was discovered, and a number of wells were drilled and abandoned without plugging. There were no requirements that wells be plugged until Michigan passed an oil and gas conservation law in 1939. As of 1965, an untold number of wells drilled for salt, coal, and oil in the Saginaw area remained unplugged, allowing deep, salty water to move upwards contaminating near-surface aquifers.

9.2.3. Disposal of Waste Saline Water.

One of the largest sources of waste saline water is brines from oil fields. In the state of Texas in 1961, 2.2 billion barrels (356 billion liters) of saltwater were produced from 67,000 oil and gas wells (McMillion, 1965). In increasing order of magnitude, there are three methods that have been used for disposal of oil field brines: discharge into streams, surface pits, and deep injection. In the state of Texas, in 1961, 10.1% of oil field brines were discharged into streams, 20.6% were placed into open surface pits, and 68.7% were disposed through deep-well injection. A remaining 0.6% were disposed of by miscellaneous methods, such as spraying on highways (McMillion, 1965).

The quantity of brine that can be disposed of through discharge into streams is usually very small. The amount of salt pollution that can be tolerated in most streams is not large, especially if municipal water for drinking is extracted downstream. Surface pits are known under various names, including retention pits, brine-storage pits, and evaporation pits. The term *evaporation pit* is usually an egregious misnomer, because the water level in such pits is reduced not by evaporation, but by seepage. Knowles (1965) suggested that a better name than *evaporation pit* was *infiltration pond.* Surface pits are constructed by excavating a hole in the surficial layer of soil. Many such pits are unlined, and have the potential to create serious salt-contamination of groundwater supplies. Brines leak from the surface pits and can be carried directly into either freshwater aquifers or surface drainage ways. If the soil or bedrock underneath

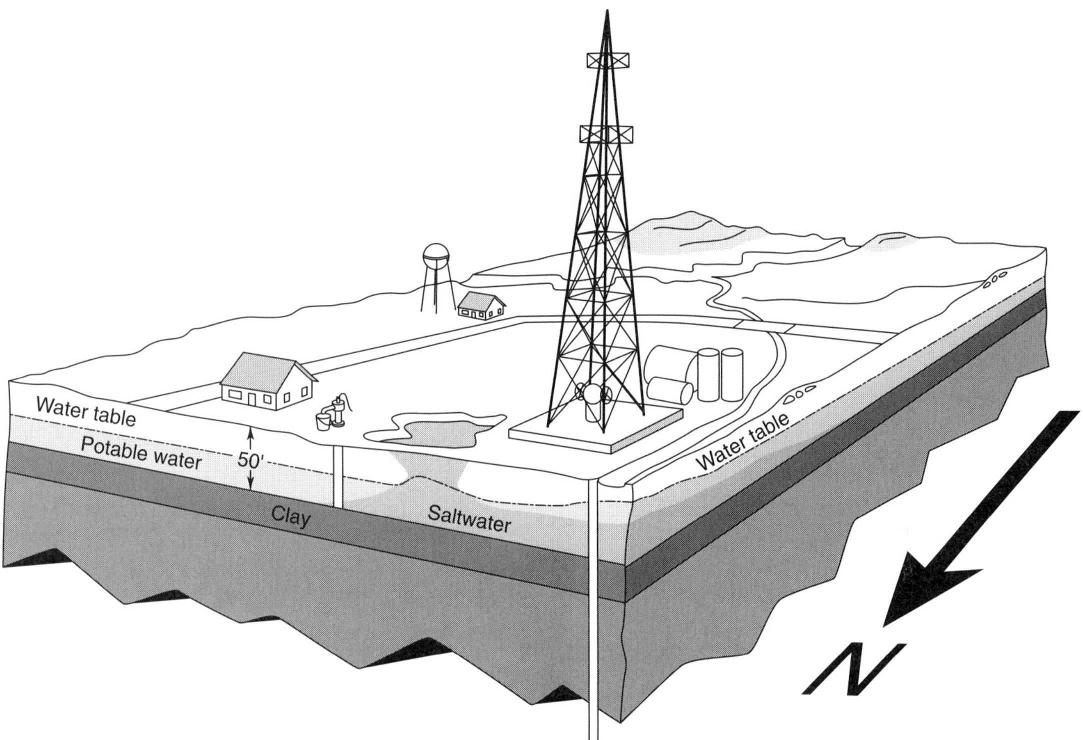

Figure 9.13 Infiltration of oilfield brines from a surface pit in the Pollard Oil Field in Alabama. Note lateral movement of brine in shallow aquifer underlain by relatively impermeable clay layer.
(From Knowles, 1965, p. 25.)

the pit contains a large amount of clay or shale, downward flow is inhibited by low permeability and substantial lateral migration of contaminant may occur. McMillion (1965) reported that vast areas of north- and west-central Texas had experienced soil salinization and water-quality deterioration as a result of leakage from surface pits. Several municipal well fields supplying water to towns and cities in parts of Texas were affected. Knowles (1965) described saltwater contamination at the Pollard oil field near the southern border of the state of Alabama (Figure 9.13). From 1944 through 1965, about 55 million barrels (8.7 billion liters) of brines were produced. Pollution resulted from leaking and overflowing surface pits, leaks from pipelines and oil well-heads, and leaks from improperly constructed oil and saltwater disposal wells. Brines from overflowing surface pits moved to nearby streams, with the result that baseflow to many streams in the area became highly mineralized. Problems in the Pollard field were exacerbated by the local geology. A 50-foot-thick section (15 meters) of highly permeable sand and gravel beds at and near the surface facilitated the spread of contaminants. Near-surface sands and gravels were underlain by relatively impermeable beds of clay that inhibited the downward migration of contaminants and facilitated the lateral spread of salt through what would have been a freshwater aquifer.

If properly done, the most effective and environmentally sound method for the disposal of oilfield brines is deep-well injection. McWilliams (1972) described the case history of the East Texas oil field as an exemplary case where successful

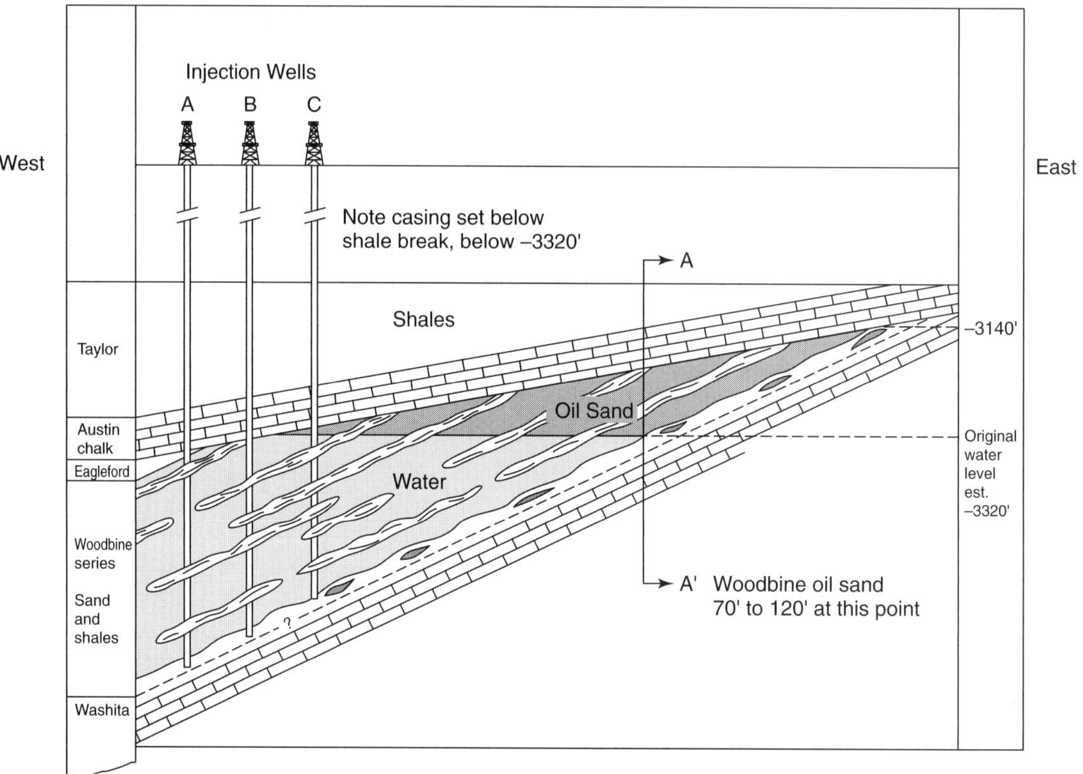

Figure 9.14 Generalized west-east geologic cross-section through the East Texas Oil Field showing the disposal of oil-field brines by downdip injection.
(From McWilliams, 1972, p. 332.)

disposal of oilfield brines not only alleviated surface pollution, but also facilitated enhanced recovery of oil by increasing reservoir pressure. The East Texas oil field was discovered in September, 1930, and is the largest known oil field in the conterminous United States. The field covers an area of more than 200 square miles (520 square kilometers). At its peak, the field contained 26,000 producing oil wells. As of 1972, more than 4 billion barrels (635 billion liters) of oil had been produced. By late 1937, 100,000 barrels (15.9 million liters) of saltwater were being produced from wells in the East Texas field each day. Saltwater production exceeded oil production, and reservoir pressure had declined by 31%. There was an obvious need to find a way to dispose of saltwater and also keep reservoir pressure levels from declining further. The decision was made to collect, process, and inject waste saltwater into the primary reservoir rock of the East Texas field, the Woodbine Formation, however, injection was to take place on the west side of the field, downdip from oil accumulations (Figure 9.14). The return of saltwater to the Woodbine Formation began in 1938 and resulted in the pressure in the reservoir leveling off at about 65% of the pre-development level. As of 1972, virtually 100% of the brines produced in the East Texas field were being disposed of through the daily injection of 500,000 barrels (79 million liters) of brine through 84 wells. McWilliams (1972) estimated that the cost of the brine disposal to operators was 2.3 cents per barrel (0.175 cents per liter). At this time, crude oil sold for about two to three dollars per barrel.

9.2.4. Control and Remediation Methods.

The control and remediation of saltwater intrusion is dependent upon the geologic circumstances in each case. A natural precursor to adopting any control or remediation strategy is a thorough geologic study to identify the causes of the saltwater intrusion and the mechanism(s) by which it is occurring. In the case of southern Alameda County in the San Francisco Bay area, groundwater is being contaminated by saltwater intrusion through a combination of four different mechanisms (EPA, 1973).

In general there are two approaches to controlling saltwater intrusion: (1) construct barriers, and (2) change the distribution of head. For those cases in which saltwater is flowing up abandoned wells that were never plugged, the solution is to locate and plug the abandoned wells by injecting concrete. In the case of seawater intrusion, head in the freshwater section of the aquifer must be increased and/or head in the saltwater section decreased. In the Los Angeles area, saltwater intrusion was reversed by injection of freshwater through a line of wells along the coastline (EPA, 1977). In the case of the Biscayne Aquifer in southeastern Florida (see Section 9.2.5), saltwater intrusion was stopped (but not reversed) by raising the freshwater level in drainage canals and installing dams that prevented saltwater from flowing inland during times of high tide. If upconing is occurring, the only solution may be to reduce the rate of pumping.

No matter what remediation is implemented, a monitoring network should be installed to verify the effectiveness of the method. In most cases, the network will consist of observation wells completed in the freshwater zone with regular withdrawals and measurement of salinity. In general, observation wells should penetrate the deepest part of the aquifer so as to detect the first presence of an advancing saltwater wedge. Regular measurements of pumping rates and changes in head will aid the recognition of the causes of saltwater intrusion (EPA, 1973).

9.2.5. Case Study: The Biscayne Aquifer, Southeastern Florida.

Perhaps the best known example of seawater intrusion into a freshwater aquifer is the case of the Biscayne Aquifer in southeastern Florida. The Biscayne Aquifer is an unconfined aquifer that underlies the entire eastern margin of Florida; it extends from the surface to depths of about 75 to 125 feet (23 to 38 meters). It is composed mainly of sandy limestones and calcareous sandstone with beds and pockets of quartz sand. Limestone in the upper part of the Biscayne Aquifer is remarkably soft. Water wells can be completed by driving steel pipe casing into the ground with a sledgehammer. The casing is then removed and the rock material extracted from it. After emptying, the casing can be reinserted and driven further down. The process is repeated until a well of desired depth is achieved. Parker et al. (1955) noted that a well of 20 to 40 feet depth could be constructed in a day without "undue hardship." In the nineteenth century, most wells were completed in this manner. The first large well was drilled in 1896 to a depth of 60 feet using a steam pump. The well was 4 inches (10 centimeters) in diameter, and supplied water to the Miami Hotel.

The average permeability of the Biscayne Aquifer is 1.7×10^{-9} m^2 (Parker et al., 1944). Parker et al. (1955) described it as "one of the most permeable aquifers ever investigated by the U.S. Geological Survey," equating its ability to transmit water to that of clean, well-sorted gravel. Recharge to the Biscayne Aquifer occurs primarily through rainfall. Annual rainfall in southeastern Florida is about 60 inches (152 centimeters). About 63% of the rainfall enters the aquifer through infiltration, the remaining 37% is lost through evapotranspiration. Direct runoff constitutes a negligible amount of precipitation.

Prior to twentieth-century development, a tremendous amount of freshwater was ponded in the Everglades west of Miami a short distance from the eastern shoreline. Shaler (1890) reported that the mean water level only three miles inland was 16 feet (5 meters) above high tide, or about 10 feet (3 meters) above mean sea level. Parker et al. (1955) reviewed several historical records of abundant freshwater springs near the shoreline. Griswold (1896) mentioned "great springs of constant flow" that emerged "in large numbers along the shore." Fuller (1904) related the observation of one spring that flowed 100 gallons per minute (379 liters per minute) from the base of a limestone bluff. Between 1907 and 1928, 10 canals were constructed in the Miami area to drain away ponded water and reclaim land for development. The Miami River was deepened by dredging. By 1910, dredging of the Miami River to a depth of 10 feet had been extended a distance of 4.25 miles (6.8 kilometers) into the Everglades. Water that was formerly ponded in the Everglades behind the coastal ridge was now free to flow into the river. Canal construction was very successful in reclaiming land for development. The water table dropped 6 feet (1.8 meters), and land on either side of the canals was reclaimed for a distance of 1.0 miles (1.6 kilometers). In a letter dated 1907 (quoted in Parker et al., 1955), W. S. Jennings described the effects of canal construction:

> The result is that the reclamation of the land is fully demonstrated. We walked for a distance of one-half mile or more along an Indian trail or canoe route through the saw grass, where 20 days ago the Indians traveled with their boats and canoes, the water having all been drawn off from this territory by the cutting of the canal, thus lowering the water level.

In 1908, the Miami water supply was derived from four wells located in a swale partly surrounded by higher land behind which the waters of the Everglades were impounded. Twelve years later, in 1920, the deeper part of the wells had to be plugged to avoid drawing up saltwater.

Saltwater intrusion occurred in two ways. First, the position of the saltwater-freshwater interface moved landward because head in the Biscayne Aquifer had been lowered by draining and pumping. Second, the construction of canals and river dredging allowed seawater during times of high tides and low seaward flow to move upstream. Pumping water wells in the vicinity of the canals had created cones of depression that touched upon the canal and rivers. This allowed saltwater to infiltrate directly into the Biscayne Aquifer.

Saltwater intrusion first became a significant problem during the drought of 1938–39. Freshwater flow through the canals to the Atlantic Ocean became diminished. Flow reversals in tidal canals occurred for periods of 2 to 5 hours during each tidal cycle, and saltwater migrated farther and farther inland. In some canals, saltwater intruded inland a distance of 10 miles. The most apparent sign of the problem was the appearance in 1939 of saltwater in the Miami well field that supplied municipal water for the city of Miami. The Miami well field was located adjacent to the Miami Canal, about 6.6 miles (10.5 kilometers) inland from Biscayne Bay. An investigation of the intrusion was begun by the U.S. Geological Survey. Parker et al. (1944) related the scope of their investigation:

> In order to carry out the investigation effectively, it was necessary to obtain basic geologic, hydrologic, and chemical data. Specifically, it was necessary to determine the depth, thickness, and areal extent of the water-bearing rock formations; the capacity of these formations to transmit and yield water; the areas and rates of recharge and discharge, the quality and quantity of water in the different parts of the water-bearing rocks; the source and approximate movement of saltwater at all depths in the water-bearing rocks, factors controlling saltwater encroachment; the height of the water table and the direction of flow of the ground water at different times of the year; the stage and discharge of the canals throughout the year; the periods when the canals drained the nearby land areas and when they fed water to these areas; and the approximate quantities of water involved.

Were this sort of investigation to be carried out today, we could add the use of geophysical methods and numerical modeling.

Up until 1943, the average rate at which saltwater was advancing landward was 235 feet (72 meters) per year, however, a severe drought occurred from 1943 through 1946. In 1945, water levels fell to all-time lows. In a 27-month period, the saltwater front advanced a total of 2,000 feet (610 meters) at an average rate of 890 feet (271 meters) per year (Figure 9.15). By 1968, judicious regulation of canal flows had stabilized the position of the saltwater front a distance from 2 to 8 miles (3 to 13 km) inland.

Saltwater intrusion into the Biscayne Aquifer was brought under control through the regulation of canal flow. Dams were constructed to (1) halt the upstream movement of seawater during times of high tides, and (2) hold higher freshwater levels so as to avoid excessive drainage of the Biscayne Aquifer.

9.3. GROUNDWATER CONTAMINATION.

Groundwater contamination is the introduction of any substance into groundwater that impairs its utility for human purposes, including drinking. Contamination of groundwater is an important environmental problem in the United States, because slightly more than half of the population relies upon groundwater as their primary household source of water. In rural areas, people rely almost exclusively on wells.

In 1984, the National Research Council (NRC) of the United States estimated that between 1 and 2% of the nation's groundwater was contaminated (NRC, 1984). However, most contaminated aquifers are in areas of dense populations, thus the percentage of the population affected is larger than 1 to 2%. The NRC (1984) also speculated that the area of the United States affected by groundwater contamination may actually be larger than 1 to 2%. Unlike surface pollution, the contamination of aquifers and underground water happens out of sight. As groundwater moves very slowly in comparison to surface water, pollution may occur over years or decades before contaminants reach a water well. Even if well water is contaminated, the level of harmful contamination may be odorless, colorless, tasteless, and therefore undetectable unless tests are run. The total number of possible contaminants may number in the thousands; however, drinking water in the United States is only tested routinely for contamination by bacteria. Comprehensive testing of a single water sample for a range of common contaminants can cost hundreds of dollars.

In the United States today, we are fighting a battle of attrition against groundwater contamination. In 1984, the chief administrator of the U.S. Environmental Protection Agency, William Ruckelshaus, said "At present, we simply do not know how to clean up most groundwater pollution." Although remediation technology developed rapidly in the 1990s and continues to develop today, Ruckelshaus' statement is still substantially true. In 1961, geologist Theodore Walker in reference to groundwater contamination at the Rocky Mountain Arsenal near Denver wrote "The only reasonable cure for such problems is the prevention of contamination" (Walker, 1961).

9.3.1. History of Groundwater Contamination.

The ancient Romans found that shallow water wells quickly became contaminated with sewage wastes from open-hole latrines and privies. They concluded that groundwater was an unreliable source of drinking water, and built an aqueduct system to supply Rome with water (Chapelle, 1997). The Roman Aqueduct system is one of the engineering wonders of the ancient world. Over a period of 500 years, the Romans built 11 major aqueducts that brought water into Rome from as far away as 57 miles (92 kilometers). Some of the Roman aqueducts are still in use.

One of the first recorded instances of groundwater contamination was the discovery by a nineteenth-century doctor in London that a cholera epidemic originated in contaminated well water. In August of 1854, London was in the grip of a cholera epidemic. Cholera is an acute bacterial infection of the small intestine; sufferers undergo massive diarrhea with rapid and severe dehydration. Death may follow in a few days if the dehydration is not treated. At that

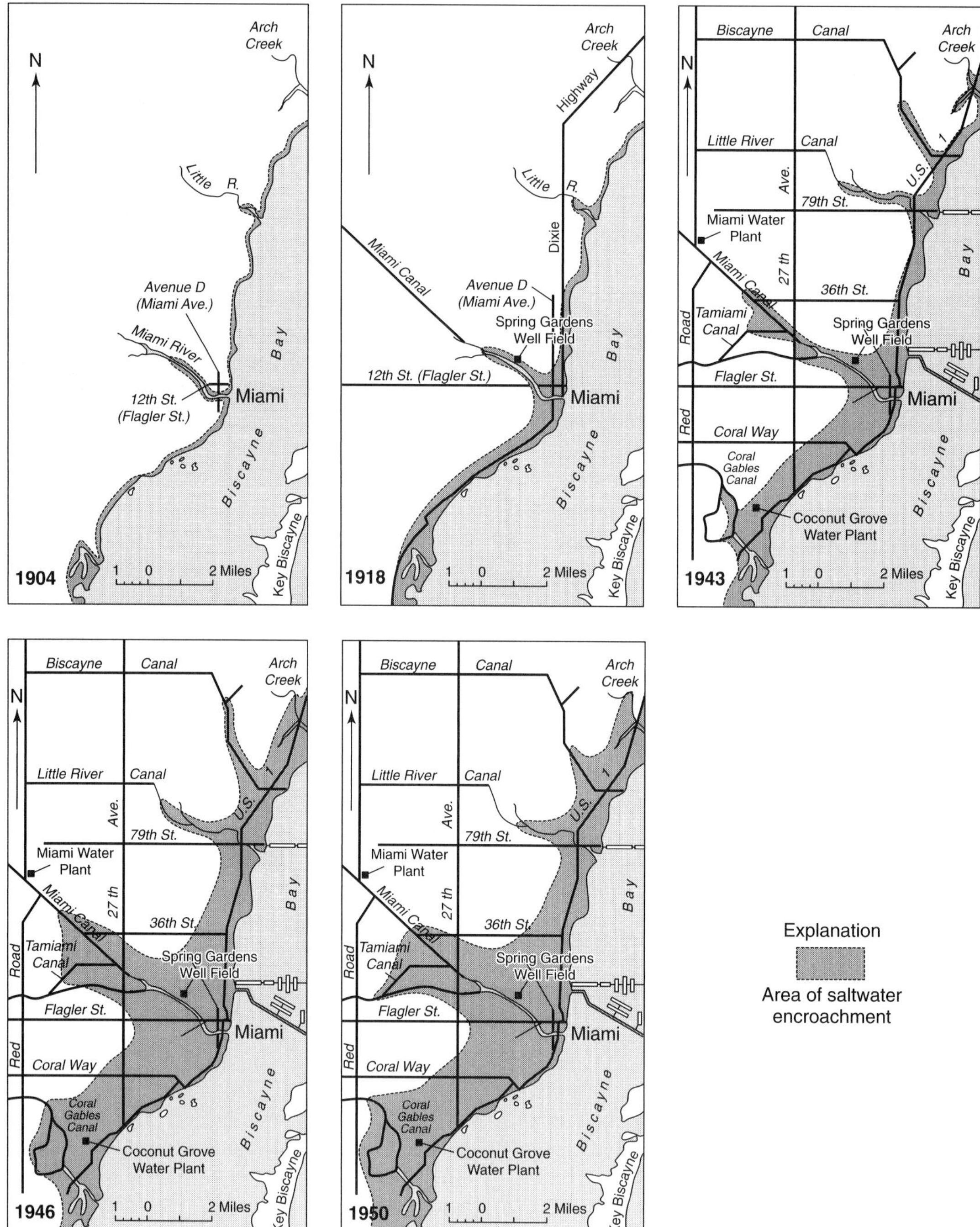

Figure 9.15 Progressive encroachment of salt water into the Biscayne Aquifer, 1904–1950.
(From Parker et al, 1955, p. 589.)

time, the cause of cholera was unknown. It was thought most likely that the disease was transmitted through the air. Dr. John Snow made a map and plotted where cholera sufferers lived. He noticed that the outbreak was centered around a particular well. On September 7, 1854, Dr. Snow demanded a meeting with the local Board of Guardians. At the meeting, he insisted that the pump handle be removed from the contaminated well. Not a single physician or resident of the affected area believed that Snow was right. Nevertheless, Dr. Snow prevailed. The pump handle was removed and the plague was abated.

As the most important source of groundwater contamination in the United States today is hazardous chemicals, the history of groundwater contamination parallels the growth of the chemical industry. The history of the modern chemical industry originates with man's ancient fascination with color and the search for dyes. In 1854 the English chemist, William Perkin, accidentally discovered that the first synthetic dye, mauve, could be manufactured from coal tar, the liquid byproduct that results when coal is heated in the absence of air to manufacture coke. Proceeding largely by trial-and-error, other chemists began to synthesize a variety of other dyes. Although the first dye was synthesized in England, the industry became established in Germany. By 1914, 90% of the world's dyes were manufactured in Germany.

With the onset of World War I in 1914, the English and United States governments found themselves at a disadvantage and were forced to develop their own chemical industries. By the end of the war in 1918, leadership in organic chemistry had begun to shift to Switzerland, England, and the United States. The chemical industry grew through the 1920s and 1930s and started to find ways to utilize petroleum as a raw material for the synthesis of organic chemicals. In October of 1938, DuPont Corporation announced the creation of the first totally synthetic fiber, nylon. Nylon arrived in the United States just as Japan cut off the import of silk and rubber from east Asia in 1939. In 1939, nylon stockings went on sale and were well received by the public. During World War II, the United States had to rely upon synthetic rubber provided by the chemical industry.

In the decades following World War II, American consumers were eager to purchase manufactured goods composed of what seemed like an endless variety of new synthetic chemicals created by the chemical industry. By the end of the twentieth century, the chemical industry was manufacturing more than 70,000 products, ranging from plastics, to soaps, drugs, pesticides, paints, and a long list of other types of products. Many of the materials manufactured by the chemical industry are used as raw materials by other industries. In many ways the chemical industry is the largest and most important sector of the United States economy.

The first widespread notice of environmental pollution due to hazardous chemicals was the 1962 publication of the book *Silent Spring* by Rachel Carson (1907–1964). Carson documented how the indiscriminate use of pesticides such as DDT was responsible for the deaths of birds and fish, and could have adverse health effects on humans. She speculated about the possibility of groundwater contamination and pointed out that the introduction of hazardous chemicals into the environment was upsetting natural ecosystems with unknown and possibly dangerous consequences. With environmental issues receiving more public attention, the United States Congress passed a number of environmental laws. These included the Water Quality Act of 1965, the Clean Air Act of 1970, the Safe Drinking Water Act of 1974, the Resource Conservation & Recovery Act of 1976, and the Toxic Substances Control Act of 1976. In 1970, President Richard Nixon created the U.S. Environmental Protection Agency to enforce these new laws by combining responsibilities formerly divided amongst other agencies.

Before the late 1970s, national attention in the United States was focused on the pollution of air and surface water. Air pollution from burning coal and wood was once so bad that in the cities of Pittsburgh and St. Louis drivers had to turn on their headlights during the day. The Cuyahoga River running through Cleveland, Ohio, was so polluted that it caught fire once in 1936, twice again during the 1950s, and once in 1969. It was long believed that groundwater was largely immune from contamination because most pollutants would be filtered out by the unsaturated zone and never reach the water table. This changed in 1978 when the infamous case of the Love Canal burst on the national scene. Love Canal was an abandoned canal in New York state that had been used for decades as a dumping ground for 22,000 tons of hazardous industrial waste chemicals. In 1953, Hooker Chemical Corporation covered the dump site with dirt and sold it to the local board of education, which promptly built an elementary school on top of the dump. Home-building in the area followed. From the late 1950s onward, people living in the area complained of strange odors and chemicals leaking into their basements. In 1978, the New York State Department of Health began a study that found high levels of chemical contaminants in the area as well as high miscarriage and birth defect rates. In 1978 and then again in 1980, U.S. President Jimmy Carter declared the Love Canal site to be an environmental emergency. Nine-hundred and fifty families living in the area were evacuated and relocated, and an extensive remediation effort began.

Following the unprecedented publicity associated with the Love Canal incident, the U.S. Congress passed the "Comprehensive Environmental Response, Compensation, and Liability Act" (CERCLA), which became more commonly known as the "Superfund" Act. This law appropriated $1.6 billion to pay for cleaning up abandoned hazardous waste sites. By the mid-1980s, the estimated number of hazardous waste sites needing to be cleaned up rose by the thousands and the cost increased to $100 billion. In 1991, the Wall Street Journal reported that the number of sites needing remediation exceeded 20,000 with an estimated cost of $600 billion to be divided between the U.S. government and industry.

The Superfund act required that contaminated groundwater be cleaned up until it met drinking-water standards. At first, it was supposed that cleanup would be straightforward using a method called "pump-and-treat." Contaminated groundwater could be pumped out of the ground, treated to remove contaminants, and then either injected back into the ground or disposed of elsewhere after treatment. For example, many contaminants can be removed by filtering water through activated charcoal.

When the Superfund law was passed in 1980, geologists and engineers had little experience with remediating groundwater contamination. Most of the early groundwater cleanup efforts were a series of uncontrolled experiments in testing whether existing technology was capable of removing contaminants from aquifers. In 1989 and 1992 the EPA published evaluations that concluded that most cleanup efforts had failed to reach their goals. When the pump-and-treat method was used, contaminant levels at first dropped rapidly, but the rate of decrease then declined. In other circumstances, contaminant levels dropped to safe levels but then mysteriously rose again months after treatment efforts had ceased. The failure was highlighted by a 1991 article in the *Wall Street Journal* titled "Throwing Good Money at Bad Water Yields Scant Improvement" (Stipp, 1991). The article described the case of IBM's Dayton facility in New Jersey. Toxic solvents had been discovered in one of the water wells supplying water for South Brunswick Township. IBM spent $10 million over six years on a pump-and-treat program. At the end of this time, the contaminant level in the South Brunswick well was in line with drinking-water standards, however, three years later the contaminant levels were once again at unsafe concentrations. At one location, the concentration of toxic solvents in the groundwater was twice as high as it was before the six-year remediation effort started (NRC, 1994). The *Wall Street Journal* quoted John A. Cherry, a leading expert on groundwater contamination, as declaring that the condition of groundwater near most major toxic-waste spills was comparable to a patient with terminal cancer. It was safe to say that everyone was unhappy. Environmentalists were unhappy, because the toxic waste dumps on the Superfund list had not been cleaned up. Businesses were unhappy, because they had spent millions of dollars on cleanup efforts that had not worked. By 1992, after 12 years and $11 billion expended, just 84 of 1,245 high-priority sites on the Superfund list had been cleaned up. The nation's leading hydrogeologist, John Bredehoeft, wrote an editorial for the journal *Ground Water* titled "Much Contaminated Ground Water Can't Be Cleaned Up" and warned that the public was being misled by unrealistic expectations (Bredehoeft, 1992).

In addition to the technical problems of remediation, the Superfund Law had created a nightmarish tangle of litigation in the courts. In 1992, the magazine *Business Week* reported that major U.S. automakers were presented with a bill for $40 million to clean up a toxic-waste dump in Metamora, Michigan. The automakers promptly sent their attorneys after 200 other parties who had also used the dump. Amongst these was a Michigan Girl Scout troop whose summer camp had sent garbage to the landfill (Hong and Galen, 1992). Unrepentant, the Girl Scouts refused to pay.

Following the Superfund debacle of the late 1980s and early 1990s, the National Academy of Sciences studied the issue of groundwater cleanup and issued a report in 1994 (NRC, 1994). They concluded that current laws and policies on groundwater cleanup were based on the assumption that restoring contaminated groundwater was technically straightforward and feasible—which it was not. In their words,

> In setting ground water cleanup goals, government regulators have only rarely considered whether existing technology is capable of meeting those goals.

During the 1990s, technical progress was made as new technologies for groundwater remediation were developed and tested. However, the United States has still not developed a rational strategy for waste disposal that provides long-term protection to both society and the environment.

9.3.2. Case Study: Love Canal.

Love Canal (Figure 9.16) is the most famous toxic waste dump in the world. It was the Love Canal case that made the public aware of groundwater contamination and the hazards of abandoned wasted dumps. Love Canal was also directly responsible for the passage of the 1980 Superfund law by the U.S. Congress.

Figure 9.16 Aerial view of Love Canal area. (Courtesy of U.S. EPA.)

The story of Love Canal starts in the year 1892 when William T. Love proposed the canal as a way of connecting the upper and lower branches of the Niagara River in New York state. There is a drop in elevation of 280 feet (85 meters) between the two, and Mr. Love hoped to harness the power of water flowing through the canal to generate electricity. The project was abandoned before it was completed, leaving a trench 60 feet (18 meters) wide by 3000 feet (914 meters) long. In 1920, the land was sold to the Hooker Chemical Corporation, a subsidiary of Occidental Petroleum. Starting in 1942, the incomplete canal was used by Hooker Chemical as a dumping ground for 22,000 tons (20 million kilograms) of toxic waste. The city of Niagara Falls and the U.S. Army also used the canal for waste disposal. Some of the chemicals stored in the Love Canal included dioxin, pesticides, halogenated organic compounds, and chlorobenzenes. Dioxin is an undesirable byproduct of the manufacture of herbicides and some other products. It is an extremely stable, fat-soluble compound that enters fatty tissue in the human body if ingested. Dioxin acquired a very bad reputation during the early 1980s when early animal experiments led some toxicologists to mistakenly conclude that dioxin was one of the most toxic substances on Earth. Later epidemiological studies on workers exposed to dioxin over many years showed it to be weakly carcinogenic at high exposures, with no carcinogenic effect at all for low exposures.

In 1953, Hooker Chemical covered the canal with dirt and sold it to the local board of education for one dollar. Included in the deed transfer was a warning concerning the chemical wastes buried on the site and a disclaimer absolving Hooker Chemical of any possible future liability. The Board of Education promptly built an elementary school on top of the former toxic waste dump. The school opened its doors to 400 students in 1955. Homebuilding in the area began; by 1978 there were about 800 single-family homes immediately surrounding the canal area. Throughout the late 1950s and into the 1970s, there were scattered complaints from residents of foul odors and strange substances appearing at the ground surface and in basements. In 1976, the city of Niagara Falls hired a consulting firm to evaluate the problems. The consulting firm found the presence of buried chemicals, and recommended that the canal be covered with clay and a drainage system installed to intercept leachate leaking from the canal, however, no actions were taken immediately. In 1978, the New York State Department of Health (NYSDH) began studying the area and found evidence of health problems amongst residents. On August 2, 1978, the NYSDH ordered the elementary school near the canal to be closed. A few days later, the state of New York announced that it would purchase the 239 homes located closest to the canal from their owners at fair market value.

Many of the residents of the Love Canal area who had not been included in the New York state buy-out were unhappy. They believed themselves to also be affected by the presence of hazardous waste, and thought the decision to evacuate only the closest 239 homes to be arbitrary. Community residents began to bring political pressure on the state of New York and the government of the United States. They marched in the streets, carried coffins to the New York State Capitol, and picketed at the canal site. In October of 1980, U.S. President Jimmy Carter ordered a total evacuation of 350 acres (1.4 square kilometers) adjacent to the canal.

Remediation of the Love Canal site included containment of leachate from the hazardous waste in the canal, groundwater monitoring, and the removal of contaminated soil and sewer and creek sediments. The 239 homes closest to the canal were demolished. Other homes were rehabilitated and eventually put up for sale. The toxic wastes located in the canal itself were never removed. Rather, they were contained. The ground overlying the canal was covered with a synthetic material to keep rain from infiltrating into the ground and contacting the buried wastes. The canal was then surrounded with a barrier and drain system to collect what leachate came off the landfill. Sewers were cleaned, and contaminated soil in creeks was collected by dredging and burned. The waste materials collected and processed at Love Canal included 32,500 cubic meters of contaminated sediments, 72,000 liters of DNAPLs, and 109,000 kilograms of carbon-filter wastes. In 1989, the EPA declared that parts of the affected area were now safe for residential occupancy, while other areas were more suitable for industrial purposes. By 1999, 238 formerly abandoned homes had been sold to new occupants.

9.3.3. Risk Assessment.

The production of wastes is an unavoidable byproduct of an industrial and technological civilization. The most effective strategies for disposing of these wastes are those that reduce pollution and groundwater contamination to the maximum amount possible for the least amount of money. **Risk assessment** is a way of ranking risks so that they may be avoided, reduced, or managed in the most efficient manner possible (Wilson and Crouch, 1987).

Almost all activities in life have risks associated with them. It is desirable to reduce risks, however, the idea that it is always desirable or feasible to reduce risks to the zero level is a false and pernicious notion that has no basis in fact or logic. A common expression of this sentiment is "if it saves just one life, its worth it." Misguided attempts to reduce risk to the zero level are likely to result in statistical murder. **Statistical murder** is the occurrence of unnecessary deaths because risks have not been reduced in the most efficient manner possible. The failure to properly assess risks and act accordingly can have very real and fatal consequences. For much of the last 100 years, chlorine has been added to public water supplies in the United States as a way of preventing bacterial growth and preventing the transmission of diseases such as typhoid fever, cholera, and dysentery. There is, however, a small risk of cancer associated with drinking chlorinated water. Wilson and Crouch (1987) estimated the lifetime risk as about 1 in 21,000. Based on the risk of cancer, in 1991 Peruvian government officials decided to not chlorinate the public water supply. The result was an outbreak of cholera that killed 4000 people.

The risk of developing cancer from ingestion of any carcinogenic substance is usually estimated by feeding laboratory rats very high amounts of a carcinogen and then linearly extrapolating to the smaller amounts that might be found in the human diet. It is questionable if this methodology is scientifically sound, because the toxicity of any substance depends upon dosage. The principle that toxicity is determined by dosage was first clearly enunciated by the medieval physician, Paracelsus (1493–1541). In general, there are three different models for the response of an organism to a toxin (Figure 9.17). The first possibility is that the nature of dose-response is linear. That is, if large amounts of a carcinogen in the diet result in large cancer rates, then small amounts of a carcinogen will result in low rates of cancer. A corollary of the linear model is that any exposure to a carcinogen, no matter how small, contains some risk of cancer. Acceptance of this model leads directly to the goal of reducing carcinogen levels in water and food to zero. The second possibility is that there is a threshold, or a minimum level of carcinogenic intake, below which no negative or positive response occurs. The third possibility is that hormesis occurs. **Hormesis** is a positive response by an organism to the limited exposure of a toxic substance (Luckey, 1992). The concept of hormesis is well illustrated by the mineral selenium. Selenium is profoundly toxic in large doses. However, it is also essential for human life and people in the United States today commonly obtain inadequate amounts for optimal health. The shelves of health-food stores in the United States are lined with bottles of selenium supplements, and a recent study found that a daily ingestion of a 200 micrograms of selenium reduced the risk of dying from one type of cancer by 50% (Clark et al., 1996).

The approach that has often been used in assessing the risk from contaminants in water and food is the linear dose-response model. The scientific validity of this approach is questionable (Goldman, 1996). Absurd conclusions can follow from the extreme linear extrapolations used in assessing cancer risks from chemicals. The background cosmic radiation Earth receives from outer space is a natural carcinogen, and the amount of radiation a person receives increases with altitude, because the Earth's atmosphere is a natural filter. Following the logic of the linear dose-response model, if everyone on Earth added one inch (2.5 centimeters) of additional height to their shoes, the increase in radiation received would kill about 1500 people over the next 50 years (Goldman, 1996). But, people living in high-altitude areas tend to have lower cancer rates, apparently due to hormesis (Luckey, 1992). Most people are unaware that plant foods and other natural substances contain natural carcinogens. Apples, strawberries, cauliflower, cabbage, peaches, celery, potatoes, bananas, as well as coffee and tea contain carcinogens in quantities that are sometimes remarkable. Plants produce these chemicals as part of a natural defense mechanism to repel predators and molds. The human body also possesses biochemical mechanisms for neutralizing these same compounds (Ross and Skinner, 1994). If the same linear dose-response model is applied to the natural carcinogens that occur in foods, we find that a wide variety of substances in our diet are more potent carcinogens than substances such as

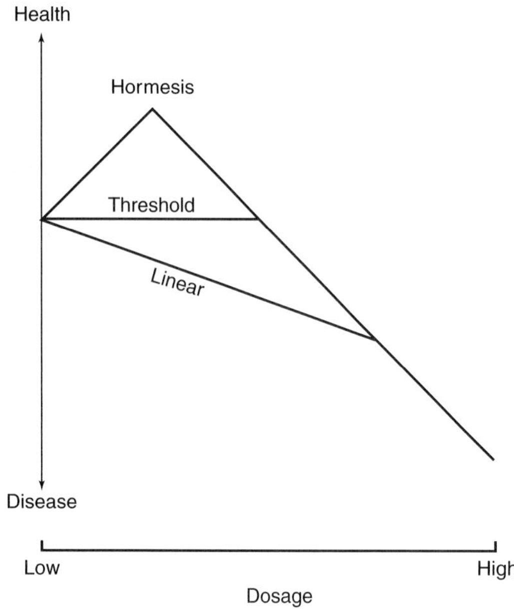

Figure 9.17 Dose-response models.
(After Luckey, 1992.)

pesticides (Ames et al., 1987). Peanut butter is often contaminated with the natural toxin aflatoxin. The risk of developing cancer from eating a peanut butter sandwich is 100 times greater than developing cancer from the residual amount of the pesticide carbaryl found in the daily average diet (Fetter, 1999). Lindane is a pesticide in the same chemical family as DDT. It is persistent in the environment, and is toxic to both humans and wildlife, however, the amount of ethyl alcohol found in one 12-ounce can of beer poses a risk of cancer that is 2.8 million times greater than the risk associated with an average daily ingestion of Lindane residue on food (Fetter, 1999).

In general, the public at large is incapable of logically and rationally assessing risks; this ineptitude is perpetuated by irresponsible media coverage (Brody, 1999). An example of a hysterical reaction founded in ignorance and compounded by yellow journalism was the Alar scare of 1989. Alar is a chemical that was used to delay ripening in apples. In the 1970s, a set of limited studies showed that a byproduct of Alar, known by the acronym UDMH, was carcinogenic. A researcher had found that when UDMH was fed to rats at a dose of 29 milligrams per kilogram of body weight per day some rodents developed tumors. The dosage in this experiment was 617,000 times higher than the amount of UDMH consumed each day in an average human diet. Later experiments based on improved methodologies revised the cancer risk down by a factor of ten, and then down again by a factor of two (Marshall, 1991). The EPA finally estimated the risk of developing cancer from Alar ingestion was 45 ten-thousandths of 1% (0.000045). In comparison, cigarette smoking increases the risk of cancer by a factor of 20 (Brody, 1999). In February of 1989, while the EPA was still conducting animal tests on Alar, an environmental advocacy group, the Natural Resources Defense Council, released a report titled *Intolerable Risk: Pesticides in Our Children's Food.* The report said that Alar was a "potent carcinogen," and was by far the biggest threat among 23 chemicals in food that the organization had studied. A copy of the report was released to CBS News and publicity was arranged with the help of a public relations firm. Actress Meryl Streep began warning of the dangers of Alar at press conferences. On February 26, 1989, the CBS News Program *60 Minutes* did a report on Alar against a backdrop of a giant apple marked with a skull and crossbones. The *60 Minutes* report labeled Alar as the most potent cancer-causing agent in the food supply and called it a cause of childhood cancer. The public reaction was mass hysteria. One person reportedly called the International Apple Institute and asked if it was safe to pour apple juice down the sink or was it necessary to take it to a toxic waste dump? A parent sent state troopers to chase down a school bus because her child had an apple in their lunch box. The hysteria quickly subsided, but that year's income for apple growers was about $120 million lower than it otherwise would have been. Both the British government and the United Nations concluded that Alar was not a cancer risk. When mice were fed UDMH at a dosage 83,000 times greater than the amount found in the average American diet, no carcinogenic activity was found (Marshall, 1991). Had nineteenth-century British writer Charles Mackay been alive, he might well have included the Alar scare as a case history in his classic book published in 1841, *Memoirs of Extraordinary Popular Delusions*.

The concept of risk analysis is correlated with the concept of a cost-benefit analysis. A **cost-benefit analysis** is an analysis that attempts to quantitatively estimate both the cost and the benefit to be obtained from any action. In our everyday lives, we commonly make cost-benefit analyses. For example, how many of us would hire a team of security guards to watch our homes while we were at work each day? The team of security guards would likely reduce the risk of burglary to near zero, however, the cost would be greater than the benefit. The maximum increase to the quality of our environment comes from obtaining the greatest benefit at the least cost. Society does not have enough resources to solve all environmental problems. Any proposed action should therefore be carefully considered and weighed in view of the costs involved. Fetter (1999) described a case history where $7.5 million was spent to prevent a toxic waste plume from reaching a single private well about 100 feet (30 meters) distant. The cost was justified on the basis of the questionable assumption that the toxic waste dump at some time in the future would be the site of residential development, even though the population in the area in question was declining.

Environmental actions are also subject to the law of unintended consequences. The **law of unintended consequences** is that any action designed to improve environmental quality or reduce risk may have the opposite effect. Much groundwater pollution may have occurred as a result of laws that limited the ability of industries to dispose of wastes in the atmosphere and surface waters. When surface options were closed, burial in the ground became a more attractive option (Bredehoeft and Usselman, 1984). Patterson (1989) concluded that the major result of environmental protection efforts in the United States since 1950 was the transfer of toxic pollutants from one medium to another. An example is provided by efforts to control air pollution. The discharge of pollutants to the air was reduced by installing devices called "scrubbers" that collect waste materials from gasses discharged from smokestacks. The toxic sludge removed from the air by the scrubber ended up in landfills. All landfills eventually leak, and have the potential to contaminate groundwater.

There are no quick and easy answers to solving the long-term problem of waste disposal. Any remedies must be based on realistic assessments of risk coupled with the understanding and evaluation of tradeoffs and the possibility of unintended consequences. The subsurface can be safely used for waste disposal if sites are selected, designed, and engineered properly (Bredehoeft and Usselman, 1984).

9.3.4. Sources of Groundwater Contamination.

Sources of groundwater contamination (Figure 9.18) may be either point sources or diffuse sources. A **point source** is a localized source of groundwater contamination, such as a single hazardous waste dump or landfill. A **diffuse source** of groundwater contamination refers to the aggregate effect of widespread activity. Diffuse sources include the application of agricultural fertilizers, homeowners using pesticides, and septic tanks.

A complete list of the possible sources of groundwater contamination would contain perhaps hundreds of different sources. A sensible approach is to consider the most important sources of contamination, while keeping in mind that other sources are possible. In the 1980s, the U.S. EPA asked the states and territories to report major sources of groundwater pollution (EPA, 1990a). The most frequently listed sources were underground storage tanks, septic tanks, agricultural activity, municipal landfills, surface impoundments, abandoned hazardous waste sites, and industrial landfills (Figure 9.19). The sources of contamination that were given the highest priority ratings for cleanup were underground storage tanks, abandoned hazardous waste sites, agricultural activity, septic tanks,

Paracelsus: "The Dose Makes the Poison"

The medieval physician who called himself Paracelsus (1493–1541) was one of the strangest figures in all of recorded history. Genius, rogue, alchemist, physician, reformer, and drunk: Paracelsus was all of these, and more besides. Belligerent, arrogant, and boastful, Paracelsus was also humble and compassionate to the sick and poor. He made so many enemies that more than once he had to flee in the middle of the night to save his own life. Even today, the final judgment on his life and work is not in.

Paracelsus was born Philippus Aureolus Theophrastus Bombastus von Hohenheim in 1493 in Einsiedeln, Switzerland, the only son of a poor German physician and chemist. Paracelsus obtained his early instruction from his father. In 1507, at the age of 14, he began to roam Europe, attending one university after another. He attended the universities of Basel, Tübingen, Vienna, Wittenberg, Leipzig, Heidelberg, and Cologne, and was disappointed in all of them. He finally graduated from the University of Vienna in 1510 at the age of 17 with a baccalaureate degree in medicine. He claimed to have subsequently received a doctorate from the University of Ferrara in 1516, but the university records for that year are missing. About this time, the youth born with the surname Hohenheim began referring to himself simply as *Paracelsus,* the name implying that he was equal to or above the legendary Roman physician of the first century AD, Celsus.

Paracelsus (1493–1541)

Paracelsus had a low opinion of professors and universities. He later wrote that he wondered how "the high colleges managed to produce so many high asses." Paracelsus believed that knowledge must come from personal experience of the world. He wrote,

> The universities do not teach all things, so a doctor must seek out old wives, gipsies, sorcerers, wandering tribes, old robbers, and such outlaws and take lessons from them. A doctor must be a traveler . . . knowledge is experience.

True to his personal philosophy, Paracelsus spent the next 10 years wandering through most of the known world. He visited England, Ireland, Scotland, Russia, Lithuania, Hungary, Italy, Egypt, Arabia, and Palestine. During part of this time he worked as an army surgeon, was held captive by the opposing army and escaped. Paracelsus' goal during his wandering was not only to find the most effective medical practices, but to discover and harness what he called the "latent forces of Nature." He wrote:

> He who is born in imagination discovers the latent forces of Nature. . . . Besides the stars that are established, there is yet another—Imagination—that begets a new star and a new heaven.

Paracelsus also believed that man's body, mind, and soul were linked. He wrote:

> Man is not body. The heart, the spirit, is man. And this spirit is an entire star, out of which, he is built.

> If therefore a man is perfect in his heart, nothing in the whole light of nature is hidden from him.

After 10 years of wandering, Paracelsus returned home at the age of 32 and was appointed town physician and lecturer in medicine at the University of Basel in Switzerland. Students flocked from all over Europe to hear his lectures. They were not disappointed. To the delight of his students and disciples, Paracelsus mercilessly attacked the questionable medical practices of the time. The prevailing theory of the time was that illness was caused by an imbalance in bodily fluids, and the most common method of treatment was to bleed a patient. The society in which Paracelsus lived was highly authoritative. The scientific method was not yet recognized; knowledge was based on authority and tradition. The accepted practice for university professors was to read aloud from books that were hundreds of years old; scholarship in medicine was synonymous with memorization of ancient texts. Paracelsus' contemporaries considered the greatest medical authorities to be the Greek, Galen (AD 129–216), and the Arab, Avicenna (AD 980–1037); the slightest deviation from their teachings was regarded as heresy. On June 24, 1527, surrounded by a crowd of cheering students, Paracelsus stood in front of the university and burned Galen's and Avicenna's books. As the books burned, Paracelsus denounced the revered physicians of the past:

> Follow after me Galen, Avicenna, Rhasis . . . not one of you will survive, even in the most distant corner where even the dogs will not piss. I shall be monarch of physicians and mine will be the monarchy.

Paracelsus attacked his fellow physicians, calling them quacks and impostors. The fact that Paracelsus' cures were successful only made the medical faculty that much madder. While his colleagues would spend their time in libraries, Paracelsus spent his time in taverns where he would challenge peasants to drinking bouts and win. To the outrage of the rich and influential townspeople in Basel, Paracelsus had a sliding scale for his medical services. Fancying himself a medical Robin Hood, he would treat the poor for free, while charging the rich a king's ransom for the same services.

Predictably enough, the man who had criticized universities for being full of "high asses" did not last long in one himself. After only a year at the University of Basel, he had made so many enemies that he had to leave abruptly in the middle of the night to save his life. Paracelsus spent the rest of his life wandering. In 1536, he published his great book on surgery, *Der Grossen Wundartzney*, which once again established his preeminence as a medical authority. He became wealthy and his services were sought after by royalty. In 1541, Paracelsus died under mysterious circumstances at the White Horse Inn at Salzburg in Austria. Some people say that he diagnosed himself with an incurable illness and retired to die quietly. Others say that he was killed in a tavern brawl, or that jealous doctors hired assassins to do away with him. We will probably never know for sure.

Paracelsus made many contributions to medical science; some of them were hundreds of years ahead of his time. A short list of some of the more important contributions include:

- Origination of the concept of hormesis, the idea that small doses of toxic substances could have beneficial effects. Paracelsus prescribed administration of the toxic element mercury as treatment for syphilis. Put more elegantly, Paracelsus is best remembered for "the dose makes the poison."
- Paracelsus was the first to advocate clean, near-aseptic conditions during surgery.
- Paracelsus was the world's first biochemist and father of modern pharmacology. He introduced the idea that the key to successful medicine was harnessing the science of chemistry, or alchemy as it was called at that time.
- Invention of the painkiller laudanum by dissolving opium in alcohol. Laudanum remained the most effective treatment for pain until morphine was first isolated in 1806.

Paracelsus: "The Dose Makes the Poison" *(continued)*

- Paracelsus was the first physician to practice holistic medicine. He understood the link between the mind and the body and emphasized that the patient's total health must be considered.
- Paracelsus was the first physician to understand that disease is caused by external factors, not an imbalance in the human body, and the first to understand that the body's own immune system enables it to heal itself.

Properly understood, Paracelsus greatest contribution was as one of the inventors of the scientific method. His recognition that knowledge comes from experience and observation, rather than tradition and authority, is the very essence of science.

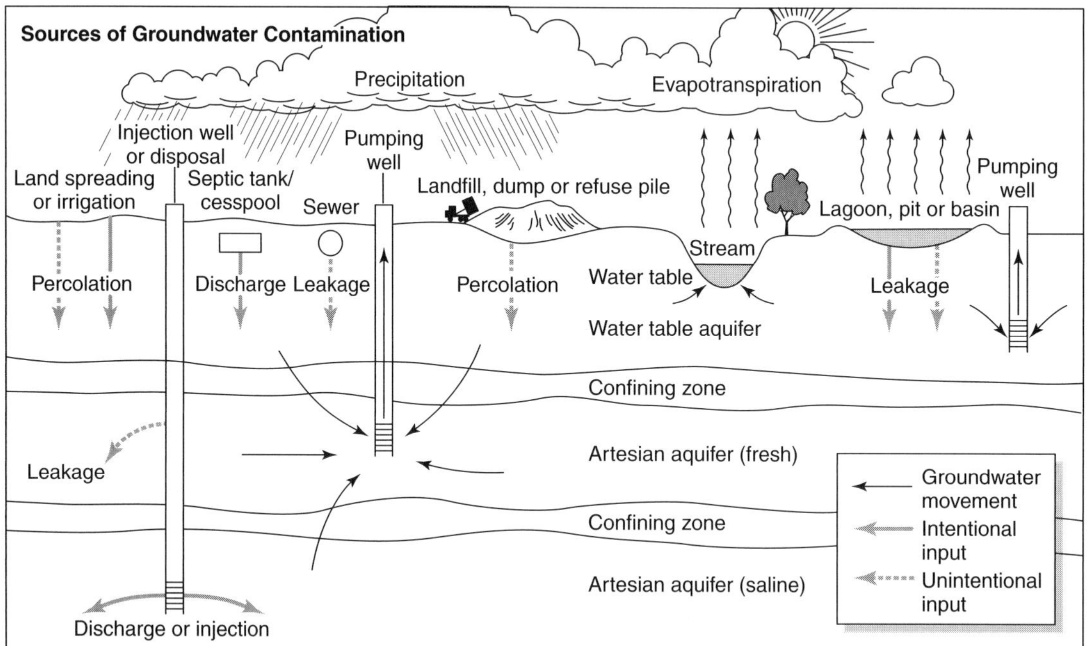

Figure 9.18 Sources of groundwater contamination. (From EPA, 1984, p. 18.)

surface impoundments, municipal landfills, and oil and gas brine pits (Figure 9.20).

9.3.4.a. Underground Storage Tanks.

In 1984, the U.S. Office of Technology Assessment (OTA, 1984) estimated there were 2.5 million underground storage tanks in the United States, counting both tanks in use and those abandoned. Underground storage tanks are primarily associated with gasoline or service stations, however they are also used by homeowners for storing fuel oil and by a variety of industries and government agencies for the storage of fuels, oils, and

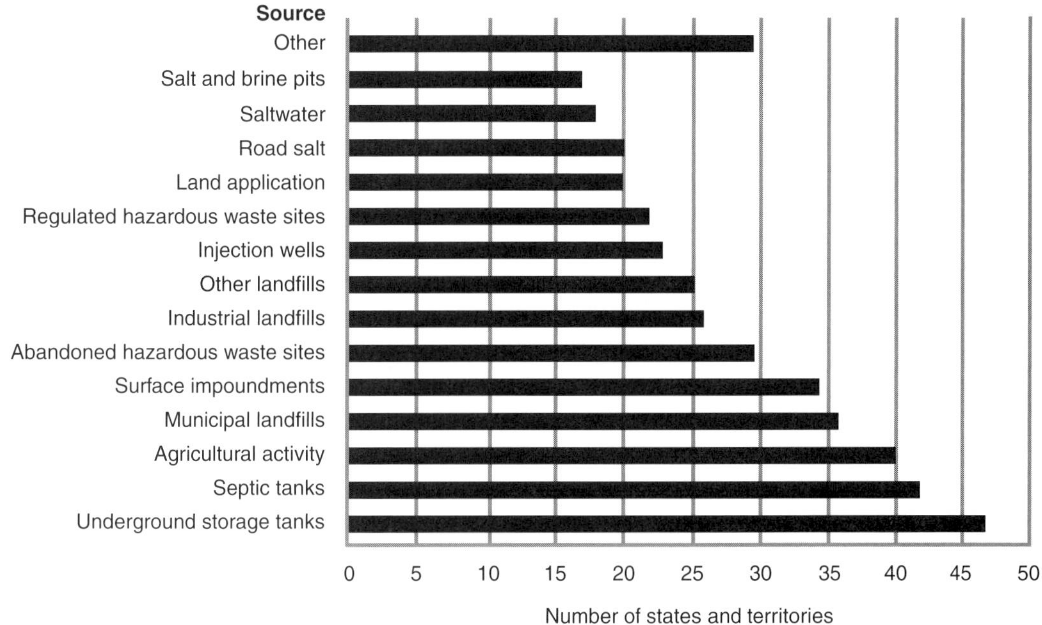

Figure 9.19 Sources of contamination ranked by states and territories of the United States as threats to groundwater.
(From EPA, 1990a.)

hazardous chemicals. As of 1988, 84% of the underground storage tanks in the United States were made of unprotected steel subject to erosion. In 1987, the EPA implemented a 10 year program for testing, repairing, and replacing underground storage tanks throughout the United States. The EPA estimated that the total number of tanks numbered around 2 million, and that half of these were leaking. The EPA rules applied to tanks holding more than 1100 gallons (4,164 liters), estimated to number 1.4 million. Of those 1.4 million tanks, 47% held petroleum products. Under EPA rules, all tanks and associated piping had to have corrosion protection and overfill prevention devices. Acceptable methods of preventing corrosion included fiberglass construction, steel coated with a corrosion-resistant material, or a cathodic protection system. Most older tanks had to be replaced; in many instances the owners were not able to afford replacement containers and went out of business. By 1998, the total number of underground storage tanks in the United States had been reduced to about 1 million, with 50% of these at gasoline stations. Even more stringent regulations governing underground storage tanks were passed by the state of Florida in 1984, requiring that all tanks be double-walled. From 1984 through 1999, Florida spent $1.0 billion cleaning up sites contaminated from leaking underground storage tanks.

9.3.4.b. Septic Tank Systems.

A **septic tank system** is a sewage disposal system used primarily for single residences (Figure 9.21). The main components of a septic tank system are a settling tank and a drainage field. Solids in the settling tank are decomposed by anaerobic bacteria; overflow of liquids from the tank goes into the drain field. Every few years the solid material that builds up in the settling tank must be pumped out and disposed of elsewhere. As of 1980, about one-third of

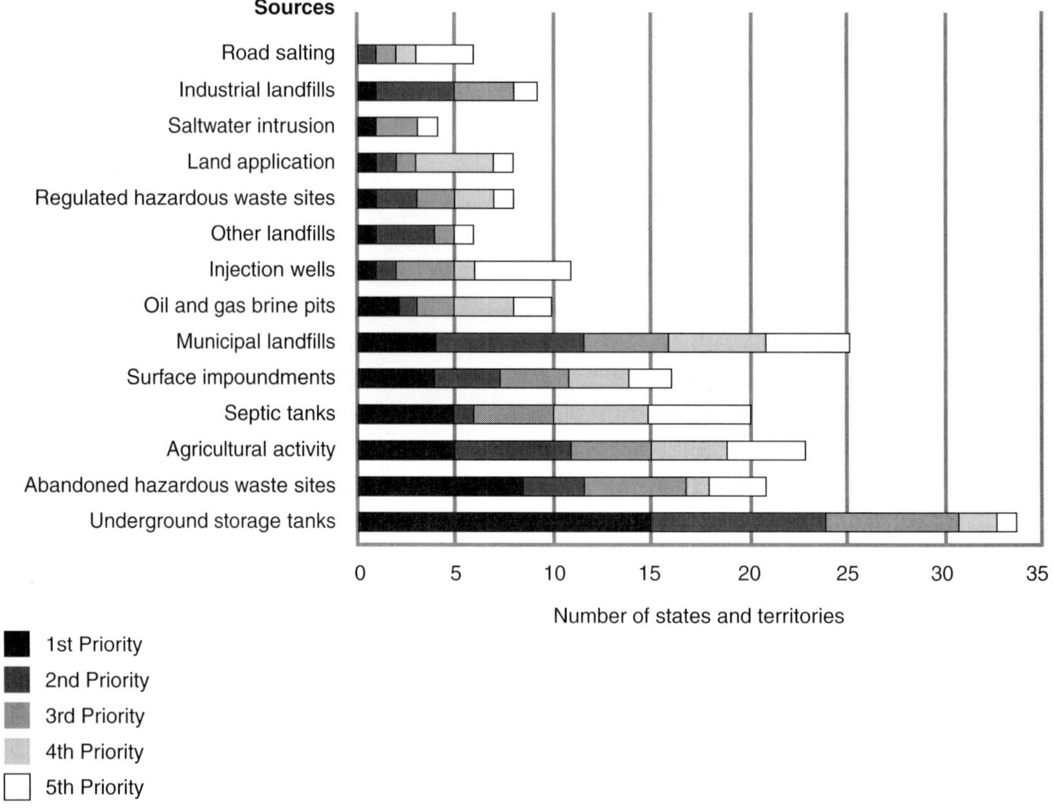

Figure 9.20 Sources of groundwater contamination ranked in terms of priority by states and territories of the United States.
(From EPA, 1990a.)

the population of the United States was served by 22 million septic tank systems discharging one trillion gallons (3.78 trillion liters) of effluent into the ground every year (EPA, 1986).

Properly designed, used, and maintained, septic tank systems are a good method for disposal of household wastes. Four factors are necessary for the proper performance of a septic system. The soil must be neither too permeable nor impermeable. Soils of low permeability will not allow septic systems to drain efficiently, while those of very high permeability will allow septic effluent to reach the saturated zone before natural processes can purify it. The unsaturated zone must be thick enough so that septic effluent is purified before it can reach groundwater in the saturated zone. The greater the thickness of the unsaturated zone, the greater the filtration capacity. Septic systems must be dispersed so as not to overload the natural capacity of the soil. In most parts of the United States, the minimum area for a septic system established by zoning regulations is 0.5 acres (2023 square meters). The fourth factor needed for good septic system performance is routine maintenance. If the sludge and scum that accumulate in the tank are not pumped out every few years, the tank will overflow and clog the soil absorption system.

The primary constituents of septic system effluent are nitrogen, phosphates, and a variety of pathogenic bacteria and viruses. These pollutants are

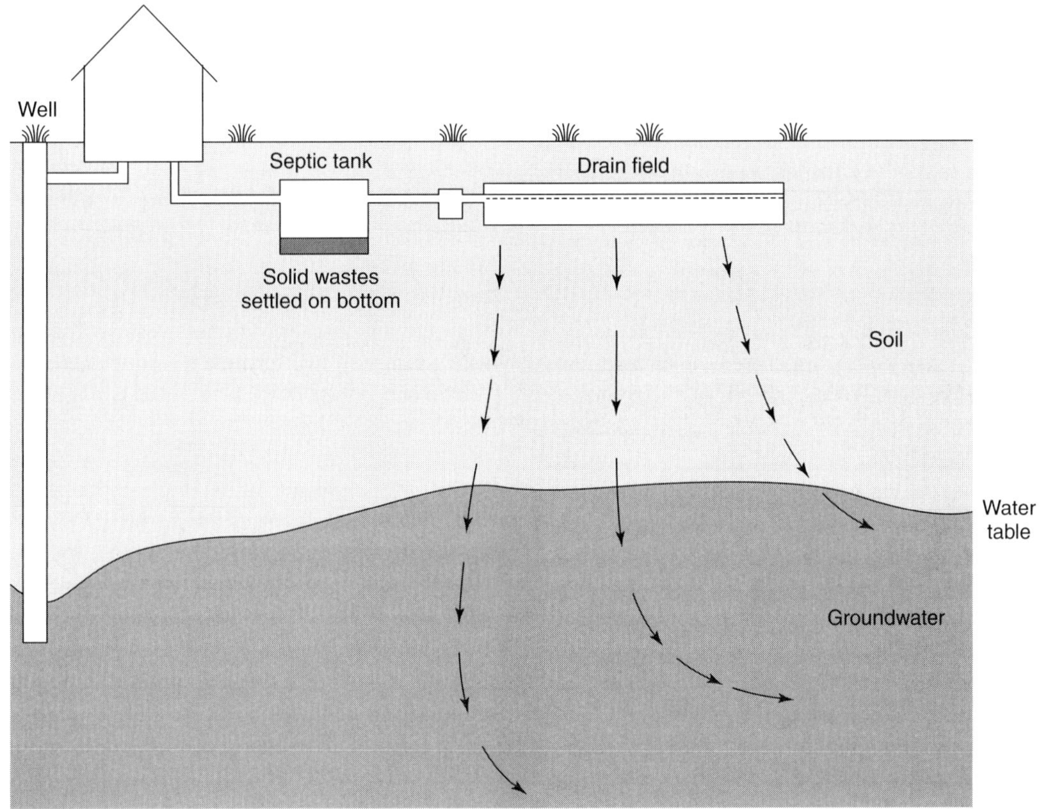

Figure 9.21 Typical household septic tank system.
(From EPA, 1986, p. 4.)

purified in soil by a combination of physical and chemical processes. Most nitrogen in the septic tank is in the form of ammonia. When the ammonia enters the soil it is quickly oxidized to nitrate. Both nitrate and phosphate are fertilizers and may be taken up by plant growth. Nitrate can also be subject to the process of denitrification where bacteria break it down to nitrogen gas that is released to the atmosphere. Bacteria and viruses die off gradually over time. Viruses typically survive from seven days to as long as 6 months (Viraraghavan, 1982).

Septic tanks are the most frequently reported cause of groundwater contamination (Yates, 1985). Consumption of contaminated groundwater was a cause of half the outbreaks of waterborne disease in the United States from 1971 through 1979. The vast majority of disease was caused by pathogenic microorganisms, not toxic chemicals. Overflow or seepage from septic systems was responsible for 63% of illnesses caused by contaminated groundwater.

Some short case histories of illness caused by contamination of groundwater by septic tanks were described by Craun (1979). Between January 17 and March 15 of 1974, approximately 1200 cases of acute gastrointestinal illness occurred amongst the populace of Richmond Heights, Florida, a town of 6500. The source of the infection was traced to one of two water wells that supplied the community's water. The contaminated well-water contained high levels of fecal coliforms. Dye studies were used to trace the source of the infection to the septic tank of a church and day care center located about 150 feet (46 meters) from the affected well. In this case, two factors conspired to produce the epidemic. In addition to the groundwater contamination by the septic tank, the town's chlorination system had broken down temporarily enabling 1 million gallons (3.78 million liters) of untreated water to enter the town's system about 48 hours before the epidemic began. In Polk County, Arkansas, there was an outbreak of 98 cases of viral hepatitis in 1971. The source of the disease was traced to commercially manufactured pellet ice. The well that supplied the water from which the ice was made had heavy coliform contamination. Dye studies revealed that high-permeability sedimentary rocks in the area had allowed effluent from a septic tank to reach groundwater. The septic tank in question was used by a family who had suffered from infectious hepatitis six weeks earlier. Between April 4 and May 22, 1972, five cases of typhoid broke out in an area of Yakima, Washington, where residents had both septic tanks and wells. A typhoid carrier in the area was identified and dye injected into his septic system. Within 36 hours, the dye had appeared in numerous wells in the area. The rapid movement of septic effluent had been facilitated by high-permeability gravel and a high water table near its seasonal peak.

The city of Charleston, South Carolina, once relied upon a shallow aquifer for an abundant supply of fresh, clean water. The entire Charlestown peninsula is underlain by a layer of quartz-rich sand about 6 meters thick. A relatively impermeable clay layer underneath the sand aquifer facilitated a high water table and the accumulation of groundwater in the shallow sand. Charlestown was founded in 1678, and for the first 50 years residents could find a ready supply of water by digging wells just a few meters deep. By the middle of the seventeenth century, though, the aquifer had become hopelessly contaminated from untreated human sewage leaking out of hundreds of outhouses. The natural ability of the aquifer to assimilate and process the waste had been overwhelmed. By 1800, most wells in Charlestown had been abandoned, replaced by rain-fed cisterns (Chapelle, 1997).

9.3.4.c. Agricultural Activities.

Agricultural activities have the potential to contaminate groundwater through the application of pesticides and fertilizers, irrigation, and the generation and disposal of animal wastes. The three basic constituents of fertilizers are nitrogen, phosphorus, and potassium. Phosphorus is not mobile; it is adsorbed by soils and generally does not reach the water table. From 1940 through 1970, laundry detergent was a major source of phosphorus added to the environment. Starting in the 1970s, state bans and voluntary actions by detergent manufacturers resulted in the removal of phosphorus from laundry products. Potassium is mobile, but there is little indication in the literature that potassium contamination is a threat to groundwater. Nitrogen is the largest component of most fertilizers. It is highly mobile, and is the single contaminant most often considered to be a major threat to groundwater quality. About 90% of nitrate contamination originates from diffuse sources. In 1999, the U.S. Geological Survey reported that 12 million tons (11 billion kilograms) of nitrogen and 2 million tons (1.8 billion kilograms) of phosphorus are applied each year in the United States in the form of fertilizer (Figure 9.22). Another 7 million tons (6.4 billion kilograms) of nitrogen and 2 million tons (1.8 billion kilograms) of phosphorus are applied in the form of manure. Excessive intake of nitrate in drinking water can result in "blue baby syndrome" where oxygen levels in the blood of infants are low, sometimes fatally so. The U.S. Geological Survey (USGS, 1999) has identified four factors that tend to increase the potential for nitrate to enter groundwater: (1) high rates of precipitation following recent application of fertilizer, (2) well-drained and permeable soils that allow the rapid downward movement of water, (3) crop management that results in slow runoff and allows more time for water to infiltrate into the ground,

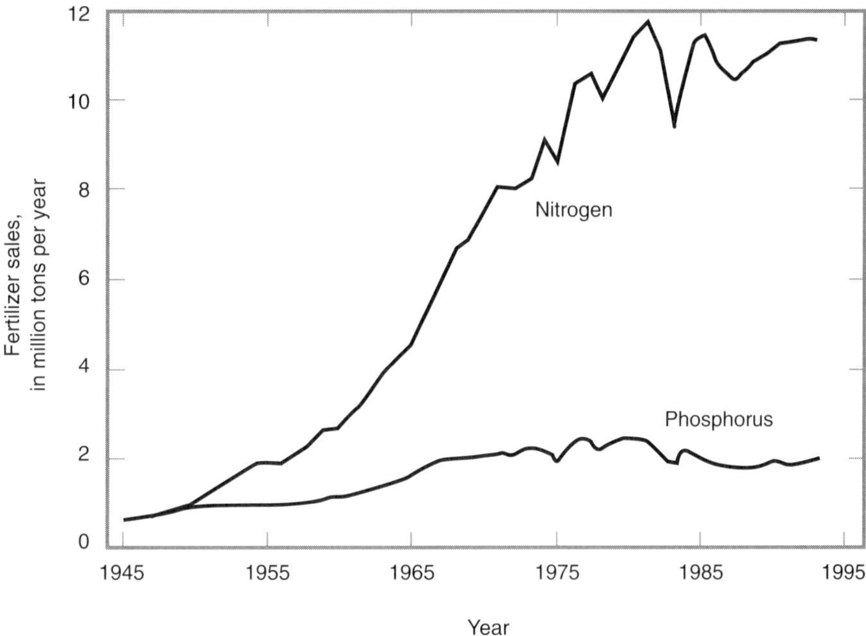

Figure 9.22 Nitrogen and phosphorus fertilizers applied to soils in the United States as estimated from fertilizer sales, 1945 to 1995.
(From USGS, 1999, p. 29.)

and (4) low organic-matter content and high levels of dissolved oxygen that can delay the chemical breakdown of nitrate to other forms of nitrogen. In 1999, the U.S. Geological Survey reported on a nationwide study of nitrate contamination of groundwater (USGS, 1999). They found that nitrate concentrations were above the background level in 53% of shallow groundwater samples obtained mostly from private wells. Major aquifers were found to be better isolated from nitrate contamination, with only 3 of 33 aquifers studied having nitrate concentrations above background levels. About 15% of water samples from wells less than 100 feet (30 meters) deep had nitrate levels above EPA standards. For wells between 100 and 200 feet (30 to 60 meters) deep, about 8% exhibited unacceptable levels of contamination. No wells deeper than 200 feet (60 meters) were found to have nitrate levels above EPA standards. The absence of nitrate in deeper wells reflects the time needed for surface contamination to reach deeper levels. The amount of nitrate contamination in groundwater is not necessarily controlled by the amount of nitrate applied as fertilizer at the surface. Areas underlain by high-permeability strata (karst, fractured rock, sand and gravel) may have high nitrate concentrations in groundwater because of the ease with which it is able to move downward. In contrast, the White River Basin in the upper midwest has some of the lowest nitrate levels in groundwater even though it has some of the highest nitrogen fertilization rates in the United States. In the White River Basin, poorly drained and relatively impermeable glacial deposits of clay and silt impede the flow of surface water and nitrate to the water table. Simultaneously, two factors act to reduce nitrate concentrations. Nitrate in surface water is intercepted by ditches that transport it to streams before it can infiltrate to the water table. Nitrate also undergoes denitrification, a bacterial process that converts nitrate to nitrogen gas (USGS, 1999).

In 1999, the USGS reported that pesticides were detectable in about 50% of the wells they sampled across the nation (1999, USGS). The highest occurrences were in shallow groundwater located in both agricultural and urban areas. The lowest frequency of pesticide contamination was for major aquifers at deeper levels. The most common pesticides found in groundwater were herbicides. Although herbicides were found in more than 50% of wells sampled, only one well of 1000 sampled had a level of contamination that exceeded EPA standards for drinking water.

The single compound found in the highest levels in shallow groundwater was atrazine and associated breakdown products. Atrazine is a herbicide used to control grasses and broadleaf weeds. It is broken down by soil microorganisms, but tends to persist and moves easily through soils. Atrazine is typically used where corn is grown; conventional municipal water treatment is ineffective at removing herbicides like Atrazine. Several herbicides are used in considerably greater quantities than Atrazine but are found in lower concentrations in groundwater due to the fact that they break down much more rapidly.

As of 1999, currently used insecticides are rarely detected in groundwater in either urban or agricultural areas. Compared to herbicides, insecticides approved for use in the United States today have lower application rates and tend to break down faster. Many insecticides used from 1940 through 1970 were composed of chlorinated hydrocarbons; these insecticides are especially persistent in the environment and slow to breakdown. Their use has been banned or restricted in the United States since the 1970s. The best known insecticide of this class is DDT, whose insecticidal properties were first discovered in 1939. The most commonly detected insecticides in urban groundwater are dieldrin and diazinon, each of which were found by the USGS (1999) to be present in about 1 to 2% of sampled wells. Dieldrin is a chlorinated hydrocarbon, and although not very mobile, it is persistent and has a historical record of heavy use. Diazinon is an organophosphate. Organophosphates tend to be more toxic than chlorinated hydrocarbons, but break down much more quickly and have little residual activity.

Although insecticides are less common than herbicides in groundwater, the USGS (1999) found they were more likely to exceed EPA drinking-water standards. The wells that exceeded EPA standards were mainly wells tapping shallow groundwater that was not used for human consumption. In all but one of the 1000 wells sampled by the USGS (1999), the offending insecticide was dieldrin. Dieldrin was widely used in the United States from the 1950s through the early 1970s. In 1975, most uses of dieldrin were banned, and it is no longer produced or imported into the United States.

After release into the environment, pesticides undergo a series of chemical and biological reactions whereby the original product breaks down first into intermediate compounds and eventually into carbon dioxide and other harmless substances. Some breakdown products are short-lived, while others are more persistent. Little is known about the occurrence of breakdown products in groundwater or their effects on human health. Although pesticides do not appear to be a significant contaminant of groundwater in the United States as of 1999, the cumulative effect of long-term exposure to a mixture of substances at low exposure levels is unknown.

Irrigation is an ancient practice; the Egyptians were irrigating fields with water from the Nile River by as early as 5000 BC. Irrigation frees farmers from the vicissitudes of natural precipitation by supplying water at a uniform rate, however, irrigation poses a number of potential threats to groundwater. When crops are irrigated, more water is added to the land than is needed for plant growth. The excess water percolates through the soil and eventually reaches the water table. The infiltration of irrigation water may mobilize agricultural chemicals such as fertilizers and pesticides, providing them with a means of reaching groundwater in the subsurface. Irrigation water may also contain dissolved salts that can be concentrated in soils by evaporation. Irrigation waters can mobilize and concentrate toxic elements that are normally present in low concentrations in some soils. Farmers in the arid Central Valley of California rely upon irrigation to grow a variety of crops. In the western part of the Central Valley where surface soils are underlain by a low-permeability clay layer, irrigation water must be drained or the water table would rise to the point that crops would literally drown. In the 1970s, farmers hit upon what seemed like an ideal solution. Excess irrigation water would be drained into basins, creating marshlands and wildlife refuges. Both agriculture and wildlife would benefit. With the help of the government, 42,000 acres (170 square kilometers) of land in Fresno and Merced counties were drained into Kesterson national wildlife refuge. Millions of migratory birds were attracted to the clear blue waters of the refuge. Then, in the early 1980s, researchers found high levels of deformed bird embryos in the Kesterson refuge. The excess irrigation water draining into the wildlife refuge had leached selenium out of the Central Valley soil. As the water in the drainage area evaporated, selenium levels increased to toxic concentrations. The refuge had to be closed by constantly firing propane cannons to scare away aquatic birds.

Animal wastes are a potential hazard for groundwater contamination, especially in the case of animal feeding operations. An **animal feeding operation (AFO)** is an agricultural enterprise where animals are kept in confined situations. Feed is brought to the animals instead of the animal grazing or feeding in pastures. In 1999, the U.S. EPA estimated there were 450,000 AFOs in the United States. A **concentrated animal feeding operation (CAFO)** is an animal feeding operation that houses a large number of animals. Some CAFOs that house hogs may have as many as 100,000 animals. AFOs and CAFOs grew rapidly in the United States during the 1990s as corporate farms sought more efficient ways to produce animal products. In 1998, 10 companies produced 92% of the poultry consumed in the United States. Hog production was also beginning to become dominated by a small number of large companies. In 1994, there were 800 farms in the state of Utah that raised 44,000 hogs. By 1997, the number of farms in Utah raising hogs had dropped to 500, but the total number of swine had grown to 295,000 (Kratz, 1998). Managed properly, animal waste can be a valuable resource that reduces the need for fertilizer, however, if mismanaged, animal waste has the potential to pollute groundwater with nitrogen, phosphorus, and pathogenic organisms, as well as the hormones and antibiotics that may be added to animal feed.

In 1992, the residents of Guymon, a town in the panhandle of Oklahoma, overwhelmingly passed a one-cent sales tax increase to attract a multimillion-dollar pork processing plant to their town. It was anticipated that the plant would add 700 jobs to the local economy and possibly another 800 in a few years. The town's mayor was quoted as stating "this is a turning point for our community and our area." The pork producer in turn pledged to contribute $175,000 each year to the town's public school for the next 25 years (McNutt, 1992). Seven years later, the mood in the Oklahoma panhandle was different. Oklahoma's hog inventory had grown from 190,000 in 1991 to 1.64 million in 1998, largely due to the development of CAFOs in the panhandle region. The amount of waste produced was enormous. Up to 900 hogs were kept in confinement barns, with some sites containing as many as 16 barns. Hogs produce 10 times the fecal mass of humans. Hog feces dropped through slats in the floor of barns and were flushed into lagoons that could be as large as football fields. There were allegations that corporate hog farming had polluted communities with odors and that waste lagoons were leaking (Lindley, 1999). In reference to the odor associated with hog waste, an editorial in the Oklahoma City newspaper (November 17, 1999) described standing next to a waste lagoon as "an unforgettable experience."

9.3.4.d. Landfills.

A **landfill** is a hole in the ground that receives waste. Older landfills were nothing more than large holes in the ground filled with waste products and then covered with dirt. Modern landfills are engineered to contain wastes and minimize the potential for leakage and groundwater contamination. A **dump** is an uncovered landfill.

The primary waste received by landfills is municipal solid waste. The largest single component is paper; municipal solid waste also contains food wastes, yard wastes, glass, metals, plastics, and liquid wastes. The U.S. EPA estimated that by the year 2000 Americans would annually generate 220 million tons (200 billion kilograms) of municipal solid waste. In 1990, 67% of municipal solid waste was sent to landfills. The trend in the United States after 1970 was to place waste in a smaller number of larger landfills. In 1973, there were an estimated 14,000 operating landfills in the United States (Cartwright, 1984). By 1997, the U.S. EPA estimated that the number had been reduced to 3091 active landfills in the United States. Landfills are widely used partly because they are a relatively low-cost manner of waste disposal. The desire to eliminate wastes going into the air or surface water has also led to an increased reliance upon burial in the ground. The U.S. EPA is trying to reduce the amount of waste taken to landfills by encouraging recycling. However, there is a limit to the amount and type of materials that can be recycled with reasonable efficiency. Burning is an alternative way of disposing of solid waste, but burning can cause air pollution.

The primary mechanism by which landfills contaminate groundwater is through the generation of leachate and its infiltration below the water table. **Leachate** is the liquid created by water infiltrating into landfills and leaching out substances from the wastes contained therein. The potential for the generation of leachate and subsequent groundwater contamination is greatest in humid climates that have high rainfalls. It is difficult to totally isolate landfill contents from underlying groundwater, however, a properly designed modern landfill system can greatly attenuate the potential for groundwater contamination. Modern landfills have liners, covers, leachate collection systems, groundwater monitoring systems, and are constructed in areas that are geologically suitable. Liners may be plastic sheets or clay layers and are designed to reduce the escape of leachate from the bottom of the landfill. The leachate that does escape is gathered by a collection system. Because the liner does not allow water to escape from the bottom of the landfill, water is prevented from draining and may fill the landfill and run over the sides. This is called the "bathtub" effect. Covers, which may be made of compressed clay, can minimize infiltration and prevent spillover. Wells surrounding a landfill allow water samples to be collected as part of a routine monitoring process to ensure groundwater contamination is not taking place. The selection of a suitable landfill site necessitates avoiding floodplains, wetlands, and unstable areas subject to landslides or sinkholes. The hydrogeology of each area is unique and must be evaluated on a case-by-case basis. In general, the more desirable sites are those situated above low-permeability media and those where the soil and/or bedrock has the capability of attenuating leachate contamination through natural processes such as adsorption.

Many older landfills received toxic industrial waste and have created serious problems. The most famous of these is the Love Canal. One of the most famous and intensively studied examples of groundwater contamination by a landfill is the case of the Borden landfill in Canada. The Borden landfill is located on a Canadian military base about 80 kilometers northwest of Toronto, and received waste from 1940 through 1976. The landfill is situated above a sand aquifer, and a plume of contaminated groundwater is slowing migrating through the aquifer. As the aquifer is not used for drinking water and there is no threat of the contaminant plume reaching surface water, no remedial action has been taken to clean up the Borden landfill site. Instead, the site was used for a long-term study of contaminant transport starting in the late 1970s. The *Journal of Hydrology* reported some of the research results in a special issue published in 1983 (Cherry, 1983).

9.3.4.e. Impoundments.

An **impoundment** is a pond or lagoon used to hold liquid wastes. Depending on the use to which they are put, impoundments may be referred to as lagoons, basins, pits, or ponds. Impoundments vary widely in their size and design; there is no standard size or design. They may be unlined, or lined with clay, asphalt, or plastic sheets to prevent leakage.

Prior to 1980, there was little to no information available on the total number of impoundments in the United States. Between 1978 and 1980, the EPA collected data on surface impoundments and published their conclusions in 1983. The EPA (1983) estimated there were a total of 180,973 surface impoundments in the United States. Of these, 36% were oil and gas brine pits, 21% were used for municipal waste, 15% for industrial waste, 14% for mining waste, 11% for agricultural waste, and 3% were used for other purposes.

The EPA found considerable potential for significant groundwater pollution from surface impoundments. Nearly all impoundments had been located without any consideration for the protection of groundwater. About half of the 180,973 identified impoundments were located in areas where the unsaturated zone was either very thin or very permeable, offering little protection to underlying groundwater. About 70% of the impoundments were located over high-quality aquifers. Approximately 15% of all sites contained waste that the EPA considered to be hazardous. Miller (1980) estimated that industries in the United States pump about 1700 billion gallons (6.4 trillion liters) of liquid wastes to impoundments each year, of which 100 billion gallons (379 billion liters) enter groundwater systems. The Resource Conservation and Recovery Act of 1976 gave the EPA the power to control the disposal of hazardous waste in the United States. After the RCRA Act was passed, it was no longer legal to dispose of hazardous waste in unlined impoundments.

Miller (1980) briefly reviewed 57 case histories where leakage from impoundment wastes had contaminated groundwater in the northeastern United States. Of these 57 cases, the most frequent sources were the chemical industry, metal plating or processing facilities, and electronics firms. The worst case of groundwater contamination involved the dumping of arsenic compounds in unlined surface pits by a chemical company. The arsenic entered groundwater and a contaminated plume migrated to a nearby stream. The site was remediated by pump-and-treat, but after two and a half years of treatment the situation was still dangerous. The most famous case of groundwater contamination from impoundments is the U.S. Army's Rocky Mountain Arsenal near Denver where chemicals used to manufacture nerve gas and pesticides were dumped into unlined impoundments from 1942 through 1956.

9.3.5. Case Study: Rocky Mountain Arsenal, Denver.

On December 7, 1941, two hundred Japanese bombers made a surprise attack on the U.S. Naval Base at Pearl Harbor, Hawaii. The attack was an overwhelming victory for the Japanese. In a little more than 30 minutes, the U.S. naval and air forces in the Pacific Ocean were severely crippled, and more than 2300 Americans were killed. U.S. President Franklin D. Roosevelt declared that the date of December 7, 1941, would "live in infamy." The next day, the U.S. Congress declared war on Japan. Prior to the attack on Pearl Harbor, the American people had been divided on the question of American entry into World War II. Overnight, the attack on Pearl Harbor unified the American people in support of the war effort.

On May 2, 1942, the U.S. government announced that it needed 19,883 acres (81 square kilometers) of farmland just north of Denver for a chemical weapons plant. Condemnation and seizure of the land from farmers met little resistance; the spectre of Pearl Harbor was close at hand and sections of the west coast actually thought that a land invasion by Japanese troops was imminent. The decision to build a chemical weapons factory was influenced by the fear that Germany and Japan would use chemical weapons. Only 27 years had passed since 1915 when Germany introduced the use of mustard and chlorine gas in World War I. The United States never used chemical weapons in World War II; it was thought, however, that the ability to deploy them would forestall their use by the enemy. Construction of the Rocky Mountain Arsenal (RMA) began on June 30, 1942. Six months later, the first production began. Three major chemicals were made at the RMA: mustard gas, Lewisite, and chlorine gas. Lewisite was better known as "blister gas;" it is a liquid whose vapor is highly toxic. Any part of a human body that is contacted by the liquid or vapor suffers inflammation, burns, and tissue damage. Lewisite had been originally developed in retaliation for German gas attacks during World War I; it was never used by the United States. The RMA also produced napalm, an explosive type of jellied gasoline that later gained notoriety during the Vietnam War. The firebombing of Tokyo with napalm on March 9 and 10, 1945, killed an estimated 100,000 people.

In May of 1947, the RMA went on stand-by status, however, tensions started to rise between the communist Soviet Union and the United States. On March 5, 1946, Winston Churchill in a speech at Fulton, Missouri, said of the communist states, "From Stettin in the Baltic to Trieste in the Adriatic, an iron curtain has descended across the Continent." The phrase "iron curtain" referred to the political and military barriers erected by the Soviet Union after World War II to seal off itself and its eastern European allies from contact with the West. On June 25, 1950, the Korean War began when North Korea invaded South Korea. On October 30, 1950, the RMA went back on wartime status. During the Korean War, the RMA manufactured incendiary munitions. Meanwhile, tensions continued to rise between the Soviet Union and the United States. In 1954, the RMA began producing a deadly new chemical weapon: Sarin nerve gas. Sarin works by interfering with a body enzyme that controls motion. Exposure results in tremors, vomiting, and breathing problems, with death coming by suffocation.

The U.S. government was not the only generator of toxic waste at the RMA. Shell Oil Company leased RMA facilities and used them to manufacture agricultural pesticides between 1952 and 1982. Between 1943 and 1957, toxic waste chemicals from manufacturing activities at the RMA were dumped into open, unlined pits. No attempt was made to prevent infiltration of toxic substances into the ground, and chemicals reached a shallow and highly permeable aquifer that was widely used in the area for irrigation and the watering of livestock. The aquifer is composed of alluvium, sand, and silt of Pleistocene to recent age, underlain by Tertiary and Cretaceous bedrock consisting of sandstone, siltstone, and other clastic sedimentary rocks. The average thickness of the surficial aquifer is about 40 feet (12 meters), and both unconsolidated and lithified sediments function as a single hydrologic unit.

Problems were first noticed when crops were damaged by contaminated irrigation water. The first reports of crop damage were made in 1951. In 1954, groundwater use for irrigation was much heavier than normal due to a drought, and crop damage was severe. The extent of the contaminated plume could be traced simply by mapping the areas of dead plant life. In 1955, steps were taken to prevent further contamination. The amount of waste chemicals dumped was reduced, and a new pond was constructed for disposal of liquid wastes. The new disposal pit was lined with asphalt to limit leakage. Although woefully inadequate by modern standards, the new impoundment was the state of the art in 1955. The asphalt liner eventually failed, but contamination problems were already severe.

By 1961, the front of the toxic plume was moving at a rate of about 3 feet (0.9 meters) per day, and had migrated 5 miles (8 kilometers) from the waste disposal sites. The total area affected was about 6.5 square miles (17 square kilometers). Due to the loss of usable groundwater, land values were reduced by half in value. In the early 1960s, the RMA tried to dispose of liquid toxic wastes by injecting them into deep wells. This led to the discovery that earthquakes could be triggered by fluid injections into the Earth's crust (see Chapter 12). In 1973 and 1974, there were new claims of crop and livestock damage. An investigation conducted by the Colorado Department of Health found a toxic nerve-gas by-product in a well 8 miles (12.9 kilometers) downgradient from the RMA waste disposal ponds. Several other toxic substances were found in measurable quantities in nearby wells. In 1975, the Colorado Department of Health issued a "Cease and Desist Order" to the U.S. Army and Shell Chemical Company, which called for them to stop discharging toxic chemicals into surface and groundwaters. In 1978, a containment operation began that was designed to stop further migration of contaminated groundwater off the RMA. A barrier was constructed by filling a 3-foot (0.9 meters) wide trench 1500 feet (457 meters) long with a mixture of soil and bentonite clay. Eight dewatering wells were installed on the upgradient side of the barrier. For three years, contaminated water was pumped out of the RMA aquifer at a rate of 10,000 gallons (37,854 liters) per hour. Toxins were removed by passing the water through activated charcoal filters. The cleaned water was then pumped back into the aquifer by twelve injection wells located on the downgradient side of the containment wall. In 1981, the system was expanded to a barrier 6800 feet (2073 meters) long with 54 withdrawal wells and 38 injection wells (Konikow and Thompson, 1984).

All manufacturing activities at the RMA stopped in 1982. In 1989, the Cold War symbolically ended when the Berlin Wall was torn down, and the U.S. Army began devoting much of its attention to cleaning up the RMA. Remediation actions at the RMA are extensive in their scope. In addition to the pumping and treating of contaminated groundwater, 10.5 million gallons (40 million liters) of hazardous liquid waste were removed from one of the pits and burned. Some 76,000 drums of hazardous salts were dug up and removed, plus 530 buildings and 250 electrical substations were demolished. As of the year 2000, groundwater pump-and-treat systems were still operational. In 1992, the U.S. Congress passed legislation designating RMA as a National Wildlife Refuge upon completion of the environmental cleanup. Remediation of the RMA is not expected to be complete until the year 2011. The lesson of the RMA is perhaps best illustrated by a statement made by one of the first geologists to study the problem. Theodore Walker (1961) said, "the only reasonable cure for such problems is the prevention of contamination."

9.3.6. Common Groundwater Contaminants.

Groundwater contaminants can be divided into three categories: organic chemicals, inorganic elements and compounds, and biological agents. The contamination potential of any substance is determined by three factors: toxicity, persistence, and mobility. If we were to make a very broad generalization, the greatest danger to groundwater comes from organic chemicals (Figure 9.23). Many organic chemicals are persistent on a human time frame, highly toxic, and mobile. Biological contaminants may be mobile and toxic, but generally are not persistent. Inorganic compounds and elements are both toxic and persistent, but, with two exceptions, are not mobile.

Generally speaking, most inorganic elements are not a serious problem in terms of groundwater contamination due to limited mobility. Under the pH and oxidation-reduction (redox) conditions normally found in groundwater, they have low solubilities and are subject to adsorption. There are two important exceptions: nitrate and chromium. Nitrate is common because of its widespread use in fertilizers and its occurrence in human and animal waste. Nitrate is also highly soluble and is unretarded by adsorption. Unconfined, near-surface aquifers are especially susceptible to nitrate contamination. The concentration of nitrate in these aquifers is likely to continue to increase in the future as nitrate use remains widespread. The second inorganic substance that is a significant groundwater contaminant is chromium. Chromium is widely used in many manufacturing processes, and is also found in sewage sludge. Like nitrate, chromium's contamination potential is due to its high solubility and its resistance to adsorption (Cherry, 1984a).

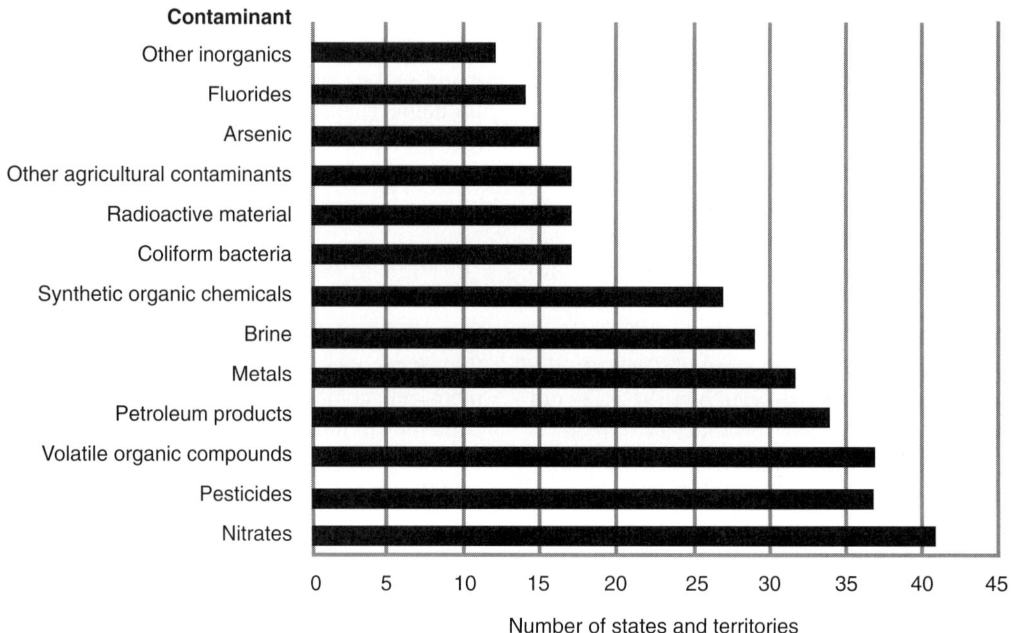

Figure 9.23 Frequency of groundwater contaminants judged by states and territories of the United States to be major threats to groundwater quality.
(From EPA, 1990a.)

The acidity or alkalinity of groundwater exerts a dominant control on solubility. Thus many inorganic elements and compounds can become mobile and contaminate groundwater in areas where the pH is unusually high or low. One place this can happen is in the vicinity of mine tailings. Mine tailings are the waste rock left over after the ore has been processed; they often contain sulfur minerals such as pyrite. When these sulfur minerals are exposed to oxidizing conditions at the surface, they combine with rainwater to form sulfuric acid. The resulting runoff is highly acidic water with a low pH that can enter the ground through infiltration. On the other extreme, if rainwater filters through the ash byproduct of coal-fired power plants, water of extremely high pH can result. Groundwater that has a pH less than about 4.5 (7.0 is neutral), usually has high concentrations of several different metals, which are rarely found in significant quantities in neutral pH waters. Contaminant plumes of acid waters are usually self-contained because of reactions with the porous medium within which they may be migrating. For example, acid tends to be neutralized by reactions with calcite cement (Cherry, 1984a).

Biological contaminants are either bacteria or viruses. Although groundwater contamination by pathogenic organisms is a documented problem, it can be corrected by proper chlorination of drinking water and remediation of the source. For example, Viraraghavan (1982) noted that viruses cannot reproduce in soils and die off between 7 days and 6 months.

The contamination of groundwater by organic chemicals is a major problem. There are tens of thousands of different types of organic chemicals manufactured and used throughout the United States. Little is known about the toxicity of many of these compounds or their breakdown products. Yet, we do know that many organic chemicals are very toxic in very small concentrations. There is no single test that can detect the presence of all organic contaminants. Organic chemicals may enter into an unknown number of reactions with microorganisms in the soil.

In the 1990s, it became apparent that the single largest problem in remediating several contaminated groundwater sites was the presence of a class of organic chemicals called non-aqueous phase liquids (NAPLs). A **NAPL** is a liquid organic compound in the subsurface that has low solubility in water and therefore generally exists as a separate phase. There are two categories of NAPLs: LNAPLs and DNAPLs. Light non-aqueous phase liquids **(LNAPLs)** are those compounds whose density is lower than water. Dense non-aqueous phase liquids **(DNAPLs)** are NAPLs that are denser than water. Common LNAPLs are hydrocarbons such as gasoline, kerosene, and diesel fuel. Gasoline has a density of about 800 kg-m^{-3}. DNAPLs are commonly composed of chlorinated solvents. A typical DNAPL is trichloroethylene, which is a colorless, somewhat toxic volatile liquid used for degreasing of metal, dry cleaning, and to remove caffeine from coffee beans. Other common DNAPLs are methylene chloride, the primary ingredient of many paint strippers, and carbon tetrachloride. Carbon tetrachloride is a dense (1580 kg-m^{-3}), colorless, volatile, and nonflammable liquid. Prior to the 1960s, it was widely used for the dry cleaning of clothes. In the 1950s it was discovered that continuous exposure to carbon tetrachloride caused diseases such as liver cancer and its use in the dry cleaning industry was discontinued.

Because they exist as a separate phase, the behavior of NAPLs in the subsurface can be very different from that of contaminants dissolved in groundwater. The density and viscosity of NAPLs is not the same as groundwater, and their behavior is affected by capillary forces. In most porous media, water is the wetting phase compared to NAPLs, while NAPLs are the wetting phase compared to air. As a consequence, when entering a porous medium NAPLs will first fill the largest pores and then the smaller ones, depending on the relative importance of the capillary entry pressure and the head gradient driving NAPL migration. Similarly, when leaving a porous medium, capillary forces will hold some residual NAPLs in place. The amount of NAPLs left in place under ambient conditions is called the residual saturation (see below). The residual saturation of a porous medium makes remediation of NAPL-contaminated zones very difficult using conventional pump-and-treat methods. A year or two of pump-and-treat will remove a contaminated groundwater plume, however, the residual saturation of a NAPL-contaminated source area may remain behind and start the contamination process all over again. The difficulty is that NAPLs tend to be slightly soluble in water. The solubility is high enough to cause contamination problems, but not so high enough as to facilitate the dissolution and removal of the NAPL contamination source. The NAPL that is left behind acts as a continuing source of groundwater contamination.

The capillary forces that make it difficult to remove NAPL contamination from the near-surface are the same forces that the petroleum industry has battled for 100 years (Bredehoeft, 1992). As long as oil has been produced, it has been found that pumping will remove only about a third of the petroleum in a reservoir rock. If enhanced recovery techniques are used, the percentage may rise to 50 or even 80%. However, successful remediation of aquifers requires removal of more than 99% of NAPL contaminants.

A key to understanding the behavior of NAPLs in the subsurface is the concept of relative permeability. **Relative permeability** is the ratio of the effective permeability of a porous medium for a given fluid at a certain saturation level, to the intrinsic permeability of that same medium when 100% saturated with that same fluid. Unlike single-phase flow, in multiphase flow the permeability of a porous medium is partly determined by the nature of the fluid saturating it and the degree of saturation. The physical origin of relative permeability lies in the fact that when two fluids compete for space in a porous medium, not all of the pore spaces, channels, and fractures are available for flow by each fluid. Thus as one fluid begins to occupy the space formerly occupied by

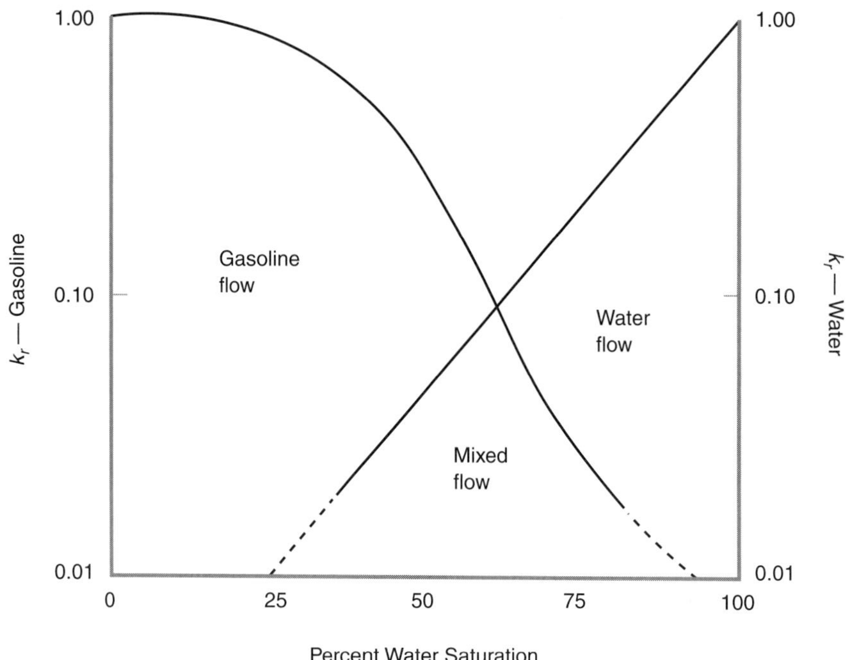

Figure 9.24 Relative permeabilities (k_r) for a gasoline-water system as a function of water saturation. (After Williams and Wilder, 1971.)

another fluid, the permeability of the medium with respect to the fluid being displaced is reduced. When two immiscible fluids compete for space in a porous medium, each has an irreducible or residual saturation. The **residual saturation** is the percent of void or pore space retained by the inhabiting fluid that cannot be replaced by the invading fluid; it is the saturation at which a fluid becomes discontinuous and immobile due to capillary forces. The existence of a residual saturation threshold implies that flow for a given fluid through a given volume of space cannot take place until the residual saturation for that space has been exceeded. It also implies that there is a residual amount of a NAPL contaminant that cannot be removed. The first part of fluid that moves into a porous medium is held there by capillary forces. Only when the residual saturation is exceeded is the fluid able to achieve through-going flow. The situation is illustrated for the gasoline-water system in (Figure 9.24). When gasoline (a LNAPL) saturation exceeds about 75%, no water flow is possible. Conversely, when water saturation exceeds 90%, no gasoline flow is possible. At intermediate saturations, the flow of both gasoline and water is possible, but at reduced permeabilities for both. Relative permeabilities in a two-phase system do not have to add up to unity.

LNAPLs, being lighter than water, typically migrate down through the unsaturated zone until they come into contact with the water table where they form a floating pool. If groundwater is moving past the LNAPL pool, it may exsolve chemicals such as benzene and toluene into the groundwater forming a contaminant plume. Because LNAPLs stop their downward movement at the water table, they are somewhat easier to clean up than DNAPLs. The hydrocarbon compounds that make up most LNAPLs are also subject to biodegradation.

DNAPLs are probably the single biggest problem in groundwater contamination. Being denser than water, they will continue to move down

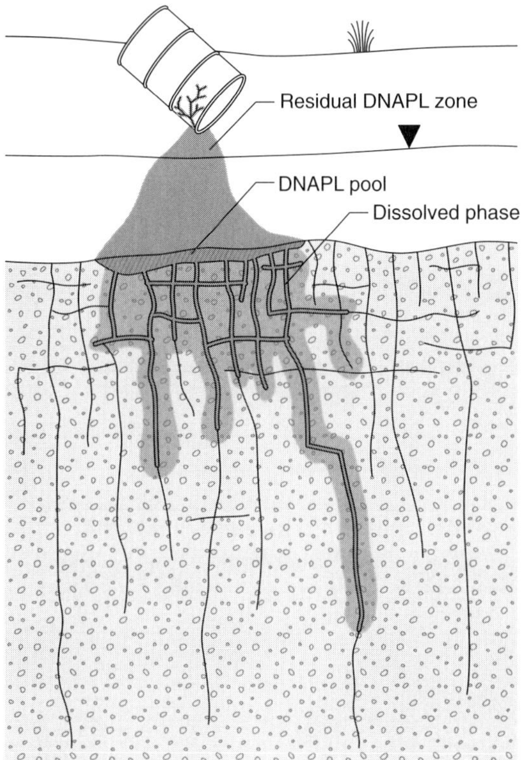

Figure 9.25 Contamination of bedrock fractures by DNAPLs.
(From Parker et al, 1994, p. 807.)

through the saturated zone until their migration is impeded by a layer of low permeability. The migration of DNAPLs below the water table generally makes their cleanup and removal much more difficult than for LNAPLs. DNAPLs move downward through high-permeability pathways such as fractures or coarse-grained sedimentary layers. It is easier for a DNAPL to displace water from larger pore channels and fractures that have lower capillary entry-pressures. Under the competing influences of buoyancy forces and hydraulic gradients, complications are possible. For example, DNAPLs may migrate downdip under the influence of gravity, while groundwater moves updip. Once a DNAPL pool forms in the subsurface, groundwater moving past it may dissolve small amounts of the contaminant for essentially forever on a human time frame. Because DNAPL levels as low as parts per billion are of concern, the contaminant potential for even small subsurface pools of DNAPLs is tremendous. One of the situations that is most intractable to remediation is the infiltration of DNAPLs into fractured bedrock and clay layers (Figure 9.25). DNAPLs can migrate through fractures with widths as small as 2×10^{-5} meters (0.02 millimeters). Because fractures typically make up a very small volume of any porous medium, the migration and concentration of DNAPLs in fractures allows them to effectively contaminate a much greater volume than they would be otherwise be capable of (Parker et al., 1994). Clay layers, which appear to be relatively impervious, may accommodate DNAPL migration through invisible networks of microscopic fractures.

9.3.7. Physical Processes.

Contaminants are transported in groundwater by three different physical processes: advection, diffusion, and dispersion. **Diffusion** is the transport of contaminants from regions of high concentration to regions of low concentration in the absence of fluid movement. Diffusion in a fluid such as water (as opposed to a porous medium) is described by **Fick's First Law.** In one spatial dimension, Fick's First Law is

$$F = -D\frac{dC}{dx} \tag{9.10}$$

where F is the mass flux of contaminant per unit area per unit time (kg-m^{-2}-s^{-1}), D is the diffusion coefficient (m^2-s^{-1}), and C is contaminant concentration (kg-m^{-3}). The **diffusion coefficient** (D) is about 1-2 $\times$ 10^{-9} m^2-s^{-1} for major anions/cations in water. In porous media, diffusion is retarded. In porous media, an **effective diffusion coefficient** D^* is defined as

$$D^* = WD \tag{9.11}$$

W is a dimensionless constant determined empirically, and is usually ~0.01-0.5 for non-reactive species. The effective diffusion coefficient for

Max Leggette: Pioneer of Hydrogeology Consultants

by Russell G. Slayback

The history of any science or profession is the story of innovators, of key discoveries, and of people who blazed new trails. Max Leggette (1899–1988) was a pioneer of another sort, the man who demonstrated, long before the era of public concern about water contamination, that a living could be earned by the full-time practice of consulting in the speciality of groundwater geology.

Today, when one can open the "yellow pages" of any city in the United States and find hydrogeologic consulting services advertised, that may seem unremarkable. It may be surprising to recall that the year 2000 marked only the 65th anniversary of the publication by C. V. Theis of the non-equilibrium formula that revolutionized quantitative hydrogeology. Moreover, 1944, only 56 years prior, marked the start of the speciality consulting business in groundwater geology by Max Leggette.

Ralph Maxwell Leggette was born on St. Valentines Day in 1899 in Ontario, Canada. He became a United States citizen in 1915, received a baccalaureate in geology from the University of Chicago in 1923, and completed three years of graduate study at Chicago.

He taught briefly at Ohio State and joined the Ground Water Division of the U.S. Geological Survey in 1928. At that time, the total professional staff of the Division of Ground Water was 8; 6 geologists and 2 engineers; indeed the infancy of that group that now employs more than 500 groundwater scientists. The Geologist in Charge was the legendary Oscar E. Meinzer and others of his early co-workers were Nelson Sayre, Lee Wenzel, Arthur Piper, Henry Barksdale, David Thompson, Harold Stearns, George Taylor, Stanley Lohman, Vic Stringfield, Al Fiedler, Vic Fishel, Harold Thomas, and C. V. Theis among the better-known scientists—a truly remarkable group.

Max's survey career started in northwestern Pennsylvania, and resulted in his first major publication, aptly titled *Ground Water in Northwestern Pennsylvania*. Leggette is credited with conducting the first pumping test by the Thiem method in Meadville, Pennsylvania (Lohman, 1979). He then worked with Stan Lohman on his study of the groundwater resources of northeastern Pennsylvania, with Lee Wenzel on the Platte River aquifer in Nebraska, and performed a groundwater investigation at Fort Leavenworth, Kansas, for the Federal Bureau of Prisons.

Max Leggette (1899–1988)

The year 1932 brought Max Leggette to Utah and one of the key stages of career and life. He performed groundwater studies with George Taylor of the Ogden Valley and of the Jordan valley near Salt Lake City, resulting in Water Supply Papers 796-D and 1029. He also met the charming Mildred Heist of Salt Lake City and married her in October. Like so many of his professional decisions, this was a wise choice that endured to the benefit of all of their associates.

In 1934, Oscar Meinzer appointed Max to be Chairman of the Survey Committee on Observation

Wells, charged with developing a nationwide program of systematic water-level observations. This assignment led to fieldwork in about 10 states, from North Carolina to Washington state, and from Texas to Wisconsin. The program led to the publication of Water-Supply Paper 777, *Water levels and artesian pressure in observation wells in the United States in 1935*, the first of a long series of basis data reports that were the precursors to the computerized data base we now use.

In 1935, Max became District Geologist in charge of groundwater investigations in New York and the New England states, a position he held for 8 years. This Survey office was based in Jamaica, Queens, on Long Island, one of the most prolific groundwater areas of the nation. During these years he continued his interest in the collection and dissemination of basis data, publishing a series of reports consisting of well logs for the four Long Island counties that are still one of the starting points for new hydrogeologic studies. His staff included C. E. Jacob, one of the great innovators of our science, and M. L. Brashears, who succeeded Max as District Geologist and was later to become his partner in the consulting business. An observation of drawdown in observation wells completed in the Lloyd Sand Member of the Raritan Formation caused by a pumping well more than 7 miles distant became Leggette's most common citation in the literature of hydrogeology.

Max left Long Island and his family in 1942 to serve in World War II with the Army Corps of Engineers. He served as a Water Supply Officer in Algeria and French Morocco during the North African campaign. Many years afterwards, we discovered Max's military service when a client who was also a World War II officer came in and asked for this most unmilitary of men as Major Leggette. Mustered out in 1944, Max returned to private life at age 45 and decided to see if he could support himself and his family as a consulting groundwater geologist. Today that seems a rather mundane choice but it was, in its time, a risky and trail-blazing move.

Max took office space on Fifth Avenue in New York City. A prestige address remained a part of his formula for conveying an aura of success throughout his consulting career. He consulted his Survey friends about what to charge and settled on a fee of $100 per day. His good friend, Henry Barksdale, told Max that no one was worth $100 a day. Of course, some hydrogeologists now charge $200 per hour, but that is another story. Max shared his first office with several fledgling businessmen. After some costly experiences, he learned to lock his telephone inside his desk drawer when he was out in the field.

The first big job, starting in March 1945, was a water-supply investigation for a dairy and poultry enterprise on the Island of Eleuthera in the Bahamas, a delicate salt-water intrusion problem. Later that year, Max landed an annual retainer from the Suffolk County Water Authority, a newly formed water-supply agency for the most easterly county on Long Island, an area that was destined for considerable growth in the post-War years. In 1950, he traveled across the country as the AAPG "Distinguished Lecturer," giving talks on "The Fundamentals of Groundwater Hydrology."

After operating as a sole proprietorship for 8 years, Max found, in 1952, that the demand for speciality consulting in hydrogeology had grown beyond his individual capabilities to provide. In a relatively short time, he took in M. L. Brashears as a partner, hired a secretary-bookkeeper, Miriam D. Miller, who was to be the backbone of the firm for 29 years, and hired two professional staff geologists; Ed Simmons, a later President of the Firm, and David W. Miller. Brash brought in new foreign and mine-dewatering work, and the firm prospered. Gene Hickok, Sid Fox, and Jim Geraghty joined the professional staff. In 1955, Jack Graham, another Survey "graduate," brought strength to the partnership in water quality and broadened the regional base with extensive contacts in Pennsylvania. Many will recognize that Geraghty & Miller and Gene Hickok moved on to form their own successful consulting firms, based in Long Island and Minnesota respectively.

The firm became consultants to water-supply companies and authorities; developed large well-water supplies for the paper, metals and brewing industries, dewatered red mines in Canada and Africa; and performed some of the earliest consulting studies involving contaminated groundwater.

Max Leggette: Pioneer of Hydrogeology Consultants *(continued)*

As a consultant, Max Leggette used his USGS training to good advantage. His broad travels and diverse experience had made him a master of sizing up a problem, focusing his efforts on the key points, and yet thoroughly examining all pertinent data. He was driven by a concern for accuracy, major technical content, minor details, and the report language that conveyed his results to the clients. We junior geologists did not look forward to the days when our report drafts were to be reviewed in head-to-head and word-by-word sessions with Max but, in retrospect, the discipline gave us excellent training.

Max Leggette's achievements as a pioneering consultant were many. Those that stand out in my memory include:

1. Drill-Stem Sampling
   When the Suffolk County Water Authority complained in 1962 of the high cost of treating some of their well water for excessive iron, Max drew on the oil business, drill-stem test technology. On his recommendation, each new production well site was drilled first by a pilot hole, in which temporary screens allowed water samples to be taken from several depth zones, which commonly varied considerably in iron content. Over a period of years it was found that such short-duration sampling was an exceptionally reliable indicator of the long-term iron content of a production well completed at a given depth. This pre-sampling program is still in use in Suffolk County in areas where iron content is a problem, and is further used to screen for nitrates and pesticides resulting from agricultural practices.
2. Expert Witness Testimony
   When the New Jersey Water Policy and Control Council established the first protected groundwater areas in 1947 and established a permit procedure for new groundwater withdrawals exceeding 100,000 gallons (378,540 liters) daily, Max became the first hydrogeologist to be accepted as an expert witness. Working before a lay council, his testimony was drafted to educate as well as set forth the facts of an application. By exceptional thoroughness, he literally set the standard for such expert testimony in New Jersey.
3. Long Island Water Balance
   When Nassau County on Long Island first began the systematic installation of sanitary sewers in 1952, with discharge of treated effluent to the ocean, Leggette's experience with the Long Island observation-well program and understanding of the pumpage and recharge balance of the area led him to conclude that a serious imbalance would occur and groundwater levels would progressively decline. As he did in so many other matters, Max devised a simple average water level comparison between sewered Nassau County and unsewered Suffolk County to convey his concern to the local government officials. First presented in 1954 and updated annually, this comparison received considerable attention during the record drought

porous media can also be related to the porosity and tortuosity (Berner, 1971),

$$D^* = D\frac{\phi}{\tau^2} \tag{9.12}$$

where ϕ is porosity and τ is tortuosity. **Tortuosity** is a dimensionless number defined as the flow path length for a water molecule divided by the straight-line distance. Besides tortuosity, other factors may retard diffusion in porous media. Ions must maintain electrical neutrality during diffusion, and diffusion may be slowed by adsorption.

Advection is the process by which a contaminant or any solute is carried by moving groundwater. The relevant groundwater velocity for transportation purposes is the linear velocity, not of the mid 1960s when there was considerable concern about the water-table decline in Nassau.

Max Leggette also began lobbying for controlled experiments with artificial recharge, using treated sewage effluent to set up a barrier to saltwater encroachment or otherwise restore the recharge-pumpage balance. Those efforts were partially responsible for the classic deep-well injection experiments conducted by the USGS at the Bay Park Sewage Treatment Plant, and go on today as scientists study the efficacy of maintaining the water table and base streamflow by return of treated sewage effluent to the upland recharge area in central Long Island.

4. Professional Status

After several years of arguing with the city and state of New York regarding the tax status of the consulting firm, Max went to court and succeeded in establishing that groundwater geology was a profession in the eyes of the law. The unanimous decision, rendered on March 4, 1958, noted that

> . . . the occupational activities of the taxpayer as a consulting geologist . . . constitutes an occupation or vocation in which a professed knowledge of the science of geology was used by its practical application to the affairs of others and in servicing their interest or welfare in the practice of an art founded on such knowledge. . . . That the activities of the taxpayer as a geologist . . . constituted the practices of a profession with the meaning . . . of the tax law.

As noted in the October, 1958, issue of *Geotimes*, "Earth scientists owe a debt of gratitude to the defendant for carrying this legal fight to a correct conclusion."

In 1969, at age 70, Max retired from his consulting firm. He passed away in February 1988. Leggette, Brashears & Graham has endured and grown. The consulting business in hydrogeology has changed tremendously in the last quarter of the twentieth century, especially with respect to the competition for work involving contaminated ground water. We have added to our tools and knowledge, and have learned how to market professional services in an era of the cattle-call Request for Proposal. Max hung out his shingle, put his professional card in a few journals, and waited for the phone to ring, which it did often enough. And yet, for all of the changes, the firm Max Leggette created retains very much of the personality and professional image of its founder.

Following Leggette's example, there are now several hundred firms that offer hydrogeologic consulting services, as either their sole business or as part of broad engineering services, and they provide employment for a few thousand hydrogeologists, hydrologists, geophysicists, and geochemists.

FOR ADDITIONAL READING

Lohman, S. W. 1979. Ground-Water Hydraulics, *U.S. Geological Survey Professional Paper 708,* p. 12.

Slayback, R. G. 1987. "Max Leggette, Pioneer of Hydrogeology Consultants" in Landa, E. R., and Ince, S., (eds), *History of Geophysics, Volume 3, The History of Hydrology,* p. 119–122, American Geophysical Union, Washington, DC.

the Darcy velocity. **Dispersion** is mechanical mixing. Dispersion along a flow path is **longitudinal dispersion.** Dispersion perpendicular to a flow path is **transverse** or **lateral dispersion.** Dispersion occurs due to both microscopic and macroscopic features. On the scale of individual solid grains in a porous medium, flow paths have branches (Figure 9.26). Flow velocity in a porous medium is also not constant. Fluid velocity in the center of a pore-throat is higher than on the sides, and larger pores have higher flow velocities. On a larger scale, heterogeneities in permeability due to fractures, lithology changes, and other geologic factors can lead to dispersion.

The first person to discover that dispersion occurs in groundwater flow was Charles Slichter

(1905). Slichter reported that his first supposition was that any spreading in the solute plume he observed would be due to diffusion. The results of his experiments caused him to reject his preconceived notion and he correctly described dispersion occurring due to pore-scale irregularities in flow velocities.

Mechanical dispersion in one dimension (x) can be defined mathematically as the product of the linear velocity (v_x, m-s^{-1}) and the **longitudinal dispersivity** (α_L, m). The processes of diffusion and dispersion can then be combined to define a **longitudinal coefficient of hydrodynamic dispersion** (D_L, m^2-s^{-1}),

$$D_L = \alpha_L v_x + D^* \qquad \textbf{(9.13)}$$

Note that the coefficient of hydrodynamic dispersion depends on properties of both the fluid and the porous medium.

An example of longitudinal hydrodynamic dispersion can be provided by a simple one-dimensional example. Consider a pipe filled with sand through which water is flowing from left to right (Figure 9.27). At time t_0, we start injecting a dye near the left end of the pipe. Assuming that we have the capability to instantaneously inject the dye throughout the diameter of the pipe, at time t_0 the relative dye concentration at the injection point is 100%, but 0% for all points to the right. At later times $t_1 < t_2 < t_3$, the dye front migrates to the right as it is carried along by advection, however, dispersion causes the formerly sharply defined front to become smoothed. Dispersion also causes the spreading of contaminant in directions transverse to the flow path as well as in the longitudinal flow direction (Figure 9.28).

The longitudinal dispersivity (α_L) of homogeneous, sandy materials measured in the lab ranges from 0.1 to 10 millimeters. The **transverse**

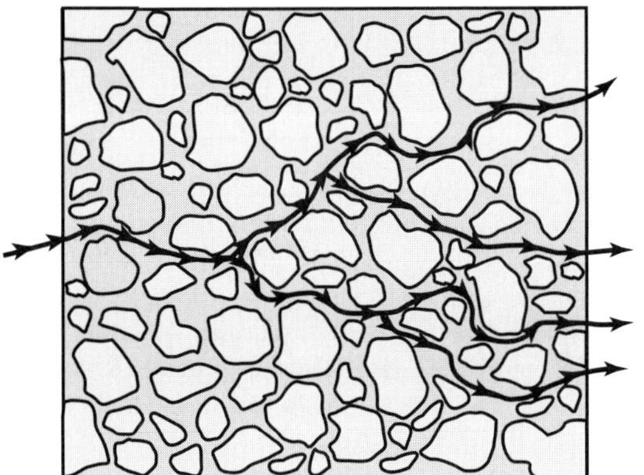

Figure 9.26 Dispersion on a microscopic scale caused by branching flow paths.

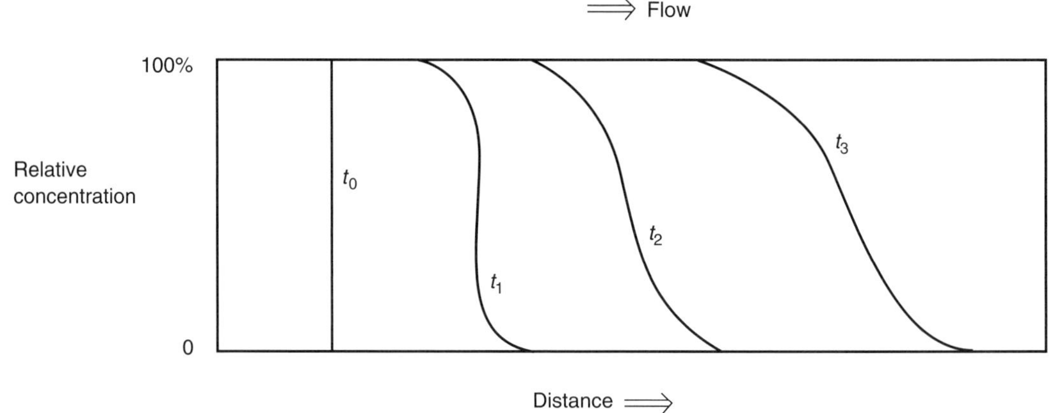

Figure 9.27 Relative contaminant concentration as a function of distance and time for one-dimensional flow. Contaminant introduction takes place at time t_0 on left. Concentration profiles at later times ($t_3 > t_2 > t_1 > t_0$) show how the initially sharp front is smoothed with passing time by dispersion.

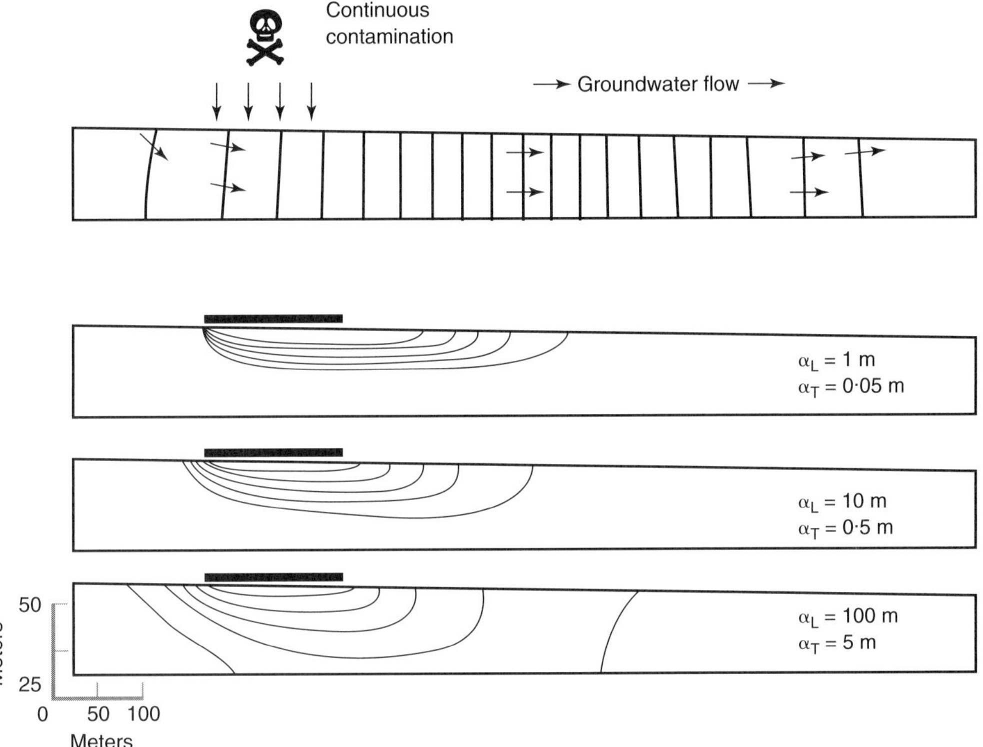

Figure 9.28 The effect of different longitudinal (α_L) and transverse (α_T) dispersivities on contaminant concentration contours.
(After Pickens and Lennox, 1976.)

dispersivity (α_T) is lower by a factor of 5 to 20. Dispersivity increases with scale. Values inferred from tracer tests in heterogeneous media are typically of the order of 0.1 to 100 meters. Near the source, dispersivity values inferred from laboratory experiments may be valid. But, as a contaminant plume spreads and encounters geologic heterogeneities, larger dispersivity values are more accurate descriptions of reality (Anderson, 1984).

Contaminant transport in porous media is described by the advection-dispersion equation. The derivation of this equation in one spatial dimension (x) is relatively straightforward. Consider a control volume with total volume V (m^3) $= A_x\Delta x$, where Δx (m) is the length of the side parallel to the x-direction, and A_x (m^2) is the cross-sectional area perpendicular to the x-direction (Figure 9.29). Fluid is moving through the control volume from left to right with Darcy velocity q (m-s^{-1}). The concentration of contaminant in the moving fluid is C (kg-m^{-3}), and the total contaminant mass inside the control volume is M (kg).

Assuming conservation of mass, the time rate of change of contaminant mass inside the control volume is the sum of the time rate of change of contaminant mass due to three factors: dispersion and diffusion, advection, and any sources or sinks that might transfer mass between the matrix and the fluid,

$$\begin{matrix} (1) & = & (2) & + & (3) & + & (4) \\ \dfrac{dM}{dt}_{\text{total}} & = & \dfrac{dM}{dt}_{\text{disp/diff}} & + & \dfrac{dM}{dt}_{\text{advect}} & + & \dfrac{dM}{dt}_{\text{source}} \end{matrix} \quad \textbf{(9.14)}$$

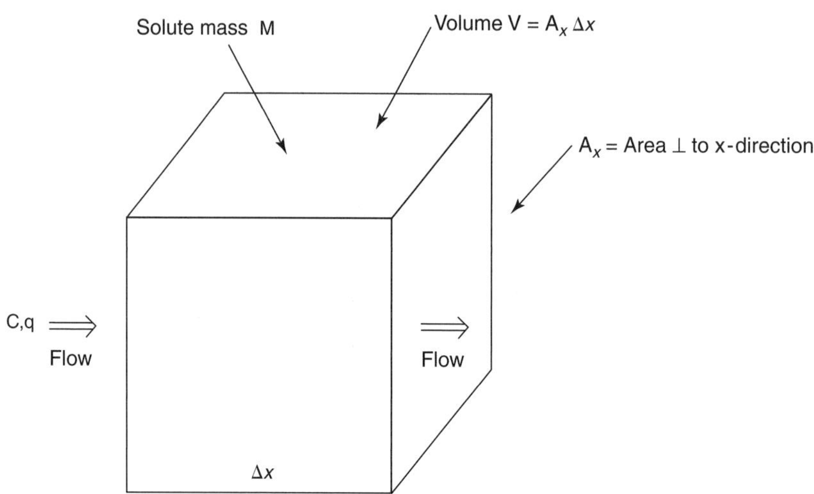

Figure 9.29 Control volume for derivation of advection-dispersion equation.

It is convenient to number each of the above terms (1, 2, 3, and 4) and expand each one at a time. Starting with term (1), the total rate at which contaminant mass inside the control volume changes is

$$\frac{dM}{dt}_{\text{total}} = \frac{d(\phi VC)}{dt} \quad \textbf{(9.15)}$$

For term (2), the change of contaminant mass associated with dispersion and diffusion is

$$\frac{dM}{dt}_{\text{disp/diff}} = \frac{\text{mass in}}{\text{unit time}} - \frac{\text{mass out}}{\text{unit time}} \quad \textbf{(9.16)}$$

Assume that contaminant flux is proportional to the concentration gradient as in Fick's First Law (equation 9.10). Because the mass **flux** is defined as "amount per unit area per unit time," the total rate at which mass enters the control volume is the product of the flux and the cross-sectional area perpendicular to flow. By analogy to Fick's First Law,

$$\frac{\text{mass in}}{\text{unit time}} = -D_x \frac{dC}{dx} \phi A_x \quad \textbf{(9.17)}$$

where D_x is the coefficient of hydrodynamic dispersion in the x-direction as defined by equation 9.13,

$$D_x = \alpha_x v_x + D^* \quad \textbf{(9.18)}$$

dC/dx is the concentration gradient, and ϕA_x is the fluid area perpendicular to flow (there can be no dispersion or diffusion through the matrix grains). The minus sign in equation 9.17 indicates that the direction of both diffusion and dispersion is from high to low concentration.

The rate at which mass leaves the control volume due to dispersion and diffusion is equal to the rate at which mass enters by dispersion and diffusion, plus any change that takes place over the length of the control volume, Δx. The total change that takes place over the length of the control volume is equal to the rate of change with respect to distance (x) multiplied by the total distance, Δx. So,

$$\frac{\text{mass out}}{\text{unit time}} = -D_x \frac{dC}{dx} \phi A_x + \frac{d\left[-D_x \frac{dC}{dx} \phi A_x\right]}{dx} \Delta x \quad \textbf{(9.19)}$$

Substituting equations 9.17 and 9.19 into equation 9.16, we obtain

$$\frac{dM}{dt}_{\text{disp/diff}} = \frac{d\left[D_x \frac{dC}{dx} \phi A_x\right]}{dx} \Delta x \quad \textbf{(9.20)}$$

Assume that D_x, ϕ, and A_x are all constant with respect to x. Then,

$$\frac{dM}{dt}_{\text{disp/diff}} = \phi D_x \frac{d^2C}{dx^2} A_x \Delta x \quad \textbf{(9.21)}$$

The product $A_x\Delta x$ is the total volume, V. Thus the second term of equation 9.14 is

$$\frac{dM}{dt}_{\text{disp/diff}} = \phi D_x \frac{d^2C}{dx^2} V \quad \textbf{(9.22)}$$

The third term is the time rate of contaminant mass change due to advective transport,

$$\frac{dM}{dt}_{\text{advect}} = \frac{\text{mass in}}{\text{unit time}} - \frac{\text{mass out}}{\text{unit time}} \quad \textbf{(9.23)}$$

The rate at which contaminant mass enters the left side of the control volume is the product of the Darcy velocity, area perpendicular to flow, and the contaminant concentration,

$$\frac{\text{mass in}}{\text{unit time}} = q_x A_x C \quad \textbf{(9.24)}$$

The rate at which contaminant mass leaves the right side of the control volume is equal to the rate at which contaminant mass enters the left side, plus the total change in mass flux that takes place over the length of the control volume. The total change that takes place over the length of the control volume is equal to the product of the rate of change with respect to distance (x) and the total distance (Δx). Thus

$$\frac{\text{mass out}}{\text{unit time}} = q_x A_x C + \frac{d(q_x A_x C)}{dx} \Delta x \quad \textbf{(9.25)}$$

Assume that q_x and A_x are constant with respect to x,

$$\frac{\text{mass out}}{\text{unit time}} = q_x A_x C + q_x A_x \Delta x \frac{dC}{dx} \quad \textbf{(9.26)}$$

The product $A_x\Delta x$ is the volume, thus

$$\frac{\text{mass out}}{\text{unit time}} = q_x A_x C + q_x \frac{dC}{dx} V \quad \textbf{(9.27)}$$

Substituting equations 9.24 and 9.27 into equation 9.23, our final expression for term (3) is

$$\frac{dM}{dt}_{\text{advect}} = -q_x \frac{dC}{dx} V \quad \textbf{(9.28)}$$

The fourth term represents the rate of contaminant mass loss/gain. Mass can be lost and/or gained from a fluid as a result of adsorption or chemical reactions with matrix grains, to name two possible mechanisms. Let S (kg-m^{-3}s^{-1}) be the rate at which the contaminant concentration (C, kg-m^{-3}) is changing. Then the total rate at which contaminant mass is added or lost from the control volume fluid is the product of S and the total fluid volume,

$$\frac{dM}{dt}_{\text{source}} = SV\phi \quad \textbf{(9.29)}$$

where the total fluid volume ($V\phi$) is the product of the control volume and the porosity. We are now ready to assemble terms (1), (2), (3), and (4) into the complete advection-dispersion equation. Substituting equations 9.15, 9.22, 9.28, and 9.29, into equation 9.14, we obtain

$$\frac{d(\phi VC)}{dt} = \phi D_x \frac{d^2C}{dx^2} V - q_x \frac{dC}{dx} V + SV\phi \quad \textbf{(9.30)}$$

Assume that V and ϕ are constant with respect to time. These terms can then be moved outside of the derivative sign, and V cancels throughout equation 9.30. Note that this should be expected. If we derive a generalized expression for contaminant transport, it should not depend on the specific size of the control volume we choose. Simplifying,

$$\phi \frac{dC}{dt} = \phi D_x \frac{d^2C}{dx^2} - q_x \frac{dC}{dx} + S\phi \quad \textbf{(9.31)}$$

The Darcy velocity (q_x) is the linear velocity (v) multiplied by the porosity (ϕ). Dividing equation 9.31 by ϕ, we obtain

$$\frac{dC}{dt} = D_x \frac{d^2C}{dx^2} - v_x \frac{dC}{dx} + S \quad \textbf{(9.32)}$$

Equation 9.32 describes the change of contaminant or solute concentration with time due to hydrodynamic dispersion, advection, and sources/sinks. The left side of equation 9.32 is the rate of change of contaminant concentration. The first term on the right side of equation 9.32 is the rate at which contaminant concentration changes due to mechanical dispersion and diffusion. Note that if the advection and source terms are left off, equation 9.32 is identical (except for the constants) to the diffusion equation (equation 5.35). The second term on the right side of equation 9.32 is the rate of contaminant concentration change due to advection. Note that the rate of change is directly proportional to both the linear velocity and the concentration gradient. If the concentration gradient is zero, there is no net transport of contaminant.

The advection-dispersion equation (9.32) can be solved by numerical means for domains of arbitrary complexity. Analytical solutions exist for several cases characterized by simple geometries and homogeneous physical properties. Consider a one-dimensional porous medium such as a pipe filled with sand. Fluid moves through the pipe with linear velocity v. The initial condition is that the contaminant concentration $C = 0$ for all x at time $t = 0$. At all times $t > 0$, a dye of concentration C_0 is injected at $x = 0$ such that the boundary condition at $x = 0$ is $C = C_0$. The boundary condition at the distant end of the pipe is that as $x \rightarrow \infty$, $C = 0$ for all times t. There are no sources or sinks. For these initial and boundary conditions, the solution to equation 9.32 is (Freeze and Cherry, 1979, p. 391; after Ogata, 1970)

$$\frac{C}{C_0} = \frac{1}{2}\left\{\operatorname{erfc}\frac{x - vt}{2\sqrt{D_x t}} + \exp\frac{vx}{D_x}\operatorname{erfc}\frac{x + vt}{2\sqrt{D_x t}}\right\} \qquad \mathbf{(9.33)}$$

where v is the average linear velocity, D_x is the longitudinal coefficient of hydrodynamic dispersion, erfc is the complementary error function, x is longitudinal distance, and t is time.

An analytical solution can also be found to describe the spread of a contaminant plume in three spatial dimensions (x, y, z) following injection at a point. Consider a three-dimensional porous medium with homogeneous physical properties. A point source of contaminant is injected instantaneously at the point $x = 0$, $y = 0$, $z = 0$, at time $t = 0$. There is fluid flow in the x-direction at an average linear velocity v. The mass of contaminant then spreads laterally in the y- and z-directions through dispersion and diffusion as it moves downstream in the x-direction (Figure 9.30). For these initial and boundary conditions, the solution to the three-dimensional form of equation 9.32 is (Freeze and Cherry, 1979, p. 395)

$$C(x, y, z, t) = \frac{M}{8(\Pi t)^{3/2}\sqrt{D_x D_y D_z}} \exp\left[\frac{vt - x}{4D_x t} + \frac{-\mathrm{y}}{4\mathrm{D_y t}} + \frac{-\mathrm{z}}{4\mathrm{D_z t}}\right] \qquad \mathbf{(9.34)}$$

where M is the total mass of contaminant injected at $t = 0$, D_x, D_y, and D_z are the coefficients of hydrodynamic dispersion in the x, y, and z-directions, respectively, t is time, and v is the average linear velocity.

9.3.8. Chemical and Biological Processes.

The chemistry of groundwater contamination can be complex. The physics of groundwater contamination is governed by a few processes: advection, dispersion, and diffusion. These processes occur in all settings, although the relative importance varies. In contrast, the number of different chemical interactions amongst contaminants and geologic media is vast. The Earth's crust contains more than 2000 minerals, each of which may act differently under different chemical conditions. There are a number of inorganic compounds and elements, which are found as groundwater contaminants (arsenic, chromium, selenium, uranium, etc.), and these inorganic elements can exist in more than one molecular and ionic form. These different forms may have different valences and different affinities for adsorption and solubility. Thus, one contaminant may be capable of assuming different forms, some of which may be mobile while others are immobile. More than 40,000 different organic compounds are manufactured, each of which is a potential contaminant with its own chemical properties. Little is usually known about

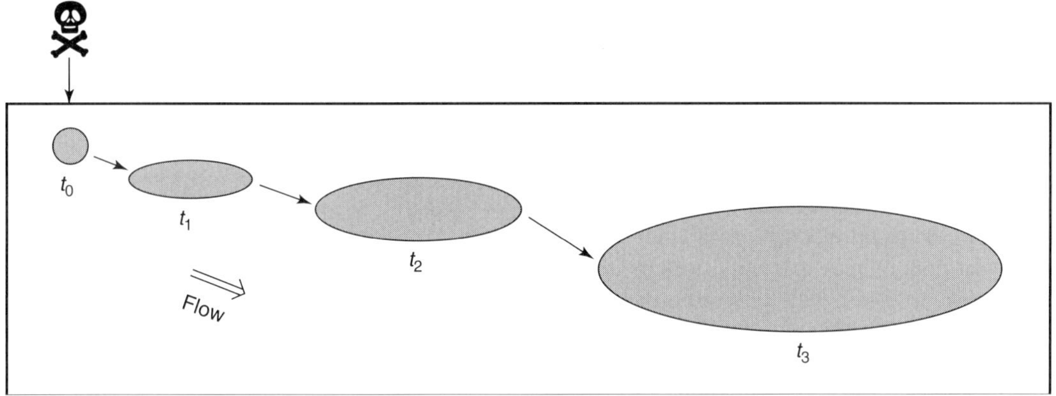

Figure 9.30 Spreading of a contaminant cloud due to dispersion following an instantaneous injection (time t_0). With increasing time $t_3 > t_2 > t_1 > t_0$, dispersion causes the volume of contaminated area to increase while the concentration of contaminant decreases.
(After Freeze and Cherry, 1979, p. 394.)

the identities and quantities of reactive minerals in many groundwater systems. The biology of groundwater contamination is also highly complex. The conditions under which biodegradation of contaminants occurs are poorly understood in many cases. Accordingly, it is often impossible to predict whether or not biodegradation will occur (Cherry, 1984b).

Although each geologic situation is unique, there are four important classes of chemical reactions important to groundwater contamination. These are acid-base reactions, precipitation and dissolution of minerals, adsorption and ion exchange, and oxidation-reduction (redox) reactions (Winter et al., 1998). Acid-base reactions involve the transfer of hydrogen ions (H^+). pH is the negative logarithm of the concentration of hydrogen ions in water. A pH of 7.0 is considered to be neutral. pH values lower than 7.0 are acid, those above 7.0 are alkaline. Under low pH conditions many metals are soluble and therefore mobile. At higher pH values, most metals will precipitate from solution.

Precipitation and dissolution are common chemical processes that occur in groundwater. When iron in solution in anaerobic groundwater comes into contact with oxygen-rich surface water, it forms iron hydroxide, which precipitates from solution. A common example of dissolution is the dissolution of limestone to form calcium (Ca^{++}) and bicarbonate (HCO_3^-) ions.

Adsorption is the process in which ions or molecules of a dissolved substance become attached to solid materials. Adsorption can be confused with absorption. Absorption is the process whereby a dissolve substance is not only attracted and held to the surface of solid, but actually penetrates into its interior. Sometimes the word sorption is used as a general classification of these phenomena. One of the best known applications of the phenomenon of adsorption is the use of activated charcoal to filter out contaminants. Charcoal is an impure form of carbon that is formed by heating a carbonaceous material (e.g., wood) in the absence of oxygen. Activated charcoal is a highly porous charcoal that has a large surface area. It is widely used in household water filters to purify drinking water and in gas masks to remove toxic gasses. The purification action is through adsorption; thus charcoal filters have to be periodically replaced as available surface area becomes covered with adsorbed contaminants. Adsorption is an important natural process for the retardation and attenuation of a plume of contaminated groundwater. In most cases, the amount of adsorption that occurs increases with the concentration of the solute being adsorbed, however, the relationship is not always linear. Clay minerals are particularly apt to act as adsorbers because of their large surface areas and uneven distribution of surface charge. Cations (positively charged ions) tend to be more strongly adsorbed than anions. Inorganic elements that tend to be strongly adsorbed by clay minerals include lead, cadmium, mercury and zinc. Potassium, ammonia, magnesium, silicon, and iron are less strongly held, while sodium and chloride ions are only weakly adsorbed (Fetter, 1994). Many organic contaminants are also readily adsorbed by soils and rocks. The chief adsorbing agent appears to be particulate organic matter found in sediments and soils. Pesticides are an example of organic chemical that tends to be readily adsorbed in most soils and therefore immobile (Cherry, 1984b).

Ion exchange is the replacement of an ion attached to the surface of a solid by another ion in solution; it is the process by which water softeners work. In a water softener, the ions responsible for hard water, calcium and magnesium, are exchanged for sodium on a solid surface. Thus the amount of calcium and magnesium in the processed water declines, while the amount of sodium increases. The Coastal Plain Aquifer on the Eastern Coast of the United States is a fascinating case of natural ion exchange that may have led to increased rates of heart disease amongst coastal inhabitants (Chapelle, 1997). The general direction of groundwater flow in the Coastal Plain Aquifer is eastward, from inland recharge areas in the west towards the Atlantic Ocean. Freshly recharged water in the aquifer has high levels of calcium and low levels of sodium. But, as the water flows eastward, the level of sodium in the groundwater increases as does the incidence of heart disease in those who live above the aquifer and drink its water. Although generally sandy, the coastal plain aquifer contains clay minerals originally deposited in a marine environment. These clay minerals have sodium ions adsorbed onto their surfaces. When surface water recharges the Coastal Plain Aquifer, it initially contains dissolved carbon dioxide as all near-surface water does. When the dissolved carbon dioxide encounters calcium carbonate as remnants of shell material in the aquifer sands, a chemical reaction takes place, which releases calcium ions,

$$CaCO_3 + H_2O + CO_2 \rightarrow Ca^{++} + 2HCO_3^- \quad \textbf{(9.35)}$$

The calcium ions, having twice the charge of the sodium ions, then preferentially replace the sodium ions on the clay minerals and release sodium ions to the groundwater,

$$Ca^{++} + 2Na^+ \text{ (clay)} \rightarrow 2Na^+ + Ca^{++} \text{ (clay)} \quad \textbf{(9.36)}$$

The process of ion exchange that takes place in the Coastal Plain Aquifer is virtually identical to that found in commerical water softeners and results in marvelous soft water ideal for washing clothes, however, high levels of sodium in the diet have been implicated as a risk factor for heart disease.

Oxidation-reduction (redox) reactions are those which involve the exchange of electrons. Oxidation is defined as the loss of electrons; reduction as the gain of electrons. As electrical charge is conserved, in order for one substance to be oxidized, another must be reduced. Redox processes can cause changes in the mobility of many inorganic contaminants. Nine common inorganic elements whose occurrence in drinking water is regulated have more than one possible oxidation state in groundwater. These are arsenic, chromium, iron, mercury, manganese, selenium, uranium, nitrogen, and sulfur. The solubility and mobility of these elements is largely controlled by the redox potential and the pH. Lead is a toxic element that is largely immobile because of prevalent pH and redox conditions (Figure 9.31). At low redox conditions, lead forms an insoluble sulfide mineral. At pH values of 7 or above and higher redox conditions, lead forms an insoluble carbonate mineral (Cherry, 1984b). Every year in the United States, a tremendous amount of lead is added to the ground at outdoor shooting ranges, however, the potential for groundwater contamination is usually very low due to lead's immobility at most pH/redox conditions.

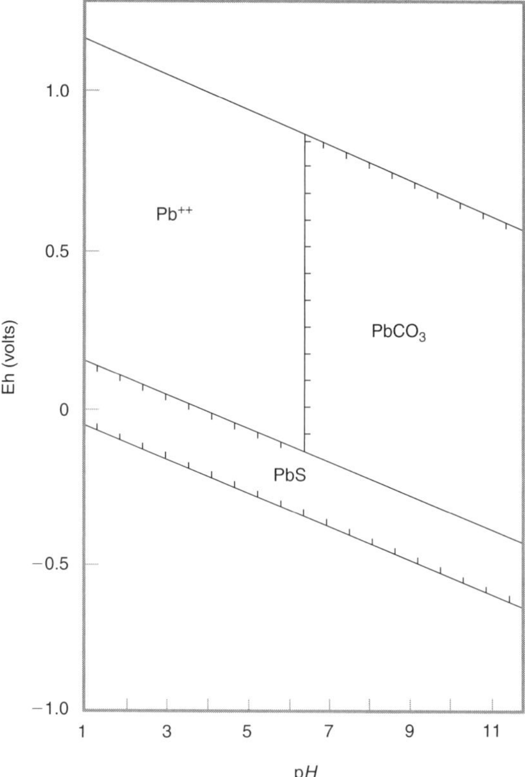

Figure 9.31 Eh-pH diagram for lead (Eh is the redox potential).
(After Cherry, 1984b.)

Biodegradation is the decomposition of organic chemicals by living organisms. Most microorganisms engage in biodegradation for the purpose of obtaining energy and carbon as food; the dominant organisms involved are bacteria. Although human interest in bacteria usually focuses on their harmful effects, bacteria have many beneficial functions. One of the most vital is as the master recyclers of the terrestrial biosphere. Without bacteria to breakdown organic wastes, the natural chemical cycles indispensable for life on Earth would grind to an eventual halt. Contaminated groundwater systems may be microbially active. Cherry (1994b) reports counting 10^6 bacteria in each gram of solid material in a number of contaminated aquifer systems. Williams and Wilder (1971) reported the case of a pipeline near Los Angeles leaking 250,000 gallons (946,350 liters) of gasoline into an aquifer. Because capillary forces made it difficult to completely remove gasoline, one of the most promising techniques for remediation in this setting was biodegradation of the gasoline by bacteria. The process of bacterial degradation in this instance was limited by the availability of oxygen and mineral nutrients. Therefore, natural biodegradation was enhanced by stimulating bacterial growth through the introduction of oxygen and nutrients. Biodegradation can result in the complete breakdown of complex organic chemicals to carbon dioxide and water, or can result in intermediate products. Whether or not biodegradation occurs depends upon the type of microorganisms present, the material under consideration, and the environmental conditions. Chlorinated hydrocarbons were once thought to be immune to biodegradation, however, we now know that these substances can be broken down by anaerobic bacteria below the water table in certain circumstances.

9.3.9. Remediation.

Remediation of groundwater contamination is any action that mitigates the undesirable effects of groundwater contamination. Remediation may involve a complete cleanup, or it may consist of actions that contain the source of contamination. In other instances, the best course of remedial action may be to do nothing at all and rely upon natural attenuation. In general, remediation tends to be complex, time consuming, and expensive (Lindorff and Cartwright, 1977). Since the initial failure of the pump-and-treat methodology at many contaminated sites was recognized in the late 1980s, it has become apparent that in many instances the successful remediation of contaminated groundwater must separately address two problems: remediation of the source, and remediation of the plume of contaminated groundwater. The greatest challenge in remediation is to clean up pools of organic chemicals such as NAPLS that act as ongoing subsurface sources of contamination (Mackay and Cherry, 1989). Remediation technology is rapidly changing, and many new methods are currently being tested and developed. In the discussion that follows, we make no attempt to comprehensively review all known technologies. Rather, a few of the more important and innovative methods are presented as an introduction to the subject.

9.3.9.a. The Site Investigation.

The first step in any remedial action is to conduct a hydrogeologic site investigation. The goal of the site investigation is to assemble an infinite amount of information with a limited amount of money and resources. The geology, chemistry, and physics of every site is unique and complex, and the tools of the hydrogeologist are limited and imprecise. Designing and implementing a site investigation is an art requiring knowledge, insight, and intuition. Each investigation must be handled as a separate problem; it is impossible to devise a cookbook approach.

The first step in any site investigation is to assemble all known information from the literature regarding the geology and hydrogeology of the affected area. Typically, the next step is to then drill a number of boreholes to access the subsurface. If the contamination source and general direction of groundwater flow are known, the first boreholes that are drilled may be downgradient from the source area so as to assess the distance that any contaminant plume may have traveled. The width of the plume can then be estimated by drilling boreholes transverse to the main direction of travel. Core samples will be collected during the drilling and submitted for permeability measurements. Permeability on a larger scale can be measured using slug tests or other quantitative methods that relate the change of head with time to permeability. It is often advisable to conduct well-scale permeability tests, as permeability is scale dependent due to the presence of geologic heterogeneities. Water levels in the investigatory boreholes are indicators of both the depth to the water table and the potentiometric surface. Water samples are collected and submitted to analytical laboratories for the purpose of detecting and measuring contaminant levels. This step may be subject to uncertainties. Analysis of water samples is technically complex. Not all known toxic substances can be detected at trace levels using routine water tests. In order to detect a contaminant, we often have to know what we are looking for before we begin to test for it.

Contour maps of the potentiometric surface will generally define the direction of groundwater flow, although this may be complicated considerably if the permeability structure is complex. If the contaminant is known and can be measured, a contour map of contaminant levels can also be made to define the size of the affected area.

The most difficult part of the hydrogeologic investigation is to assess the chemical, biological, and physical processes that affect contaminant behavior and mobility in the subsurface. How much dispersion will occur? Will adsorption occur? How will possible oxidation-reduction, acid-base, solution-precipitation, and other chemical processes affect contaminant concentrations? What about biodegradation? If these processes are present, will they be sufficient to naturally attenuate the contamination? It may be difficult to answer these questions.

9.3.9.b. Pump-and-Treat.

The classic method that has been used for many years to remediate contaminated groundwater problems is pump-and-treat. **Pump-and-treat** is a method for remediating contaminated groundwater in which the affected water is pumped out of the ground and treated at the surface to remove the contaminant (Figure 9.32). The treated water is then either injected back into the ground or disposed of somewhere else. A typical procedure is to pump extracted water through a medium such as activated charcoal that will adsorb the contaminant.

Pump-and-treat technology works well in some instances to treat contaminated plumes of groundwater. In many cases, however, it does not. In 1994, the National Research Council (NRC, 1994) surveyed 77 sites where the pump-and-treat

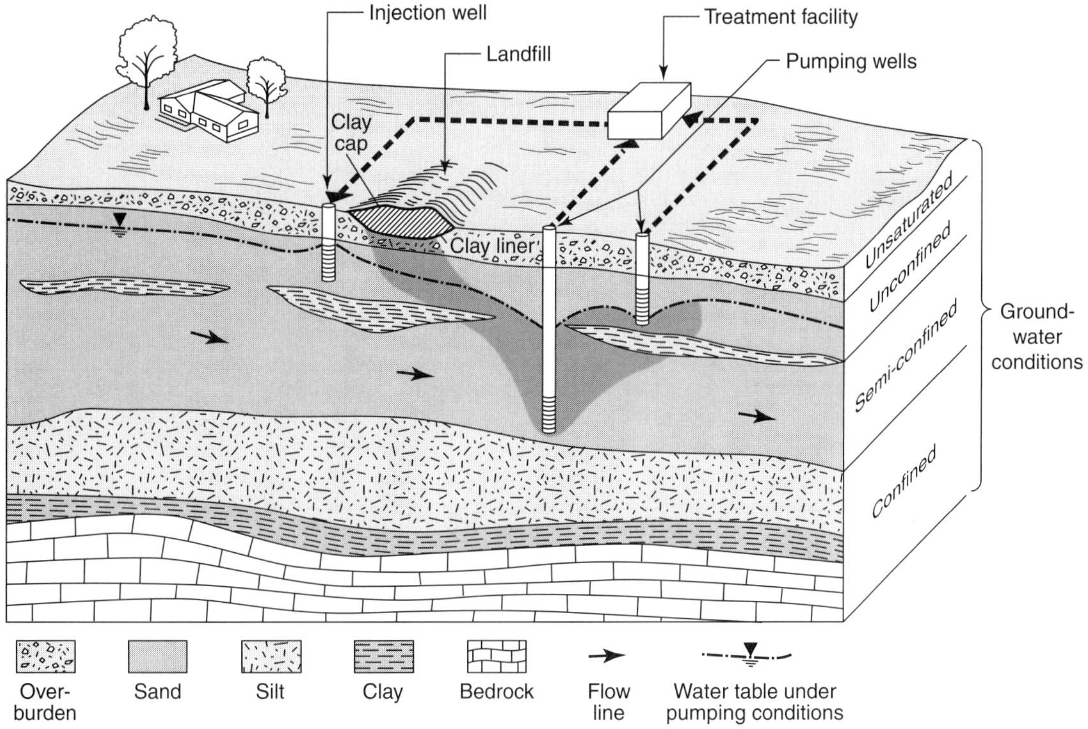

Figure 9.32 Pump-and-treat system for groundwater remediation.
(From EPA, 1990b, p. 2.)

method was either in use or had been used in the past. Of these 77 sites, only 7 had achieved their cleanup goals. In each case, pump-and-treat had reduced contaminant levels, but not enough. When the pumping began, contaminant levels fell rapidly, however, instead of continuing their rapid fall, it was found that tailing occurred. **Tailing** is a decline in the rate of contaminant removal (Figure 9.33). If the cleanup goal was reached, when the pumps were shut off the concentration of contaminants sometimes mysteriously rebounded. The existence of tailing implied that a complete cleanup to drinking water standards in some cases might take decades or even centuries. Today, we understand that the phenomena of tailing and rebound occur in cases where the contaminant has a separate source area that continuously feeds the groundwater. In general, the pump-and-treat method does not work well in the following three cases (EPA, 1990b):

1. Heterogeneous aquifer conditions where the presence of low-permeability zones restricts the flow of contaminants towards extraction wells.
2. Conditions where contaminants have adsorbed onto sediments or rocks and slowly desorb in response to the treatment process.
3. If immobile NAPLs are present and feed contaminant plumes through ongoing and prolonged dissolution of miscible components.

Even though pump-and-treat may not be able to completely cleanup some of the most serious contamination problems, the technology can nevertheless be used to contain the contaminated zone and

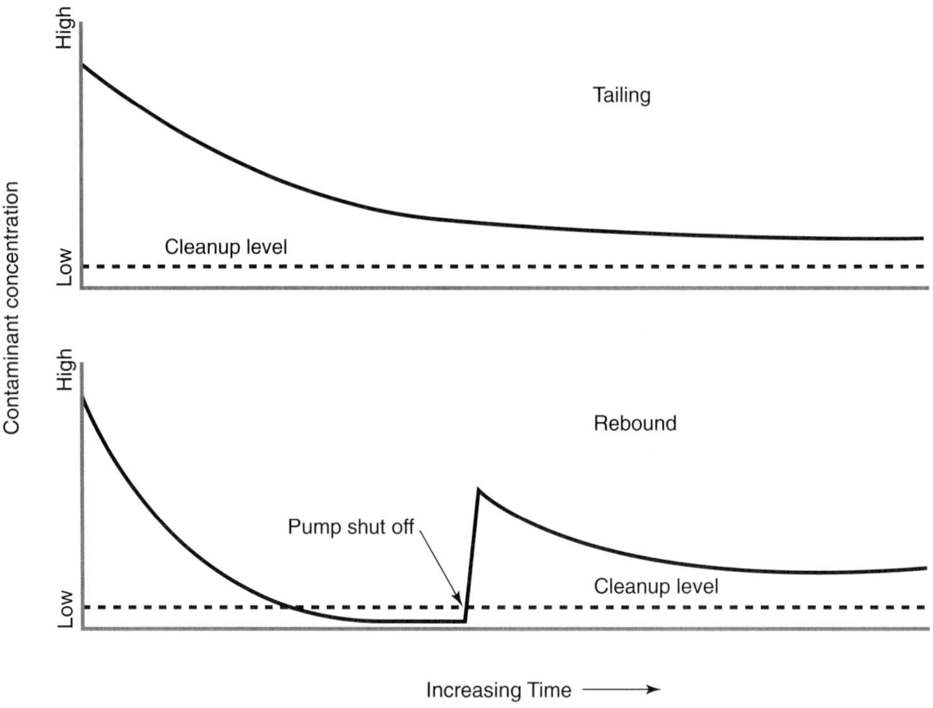

Figure 9.33 Tailing and rebound phenomena often found in pump-and-treat systems.
(After Wiedemeier et al, 1999, p. 5.)

prevent it from spreading. There are also ongoing efforts to enhance pump-and-treat technology by injecting chemicals that enhance water's ability to remove contaminants. One class of additives that have been used are surfactants. A **surfactant** is a substance that, when added to a liquid, reduces its surface tension and increases its wetting properties. Soaps and detergents are surfactants.

9.3.9.c. Natural Attenuation.

Natural attenuation is any reduction of contaminant mass, toxicity, mobility, or concentration that occurs without human intervention. A variety of natural biological and chemical processes can lead to natural attenuation. Some of these processes are destructive, while others are not. An example of a destructive process is the biodegradation of hydrocarbons by microorganisms. Almost all types of hydrocarbons will undergo biodegradation under aerobic conditions, producing carbon dioxide and water (Wiedemeier et al., 1999). There are also certain types of non-biological destructive processes; however, abiotic processes generally operate at much slower rates. Examples of nondestructive attenuation processes include adsorption, dispersion, and dilution.

One of the most striking proofs of the power of natural attenuation comes from studies of the lengths of contaminant plumes from leaking tanks containing hydrocarbon fuels. Microorganisms that break down hydrocarbons are found virtually everywhere in the subsurface. A study of 271 sites in California found that the lengths of contaminant plumes usually did not exceed 250 feet (76 meters). Of those 271 plumes, 59% were found to be stable, 33% were shrinking, and only 8% were growing. Similar results were found in a study of 217 benzene plumes in Texas by the Texas Bureau of Economic Geology. These studies indicate that

85 to 90% or more of the petroleum plumes in the United States are either at steady-state or receding. In some states, it is now necessary for applicants to demonstrate why natural attenuation will not take place if they are seeking funds to cleanup a fuel spill. Although the results for chlorinated solvents are not so well-documented as for hydrocarbon fuels, some studies show that these substances may also be subject to natural attenuation through biodegradation. Unfortunately, not all hydrocarbons may be susceptible to natural attenuation through biodegradation. The gasoline additive known by the acronym MTBE appears to be resistant to breakdown by microbial action (see review by Wiedemeier et al., 1999).

9.3.9.d. Funnel and Gate Systems.

A **funnel-and-gate system** is a way of treating contaminated groundwater by funneling it through a gate that contains a reactive medium that neutralizes the contaminant (Starr and Cherry, 1994). Funnel-and-gate systems are composed of two components: cutoff walls and a reactive gate. The simplest type of funnel-and-gate system consists of two cutoff walls that funnel a plume of contaminated groundwater through a reactor gate (Figure 9.34).

The heart of the funnel-and-gate system is the reactor gate that removes contamination. Although this technology is still developing, there are at least five different types of reactor gates. The first type of reactor modifies pH or oxidation-reduction (Eh) conditions in the subsurface, causing contaminants to precipitate out of solution. Chromium becomes less soluble under reducing conditions, and some biodegradation processes proceed more readily under oxidizing conditions. The second type of reactor is one containing a material that dissolves and causes a contaminant to precipitate out of solution. Reactors containing calcium phosphate will react with lead to cause the precipitation of lead phosphate minerals. The third type of reactor removes contaminants by adsorption. Activated charcoal removes several types of organic compounds, and ion exchange resins can remove contaminants in solution as ions. Adsorption is a reversible process, so the active component in this type of gate has to be periodically replaced. A fourth type of reactor stimulates biodegradation by providing a vital nutrient whose availability limits the rate of contaminant breakdown. Ordinary sawdust can be used as a source of organic material to stimulate the breakdown of nitrate through denitrification. The fifth type of reactor is one in which a contaminant is physically removed or transformed. Some volatile compounds can be removed by air sparging. **Sparging** is the introduction of a gas into a liquid.

The type of cutoff wall that is used in a funnel-and-gate system is not critical so long as the low-permeability material that constitutes the wall does not become dislodged and clog up the reactive gate. Steel sheets are satisfactory cutoff walls because the joints between them can be sealed and they will not interfere with gate function. In contrast, if the reactive gate is surrounded by slurry walls made of clays, it may be possible for the clay minerals to enter the flow system and plug up the reactive gate.

Funnel-and-gate systems have several theoretical advantages over the conventional pump-and-treat method. Many types of reactive gates completely process contaminants in the subsurface. In pump-and-treat, contaminants are brought to the surface, encumbering the remediator with the additional burden of their processing and/or disposal. The pumps used in the pump-and-treat method run 24 hours a day, commonly for years at a time, however, many of the in situ reactors used in the funnel-and-gate technique are passive. They require no maintenance and are immune from mechanical breakdown and power outages.

The design of a funnel-and-gate system must satisfy the following criteria. The system must be designed so that all contaminated groundwater passes through the reactive gate. This is not a trivial issue. The permeability structure of many aquifers is complex and spatially variable. For example, sequences of clastic sediments commonly contain interbedded layers, fingers, and lens of sand and clays whose permeability may vary by many orders of magnitude. If an aquifer is shallow (as most contaminated aquifers are), flow velocities

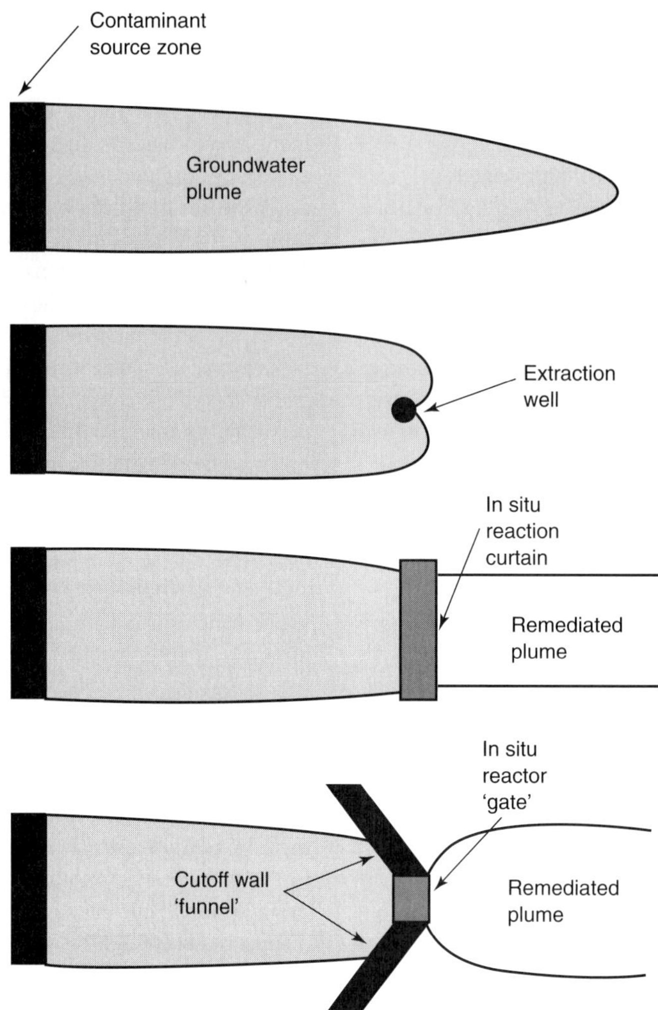

Figure 9.34 Remediation options for treatment of contaminated groundwater. (a) no treatment, (b) pump-and-treat, (c) in situ reaction curtain, and (d) funnel-and-gate system.
(From Starr and Cherry, 1994, p. 466.)

and directions may conceivably be affected by seasonal and longer term changes in precipitation. The second essential design criterion is that the residence time of the contaminated groundwater in the reactive gate must be long enough that the desired remediation is achieved. Finally, the size of the system in terms of the length of the cutoff walls and number of gates must be as small as possible in order to reduce cost. Numerical modeling can be a useful tool in the design of funnel-and-gate systems (Starr and Cherry, 1994)

9.3.10. Case Study: A Toxic Waste Dump in Louisiana.

The problem of adequately characterizing the representative elementary volume (REV) for estimation of in situ permeability is highly pertinent to

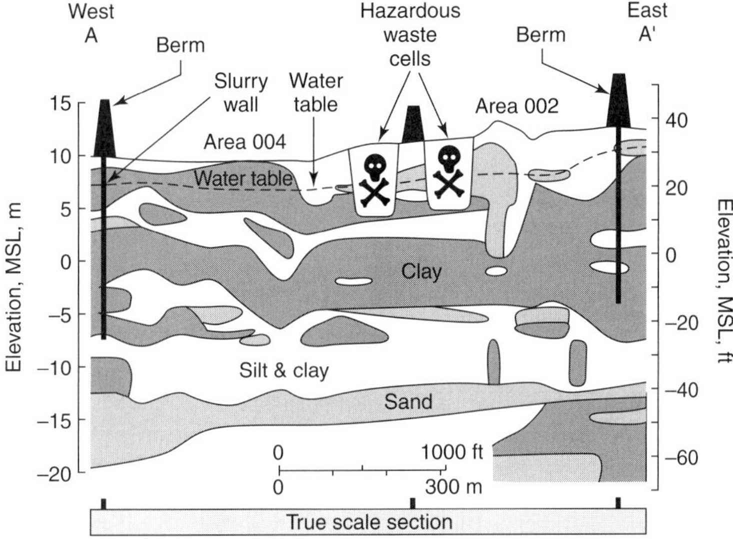

Figure 9.35 Simplified geologic cross-section through Livingston toxic waste dump in Louisiana. Vertical exaggeration is 20×.
(From Hanor, 1993, p. 3692.)

problems of groundwater contamination, as shown by Hanor's (1993) study of a toxic waste dump in Louisiana. In 1977, a site near the town of Livingston, Louisiana, was opened as a repository for hazardous wastes. The surface area of the site was approximately 1600 × 950 m. Waste was stored in landfill cells located beneath the water table (Figure 9.35). While some of the waste was buried in drums or other containers, much of it was simply poured into excavated cells and then covered. In 1985, site operators informed the Louisiana Department of Environmental Quality (LADEQ) that waste had been discovered beneath the natural clay sequence that extends underneath most of the site. Subsequently, the site was closed in 1990.

The Livingston site is located on a late Pleistocene sequence of fluvial, deltaic, and marine sands, silts, and clays (Figure 9.35). The site lies above, and in the recharge area, of a regional aquifer. At the time the Livingston site was developed, it was thought that a clay layer about 15 meters thick would act as a confining layer and prevent the leakage of toxic waste into the underlying aquifer. Additionally, a subsurface wall was placed around the entire perimeter in 1984 to prevent lateral migration of wastes. The wall was constructed by digging a trench 12 to 15 m deep and filling it with a clay slurry. Emplacement of the subsurface slurry wall was followed in late 1984 by the construction of a 5-meter-high berm or levee on the surface that completely encircled the Livingston site. The purpose of the surface levee was to control surface runoff. Drainage of the site was accomplished through sluice gates and box culverts placed around the perimeter. Monthly records of the volume of water discharged from the site were kept by the operator and filed with the LADEQ.

Annual rainfall in southern Louisiana typically exceeds annual evapotranspiration. The surplus water is lost by surface runoff to streams. Because surface runoff at the Livingston site was controlled, the absence of a standing body of water at the surface suggested to Hanor (1993) that significant vertical leakage had occurred via downward infiltration through the supposed confining layer. By estimating annual precipitation, evaporation, and site runoff, Hanor (1993) was able to show that there was substantial infiltration and downward flow. The total leakage through the confining layer was estimated to be about 1 meter over about a three and one-half year period of time. The vertical head gradient at the Livingston site was known from measurements of the water level in monitoring wells to be about 0.1. Knowing both the volumetric flux (~0.29 m-yr^{-1} $\approx 9.2 \times 10^{-9}$ m-s^{-1}) and the head gradient (0.1), Hanor (1993) calculated the in situ hydraulic conductivity from Darcy's Law as

$$K_z = \frac{q_z}{dh/dz} = \frac{9.2 \times 10^{-9}(\text{m-s}^{-1})}{0.1\ (\text{m-m}^{-1})} = 9.2 \times 10^{-8}\ \text{m-s}^{-1} \qquad \textbf{(9.37)}$$

By constructing above- and below-ground perimeter levees at the Livingston site, the operators had unintentionally converted the site to a giant permeameter. A **permeameter** is a laboratory device used to measure permeability (see chapter 3). Restricting runoff at the surface increased the residence time of surface water and the head gradient driving toxic wastes into the underlying aquifer. Construction of the perimeter walls thus not only did not help prevent contamination, the presence of the walls actually promoted the escape of toxic wastes.

In that estimates of surface runoff made from water departing the site through sluice gates and box culverts were maximum estimates, Hanor (1993) argued that both the resulting estimates of q_z and K_z were minimum estimates, and that the likely in situ vertical hydraulic conductivity was at least as high as 10^{-7} m-s^{-1}. This hydraulic conductivity is equivalent to a permeability of 10^{-14} m^2, characteristic of fine sand or silt—not clay. A comparison of the in situ estimate of the vertical conductivity with measurements on core samples (Figure 9.36) showed that the in situ conductivity was one to four orders of magnitude larger than the core measurements. The failure of the core measurements to adequately characterize the in situ permeability was due to a failure to understand the appropriate REV. The increase of conductivity with scale was due to heterogeneities larger than the core scale. These included (1) numerous sandy and silty distributary channels in the clay layers, and (2) pedogenic features in the clays. **Pedogenic** features are those related to soil formation. Sand layers in the clays appeared to be discontinuous in random, two-dimensional cross-sections (e.g., Figure 9.35), but were likely connected in three dimensions. Pedogenic features in the clays included joints, slickenslides, concretions, concentrations of organic material, and roots. A **joint** is a fracture along which no movement has occurred. A **slickenslide** is a polished and striated surface that results from movement on a fault plane. A **concretion** is a hard feature formed by the concentration of certain substances. These features are normally associated with rocks, but can also form in soils. The pervasive secondary fracture porosity represented by joints and slickenslides in the clays most likely developed as a result of alternate wetting and drying of clays after deposition. The lesson of the Livingston Toxic Waste Dump is that conventional engineering measurements alone may be inadequate for site characterization—geology cannot be ignored.

Review Questions

1. Define the following terms in the context of hydrogeology:
   - a. land subsidence
   - b. hydrocompaction
   - c. peat
   - d. sinkhole
   - e. Karst
   - f. miscible
   - g. dispersion
   - h. upconing
   - i. risk assessment
   - j. statistical murder
   - k. hormesis
   - l. cost-benefit analysis

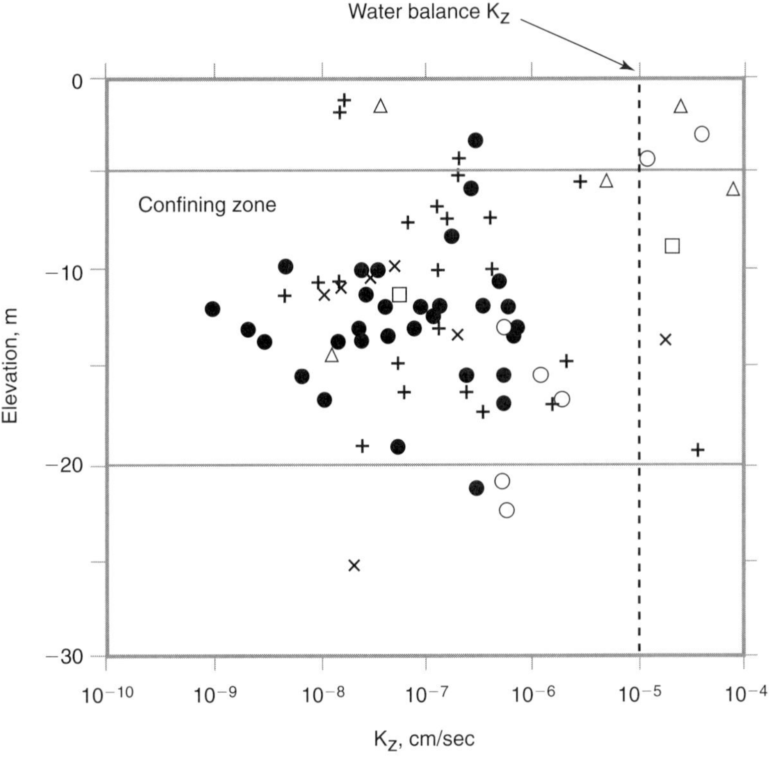

Figure 9.36 Comparison of permeability measurements on cores from the Livingston site with large-scale, in situ permeability estimated from water budget analysis.
(From Hanor, 1993, p. 3697.)

m. law of unintended consequences
n. point source
o. diffuse source
p. septic tank system
q. animal feeding operation (AFO)
r. concentrated animal feeding operation (CAFO)
s. landfill
t. dump
u. leachate
v. impoundment
w. NAPL
x. LNAPL
y. DNAPL
z. relative permeability
aa. residual saturation
bb. diffusion
cc. Fick's First Law
dd. diffusion coefficient
ee. effective diffusion coefficient

ff. tortuosity
gg. advection
hh. longitudinal dispersion
ii. transverse or lateral dispersion
jj. longitudinal dispersivity
kk. longitudinal coefficient of hydrodynamic dispersion
ll. transverse dispersivity
mm. flux
nn. adsorption
oo. ion exchange
pp. biodegradation
qq. remediation
rr. pump-and-treat
ss. tailing
tt. surfactant
uu. natural attenuation
vv. funnel-and-gate system
ww. sparging
xx. permeameter
yy. pedogenic
zz. joint
aaa. slickenslide
bbb. concretion

2. What are three mechanisms by which human actions can cause land subsidence?

3. Head declines 50 meters in a 25-meter-thick aquifer composed of sand. How much land subsidence will occur? Suppose the aquifer is composed of half sand, half clay layers. How does your answer change?

4. The maximum land subsidence in Tokyo due to groundwater extraction is 2 to 3 meters. The maximum land subsidence in the San Joaquin Valley is 9 meters. Which is a greater environmental problem, and why?

5. How has aquifer compaction affected aquifer storage capacity in the San Joaquin Valley?

6. What is the cause of land subsidence in the Everglades, and why can't it be reversed?

7. What physical and chemical processes cause sinkholes? In what two ways can human actions precipitate the formation of sinkholes?

8. What are three ways in which freshwater aquifers can become contaminated by saltwater?

9. Derive the Badon-Ghyben-Herzberg equation. Use a figure. Why did Hubbert (1940) refer to this equation as an "approximation"?

10. A tongue of saltwater is intruding into a freshwater aquifer, but the advance has been stopped and the position of the freshwater-saltwater interface stabilized. Explain how saltwater can still be actively transported into the freshwater aquifer and back to the sea.

11. Why do state regulations now require that abandoned oil and gas wells be cemented shut?

12. What are two reasons for injecting saltwater from oil wells back into underground reservoirs?

13. What caused the intrusion of saltwater into the Biscayne Aquifer in Southeastern Florida? What actions stopped the intrusion?

14. How much of the United States groundwater resources are contaminated?

15. What is the best cure for groundwater contamination?

16. Why is it usually undesirable to try to reduce risks to the zero level?

17. Explain how efforts in the United States to control air and surface water pollution led to more groundwater pollution. What general principle does this illustrate?

18. Sources of groundwater contamination can be divided into what two general categories? Give examples of each type.

19. What four factors are necessary for the proper performance of a septic tank system?

20. What is the most frequently reported cause of groundwater contamination?

21. Why did the inhabitants of Charlestown, South Carolina, have to abandon most of their water wells in the middle of the seventeenth century?

22. What are the three basic components of most fertilizers? What is the potential of each for groundwater contamination?

23. What single pesticide is found in the highest levels in groundwater? Why is this pesticide found more often than others?

24. Which compounds are more often found as a groundwater contaminant—herbicides or insecticides? Why?
25. What is the dominant type of waste buried in landfills?
26. Explain how modern landfills are designed so as to minimize the possibility of groundwater contamination.
27. The contamination potential of any substance is determined by what three factors?
28. The greatest danger to groundwater comes from what type of contaminant? Why?
29. What two inorganic chemicals pose the greatest threat to groundwater? Why?
30. What class of organic chemicals poses the greatest threat to groundwater?
31. Why is it difficult to remove NAPLS from contaminated aquifers?
32. Why are DNAPLs generally more difficult to remediate than LNAPLs?
33. What three physical processes transport contaminant?
34. What is the difference between a diffusion coefficient and an effective diffusion coefficient? By how much do they differ?
35. Why does dispersion occur? List two reasons.
36. Derive the advection-dispersion equation that describes contaminant transport in one dimension. Use a figure and be sure to state your assumptions in the appropriate places.
37. What four classes of chemical reactions are important to groundwater contamination?
38. What substance is commonly used to adsorb contaminants? In the ground, what type of minerals tend to act as adsorbers?
39. In the process of ion exchange, why do clay minerals tend to release sodium ions and adsorb calcium ions?
40. Why is the lead added to the ground at outdoor shooting ranges generally not a threat to groundwater?
41. What type of contaminant is known to be susceptible to biodegradation?
42. What is the greatest challenge in groundwater remediation?
43. Where does the pump-and-treat method work well? In what cases does it not work well?
44. List one type of destructive natural-attenuation process, and three types of nondestructive natural-attenuation processes.
45. What are the two components of a funnel-and-gate system?
46. List three of the five different types of in situ reactors that can be used in a funnel-and-gate system.
47. Why were most of the buried toxic chemicals never removed from the Love Canal?
48. How was groundwater contamination at the Rocky Mountain Arsenal originally discovered?
49. What geologic factors caused permeability at the Livingston toxic waste dump in Louisiana to be higher than measurements on core samples indicated?

SUGGESTED READING

ASCE (American Society of Civil Engineers). 1969. Saltwater intrusion in the United States. *Proceedings of the American Society of Civil Engineers, Journal of the Hydraulics Division,* 95 (HY5): 1651–1669.

Galloway, D., Jones, D. R., and Ingebritsen, S. E. (eds.). 1999. Land Subsidence in the United States. *U.S. Geological Survey Circular 1182.*

Hanor, J. S. 1993. Effective hydraulic conductivity of fractured clay beds at a hazardous waste landfill, Louisiana Gulf Coast. *Water Resources Research,* 29: 3691–3698.

Mackay, D. M., and Cherry, J. A. 1989. Groundwater contamination: pump-and-treat remediation. *Environmental Science and Technology,* 23 (6): 630–36.

Poland, J. F. 1972. Subsidence and its control in Underground Waste Management and Environmental Implications, T. D. Cook (ed.), *Am. Assoc. Petroleum Geologists Memoir* 18, Tulsa, Oklahoma, 50–71.

Starr, R. C., and Cherry, J. A. 1994. In situ remediation of contaminated ground water: the funnel-and-gate system. *Ground Water,* 32 (3): 465–476.

Notation Used in Chapter Nine

Symbol	Quantity Represented	Physical Units
α	compressibility of a porous medium	$Pa^{-1} = m\text{-}s^2\text{-}kg^{-1}$
α_L, α_T, α_x	longitudinal (L), transverse (T) dispersivity, dispersivity in x-direction	m
A_x	area perpendicular to x-direction	m^2
b, Δb	aquifer thickness, change in aquifer thickness	m
C	contaminant concentration	$kg\text{-}m^{-3}$
D	diffusion coefficient	$m^2\text{-}s^{-1}$
D^*	effective diffusion coefficient	$m^2\text{-}s^{-1}$
D_L, D_x ,D_x ,D_x	longitudinal coefficient of hydrodynamic dispersion, coefficient of hydrodynamic dispersion in x-, y-, and z-directions	$m^2\text{-}s^{-1}$
Δx	length of control volume side	m
erf, erfc	error function, complementary error function	dimensionless
F	mass flux of contaminant	$kg\text{-}m^{-2}\text{-}s^{-1}$
ϕ	porosity	dimensionless
g	acceleration due to gravity	$m\text{-}s^{-2}$
h, Δh	head, change in head	m
K	hydraulic conductivity	$m\text{-}s^{-1}$
M	mass inside control volume	kg
P, ΔP	fluid pressure, change in fluid pressure	Pascal (Pa) = $kg\text{-}m^{-1}\text{-}s^{-2}$
q, q_x, q_z	Darcy velocity, Darcy velocity in the x- and z-directions	$m\text{-}s^{-1}$
ρ	fluid density or freshwater density	$kg\text{-}m^{-3}$
ρ_s	saltwater density	$kg\text{-}m^{-3}$
S	rate of contaminant concentration change	$kg\text{-}m^{-3}s^{-1}$
σ_e, $\Delta\sigma_e$	effective stress, change in effective stress	Pascal (Pa) = $kg\text{-}m^{-1}\text{-}s^{-2}$
t	time	s
τ	tortuosity	dimensionless
V	volume	m^3
v, v_x	linear velocity, linear velocity in the x-direction	$m\text{-}s^{-1}$
V_T, ΔV_T	total (matrix plus fluid) volume of a porous medium, change in total volume	m^3
W	empirical diffusion constant	dimensionless
x, y	distance	m
z	elevation or depth	m

CHAPTER 10

Petroleum Migration

10.1 Definition and Origin of Petroleum.

Following Mackenzie and Quigley (1988), we define **petroleum** as a natural subsurface hydrocarbon material, including oil and gas. In the Earth's crust, petroleum is found in rock pores and fractures. Economic petroleum accumulations are found in reservoir rocks. A **reservoir** is a geologic unit saturated with oil or gas that is sufficiently porous so as to contain appreciable quantities of petroleum and permeable enough (usually $\geq 10^{-14}$ m^2) to allow for large quantities of petroleum to be pumped out of the reservoir.

Oil is a complex liquid-phase mixture of organic compounds. Oil consists dominantly of hydrocarbon molecules composed of various proportions of hydrogen and carbon, and smaller amounts of nonhydrocarbons such as sulfur, nitrogen, oxygen, and trace elements, including some metals. The total number of compounds that have been identified in some oils is at least several hundred and may exceed one thousand. Natural gas consists primarily of methane (CH_4), the simplest hydrocarbon, with lesser amounts of ethane, propane, and butane. Natural gas also contains lesser amounts of nonhydrocarbon gases such as hydrogen sulfide, carbon dioxide, and helium. Because oil and gas consist of complex mixtures of hydrocarbon molecules that behave differently at different temperatures and pressures, they may undergo phase changes as they are extracted from the subsurface. Oil and gas are mutually soluble in each other. Gas may exsolve from oil, and oil may condense from gas. Oils that contain large portions of waxy paraffins may be liquid at high subsurface temperatures but may solidify at the surface (Meissner, 1991).

The origin of the petroleum (literally, "rock oil") found in the Earth's crust is an old problem that has been debated for more than 100 years. There are two main schools of thought concerning the origin of petroleum. **Abiogenic theories** postulate that petroleum arises inorganically as a result of hydrocarbon outgassing from the mantle. One of the first advocates of an inorganic origin was the Russian chemist Mendeleev (1877), who created the periodic table of the elements. **Biogenic theories** postulate that petroleum forms from the transformation of organic material buried in sedimentary rocks as they are subjected to increasing pressure and temperature with increasing depth of burial.

The chief modern advocate of the abiogenic theory today is Thomas Gold (Gold, 1993; 1999). Gold (1993) cites several arguments in favor of each theory. In favor of the biogenic theory:

1. Petroleum contains groups of molecules that can be clearly identified as resulting from the breakup of common organic molecules that occur in plants and that could not have formed from nonbiological processes.
2. Petroleum molecules frequently exhibit optical activity that is a characteristic property of biological materials and is absent in materials of nonbiological origin.
3. Petroleum is found mostly in sedimentary rocks, frequently in close association with one or more organic-rich sedimentary rocks that appear to be the source of the petroleum.

In favor of the abiogenic theory, Gold (1993) lists:

1. Petroleum is often found in geographic patterns of long lines or arcs that are more related to the large-scale structure of the Earth's crust than to the smaller-scale patchwork of sedimentary deposits.
2. Areas with deposits of petroleum tend to have petroleum present at all stratigraphic levels, suggesting that the petroleum originated as a stream from the basement below.
3. Some petroleum from deeper levels lacks substantial evidence of biologic activity.
4. Methane (natural gas) is found in many areas where a biogenic origin is improbable.
5. Areas with petroleum deposits often are also characterized by relatively high rates of helium seeping up from the mantle.

The primary difficulty with abiogenic theories is the incontrovertible evidence of biologic activity in the organic molecules that constitute petroleum. Gold (1999) has proposed that the upper 3 to 6 km of the Earth's crust contains a large community of thermophilic ("heat loving") bacteria that utilize inorganic petroleum streaming up from the mantle as a food source. It is the digestion of inorganic petroleum by these bacteria in the crust that turns it into an organic product. As interesting as this idea is, it has found little acceptance. Nearly all geologists and geochemists currently favor the biogenic theory of petroleum origin; abiogenic theories are seen mostly as curiosities not to be taken seriously. As Apps and van dep Kamp (1993, p. 81) wrote in reference to the origin of natural gas, "Overwhelming evidence supports the belief that the world's natural-gas resources come from the decomposition of organic matter in sedimentary rock formations."

10.2. Biogenic Origin of Petroleum.

Organic matter originates from the process of photosynthesis, where plants use the energy of sunlight to combine carbon dioxide and water, creating hydrocarbons. When plants and animals die, most organic matter is destroyed by scavenging organisms. If organic matter escapes destruction, it may become incorporated into a sediment. An organic-rich sedimentary rock that is capable of generating oil or gas is a **source rock.** Organic sedimentary material is subject to alteration upon burial and exposure to increasing pressure and temperature. In general, the progressive alteration of sedimentary organics can be divided into three stages: diagenesis, catagenesis, and metagenesis.

In a larger context, **diagenesis** is a general term that refers to any postdepositional change that takes place in a rock. In petroleum geology, diagenesis refers to the changes that take place to organic material between the surface and the depths at which oil generation first occurs. There are two important diagenetic changes. In the first, organic material is changed into a generic petroleum precursor termed **kerogen.** Kerogen can be divided into an inert component that has no potential to produce petroleum, a **labile** fraction that has the potential to chiefly generate oil, and a **refractory** component that yields only gas (Figure 10.1). The second change that may occur is the generation of methane by bacterial action. Bacterial action can be the source for economically significant accumulations of natural gas at relatively shallow depths.

Catagenesis is the process in which kerogen is converted to oil and gas at elevated temperatures.

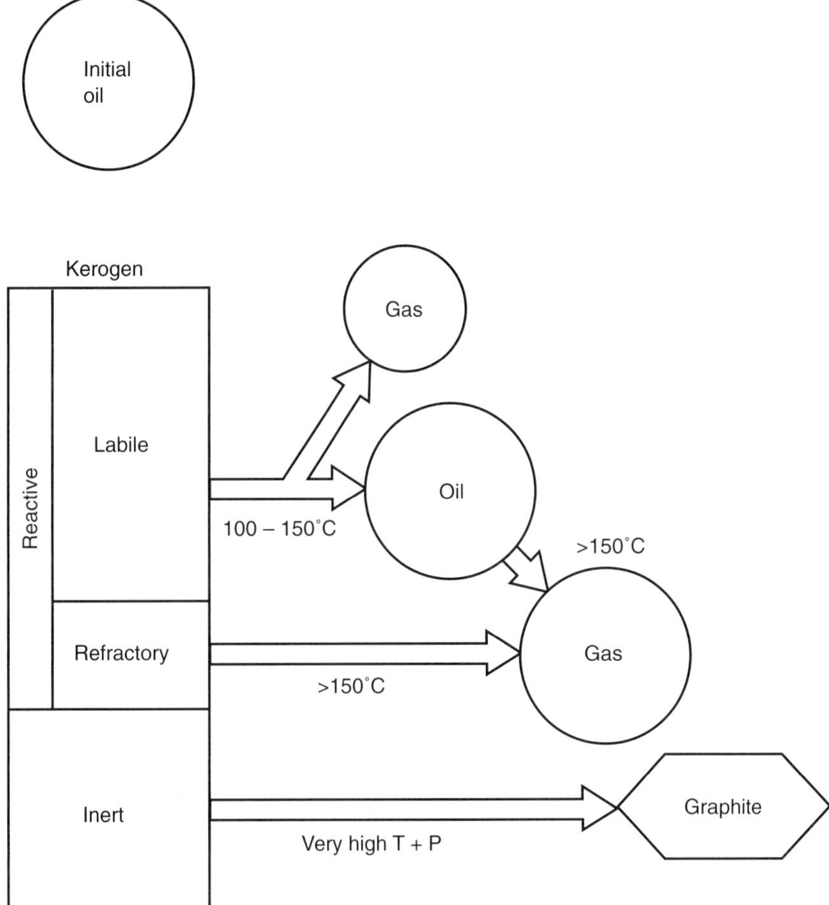

Figure 10.1 Classification and fate of organic matter in source rocks.
(From Mackenzie and Quigley, 1988, p. 402.)

Both time and temperature are important factors in the catagenic process of oil generation and the subsequent **cracking** of oil to natural gas (Waples, 1980). According to the **kinetic theory** of oil and gas formation (Quigley et al., 1987), the rate at which either oil or gas is formed is linearly dependent upon time and exponentially dependent upon temperature. The rate at which the concentration of any component changes is dependent upon the amount present

$$\frac{dC}{dt} = -k_r C \tag{10.1}$$

where C (dimensionless) is the concentration of either kerogen or oil, t (s) is time, and k_r (s^{-1}) is the reaction-rate coefficient. The reaction-rate coefficient (k_r) is given by the Arrhenius equation,

$$k_r = A \exp\left[\frac{-E_a}{RT}\right] \tag{10.2}$$

where A (s^{-1}) is a constant peculiar to an individual chemical reaction, E_a (J-mole^{-1}) is activation energy, R (J-mole^{-1}-°K^{-1}) is the universal gas constant, and T (°K) is the absolute temperature.

Most oil is generated from labile kerogen at temperatures in the range of 100 to 150°C. At

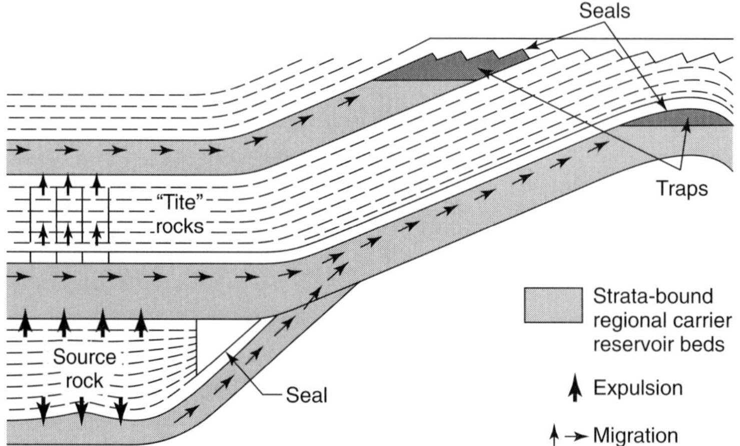

Figure 10.2 Hydrocarbon machine. The function of a hydrocarbon machine is to turn organic matter in a source rock into a hydrocarbon accumulation.
(From Meissner et al., 1984, p. 3.)

temperatures greater than 150°C, oil tends to crack to gas, and refractory kerogen begins to generate gas directly.

Metagenesis is the third stage of organic evolution. During metagenesis, residual kerogen that has not been converted to oil or gas becomes highly condensed and enriched in carbon. The ultimate product of metagenesis is pure carbon in the form of graphite.

10.3. The Petroleum System.

A **petroleum system** consists of the essential geologic components and processes that may result in an economic petroleum accumulation (Magoon and Dow, 1994); petroleum systems are usually named after a specific source rock. The formal recognition and classification of petroleum systems has its roots in Meissner's concept of hydrocarbon machines (Meissner et al., 1984a). Meissner's **hydrocarbon machine** consists of "all of the factors which affect the processes of hydrocarbon generation, migration, and accumulation" (Meissner et al., 1984a, p. 1). The function of a hydrocarbon machine is to turn organic matter in a source rock into a hydrocarbon accumulation. The elements of a hydrocarbon machine exist in interdependent, cause-and-effect relationships. In the context of hydrogeology, one important element of a hydrocarbon machine is a plumbing system that allows the movement of petroleum fluids outward from their site of generation to sites where they accumulate in traps (Figure 10.2). This movement is termed migration.

10.4. Migration.

Migration is the movement of petroleum in the subsurface. Meissner (1991) cites four types of evidence that petroleum migrates:

1. Most mature source rocks do not contain the amount of petroleum that they were capable of generating.
2. Reservoir rocks that contain petroleum are commonly incapable of having generated the oil found within them.
3. A number of oils in different locations have been geochemically correlated with distant source rocks.
4. Petroleum moves through a reservoir rock during production.

Migration can be either primary or secondary. The movement of petroleum out of the source rock in which it was generated to a reservoir is termed **primary migration.** Movement of petroleum within a reservoir is termed **secondary migration** (North,

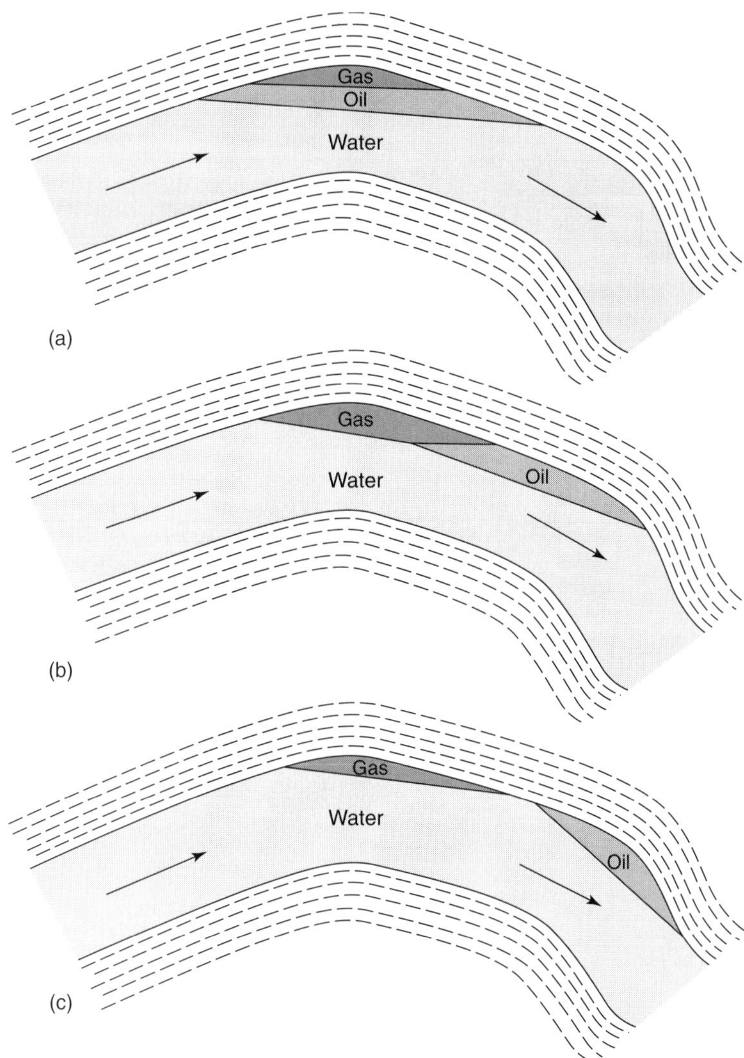

Figure 10.3 Hydrodynamic petroleum trap. Capillary forces prevent oil and gas from escaping through overlying shale. Hydrocarbons are hydrodynamically trapped whenever the dip of the oil-water (or gas-water) contact is less than the dip of the reservoir. As the hydrodynamic force increases (part a to part c), the relative importance of the buoyancy force decreases.
(From Hubbert, 1953, p. 230.)

1985). The direction of secondary migration is controlled by regional head gradients, permeability structure, and buoyancy forces. Because oil is less dense (~820 to 930 kg-m^{-3}) than most aqueous brines, it tends to migrate vertically upward. Most of the oil and gas that have been generated throughout geologic time have escaped to the surface and been destroyed. However, some small fraction of the total generation volume has become trapped in the subsurface at sites where its potential energy is a minimum. A typical situation is an anticline where petroleum in a coarse-grained reservoir rock is kept from further vertical migration upward by capillary forces in a fine-grained seal rock (Figure 10.3). Capillary force (see section 6.1) is the force that causes a wetting liquid to

M. King Hubbert: The Theory of Groundwater Motion

The contributions of M. King Hubbert (1903–1989) to hydrogeology may have been the most significant of the twentieth century. Hubbert, in fact, made so many significant and original contributions to the geologic sciences that it would not be an exaggeration to characterize him as the greatest geologist of that century.

M. K. Hubbert was born in 1903 and spent his childhood on a farm in San Saba County, Texas. Hubbert's early schooling was episodic, the length of the school year determined largely by the seasonal demands of farm work. Hubbert was dissatisfied with life on the farm and dreamed of obtaining a first-class education. He spent two years at Weatherford College in Weatherford, Texas, but wanted more than that school could offer. After consulting with the president of Weatherford, Hubbert applied to the University of Chicago and was accepted. Lacking the money to travel to Chicago, Hubbert worked his way north by following the wheat harvest. He put in 13-hour days of hard manual labor and slept at night in haystacks. Hubbert arrived in Chicago in September 1924, where he supported himself by working at various jobs while attending the university full time. Near the end of Hubbert's first year there, he was asked by his dean what he intended his major to be. His reply was that he had not intended to major in anything —he came to Chicago to obtain an education. Forced to declare a major subject, Hubbert chose an obscure double major in physics and geology with a minor in mathematics. In 1931, Hubbert became an instructor of geophysics at Columbia University, where he lobbied for the inclusion of physics, mathematics, and geophysics into the standard geology curriculum. These efforts were ahead of their time and wholly unsuccessful. Hubbert (1982b, p. 249) later explained his philosophy of education:

> It became increasingly clear that the knowledge appropriate to any domain of phenomena was dictated by the phenomena themselves. If they were chemical, a knowledge of chemistry would be required; if mechanical, a knowledge of mechanics, etc. Since the Earth represents a composite of chemical, mechanical, thermal, gravitational, electrical, magnetic, nuclear, and biological phenomena, then a very broad education in basic sciences becomes a logical necessity for an earth scientist.

M. King Hubbert (1903–1989)

In 1937, Hubbert published a paper titled "Theory of Scale Models as Applied to the Study of Geological Structures" in the *Geological Society of America Bulletin*. This paper was described by Doan (1994, p. 40) as "arguably one of the most seminal published in geology in this century." The paper was accepted by the faculty at the University of Chicago as Hubbert's Ph.D. thesis, an outcome Hubbert had not foreseen.

Toward the end of his years at Columbia, Hubbert turned his attention to hydrogeology. In 1940, he published the most significant hydrogeology paper of the twentieth century, "The Theory of Ground-Water Motion," in the *Journal of Geology*. The paper was 159 pages long and established the physical principles upon which hydrogeology was to develop. Darcy fathered the law that bears his name, but it was Hubbert who nurtured and developed it. Bredehoeft (1990, p. 1325) described Hubbert's work as "the clearest exposition of Darcy's Law." One of Hubbert's contributions was to point out that many petroleum engineers had corrupted Darcy's Law by incorrectly writing an equation that equated specific discharge as proportional to the fluid-pressure gradient instead of the head gradient. Evidently not one prone to suffer fools, in a review of Morris Muskat's book *Physical Principles of Oil Production*, Hubbert (1950, p. 659) noted that the incorrect formulation of Darcy's Law resulted in "the perpetuation of a state of confusion which has already persisted too long." The significance of Hubbert's 1940 paper was not realized for quite some time. Even in the 1950s, there were those in the groundwater community who had not read it. John Bredehoeft asked C. V. Theis why it took so long for the importance of Hubbert's work to be recognized. Theis was not sure; but commented, "I was having my own problems getting my ideas accepted as was Hubbert."

Although no one knows exactly why, Hubbert failed to obtain tenure at Columbia. Perhaps the well-bred New Yorkers on the Columbia faculty resented being lectured to by a Texas farm boy. Hubbert's rejection by Columbia created a wound in him that never healed. This failure, however, so drove him for the rest of his career that his productivity was probably greater than it otherwise would have been.

Following the end of World War II, many of the large oil companies formed research divisions. Foremost among these was Shell Development Company. The industrial scientists at Shell fully recognized the brilliance of Hubbert's work. Willy Hafner, a geophysicist, persuaded Hubbert to join him at Shell. At Shell, Hubbert was given free rein to work on whatever he wanted to without pressure from management. The results were spectacular. In 1951, Hubbert and Hafner published separate papers that forever changed the entire science of structural geology. Hubbert's "Mechanical Basis for Certain Familar Geologic Structures" and Hafner's "Stress Distributions and Faulting" were published back-to-back in the *Geological Society of America Bulletin*. In 1951, most geologists were largely ignorant of physics. The Hubbert and Hafner papers were considered by most geologists to be too highly mathematical to be of any practical value. It took a new generation of geologists to adopt, teach, and foster these ideas; the older generation took their old ideas to their graves.

In 1953, Hubbert published "Entrapment of Petroleum under Hydrodynamic Conditions" in the *American Association of Petroleum Geologists Bulletin*. In this paper, Hubbert established the hydrodynamic theory of petroleum migration that explained why the oil-water contact in reservoirs could be tilted. Hubbert further showed how this tilt could be used to understand the process of oil migration and thus forever altered the science and art of petroleum exploration.

The origin of Hubbert's hydrodynamic theory lay in his work on saltwater-freshwater interfaces (pers. comm., F. Meissner, 2001). When Hubbert came to work at Shell Development, he was asked to give an illustrated talk to the technical group working there. He showed a diagram very similar to one that he eventually published as Figure 23 (1953, p. 1994) in "Entrapment of Petroleum under Hydrodynamic Conditions." This diagram depicted the displacement of a slug of brine in a freshwater aquifer. Because of its greater density, the salt-water mass rested at the bottom of a syncline. Under static conditions, the interface between the salt water and the overlying freshwater would be horizontal; however, under hydrodynamic conditions within the freshwater aquifer, the static slug of denser brine would be shifted toward the direction of flow and the interface would be inclined and nonhorizontal. When Hubbert showed this illustration, someone in the group asked him:

> What would happen if the diagram were turned upside down, and the syncline became an anticline and the dense saltwater body were replaced by a body of oil or gas having a density less than that of fresh water? Under hydrodynamic conditions, would the low density hydrocarbon accumulation have a non-horizontal contact tilted in the direction of groundwater flow?

M. King Hubbert: The Theory of Groundwater Motion *(continued)*

As with Hubbert's ideas in structural geology, there was resistance to the adoption of hydrodynamic principles. When management ordered the implementation of hydrodynamic analyses, the new method became known by its reluctant practitioners as "hydrogoddamits." David Stearns (pers. comm., 2000) later ventured that the primary reason for the resistance to Hubbert's ideas was that it was easier to resist them than to learn the mathematics and physics necessary to apply them.

In collaboration with W. W. Rubey, in 1959 Hubbert published two papers in the *Geological Society of America Bulletin* that demonstrated that the state of stress in the Earth depended on the pore-fluid pressure. The immediate application suggested by Hubbert and Rubey was to provide an explanation for the long-standing problem of the apparent mechanical impossibility of large overthrusts ("Smoluchowski's Dilemma"). In a larger sense, earth scientists realized the application of the principle to all types of faulting within the Earth. Subsequent earthquakes near Denver and controlled experiments at Rangely, Colorado, demonstrated convincingly that movement on faults was dependent on the ambient fluid pressure (Bredehoeft, 1990).

Hubbert was evangelical in his primary cause—the application of physics and mathematics to geology. His contemporaries found this zeal irritating; but without it, the geological sciences would not have advanced. Simultaneously sensitive and abrasive, Hubbert inspired in his disciples the dual emotions of awe and sorrow. Sorrow, because Hubbert was egotistical, had a short temper, and was totally incapable of accepting criticism of any kind. Awe, because despite these personality flaws, Hubbert was a true genius who made gargantuan contributions.

As a young, junior geologist, Fred Meissner worked with Hubbert at Shell Oil in the 1950s. He recalls:

> Hubbert was arrogant, egotistical, dogmatic and intolerant of work he perceived to be incorrect. But above all he was a great scientist who demanded sound thinking and excellence. When he was lecturing in a classroom-type setting, he would get a "student" up to the blackboard, ask him questions, have him write equations, draw diagrams, etc. until the "student" started to fail. He pushed people to the limit, and often berated them for failure. Hubbert was the greatest blackboard lecturer I have ever seen. He could talk for hours without referring to a single note and could draw complicated diagrams and write legible text statements, equations, etc. with ease. It was very easy to take notes from his lectures and talks.

Raleigh (1982) noted that Hubbert's influence was attributable to two factors. First, Hubbert chose to work on problems of fundamental importance. Second, his papers were models of clarity in their exposition and

ascend a small ($<$ 0.5-mm diameter) tube against the force of gravity. Some liquids are more wetting than others. In the oil-water-rock system, oil is the nonwetting phase relative to water when rock pores are already saturated with water (North, 1985, p. 242). It thus takes an external force for a nonwetting liquid such as oil to displace a wetting-liquid such as water from the pores of a rock. The magnitude of the force involved is inversely proportional to the radius of the pore throat (Downey, 1994, p. 160). Thus, unless a buoyancy or other external driving force large enough to exceed capillary forces is present, a water-saturated, fine-grained rock such as a shale may constitute a zero-permeability seal to oil migration.

10.5. Hydrodynamic Theory of Oil Migration.

The hydrodynamic theory of oil migration was first postulated by M. K. Hubbert in 1953 (Hubbert, 1953). The theory essentially states that oil is affected by two forces: a hydrodynamic force related to head gradients in surrounding groundwater are accessible to anyone with a basic undergraduate education in physics and mathematics. Succinctly, Raleigh (1982) described Hubbert's contributions as "elegant essays in the clear and logical application of fundamental physical principles to a problem. A careful and thorough understanding of one of his papers is an educational experience that leaves the student forever well equipped in the subject and a better scientist as well."

Over the course of his career, Hubbert was the recipient of numerous honors, the complete listing of which would fill most of a page. The highest honor Hubbert received was the Vetlesen Prize (Raleigh, 1982), which is awarded by Columbia University for fundamental contributions to the understanding of the Earth. Doan (1994) described the Vetlesen Prize as the premier award in the earth sciences, comparable to a Nobel Prize, but awarded less often. In his response to the Vetlesen award, Hubbert warned against the "trap of specialization" and noted that his productive investigations had shared two characteristics: (1) a desire to understand the mechanism of some puzzling phenomenon, and (2) an almost total initial ignorance of the problem to be investigated.

FOR ADDITIONAL READING

Bredehoeft, J. D. 1990. M. King Hubbert: Contributions to the Study of Fluids in the Crust. *EOS, Transactions, American Geophysical Union* 71 (43): 1325.

Doan, D. B. 1994. Memorial to M. King Hubbert. *Geological Society of America Memorials* 24: 39–46.

Hubbert, M. K. 1940. The Theory of Ground-Water Motion. *Journal of Geology* 48: 785–944.

Hubbert, M. K. 1950. Review and Discussion, Physical Principles of Oil Production by Morris Muskat. *Journal of Geology* 58: 655–60.

Hubbert, M. K. 1963. Are We Retrogressing in Science? *Geological Society of America Bulletin* 74: 365–78.

Hubbert, M. K. 1982a. Avoid the Trap of Specialization. *AAPG Explorer* (9): 32–33.

Hubbert, M. K. 1982b. Response of M. King Hubbert. *EOS, Transactions, American Geophysical Union* 63 (17): 249–50.

Hubbert, M. K., and Rubey, W. W. 1959. Role of fluid pressure in mechanics of overthrust faulting. I. Mechanics of fluid-filled porous solids and its application to overthrust faulting. *Geological Society of America Bulletin* 70: 115–66.

Raleigh, C. B. 1982. The Vetlesen Prize to M. King Hubbert *EOS, Transactions, American Geophysical Union* 63 (17): 249–50.

and a buoyancy force due to the density contrast between oil or gas and the aqueous brines commonly found in the subsurface. The key thesis of the hydrodynamic theory is that oil and/or gas accumulate in areas that represent minima in potential energy. These minima are not necessarily found on structural highs, as in the classic anticlinal oil trap. In areas of active groundwater flow, the hydrodynamic theory of oil migration predicts the existence of hydrodynamic oil traps. These traps can be found only through the knowledge and application of Hubbert's theory.

In the following section, we first define general terms for the potential energy per unit mass (Φ, J-kg^{-1}) and forces per unit mass (E, N-kg^{-1} or m-s^{-2}) that act upon a fluid. Equations for both oil (Φ_0, E_0) and water (Φ_w, E_w) are developed. Substitution of terms then allows us to express the vectors that describe the direction in which oil migrates in terms of the regional head gradients and the density contrast between oil and/or gas and groundwater.

Following Hubbert (1953), we define the potential energy per unit mass of fluid (Φ) to be (equation 2.40)

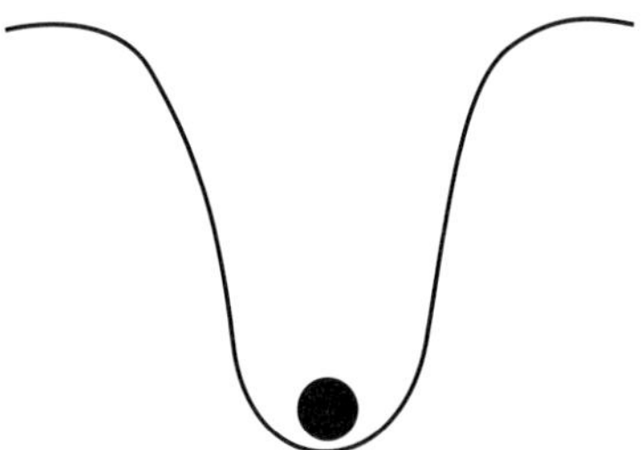

Figure 10.4 Ball trapped in a well exists at a state of minimum potential energy.

$$\Phi = gz + \frac{P - P_0}{\rho} \quad \textbf{(10.3)}$$

where g (m-s^{-2}) is the acceleration due to gravity, z (m) is elevation, P (kg-m^{-1}-s^{-2}) is fluid pressure, P_0 (kg-m^{-1}-s^{-2}) is a reference fluid pressure, and ρ (kg-m^{-3}) is fluid density. Let $P_0 = 0$, such that P is gauge pressure. Then

$$\Phi = gz + \frac{P}{\rho} \quad \textbf{(10.4)}$$

or

$$\frac{\Phi}{g} = z + \frac{P}{\rho g} = h \quad \textbf{(10.5)}$$

where h (m) is head. According to the hydrodynamic theory of oil migration as developed by Hubbert (1953), oil becomes trapped in a region of low Φ because fluids move from regions of high to low Φ. The situation is analogous to a ball trapped in a well (Figure 10.4). The ball rolls down the side of the well because it is moving from a region of high potential energy to one of low potential energy. Once the ball arrives at the bottom of the well, it is trapped because any movement up the sides would require an increase in potential energy. Until an external energy source becomes available, the ball must remain at the bottom of the well.

An isopotential is an imaginary line in space along which the potential energy of a fluid (or anything else) is constant. Consider two isopotential lines separated by a distance Δs (m), where s (m) is the direction perpendicular to the isopotentials (Figure 10.5). Let $\vec{E}$ (N-kg^{-1} or m-s^{-2}) be a vector representing force per unit mass that points in the positive s-direction. Then

$$\Delta\Phi = -\vec{E} \cdot \vec{\Delta s} \quad \textbf{(10.6)}$$

where $\vec{\Delta s}$ (m) is a vector in the s-direction of length Δs. Equation 10.6 follows from a consideration that energy = force $\times$ distance. If a force (E) is applied over a distance (Δs), the result is a change in potential energy ($\Delta\Phi$). A simple analogy would be picking up a book off the floor and placing it on a shelf. Energy is applied over a distance to place the book in a region where its potential energy is higher. The higher the shelf, the greater the distance (Δs) and the greater the increase in potential energy. Note that we must be careful to use vectors to specify the direction of motion as well as the total distance involved. For example, if we moved our book laterally on the floor, there would be no change in potential energy no matter how great the distance involved. The negative sign in equation 10.6 indicates that it takes negative energy to move "downhill" from a region of higher potential energy ($\Phi + \Delta\Phi$) to a region of lower potential energy (Φ) as shown in Figure 10.5. Equivalently, we can write

$$\vec{E} = \frac{-\Delta\Phi}{\Delta s}\hat{E} = -\vec{\nabla}\Phi \quad \textbf{(10.7)}$$

where $\hat{E}$ is a unit vector in the $\vec{E}$ direction, $\vec{\nabla}$ is the gradient operator (m^{-1}) and $\vec{\nabla}\Phi$ is the gradient of the fluid potential energy per unit mass. Recall that the potential energy per unit mass of a fluid (Φ) is related to its elevation (z) and pressure (P) by the following equation

$$\Phi = gz + \frac{P}{\rho} \quad \textbf{(10.8)}$$

where ρ is fluid density. Substituting equation 10.8 into equation 10.7,

$$\vec{E} = -\vec{\nabla}\Phi = \vec{g} - \vec{\nabla}\frac{P}{\rho} \quad \textbf{(10.9)}$$

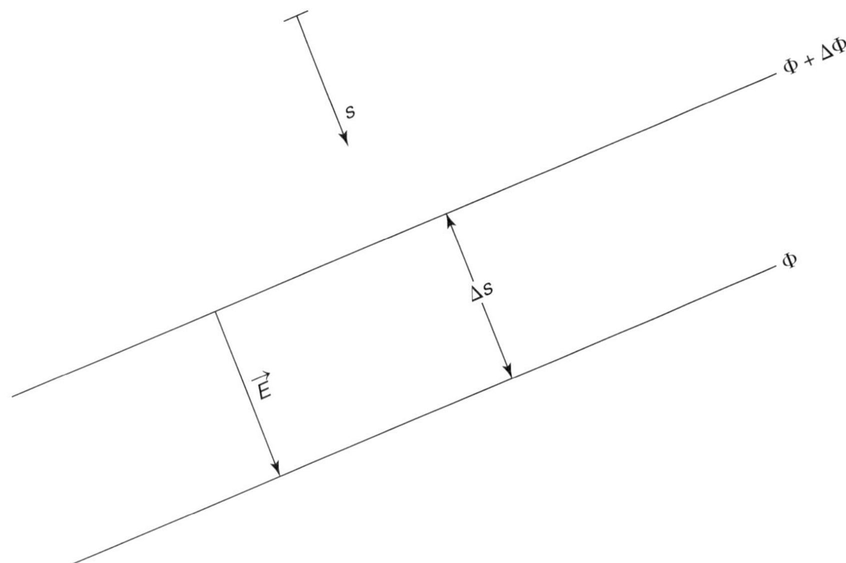

Figure 10.5 Two isopotential lines separated by a distance Δs.

The gravity vector $\vec{g}$ in equation 10.9 is positive, because it points in the "$-z$" direction. The gradient operator ($\vec{\nabla}$) acts upon a scalar to produce a vector. For example,

$$\vec{\nabla} u = \frac{du}{dx}\hat{x} + \frac{du}{dy}\hat{y} + \frac{du}{dz}\hat{z} \qquad \textbf{(10.10)}$$

where u is any scalar field, and $\hat{x}$ is a unit vector in the x-direction, etc. If the x- and y-directions are mutually perpendicular to elevation (z)

$$\frac{d(gz)}{dx}\hat{x} = \frac{d(gz)}{dy}\hat{y} = 0 \qquad \textbf{(10.11)}$$

and

$$\frac{d(gz)}{dz}\hat{z} = g(-\hat{z}) = -\vec{g} \qquad \textbf{(10.12)}$$

Recall that we have defined the coordinate direction z to be *elevation*, not depth. The minus sign in equation 10.12 indicates the gravity vector $\vec{g}$ points in the minus *elevation* direction and cancels the minus sign in equation 10.7, thus resulting in a positive $\vec{g}$ in equation 10.9.

Equation 10.9 implies that at any point in space, a fluid will be acted upon by two forces: gravity and the negative gradient of the ambient fluid pressure divided by fluid density. In general, these forces are *not* in the same direction. Note that equation 10.9 implies that the direction in which a fluid moves is not controlled solely by the fluid-pressure gradient.

Now let us develop potential energy and force vector expressions for both water and oil. We will denote the water terms by the subscript "w" and the oil terms by the subscript "o". The potential energy per unit mass of oil (Φ_o) is not the same as for water, because the density of oil is not the same. For oil,

$$\Phi_o = gz + \frac{P}{\rho_o} \qquad \textbf{(10.13)}$$

where ρ_o (kg-m^{-3}) is the density of oil and P is the ambient fluid pressure (oil or water). The density of oil is usually in the range $820 \leq \rho_o \leq 930$ kg-m^{-3}. The density of gas is about 120 kg-m^{-3}.

By analogy to equations 10.7 and 10.9,

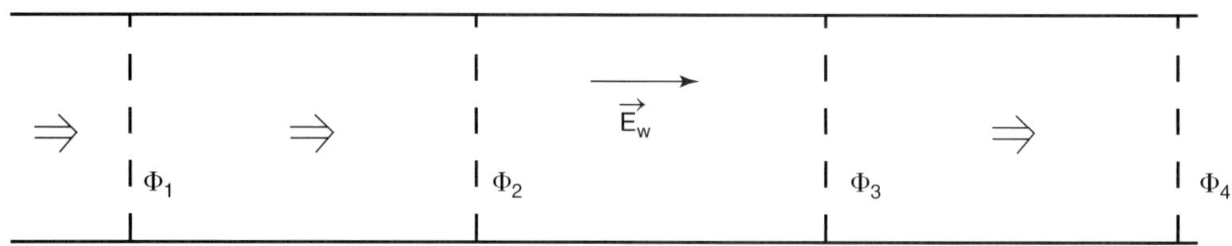

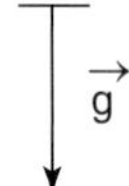

Figure 10.6 Isopotentials in a horizontal reservoir.

$$\vec{E}_o = -\vec{\nabla}\Phi_o \qquad \textbf{(10.14)}$$

and

$$\vec{E}_o = \vec{g} - \vec{\nabla}\frac{P}{\rho_o} \qquad \textbf{(10.15)}$$

From equation 10.8,

$$\Phi_w = gz + \frac{P}{\rho_w} \qquad \textbf{(10.16)}$$

and from equations 10.7 and 10.9,

$$\vec{E}_w = -\vec{\nabla}\Phi_w \qquad \textbf{(10.17)}$$

and

$$\vec{E}_w = \vec{g} - \vec{\nabla}\frac{P}{\rho_w} \qquad \textbf{(10.18)}$$

From equation 10.16,

$$P = (\Phi_w - gz)\rho_w \qquad \textbf{(10.19)}$$

Substituting equation 10.19 into equation 10.13,

$$\Phi_o = gz + \frac{(\Phi_w - gz)\rho_w}{\rho_o} \qquad \textbf{(10.20)}$$

$$\Phi_o = gz + \frac{\rho_w}{\rho_o}\Phi_w - \frac{\rho_w}{\rho_o}gz \qquad \textbf{(10.21)}$$

$$\Phi_o = \frac{\rho_w}{\rho_o}\Phi_w + gz\left[1 - \frac{\rho_w}{\rho_o}\right] \qquad \textbf{(10.22)}$$

$$\Phi_o = \frac{\rho_w}{\rho_o}\Phi_w + gz\left[\frac{\rho_o - \rho_w}{\rho_o}\right] \qquad \textbf{(10.23)}$$

$$\Phi_o = \frac{\rho_w}{\rho_o}\Phi_w - gz\left[\frac{\rho_w - \rho_o}{\rho_o}\right] \qquad \textbf{(10.24)}$$

Equation 10.24 relates the isopotentials for oil (Φ_o) to the isopotentials for water (Φ_w). Re-arranging equation 10.18,

$$-\vec{\nabla}P = \rho_w(\vec{E}_w - \vec{g}) \qquad \textbf{(10.25)}$$

Substituting equation 10.25 into equation 10.15,

$$\vec{E}_o = \vec{g} + \frac{\rho_w(\vec{E}_w - \vec{g})}{\rho_o} \qquad \textbf{(10.26)}$$

Equation 10.26 gives the impelling force per unit mass for oil migration ($\vec{E}_o$) in terms of a hydrodynamic force ($\vec{E}_w$) related to head gradients and a buoyancy force related to the fluid-density contrast between water and oil. Note that if $\rho_o = \rho_w$, $\vec{E}_w = \vec{E}_o$.

Now, suppose we consider an aquifer or petroleum reservoir through which secondary migration may occur (Figure 10.6). Let $\vec{E}_w$ be in the $+x$-direction as shown in Figure 10.7. Because the gravity vector points "downward," $-\vec{g}$ is in the "upward" direction. By adding the vectors in equation 10.26 head-to-tail, we find that $\vec{E}_o$, the impelling force per unit mass for oil migration, is in the same direction as $\vec{E}_w$ but is tilted upward by a buoyancy force as $\rho_o < \rho_w$.

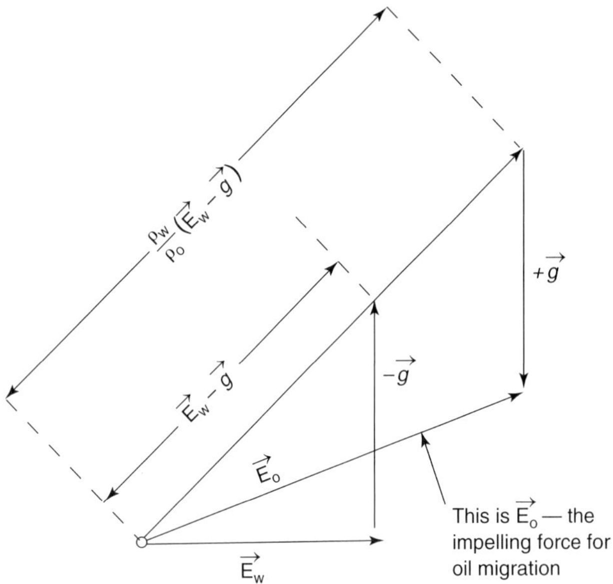

Figure 10.7 The impelling force vector for oil migration is found by head-to-tail addition of vectors representing a hydrodynamic force and a buoyancy force.

A typical hydrodynamic trap occurs when oil is trapped in a reservoir at some point where the dip suddenly increases (Figure 10.8). Note that if the hydrodynamic force were not present, the oil would simply migrate updip in the direction dictated by its buoyancy. Instead, the petroleum fluid moves downdip. However, if the dip of the formation suddenly increases, the hydrodynamic force that was sufficiently large enough to drive the petroleum fluid down the shallow dip of the reservoir is now insufficient to force the same low-density fluid down the steep wall of the reservoir. At the same time, the hydrodynamic force prevents the oil and/or gas from moving upstream. The petroleum fluid is trapped in a region of minimum potential energy. If we did not have knowledge of the hydrodynamic regime, we would not predict an oil trap in this situation.

It is important to note that the oil-water contact lies along a Φ_o isopotential. This must be true. If the oil-water contact were not a Φ_o isopotential, then there would exist a nonzero potential energy gradient in the oil, and the oil would migrate until the oil-water contact was parallel to the Φ_o isopotentials (Figure 10.9).

Under the situation portrayed in Figure 10.8, oil cannot escape through the overlying seal rock, because it is held in place by capillary forces. The preferential wetting of pore spaces in the shale by water excludes oil. The seal rock in this instance thus has a true zero permeability (to the oil) *unless* the capillary force is exceeded.

Note that if $\vec{E}_w = 0$, there is no impelling force for water flow through a reservoir, and

$$\vec{E}_o = \vec{g}\left[1 - \frac{\rho_w}{\rho_o}\right] \tag{10.27}$$

In this instance, $\vec{E}_o$ is in the $-\vec{g}$ direction (because $\rho_w > \rho_o$ in equation 10.27) and the oil-water

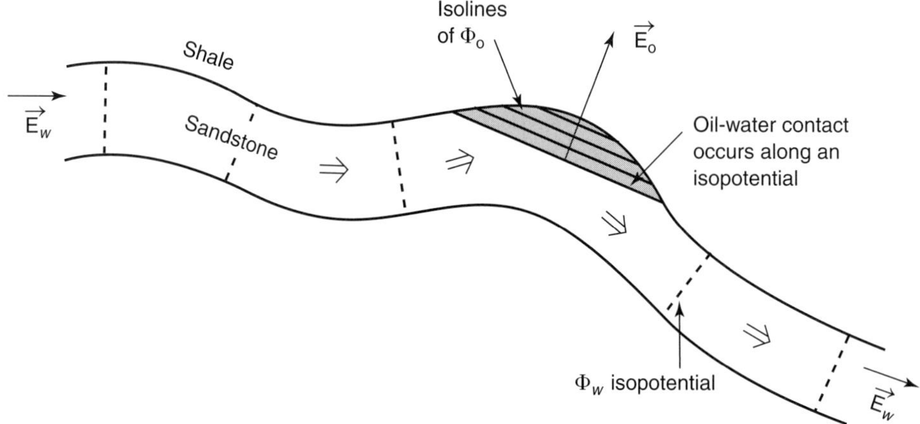

Figure 10.8 Oil trapped in a gently folded reservoir by a sudden increase in dip angle.

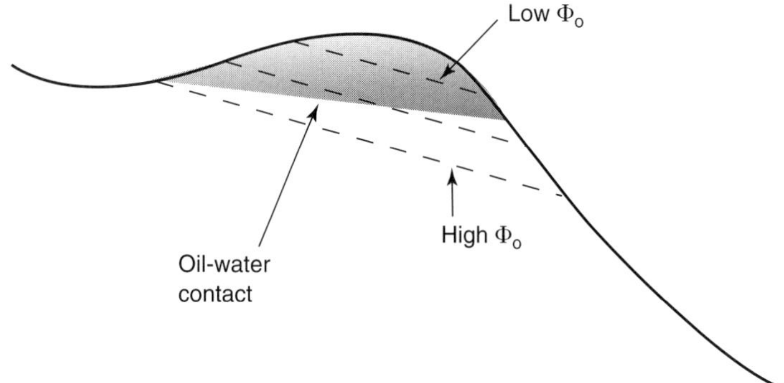

Figure 10.9 If the oil-water contact is not parallel to the oil isopotentials, oil will flow from regions of high potential to low potential until the contact is parallel.

contact is perpendicular to the gravity vector, i.e., it is *not* tilted. In other words, a tilted oil-water contact implies active groundwater flow.

Let us first consider the simple case of a horizontal reservoir. In this case, the tilt of the oil-water contact is controlled by the head gradient in the reservoir. From equations 10.17 and 2.44,

$$\vec{E}_{\mathrm{w}} = -\vec{\nabla}\Phi_{\mathrm{w}} = -\vec{\nabla}(gh) \qquad \textbf{(10.28)}$$

where h (m) is hydraulic head and g (m-s^{-2}) is a scalar equal to the absolute value of the vector $\vec{g}$. If flow in the reservoir is purely horizontal (x-direction), then $dh/dz = 0$ and

$$\vec{E}_{\mathrm{w}} = -g\frac{dh}{dx}\hat{x} \qquad \textbf{(10.29)}$$

where $\hat{x}$ is a unit vector in the x or horizontal direction. Substituting equation 10.29 into equation 10.26,

$$\vec{E}_{\mathrm{o}} = \vec{g} + \frac{\rho_{\mathrm{w}}}{\rho_{\mathrm{o}}}\left[-g\frac{dh}{dx}\hat{x} - \vec{g}\right] \qquad \textbf{(10.30)}$$

Breaking the components of equation 10.30 into two mutually orthogonal vector components (vertical and horizontal),

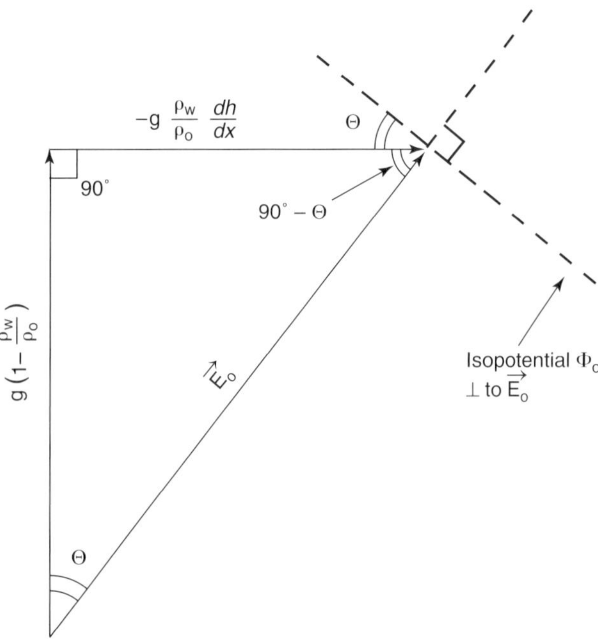

Figure 10.10 The dip of the oil-water contact (Θ) can be calculated as the angle whose tangent is described by the ratio of the horizontal to vertical components of the force vector ($\vec{E_o}$), which points in the direction of oil migration. Θ is also the angle by which the force vector for oil migration ($\vec{E_o}$) is deflected from the vertical.

$$\vec{E_o} = \vec{g}\left[1 - \frac{\rho_w}{\rho_o}\right] - g\,\frac{\rho_w}{\rho_o}\frac{dh}{dx}\,\hat{x} \qquad \textbf{(10.31)}$$

Note that $\vec{g}$ in equation 10.31 is a vector that points "downward," while g is a scalar. The first term on the right side of equation 10.31 is negative, because $\rho_w > \rho_o$. As a consequence, the vector that represents the first term on the right side of equation 10.31 must point "upward." If flow is in the $+x$-direction, then (dh/dx) in equation 10.31 must be negative. Thus, the vector that represents the second term on the right side of equation 10.31 must point in the $+x$-direction (Figure 10.10).

Drawing the two vector components of $\vec{E_o}$ defines an angle Θ. The angle Θ is the angle that the oil-water contact makes with the horizontal, as well as the angle by which the vector $\vec{E_o}$ is deflected from the vertical (Figure 10.10).

$$\text{Tan}\,\Theta = \frac{-g\,\frac{\rho_w}{\rho_o}\frac{dh}{dx}}{g\left[1 - \frac{\rho_w}{\rho_o}\right]} \qquad \textbf{(10.32)}$$

$$\text{Tan}\,\Theta = \frac{\rho_w}{\rho_w - \rho_o}\frac{dh}{dx} \qquad \textbf{(10.33)}$$

$$\Theta = \text{Tan}^{-1}\left\{\frac{\rho_w}{\rho_w - \rho_o}\frac{dh}{dx}\right\} \qquad \textbf{(10.34)}$$

Consider a typical case where $\rho_w = 1000$ kg-m^{-3}, and $\rho_o = 900$ kg-m^{-3}. Then

$$\Theta = \text{Tan}^{-1}\left\{10\,\frac{dh}{dx}\right\} \qquad \textbf{(10.35)}$$

What is a typical value for dh/dx? In the case of regional topographically driven flow, we might find a head drop of 1 km over 100 km with $dh/dx = 0.01$.

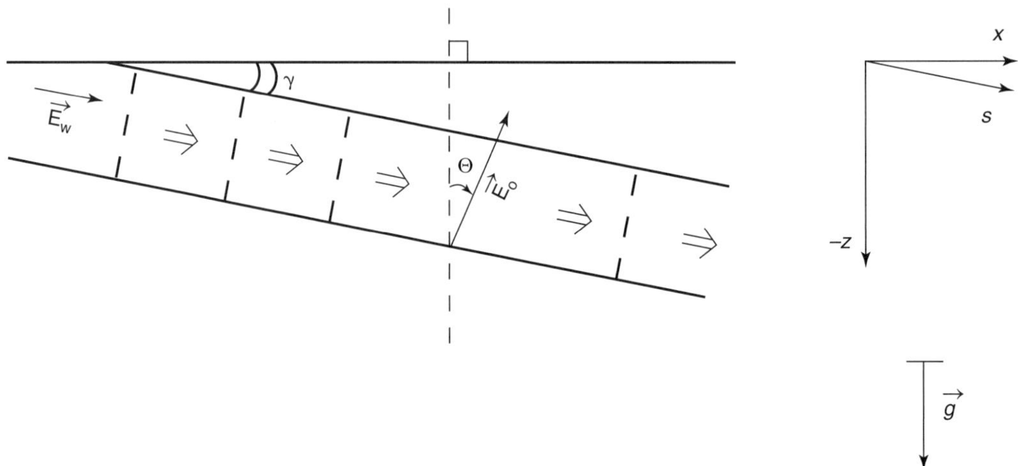

Figure 10.11 Calculation of the dip of the oil-water contact (Θ) in a reservoir with a dip angle γ.

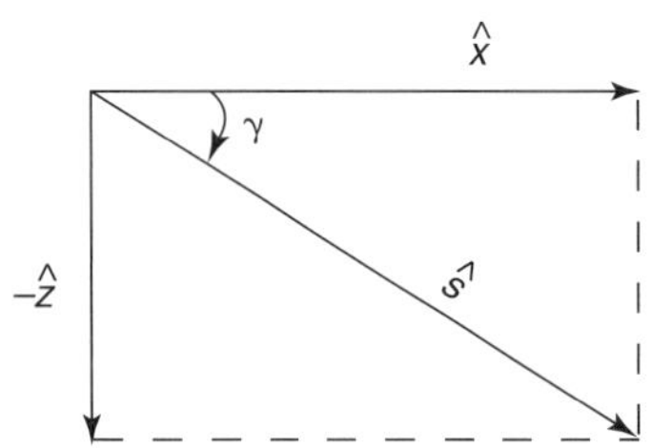

Figure 10.12 The unit vector s in the direction of the dipping reservoir (s-direction) can be resolved into horizontal ($\hat{x}$) and vertical ($\hat{z}$) components.

For $dh/dx = 0.01$, $\Theta = 5.7°$; for $dh/dx = 0.1$, $\Theta = 45°$; and for $dh/dx = 0.001$, $\Theta = 0.57°$.

Consider now the more general problem of finding Θ for a reservoir that dips at an angle γ from the horizontal (Figure 10.11). Let the direction parallel to the reservoir be s, and start with equation 10.26,

$$\vec{E}_o = \vec{g} + \frac{\rho_w}{\rho_o}(\vec{E}_w - \vec{g}) \qquad \textbf{(10.36)}$$

$$\vec{E}_o = -g\hat{z} + \frac{\rho_w}{\rho_o}(\vec{E}_w + g\hat{z}) \qquad \textbf{(10.37)}$$

where z is elevation and $\hat{z}$ is a unit vector in the $-\vec{g}$ direction (up).

If flow in the reservoir is parallel to the dip of the bed, then from equation 10.28,

$$\vec{E}_w = -g\vec{\nabla}h = -g\frac{dh}{dx}\hat{s} \qquad \textbf{(10.38)}$$

where $\hat{s}$ is a unit vector in the s-direction, parallel to reservoir bedding. Any vector in the s-direction can be resolved into its x- and z-components (Figure 10.12), thus

$$\vec{E}_w = -g\left[\frac{dh}{ds}\cos\gamma\,\hat{x} - \frac{dh}{ds}\sin\gamma\,\hat{z}\right] \qquad \textbf{(10.39)}$$

Substituting equation 10.39 into equation 10.37,

$$\vec{E}_o = -g\hat{z} + \frac{\rho_w}{\rho_o}\left[-g\frac{dh}{ds}\cos\gamma\hat{x} + g\frac{dh}{ds}\sin\gamma\hat{z} + g\hat{z}\right] \qquad \textbf{(10.40)}$$

Note that g in equation 10.40 is a scalar. Thus,

$$\frac{\vec{E}_o}{g} = -\hat{z}\left\{1 - \frac{\rho_w}{\rho_o}\left[\frac{dh}{ds}\sin\gamma + 1\right]\right\} + \hat{x}\left[\frac{\rho_w}{\rho_o}\frac{-dh}{ds}\cos\gamma\right] \qquad \textbf{(10.41)}$$

Note that for flow "downdip," dh/ds is negative (Figure 10.11). Having resolved $\vec{E}_o$ into its $\hat{x}$ and $\hat{z}$ components, we can now solve for the angle Θ

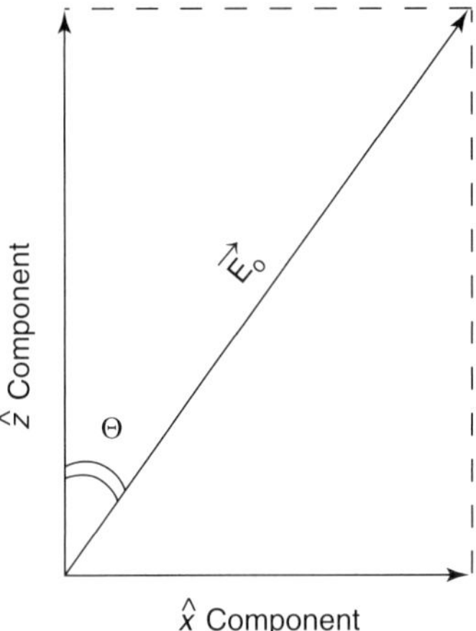

Figure 10.13 The angle by which the vector ($\vec{E}_o$) is deflected from the horizontal can be calculated as the arctangent of the ratio of its horizontal component to its vertical component.

that defines the orientation of the $\vec{E}_o$ vector and the oil-water contact (Figure 10.13).

$$\Theta = \mathrm{Tan}^{-1}\left\{\frac{\hat{x}\text{ component}}{\hat{z}\text{ component}}\right\} \quad \textbf{(10.42)}$$

$$\Theta = \mathrm{Tan}^{-1}\left\{\frac{\frac{\rho_w}{\rho_o}\left[\frac{-dh}{dx}\cos\gamma\right]}{\frac{\rho_w}{\rho_o}\left[\frac{dh}{ds}\sin\gamma + 1\right] - 1}\right\} \quad \textbf{(10.43)}$$

$$\Theta = \mathrm{Tan}^{-1}\left\{\frac{\frac{-dh}{ds}\cos\gamma}{\left[\frac{dh}{ds}\sin\gamma + 1\right] - \frac{\rho_o}{\rho_w}}\right\} \quad \textbf{(10.44)}$$

If the angle Θ is greater than the dip angle γ, the hydrodynamic force is greater than the buoyancy force, and the oil (or gas) will migrate downdip. If the angle Θ is less than the dip angle γ, the hydrodynamic force is lower than the buoyancy force, and the oil (or gas) will migrate updip (see Figure 10.11). Oil can become trapped if the local dip angle exceeds Θ, while the "upstream" dip angle is less than Θ.

Problem: A reservoir has a dip angle γ of 1°. The head gradient in the reservoir is −0.01. The reservoir contains oil with density 900 kg-m^{-3}. In what direction is the impelling force for oil migration directed? Will the oil migrate updip or downdip? What if the petroleum fluid were gas with a density of 120 kg-m^{-3}? Will the gas migrate updip or downdip?

Assume a water density of 1000 kg-m^{-3}. Substituting into equation 10.44,

$$\Theta = \mathrm{Tan}^{-1}\left\{\frac{(0.01)\cos(1°)}{[(0.01)\sin(1°) + 1] - (900/1000)}\right\} \quad \textbf{(10.45)}$$

$$\Theta = \mathrm{Tan}^{-1}\left\{\frac{(0.01)(0.99985)}{[(0.01)(0.017445) + 1] - 0.9}\right\} \quad \textbf{(10.46)}$$

$$\Theta = \mathrm{Tan}^{-1}\left\{\frac{0.01}{1.00017 - 0.9}\right\} \quad \textbf{(10.47)}$$

$$\Theta = \mathrm{Tan}^{-1}\{0.09981\} \quad \textbf{(10.48)}$$

$$\Theta = 5.7° \quad \textbf{(10.49)}$$

As Θ = 5.7° > γ = 1°, the oil will migrate downdip. Note that if $dh/ds = 0$, we would find $\Theta = \mathrm{Tan}^{-1}(0°) = 0°$, and the impelling force for oil migration would be directed upward as appropriate for a buoyancy force only.

If we had gas (density 120 kg-m^{-3}) instead of oil (density 900 kg-m^{-3}), the angle would be

$$\Theta = \mathrm{Tan}^{-1}\left\{\frac{0.01}{1.00017 - 0.12}\right\} \quad \textbf{(10.50)}$$

$$\Theta = \mathrm{Tan}^{-1}\{0.01136\} \quad \textbf{(10.51)}$$

$$\Theta = 0.65° \quad \textbf{(10.52)}$$

A relatively greater buoyancy force for the lower-density gas results in less deflection of $\vec{E}_o$ from the vertical. In this case, Θ = 0.65° < γ = 1°, and the gas will migrate updip.

Experiences of a Petroleum Hydrogeologist

by

Fred F. Meissner

Fred F. Meissner

I first became aware of the significance of groundwater flow in affecting the migration and accumulation of petroleum in 1953, when I was a graduate student at the Colorado School of Mines. I was taking a seminar course in which each of the students was asked to orally review a paper of his choosing from a recent edition of the *American Association of Petroleum Geologists Bulletin*. I recalled having looked casually at a paper in the August 1953 issue by M. K. Hubbert entitled "*Accumulation of Petroleum under Hydrodynamic Conditions*." I thought it sounded appropriate for review, so I reread the article in more detail and discussed it before my professors and fellow students. The paper was full of physics and mathematics and was difficult for many petroleum geologists to understand. I happened to be an engineer, a background that was of benefit in deciphering Hubbert's paper.

After I had reviewed the paper, I promptly forgot about it for several years. In 1957, I was employed by Shell Oil Company as a junior geologist in its Exploration Department office in Roswell, New Mexico. I was assigned the task of studying and developing petroleum prospects in the Delaware Basin portion of the larger Permian Basin. A series of oil fields was present in the uppermost Delaware Sandstone, and new discoveries were being made. The Delaware Sandstone of Upper Permian Guadalupian age is a deep-water deposit with turbidite or density-current channels that contain the oil reservoirs. The channels may be traced and projected for many miles; they are generally linear but also contain meanders. The geometry of the meanders does not provide structural/stratigraphic "closure" against regional dip. The west side of the Permian Basin is bounded by mountainous uplifts associated with the Rio Grande Rift. However, the east side in west-central Texas has much lower elevations and is associated with large salt-water springs that discharge from rocks of Permian age. This situation indicates the presence of easterly to northeasterly flowing groundwater across the Delaware Basin. In studying the existing oil fields, I found that the accumulations contained tilted oil-water contacts and were not controlled solely by structure or stratigraphy. At this point, I remembered Hubbert's paper on hydrodynamically controlled accumulations. I read it once again, and applied Hubbert's concepts to the setting of oil fields in the

Delaware Basin. I worked largely alone and unsupervised for a period of time and finally showed the results of my work to the Division Exploration manager. He was a crusty, salty man of about 60 years with many years of experience in the oil and gas business. When I talked to him about hydrodynamics and tilted oil-water contacts, he made some comments to the effect that:

> Oil-water contacts are flat, sonny. The whole anticlinal theory of density segregation and oil exploration is based on that fact. If you bring me any more wild ideas like this, you had better look for another place of employment.

I went back to my office with my tail between my legs and proceeded to do other things. However, my manager retired a few months later and was replaced by a younger, more aggressive man. He came around to my office and asked me what I had been doing. I showed him my work on the Delaware Sand, and he liked it. Shell took leases on some of my prospects. Two of them were subsequently drilled through "farm-out" arrangements made to small, independent operators and resulted in the discovery of the Screwbean and Meridian oil fields.

After my initial work on the Delaware Sand, I was sent on a short assignment to the Shell Development Company laboratory in Houston, Texas, where I took an in-house training course from M. King Hubbert, who at that time was employed by Shell. This was a tremendous opportunity, and what I learned greatly influenced my career.

After being exposed to Hubbert's science, insight, and wisdom, I returned to Shell's Roswell office and was assigned the task of developing prospects in the San Andres Formation of Permian lower Guadalupian-upper Leonardian age. The principal oil-productive reservoir in the San Andres formation is a dolomite zone characterizing shelf margin and shallow marine to intertidal backshelf facies. The backshelf facies is overlain by anhydrites and grades northward into a series of east-west trending tidal-flat facies porosity pinchouts where dolomite is replaced by anhydrite. Regional eastward groundwater flow through the dolomite zone has created a series of hydrodynamically controlled oil fields along structural noses and monoclines, as well as along stratigraphic porosity/permeability pinchouts. Most of the oil fields are characterized by eastwardly tilted oil-water contacts caused by groundwater flowing from elevated outcrops on the west side of the Permian Basin in New Mexico toward distant low-elevation outcrops on the east side of the basin in Texas. A series of three tidal-flat porosity pinchouts combined with a monoclinal flexure and an easterly-tilted oil-water contact localize the giant Slaughter-Leveland oil field. I used subsurface well control to map porosity pinchouts and structural attitude along the "Slaughter-Leveland trend" for a distance of 100 miles westward to the San Andres outcrop. This mapping, when combined with potentiometric surface data, led to the delineation of several hydrodynamic prospects on which Shell acquired strong lease positions. Subsequent drilling by farmout promotion and nearby development resulted in the discovery of a number of economically significant oil fields, including the Cato and Chaveroo oil fields.

Ever since my experience of applying Hubbert's principles to exploration in the Permian Basin, I have been aware of the importance of hydrodynamics in controlling oil and gas accumulation in other areas—particularly in Rocky Mountain basins that are characterized by mountainous terrain and differential outcrop elevation. Hydrodynamic control may not be significant in many fields or prospects, but its effects should always be considered in areas where groundwater is in a substantial state of motion. The application of hydrodynamic principles is one of the key instruments in the explorationist's tool box.

CASE STUDY 10.6 The Bakken-Madison Petroleum System, Williston Basin

The Bakken-Madison Petroleum System is located in the Williston Basin, an intracratonic basin in the northcentral United States and southcentral Canada (Figure 10.14). The Bakken Formation is a Mississipian-age black shale believed to be the dominant source rock within the Williston Basin.

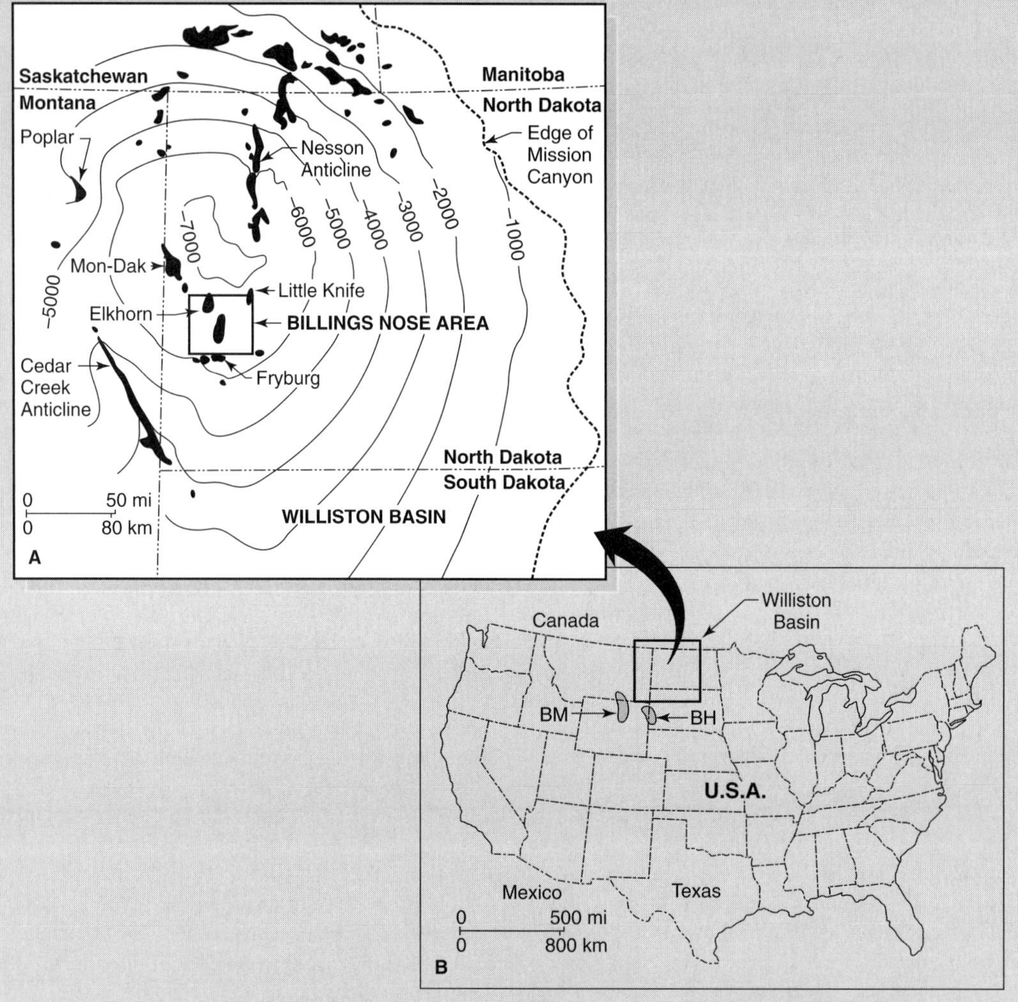

Figure 10.14 Location map for the Williston Basin and the Billings Nose area of the Williston Basin. Contour lines in (A) show elevation of Mission Canyon Formation in feet. "BM" and "BH" in (B) refer to the Bighorn Mountains and the Black Hills, respectively, as shown in Figure 10.15.
(From Berg et al., 1994, p. 502.)

Most petroleum reservoirs are found within the directly overlying carbonate rocks of the Madison Group, including the Lodge Pole, Mission Canyon, and Charles Formations.

The Bakken Formation is relatively thin, varying from zero thickness near the basin edges to a maximum thickness of about 43 meters at the center of the Williston Basin. The effective matrix porosity and permeability of the Bakken Formation was described by Meissner (1984b) as "low to non-existent." In the context of petroleum geology, these indications are made relative to what constitutes a good reservoir rock. Measurements on Bakken cores indicate porosities of 5 to 6% and permeabilities less than 0.1 milliDarcies (lower than $10^{-16}m^2$).

In the deep, central part of the Williston Basin, oil shows are universal, but economic production from the Bakken is uncommon. No water is ever recovered from the Bakken shale during drill-stem tests or initial well completions; petroleum appears to be the only mobile fluid found in the Bakken. Those few wells that produce petroleum directly from the Bakken do so at rates that could not be sustained by the low permeabilities measured on Bakken core samples. At these sites where oil production is possible, it appears that the Bakken contains a pervasive fracture system that provides permeability of reservoir quality.

Persuasive geochemical evidence that the Bakken is the source rock for oils found in overlying Madison Group reservoirs was first provided in studies reported by Dow (1974) and Williams (1974). Dow estimated that, with an average organic richness of about 4%, the Bakken had generated about 10 billion barrels of oil in the deeper parts of the Williston Basin.

In the sections of the Williston Basin where the Bakken shale is deeper than about 1900 to 2500 m, electrical logs show a dramatic decrease in electrical conductivity. More precisely, Meissner (1984b) pointed out that the change in electrical conductivity was more correlated with temperature than with depth. At temperatures near the point where kinetic models predict oil generation begins (around 71°C in this case), Bakken conductivity dramatically decreases. Meissner hypothesized that the decrease in electrical conductivity was due to relatively conductive aqueous brines being driven out by nonconductive thermally generated oils. Meissner also estimated fluid pressures in the Bakken from sonic logs and found that anomalously high fluid pressures correlated with the presence of oil as indicated by electrical logs. The simultaneous occurrence of abnormally high fluid pressures with hydrocarbon generation suggested a cause-and-effect relationship between the two, with oil generation creating overpressuring. The generation of lower-density liquid oil from higher-density solid kerogen leads to overpressuring as the more voluminous oil tries to occupy the smaller space formerly occupied by its precursor.

The picture that emerges is a mechanism that simultaneously provides for both the generation of petroleum from the Bakken formation and its primary migration. In areas where the Bakken is thermally mature, the rock appears to be completely oil-saturated and overpressured. The process of oil generation thus appears to create the high fluid pressures, which drive the oil out of the source rock. An unanswered question is precisely how Bakken oils migrate vertically to reservoirs in the Mission Canyon formation. The bottom formation in the Madison Group, the Lodgepole Formation, is of low permeability. In petroleum geology parlance, the word that is used is *tight.* Dow (1974) speculated that the movement takes place through vertical fractures in the Lodgepole Formation. Meissner (1984b) further postulated that the vertical fracturing was caused by the same generative processes that had fractured the Bakken.

Most of the oil produced in the Williston Basin comes from reservoirs in limestones and dolomites of the Mission Canyon Formation. The location and distribution of oil within the Mission Canyon Formation appears to be affected by active groundwater flow, precisely as predicted by Hubbert's (1953) hydrodynamic theory of oil migration.

Although Berg et al. (1994) list at least four periods of geologic time during which hydrodynamic flow could have affected oil migration and accumulation in the Williston Basin, the present hydrodynamic regime is believed to have begun with

CASE STUDY 10.6 *(continued)*

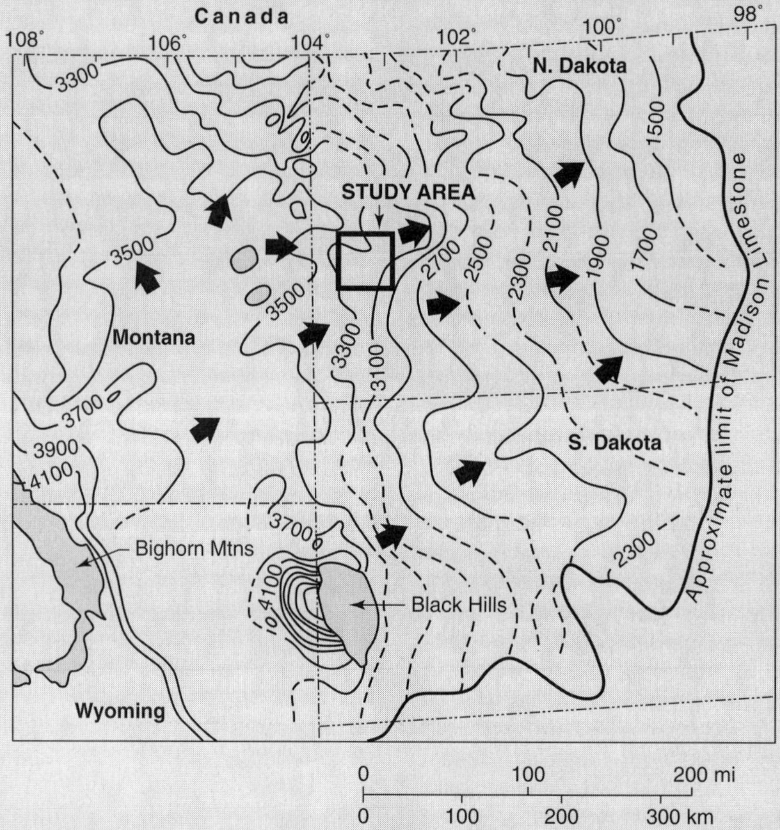

Figure 10.15 Potentiometric surface (feet) in the Mississippian Madison Group, Williston Basin. Arrows show inferred directions of topographically-driven groundwater flow. "Study Area" refers to Billings Nose area. (From Berg et al., 1994, p. 504.)

regional uplift of the Rocky Mountains and Great Plains during the late Pliocene (about 2 million years ago). At this time, the entire stratigraphic section of the Williston Basin was exposed around the margins of the Black Hills and the mountain ranges to the west and southwest. The Mission Canyon and other stratigraphic units became open to infiltration and recharge, and topographically driven flow downdip in a northeast direction began (Figure 10.15).

In general, three types of oil traps exist in the Billings Nose area of the Williston Basin: stratigraphic, stratigraphic-hydrodynamic, and pure hydrodynamic (Figure 10.16). The pure stratigraphic trap occurs when porosity abruptly decreases, resulting in a capillary-force barrier that prevents oil from being flushed downdip by the regional groundwater flow. In the stratigraphic-hydrodynamic trap, oil is held in place by a combination of hydrodynamic forces and capillary barriers. In a pure hydrodynamic trap, oil is trapped by a slight increase in structural dip in the direction of flow. The increase in dip is sufficiently large enough so that the oil's buoyancy

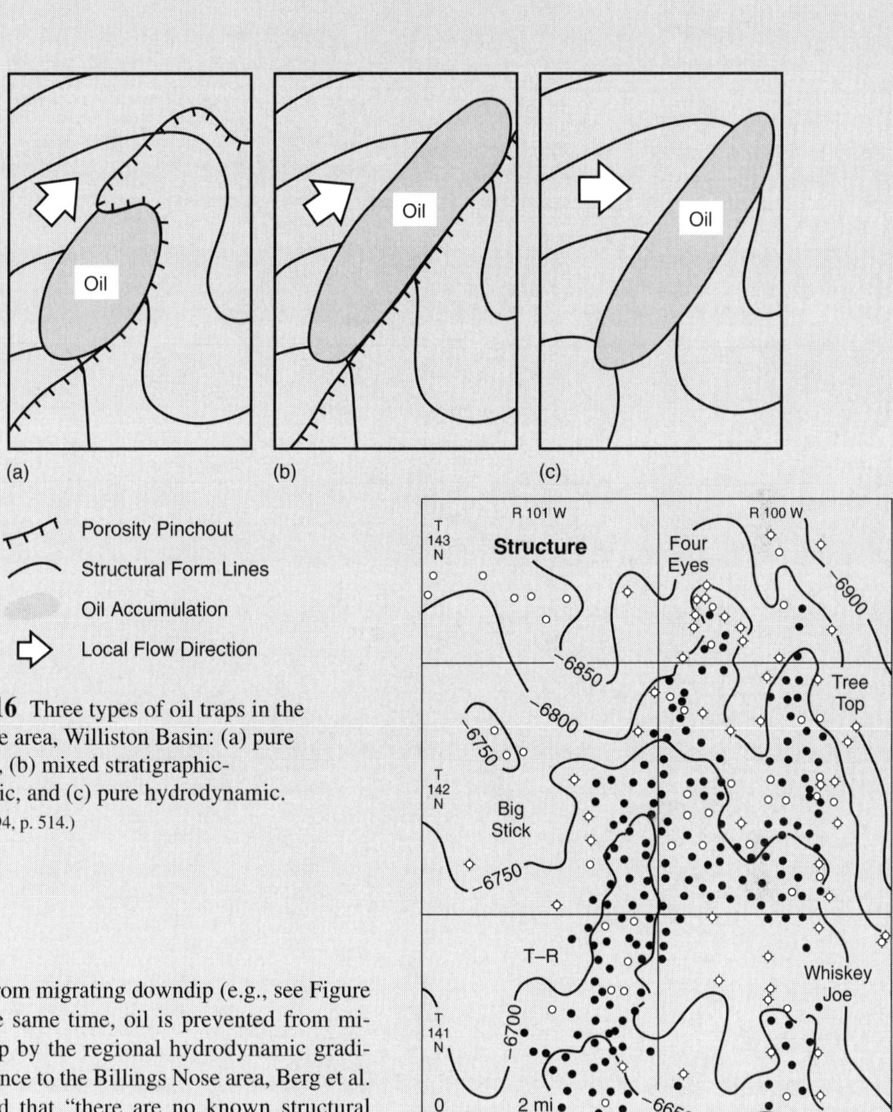

Figure 10.16 Three types of oil traps in the Billings Nose area, Williston Basin: (a) pure stratigraphic, (b) mixed stratigraphic-hydrodynamic, and (c) pure hydrodynamic.
(From Berg, 1994, p. 514.)

Figure 10.17 Elevation (feet) of the top of the Mission Canyon Formation, Billings Nose area, Williston Basin. Plot of producing oil wells (filled circles) versus non-producing wells (hollow circles) indicates oil accumulation has been displaced from the structural high point to the northeast.
(From Berg et al., 1994, p. 508.)

prevents it from migrating downdip (e.g., see Figure 10.8). At the same time, oil is prevented from migrating updip by the regional hydrodynamic gradient. In reference to the Billings Nose area, Berg et al. (1994) stated that "there are no known structural traps in which the present accumulation is determined solely by structure." In this setting, a theoretical knowledge of hydrodynamics is completely necessary if any endeavor undertaken to find oil is to be successful.

A structure map of the Mission Canyon Formation in the Billings Nose area (Figure 10. 17) shows that oil production (and accumulation) tends to be displaced from the structural high to the northeast—in the direction of regional groundwater flow. Head

Case Study 10.6 *(continued)*

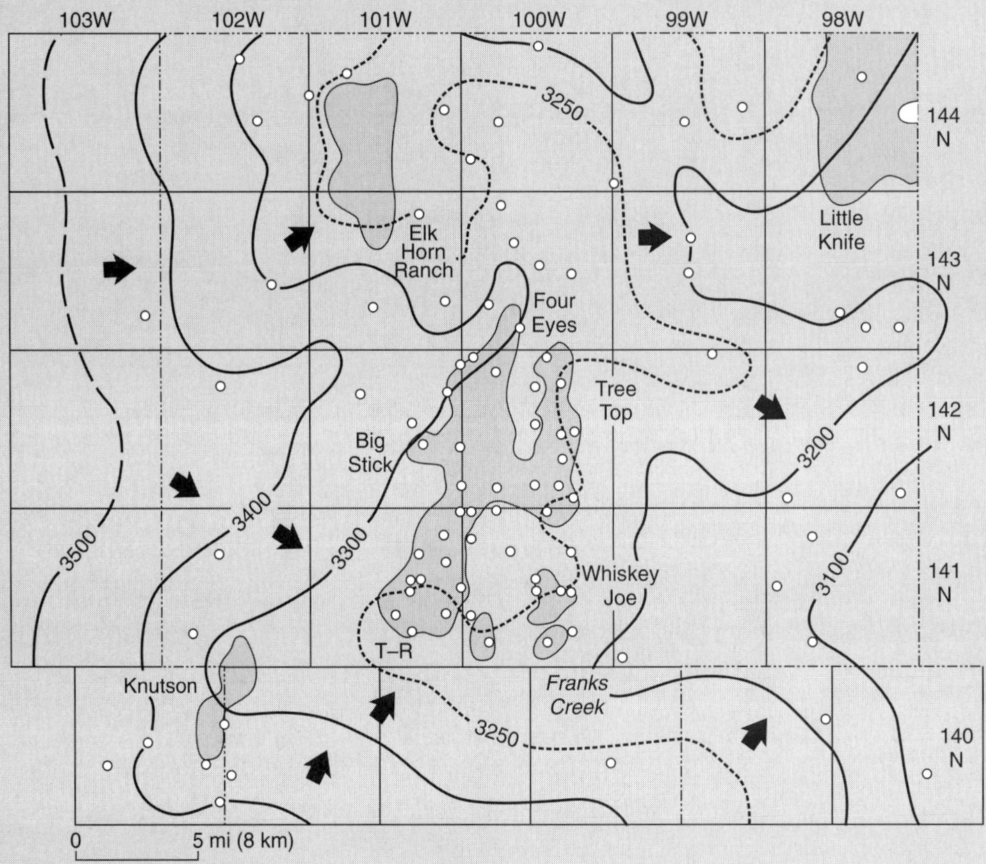

Figure 10.18 Freshwater potentiometric surface (feet) in the Mission Canyon Formation, Billings Nose area.
(From Berg et al., 1994, p. 507.)

or potentiometric gradients in the Mission Canyon Formation in the Billings Nose area range from 0.002 to 0.006 (Figure 10.18). The dip of the oil-water contact is about 25 ft-mile^{-1} (4.73 m-km^{-1}) (Figure 10.19), implying the oil-water contact dips northeastward about 0.27° ($\Theta = \text{Tan}^{-1}$ [25/5280] = 0.27°).

Because the Mission Canyon Formation is nearly horizontal, the angle of the oil-water contact can be estimated using equation 10.33,

$$\text{Tan } \Theta = \frac{\rho_w}{\rho_w - \rho_o} \frac{dh}{dx} \quad \textbf{(10.53)}$$

The density of Mission Canyon oils is 625 kg-m^{-3}; the density of the saline groundwater is 1030

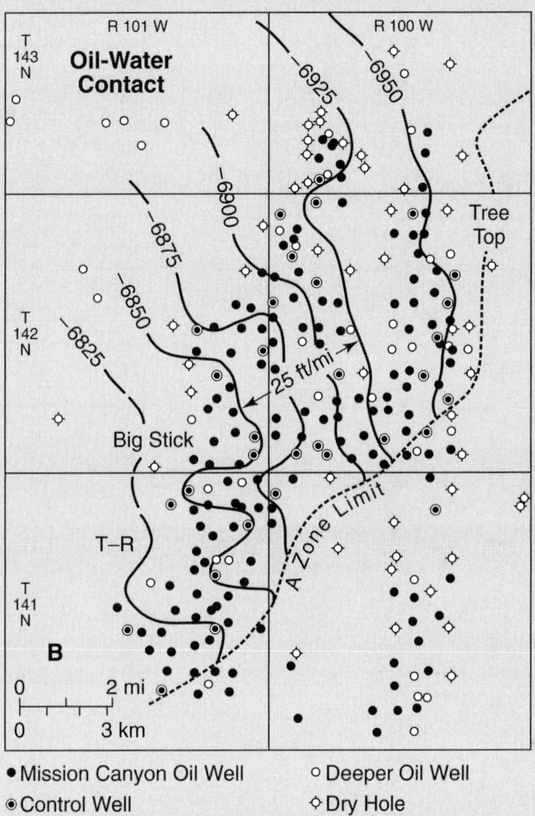

Figure 10.19 Elevation (feet) of the oil-water contact in the Billings Nose area. Average dip is about 25 ft/mile (~0.56°).
(From Berg et al., 1994, p. 509.)

kg-m^{-3}. For these densities and a regional hydrodynamic gradient (dh/dx) of 0.002, the predicted angle Θ is 0.29°, in good agreement with the observed dip angle of 0.27°.

The regional structural dip (Figure 10.18) in the Billings Nose area is about 0.1° to 0.7°, averaging around 0.2°. If the regional average dip is 0.2° while the observed dip angle Θ of the oil-water contact is 0.27°, then oil will be forced downstream by the regional hydrodynamic gradient. However, in those sections of the Mission Canyon Formation where the dip locally exceeds 0.27°, the natural buoyancy of the oil will prevent it from migrating downstream, and it will be retained in a purely hydrodynamic trap.

Exploration strategies in the Billings Nose area that rely solely on structural closure are flawed. As the above discussion indicates, it is possible to hydrodynamically trap petroleum in the absence of structural closure.

Review Questions

1. Define the following terms in the context of hydrogeology:
    a. petroleum
    b. reservoir
    c. oil
    d. abiogenic theory
    e. biogenic theory
    f. source rock
    g. diagenesis
    h. kerogen
    i. labile
    j. refractory
    k. catagenesis
    l. cracking
    m. kinetic theory
    n. metagenesis
    o. petroleum system
    p. hydrocarbon machine
    q. migration
    r. primary migration
    s. secondary migration
2. In Figure 10.3b, what is the dip of the oil-water contact, Θ? What is the dip of the reservoir section (Figure 10.3b) in which the oil is trapped? If the density of the oil is 900 kg-m^3, what is the head gradient? Now refer to Figure 10.3c. If the reservoir permeability is 10^{-13} m^2, what groundwater flow velocity would result in the oil in Figure 10.3c being displaced downstream, out of the figure?

Suggested Reading

Berg, R. R., DeMis, W. D., and Mitsdarffer, A. R. 1994. Hydrodynamic effects on Mission Canyon (Mississipian) oil accumulations, Billings Nose Area, North Dakota. *American Association of Petroleum Geologists Bulletin*. 78: 501–18.

Hubbert, M. K. 1953. Entrapment of petroleum under hydrodynamic conditions. *American Association of Petroleum Geologists Bulletin* 37: 1954–2026.

Meissner, F. F. 1991. Origin and migration of oil and gas. In Gluskoter, H. J., Rice, D. D., and Taylor, R. B., eds. *Economic Geology*, United States, Volume 2, *The Geology of North America*, pp. 225–40. Boulder, Colo.: Geological Society of America.

Notation Used in Chapter Ten

Symbol	Quantity Represented	Physical Units
A	Arrhenius constant	s^{-1}
C	concentration of kerogen or oil	dimensionless
Δs, $\overrightarrow{\Delta s}$	distance interval in the s-direction, vector of length Δs in s-direction	m
E_a	activation energy for hydrocarbon reactions	$J\text{-mole}^{-1}$
E, $\vec{E}$	force per unit mass, force per unit mass vector	$N\text{-kg}^{-1} = m\text{-s}^{-2}$
$\vec{E}_o$	force per unit mass vector in oil	$N\text{-kg}^{-1} = m\text{-s}^{-2}$
$\vec{E}_w$	force per unit mass vector in water	$N\text{-kg}^{-1} = m\text{-s}^{-2}$
Φ, $\Delta\Phi$	fluid potential energy per unit mass, change in fluid potential energy per unit mass	$J\text{-kg}^{-1}$
Φ_o	oil potential energy per unit mass	$J\text{-kg}^{-1}$
γ	reservoir dip (bedding angle measured from the horizontal)	dimensionless
g, $\vec{g}$	acceleration due to gravity, gravity vector	$m\text{-s}^{-2}$
h	head	m
k_r	reaction-rate coefficient	s^{-1}
P, P_o	fluid pressure, reference fluid pressure	Pascal (Pa) = $kg\text{-m}^{-1}\text{-s}^{-2}$
Θ	angle the oil-water contact makes with the horizontal	dimensionless
R	universal gas constant	$J\text{-mole}^{-1}\text{-}^{\circ}K^{-1}$
ρ	fluid density	$kg\text{-m}^{-3}$
ρ_o	oil density	$kg\text{-m}^{-3}$
ρ_w	density of water	$kg\text{-m}^{-3}$
$\hat{s}$	unit vector in the s-direction	dimensionless
T	temperature or absolute temperature	$^{\circ}K$
t	time	s
u	any scalar field	varies
x, y, z	distance	m
$\hat{x}$, $\hat{y}$, $\hat{z}$	unit vectors in the x-, y-, and z-directions	dimensionless
z	elevation	m
$\vec{\nabla}$	gradient operator	m^{-1}

CHAPTER 11

HEAT TRANSPORT

11.1 TERRESTRIAL HEAT FLOW.

The flow of heat from the Earth's interior is the ultimate source of the Earth's internal mechanical energy. The Earth is a heat engine, a machine that converts thermal energy to mechanical energy. Mechanical energy manifests itself in the form of earthquakes, movement of the tectonic plates, uplift, deformation, and orogeny. While heat flow provides the energy that creates topographic relief on the Earth's surface, erosion continually acts to reduce the continents to featureless peneplains. Erosional rates are largely determined by climate acting in concert with gravity. The terrestrial climate system is powered by solar energy.

Heat in both the continental and oceanic crust is transported by moving fluids, with important consequences for many geologic processes. The rate at which magmatic intrusions cool is affected by hydrothermal circulation. Nearly all chemical reactions are temperature dependent; as moving fluids transport heat and change crustal temperatures, they may also affect processes such as metamorphism and diagenesis. The very existence of ore deposits implies concentration of ores by moving fluids, and temperature changes may play an important role in the sometimes complex chemical reactions involved in the precipitation of ore minerals. The formation of oil and gas in sedimentary basins through the thermal maturation of organic-rich source rocks has an exponential dependence upon temperature. While temperature changes as small as 5°C can double or halve the rate at which hydrocarbons are formed, regional groundwater flow systems have the potential to change basin temperatures by 50°C. In the oceanic crust, circulating water cools the newly formed rocks near spreading centers, forms ore deposits, and chemically reacts with minerals in the basaltic rocks to form new minerals.

Heat is transported by conduction, convection, and radiation. **Conduction** is the transport of heat through a substance by molecular collisions; it is the dominant mode of heat transport in the Earth's crust and lithosphere. **Convection** is the transport of heat by the movement of a solid, liquid, or gas. **Radiation** is the transport of heat by the emission and absorption of electromagnetic energy. The relative importance of radiative heat transfer increases as temperature increases.

The Earth's mean air temperature is determined by the flux of heat from the sun and the capacity of the planet to reflect and retain heat. The mean planetary air temperature is about 15°C. Ground temperatures tend to be warmer than air temperatures by 2 to 3°C, because the ground absorbs heat from the sun and then radiates heat to the atmosphere. Ground temperatures near the Earth's surface are thus determined largely by climate.

Temperature in the Earth increases with increasing depth; the rate of increase is the **geothermal gradient.** Geothermal gradients commonly vary from 10 to 60°C-km^{-1}; a nominal value for the average geothermal gradient in the crust is 25°C-km^{-1}. In areas such as subduction zones or volcanic provinces, the geothermal gradient may take on extreme values.

Heat flows from the Earth's hot interior upward to its comparatively cool surface. This flow of heat is measured in terms of energy (Joules) per unit area (m^2) per unit time (s). A Joule per second (J-s^{-1}) is a **Watt** (W), which is a unit of power. **Power** is energy per unit time. The common units for terrestrial heat flow are milliWatts per square meter (mW-m^{-2}). The mean planetary heat flow is 87 mW-m^{-2}. The average flow of heat from the oceanic crust is 101 mW-m^{-2}, while the average heat loss from the continents is somewhat lower (65 mW-m^{-2}) (Pollack et al., 1993). The known sources of the Earth's heat flow are the decay of radioactive elements and the original heat of accretion associated with the formation of the Earth.

11.2. FOURIER'S LAW OF HEAT CONDUCTION.

Empirically, the amount of heat energy transported by conduction per unit time per unit area (the flux) is proportional to the thermal gradient and a material property termed the thermal conductivity. This relationship is **Fourier's Law of Heat Conduction,**

$$\vec{q}_{\mathrm{h}} = -\lambda \vec{\nabla} T \quad \textbf{(11.1)}$$

where $\vec{q}_{\mathrm{h}}$ is the conductive heat flow (J-s^{-1}-m^{-2} or W-m^{-2}), λ is the thermal conductivity (W-m^{-1}-°K^{-1}), T is temperature (°K or °C), and $\vec{\nabla} T$ is the temperature gradient (°C-m^{-1} or °K-m^{-1}).

TABLE 11.1 Thermal Conductivity of Common Materials and Rocks.

Material	Thermal Conductivity (W-m^{-1}-°K^{-1})
Diamond	1489
Silver	410
Copper	385
Aluminum	202
Lead	35
Steel	20–40
Water	0.6
Air	0.024
Rocks and minerals	1–7
Shale	1–2
Sandstone	2–4.5
Limestone	2–4
Granite	3–4
Basalt	1.5–2.5
Quartz	7–8
Coal	0.2–0.5

Less formally, if the vector notation is dropped, Fourier's Law in one dimension (x) can be written

$$q_{\mathrm{hx}} = -\lambda_{\mathrm{x}} \frac{dT}{dx} \quad \textbf{(11.2)}$$

The material with the highest known thermal conductivity (1489 W-m^{-1}-°K^{-1}) is diamond. Most metals are also highly conductive, with conductivities ranging from 410 W-m^{-1}-°K^{-1} for silver to 20-40 W-m^{-1}-°K^{-1} for steel. Rocks tend to be neither good conductors nor good insulators. The thermal conductivity of rock commonly varies between 1 and 7 W-m^{-1}-°K^{-1}, depending on composition, porosity, temperature, and other factors. An average thermal conductivity for rock is about 2.5 W-m^{-1}-°K^{-1}. Water has a relatively low thermal conductivity of 0.6 W-m^{-1}-°K^{-1}. Most insulating materials rely on trapping dead air spaces, because air has a relatively low thermal conductivity of 0.02 W-m^{-1}-°K^{-1} (Table 11.1).

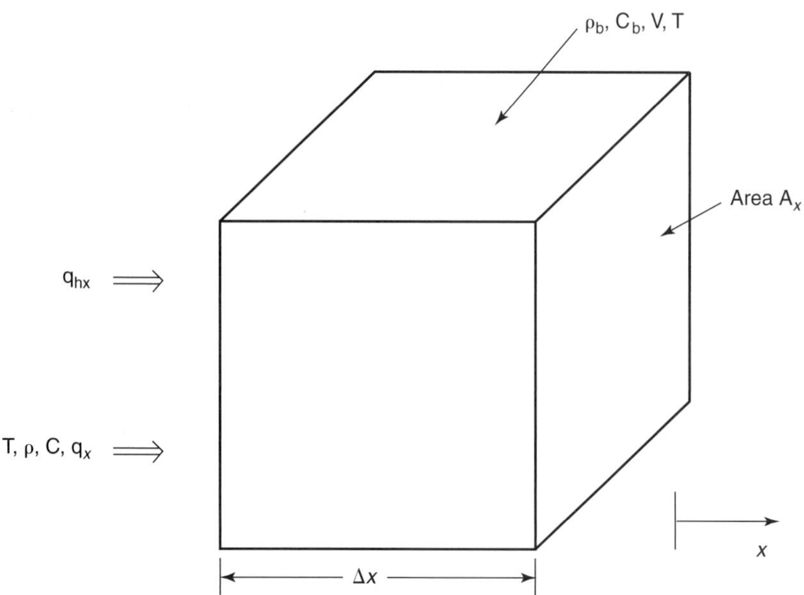

Figure 11.1 Control volume for derivation of the heat equation.

In the discussion that follows we distinguish the conductive heat flow from the total heat or energy transport. For example, heat may be transported through convection by the movement of groundwater. This heat transport represents a flow of heat but should not be confused with the heat flow that takes place through the separate process of conduction.

11.3. The Heat Equation.

By assuming conservation of energy, an equation that describes the transport of heat by conduction and convection in a porous medium can be derived. We proceed to derive the heat equation in one dimension (x).

Consider a control volume of width Δx (Figure 11.1). The control volume has volume V (m^3), bulk density ρ_b (kg-m^{-3}), specific heat capacity C_b (J-kg^{-1}-°K^{-1}) and temperature T (°C or °K). Because we are dealing with a porous medium, we need to distinguish between physical properties of the porous medium (bulk properties) and distinct physical properties of the fluid or matrix. Therefore, we use the subscript "b" (for "bulk") to indicate quantities that are characteristic of the saturated porous medium. Bulk properties are averages of the properties of the solid matrix and the pore fluid.

The conductive heat flow in the x-direction as defined by Fourier's Law is q_{hx} (J-s^{-1}-m^{-2}); the cross-sectional area perpendicular to heat flow is A_x (m^2). Fluid moves through the control volume in the $+x$-direction (Figure 11.1). The moving fluid is in thermal equilibrium with the porous medium that constitutes the control volume: both have temperature T. The fluid has density ρ (kg-m^{-3}), and specific heat capacity C (J-kg^{-1}-°K^{-1}), and is moving with Darcy velocity q_x (m-s^{-1}).

Assuming conservation of energy, the rate at which the total internal thermal energy of the control volume changes is equal to the sum of the individual rates of change due to conduction, convection, and internal heat generation,

$$\begin{array}{ccccccc} (1) & = & (2) & + & (3) & + & (4) \\ \left[\dfrac{dH}{dt}\right]_{\text{total}} & = & \left[\dfrac{dH}{dt}\right]_{\text{cond}} & + & \left[\dfrac{dH}{dt}\right]_{\text{conv}} & + & \left[\dfrac{dH}{dt}\right]_{\text{internal}} \end{array} \qquad \textbf{(11.3)}$$

It is convenient to number each of the above terms (1), (2), (3), and (4), and expand each in turn.

For the first term, the rate at which the total internal heat is changing is equal to the rate at which the product of specific heat capacity, temperature, and mass is changing,

$$\left[\frac{dH}{dt}\right]_{\text{total}} = \frac{d[\rho_b V C_b T]}{dt} \tag{11.4}$$

where the total mass is equal to the product of the bulk density (ρ_b) and the volume (V). We can convince ourselves that equation 11.4 is correct by checking the dimensions,

$$\frac{\text{J}}{\text{s}} = \frac{(\text{kg-m}^{-3})(\text{m}^3)(\text{J-kg}^{-1}{}^\circ\text{K}^{-1})(^\circ\text{K})}{\text{s}} \tag{11.5}$$

Both sides of equation 11.4 have physical units of energy per unit time (J-s^{-1}).

The rate of heat change due to heat conduction (term 2 in equation 11.3) is

$$\left[\frac{dH}{dt}\right]_{\text{cond}} = \frac{\text{heat in}}{\text{unit time}} - \frac{\text{heat out}}{\text{unit time}} \tag{11.6}$$

where "heat in" and "heat out" refer exclusively to heat transported by conduction. The heat energy in per unit time is the heat energy per unit area per unit time (the heat flow q_{hx}) multiplied by the cross-sectional area (A_x). The heat out per unit time is equal to the heat in per unit time, plus any change that takes place over the length of the control volume Δx. The total change that takes place over the length of the control volume Δx is the rate of change with respect to distance multiplied by the total distance Δx. Thus,

$$\left[\frac{dH}{dt}\right]_{\text{cond}} = q_{\text{hx}}A_x - \left[q_{\text{hx}}A_x + \frac{d(q_{\text{hx}}A_x)}{dx}\Delta x\right] \tag{11.7}$$

$$\left[\frac{dH}{dt}\right]_{\text{cond}} = \frac{-d(q_{\text{hx}}A_x)}{dx}\Delta x \tag{11.8}$$

$$\left[\frac{dH}{dt}\right]_{\text{cond}} = \frac{-d(q_{\text{hx}})}{dx}V \tag{11.9}$$

Where $A_x\,\Delta x = V$, and A_x can be taken outside of the derivative because it is constant with respect to x. Substituting equation 11.1 (Fourier's Law) in equation 11.9,

$$\left[\frac{dH}{dt}\right]_{\text{cond}} = \frac{-d\left[-\lambda_x \dfrac{dT}{dx}\right]}{dx}V \tag{11.10}$$

or

$$\left[\frac{dH}{dt}\right]_{\text{cond}} = \frac{d\left[\lambda_x \dfrac{dT}{dx}\right]}{dx}V \tag{11.11}$$

where λ_x is not necessarily constant with respect to x.

The third term of equation 11.3, which represents the convective component, can be written

$$\left[\frac{dH}{dt}\right]_{\text{conv}} = \frac{\text{heat in}}{\text{unit time}} - \frac{\text{heat out}}{\text{unit time}} \tag{11.12}$$

where "heat in" and "heat out" now refer exclusively to heat transported by convection. Expanding the first term on the right side of equation 11.12,

$$\frac{\text{heat in}}{\text{unit time}} = \frac{\text{mass in}}{\text{unit time}} \times \frac{\text{heat energy}}{\text{unit mass}} \tag{11.13}$$

Mass in per unit time is equal to the product of the Darcy velocity, cross-sectional area perpendicular to flow, and fluid density. Heat energy per unit mass is equal to the product of temperature and specific heat capacity. Thus,

$$\frac{\text{heat in}}{\text{unit time}} = (q_x A_x \rho)(CT) \tag{11.14}$$

Checking the physical dimensions of equation 11.14,

$$\frac{\text{J}}{\text{s}} = (\text{m-s}^{-1})\,(\text{m}^2)\,(\text{kg-m}^{-3})\,(\text{J-kg}^{-1}\text{-}^\circ\text{K}^{-1})\,(^\circ\text{K}) \tag{11.15}$$

Applying the same reasoning we used to arrive at equations 11.7 and 11.8,

$$\left[\frac{dH}{dt}\right]_{\text{conv}} = q_x A_x \rho C T - \left[q_x A_x \rho C T + \frac{d(q_x A_x \rho C T)}{dx}\Delta x\right] \tag{11.16}$$

Again, A_x is constant with respect to x, and $A_x\,\Delta x = V$.

$$\left[\frac{dH}{dt}\right]_{\text{conv}} = \frac{-d(q_x \rho C T)}{dx} V \qquad \textbf{(11.17)}$$

Lastly, the fourth term (internal heat generation) can be characterized simply as

$$\left[\frac{dH}{dt}\right]_{\text{internal}} = AV \qquad \textbf{(11.18)}$$

Where A (J-m^{-3}-s^{-1} or W-m^{-3}) is the heat generation rate per unit volume. Most internal heat generation in crustal rocks is due to radioactivity. An average value of radioactive heat generation for crystalline rocks in the upper continental crust is 2.5×10^{-6} W-m^{-3}.

Assembling the four terms introduced in equation 11.3,

$$\frac{d[\rho_b V C_b T]}{dt} = \frac{d\left[\lambda_x \frac{dT}{dx}\right]}{dx} V + \frac{-d(q_x \rho C T)}{dx} V + AV \qquad \textbf{(11.19)}$$

Assume that the volume (V) is constant with respect to time. Further assume that the following quantities are constant with respect to distance (x): fluid density (ρ), fluid heat capacity (C), thermal conductivity (λ_x), fluid Darcy velocity (q_x), and the density (ρ_b) and specific heat capacity (C_b) of the porous medium. We then obtain

$$\rho_b C_b \frac{dT}{dt} = \lambda_x \frac{d^2T}{dx^2} - q_x \rho C \frac{dT}{dx} + A \qquad \textbf{(11.20)}$$

The left-side of equation 11.20 represents the change of temperature with respect to time. Under steady-state conditions, $dT/dt = 0$. The first term on the right-side of equation 11.20 contains the second spatial derivative, or the curvature of the temperature field. Thus, the change of temperature with respect to time due to conductive heat flow is proportional to the curvature of the temperature field. The second term on the right-side of equation 11.20 represents advective heat transport by a fluid moving through a porous medium. Note that the total convective heat transport is proportional to both the Darcy velocity (q_x) and the thermal gradient (dT/dx). If the thermal gradient is zero, there is no net heat transport. In summary, equation 11.20 defines the change of temperature with time as depending on three factors: the curvature of the temperature field, convection, and internal heat generation.

11.4. Effect of Fluid Flow on Conductive Heat Flow.

Following Lachenbruch and Sass (1977), let q_{hz} be the conductive heat flow in the z-direction (J-s^{-1}-m^{-2} or W-m^{-2}), where z is depth (m), and q_z be the Darcy velocity in the z-direction (m-s^{-1}). Assume steady-state conditions ($dT/dt = 0$) and no internal heat generation ($A = 0$). Equation 11.20 then reduces to

$$\lambda_z \frac{d^2T}{dz^2} = q_z \rho C \frac{dT}{dz} \qquad \textbf{(11.21)}$$

We can move the thermal conductivity (λ_z) back inside the derivative,

$$\frac{d\left[\lambda_z \frac{dT}{dz}\right]}{dz} = q_z \rho C \frac{dT}{dz} \qquad \textbf{(11.22)}$$

Multiply the right side of equation 11.22 by the number one in the form $(-\lambda_z/-\lambda_z)$,

$$\frac{d\left[\lambda_z \frac{dT}{dz}\right]}{dz} = \frac{-q_z \rho C}{\lambda_z}\left[-\lambda_z \frac{dT}{dz}\right] \qquad \textbf{(11.23)}$$

We can now substitute Fourier's Law (equation 11.2) into equation 11.23,

$$\frac{-d(q_{hz})}{dz} = \frac{-q_z \rho C}{\lambda_z} q_{hz} \qquad \textbf{(11.24)}$$

or

$$\frac{d(q_{hz})}{q_{hz}} = \frac{q_z \rho C}{\lambda_z} dz \qquad \textbf{(11.25)}$$

Integrating both sides of equation 11.25,

$$\log_e(q_{hz}) = \frac{q_z \rho C z}{\lambda_z} + C_1 \qquad \textbf{(11.26)}$$

where C_1 is a dimensionless constant of integration. Because $e^{(a+b)} = e^a e^b$,

Charles Sumner Slichter: The Wills of Homer and Shakespeare

Charles Sumner Slichter (1864–1946) was born in St. Paul, Minnesota, on April 16, 1864. His family moved to Chicago in 1869. He was the youngest of nine children, the son of descendants of Swiss Mennonites who had immigrated to the United States in search of religious liberty. Slichter later wrote an historical account of his family, *Two Hundred Years of Pioneering,* in which he recounted a summer-long visit to relatives in Canada:

> I have a high respect for those simple people. They got down on their knees every morning in family groups for prayer and for a few verses of scripture. They worked hard and laid up money. They lived well, for the earth yielded its abundance, and their gardens and orchards seemed to know that they were in the hands of the favored of the gods.

Slichter attended Northwestern University and graduated with special honors in 1885 with a bachelor of science degree. In 1888, Slichter received a master's degree in science followed by an honorary degree of doctor of science in 1916. In 1889, Slichter joined the mathematics department at the University of Wisconsin as an assistant professor. Three years later, in 1882, Slichter was appointed to a professorship. He became chairman of the mathematics department in 1902 and dean of the Graduate School in 1920. Altogether, Slichter spent 45 years as a faculty member at Wisconsin, retiring as dean in 1934.

Charles Sumner Slichter (1864–1946)

Slichter became interested in groundwater in 1894 when he began a collaboration with Wisconsin professor Franklin H. King on quantifying the conditions under which groundwater moved. The results of Slichter's investigations were published in the *19th Annual Report of the U.S. Geological Survey* in 1899. Slichter's 84-page paper was titled "Theoretical Investigation of the Motion of Ground Waters." Slichter was the first to show that the movement of groundwater was analogous to the flow of heat or electricity and could be described by potential-field theory. He derived Laplace's equation and wrote:

> It seems remarkable that the fact that the solution of any problem in the motion of ground waters depends upon the solution of the differential equation (5) has not been pointed out before.

Slichter applied the results of his theory to predict how pumping wells would interfere with each other and the effects of irrigation ditches on groundwater movement. Although Slichter clearly understood the influence of gravity on vertical flow, his nomenclature could be criticized for referring to a "pressure" potential instead of clearly differentiating between pressure and head. Nevertheless, Slichter laid the groundwork for important theoretical advances that would be made in the twentieth century by Theis, Hubbert, and Jacob.

Slichter also conducted field experiments as a means of checking his theoretical work. In 1904, he published a paper titled "Measurement of Underflow Streams in Western California" in the journal *Western Society of Engineers*. This was followed in 1905

Charles Sumner Slichter: The Wills of Homer and Shakespeare *(continued)*

by U.S. Geological Survey Water-Supply Paper 140, "Field Measurements of the Rate of Movement of Underground Waters," which described field work conducted in Kansas, California, and Long Island, New York. In his introduction to Water-Supply Paper 140, Slichter claimed that "These measurements, it is believed, constituted the first direct determinations of the rate of flow of ground water that had been made in this country." Slichter measured flow velocities by pouring salt into one well and measuring the time it took to be carried to a nearby well. The appearance of salt water at the second well was detected by measuring changes in the electrical conductivity of the groundwater. In addition to his field work, Slichter carried out laboratory experiments designed to verify both his theoretical calculations and his electrolytic method of measuring groundwater flow velocities. In Water-Supply Paper 140, he described experiments carried out with the vertical and horizontal flow of water through sand-filled tanks. Slichter's experiments led him to correctly infer that dispersion, not diffusion, was the primary mechanism by which contaminant plumes spread. Wang (1987) concluded that Slichter was the first person to conduct research on contaminant transport by groundwater movment.

Following the publication of his influential theoretical paper of 1899, Slichter was employed as a consultant on a variety of applied hydrologic studies involving the survey of water resources throughout the United States. In 1912, he estimated that his government work alone had resulted in 30,000 pages of manuscripts and notes.

Slichter had a mischievous side to his character. He claimed that the card game of poker had great educational value and recommended that everyone learn it. In his undergraduate days, he related how he would often play poker in the room of a theological student, since that site was considered by the campus authorities to be above suspicion. The theological student with whom Slichter played eventually became a bishop, and Slichter offered his opinon that he was "a damn sight better bishop for playing poker." In 1922, Slichter was an expert witness in a lawsuit in Seattle. He described the courtroom events in a letter to his son (quoted in Ingraham, 1972):

> I finally got my testimony in and left this morning. The case will run two weeks more. I can not tell the outcome. I had a fine time on the stand and had a lot of fun with the lawyers because they were so ignorant of underground waters. The judge is a fine able fellow and enjoys a joke very much, and I had a good many chances to cheer him up. The jury had not heard a joke in three weeks. The lawyer on cross examination asked me the ultimate origin of all waters—rivers, lakes, and springs. He wanted me to say "the rain." I said "the seas." He then said, "Isn't the ultimate source the rainfall?" I then quoted: "God said, let the waters be gathered together in one place, and the gathering together of the waters called He the sea; and the morning and the evening were the third day." The jury roared. He was then foolish enough to read from one of my publications where I said that the source of all surface and subterrane waters was the rainfall. He

$$q_{hz} = e^{C_1} \exp\left[\frac{q_z \rho C z}{\lambda_z}\right] \quad \textbf{(11.27)}$$

Let the conductive heat flow at the surface ($z = 0$) be q_{h1}. Then $q_{h1} = e^{C_1}$, and

$$q_{hz} = q_{h1} \exp\left[\frac{q_z \rho C z}{\lambda_z}\right] \quad \textbf{(11.28)}$$

Equation 11.28 relates the conductive heat flow at the surface (q_{h1}) to the heat flow at any depth z (q_{hz}).

Consider the change in conductive heat flow that takes place from the surface ($z = 0$) to some depth $z = L$ (Figure 11.2). Let the conductive heat flow at the depth $z = L$ be q_{h2}. Then

$$\frac{q_{h2}}{q_{h1}} = \exp\left[\frac{q_z \rho C L}{\lambda_z}\right] \quad \textbf{(11.29)}$$

Equation 11.29 relates the change in conductive heat flow (q_h) that takes place at different depths asked me if that were not true. I said "Who said that." He said, very decisively, "Slichter." I then said, "I have quoted what God said. You have quoted what Slichter said. There is a conflict of authority." I waited a minute, until they had laughed several minutes, and then said, "Slichter is right." I had no trouble with the jury after that.

Slichter was a great teacher. He taught mathematics as a means of enriching people's lives, giving his students insights into the relation of mathematics to the entire range of science and to the place of science in human thought. Slichter's philosophy of teaching is illustrated by the following quote from an address he delivered on September 5, 1931, "The Self-Training of a Teacher":

> We are all mentioned in the wills of Homer and Shakespeare and of all the great masters of letters; we do not share merely in part, but each of us inherits in full all their rich chattels. You may be assured that great literature, great history, and great biography belong to us—even to mathematicians—and may be claimed and cultivated as our own. Hence the students have a right to find something more in the classroom than the narrow mechanics of a scientific machine. Instead of twenty times at the movies, better the guest of Homer twenty times. What has Homer to do, you will ask, with the teaching of mathematics? I use Homer, of course, merely as an example. His heavy-breasted heroes contending in the sweat of their primal passions are reacting to the same emotions that sway men everywhere, and to motives that rule our own destiny. As you follow the story of their strife the fibers of your being will vibrate with the ambitions and angers universal among humanity; your personality will quicken to a new appreciation of the facts of life. In other words, the ingredients necessary to compound a man can hardly be omitted from the recipe for making a teacher. You need not be afraid, therefore, to enrich your life by the cultivation of letters and by indulgence at the feast of the humanities.

When Charles Sumner Slichter passed away in 1946, he was eulogized by the faculty at the University of Wisconsin. Their eulogy concluded:

> He was reverent in the deepest sense; reverent before the unities of nature, reverent before all great expressions of the human spirit and of the human intellect and especially reverent before the possibilities of youth.

FOR ADDITIONAL READING

Ingraham, M. H. 1972. *Charles Sumner Slichter: The Golden Vector*. Madison: University of Wisconsin Press; 316 pp.

Slichter, C. S. 1958. *Science in a Tavern: Essays and Diversions on Science in the Making*, 2d ed. Madison: University of Wisconsin Press, 206 pp.

Wang, H. F. 1987. Charles Sumner Slichter—An Engineer in Mathematician's Clothing. In, Landa, E. R., and Ince, S., eds. *History of Geophysics, Volume 3, The History of Hydrology*, pp. 103–12, Washington, D.C.: American Geophysical Union, 122 pp.

due to vertical fluid movement. If the fluid flow is downwards in the $+z$-direction, the conductive heat flow at the surface (q_{h1}) will be lower than the conductive heat flow (q_{h2}) at depth $z = L$. Conversely, if the fluid flow is upward in the $-z$-direction, the conductive heat flow at the surface (q_{h1}) will be higher than the conductive heat flow (q_{h2}) at depth $z = L$. In each case, the total heat transport due to both conduction and convection is constant with depth. Upward-moving fluids elevate the near-surface conductive heat flow and geothermal gradient, while downward-moving fluids depress the near-surface conductive heat flow and geothermal gradient (Figures 11.3 and 11.4).

Note that the extent to which the conductive heat flow is affected by fluid flow depends exponentially on the Darcy velocity (q_z), the depth of circulation (L), the thermal conductivity (λ_z), and

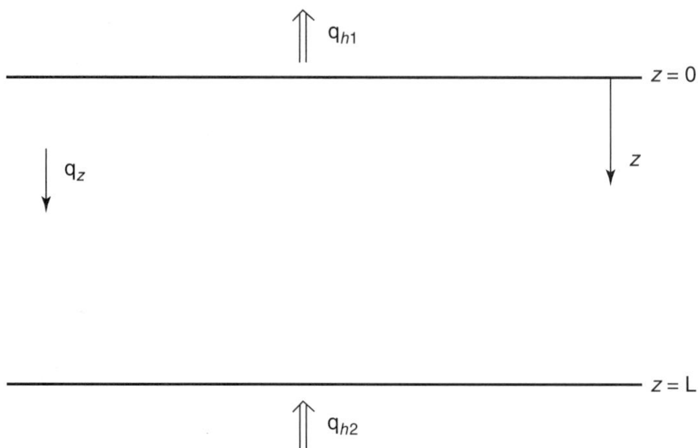

Figure 11.2 Boundary conditions for derivation of equation 11.29. Conductive heat flow at the surface (q_{h1}) is lower than conductive heat flow at depth $z = L$ (q_{h2}) due to downward fluid flow with Darcy velocity q_z.

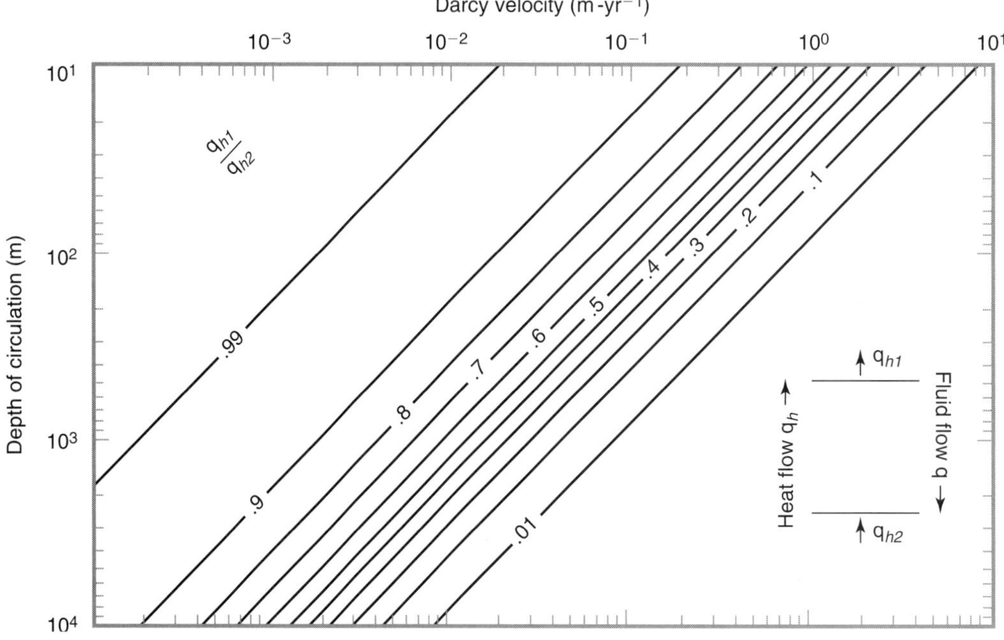

Figure 11.3 Effect of fluid flow on conductive heat flow. The lines show the ratio of conductive heat flow at the surface (q_{h1}) to conductive heat flow at depth (q_{h2}). Surface heat flow is reduced by both high flow velocities and increased depth of circulation.

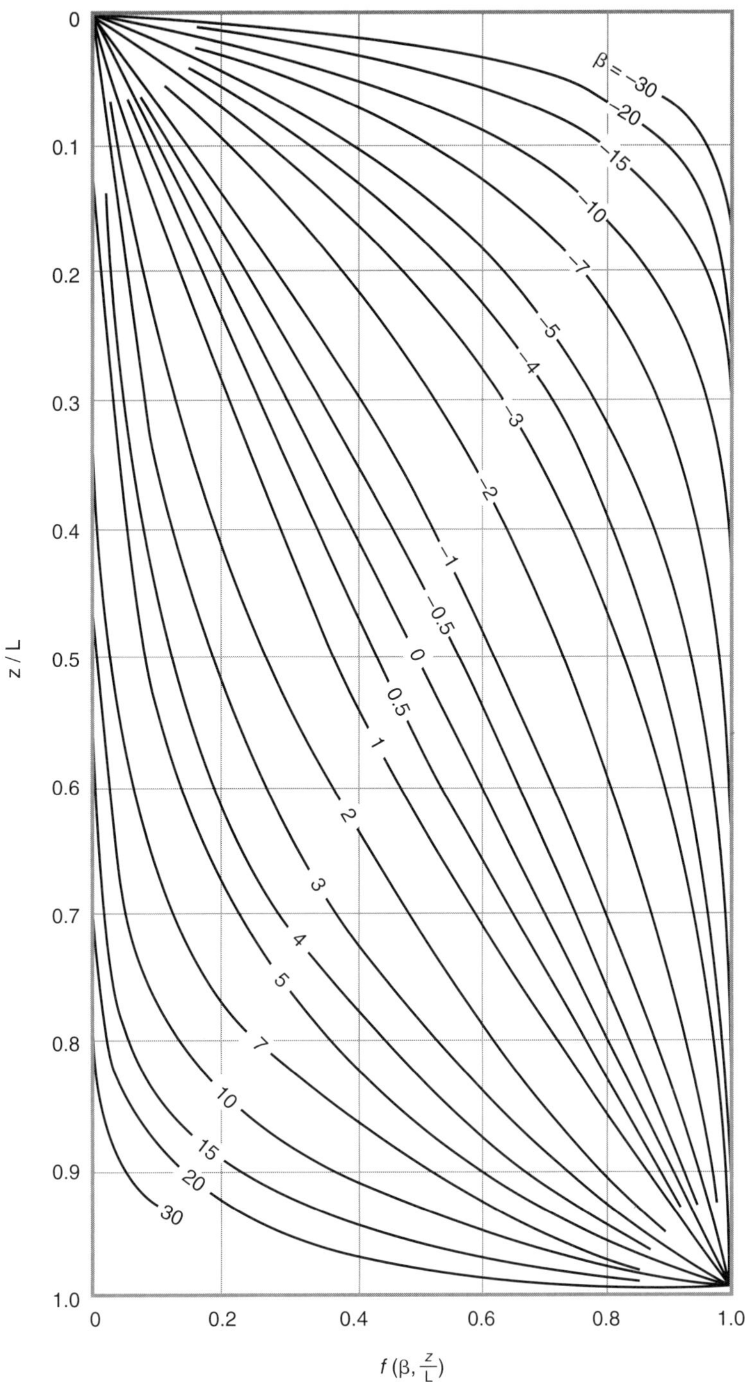

Figure 11.4 Effect of upward (B < 0) fluid flow and downward (B > 0) fluid flow on temperature. (From Bredehoeft and Papadopolous, 1965, p. 326).

the density (ρ) and heat capacity (C) of the moving fluid.

The extent to which fluid movement perturbs the ambient geothermal regime in the Earth's crust depends on the fluid velocity and the depth of circulation. Let $\rho = 1000$ kg-m^{-3}, $C = 4200$ J-kg^{-1}°K^{-1}, and $\lambda = 2.5$ W-m^{-1}°K^{-1}. Substituting into equation 11.29,

$$\frac{q_{h2}}{q_{h1}} = \exp\,(0.0532 q_z L) \tag{11.30}$$

where L is the depth of circulation in meters, and q_z is in the convenient units of m-year^{-1} (as opposed to m-s^{-1}). Or, turning each side of equation 11.30 upside down,

$$\frac{q_{h1}}{q_{h2}} = \exp\,[-0.0532(q_z L)] \tag{11.31}$$

for L in meters and q_z in m-year^{-1}. Consider the downward percolation of groundwater in a recharge area. If the infiltration Darcy velocity is a centimeter per year (0.01 m-year^{-1}) and the total depth of circulation is 100 m, the surface heat flow is depressed by 5%. However, if the depth of circulation is 8000 m for the same Darcy velocity (0.01 m-year^{-1}), the surface heat flow will be depressed by nearly 99% (Figure 11.3).

11.5. Effect of Fluid Flow on Temperature.

Following Bredehoeft and Papadopolous (1965), we can find the effect of moving groundwater on temperature by solving equation 11.20. In section 11.4, the boundary conditions were fixed heat flow at $z = 0$ and $z = L$. If the boundary conditions are fixed temperature instead of heat flow, the solution can be found in terms of temperature. Let the temperature at $z = 0$ be T_0 (°C or °K), and the temperature at $z = L$ be $T_0 + \Delta T$. Temperature as a function of depth is then given by

$$T(z) = T_0 + \Delta T \frac{\exp(Bz/L) - 1}{e^B - 1} \tag{11.32}$$

where

$$B = \frac{C\rho q_z L}{\lambda_b} \tag{11.33}$$

and q_z is the Darcy velocity in the z-direction, L is the depth of fluid circulation, λ_b is the thermal conductivity of the porous medium, and C and ρ are, respectively, the specific heat capacity and density of the moving fluid.

The effect of moving groundwater is to introduce curvature into the geothermal gradient (Figure 11.4). Upward-moving groundwater increases the near-surface geothermal gradient (and conductive heat flow), while downward-moving groundwater decreases the near-surface geothermal gradient (and conductive heat flow).

Because moving groundwater introduces curvature into the geothermal gradient, a linear geotherm is commonly invoked as evidence for the lack of a hydrologic disturbance. However, this is true only if the depth of fluid circulation is known—and it usually isn't. Differentiating equation 11.32 twice with respect to depth (z),

$$\frac{d^2T}{dz^2} = \frac{\Delta T}{e^B - 1}\frac{B^2}{L^2}e^{Bz/L} \tag{11.34}$$

Thus, the curvature of the geotherm (d^2T/dz^2) increases exponentially with increasing depth (z). Curvature of a geotherm may also be present due to changes in thermal conductivity. It is thus difficult to uniquely interpret curvature in a geotherm as evidence either for or against the presence of fluid circulation.

11.6. Heat Transport in Sedimentary Basins.

Domenico and Palciauskas (1973) considered steady-state heat and fluid flow in an idealized, two-dimensional system with homogeneous and isotropic physical properties (Figure 11.5). Thermal boundary conditions were insulating lateral boundaries, fixed temperature at the surface, and a fixed thermal gradient (or heat flow) at the base.

Balneology, Hot Springs, and Spas

Balneology is the practice of using mineral-rich spring water for the treatment of disease. A **spring** is a place where there is a natural flow of groundwater at the surface. A **hot spring** is a spring whose water has a higher temperature than that of the human body (98.6°F or 37°C). A **spa** is usually defined as a fashionable hotel or resort that has a mineral spring. People visit spas for therapeutic reasons, as well as for relaxation and leisure.

In general, hot springs are found in two types of environments: volcanic areas, where there is a near-surface heat source, and areas where significant topographic relief combined with high-permeability avenues for deep circulation result in hot springs. The best-known example of hot springs in North America related to volcanism is Yellowstone Park in the northwest corner of Wyoming. Yellowstone lies over a hot spot and has had three major volcanic eruptions in the last 2 million years. The park contains 10,000 hot springs in the form of steam vents, fumaroles, mud and water pools, and geysers. After a day of hiking through woods populated by bears, a visitor can relax tired muscles in a bath of hot mineral water. The older section of the Old Faithful Inn has a few private bathtubs whose plumbing is connected directly to the geothermal waters of the park. The geothermal water that fills the old-fashioned, high-sided bathtubs of the lodge has a yellow tinge. The hot mineral water somehow seems hotter than ordinary hot water and has an amazing ability to ease aching muscles.

The warm and hot springs of the Appalachian Mountains in the eastern United States are an example of springs that acquire their heat from the background heat flow of the Earth. Most Appalachian springs discharge in the central valleys of anticlinal structures formed by folded and faulted sandstone and carbonate rocks (see box figure). Water flow is driven by topographic relief and follows subterranean paths along bedding planes, faults, and other high-permeability

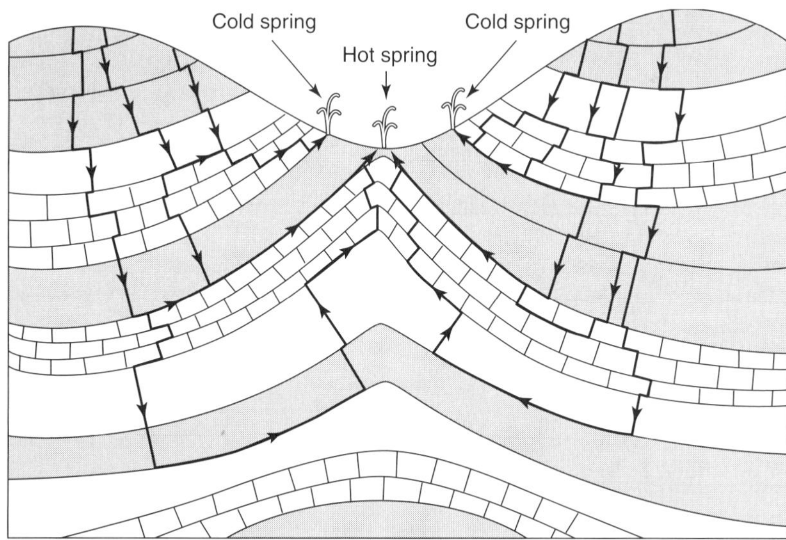

Conceptual model of how groundwater moves through faulted and folded limestone and sandstone beds in the Appalachian Mountains to form both cold and hot springs. The primary routes for water movement are fractures and bedding planes. (After Hobba et al., 1979, p. 11.)

Balneology, Hot Springs, and Spas *(continued)*

features. It is not uncommon to find cold, warm, and hot springs discharging in the same area. The temperature of each spring is controlled by the depth of circulation. Deeper-circulating waters have higher temperatures. The temperature of deep-sourced springs also tends to remain constant throughout the year, whereas the temperature of springs formed by a mixture of deep and shallow-circulating cold water tends to vary. The circulation path of spring waters also controls their chemistry and medicinal properties. Water that discharges from Warm Springs, Virginia, has a constant temperature of 98°F (37°C), and is believed to have circulated to depths of about 4800 ft (1463 m). The Warm Springs water contains a high concentration of sulfate due to the progressive dissolution of gypsum ($CaSO_4$) found in the limestone it passes through. Sulfate in the Warm Springs water makes the water a good laxative. Water from a cold spring in the Appalachians known as the Montgomery White Sulphur Springs has an entirely different chemical composition. The White Sulphur Springs water circulates to depths of only a few hundred meters through fractures in limestone bedrock and is enriched in bicarbonate (HCO_3) due to dissolution of calcite ($CaCO_3$). Calcite is more soluble in cold water than hot; thus, the cold spring water from White Sulphur Springs contains more bicarbonate than hot mineral waters and is a good remedy for indigestion (Chapelle, 1997). In the nineteenth century, patients commonly toured several springs and associated spas in Virginia, seeking the mineral water that might be a specific cure for their ailment. Although many of these spas have vanished, some remain. White Sulphur Springs in West Virginia is now occupied by the Greenbrier Hotel, a five-star resort. In addition to the traditional mineral baths, the Greenbrier offers golf, tennis, horseback riding, and other attractions.

The practice of balneology extends into the indefinite past. Archeological evidence exists that mineral water was used for bathing at least as early as 5000 years ago. In the ancient world, the most ardent devotees of natural hot-water baths were the Romans. They built great public baths called *thermae*. The largest of these could accommodate more than 3000 people at one time. The *thermae* were complex and ornate structures with separate hot, tepid, and cold baths. Slaves moved through underground corridors and anointed the bodies of the bathers with oils. Floors were typically made of marble, and the walls were sheathed with marble to a considerable height. Both walls and floors were decorated with mosaics, and the Romans used an ingenious system of clerestory windows to light the baths.

Altogether, the Romans founded more than one hundred bathing places throughout their empire. Included in these was the city of Bath in England. The Romans built a large structure over the hot (120°F or 49°C) mineral springs at Bath and named it *Aque Sulis*, after the god Sul, also known as Minerva. The Roman *thermae* at Bath included a *frigidarium* (cold bath), *tepidarium* (warm bath), *calidarium* (hot bath), and an exercise court. Like other peoples, the Romans believed that springs and wells were the abodes of gods and nymphs. Later excavation of the Roman *thermae* at Bath found thousands of coins that had been tossed into pools of water as votive offerings. Not all of these offerings were benign. Some of the Romans wrote curses on pieces of pewter and tossed them into the water. One ancient curse reads:

> May he who stole my cloak, whether he be man or woman, boy or girl, freedman or slave, become impotent and die.

The most famous of the Roman resorts built around a hot spring was the city of Baiae, located about 10 mi (16 km) from Naples. Baiae possessed sulphur hot springs, considered to have considerable curative powers. The hot springs, mild climate, and luxuriant vegetation in the region made it a popular resort. Many wealthy Romans, including Nero and Julius Caesar, built magnificent villas in Baiae to which they would retire in the summer months.

In England, the facilities at Bath fell into disrepair and disuse when the Romans left. Interest was revived in the sixteenth century, when England underwent a cultural renaissance under the leadership of Queen Elizabeth I (1533–1603). One of the most remarkable women who has ever lived, Elizabeth reigned for 44 years, from 1558 to her death in 1603. The French ambassador said of her, "She is a very great Princess, who knows everything." In May 1588, England faced imminent invasion by the Spanish Armada. Queen Elizabeth rode out upon a white horse and addressed her troops:

> And therefore I am come amongst you all, as you see at this time, not for my recreation and disport, but being resolved, in the midst and heat of the battle, to live or die amongst you all; to lay down for my God, and for my kingdom, and for my people, my honour and my blood even in the dust. I know I have the body of a weak and feeble woman, but I have the heart and stomach of a king, and a king of England too.

The English fleet convincingly defeated the Spanish in one of the greatest sea battles of all time. The stage was set for England to build the empire upon which the sun never set.

Elizabeth's father, King Henry VIII, had closed many of the holy wells and sulfurous hot springs in England, as the superstitious practices associated with these places were considered to be incompatible with the new Protestant regime. Elizabeth fostered a secular approach that allowed the bathing and drinking of mineral waters for medicinal purposes. Following Elizabeth's ascension to the throne in 1558, Dr. William Turner published the first book on the springs at Bath, *A Book of the Natures and Properties as well of the Baths in England as of Other Baths in Germany and Italy*. The nobility and gentry of England soon found that a week or two spent at Bath was not only a possible cure for any disease from which they might be suffering but also offered a pleasant relaxation from their responsibilities. The use of mineral waters for medicinal purposes acquired a new respectability, and the tradition of the English secular holiday was born.

The therapeutic powers of bathing in or ingesting mineral waters have been promulgated for thousands of years, sometimes by earnest physicians, other times by quacks and hustlers. Modern medicine does not recognize any physiological benefit from bathing in hot mineral waters; any benefit that is derived is believed to be purely psychological, which is not to say that it is necessarily inconsequential. The zenith of balneology and hot spring resorts was the nineteenth century. In 1900, there were more than 2000 spas in the United States. By 1984, the number had fallen to 150. In 1996, there were 115 major geothermal spas in the United States (Lund, 1996).

Several factors conspired to kill the spa industry in the United States. The advent of modern medicine with specific and proven drugs such as antibiotics ended balneological prescriptions. Many of the medicinal claims that had been made for bathing in mineral waters were outlawed. A common inducement to head to country resorts in the nineteenth century was the tendency for epidemics of cholera, malaria, yellow fever, and other diseases to break out in major population centers during the summers. When these epidemics were brought under control by modern medicine, one reason to make an annual migration to the mountain resorts was lost. With the advent of the automobile, people were able to travel to several places instead of journeying to a single spa for their vacation.

Spas remain popular in Europe. The region that was formerly Czechoslovakia has 52 mineral-water spas and more than 1900 mineral springs. Budapest, Hungary, is the spa capital of Europe. Hungary has more than a hundred sources for hot mineral baths, and each year, 2 million citizens of Budapest take balneological cures. In 1996, a London newspaper reported that more than 15 million Europeans annually spend up to four weeks a year at geothermal spas with the entire treatment paid for by either government or private health insurance (Carter, 1996).

In Japan, bathing in hot springs is perhaps more popular than anywhere else on Earth (see photograph). The islands of Japan are a volcanic arc formed by the subduction of the Pacific Plate beneath the Asian Plate. There are 50 to 60 volcanoes on the Japanese islands that have been active in historical times. Mountain ranges cover more than four-fifths of Japan's land surface, and the average annual precipitation is greater

Balneology, Hot Springs, and Spas *(continued)*

Japanese hot-spring bathers.
(Photograph by Anatol Filin)

than 60 in. (152 cm). The combination of high relief, abundant rainwater, and volcanic heat has produced about 1000 springs throughout Japan, five-sixths of which are hot springs. The Japanese prefer to take very hot baths, emerging with their bodies as red as a boiled octopus. According to Buddhist teachings, bathing is synonymous with ablution and purification. It brings sevenfold luck and removes the seven diseases to which the flesh is heir. Japanese hot springs have a variety of chemical compositions. Some are alkaline and make the skin feel slimy. Others are iron-rich and have a reddish color. It is common in Japanese folklore for springs to have been discovered through divine revelation. A typical story is that a person who suffers from a lingering illness prays for recovery. In a dream, an oracle tells the sufferer to journey to a certain spring. The spring is discovered, and the patient cured. Other legends attribute the discovery of some hot springs to Buddhist priests. In the olden days, these priests would make pilgrimages to lofty peaks that they regarded as sacred places. As they wandered up and down these mountains, it was only natural that they should have occasionally discovered mineral springs.

For Additional Reading

Carter, R. 1996. Return of the water cure. *The Independent* (London), 23 July, 1996, p. 6.

Chapelle, F. H. 1997. *The Hidden Sea.* Tucson, Ariz.: Geoscience Press, 238 pp.

Crook, J. K. 1899. *The Mineral Waters of the United States and Their Therapeutic Uses.* New York: Lea Brothers & Co., 588 pp.

Hembry, P. 1990. *The English Spa, 1560–1815: A Social History.* London: Athlone Press, 401 pp.

Hobba, W. A., Jr., Fisher, D. W., Pearson, F. J., Jr., and Chemerys, J. C. 1979. Hydrology and geochemistry of thermal springs of the Appalachians. *U.S. Geological Survey Professional Paper 1044-E.*

Lund, J. W. 1996. Balneological use of thermal and mineral waters in the U.S.A. *Geothermics* 25 (1): 103–47.

Valenza, J. M. 2000. *Taking the Waters in Texas.* Austin: University of Texas Press, 265 pp.

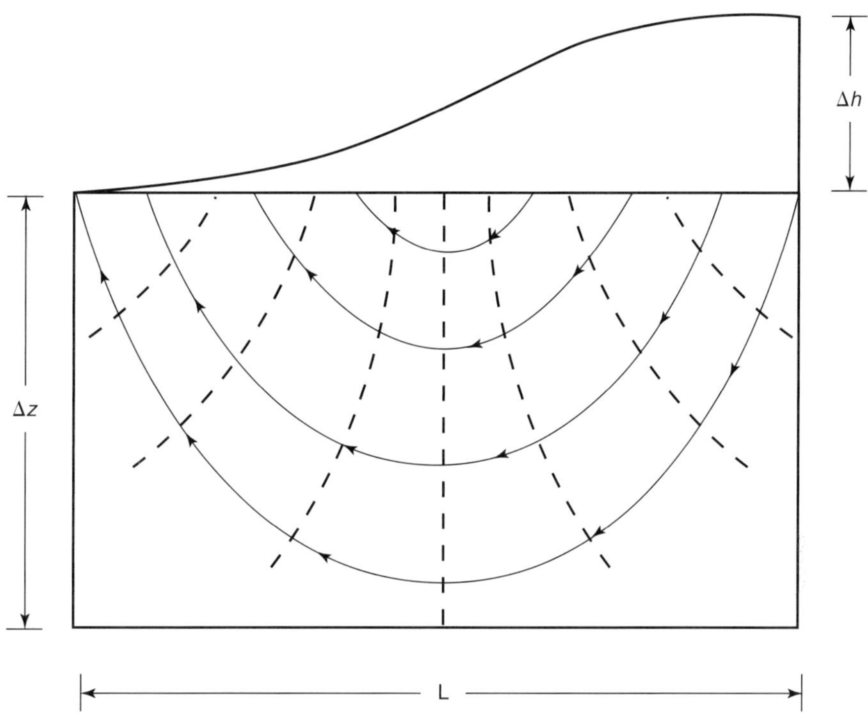

Figure 11.5 Idealized basin as analyzed by Domenico and Palciauskas (1973). Head contours are dashed lines; streamlines are solid lines.
(After Domenico and Palciauskas, 1973, p. 3805)

Hydrologic boundary conditions were no-flow lateral boundaries, fixed head at the surface, and a no-flow bottom boundary. A relative gauge of the magnitude of convective to conductive heat transport (**Peclet number**) under these circumstances is given by the dimensionless group

$$P_e = \frac{\Delta h K \Delta z \rho C}{2L\lambda_b} \quad \textbf{(11.35)}$$

where Δh is the total head drop (m) across a basin of length L (m) and depth Δz (m), K is hydraulic conductivity (m-s^{-1}), ρ is fluid density (kg-m^{-3}), C is fluid specific heat (J-kg^{-1}-°K^{-1}), and λ_b is the bulk thermal conductivity (W-m^{-1}-°K^{-1}) of a porous medium. By taking the average head gradient to be $\Delta h/L$, equation 11.35 can be written as

$$P_e = \frac{q\Delta z \rho C}{2\lambda} \quad \textbf{(11.36)}$$

where q is the Darcy velocity (m-s^{-1}), and the familiar dimensionless group ($C\rho q_z L/\lambda_b$) from equations 11.28 and 11.33 appears again. For nominal values of $k = 10^{-15}$ m^2, $\rho = 1000$ kg-m^{-3}, $g = 9.8$ m-s^{-2}, $C = 4200$ J-kg^{-1}-°K^{-1}, $\lambda = 2.5$ W-m^{-1}-°K^{-1}, $\mu = 5 \times 10^{-4}$ kg-m^{-1}-s^{-1}, $\Delta h/L = 0.01$, and $\Delta z = 5000$ m, the Peclet number is approximately equal to one ($P_e \approx 1$), and it is apparent that thermal regimes in sedimentary basins with head gradients greater than 0.01 and average permeabilities of the order of 10^{-15} m^2 will be perturbed by topographically driven groundwater flow. If the average permeability is an order of magnitude lower (10^{-16} m^2), $P_e \sim 0.1$ and the magnitude of advective heat transport will still be significant. Only at average permeabilities of the order of 10^{-17} m^2 and lower will topographically driven fluid flow fail to be an important mechanism for heat transport in sedimentary basins.

Case Study 11.7 North Slope Basin, Alaska.

The North Slope Basin in Alaska (Figure 11.6) is a well-documented case of thermal anomalies in a sedimentary basin that are apparently due to subsurface fluid flow driven by regional topographic gradients (Deming et al., 1992). From 1977 to 1984, the U. S. Geological Survey made repeated temperature measurements in the upper sections (< 900 m depth) of 21 boreholes drilled for petroleum exploration in the North Slope Basin (Lachenbruch et al., 1987; 1988). The repetition of these temperature measurements over a period of several years enabled extrapolation to a thermal state that was near equilibrium (±0.1°C). Plots of temperature versus depth (Figure 11.7) showed that thermal gradients on the coastal plain to the north were systematically higher than those measured in wells located in the foothills of the Brooks Mountain Range to the south. The maximum thermal gradient in the coastal plain was 53°C-km^{-1} (well NKP, Figure 11.7); the lowest thermal gradient in the foothills province was 22°C-km^{-1} (well LBN, Figure 11.7). Bottom-hole temperature data (Figure 11.8) measured at the bottom of these same wells during drilling and geophysical logging were also collected and used to estimate temperature and thermal gradients at greater depths (Blanchard and Tailleur 1982; Deming et al. 1992).

Thermal gradients estimated from both equilibrium temperature logs and bottom-hole temperature logs were combined with thermal conductivity measurements to estimate heat flow (Figure 11.8). It was found that the near-surface conductive heat flow varied systematically from south to north, just as the thermal gradients did. Heat flow increased from a low of 27 ±6 mW-m^{-2} in the Foothills province to the south, to a high of 90 ±27 mW-m^{-2} on the coastal plain to the north. Similarly, at a depth of 3 km, the

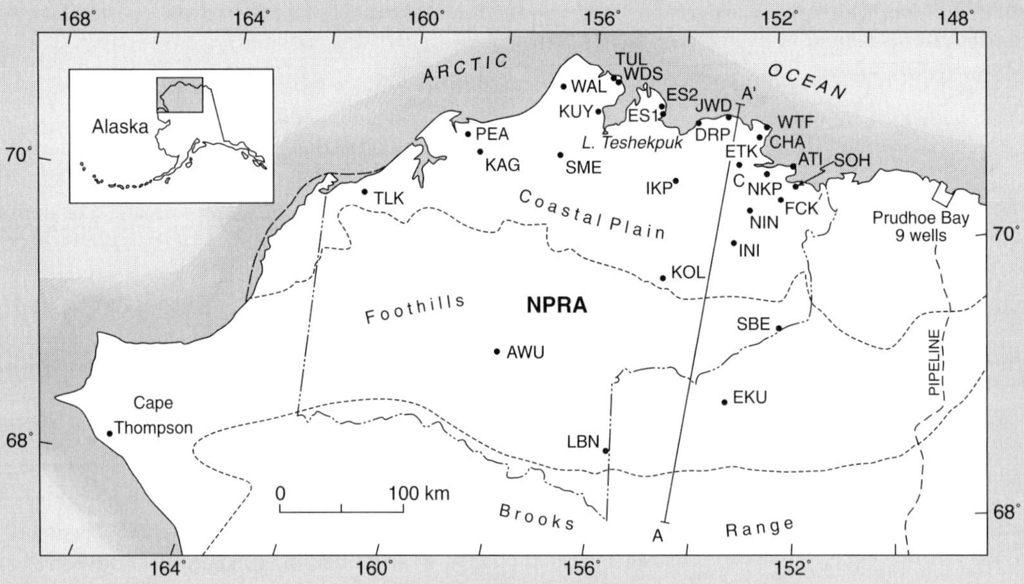

Figure 11.6 Location map for North Slope sedimentary basin, North Slope of Alaska. Three-letter acronyms refer to well names; see Figure 11.7 for temperature-depth plots from these wells.
(From Deming et al., 1992, p. 529.)

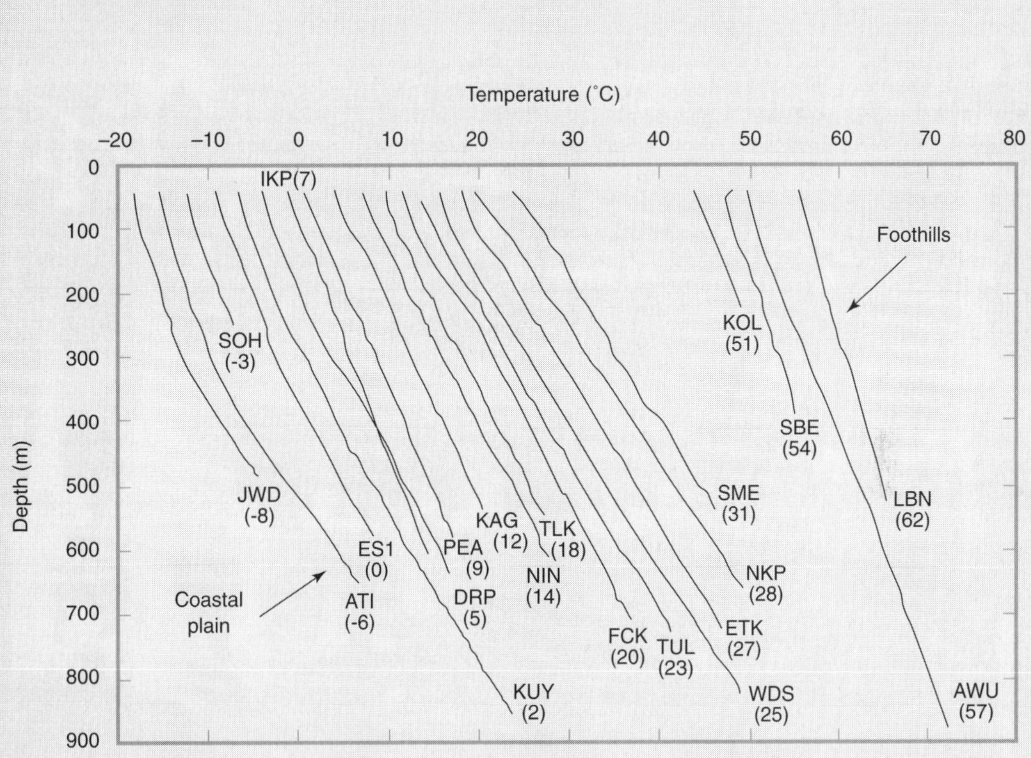

Figure 11.7 Equilibrium temperature logs from the North Slope Basin, Alaska. Note segregation of logs by physiographic province (coastal plain versus foothills province).
(From Deming et al., 1992, p. 532.)

maximum difference in equilibrium formation temperatures estimated from bottom-hole temperatures was 60°C, about 20 times the estimated level of noise in the data (Figure 11.8). Deming et al. (1992) considered a number of hypotheses to explain the geothermal pattern observed in the North Slope Basin. Possible causes included lateral changes in radioactive heat generation, thermal transients related to tectonic factors such as rifting or sedimentation, migration of bodies of water over the land surface, conductive heat refraction, and a change in heat flux at the base of the lithosphere due to some unknown cause. Although each of these mechanisms could, and probably did, contribute to scatter in the data, Deming et al. (1992) concluded that none of them could reasonably explain the observed systematic changes in heat flow and temperature by itself. The systematic variation of heat flow from south to north is consistent with the probable existence of a regional scale (~330 km) groundwater flow system that transports heat by advection from regions of high elevation in the Brooks Range and its foothills to lower elevations on the Arctic coastal plain (Figure 11.9). The hydrologic hypothesis is simple and elegant, relying on a physical mechanism (regional groundwater flow) whose existence has been documented for several sedimentary basins from measurements of hydraulic head and permeability (Bredehoeft et al., 1982).

The integrated flux of groundwater through the North Slope basin can be estimated by equating the

CASE STUDY 11.7 *(continued)*

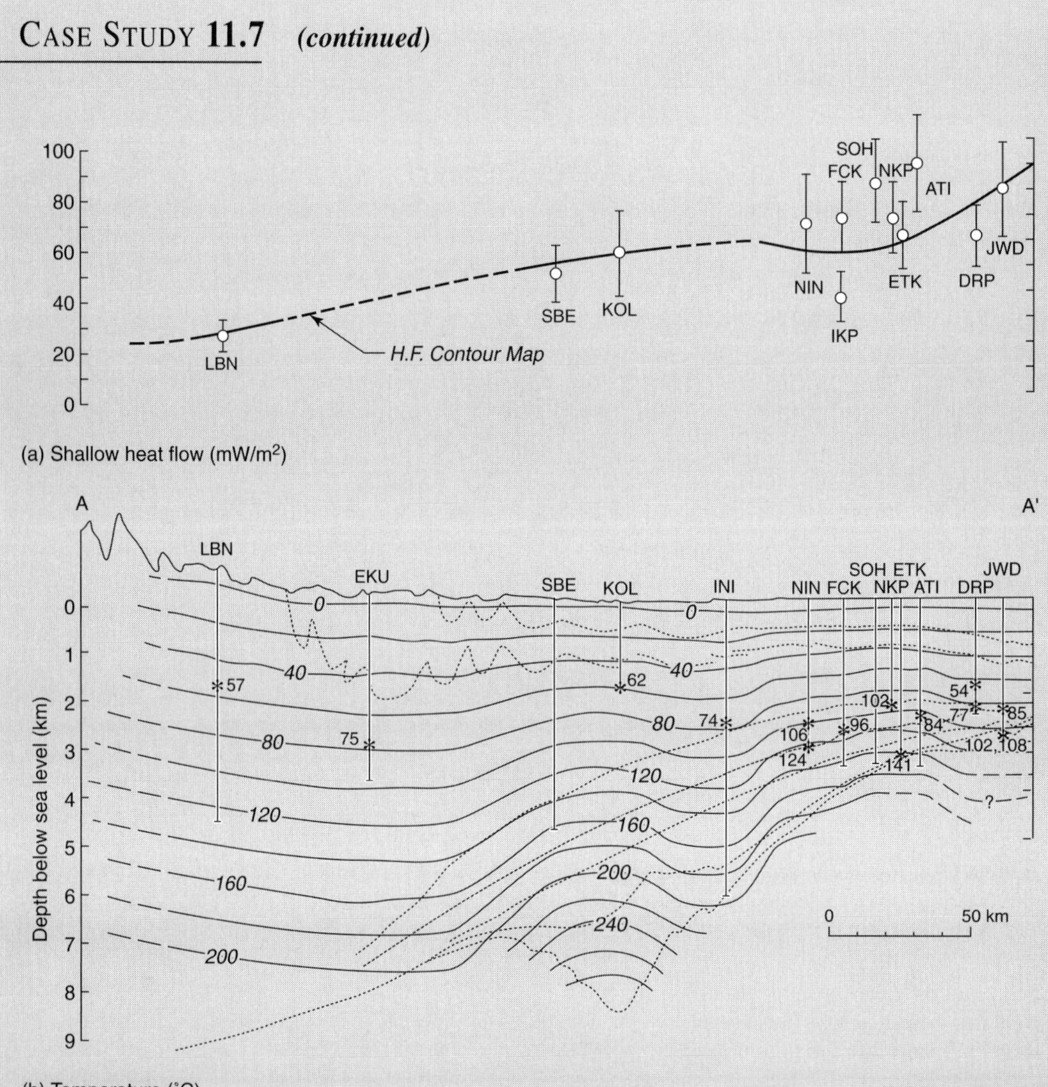

Figure 11.8 (a) Near-surface heat flow (mW/m^2) and (b) subsurface temperature along a north-south cross section through the North Slope sedimentary basin, Alaska. In (a), open circles with error bars represent heat flow estimates at individual wells. Solid (and dashed) line represents generalized heat flow trend along a cross section from south to north. In (b), dotted lines represent stratigraphic boundaries, solid lines represent estimated temperature in °C, and asterisks represent location of bottom-hole temperatures with estimated equilibrium temperatures shown in °C.

(From Deming et al., 1992, p. 538.)

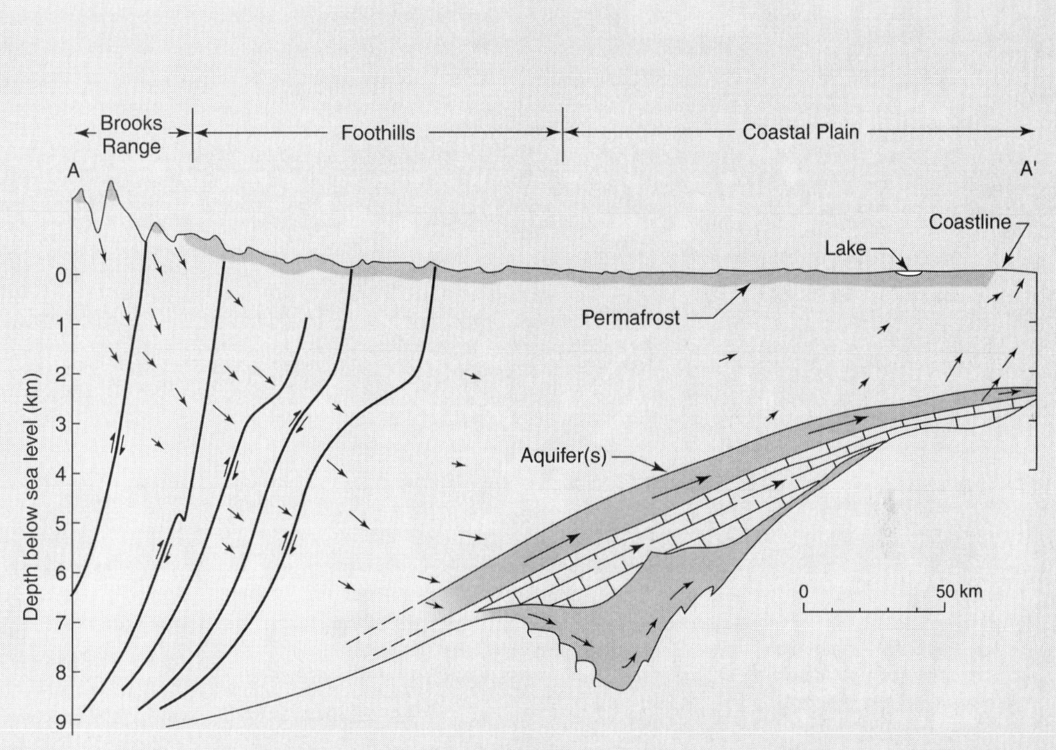

Figure 11.9 Conceptual model of regional groundwater flow in North Slope sedimentary basin, Alaska. Thick (~300 m) layer of surface permafrost across most of foothills and coastal plain province must prevent both infiltration and discharge (see discussion in Deming et al., 1992).
(From Deming et al., 1992, p. 539.)

thermal energy redistributed by the flow system to the apparent loss and gain in heat flow at the surface. Simple assumptions about the geometry of the groundwater flow paths can then be invoked to arrive at an estimate of the average Darcy velocity. Perhaps the simplest model is a two-dimensional v-shaped channel (Figure 11.10), where flow is confined by permafrost on the top and relatively impermeable basement rocks on the bottom. Over the left half of the model (descending channel of length l_1) the heat flow that is lost to the downward flow of groundwater is Δq_h. The total heat loss rate (per unit channel width perpendicular to the flow direction) is then the product of the heat flow loss and the channel length ($\Delta q_h\, l_1$). If energy is to be conserved, the heat loss at the surface must be equal to the heat transported through and out of the channel by moving groundwater. The heat loss rate due to moving groundwater is equal to the mass flux (per unit channel width perpendicular to the flow direction) multiplied by the heat content per unit mass. Thus

$$\Delta q_h l_1 = q\rho C(T_2 - T_1)b \qquad \textbf{(11.37)}$$

where Δq_h is the average heat flow deficit (W-m^{-2}) over the horizontal length of the descending channel (l_1), q is the Darcy velocity (m-s^{-1}) or specific discharge, ρ is fluid density (kg-m^{-3}), C is fluid specific heat (J-kg^{-1}-°K^{-1}), T_1 and T_2 (°C or °K) are fluid temperatures at the ends of the channel, and b (m) is channel height. If we take the hingeline of the flow system to be near the north-south midpoint of the basin where the surface heat flow is approximately equal to the apparent background heat flow (see Figure 11.8), then l_1 = 170,000 m. If we further take

CASE STUDY 11.7 *(continued)*

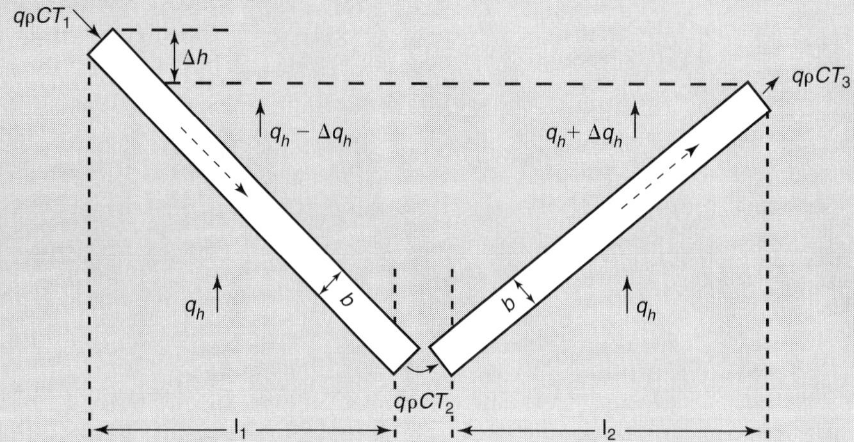

Figure 11.10 Channel model of fluid flow in the North Slope sedimentary basin, Alaska. Notation is as follows: q_h = background conductive heat flow, q = groundwater Darcy velocity, ρ = groundwater density, C = groundwater specific heat capacity, Δh = total head drop across basin, Δq_h = apparent deficit/surplus in conductive heat flow at the surface due to regional groundwater flow, b = channel height, T_1, T_2, T_3 = temperature at points shown.

(From Deming et al., 1992, p. 540.)

Δq_h = 0.02 W-m^{-2}, ρ = 1000 kg-m^{-3}, C = 4200 J-kg^{-1}-°K^{-1}, T_1 = 0°C, T_2 = 180°C (see Figure 11.8), and b = 2000 m, then q = 0.07 m-year^{-1}.

The effective basin-scale permeability can also be estimated by applying Darcy's law. Let s be a distance coordinate along the channel (Figure 11.10). The average head gradient (dh/ds) in the channel can then be approximated by the total drop in head (Δh = 1500 m) across the basin divided by the length of the basin (Δs = 340,000 m). By Darcy's law the permeability k(m^2) is then given by

$$k = \frac{q \Delta s \mu}{\rho g \Delta h} \qquad \textbf{(11.38)}$$

where q is the Darcy velocity of water moving through the channel, μ is water dynamic viscosity, ρ is fluid density, and g is the acceleration due to gravity. If we take q = 0.07 m-year^{-1}, μ = 5 × 10^{-4} kg-m^{-1}-s^{-1} (pure water at 50°C), ρ = 1000 kg-m^{-3}, and g = 9.8 m-s^{-2}, then $k = 2.6 \times 10^{-14}$ m^2.

Note that these estimates of the average Darcy velocity and basin-scale permeability depend on an assumption regarding the thickness (b) of the channel model. For example, if we take the thickness of the channel (b) to be 100 m (instead of 2000 m), our estimate of both the Darcy velocity and permeability increases by a factor of 20. By themselves, the thermal data are only sufficient to discern the integrated flux of groundwater through the basin; they do not allow us to discern if the flow takes place relatively slowly through a thick unit or quickly through a thinner layer of higher permeability.

REVIEW QUESTIONS

1. Define the following terms in the context of hydrogeology:
   a. conduction
   b. convection
   c. radiation
   d. geothermal gradient
   e. Watt
   f. power
   g. Fourier's Law
   h. balneology
   i. spring
   j. hot spring
   k. spa
   l. Peclet number
2. What is the mean air temperature on Earth?
3. What is the mean global heat flow? The mean oceanic heat flow? The mean continental heat flow?
4. Write down Fourier's Law of Heat Conduction. Define each term and give its physical units.
5. Derive the equation in one dimension (x) that describes the transport of heat by conduction and convection. Include internal heat generation.
6. Derive equation 11.29.
7. Apply equation 11.29 to model the apparent depression of heat flow in the recharge area of the North Slope Basin. What Darcy velocity q_z will give the apparent reduction in surface heat flow observed? How does this value compare to the average Darcy velocity estimated by Deming et al. (1992)? If they are different, how can the two different values be reconciled?
8. The surface temperature T_0 is 20°C, and the temperature at 5 km depth is 145°C. If groundwater is moving upward at a Darcy velocity of 1 cm-year^{-1}, what is the geothermal gradient at a depth of 2.5 km? If groundwater is moving downward instead of upward, how does your answer change? (Hint: Differentiate equation 11.32.)

SUGGESTED READING

Bredehoeft, J. D., and Papadopulos I. S. 1965. Rates of vertical groundwater movement estimated from the Earth's thermal profile. *Water Resources Research* 1: 325–28.

Deming, D., Sass, J. H., Lachenbruch, A. H., and De Rito, R. F. 1992. Heat flow and subsurface temperature as evidence for basin-scale groundwater flow. *Geological Society of America Bulletin* 104: 528–42.

Notation Used in Chapter Eleven

Symbol	Quantity Represented	Physical Units
A	internal heat generation rate	$J\text{-}m^{-3}\text{-}s^{-1}$ or $W\text{-}m^{-3}$
A_x	cross-sectional area perpendicular to the x-direction	m^2
B	dimensionless group	dimensionless
b	channel height	m
C	specific heat capacity of a fluid	$J\text{-}kg^{-1}\text{-}°K^{-1}$
C_1	constant of integration	dimensionless
C_b	bulk specific heat capacity of a saturated porous medium	$J\text{-}kg^{-1}\text{-}°K^{-1}$
Δx	length of control volume side	m
g	acceleration due to gravity	$m\text{-}s^{-2}$
H	internal heat energy	J
$h, \Delta h$	head, change in head	m
K	hydraulic conductivity	$m\text{-}s^{-1}$
k	permeability	m^2
L	fixed distance or depth	m
$\lambda, \lambda_x, \lambda_z$	thermal conductivity, thermal conductivity in the x- and z-direction	$W\text{-}m^{-1}\text{-}°K^{-1}$
λ_b	bulk thermal conductivity of a saturated porous medium	$W\text{-}m^{-1}\text{-}°K^{-1}$
l_1	channel length	m
μ	fluid viscosity	$Pa\text{-}s = kg\text{-}m^{-1}\text{-}s^{-1}$
P_e	Peclet number	dimensionless
q, q_x, q_z	Darcy velocity, Darcy velocity in the x- and z-direction	$m\text{-}s^{-1}$
$q_h, \Delta q_h, q_{hx}, q_{hz}$	conductive heat flow, change in conductive heat flow, conductive heat flow in the x- and z-direction	$J\text{-}s^{-1}\text{-}m^{-2}$ or $W\text{-}m^{-2}$
q_{h1}, q_{h2}	conductive heat flow at depths z=0, z=L	$J\text{-}s^{-1}\text{-}m^{-2}$ or $W\text{-}m^{-2}$
ρ	fluid density	$kg\text{-}m^{-3}$
ρ_b	bulk density of a saturated porous medium	$kg\text{-}m^{-3}$
s	distance along channel	m
t	time	s
$T, T_0, \Delta T$	temperature, surface temperature, change in temperature	°C or °K
V	volume	m^3
x	distance	m
$z, \Delta z$	depth, fixed depth	m
∇	gradient operator	m^{-1}

CHAPTER 12

EARTHQUAKES, STRESS, AND FLUIDS

12.1. HYDROLOGICAL EFFECTS OF EARTHQUAKES.

A variety of reports have been made of various hydrologic disturbances both preceding and following earthquakes. The existence of these hydrologic disturbances is unequivocal evidence that fluids in the crust respond to earthquakes and changes in the state of crustal stress. Typical hydrologic disturbances include changes in water levels in wells (hydroseisms), oscillatory changes in the level of surface bodies of water (seismic seiches), changes in streamflow or spring discharge, water-sediment ejections, and liquefaction.

12.1.1. Hydroseisms and Seismic Seiches.

A **hydroseism** is a seismically induced water-level fluctuation in a well; it is one of the most commonly reported hydrologic disturbances associated with earthquakes. Roeloffs (1988) cites several examples of water-level changes that preceded large earthquakes. These include hydrologic disturbances associated with the 1978 Izu-Oshima-Kinkai earthquake in Japan, the 1975 Haicheng and 1976 Tangshan earthquakes in China, and two earthquakes in southern California in 1980 and 1982. One of the most remarkable aspects of these hydrologic changes is that many of them occurred at distances of several hundred kilometers from earthquake epicenters. Before the 1976 Tangshan earthquake (magnitude 7.8), groundwater anomalies were observed at distances of up to 200 km from the epicenter (Roeloffs, 1988). The most common change that takes place before an earthquake is a drop in water level. But increases have also been reported. Water-level changes may also propagate spatially over time, finally converging on the epicentral area. It is possible that these migrating hydrologic disturbances may record the passage of strain through the crust, finally accumulating in the earthquake zone.

Some of the most dramatic examples of how changes in stress can have hydrological effects were provided by the great Alaskan earthquake of 1964. The March 27, 1964, earthquake in Prince William Sound was one of unusual severity, not only in terms of magnitude but also in its duration and coverage. It had a Richter-scale magnitude of 8.5, the ground shaking lasted for 3 to 4 minutes, and the damage zone covered 50,000 mi^2 (129,499 km^2) (Krauskopf, 1968). Immediately following the Alaskan earthquake, hydroseisms occurred in wells not only in the United States but throughout

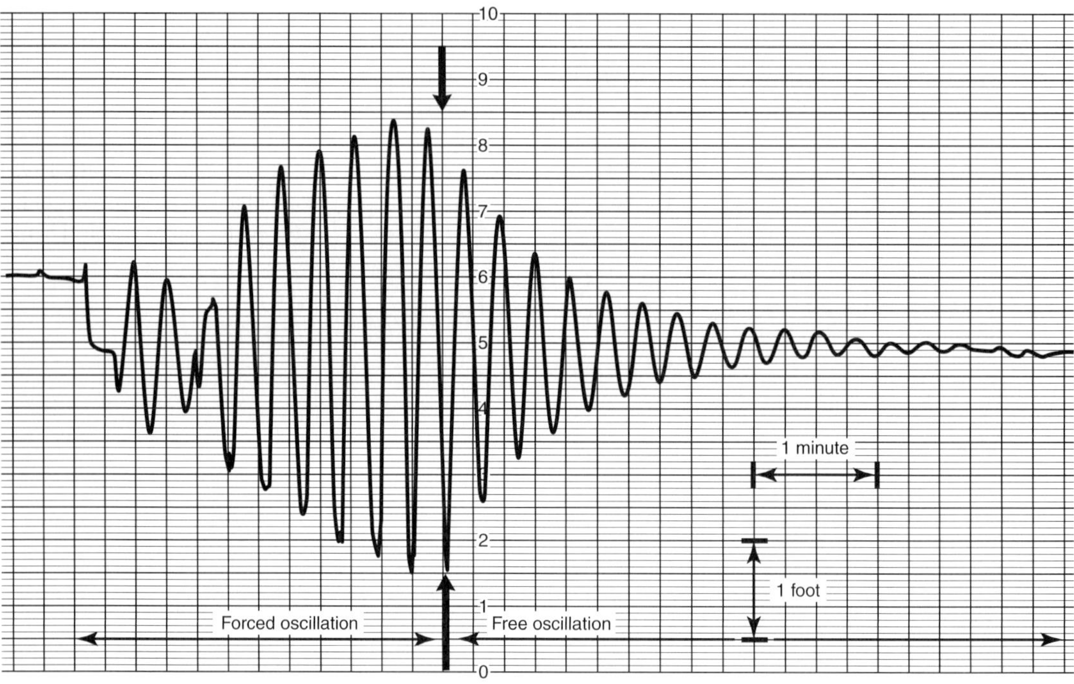

Figure 12.1 Oscillation of the water level in a well near Perry, Florida, in response to the great Alaskan earthquake of March 27, 1964.
(From Cooper et al., 1965.)

the world. The largest recorded hydroseism that occurred outside Alaska was one observed in a well in South Dakota at the northwest edge of the Black Hills. The maximum change in head observed in this well was 23 ft (7.0 m). A well recorder located as far away as Australia recorded a maximum water-level change of 2.25 ft (0.7 m).

One of the fascinating aspects of hydroseisms is the observation that they occur in some wells, but not in others, even in cases where wells penetrate the same aquifer. Cooper et al. (1968) suggested that well-aquifer systems have natural resonant frequencies just as seismographs consisting of coupled spring-mass systems do (Figure 12.1). Through the phenomenon of resonance, the amplitude of water-level fluctuations in a well can be a thousand times greater than the amplitude of the displacement in the solid Earth induced by seismic waves. **Resonance** is technically defined as the frequency at which forced oscillations have their maximum amplitude. A simple and intuitive example of resonance is pushing someone sitting in a swing. If we push the swing at its natural resonant frequency, a very small effort on our part can result in a large displacement of the swing. To time the frequency of our pushes with the swing's natural resonance, we push forward when the swing reaches its maximum backward travel and is just starting to move forward again. In this way, a series of very small pushes occurring at precisely the right times maintains cyclic motion of relatively large amplitude. But if we time our pushes randomly, with some forward pushes occurring when the swing is moving backward, we would be frustrated in our attempts to maintain the swing's motion, even if we exert considerable force. The phenomenon of resonance has important applications throughout the natural world. One of the most famous examples of resonance occurred on November 7, 1940, when the newly constructed Tacoma Narrows Bridge in Puget Sound, Washington, collapsed because of wind-induced resonant vibrations.

A **seismic seiche** is an oscillation in a body of surface water (e.g., a lake) caused by seismic waves. The Alaskan earthquake of 1964 caused seismic seiches with amplitudes of 1.83 ft (0.6 m) on a reservoir in Michigan and 1.45 ft (0.4 m) on Lake Ouachita in Arkansas (Vorhis, 1986).

12.1.2. Changes in Streamflow and Spring Flow.

Dramatic changes in stream and spring flow may occur immediately after an earthquake (review section 7.4.2). In general, two hypotheses seek to explain these changes. One possibility is that the solid earth responds to changes in stress. Just as a sponge releases water when it is squeezed or absorbs water when an external stress is decreased, the solid earth acts as an elastic porous medium, taking up or releasing water as tectonic stress changes. To isolate the influence of tectonic strains, it is necessary to consider the influence of other forces and factors that can also change groundwater levels. These include tidal strains, changes in barometric pressure, and rainfall events. Each of these can produce a water-level change in a well. In the case of rainfall, a head change may be delayed by weeks to months, depending on the hydraulic diffusivity of the crust at a specific site.

The other hypothesis that attempts to explain the apparent influence of earthquakes on water levels is the possibility that faulting and strain increase the permeability and porosity of the crust in the vicinity of the fault zone. Major faults (e.g., the San Andreas Fault in California) do not consist solely of a single fault plane. Rather, a wide area surrounding the fault trace is subject to extensive fracturing. A result of permeability increases is that flow to the area surrounding the fault zone temporarily increases. Baseflow to streams is higher, and the water level in wells rises for a period of time. However, the water level ultimately adjusts to new and lower equilibrium levels. Considerable support for the increased permeability hypothesis is provided by a study of hydrologic changes that occurred in central California following the 1989 Loma Prieta earthquake (Rojstaczer et al., 1995). Three pertinent types of changes took place. (1) Immediately following the earthquake (within 15 minutes), streamflow increased. (2) The amount of dissolved solids in the streams increased, but the relative proportions of the different ionic species remained unchanged. (3) Within weeks to months following the earthquake, the water table dropped. Rojstaczer et al. (1995) argued that the pattern of these changes suggested the likely mechanism responsible for the hydrologic phenomena was an increase in crustal permeability through faulting associated with the earthquake. The increased concentration of dissolved solids in streams was interpreted to represent an increase in the relative amount of baseflow to the streams. However, the constancy of the composition of the added material suggested that there were no new or unusual fluid sources, as might have occurred if seismic strain had squeezed fluids up from deep within the crust. There was also no noticeable correlation between the degree to which streamflow increased and the magnitude or type of seismic deformation. For example, areas undergoing extension might have been expected to take up fluids. However, baseflow to streams in these areas apparently increased. The permeability enhancement inferred by Rojstaczer et al. (1995) was estimated to be approximately an order of magnitude and appeared to be persistent, at least for 3 to 4 years after the earthquake.

Although the hydrologic changes following the 1989 Loma Prieta earthquake strongly suggest permeability changes as the mechanism that links seismic activity with crustal fluids, the evidence is not unequivocal for other events. Both rising and falling water levels have been observed before and following earthquakes, suggesting that the crust is responding as an elastic porous medium to extensional and compressional strains. At the present time, it is not clear if a single hypothesis can adequately explain all of the data.

12.1.3. Water-Sediment Ejection.

Another hydrologic phenomenon associated with earthquakes is water-sediment ejection. Half of the documented records of strong earthquakes contain records of ejections. Water-sediment ejections caused by earthquakes may take various forms. In the literature, they have been described as sand and mud spouts or fountains, mud volcanoes, sand boils, mud or sand vents, craters, and various other terms. Ejections typically commence after seismic motion and may continue for up to 24 hours after the cessation of strong ground motion. Ejecta range from clear water, to mud, to water containing coarse gravel, and ejections are usually observed to occur in a pulsating manner. In all cases, water-sediment ejections are associated with the near-surface presence of water-saturated unconsolidated sediments and appear to be a response to stress changes in the crust caused by an earthquake. In the case of the Alaskan earthquake of 1964, water-sediment ejections occurred at distances up to 250 mi (402 km) from the epicenter. Eyewitnesses to the 1964 Alaskan earthquake reported that eruptions were as high as 100 ft (30 m). Water-sediment ejections typically occur in locations where the water table is high. These include the toe of alluvial fans, stream valleys, floodplains, and near lakes, swamps, and coastal plains (Waller, 1968).

12.1.4. Liquefaction.

Liquefaction is the loss of shear strength a water-saturated sediment undergoes when shaken. Consider a container of water-saturated sand. The sand maintains its structural integrity and resistance to shear stress through grain-to-grain contacts. Each individual sand grain is supported by adjacent grains. However, if we were to stir or violently shake the container of water-saturated sand, some of the grain-to-grain contacts would be lost. A sand grain suspended in fluid has no ability to resist shearing stress. In effect, the material has been liquefied.

The stress borne by the matrix grains in a porous medium is the effective stress (see section 3.4.2). When grain-to-grain contacts are disturbed, the weight of these grains must be borne by the fluid. Thus, the effective stress must decrease and the fluid pressure must increase. The degree to which a material has undergone liquefaction depends on the number of grain-to-grain contacts that have been lost. If all contacts are lost, the average effective stress goes to zero. The decrease in effective stress and subsequent increase in fluid pressure are effects of liquefaction, not causes.

Figure 12.2 These earthquake-resistant apartment buildings in Niigata, Japan, survived the June 16, 1964 earthquate intact. However, liquefaction of the underlying soil caused the buildings to tilt and fall over. (Photograph courtesy of the National Geophysical Data Center.)

Damage to structures by liquefaction is of two types. In the first case, structures resting on liquefied soils that are otherwise earthquake-resistant may be tilted by differential subsidence (Figure 12.2). In the second case, liquefied soil may be displaced and act as a disruptive force, damaging airport runways, buried pipes, and pile foundations for bridges and buildings.

12.2. Smoluchowski's Dilemma.

In several parts of the world, there are large areas where older rocks have been thrust up on top of younger rocks. Structural relationships indicate that the older rocks were superposed over their younger cousins by upward movement along thrust faults. These large-scale thrust faults are called overthrusts. An **overthrust fault** is a reverse

Quicksand

Quicksand is a mixture of sand and water that has a low shear strength due to flowing water. Like a sediment that has undergone liquefaction, quicksand owes its low resistance to shearing to a loss of grain-to-grain contacts. However, in the case of quicksand, the disruption of grain contacts occurs not by shaking but by water flowing up through a sand-water mixture. The genesis of quicksand can be demonstrated through the following experiment (Matthes, 1953). An old-fashioned flatiron is placed on top of a barrel of water-saturated sand (see box figure). A pipe is inserted into the base of the barrel through which a flow of water may be admitted. So long as the water is turned off, the water-saturated sand is able to support the weight of the flatiron. However, if the water is turned on, the flow of water from the base of the sand pile to its top disrupts the grain-to-grain contacts among the sand particles. The sand loses its resistance to shearing, and the flatiron falls to the bottom of the barrel. If the water is subsequently turned off, the sand settles and the excess water collects at the top. If a second flatiron is then placed on top of the sand, its weight will be supported. Thus, the key to the occurrence of quicksand is not the amount of water present, but the flow of water.

Much of our understanding of quicksand comes from the studies of Ernest Rice Smith (1891–1952), who was a professor of geology at DePauw University in Indiana. Smith reportedly obtained his understanding that the key requirement for the formation of quicksand was an upward flow of water by observing quicksand beds in a pasture near Greencastle, Indiana. After much frustration, Smith was unable to understand why some wet sand was "quick" while other sand was firm. The answer was provided to Smith by a farmer. The farmer told Smith that sand in his pasture was sometimes quick and sometimes firm. The farmer said, "Come back in August and you'll be able to square dance on it." August was the month of the year in which springflow was the lowest. Immediately, Smith realized that the key was not the amount of water but the flow.

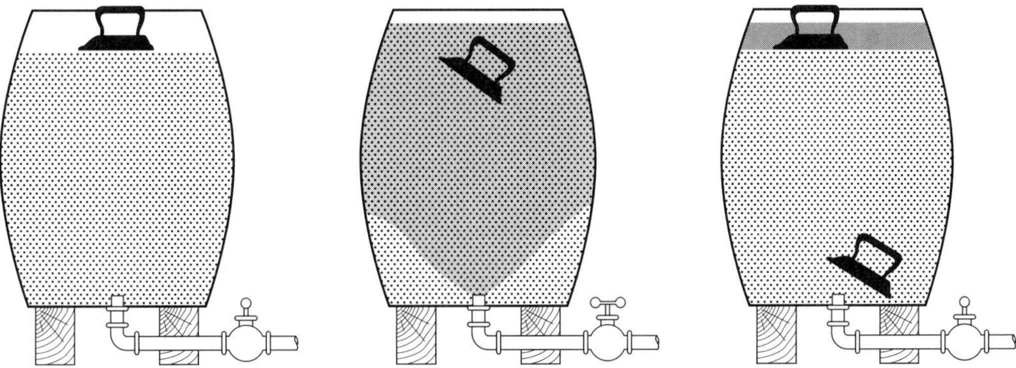

Experiment demonstrating that quicksand is formed by the upward flow of water through sand. In the picture on the left, a flatiron rests on sand in a barrel. The sand is saturated with water, but able to support the weight of the flatiron. In the middle picture, a valve has been turned, allowing water to flow from the pipe in the bottom of the barrel up through the sand particles. The upward flow of water disturbs the grain-to-grain contacts, and the sand becomes "quick". In the picture on the right, the flow of water has been turned off. The sand settles, and the excess water collects near the top of the barrel. A second flatiron is placed on the surface and its weight is supported by the sand. The fact that the second flatiron does not sink is evidence that the primary factor needed for the formation of quicksand is not the amount of water, but its flow.
(From Matthes, 1953, p. 97.)

Quicksand *(continued)*

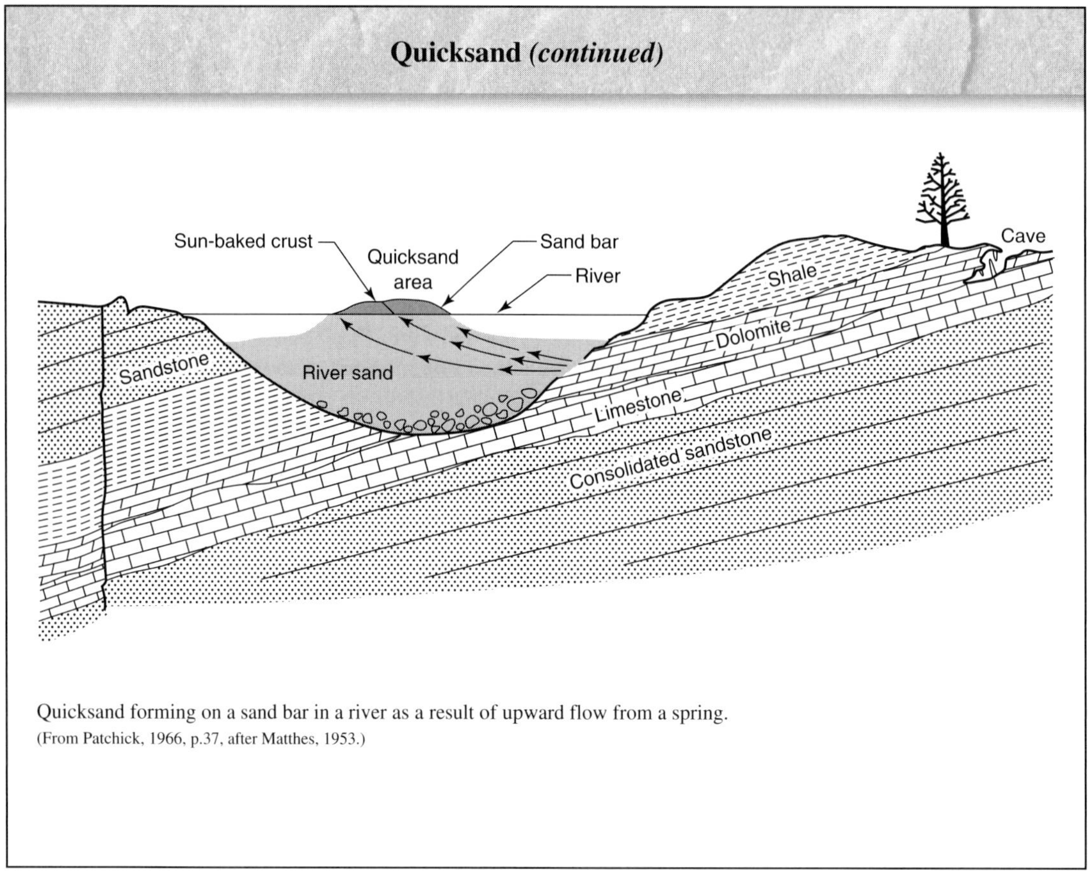

Quicksand forming on a sand bar in a river as a result of upward flow from a spring.
(From Patchick, 1966, p.37, after Matthes, 1953.)

fault in which older strata have been superimposed on younger strata (Figure 12.3). The total displacement on overthrusts is in the range of 10 to 100 km, and the thickness of the older rocks that have been overthrust onto the younger strata (the hanging wall) may exceed several kilometers. Some examples include the Heart Mountain overthrust in Wyoming, with an inferred displacement of 55 km; the Roberts Mountain overthrust in Nevada, whose total movement exceeded 80 km; and the Taconic thrust sheet near Albany, New York, which has overridden younger sediments by 80 km or more (Hubbert and Rubey, 1959).

Understanding the forces by which such huge blocks of rock are moved has long been one of the classic problems of structural geology. The primary difficulty in explaining the movement of overthrust blocks is that the forces required to push the rocks along the thrust fault apparently exceed the crushing strength of the rocks involved, therefore implying that the movement was mechanically impossible. Yet it occurred.

The situation can be analyzed by modeling the movement of the overthrust sheet as a block on a ramp (Figure 12.4). Although the geologic relationships in most overthrusts indicate that the hanging wall strata were pushed uphill, let us be optimistic and consider a case where the ramp is horizontal so as to minimize the needed force. The force needed to push the block along the ramp is the minimum needed to overcome the frictional resistance to sliding. The frictional resistance to sliding the block along the ramp is the product of the weight of the block and the coefficient of friction. Thus

Matthes (1953) described quicksand as a "malevolent phenomenon" that figured prominently in the landscapes of nineteenth-century romance novels, where it was frequently used as a literary device for the disposal of unwanted characters. Although quicksand is less frequently encountered in the age of paved roads and travel by automobile, it is as widespread as ever and is probably more dangerous because it is less familiar (Patchick, 1966). Typically, quicksand is found in hilly terrains, especially where the bedrock is composed of cavernous carbonate rocks that give rise to numerous groundwater springs. Where it occurs, quicksand is confined to a relatively small area where an underlying spring maintains an upward flow of water (see box figure). Matthes (1953) suggested that it was possible for a person trapped in quicksand to escape if they had a pole with them and were careful to follow this procedure. First, they must not try to force their way to solid ground; that action will simply hasten sinking into the morass. Standing still and yelling for help is also likely to prove useless unless help arrives very quickly. The first move in the recommended escape procedure is for the trapped person to drop their pole behind them and fall back on it while sticking their arms out at right angles to their body. In this way, they can float on the surface of the quicksand and call for help. If help is not forthcoming, it is possible to escape if the right actions are taken. The first step is to move the pole until it is at right angles to the body and underneath the hips. Once this is done, the victim may pull their legs out one at a time. This should be done slowly, with frequent rests. Once both feet have been extracted from the quicksand, the floating prisoner of the morass should look around and find the shortest route to solid ground. The escape is then done by rolling to the edge of the quicksand pit. Indeed, rolling is the only way to escape. The author has not tested this procedure, and readers are advised to proceed at their own risk.

For Additional Reading

Matthes, G. H. 1953, Quicksand. *Scientific American* 188, (6): 97–102.

Maxwell, J. C. 1953, Memorial to Ernest Rice Smith (1891–1952). *Proceedings of the Geological Society of America*, 139–42.

Patchick, P. F. 1966. Quicksand and water wells. *Groundwater* 4, (2): 32–46.

$$F = mg\mu_f \quad \textbf{(12.1)}$$

where F (kg-m-s^{-2}) is the force needed to move the block, m (kg) is the mass of the block, g is the acceleration due to gravity (9.8 m-s^{-2}), and μ_f (dimensionless) is the coefficient of friction. Denote the height of the block by h (m), the length of the block parallel to F as b (m), and the width of the block perpendicular to F as w (m) (Figure 12.4). The mass of the block is equal to the product of its density (ρ_b, kg-m^{-3}) and volume (bwh, m^3). Thus

$$F = \rho_b bwhg\mu_f \quad \textbf{(12.2)}$$

The definition of stress is force per unit area. If the force F is applied equally over the left side of the block (Figure 12.4), that end of the block is subject to an applied stress σ_a (kg-m^{-1}-s^{-2}),

$$\frac{F}{wh} = \sigma_a = \rho_b bg\mu_f \quad \textbf{(12.3)}$$

The normal stress (σ_n, kg-m^{-1}-s^{-2}), or stress perpendicular to the ramp, is the weight of the block ($mg = \rho_b whg$, kg-m-s^{-2}) divided by the area of the bottom of the block (bw, m^2),

$$\sigma_n = \rho_b hg \quad \textbf{(12.4)}$$

Thus, equation 12.3 can be rewritten as

$$\sigma_a = \sigma_n \mu_f \frac{b}{h} \quad \textbf{(12.5)}$$

The left side of equation 12.5 is the minimum applied stress needed to move the block; the right side of equation 12.5 is the frictional resistance to movement. The applied stress needed to move a

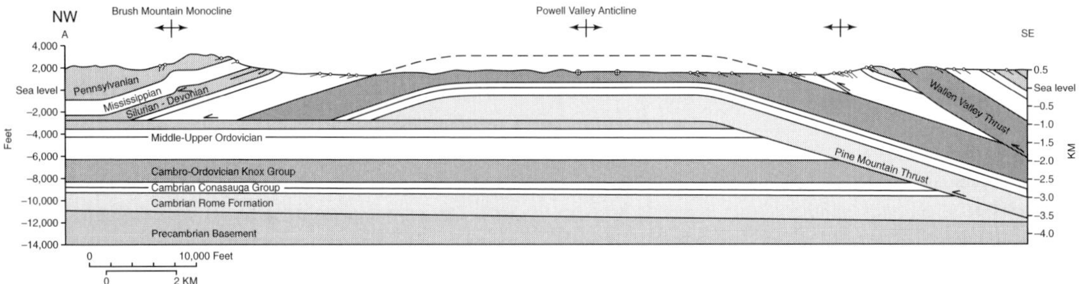

Figure 12.3 Overthrust faults in the Pine Mountain thrust system, southern Appalachian fold-and-thrust belt. (From Mitra, 1988, p. 78.)

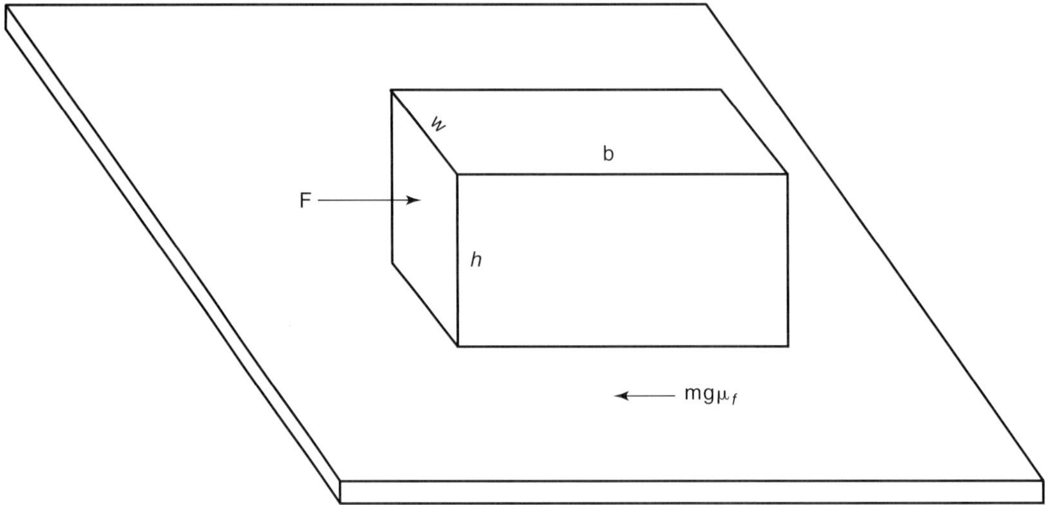

Figure 12.4 Forces acting on a block on a plane.

block is proportional to its length (b), the coefficient of friction (μ_f), and the normal stress. Although a cursory inspection of equation 12.5 implies that is easier to move taller blocks (because the height, h, appears in the denominator), this is not true. The normal stress ($\sigma_n = \rho hg$) is also proportional to block height (h). Block height has no effect on the relative ease or difficulty of movement, because both the normal stress (σ_n) and the applied stress (σ_a) increase as block height increases.

If the block of rock had infinite strength, an arbitrarily large stress σ_a could be applied to the left end of the block to move blocks of any length. However, rocks have a finite compressive strength. A column of rock of infinite height cannot be constructed, because the rock at the bottom of the column would be crushed by the weight of the overlying rock. The maximum compressive stress that can be borne by most rocks is in the range of 10 to 200 $\times$ 10^6 Pa. A Pascal (Pa) is unit of stress and is equal to one Newton (N) per square meter (1 Pa = 1 N-m^{-2} = kg-m^{-1}-s^{-2}). A Newton (N) is a unit of force (1 N = 1 kg-m^{-1}-s^{-2}).

If we assume that the applied stress (σ_a) can be no greater than an average value for rock strength (10^8 Pa), we can calculate the maximum length of a block that can be moved by pushing it from an end. Solving equation 12.5 for block length (b),

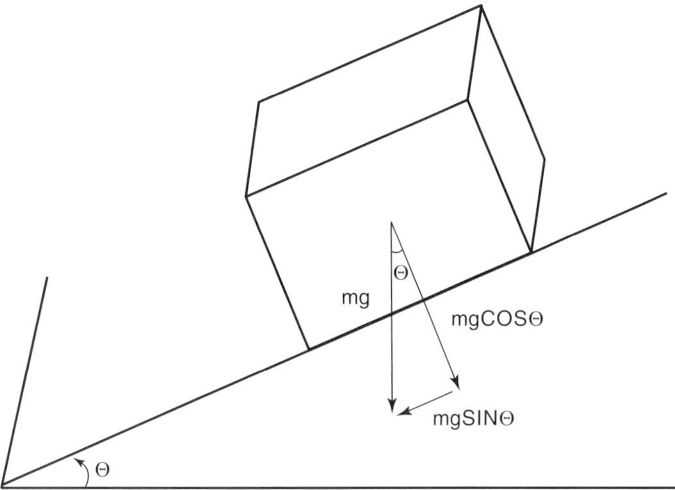

Figure 12.5 Forces acting on a block on a ramp inclined at an angle Θ to the horizontal.

$$b = \frac{h\sigma_a}{\sigma_n \mu_f} \quad \textbf{(12.6)}$$

But $\sigma_n = \rho_b hg$, so

$$b = \frac{h\sigma_a}{\rho_b hg\mu_f} = \frac{\sigma_a}{\rho_b g\mu_f} \quad \textbf{(12.7)}$$
$$= \frac{10^8 \text{ (kg-m}^{-1}\text{-s}^{-2})}{2650 \text{ (kg-m}^{-3}) \times 9.8 \text{ (m-s}^{-2}) \times 0.6}$$
$$= 6418 \text{ m}$$

where we have assumed average numbers for rock density (2650 kg-m^{-3}), the acceleration due to gravity (9.8 m-s^{-2}), and the coefficient of friction for rock-on-rock (0.6).

Under the most optimistic assumptions possible (horizontal surface and maximum applied stress), it is thus impossible to move blocks longer than about 6 km because the necessary force would be of such magnitude that the rock would simply be crushed instead of sliding. Yet geological studies throughout the world show irrefutably that blocks up to 100 km long have been moved for considerable distances. This is **Smoluchowski's dilemma** (Smoluchowski, 1909).

One possible reconciliation is the possibility that overthrust blocks may not have been pushed up ramps by tectonic forces but rather simply slid down inclined slopes. Over geologic time, differential uplift may have reversed the relative dip, giving the impression that blocks that slid downhill were pushed uphill. In this case, we conveniently avoid the problem of crushing the end of the block with a large applied force. The force needed to move the thrust sheet is gravity, which is a body force that is applied equally over the entire volume of the thrust sheet. This appears to be a reasonable hypothesis. However, unreasonably high dip angles are apparently necessary before blocks of rock will slide downhill.

Consider a block on a ramp again (Figure 12.5). The ramp is inclined at an angle Θ to the horizontal. The force (mg, kg-m-s^{-2}) acting on the block is the product of mass (m, kg) and the acceleration due to gravity (g, m-s^{-2}). This force can be resolved into two components. One component ($mg \sin \Theta$) acts in a direction parallel to the ramp. The other component ($mg \cos \Theta$) is in a direction perpendicular to the ramp surface. The block will slide downhill when the propelling force ($mg \sin \Theta$) is larger than the frictional force resisting movement. The frictional force resisting the downhill slide is equal to the product of the normal force ($mg \cos \Theta$) and the coefficient of friction (μ_f).

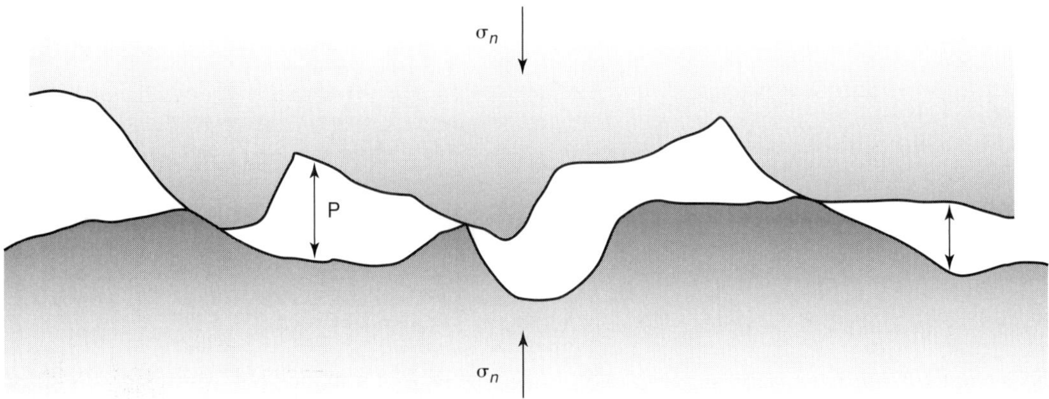

Figure 12.6 Microscopic view of opposite surfaces in contact across a fault plane subject to normal stress σ_n and a fluid pressure P.

Thus, the minimum criterion for downhill movement is

$$mg \sin \Theta = \mu_f \, mg \cos \Theta \qquad \textbf{(12.8)}$$

Solving for Θ,

$$\frac{\sin \Theta}{\cos \Theta} = \tan(\Theta) = \mu_f \qquad \textbf{(12.9)}$$

or

$$\Theta = \tan^{-1}(\mu_f) \qquad \textbf{(12.10)}$$

For $\mu_f = 0.6$, $\Theta = 31°$. In view of the existing low angles that commonly characterize overthrusts (about 10° or less), it seems unlikely that such steep reversals could have occurred in the past. Gravity sliding does not, therefore, appear to be an attractive hypothesis.

The first plausible explanation for Smoluchowski's dilemma was provided by the American geologists M. K. Hubbert and W. W. Rubey. In 1959, they proposed that thrust faulting and block movement was facilitated by high pore-fluid pressures.

12.3. Role of Fluid Pressure in Earthquakes and Fault Movement.

Fluid pressure in the Earth's crust may affect the state of stress and therefore faulting, earthquakes, and the occurrence of landslides. A knowledge of the importance of fluid pressure on mechanical phenomena in the Earth's crust was first developed in the Hubbert-Rubey pore pressure hypothesis.

12.3.1. Hubbert-Rubey Pore Pressure Hypothesis.

Hubbert and Rubey (1959) suggested that a solution to Smoluchowski's dilemma could be provided by a consideration of the possibility of high fluid pressures in the Earth. During times of active tectonism and rapid sediment accumulation, such as are found today in the Gulf Coast Basin of the southeast United States, fluid pressures approaching some appreciable fraction of lithostatic are common (review Chapter 8). If such conditions prevailed at the times when overthrust movements occurred, the amount of applied stress necessary for fault movement may have been substantially lower than that required under hydrostatic conditions. Fault movement facilitated by high fluid pressures is the **Hubbert-Rubey pore pressure hypothesis.**

Consider a microscopic view of a fault surface (Figure 12.6). The interface between the two sides of the fault is not perfectly smooth but contains both macroscopic and microscopic irregularities. The rock surfaces are in contact at certain points but in other areas are separated by a void normally filled with fluids. A stress normal to the fault surface tends to hold the two surfaces together. Depending on the orientation of the fault plane and of tectonic stresses, the normal stress may be due to overburden weight or horizontal stresses arising from tectonic factors. However, fluid pressure (P, kg-m^{-1}-s^{-2}) acts in the opposite direction to push the two sides apart. As the fluid pressure increases, the stress normal to the fault plane (σ_n, kg-m^{-1}-s^{-2}) decreases, even though the two sides of the fault do not necessarily move apart. The reduction of normal stress across a fault plane by abnormally high fluid pressures may thus facilitate movement on the fault by reducing the net normal stress across the fault plane.

Recall (equation 12.5) that the frictional resistance to block movement depends not just on the coefficient of friction (μ_f, dimensionless) but also on the normal stress across the interface between the block and the ramp ($\sigma_n = \rho hg$). In the case of a porous medium, the stress across the fault plane is the effective stress (σ_e, kg-m^{-1}-s^{-2}) (review section 3.4.2.). Thus, the length (b, m) of a block that can be moved by an applied force is

$$b = \frac{h\sigma_a}{\sigma_e \mu_f} \qquad \textbf{(12.11)}$$

Recall that as the ambient fluid pressure *increases*, the effective stress *decreases* (equation 3.43)

$$\sigma_e = \sigma_n - P \qquad \textbf{(12.12)}$$

where σ_n is the total stress due to the overburden weight and P is the ambient fluid pressure. Thus,

$$b = \frac{h\sigma_a}{(\sigma_n - P)\mu_f} \qquad \textbf{(12.13)}$$

Thus, as the fluid pressure approaches lithostatic, arbitrarily large blocks of the Earth's crust may be moved by relatively small forces.

It is important to note that the role of the fluid is *not* to lubricate the fault. The coefficient of friction remains unchanged. The reduction in the frictional resistance to movement is facilitated by reduction of normal stress, not reduction of the coefficient of friction.

The principle underlying the Hubbert-Rubey pore pressure hypothesis can be demonstrated through a simple and elegant experiment involving only a sheet of glass and an empty beer can. A sheet of glass is slightly tilted at a small angle from the horizontal and wet with a detergent solution. The addition of detergent reduces the water's surface tension, allowing it to spread uniformly over the glass surface. An empty beer can is placed on the glass with an open end downward; the can's other end is sealed. If the glass sheet is tilted at increasingly greater angles, the beer can will begin to slide downhill when the angle is about 17°, corresponding to a coefficient of friction $\mu_f = 0.3$ for metal moving on wet glass. However, if the metal can is first chilled, a different result can be obtained. If a cold can is placed on the glass sheet at an angle less than 17°, nothing happens. As the cold air in the can begins to heat up and expand, the increased air pressure between the base of the can and the glass surface increases, decreasing the normal stress and allowing the can to slide down a modest slope. The increase in air pressure at the interface between the beer can and glass sheet is entirely analogous to an increase in fluid pressure along a fault plane.

At the present time, it is unknown if the Hubbert-Rubey pore pressure hypothesis accurately describes the mechanism by which overthrusting took place. Other hypotheses, perhaps equally valid, remain possible. For example, Smoluchowski (1909) suggested that overthrust movement was not analogous to the frictional sliding of a block on a ramp but rather represented a type of plastic deformation where rocks subject to high pressures over geologic time flowed. Another possibility is that the net movement of large fault blocks is the result of the repeated motion of several small blocks. The demonstration that a mechanism is feasible does not prove that it occurs in nature.

12.3.2. Earthquakes from Reservoir Impounding.

The accumulation or impounding of water in a reservoir can induce earthquakes by changing the state of stress in the crust. An increase in seismicity

Case Study 12.3.3 The Denver and Rangely Earthquakes.

In 1942, at the height of World War II, the U.S. Army purchased 17,000 acres (69 km^2) of land 10 mi (16 km) northeast of Denver, Colorado, on which to manufacture chemical weapons, such as mustard gas, white phosphorus, and napalm. The site became known as the Rocky Mountain Arsenal (RMA). After the end of the war, sections of the site were leased to private industry, which produced pesticides there. In the mid-1950s, it was discovered that liquid wastes from the RMA had contaminated groundwater and caused crop damage to the north of the RMA (see Chapter 9). Efforts to contain waste products and remediate environmental contamination began soon after.

Disposal of chemical waste was a difficult problem. The ideal solution appeared to be injection of the wastes into a deep well. In 1961, a 3671-m-deep well was drilled through the sedimentary rocks of the Denver Basin, bottoming in Precambrian crystalline rocks. Injection of waste fluids at a rate of 21 million L per month (21 million L is a cube 27.6 m on a side) began in March 1962 and continued through September 1963. No fluid was injected from October 1963 through August 1964. Fluid was then allowed to flow into the well under the influence of gravity at an average rate of about 7.5 million L per month until April 1965. Forced injection then resumed at an average rate of 17 million L per month until February 1966. Injection was stopped at that time because of concern that it was precipitating a series of earthquakes in the Denver region.

In 1962, there were two seismograph stations in the Denver area. One was operated by the Colorado School of Mines, in Golden, about 10 mi (16 km) west of central Denver. The other station was in Regis College in Denver. On April 24, 1962, about a month and a half after the initiation of deep waste injection at the RMA, these stations started to record a series of earthquakes whose epicenters were northeast of Denver. A number of the larger earthquakes, with Richter-scale magnitudes between 3 and 4, were felt over a wide area, with minor damage reported near the epicenters. The sudden appearance of this previously absent seismic activity next to a major city raised concerns, and the deep waste injection program was suspended pending the results of an investigation.

One of the first questions raised was whether the magnitude 3 to 4 earthquakes felt in the Denver area were anomalous or had historical precedents. This question was difficult to answer, because the historical record was incomplete. The Colorado School of Mines seismograph station had only been in operation since 1959. The Regis College station had records dating back to 1909. However, it was located in an area of high background noise and, for most of its history, had been operated at low magnification. Thus, if a series of magnitude 3 to 4 earthquakes had occurred in the past, they might well have escaped seismographic detection.

A third seismographic station had operated in Boulder about 30 mi (48 km) northwest of central Denver, between 1954 and 1959. A search of the records from this station revealed about 13 events that might have been small-magnitude earthquakes in the Denver area. All had occurred, however, during normal daylight working hours. This suggested that they most probably represented blasting during construction or the disposal of explosives at the RMA. There appeared, therefore, to be no historical evidence for significant seismic activity in the Denver area before deep waste injection at the RMA.

In January and February of 1966, the U.S. Geological Survey established a dense network of seismic stations in the vicinity of the RMA disposal well for the purpose of obtaining the accurate locations of earthquake hypocenters. The **hypocenter** is the actual three-dimensional point in the Earth's crust at which an earthquake originates. An **epicenter** is the point on the ground surface directly above the hypocenter. During two months of operation for about 6 hours a day, the USGS seismographic network located 62 earthquakes clustered in a distribution that was elliptical with the disposal well near its center (Figure 12.7). Earlier, in November 1965, consulting geologist David Evans had established a

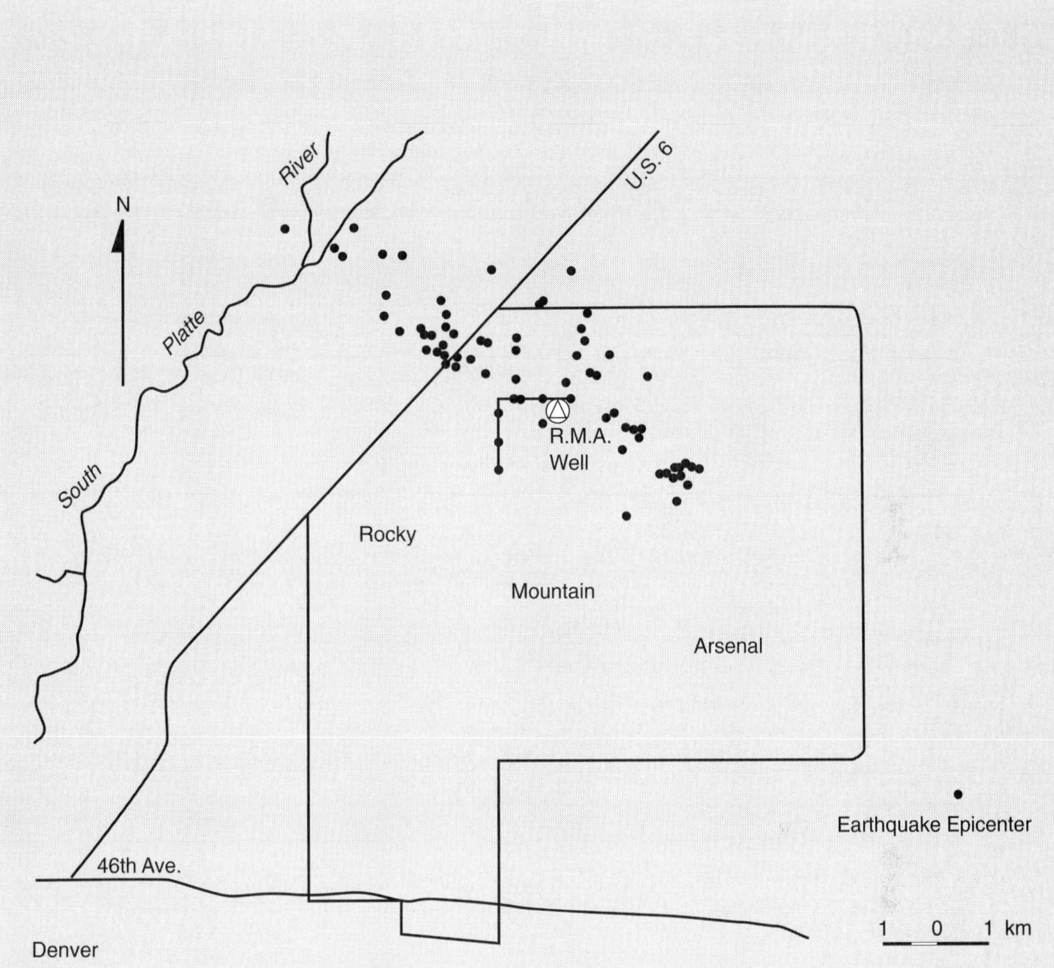

Figure 12.7 Earthquake epicenters near the Rocky Mountain Arsenal (RMA) injection well, northeast of Denver, Colorado.
(From Healy et al., 1968, p. 1303.)

temporal correlation between the volume of wastes injected at the RMA and the frequency of earthquakes (Figure 12.8). It thus became apparent that the Denver earthquakes were correlated with the RMA deep injection well both spatially and temporally. Healy et al. (1968) estimated that the probability of such correlations occurring by chance was one in 2.5 million. Not only was it improbable that the Denver earthquakes would be associated with the location and timing of waste injection by coincidence, but Healy et al. (1968) also pointed out that the region in which the earthquakes occurred showed very little evidence of folding or faulting, as might have been expected in a seismically active area. It therefore appeared nearly certain that the fluid injection had caused the earthquakes.

Deep waste injection at the RMA well stopped in February 1966, but the earthquakes did not stop.

CASE STUDY 12.3.3 *(continued)*

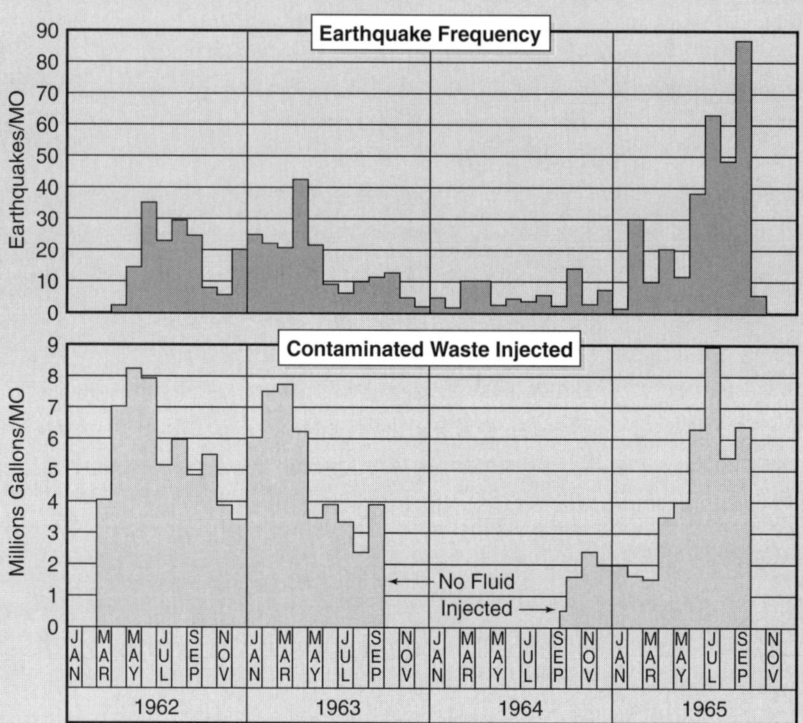

Figure 12.8 The occurrence of earthquakes in the Denver region temporally correlated with the amount of waste injected into a deep well at the Rocky Mountain Arsenal.
(From Healy et al., 1968, p. 1302.)

The three largest earthquakes of the entire sequence occurred after injection had ceased. Earthquakes with magnitudes of 5.0, 5.3–5.5, and 5.1 occurred in April, August, and November 1966. As the energy released in earthquakes increases approximately by a factor of 30 for each 1-point increase in magnitude, the three magnitude 5 earthquakes accounted for nearly all of the energy released in the entire sequence of Denver earthquakes.

If fluid injection caused the Denver earthquakes, then why had the three largest earthquakes of the sequence occurred after the injection had ceased? Healy et al. (1968) concluded that the Denver earthquakes were not so much caused by the deep waste injection as they were triggered. The general orientation of fault-plane movement suggested that the earthquakes occurred along pre-existing faults in the basement, with the direction of movement being consistent with regional tectonic stress fields. Assuming reasonable values for the coefficient of friction and strength of the basement rocks near the RMA well, Healy et al. (1968) concluded that basement rocks were stressed to near their breaking point before fluid injection. Fluid injection allowed fault movement to take place by reducing the normal stress across fault planes, according to the Hubbert-Rubey pore pressure hypothesis.

The reason that the earthquakes continued after the injection ceased was that the pressure front created by injection continued to migrate outward through a diffusive process. As longer fault segments came under the influence of high fluid pressures, movement along greater fault lengths became possible, resulting in larger earthquakes. Healy et al. (1968) suggested that it was advisable to consider drilling new wells to pump fluids out of the Denver basement and reduce the pore pressure in the hopes of preventing possible large earthquakes in the future. This was never done, however, and the Denver earthquakes eventually stopped as the high pore pressures induced by injection diffused.

The discovery that fluid injection had triggered earthquakes led to speculation that earthquakes might be controllable (Raleigh et al., 1976). If high fluid pressures from fluid injection triggered earthquakes, then perhaps the creation of low fluid pressures by withdrawals would prevent them. To study the feasibility of this scheme, in 1967 a team of scientists from the U.S. Geological Survey installed a network of seismographs in the Rangely oil field, near Vernal, Utah. Water had been injected into oil reservoirs there since 1957 to enhance oil recovery.

In the Rangely experiment, fluid injection was controlled while seismic activity was monitored. It was found that when in situ fluid pressures exceeded the values at which the Hubbert-Rubey pore pressure hypothesis predicted fault movement, earthquakes occurred. Similarly, when pore pressure was reduced by reversing the direction of pumping, the earthquakes stopped. The Rangely experiments provided strong confirmation for the Hubbert-Rubey pore pressure hypothesis.

Raleigh et al. (1976) speculated that the timing and size of major earthquakes could be controlled by drilling wells near major faults and controlling fluid pressures by injecting or withdrawing fluid from the boreholes. The goal of such a scheme would not be to prevent major earthquakes by lowering fluid pressures. That would only delay the onset of a major earthquake and ensure that an unusually large earthquake would eventually take place when the tectonic stresses eventually built up to the point that slip became inevitable. Rather, the idea would be to accommodate fault movements gradually, allowing energy to be released in a uniform manner through a series of small earthquakes rather than a destructive large earthquake.

For the San Andreas fault, Raleigh et al. (1976) described a scheme that might work as follows. The San Andreas has an average fault slip of 2 to 3 cm per year. An earthquake of magnitude 4.5 involves about 2 cm of slip along a fault length of 5 km. If fault movement could be accommodated through a yearly series of magnitude 4.5 earthquakes along the San Andreas, then perhaps the great earthquakes that occur there every 100 to 200 years could be prevented. Wells would be drilled along the fault with a 5-km spacing (Figure 12.9). Fluid would be pumped out of wells on either side of a 5-km-long section, preventing fault slip. Along one 5-km section of the fault, however, fluid would be injected, raising pore pressures and precipitating a magnitude 4.5 earthquake. After tectonic stress had been relieved along that 5-km-long section of the fault, the procedure would be repeated for the next section until movement occurred along the entire length of the fault. This procedure has never been attempted near the San Andreas or any other major fault. Although theoretically feasible, the costs involved are likely to be quite high, perhaps higher than the eventual cost of a great earthquake. There is also the possibility that fluid injection could trigger a great earthquake, despite attempts to lock up surrounding fault segments by lowering fluid pressures. Our understanding of the relationship between fluids and earthquakes is still incomplete, and the political consequences of accidentally triggering a great earthquake would be enormous.

CASE STUDY 12.3.3 *(continued)*

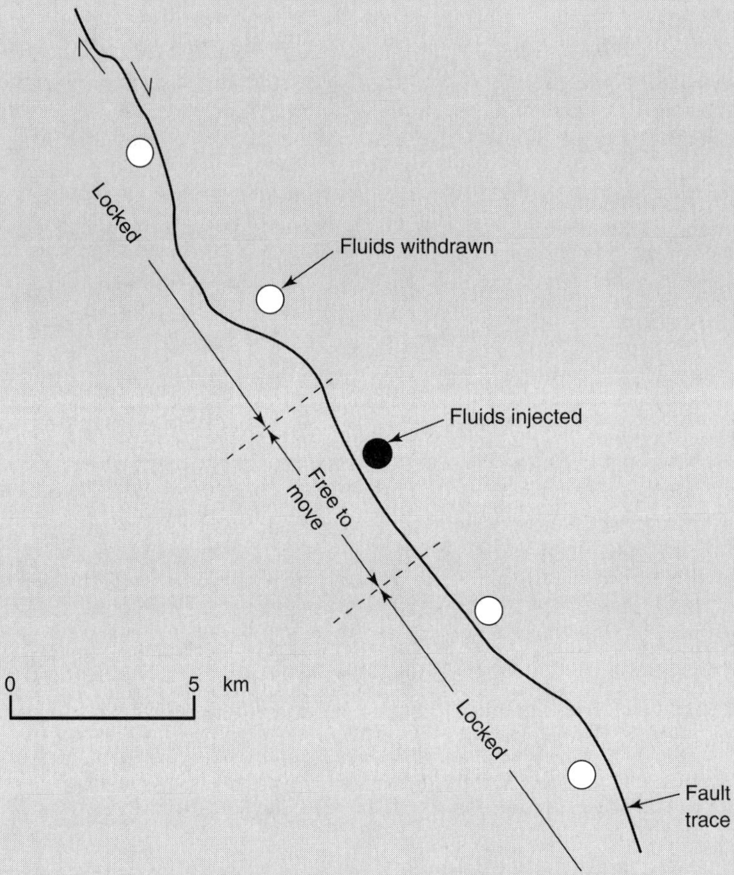

Figure 12.9 Scheme proposed by Raleigh et al. (1976) for controlling the release of earthquake energy on the San Andreas Fault in California. Earthquakes on sections of the fault about 5 km long would be induced by injecting fluids. Simultaneously, surrounding fault segments would be locked up by withdrawing fluids. After relieving stress on one 5-km-long section of the fault, the procedure would be repeated until the entire length of the fault was treated. In this way, great earthquakes could be prevented by gradually releasing tectonic strain through hundreds of smaller earthquakes.

CASE STUDY 12.3.4 The Vaiont Dam Disaster.

The role of pore-fluid pressure in influencing mechanical movement was dramatically illustrated in Italy in 1963, when a landslide likely precipitated by rising pore-fluid pressures killed 2600 people. On the evening of October 9, 1963, 240 million m^3 of bedrock and soil slid into the Vaiont Reservoir. Within 60 seconds, 1.8 km of the reservoir was filled with material that obtained heights of up to 150 m above previous water levels. The displacement of air and water created an enormously destructive flood wave and blast of compressed air. A 70-m-high flood wave swept over the Vaiont Dam and traveled downstream, destroying everything in its path. No one who witnessed the landslide survived the destructive flood wave. The nearest survivor witness lived in a house 260 m above the reservoir and on the side opposite of that on which the landslide occurred. He was awoken at 10:40 P.M. when the roof of his house lifted off and he was inundated by a shower of water and rocks. Remarkably, the Vaiont Dam itself was left intact. However, water and compressed air penetrated all the galleries and interior areas of the dam and abutments. The force of the water and air was sufficient to twist and shear steel I-beams in the dam's powerhouse. Steel doors were torn from their hinges and bent (Kiersch, 1965, 1976).

Kiersch (1965, 1976) concluded that several adverse features of the reservoir area geology contributed to the disastrous landslide. Notably, the sedimentary strata that made up the walls of the reservoir had a dip parallel to the slope of the canyon (Figure 12.10). Some of the rock units involved were inherently weak and possessed little resistance to shearing. Before the landslide, there had been heavy rains for two weeks, and the reservoir water level had subsequently risen by 20 m. The rising water level infiltrated the banks of the reservoir, causing the water table to rise. The rising water table intercepted a plane

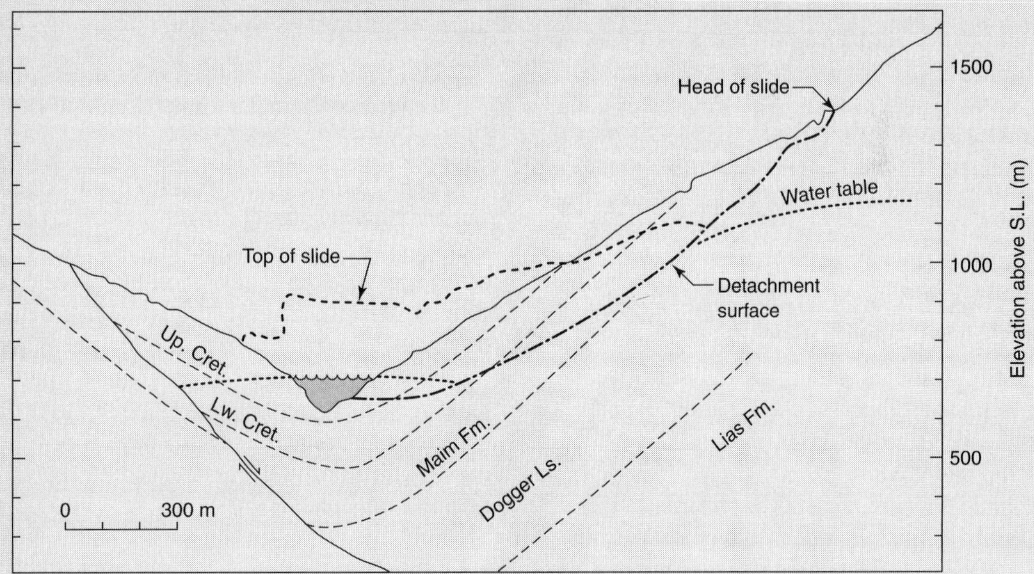

Figure 12.10 Geologic cross-section through the Vaiont Reservoir. Note coincidence of water table with much of the detachment surface.
(After Kiersch, 1965.)

CASE STUDY 12.3.4 *(continued)*

of weakness along the layered sedimentary rocks of the canyon walls, decreasing the effective stress across this plane. The frictional resistance to downward gravity sliding was no longer sufficient to retain the rocks of the canyon walls, and they precipitously failed.

Remarkably, the Vaiont Dam disaster was preceded by several warning signals. The signals were not totally ignored, but the magnitude of the threat was drastically underestimated. Three years before the disaster, in 1960, a slide of 700,000 m^3 occurred on a bank of the reservoir near the dam. Because of this slide, a network of geodetic stations was installed to monitor earth movements. During the spring and summer preceding the 1963 landslide, the slide area was observed to be creeping at a rate of 1 cm per week. Beginning September 18, the rate of creep at some geodetic stations increased to 1 cm per day. On the day before the disastrous landslide, engineers finally realized that a high rate of creep was occurring at all observation stations and that an enormous mass of rock was unstable. They responded by starting to lower the level of water in the reservoir but made no attempt to evacuate people from the area. Engineers expected that a slide would lead to a flood wave 20 m high, but the actual flood wave reached 70 m and accordingly was much more destructive than anticipated (Kiersch, 1965, 1976).

following reservoir impoundment was first noticed in conjunction with the filling of Lake Mead in the late 1930s. In the late 1960s, earthquakes larger than magnitude 5.5 had occurred at large reservoirs in China, Rhodesia, Greece, and India. The largest known earthquake triggered by reservoir impounding was a magnitude 6.5 earthquake at the Koyna Reservoir in India. The Koyna Reservoir earthquake occurred in 1967 and caused 200 deaths, 1500 injuries, and major damage to buildings in a nearby town. As of 1986, there were at least 50 recognized cases where reservoir impounding was believed to have led to changes in earthquake activity (Simpson, 1986). Impounding water in a reservoir has three main effects:

1. The solid earth undergoes an instantaneous elastic stress from the addition of the weight of the water impounded in the reservoir.
2. Fluid pressure in the rocks underneath the reservoir increases due to compaction from the added weight of the reservoir water.
3. Fluid pressure in the rocks underneath the reservoir also increases due to fluid migration and diffusion of pressure (head) transients related to fluid migration from the reservoir itself and fluids moving in response to pore collapse caused by loading.

In areas where the groundwater table had been deep before the impounding of the reservoir, a fourth effect must be considered:

4. The water table rises by flow of water from the reservoir into the formerly unsaturated zone.

In general, fluid pressure changes due to loading and diffusion operate on two different time scales. Earthquakes that immediately follow rapid changes in water level have most likely been triggered by the instantaneous propagation of a stress due to loading. In other cases, earthquakes only occur many years after reservoirs have filled. In the latter situations, earthquakes may be triggered by fluid pressure increases that have arisen from the slow diffusion of a pressure increase related to reservoir impounding.

12.4. DILATANCY.

Dilatancy is a volume increase due to deformation (Mead, 1925). A material subject to dilation cannot change its shape without changing its volume.

If volume changes are resisted, then changes of shape are prevented. Mead (1925) suggested that this principle can be illustrated through a simple experiment first described by the English engineer Osborne Reynolds (1842–1912). A balloon is filled with sand; the sand contains just enough water to maintain cohesion between the sand particles. After the sand has been shaken into a condition of dense packing, it resists deformation "to an astonishing degree" (Mead, 1925, p. 687). For the sand to deform, it must experience a volume increase (dilatancy). However, the formation of any void within the sand-packed balloon would create a vacuum, which would be resisted by the considerable atmospheric pressure on the surface of the sand-packed balloon. The situation is similar in the Earth's crust, where the ambient pressure regime is not atmospheric but usually hydrostatic. A corollary is that a contained granular material, when forced, will not fail plastically but by shearing along a discrete plane, just as a solid does. Plastic deformation of closely-packed grains is not favored because the entire mass is deformed and a larger volume increase occurs than when failure occurs along a single shear plane.

The principle of dilatancy implies that for rocks in the Earth's crust to deform, there must be a movement of fluids toward the voids created through deformation and the accompanying volumetric increase. Mead (1925) again suggests a simple experiment to demonstrate this principle. A thin-walled rubber cylinder such as a bicycle tube is filled with sand. The sand is filled with water, and the ends are capped with large corks. A glass tube is inserted into a hole in one of the end corks and bent upward at a right angle so that the water level in the tube can be observed. Observation is facilitated by coloring the water. It will be found that the sand-filled cylinder cannot be bent, squeezed, or deformed in any manner without lowering the level of water in the tube. For deformation to be accommodated, there must be some movement of water to fill the void created. If fluid cannot move into the void fast enough, deformation will be resisted. This phenomenon is termed **dilatancy hardening.** Dilatancy hardening can also be understood from an effective stress viewpoint. Recall that effective stress—the stress borne by the solid matrix of a porous medium (equation 3.41)—is equal to the total stress minus the fluid pressure. If fluids are not readily available to move into the void created by dilatancy, a tremendous drop occurs in the ambient fluid pressure. This drop in fluid pressure results in an increase in the effective stress, and the material hardens.

In 1969, seismologists in the former Soviet Union found that the ratio (V_p/V_s) of the velocity of the compressional seismic wave (V_p) to the velocity of the shear wave (V_s) decreased before a series of earthquakes. Similar results were reported for other earthquakes in different parts of the world. In each case, the ratio (V_p/V_s) returned to normal shortly before an earthquake occurred. One explanation for this observation is provided by the **dilatancy-diffusion model** (Nur, 1972; Scholz et al., 1973). In the dilatancy-diffusion model, dilatancy is produced by the formation and propagation of cracks near earthquake faults due to increases in stress. Laboratory experiments have shown that dilatancy can result in the lowering of the seismic velocity ratio (V_p/V_s) observed before earthquakes. Dilatancy causes the rock volume surrounding the fault to become undersaturated. Incomplete fluid saturation strongly reduces V_p but has little effect on V_s. The entire sequence of events culminating in an earthquake can be divided into three stages:

1. Stress and accumulated tectonic strain become large enough to produce dilatancy.
2. Dilatancy increases faster than pore fluid can diffuse into the newly created fractures. Accordingly, fluid pressure decreases, effective stress increases, and dilatancy hardening occurs. The ratio of seismic velocities (V_p/V_s) decreases as V_p decreases. Any earthquake that would otherwise occur is inhibited by dilatancy hardening.
3. The flow of fluid into the underpressured, dilatant volume near the earthquake fault exceeds the dilatancy increase. Fluid pressure rises, effective stress decreases, and both V_p

and the ratio of V_p to V_s return to normal. Because tectonic stress has continued to accumulate during the dilatant period, an earthquake is triggered by the rising fluid pressure.

The net effect of dilatancy is to delay an earthquake by lowering the ambient fluid pressure in the fault zone and then to trigger an earthquake when the fluid pressure recovers.

12.5. How Faulting Keeps the Crust Strong.

In the interior of the tectonic plates, the continental crust is everywhere stressed to the point of failure. Townend and Zoback (2000) suggest that there are three lines of evidence for this: (1) seismicity induced by reservoir impoundment, (2) earthquakes triggered by other earthquakes, and (3) in situ stress measurements in deep boreholes. The state of stress in the continental crust suggests that fault movement is resisted by friction, consistent with the results of laboratory measurements on rocks that have found friction coefficients in the neighborhood of 0.6 to 1.0. Under conditions of hydrostatic fluid pressure, the continental crust is thus relatively strong. If fluid pressures were higher than hydrostatic, the Hubbert-Rubey pore-pressure hypothesis would predict failure at lower stress differentials and lower apparent coefficients of friction.

The hydrostatic fluid pressures that result in a strong crust are maintained by average permeabilities in the upper 10 km of the continental crust that are in the range of 10^{-16} m^2 to 10^{-17} m^2. These permeabilities are typically three to four orders of magnitude higher than those measured on intact core samples from the crystalline crust (Figure 12.11). Thus, the continental crust develops virtually all of its permeability through fractures and faults. These faults and fractures maintain high permeability in the crust and keep it strong by allowing the dissipation of high fluid pressures.

The characteristic time t for diffusion of a thermal transient through the continental crust is (equation 5.86)

$$t = \frac{y^2}{D} \quad \textbf{(12.14)}$$

where y (m) is the characteristic length (thickness of the brittle crust) and D (m^2-s^{-1}) is the hydraulic diffusivity. Substituting in definitions of hydraulic diffusivity (equation 5.35), hydraulic conductivity (equation 2.56), and specific storage (equation 3.70),

$$t = \frac{y^2(\alpha + \phi B)\mu}{k} \quad \textbf{(12.15)}$$

where α (Pa^{-1} = m-s^2-kg^{-1}) is porous-medium compressibility, ϕ (dimensionless) is porosity, B (Pa^{-1} = m-s^2-kg^{-1}) is water compressibility, μ (kg-m^{-1}-s^{-1}) is water viscosity, and k (m^2) is permeability. Townend and Zoback (2000) suggest average values for the continental crust of $\alpha = 2 \times 10^{-11}$ Pa^{-1}, $\phi = 0.02$, $B = 5 \times 10^{-10}$ Pa^{-1}, and $\mu = 1.9 \times 10^{-4}$ kg-m^{-1}-s^{-1}. For these average values, equation 12.15 reduces to

$$t = \frac{y^2}{k} \times 5.7 \times 10^{-15}\ \text{s} \quad \textbf{(12.16)}$$

or,

$$t = \frac{y^2}{k} \times 1.8 \times 10^{-22}\ \text{years} \quad \textbf{(12.17)}$$

If the numerator (y) is expressed in kilometers squared while the denominator is expressed in meters squared, the numerator must be multiplied by 10^6,

$$t = \frac{y^2}{k} \times 1.8 \times 10^{-16}\ \text{years} \quad \textbf{(12.18)}$$

Taking the log of both sides of equation 12.18,

$$\log_{10}(t) = 2\log_{10}(y) - \log_{10}(k) - 16 \quad \textbf{(12.19)}$$

where we have used the properties of logarithms ($\log a/b = \log(a) - \log(b)$, $\log(ab) = \log(a) + \log(b)$, and $\log(a^b) = b\log(a)$). A plot of equation 12.19 is linear for permeability (k) in terms of log(distance) versus log(time) (Figure 12.12). For average crustal permeabilities in the range of 10^{-16} to 10^{-17} m^2, characteristic times for the diffusion of fluid transients over characteristic length scales for the upper continental crust (1 to 10 km) are in the

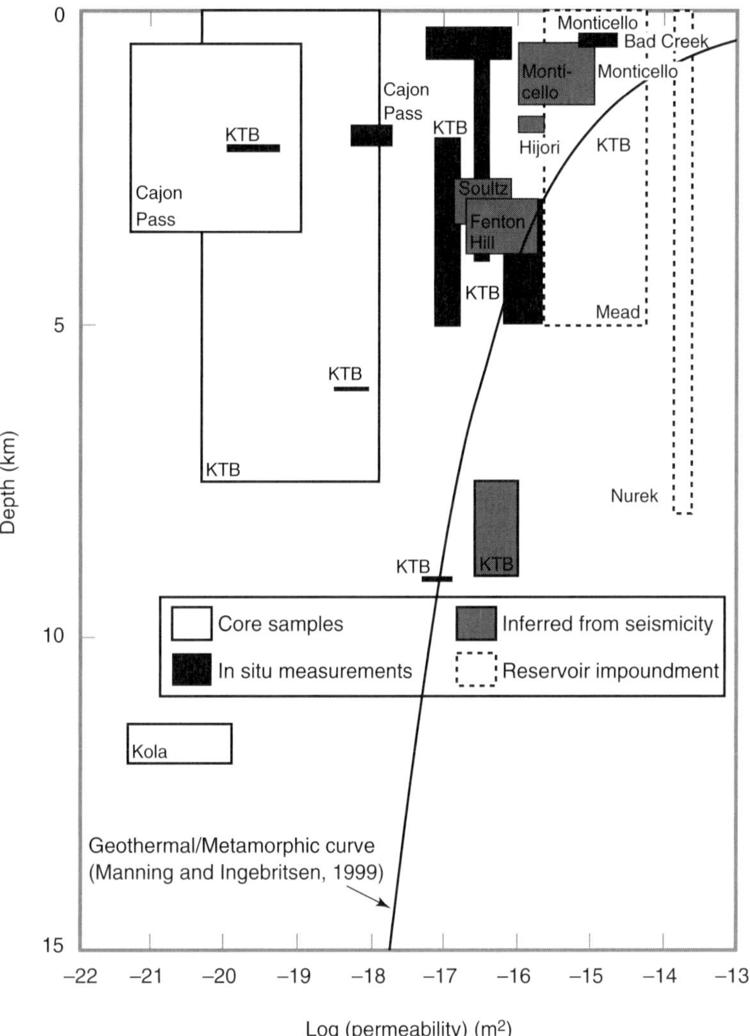

Figure 12.11 Crustal permeability data.
(From Townend and Zoback, 2000, p. 400.)

range of 10 to 1000 years. Thus, fluid pressures in the crust equilibrate over relatively short time scales, and hydrostatic fluid pressures are maintained.

12.6. The Mystery of Deep Earthquakes.

One of the yet unexplained aspects of earthquakes is the occurrence of deep earthquakes in subduction zones. Earthquakes have been observed to nucleate at depths of up to 650 km. At these great depths, the high temperatures and pressures that are present should make mechanical failure impossible. The mantle should deform by plastic flow, yet earthquakes occur.

A number of hypotheses have been proposed to explain deep earthquakes. Most of the proposed explanations involve phase changes of some type. A phase change is an alteration in the structure of a

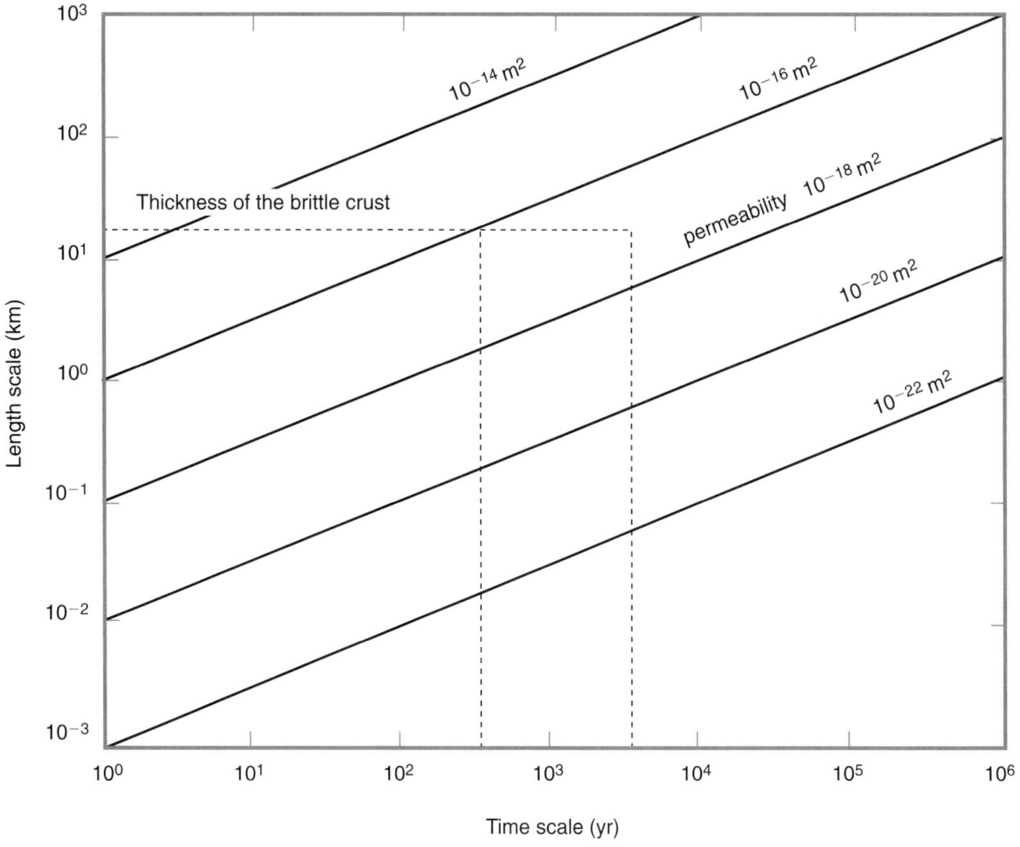

Figure 12.12 Length and time over which a hydraulic disturbance in the continental crust diffuses as a function of average permeability.
(From Townend and Zoback, 2000, p. 401.)

material with no change in composition. For example, at a depth of about 400 km, the mineral olivine transforms to a denser structure similar to the structure of the mineral spinel. Meade and Jeanloz (1991) studied the reaction of common mantle minerals to high pressures and temperatures using a diamond-anvil cell apparatus. The **diamond-anvil cell** is a device in which two gem-quality diamonds are used to apply extremely high pressures to samples. Diamond is used not only because it is extremely strong but also because its transparency to visible light and other radiation bands allows simultaneous observation during pressure application. The transparency of diamond also allows samples under pressure to be heated with a laser beam and the temperature estimated by measuring the wavelength of the thermal radiation emitted by the heated sample. The diamond-anvil cell was invented in 1958; by 1986, temperatures of thousands of degrees Celsius and pressures greater than those at the Earth's center could be achieved in the diamond-anvil cell (Jeanloz and Lay, 1993). In the earth sciences, use of the diamond-anvil cell has revolutionized our knowledge of the Earth's mantle and core.

The minerals olivine and pyroxene in their various phases are the most common minerals in the upper mantle and constitute over half the mass of a subducting slab. Meade and Jeanloz (1991) tested olivine and pyroxene in the diamond-anvil cell at high pressures, both with and without heating, but failed to observe any acoustic emissions or other seismic behaviors. But a different result was obtained when hydrous minerals were tested. In young oceanic lithosphere, the hydrothermal circulation of seawater through the oceanic crust converts the anhydrous minerals olivine and pyroxene to the mineral serpentine. Serpentine is the major hydrous mineral found in the upper part of subducting slabs; it is about 13% water by mass. Serpentine is also commonly found in libraries, museums, and other buildings as the primary constituent of the rock serpentinite, a dark green rock that is commonly cut, polished, and used as an exterior or interior facing stone. Meade and Jeanloz (1991) observed acoustic emissions from serpentinite as it dehydrated at pressures corresponding to depths ranging between 62 and 773 km. This observation suggests that deep earthquakes may be triggered by the release of water from subducting slabs. A corollary is that the upper mantle of the Earth may contain substantial amounts of water. Smyth (1994) suggested that the upper mantle may contain an amount of water equal to "several oceans." Water may thus play a pervasive global role in mantle chemistry and tectonics.

REVIEW QUESTIONS

1. Define the following terms in the context of hydrogeology:
   a hydroseism
   b. resonance
   c. seismic seiche
   d. liquefaction
   e. quicksand
   f. overthrust fault
   g. Smoluchowski's dilemma
   h. Hubbert-Rubey pore pressure hypothesis
   i. hypocenter
   j. epicenter
   k. dilatancy
   l. dilatancy hardening
   m. dilatancy-diffusion model
   n. diamond-anvil cell
2. List some of the hydrological effects of earthquakes.
3. How far from an earthquake can water-level changes in a well be observed?
4. Why do hydroseisms occur in some wells but not in others?
5. What two hypotheses have been proposed to explain changes in stream and spring flow that follow an earthquake? What observation(s) could discriminate between the two hypotheses?
6. Where do water-sediment ejections occur after an earthquake?
7. What is the cause of liquefaction?
8. Assuming a compressive strength of 10^8 Pa, a coefficient of friction $\mu_f = 0.6$, and a rock density of 2650 kg-m^{-3}, the maximum length of a block that can be pushed along a horizontal plane is about 6 km. How long of a block can be pushed up an incline of 10°?
9. What aspect of hydrogeology contributed to the Vaiont Dam disaster?
10. In what three ways can reservoir impounding cause fluid pressures (at a point of constant elevation) under the reservoir to increase?
11. For a material to deform, it must experience a ______________.
12. What is the cause of dilatancy hardening?
13. What are the three stages of the dilatancy-diffusion earthquake model?
14. Explain how faulting keeps the continental crust strong.
15. What is a possible cause of deep-focus earthquakes?
16. Why did the largest Denver earthquakes occur after fluid injection at the RMA had stopped?
17. What is the cause of quicksand?

Suggested Reading

Healy, J. H., Rubey, W. W., Griggs, D. T., and Raleigh, C. B. 1968. The Denver earthquakes. *Science* 161: 1301–10.

Hubbert, M. K., and Rubey, W. W. 1959. Role of fluid pressure in mechanics of overthrust faulting, I. Mechanism of fluid-filled porous solids and its application to overthrust faulting. *Geological Society of America Bulletin.* 70 p: 115–66.

Kiersch, G. A. 1965. Vaiont Reservoir disaster: *Geotimes* 9 (9): 9–12.

Matthes, G. H. 1953 Quicksand. *Scientific American* 188 (6): 97–102.

Mead, W. J. 1925. The geologic role of dilatancy. *Journal of Geology* 33: 685–98.

Notation Used in Chapter Twelve

Symbol	Quantity Represented	Physical Units
α	compressibility of a porous medium	$Pa^{-1} = m\text{-}s^2\text{-}kg^{-1}$
B	fluid compressibility	$Pa^{-1} = m\text{-}s^2\text{-}kg^{-1}$
b	block length	m
D	hydraulic diffusivity	$m^2\text{-}s^{-1}$
F	force	Newton (N) = $kg\text{-}m\text{-}s^{-2}$
ϕ	porosity	dimensionless
g	acceleration due to gravity	$m\text{-}s^{-2}$
h	block height	m
k	permeability	Darcy = 10^{-12} m^2
m	mass	kg
μ	water viscosity	Pa-s = $kg\text{-}m^{-1}\text{-}s^{-1}$
μ_f	coefficient of friction	dimensionless
P	fluid pressure	Pascal (Pa) = $kg\text{-}m^{-1}\text{-}s^{-2}$
Θ	angle of block inclination	dimensionless
ρ_b	density of block	$kg\text{-}m^{-3}$
σ_a	stress applied to end of block	Pascal (Pa) = $kg\text{-}m^{-1}\text{-}s^{-2}$
σ_e	effective stress	Pascal (Pa) = $kg\text{-}m^{-1}\text{-}s^{-2}$
σ_n	normal stress	Pascal (Pa) = $kg\text{-}m^{-1}\text{-}s^{-2}$
t	time	s
V_p	P-wave seismic velocity	$m\text{-}s^{-1}$
V_s	S-wave seismic velocity	$m\text{-}s^{-1}$
w	block width	m
y	length	m

CHAPTER 13

FLUIDS IN THE OCEANIC CRUST

13.1. THE OCEANIC FLUID CYCLE.

The oceanic crust, which covers 71% of the Earth's surface area, hosts an enormous volume and active flux of fluids. The **oceanic fluid cycle** is the addition of fluids to the young rocks of the oceanic crust near the sites of their creation and the return of the fluids to the surface at and near subduction zones where old oceanic crust is destroyed. The energy source that drives this cycle is the escape of the Earth's internal heat. At spreading ridges, where the plate tectonic cycle begins with the creation of the oceanic lithosphere, seawater circulates through the newly formed oceanic crust, simultaneously altering both it and the rocks of the oceanic crust. Water and volatile elements such as carbon chemically combine with minerals in the oceanic crust, and the crust becomes hydrated. More water is added to the top of the oceanic crust by the accretion of a layer of porous sediments and sedimentary rocks. At subduction zones, the mechanically and chemically bound water in the oceanic crust is released and may flow back up through the rocks of the continental margins, altering their composition and creating volcanoes at the surface. The oceanic fluid cycle thus signficantly affects the composition of both seawater and oceanic and continental rocks.

13.2. CREATION AND STRUCTURE OF THE OCEANIC CRUST.

Before the advent of plate tectonic theory and our understanding of the process of seafloor spreading, the ocean basins were thought to be the oldest and most stable parts of the Earth's crust. We now know, however, that the ocean basins are, in a geologic sense, ephemeral, subject to constant creation and destruction. The oceanic lithosphere, including the crust, is created at spreading ridges in ocean basins. As the lithosphere spreads laterally away from a central ridge, new material rises up from the mantle to take its place and is accreted onto the trailing edge of a spreading oceanic plate (Figure 13.1). As the initially hot oceanic lithosphere cools, its density increases, and it isostatically subsides. Subduction occurs when old oceanic lithosphere becomes gravitationally unstable and sinks back into the mantle or when younger oceanic lithosphere is overidden by more buoyant continental lithosphere. The entire process, from creation to destruction through subduction, typically takes 150 to

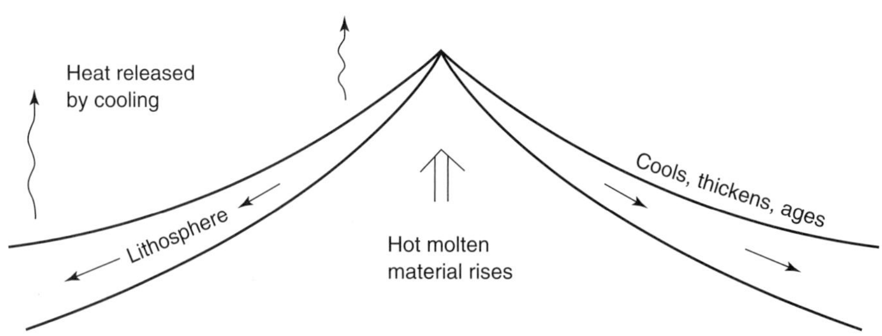

Figure 13.1 Conceptual illustration of seafloor spreading.

200 million years. The oldest known parts of the oceanic crust are on the order of 200 million years old. This is only 5% of the age of the oldest known continental rocks.

The oceanic and continental crusts form through different processes and therefore have different compositions. Continental crust is formed by accretion at continental margins. It tends to be heterogeneous, thermally stable, and old. The oceanic crust, formed by the more-or-less uniform process of seafloor spreading, tends to be relatively homogeneous, thermally unstable, and young. The oceanic crust is composed of three layers (Table 13.1). Unlike the complex and heterogeneous continental crust, these layers are present everywhere, although their thickness may vary. The top layer (layer 1) is composed of sediments or sedimentary rocks. On average, it is about 500 m thick but can vary from zero thickness on young oceanic crust in mid-ocean to as deep as 20 km near the mouth of a major river such as the Mississippi. The second layer (layer 2) is composed of volcanic rocks. Rocks in the upper part of layer 2 (layer 2a) tend to be extrusive basalts that develop high permeability, while rocks in the lower section (layer 2b) are intrusive and tend to have low permeability. The total thickness of the volcanic layer (layer 2) is about 1500 m. The bottom layer, layer 3, is composed of gabbro in the form of sheeted dikes; it averages about 5 km in thickness. The total thickness of the oceanic crust is about 7 km. It is thus much thinner than the continental crust, which averages about 40 km in thickness.

TABLE 13.1 Structure of the Oceanic Crust

Layer	Thickness (m)	Lithology	Permeability (m^2)	Porosity
1	0–20,000 avg. 500	sediments, sed. Rock	$10^{-15} - 10^{-17}$	10–60%
2a	500	extrusive basalts	$10^{-12} - 10^{-15}$	7–10%
2b	1000	intrusive basaltic dikes	$10^{-16} - 10^{-18}$	1–3%
3	5000	gabbro	low	low

The uppermost oceanic crust is distinctly layered in its most important hydraulic property, permeability (Figure 13.2). Although the permeability of all layers may vary widely between sites, the uppermost 500 m of layer 2, consisting of extrusive basalts, tends to have a higher permeability than either the underlying intrusive basaltic layer or the overlying sedimentary layer. Davis et al. (1996) reported a permeability of 10^{-15} m^2 for sediments at the top of layer 1, decreasing to 10^{-17} m^2 at a depth of 200 m. Most measurements and estimates of permeability in the top of layer 2 are in the range of 10^{-13} m^2 to 10^{-14} m^2. At depths of about 500 to 600 m below the top of the basement, permeability drops three to four orders of magni-

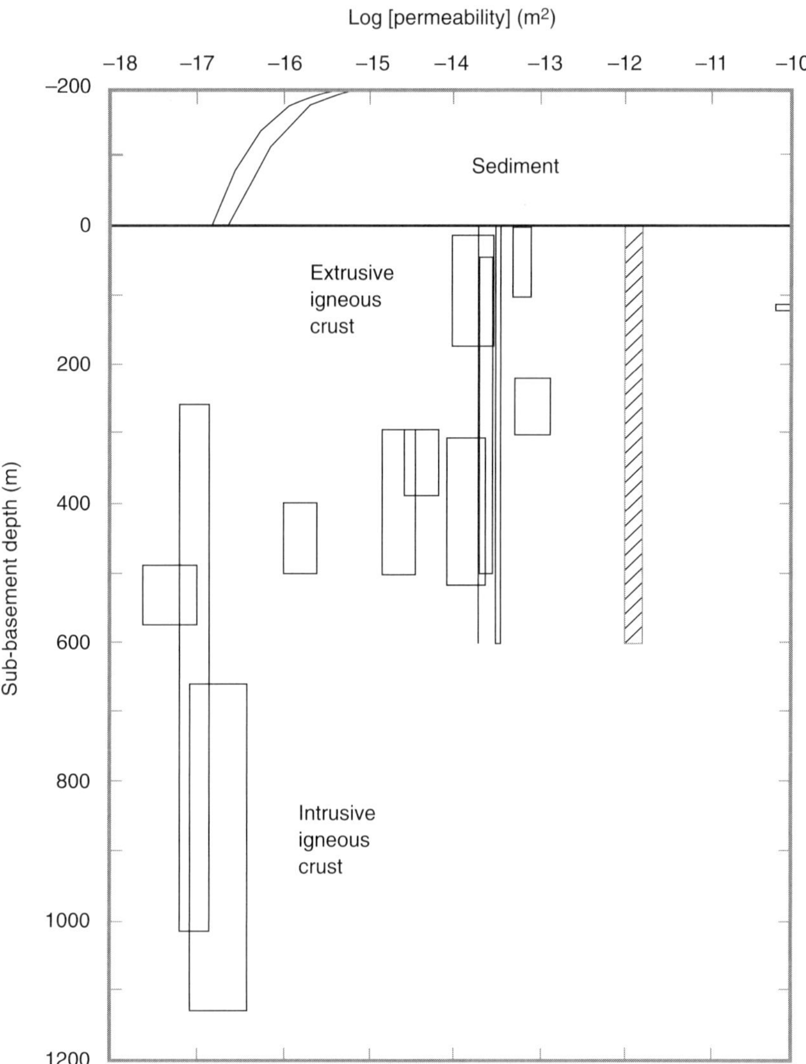

Figure 13.2 Permeability measurements and estimates for the upper oceanic crust.
(From Davis et al., 1996, p. 2938.)

tude. Permeabilities in the lower part of layer 2 are typically in the neighborhood of 10^{-17} m^2 (Davis et al., 1996; Fisher, 1998). The distinguishing hydraulic characteristic of the oceanic crust is thus a 500-m-thick layer of highly permeable fractured basalts layered between a lower-permeability layer of sediments and a relatively impermeable lower basement of intrusive igneous rocks.

Although the pattern of permeability distribution described in the preceding paragraph is generally accepted, the degree to which permeability is enhanced in the top of layer 2 is not well constrained. The extrusive basalts near the top of the oceanic basement develop high permeability during their formation. Thermal-contraction cooling and tectonic fracturing both contribute to the development of fracture permeability. Most borehole-scale measurements of permeability in this layer yield numbers that are in the range of 10^{-13} to 10^{-14} m^2. However, Davis et al. (1997) argued that thermal modeling of hydrothermal circulation near the Juan de Fuca Ridge offshore of Vancouver Island, Canada, constrained the permeability of this layer to be at least 10^{-11} m^2 and possibly as great as 10^{-9} m^2. At the present time, it is not clear if these results are due to an increase of permeability with scale or if this particular site has an anomalously high permeability.

13.3. Thermal Evolution of the Oceanic Lithosphere.

Primary clues to the transient nature of the ocean basins were first provided by studies of the flow of heat from the oceanic crust. One of the first measurements of oceanic heat flow was made by Bullard (1954), who found a heat flow of 41 mW-m^{-2} in the North Atlantic Ocean. This result compared to earlier work by Revelle and Maxwell (1952), who had found a slightly higher heat flow from the crust of the Pacific Ocean. Although these values seem very low compared to our current estimate of the average oceanic heat flow (101 mW-m^{-2}), at the time they were considered to be anomalously high because the basalt, gabbro, and other mafic rocks that make up the bulk of the oceanic crust were known to be highly depleted in radioactive heat-generating elements compared to the rocks that make up the continental crust. It was therefore quite surprising to find that the mean oceanic heat flow was virtually the same as the continental.

Although the ultimate source of high oceanic heat flow remains unknown, with the advent of seafloor spreading and plate tectonics, it was discovered that the immediate source of high oceanic heat flow was the cooling of the oceanic lithosphere. The oceanic crust and lithosphere form from the upwelling of hot mantle rock at spreading ridges. This material mixes with seawater, forming the oceanic crust. After formation at the spreading ridge, the oceanic lithosphere moves laterally away from the ridge, symmetrically spreading on both sides. As the oceanic lithosphere moves away from the site of its birth, it cools, contracts, and subsides. Eventually, it is destroyed by subduction, although a piece may occasionally escape destruction through being thrust up into the continental crust in a plate collision. A piece of the oceanic crust (and possibly upper mantle) that escapes destruction in this manner is termed an **ophiolite.** The knowledge gained from the study of ophiolites constitutes an important contribution to our understanding of the nature of the oceanic crust as it is otherwise difficult to sample the deep parts of the oceanic crust.

The relatively high heat flow from the oceanic crust is largely the result of the cooling of the oceanic lithosphere. Oceanic heat flow is thus inherently transient. In the following section, we derive a mathematical expression for the flow of heat from the oceanic crust as a function of its age and show how deviations from this theoretical formula led to the discovery of one of the most important geological processes on Earth: hydrothermal circulation of seawater in the oceanic crust.

13.4. Kelvin's Problem.

The thermal evolution of oceanic lithosphere can be modeled as the conductive cooling of an initially hot half-space. This is an old and very well-known problem in geophysics that was first solved by the British physicist William Thompson (1824–1907) around 1862. In that year, the British government gave Thompson the title of Baron Kelvin of Largs, and he is frequently referred to as Lord Kelvin, or simply "Kelvin."

Kelvin was interested in estimating the age of the Earth. He assumed the Earth formed as an initially hot sphere of temperature T_m (°C or °K) that cooled through time. Kelvin reasoned that for a given thermal diffusivity, the total amount of cooling that had taken place over geologic time would depend only on the amount of time that had passed since the Earth's formation. The geothermal gradient would

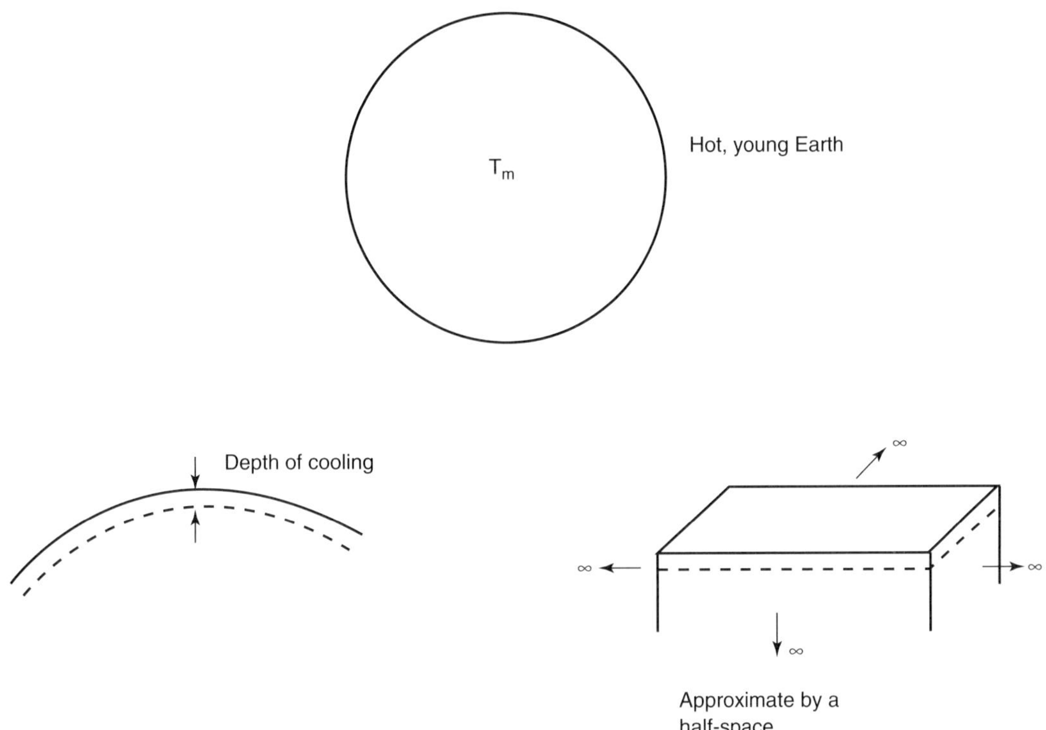

Figure 13.3 Kelvin's model of a cooling Earth approximated by a cooling half-space.

be very high early in the Earth's history but decrease through time. Thus, given the proper mathematical tools with which to attack the problem, the age of the Earth could be estimated simply by measuring the geothermal gradient.

Kelvin assumed that the cooling of the spherical Earth could be modeled as the cooling of a one-dimensional half-space, reasoning that if the cooling had not progressed a substantial distance into the interior, the physical behavior would approximate that of a half-space (Figure 13.3). Kelvin further assumed no internal heat generation inside the Earth, as radioactivity was unknown at the time. These two assumptions simplify the mathematical treatment considerably.

The problem is set up as follows (Figure 13.4). At time t_0 (s), the initial condition is constant temperature T_m everywhere. At some instantaneous increment of time later t_0^+ (s), the surface temperature cools from T_m to T_0 (°C or °K). The cooling thereafter propagates into the subsurface. At time t_1 (s), the near-surface thermal gradient is higher than at time t_2, which is higher than at time t_3, where $t_3 > t_2 > t_1 > t_0$. Should the cooling be allowed to continue indefinitely, the ultimate result at time t_∞ would be a constant temperature T_0 throughout the half-space.

The essential relevance of Kelvin's solution to the problem of a cooling Earth is that the geothermal gradient is a function of time. Thus, if we can measure the geothermal gradient, we can determine how much time has passed since the Earth formed. Note that the solution to Kelvin's problem predicts a geothermal gradient that decreases with increasing depth (Figure 13.4). This effectively takes care of the problem of obtaining unreasonably high temperatures when the average near-surface geothermal gradient is extrapolated to

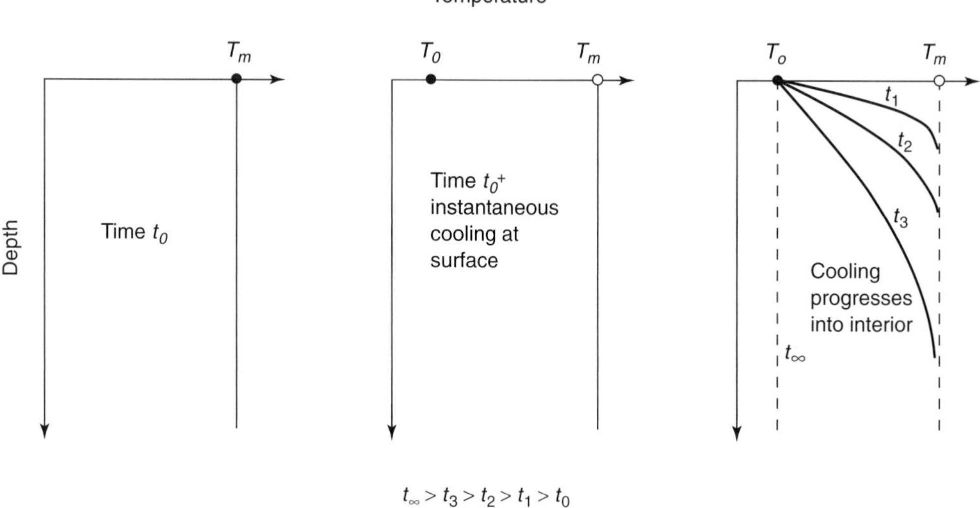

Figure 13.4 Kelvin's problem.

the center of the Earth (e.g., 25 °C-km^{-1} × 6371 km = 159,275 °C).

Thus, Kelvin's model is attractive in that it (a) explains the problem of extrapolation of the near-surface geothermal gradient to great depths and (b) offers an elegant and simple way to estimate the age of the Earth. The solution to Kelvin's problem is the most important result in all of heat conduction theory. It is (Turcotte and Schubert, 1982, p. 159)

$$T(z, t) = T_m + (T_0\text{-}T_m)\ \text{erfc}\left[\frac{z}{\sqrt{4\kappa t}}\right] \quad \textbf{(13.1)}$$

where z (m) is depth, t (s) is time, T (°C or °K) is temperature, T_m is the initial temperature at time $t_0 = 0$, T_0 is the surface temperature after cooling, κ is the thermal diffusivity, and erfc is the complementary error function. The complementary error function (erfc) is

$$\text{erfc}(x) = 1 - \text{erf}(x) \quad \textbf{(13.2)}$$

where erf(x) is the error function,

$$\text{erf}(x) = \frac{2}{\sqrt{\pi}}\ {}_0\int^x e^{-y^2} dy \quad \textbf{(13.3)}$$

The error function, like any other function, is simply a way of mapping one variable (x) to another [erf(x)] (Figure 5.10). The erf(0) = 0, erf(1) = 0.84, and erf(2) = 0.9952. The error function thus asymptotically approaches 1 very quickly. For small numbers, erf(x) ≈ x. For example, erf(0) = 0, erf(0.10) = 0.11, erf(0.50) = 0.52, etc.

Let us simplify equation 13.1 by letting $T_0 = 0$. We then obtain

$$T(z, t) = T_m\left\{1 - \text{erfc}\left[\frac{z}{\sqrt{4\kappa t}}\right]\right\} \quad \textbf{(13.4)}$$

$$T(z, t) = T_m\left\{1 - 1 + \text{erf}\left[\frac{z}{\sqrt{4\kappa t}}\right]\right\} \quad \textbf{(13.5)}$$

$$T(z, t) = T_m\ \text{erf}\left[\frac{z}{\sqrt{4\kappa t}}\right] \quad \textbf{(13.6)}$$

To estimate the age of the Earth, we need to find the near-surface geothermal gradient, dT/dz at $z = 0$. Differentiating the error function, we obtain

$$\frac{dT}{dz} = \frac{T_m d\left\{\text{erf}\left[\frac{z}{\sqrt{4\kappa t}}\right]\right\}}{dz} \quad \textbf{(13.7)}$$

$$\left.\frac{dT}{dz}\right|_{z=0} = \frac{T_m}{\sqrt{\pi\kappa t}} \tag{13.8}$$

or

$$\left.\frac{dT}{dz}\right|_{z=0} = g_0 = \frac{T_m}{\sqrt{\pi\kappa t}} \tag{13.9}$$

where g_0 (°C-m^{-1}) is the near-surface geothermal gradient. Thus, the time elapsed since cooling began (t) is

$$t = \frac{(T_m/g_0)^2}{\pi\kappa} \tag{13.10}$$

For T_m = 1000 °C, g_0 = 0.025 °C-m^{-1}, and $\kappa = 10^{-6}$ m^2-s^{-1}, t = 5.1 × 10^{14} s = 16 × 10^6 yrs. Taking uncertainties into account, Thompson estimated the age of the Earth to be in the range of 10 to 100 million years. This result was the subject of great controversy, as geologists argued that even 100 million years was far too short a time to account for the rock record or to allow for biological evolution. The reconciliation of the conflicting geological evidence and geophysical model came when radioactivity was discovered around the year 1900. With the recognition that the Earth was not just cooling from an initially hot state but could actually be heating up, the premise of Thompson's model was shown to be false. The later advent of radioactive dating yielded an age for the Earth (~4.5 × 10^9 yrs) that could be reconciled with the geologic record.

Thompson had no way of knowing that such a thing as radioactivity existed. He was furthermore careful to state that his result was conditional on the absence of heat sources within the Earth. Because Thompson stated his assumptions, his work endures as an elegant and striking example of the power of geophysical methods, even though his conclusion was false.

We now apply the results of Kelvin's problem to model oceanic heat flow as resulting from the cooling of a half-space. From equation 13.6, temperature in the oceanic lithosphere is

$$T(z,t) = T_m \operatorname{erf}\left[\frac{z}{\sqrt{4\kappa t}}\right] \tag{13.11}$$

where T is temperature as a function of depth (z) and age (t) of the oceanic lithosphere. According to Fourier's Law (equation 11.2), surface heat flow q_{ho} (W-m^{-2}) is the product of the geothermal gradient (dT/dz) at the surface (z = 0) and the thermal conductivity (λ, W-m^{-1}-°K^{-1}),

$$q_{ho} = \lambda \left.\frac{dT}{dz}\right|_{z=0} \tag{13.12}$$

where we have dropped the minus sign from Fourier's Law. Substituting equation 13.9 into equation 13.12,

$$q_{ho} = \frac{\lambda T_m}{\sqrt{\pi\kappa t}} \tag{13.13}$$

According to our theoretical model of a cooling half-space, heat flow from the oceanic crust is inversely proportional to the square root of time. It is a general characteristic of transient diffusion problems, whether hydraulic or thermal, to exhibit a dependence on the square root of time.

Let us now examine the physical significance of equation 13.13 in some detail. The model breaks down at time t = 0. This is because the model is based on a nonphysical assumption of an instantaneous temperature change and a discontinuous temperature field. The model predicts surface heat flow, which is something that can be estimated from measurements of temperature and thermal conductivity near the surface of the oceanic crust. Thus, the model can be tested. Finally, note that the model may also break down as time (t) becomes very large. The model assumes a half-space extending to infinity, but the oceanic lithosphere has a finite bottom. It is possible that at large times, the existence of this boundary may affect thermal conditions near the surface.

The values of heat flow predicted by equation 13.13 depend on the values used for thermal conductivity, initial temperature of the cooling slab (T_m), and thermal diffusivity. Stein and Stein (1992) suggest the best choices are T_m = 1425 °C, λ = 3.25 W-m^{-1}-°K^{-1}, and κ = 26.3 km^2-Ma^{-1} = 26.3 × 10^6 m^2-Ma^{-1}. We then find that

$$q_{ho} = \frac{510}{\sqrt{t}} \tag{13.14}$$

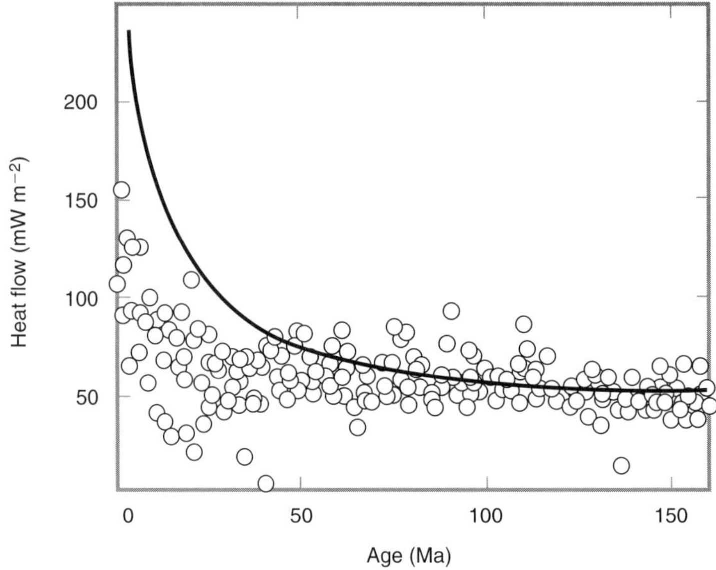

Figure 13.5 Observed heat flow from the oceanic crust (open circles) and heat flow predicted by a theoretical model of a cooling oceanic lithosphere (solid line). The failure of the observations to match the theoretical prediction for oceanic crust younger than 65 Ma is believed to indicate the presence of hydrothermal circulation. (From Alt, 1995, p. 88.)

where q_{ho} is surface heat flow in mW-m^{-2} and t is age in 10^6 years (1 Ma $= 10^6$ years). When we compare the heat flow predicted by equation 13.14 to measurements, we find a significant discrepancy for oceanic lithosphere younger than about 65 million years. The observed heat flow is much lower than that predicted by the theoretical model (Figure 13.5). It is this discrepancy that led to the discovery of hydrothermal circulation through the oceanic crust.

13.5. Hydrothermal Circulation in the Oceanic Crust.

Elder (1965) first suggested that what appeared to be anomalously low heat flow in young oceanic crust could be due to the circulation of seawater. Geophysical surveys of heat flow from the oceanic crust measure only the heat that is transported by the process of conduction. If significant movement of heat by moving fluids were taking place, the total heat flow from the cooling oceanic lithosphere would be partitioned into a conductive and advective component. Because the advective component would be missed by a conductive measurement, the measurements would indicate values considerably lower than that predicted by a theoretical model. The hypothesis of pervasive hydrothermal circulation in the oceanic crust was subsequently validated by many lines of evidence. These included the discovery of hydrothermal vents and mineralization near midocean spreading centers, studies of altered basalts from the seafloor, and analysis of geochemical anomalies in deep ocean waters.

In general, hydrothermal circulation through the oceanic crust can be divided into ridge and flank circulation. **Ridge circulation** occurs in the immediate vicinity of spreading ridges and is characterized by high temperature (350°C), rapid circulation, and the presence of hydrothermal vents termed black and white smokers (Figure 13.6). **Flank circulation** occurs away from the ridge axis

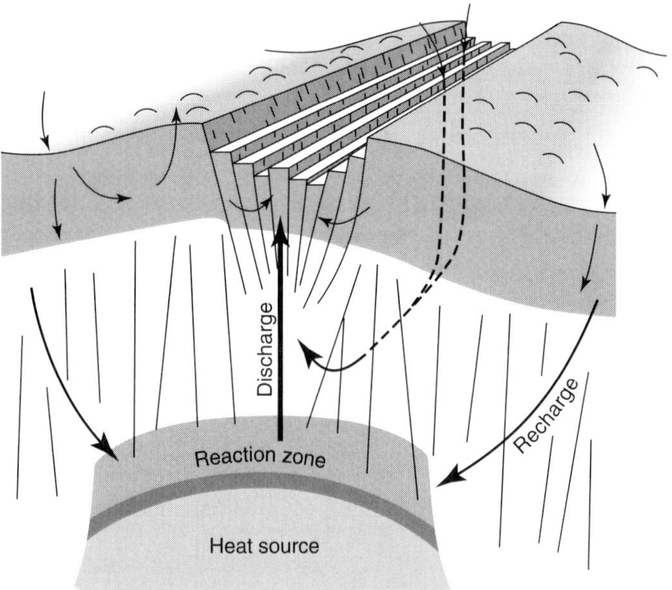

Figure 13.6 Ridge circulation in the oceanic crust.
(From Alt, 1995, p. 89.)

and is characterized by lower temperature fluids (<200 °C) and slower circulation (Figure 13.7). It is difficult to make quantitative estimates of the relative sizes of the hydrothermal fluxes that take place by ridge and flank circulation. What estimates have been made indicate that the ridge circulation is perhaps responsible for 10 to 20% of the total hydrothermal flux through the oceanic crust. Ridge and flank circulation have also been termed "active" and "passive" circulation (Lister, 1972). Active circulation is driven by the presence of an inferred magmatic heat source beneath a spreading ridge, whereas passive circulation is driven by the passive cooling of oceanic lithosphere off-axis. In the case of both ridge and flank circulation, the discharge of hydrothermal fluids tends to be focused, and recharge through seawater infiltration into the crust tends to be diffuse.

It is hardly possible to overestimate the global importance of hydrothermal circulation in the oceanic crust. Stein and Stein (1994) estimated that 34% (±12%) of the total oceanic heat loss (32 $\times 10^{12}$ W) occurs through hydrothermal circulation. This is 25% of the total terrestrial heat-loss rate of 4.4×10^{13} W. The rate at which seawater circulates through the oceanic crust is more difficult to estimate, because the average temperature of heated water that leaves the crust is poorly constrained to be in the range of 100 to 400°C. Stein and Stein (1994) estimated that the water flux was in the range of 10^{13} to 10^{15} kg-yr^{-1}. The mass of Earth's oceans is about 1.37×10^{21} kg. Using the above estimates of water-flux rates, in 10^9 years, a given volume of seawater has circulated through oceanic crust from 7 to 730 times. If the Earth is assumed to be 4 billion years old and the water inventory constant through geologic time, then the oceans have circulated through the rocks of the oceanic crust between 30 and 3000 times.

Hydrothermal circulation of seawater through the oceanic crust alters the composition and chemistry of both seawater and rock. In general, the chemical effect of seawater circulation on the basalts of the oceanic crust is to hydrate, sulfidize, and oxidize the minerals. Rocks of the oceanic crust become highly hydrated with the formation

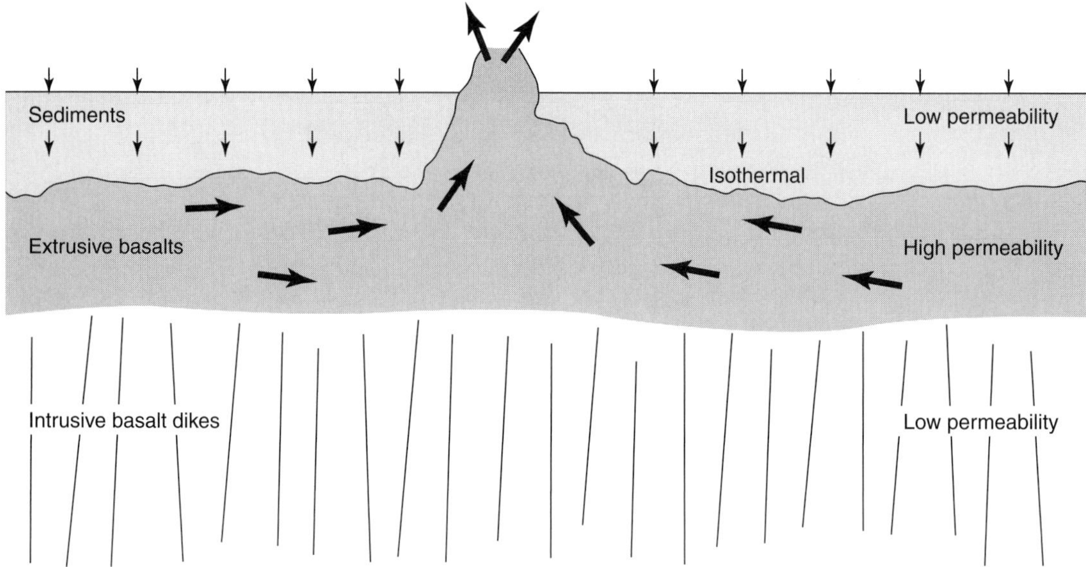

Figure 13.7 Flank circulation in the oceanic crust.

of water-bearing minerals such as zeolites, epidotes, chlorites, amphiboles, and hydrogarnets. The magnesium- and iron-rich silicates found in the oceanic crust (e.g., olivine) are converted to the mineral serpentine through the addition of water. This process is termed **serpentinization.** After alteration, the water content of the upper oceanic crust may be between 5 and 15%. In addition to water, hydrothermal circulation also adds carbon dioxide to the oceanic crust (Fyfe, 1997). Seawater circulation also adds magnesium and possibly sodium to the oceanic crust (Cathles, 1990).

13.5.1. Ridge Circulation.

Ridge circulation (Figure 13.6) is characterized by the presence of black and white smokers, plumes, biotic communities near hydrothermal-vent orifices, and massive sulfide deposits.

A **black smoker** is a hot spring located along eruptive fissures at the center of a spreading ridge that discharges hot water laden with sulfide minerals at temperatures near 350°C (Figure 13.8). The black color of the water discharged from these vents arises when the hydrothermal fluid discharged from a vent contains enough metals and sulfur to cause precipitation of sulfide particles during mixing with seawater at the vent orifice (Hannington et al., 1995). Typical sulfide minerals that are precipitated during venting include chalcopyrite, pyrrhotite, anhydrite, pyrite, and sphalerite. The rate at which energy is contributed to ocean water by the discharge of some black smokers is greater than 100 million Watts (Lowell et al., 1995). The highly concentrated nature of the energy flux from black smokers and the discovery of inactive vents implies that these hydrothermal vents are transient features. Otherwise, the rate of energy release could not be maintained over geologic time. **White smokers** are hydrothermal vents that discharge fluids at temperatures in the range of 100 to 300°C. The white color of the water discharging from these vents is due to white particles of silica, anhydrite, and barite that precipitate from solution when hot hydrothermal fluids encounter cold seawater. Both white and black smokers often occur on the same vent complex and are generally confined to a relatively narrow zone about 100 m wide directly above what is presumed

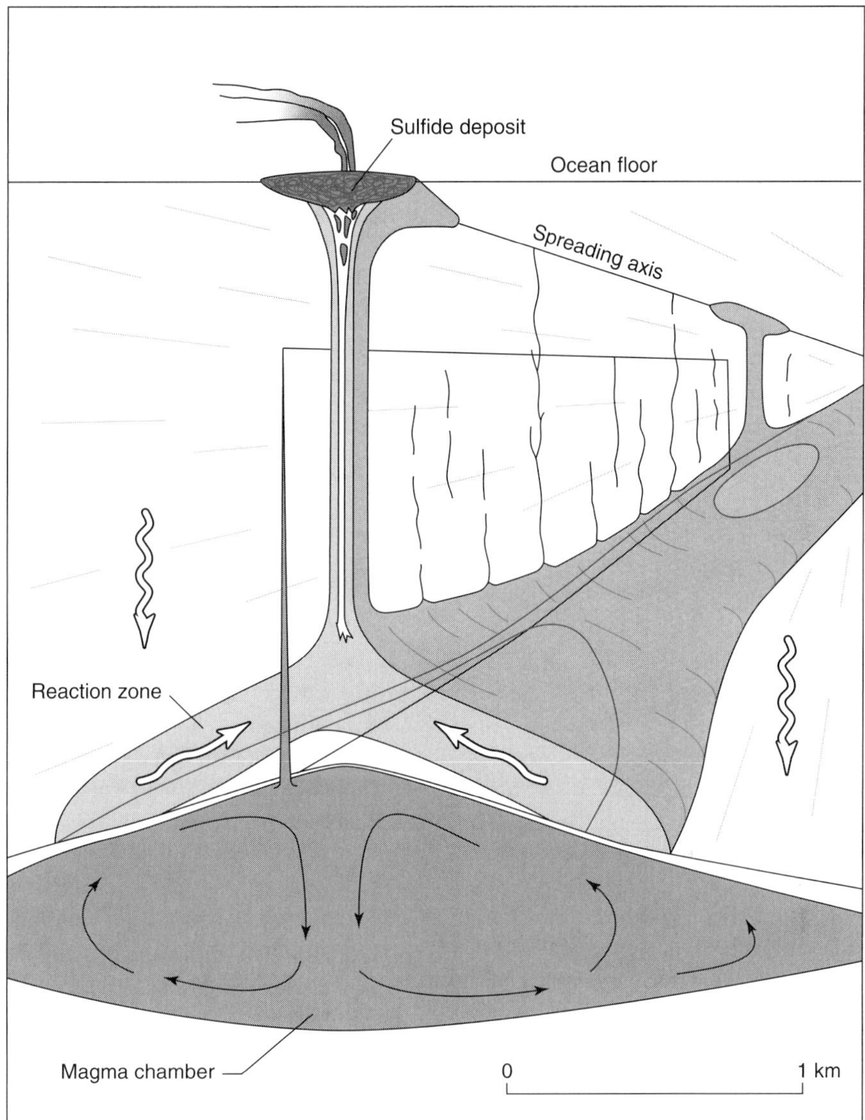

Figure 13.8 Black smoker.
(From Cann and Strens, 1989, p. 12,228.)

to be a magma chamber. Black smokers are believed to result from direct discharge of high-temperature hydrothermal fluids, whereas white smokers represent the mixing of a 350°C hydrothermal fluid with seawater.

Isotopic measurements indicate the typical residence time for a water molecule involved in ridge-axis convection is around 100 years (Cathles, 1990). The vigor of hydrothermal ridge circulation increases as the spreading rate increases. The mid-Atlantic ridge spreads at a rate of 2 to 4 cm-yr^{-1}, and is considered to be a slow-spreading ridge with lower hydrothermal activity. A well-studied example of a ridge that spreads at intermediate rates (5 to

6 cm-yr^{-1}) is the ridge between the Juan de Fuca and Pacific Plates, offshore from the northwestern United States and southwest Canada. The East Pacific Rise between 9°N and 13°N, west of Mexico and Central America, marks the boundary between the Pacific and Cocos Plates and is a fast-spreading ridge, characterized by a spreading rate of 11 cm-yr^{-1} and vigorous hydrothermal activity (Fornari and Embley, 1995).

A **plume** is formed by a sudden release of hot water from a hydrothermal system. The released water is hotter and more buoyant than surrounding seawater and thus rises above the discharge point until it dissipates and spreads laterally (Baker et al., 1995). Intense, short-lived episodes of hydrothermal activity can give rise to unusually large plumes termed **megaplumes** (Lowell, 1995). The first discovered megaplume had a thickness of about 700 m and a diameter of 20,000 m. Baker et al. (1989) estimated this megaplume formed during a hydrothermal discharge that lasted from 2 to 20 days, with a heat output of 10^{16} to 10^{17} Joules. This heat output is equivalent to the annual heat output of between 200 and 2000 black smokers. Based on an estimated average annual per capita consumption of electrical energy in the United States of 4.6×10^{10} J, the energy released during this megaplume event was equivalent to the electrical energy consumed by 1 million people over one year. The precise mechanism responsible for megaplumes is unknown. Whatever agent is ultimately responsible, the creation of a megaplume requires that hydrothermal venting rates undergo a short-lived, dramatic increase. Cathles (1993a) suggested that this could be accomplished through temporary increases in fracture permeability that might be caused by tectonic stresses resulting from magma withdrawal, thermal contraction, dike emplacement, or pulses of magmatic volatiles released from the magma chamber that underlies spreading ridges. Another possibility is that hot hydrothermal fluids are stored in chambers or compartments that are periodically breached. However, Cathles (1993a) pointed out the difficulty of sealing such a compartment in the vicinity of a ridge spreading at a rate of 1 to 10 cm-yr^{-1}.

One of the most shocking discoveries in the history of science was the discovery of active communities of living creatures near the hydrothermal vents of the ocean's spreading centers. This environment is inherently inimical to life as we commonly know and define it. Sunlight is virtually absent, and the temperature of venting fluids averages 350°C. The deep-sea hydrothermal vent communities are the only biotic communities on Earth that can exist without sunlight. The presence of hydrothermal vent communities in the deep ocean is analogous to oases in the desert. Life in the deep sea is scarce because food is scarce. Plants cannot grow without sunlight. Hydrothermal vents are able to sustain life by providing an alternative energy source in the form of reduced ions, principally sulfide. These are used by autotrophic bacteria to convert carbon dioxide, water, and nitrate into essential organic substances. Just as green plants use sunlight as an energy source for the process of photosynthesis, oxidizing bacteria near hydrothermal vents in the deep ocean use the process of chemosynthesis to make food from inorganic components. **Chemosynthesis** is a process by which the base of food chain is created through chemical processes other than photosynthesis. Most of the mass of a vent community is located near the vent opening. Animal crowding there is so great that the principal factor limiting population seems to be space rather than food. Animals present in vent communities include tubeworms, clams, mussels, shrimp, crabs, and fish (Hessler and Kaharl, 1995).

Ridge circulation is also characterized by the presence of massive sulfide mineral deposits. The primary characteristic of chemical interaction between the oceanic crust and seawater is the formation of a hydrothermal fluid that is acidic, reducing, and sulfur- and metal-rich. Metals are leached from rocks of the oceanic crust into circulating hydrothermal fluids. When primary, high-temperature (350°C) hydrothermal fluids are vented directly into the sea without dilution, a black smoker occurs. The rapid changes in temperature and chemical conditions that occur near the opening of the vent cause sulfide minerals to precipitate. These minerals accumulate on the ocean floor near black smoker vents and form massive sulfide deposits. No two sulfide deposits are exactly alike. The shape of the sulfide mounds and the places of their occurrence depend strongly on the permeability structure of the oceanic crust, which in turn depends on the structure and style of fracturing (Hannington et al., 1995).

13.5.2. Flank Circulation.

Flank circulation occurs on the older flanks of ridge axes and is characterized by lower temperature (<200°C) circulation. Flank circulation is not as impressive a process as high-temperature ridge circulation but may have a greater cumulative effect with respect to altering the thermal and chemical nature of both the oceanic crust and seawater (Davis and Becker, 1998). A distinguishing feature of flank circulation is the presence of layer 1 sediments in thicknesses that range from a few to several hundred meters. The thickness of layer 1 sediments varies according to the topography of the buried oceanic basement. In some areas, the oceanic basement is relatively flat, with little relief. However, in other areas, the basement contains ridges and peaks that may be high enough for the top of layer 2 to project through layer 1 sediments and reach the ocean floor.

A key to understanding the nature of flank circulation was the discovery (Davis et al., 1996) that heat flow on the ocean floor is inversely correlated with depth to the oceanic basement (top of layer 2). Where the basement topography is high and sediment thickness low, heat flow is high. Where basement topography is subdued and sediment thickness high, heat flow is low. Consideration of Fourier's Law suggests that this inverse correlation may be due to the sediment-basement contact being isothermal. Fourier's Law of Heat Conduction (equation 11.2) can be written

$$q_h = \lambda \frac{\Delta T}{\Delta z} \tag{13.15}$$

where q_h is the heat flow as measured near the surface of the ocean floor, λ is the thermal conductivity of the sedimentary layer, ΔT is the temperature drop between the ocean floor and top of the basement (or bottom of sediments), and Δz is the thickness of the sediment layer (or depth to basement). Suppose that sediment thermal conductivity (λ) is essentially constant. If the sediment-basement contact is isothermal, then ΔT will be constant no matter how Δz changes. Thus, the heat flow q_h is inversely proportional to sediment thickness, Δz. Under more normal circumstances, temperature would increase with depth, and ΔT would change in the same proportion as Δz, allowing heat flow to remain constant.

The only practical way to maintain an isothermal basement is through horizontal fluid flow. If some active hydrothermal process were not acting, the lateral temperature gradients would soon disappear through the horizontal transport of heat by conduction. In consideration of the permeability structure of the oceanic crust (Table 13.1, Figure 13.2), rapid horizontal flow most likely takes place in the upper 500 m or so of the oceanic basement, where permeability is one to three orders of magnitude greater than in either underlying basement rocks or overlying sediments. Flow is probably vertical downward through the sediments of layer 1. Fluid is then strongly refracted into horizontal flow through the upper basement (Figure 13.7). Horizontal flow is likely focused to discharge at locations where the sedimentary cover is thin or absent, as these areas represent the least resistance to flow. Rapid horizontal flow near the top of layer 2 may be promoted by a marked permeability anisotropy, with the horizontal permeability much greater than the vertical. Anisotropy likely results from layering produced by sequences of eruption events. Overall, the oceanic crust tends to be heterogeneous and strongly anisotropic in its permeability structure (Fisher, 1998).

As the oceanic crust ages, hydrothermal circulation tends to slow down and eventually cease. There are several reasons for this. The blanket of low-permeability sediments thickens and lithifies with increasing age. Simultaneously, fractures in layer 2 may be closed off by mineral precipitation, and the heat flow that drives hydrothermal circulation

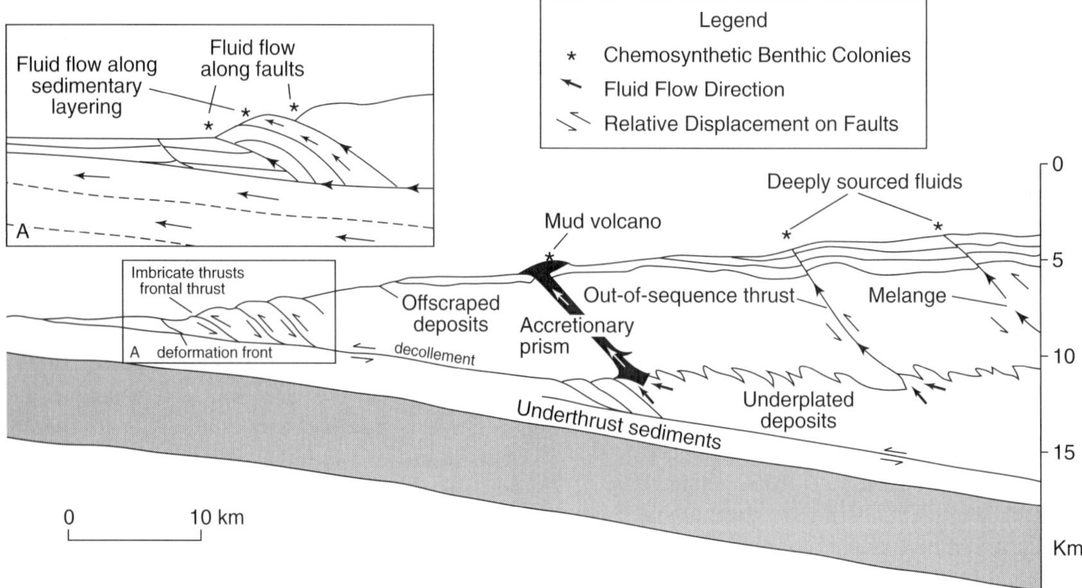

Figure 13.9 Features of a typical accretionary prism.
(From Moore and Vrolijk, 1992, p. 114.)

diminishes. The age at which flank circulation essentially ceases can be estimated by comparing observations of heat flow with theoretical cooling models, such as the half-space model of Kelvin. Stein and Stein (1994) concluded that hydrothermal circulation is absent in oceanic crust older than 65 ±10 Ma. Stein and Stein (1994) also note that the primary control on the rate of heat removal by hydrothermal circulation appears to be seafloor age, not sediment thickness. The development of thick, cemented layers of sedimentary rock in older ocean basins thus does not appear entirely sufficient by itself to choke off hydrothermal circulation in the oceanic crust.

13.6. Subduction Zones.

Subduction zones are the most dynamic structural environment on Earth, characterized by the highest strain rates and largest earthquakes (Moore and Vrolikj, 1992). As the water-logged sediments and hydrated rocks of the oceanic crust become subducted, they are exposed to increasing pressures and temperatures that tend to release the fluids adsorbed during hydrothermal interactions in younger oceanic crust as well as volatile elements (e.g., carbon) accreted onto the oceanic floor by sedimentation. The release of fluids and volatile gases in an area subject to high strain and deformation rates leads to complex interactions between hydrogeological and structural phenomena. Fluid pressures affect how rocks deform, while the manner of deformation (e.g., faulting) affects how fluids move.

13.6.1. Accretionary Prisms.

An **accretionary prism** (Figure 13.9) is a wedge-shaped accumulation of sediments and sedimentary rock that forms near a subduction zone. The key hydrological process that occurs in the accretionary prism is the dewatering of wet sediments in a structurally dynamic environment. Sediments

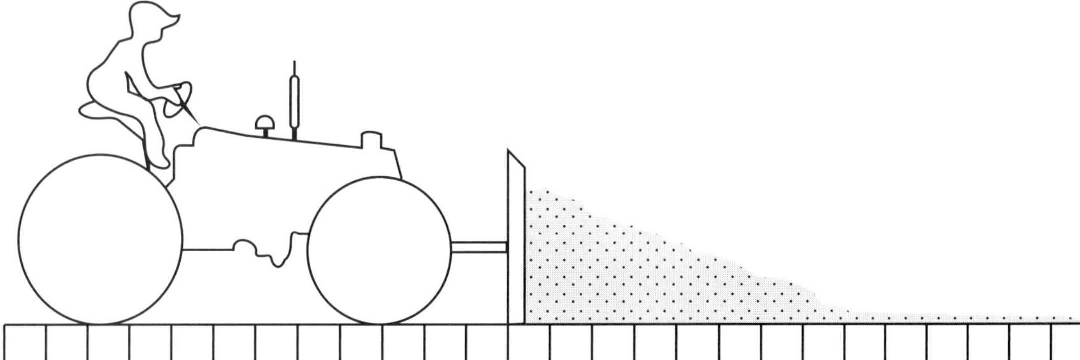

Figure 13.10 The mechanics of an accretionary prism are similar to a bulldozer pushing a pile of sand.
(From Davis et al., 1983, p. 1157.)

that accumulate in accretionary prisms are a mixture of terrigenous materials, derived from nearby volcanic island arcs or continents, and pelagic sediments scraped from a subducting plate. **Pelagic** sediments are those that form on the ocean floor in deep water. The average composition (by volume) of pelagic sediment on the sea floor is 72% carbonate ooze, 19% red clay, and 9% siliceous ooze (Peacock, 1990). **Terrigenous** sediments are those derived from the erosion of a land mass. In most cases, the flux of terrigenous sediments into an accretionary prism is greater than the flux of pelagic sediments (Peacock, 1990).

Accretionary prisms have three distinguishing features: (1) There is a basal detachment or decollement, below which deformation is minimized. A **decollement** is a detachment surface between two bodies of rock that have separate styles of deformation. Decollements are usually associated with thrust faults. The decollement at the base of accretionary prisms is believed to be a structurally weak zone, with the weakness likely resulting from high fluid pressures (see section 12.3). (2) Material above the basal decollement is subject to large horizontal compression. (3) Accretionary prisms have a characteristic wedge shape (Davis et al., 1983).

The most succesful model that explains the wedge shape of accretionary prisms and their structural evolution is the **Coulomb wedge model** of Davis et al. (1983). The Coulomb wedge model is analogous to a bulldozer pushing a pile of sand, soil, or snow (Figure 13.10). The material pushed by the bulldozer blade deforms until it takes on a wedge shape characterized by a critical taper. The **critical taper** is the shape for which the wedge is on the verge of failure under horizontal compression everywhere. Once the critical taper is reached, the wedge can be pushed by the bulldozer, but its shape remains constant. If material is added to the wedge, it becomes larger but its shape and critical taper remain constant. Coulomb theory predicts that the shear stress necessary for the failure of a brittle substance (rock) depends on the cohesive strength, the normal stress and coefficient of friction across a failure surface (fault), and the fluid pressure. In the case of accretionary prisms, the cohesive strength of the sediments is relatively unimportant. The deformation style, and thus the shape of the critical wedge, largely depends on the coefficient of friction and the fluid pressure.

The Coulomb wedge model predicts that as friction across the basal decollement increases, so does the taper. On the other hand, an increase in the internal strength of the wedge decreases the critical taper. The observed tapers of modern accretionary wedges are consistent with low cohesive strengths and coefficents of friction that range from 0.85 to 1.03. The deformation within an accretionary wedge necessary to maintain the critical taper is usually acommodated through thrust faulting that may be facilitated by high fluid pressures. The role of fluid pressure is fundamental. According to the Hubbert-Rubey pore pressure hypothesis (section 12.3.1), fluid pressure governs whether slip occurs at the decollement or if strain is transferred to thrust faults in the upper part of the accretionary prism.

A variety of evidence exists for active fluid flow in accretionary prisms. The location of fluid seeps on the ocean floor is marked by biologic communities of organisms that derive their energy through chemosynthesis. The size of the chemosynthetic community depends on the flow rate and chemistry of the vented fluids. Thus, the mapping of chemosynthetic communities provides a basis for quantitatively estimating fluid fluxes in accretionary prisms. Accretionary prisms also commonly contain mud volcanoes and mud diapirs. A **diapir** is a dome-shaped upwelling of a plastic material squeezed through overlying strata. Mud diapirs and volcanoes in accretionary prisms are believed to be the surface manisfestation of high fluid pressures in the subsurface. Overpressuring is promoted by the rapid accumulation of porous, clay-rich sediments that tend to have low permeabilities. The high rates at which plates converge (1–10 cm-yr^{-1}) substantially exceed "normal" burial rates in sedimentary basins; thus, sediments may accumulate so quickly that excess pore pressures develop.

The water expelled from accretionary prisms has two sources. One is sediment compaction. As the relatively thin layer of oceanic sediments is thrust into deeper burial within an accretionary wedge, pore fluid is expelled as sediments compact after being subjected to deeper burial with concomitant increases in effective stress. Before burial, ocean floor sediments (layer 1) have porosities that range from 10 to 50%. After burial and low-grade metamorphism, porosities have been reduced to less than 5 to 10%. Thus, 50 to 80% of the available pore water is expelled by mechanical compaction during subduction (Peacock, 1990). The highest rates of fluid expulsion from mechanical compaction tend to occur near the front of the accretionary prism, where ocean floor sediments begin rapid burial in the prism. This is a reflection of the exponential nature of porosity-depth curves that describe settings where porosity changes are mostly mechanical (as opposed to chemical). In these cases, the highest rates of compaction occur near the surface. The fluid expulsion rate also depends on the taper of the accretionary prism. In prisms with high taper angles, fluid expulsion tends to occur more quickly, as sediments are buried at a faster rate.

The second source of water is the dehydration of minerals as they are subducted and exposed to higher temperatures at greater burial depths. The occurrence of pore water that is fresher (lower in dissolved solids) than seawater near faults and permeable stratigraphic conduits is evidence for dehydration, as water released from dehydration would not normally contain dissolved salts (Moore and Vrolijk, 1992). In general, fluid production by mechanical pore collapse tends to occur at more shallow depths because the decrease of porosity with depth tends to be exponential. Fluid production by dehydration tends to occur at greater depths, as dehydration is largely triggered by the high temperatures that exist at depth. Dehydration sources may be either clay minerals in sediments and sedimentary rocks or the hydrated crystalline rocks (e.g., basalt) of the oceanic crust. Perhaps the most important clay dehydration is the conversion of smectite to illite (see section 8.3.2). The smectite-illite conversion is a kinetic reaction that depends on time and temperature. The precise nature of the reaction is poorly known, but it may occur at temperatures between 60 and 200°C, and the total amount of water released may be up to 35% of the original smectite volume. At higher temperatures and pressures, minerals in the basaltic layer begin to break down and release water. The fluid production of subducting basalts is complex, because basalts contain nine major chemical constitutents and several minor components. The chemical reactions by which hydrous minerals such as amphibole, mica, and chlorite break down are complex functions of pressure, temperature, composition, and the mobility of oxygen.

Extensive evidence exists that the dewatering of accretionary prisms largely occurs through focused flow. In addition to the presence of biologic communities at vents, mineral crusts are also found. Core samples from boreholes that have passed through faults show extensive mineralization, implying substantial fluid flow through faults.

13.6.2. Vulcanism.

Volcanoes tend to occur in arcs above subduction zones. The presence of volcanoes, with high heat flow and high temperatures near the surface, is a conundrum if the transport of heat by fluid movement is not considered. Subduction involves the transport of relatively cold material to depths where the ambient temperature is much higher. The introduction of cold, near-surface material into a hotter environment should absorb heat, leading to colder conditions above a subduction zone. Instead, volcanic activity is a dramatic manifestation of increased heat flow and higher temperatures.

The apparent discrepancy can be resolved if the transport of heat by fluids released from the subducting slab is considered. As the hydrated minerals of the oceanic crust are subjected to higher temperatures at depth, water and other volatile elements such as carbon are released. Being less dense than surrounding rocks, these elements tend to migrate upward carrying heat with them. Most fluids are released at depths of 10 to 40 km (Peacock, 1990). Even if the upward velocity of these fluids is small, the rate at which heat is transported to the surface may be large because the magnitude of heat transport by a moving fluid depends exponentially on the vertical distance transversed (see section 11.4). The introduction of water into formerly anhydrous rocks also tends to promote melting and the formation of magma. A water-saturated rock typically melts at temperatures several hundred degrees Celsius lower than a dry rock of the same composition (Figure 13.11).

Not only are the hydrated rocks of the oceanic crust carried down into the upper mantle through subduction, but sediments and sedimentary rocks may also be subducted. As the oceanic lithosphere bends near a subduction zone, the upper part cracks and forms ridge and valley-type (horst and graben) structures that may fill with sediments. Thus, low-density sedimentary materials may become tectonically trapped and carried down into the mantle (Fyfe, 1997). Peacock (1990) noted that the presence of the mineral coesite in the high-pressure metasedimentary rocks now exposed in the Alpine collision zone implies that sedimentary material had been subducted to depths greater than 90 km. **Coesite** is a type of quartz that forms at very high pressures.

Fluids released from a subducting slab may change the chemical composition of both the continental crust and the mantle wedge that overlie the slab. The movement of fluids from subducting slabs to erupting volcanoes is mostly inferred from circumstantial evidence. The precise paths taken by rising fluids are largely "unseen and unknown" (Plank, 1996). Volcanic rocks from convergent margins have a unique composition; they are enriched in alkaline elements such as rubidium and barium while simultaneously being depleted in highly charged cations such as niobium and titanium. Keppler (1996) concluded that the agent responsible for the generation of arc magmas was most likely an alkali-chloride-rich aqueous fluid. More simply put, saltwater. As the generation of arc magmas is thought to be the most important mechanism for the growth of the continental crust since the Proterozoic eon (600 to 2500 Ma), the circulation of saltwater through the mantle and crust at convergent margins appears to be a major agent in the creation of the continental crust. Almost one-third of the crust of the Americas may have been influenced by "recent" subduction events Fyfe (1997).

Rising fluids released from a subducting slab are an extremely efficient mechanism for heat and mass transport; they drain energy and mass from an overyling mantle wedge and transport them to the surface. Fyfe (1997, p. 246) estimated that "every gram of fluid introduced will probably lead to something like 100–1000 times the mass of volcanic rock."

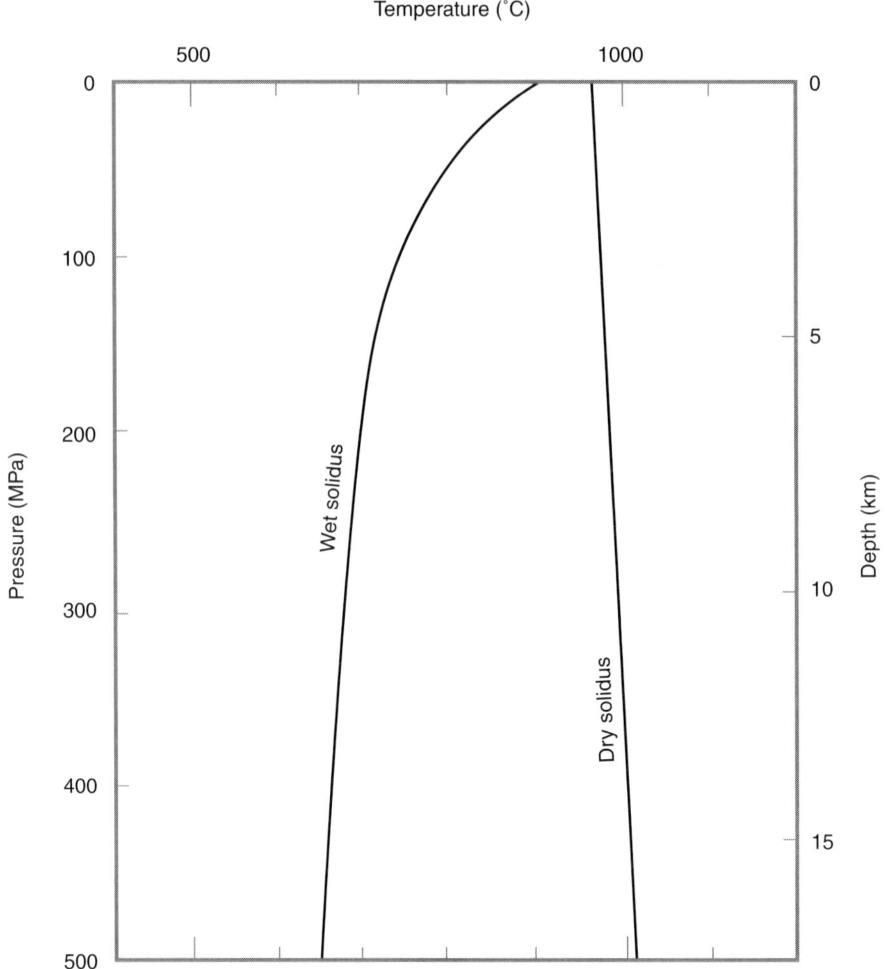

Figure 13.11 Solidus for wet and dry granodiorite. The solidus defines the pressure and temperature at which melting first begins. Granodiorite is a rock that represents the average composition of the continental crust. The addition of water lowers the melting point by several hundred degrees Celsius.
(After Wyllie, 1971.)

13.7. Origin of the Oceans.

One of the great unresolved problems in geology is the origin of the oceans. The problem was first considered in detail by Rubey (1951). Rubey (1951) pointed out that volatile compounds including water, carbon dioxide, chlorine, nitrogen, and sulfur were much too abundant near the Earth's surface to be explained as the products of rock weathering. It is also not possible for all of the volatile compounds found near the surface today to have been present early in the Earth's history in the form of a "dense primitive atmosphere." High atmospheric concentrations of carbon dioxide, nitrogen, and hydrogen sulfate in contact with seawater would have led to a highly acidic ocean with the concomitant deposition of large quantities of carbonate rocks. However, the percentage of limestone in sediments of Precambrian age is about the same as that in younger sedimentary rocks.

Rubey (1951) reasoned that carbon has been added to the atmosphere and oceans continuously through geologic time. The amount of carbon tied up in rocks is 600 times the amount of carbon present in today's atmosphere, hydrosphere, and biosphere combined. As carbon is continuously extracted from the atmosphere-ocean system by sedimentation, Rubey (1951) pointed out that if a continuous rate of supply were not present, brucite would replace calcite as a common marine sediment. Yet the geologic record contains no brucite deposits. Rubey (1951) concluded that the geologic record strongly indicated that Earth's oceans and crustal carbon inventory formed as the result of slow accumulation over geologic time from a more-or-less continuous and gradual supply mechanism. The only such mechanism known to Rubey (1951) was volcanic outgassing. Earth scientists have subsequently accepted with little question that the Earth's oceans and its near-surface inventory of volatiles such as carbon have originated from volcanic outgassing.

However, Rubey (1951) never measured the rate at which water and carbon are added to the terrestrial surface environment through volcanic outgassing. Any attempt by Rubey (1951) to make such estimates would have been badly flawed, because he was unaware of the process of subduction. Subduction carries both water and carbon down into the mantle at significant rates, and it is questionable whether the water and carbon thus removed from the near-surface environment are fully returned through outgassing.

Since the early 1980s, quantitative estimates have been made that compare the rate at which water and carbon are released by volcanic outgassing to their uptake by subduction (see review by Bebout, 1995). It turns out that the net rate at which water is added to the hydrosphere by volcanic outgassing is only 5 to 15% of the rate at which it is carried down into the mantle by subduction. Similarly, the net rate at which carbon is added to the atmosphere by volcanic outgassing is only 10 to 44% of the rate at which it is lost to the mantle. Based on all existing estimates and studies, volcanic outgassing is a grossly inadequate mechanism to explain the abundant existence of water and carbon at and near the Earth's surface.

If near-surface supplies of water and carbon did not come from volcanic outgassing, where did they originate? There are at least three possible explanations. First, it is eminently possible that substantial amounts of water and carbon return to the surface through processes other than volcanic outgassing. One possibility is updip transport back up a subducting wedge. Seafloor fluid venting occurs along trenches, and studies of accretionary complexes document updip fluid flow. On the other hand, estimates of water-loss rates are very conservative because they only include water that is mineralogically bound. Pore fluids, which account for 90% of subducted water, are not included. It is therefore possible that all accounts of seafloor venting and other evidence for shallow return near subduction zones could be explained by the return of pore fluids.

A second possibility is that the rate of volcanic outgassing was higher early in the Earth's history. This is not really an attractive hypothesis, however. High outgassing rates would have resulted in the "dense primitive atmosphere" that Rubey (1951) showed was geochemically impossible. It is also probable that high levels of volcanic activity on a young and hotter Earth would have been matched by higher subduction rates.

A third possibility was suggested by Frank et al. (1986) and Frank (1990), who proposed that the source of the Earth's water was ongoing bombardment by a unique species of "small comets" composed almost entirely of water (Figure 13.12). The **small comet hypothesis** proposes that Earth is currently accreting water at a rate of 0.2 to 1.0 $\times$ 10^{12} kg-yr^{-1} from an influx of small (mass 0.2 to 1.0 $\times$ 10^5 kg, diameter about 4 to 6 m), cometlike objects. Small comets are believed to be composed primarily of water, but a carbon mantle is required to avoid vaporization in interplanetary space. Putative estimates of carbon-mantle density and thickness made by Frank and Sigwarth (1993) imply

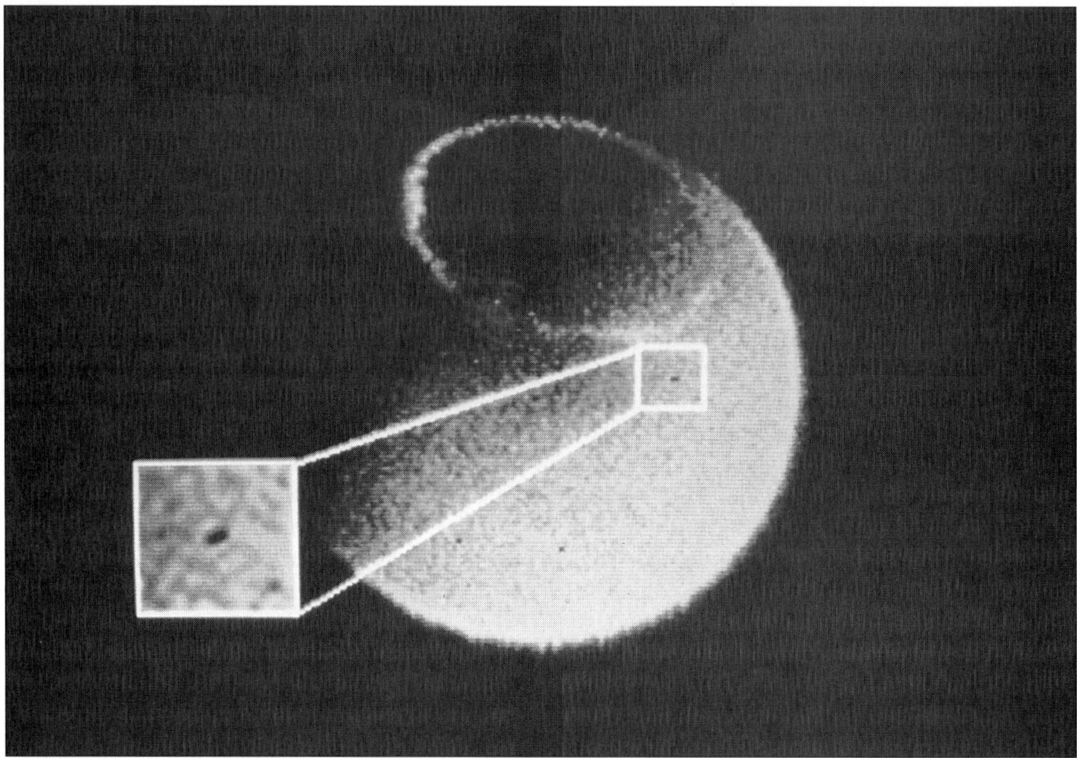

Figure 13.12 Dark spot, or "atmospheric hole," which may be due to small-comet impact. Atmospheric holes were interpreted by Frank et al. (1986) as due to the absorption of ultraviolet light by an expanding water vapor cloud from a small comet impact. Others (Dessler, 1991) believe that such atmospheric holes are instrumental artifacts and that small comets do not exist.

(Photograph courtesy Louis A. Frank, University of Iowa.)

that small comets are probably about 90% water, and 10% carbon.

The small comet hypothesis created a sensation and an immediate controversy. The accretion rates postulated by Frank et al. (1986) were 10,000 to 100,000 higher than the generally accepted estimates for the total extraterrestrial accretion rate. One of the reviewers of the original paper (Frank et al., 1986) noted that "if this is correct, we would have to burn half the contents of the libraries in the physical sciences" (Frank, 1990). A number of criticisms of the small comet hypothesis were subsequently published (see review by Dessler, 1991). In the spring of 1997, new high-resolution satellite data appeared to substantially bolster the small comet hypothesis (Frank and Sigwarth, 1997a, b, c, d). But the new observations were followed by a new round of criticisms (e.g., Parks et al., 1997, 1998) and a rebuttal (Frank and Sigwarth, 1999). As of 1999, it was not clear from a review of the literature if the small comet hypothesis was faulty or if the criticisms were based on faulty assumptions.

From a geologic standpoint, the small comet hypothesis can be put to a test of logical consistency. That is, how does the estimated rate of water and carbon accretion inferred from satellite data compare to the rate implied by geological data as necessary to create and maintain Earth's near-surface inventories of water and carbon? If it is assumed that the Earth formed without water and carbon at the surface, then the supply rate necessary to explain existing inventories can be estimated as

$$\text{supply rate} = \frac{\text{present inventory}}{\text{age of earth}} + \text{depletion rate}$$

where "depletion rate" refers to the net depletion rate (subduction minus outgassing). Based on current estimates from several studies, net depletion rates for water and carbon are in the range of 0.8 to 1.8×10^{12} kg-yr^{-1} and 0.3 to 2.5×10^{11} kg-yr^{-1}, respectively. Estimating present water and carbon inventories to be 1.37×10^{21} kg and 1.0×10^{20} kg, respectively, and the age of the Earth to be the age of the oldest known rock (4.0×10^9 yr), the water and carbon supply rates necessary to explain current inventories are in the range of 1 to 2×10^{12} kg-yr^{-1} and 0.6 to 3×10^{11} kg-yr^{-1}, respectively. These estimates are not significantly affected by the assumed amount of volatiles present early in the Earth's history, because the calculated supply rate is largely determined by the depletion rate.

Frank and Sigwarth (1993) estimated water was being added to Earth from small comets at a rate of 0.2 to 1.0×10^{12} kg-yr^{-1}. Based on putative estimates of carbon-mantle density and thickness made by Frank and Sigwarth (1993), small-comet carbon content is probably equivalent to about 10% of the total water mass. Thus, small-comet carbon-accretion rates are in the range of 0.2 to 1.0 $\times 10^{11}$ kg-yr^{-1}. In consideration of the uncertainties involved in assuming accretion, subduction, and outgassing rates constant over geologic time, these numbers are very close to those that satisfy the geologic constraints.

If water and carbon are being carried down into the mantle and not returned, substantial amounts of these elements should be stored in the Earth's interior. Presumably, most storage would take place in the upper mantle, although the degree to which the lower and upper mantle mix is poorly known. Assuming that the rates cited above have been constant over geologic time, the net amount of water carried down into the mantle over the last 4.0×10^9 yrs should be in the range of 3 to 7 $\times$ 10^{21} kg, two to five times the mass of today's oceans (1.37×10^{21} kg). Ten years ago, there was little to no evidence for substantial amounts of water storage in the upper mantle (Ahrens, 1989). However, in the past few years, it has become apparent that large amounts of water may be stored in the upper mantle in the form of **dense hydrous magnesium silicates (DHMS).** One of the most important minerals in this group is **wadsleyite,** which may contain up to 3.3% H_2O by weight (Smyth, 1994). Smyth (1994) estimated that if the mantle between the depths of 400 and 525 km were composed of 60% fully hydrated wadsleyite, the amount of water contained therein would be "more than four times the amount of H_2O currently in the Earth's hydrosphere." More recently, Bose and Navrotsky (1998) showed "there is no barrier to subducting substantial amounts of water to depths of 400–600 km in colder slabs, since the slab can remain in the stability field of hydrous phases throughout its descent." Other evidence for the presence of water in the upper mantle is provided by seismic tomography (Nolet and Zielhuis, 1994, p. 15,816), excess helium in groundwater (Torgersen et al., 1995), and deep-focus earthquakes (Meade and Jeanloz, 1991).

REVIEW QUESTIONS

1. Define the following terms in the context of hydrogeology:
   a. oceanic fluid cycle
   b. ophiolite
   c. ridge circulation
   d. flank circulation
   e. serpentinization
   f. black smoker
   g. white smoker
   h. plume
   i. megaplume
   j. chemosynthesis

k. accretionary prism
l. pelagic
m. terrigenous
n. decollement
o. Coulomb wedge model
p. critical taper
q. diapir
r. coesite
s. small comet hypothesis
t. dense hydrous magnesium silicate
u. wadsleyite

2. Compare the ages of the oldest continental and oceanic rocks.

3. Describe the process by which the oceanic crust forms, and compare it with the process or processes that create the continental crust.

4. Describe the composition of the oceanic crust. Use a figure and show the thickness of each layer, its composition, and its average permeability.

5. Which layer of the oceanic crust has the highest permeability? How did this permeability originate?

6. Why is the flow of heat from the oceanic lithosphere of the same order of magnitude as that from the continents? Where do continental and oceanic heat flow originate? Is their source different or the same?

7. Use a set of three figures to describe Kelvin's problem. Describe its relevance to the oceanic lithosphere and fluid circulation in the oceanic crust.

8. What are four types of evidence for hydrothermal circulation through the oceanic crust?

9. What are "active" and "passive" hydrothermal circulations in the oceanic crust?

10. What percentage of the total terrestrial heat-loss rate occurs through hydrothermal circulation?

11. What effect does hydrothermal circulation have on the chemical composition of the rocks of the oceanic crust? On seawater?

12. Explain why black smokers are thought to be geologically transient.

13. Outline two possible mechanisms for the formation of megaplumes.

14. What is the relationship between heat flow and the topography of the oceanic basement (top of layer 2)? What does this relationship suggest concerning hydrothermal circulation?

15. How does hydrothermal circulation change as the oceanic lithosphere ages?

16. Why are subduction zones "the most dynamic structural environment on Earth"?

17. What is the single most important hydrogeologic process that occurs in a subduction zone?

18. What are the sources of sediments that accumulate in accretionary prisms? Which source is larger?

19. What are the three distinguishing features of accretionary prisms?

20. According to the Coulomb wedge model, what two factors are most important in determining the shape of an accretionary prism?

21. What determines if strain in an accretionary prism is accommodated through movement on the base decollement or movement along overlying thrust faults?

22. What are two types of evidence for active fluid flow in accretionary prisms?

23. What are two sources for the fluids expelled from accretionary prisms?

24. What evidence is there to support the idea that fluid flow in accretionary prisms is focused?

25. What fluid may be a major agent in determining the composition of the continental crust?

26. List and describe three hypotheses to explain the origin of the oceans. Cite any evidence that supports these theories, as well as arguments against them.

27. How can water be stored in the mantle?

SUGGESTED READING

Alt, J. C. 1995. Subseafloor processes in Mid-Ocean Ridge hydrothermal systems. In Humphris, S. E., Zierenberg, R. A., Mullineaux, L. S., and Thomson, R. E., eds. *Seafloor Hydrothermal Systems, American Geophysical Union Geophysical Monograph 91*, pp. 85–114. Washington, D. C.: American Geophysical Union.

Frank, L. A. 1990. *The Big Splash.* Seacaucus, N. J.: Carol Pub. Group, 255 pp.

Hallam, A. 1983, The age of the Earth. In *Great Geological Controversies*, pp. 82–109. New York: Oxford University Press, 182 pp.

Moore, J. C. and Vrolijk, P. 1992. Fluids in accretionary prisms. *Reviews of Geophysics*, 30: 113–35.

Peacock, S. M. 1990. Fluid processes in subduction zones. *Science* 248: 329-37.

Rubey, W. W. 1951. Geologic history of sea water: An attempt to state the problem. *Geological Society of America Bulletin* 62: 1111–48.

Notation Used in Chapter Thirteen

Symbol	Quantity Represented	Physical Units
ΔT	temperature change	°C or °K
Δz	vertical distance	m
erf	error function	dimensionless
erfc	complementary error function	dimensionless
g_0	geothermal gradient at $z=0$	°C-km^{-1}
κ	thermal diffusivity	m^2-s^{-1}
λ	thermal conductivity	W-m^{-1}-°K^{-1}
q_h, q_{ho}	conductive heat flow, conductive heat flow at zero depth	J-s^{-1}-m^{-2} or W-m^{-2}
T, T_0, T_m	temperature	°C or °K
t, t_0, t_1, etc.	time	s
z	depth	m

CHAPTER 14

FLUIDS AND ORE DEPOSITS

14.1. FACTORS NECESSARY FOR THE FORMATION OF ORE DEPOSITS.

An **ore deposit** is a concentration of a mineral that is economically viable to mine. Not all mineral concentrations in the Earth's crust are ore deposits, because not all minerals are valuable enough to justify the cost of extraction. Ore deposits and mineral concentrations can be formed in many ways. Most ore deposits are formed through a process by which mass was mobilized and concentrated by moving fluids. If the fluid that formed the ore deposit was hot, the deposit is referred to as a **hydrothermal ore deposit.** Hydrothermal ore deposits (Figure 14.1) are generally recognized as the most important class of both base and precious metal deposits; they can be subdivided into hypothermal, mesothermal, and epithermal ore deposits. **Hypothermal** ore deposits form at great depths at temperatures of 300 to 500°C; **mesothermal** deposits form at intermediate depths at temperatures in the range of 200 to 300°C; and **epithermal** deposits form at relatively shallow depths at temperatures of 50 to 200°C.

Not all ore deposits form through the movement of fluids. For example, sulfur can be deposited around volcanic fumaroles, precipitating directly from a vapor to a solid phase. Evaporation of brines can concentrate evaporite minerals, such as halite (table salt), potash, and borax. The aluminum ore, bauxite, is formed not by the transport of aluminum minerals but by the de facto concentration of aluminum through the removal of other elements by extensive weathering in tropical regions.

There is a multitude of types of ore deposits; their study is accompanied by a complex and sometimes obscure jargon. Some terminology in ore geology is descriptive, referring to characteristics and properties inherent to a specific deposit or family of deposits, whereas other terms are genetic, referring to the mechanism thought to have formed a particular deposit or type of deposit. In truth, every ore deposit found in nature is unique, arising from a set of circumstances not precisely duplicated anywhere else on Earth through geologic time. Many deposits, however, have formed through similar processes, creating a spectrum of similarities. Confusion arises when attempts are made to create strict divisions and boundaries where none exist in nature.

Four factors must be present for a hydrothermal ore deposit to form (Sharp and Kyle, 1988; White, 1968):

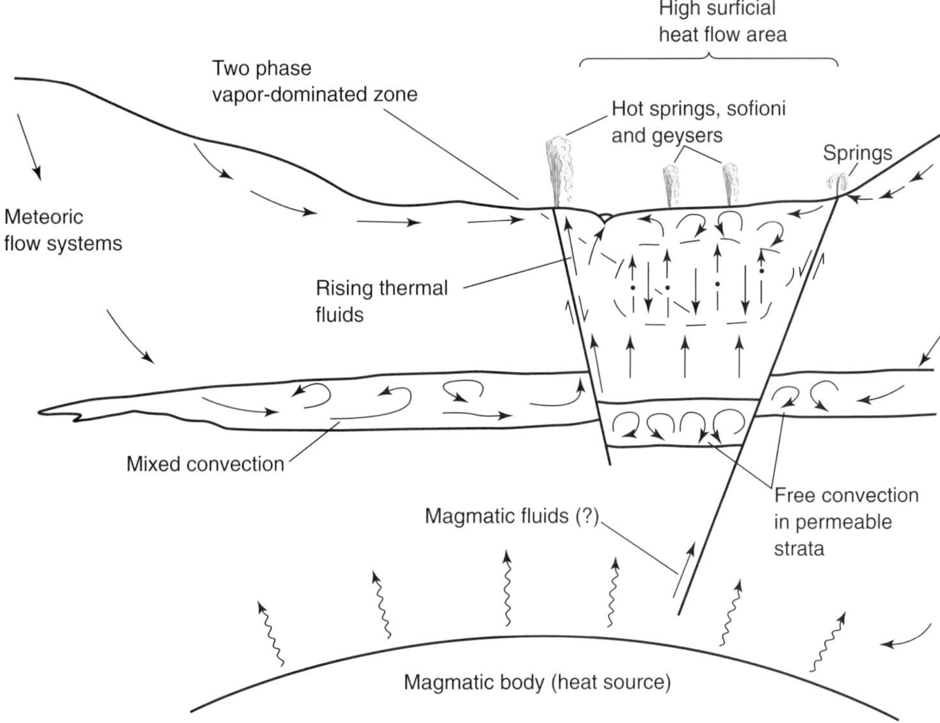

Figure 14.1 Possible flow paths and mechanisms in a hydrothermal system induced by a magmatic intrusion. (From Sharp and Kyle, 1988, p. 475.)

1. There must be a source for the mineral that forms the deposit.
2. The mineral must be dissolved in a fluid.
3. The fluid must move from the site of dissolution to the site of precipitation.
4. Precipitation must occur in response to physical and/or chemical changes in the fluid or porous medium it is traversing.

Other factors are implied, and the above list can be expanded. For example, it is implicit that permeability must be high enough for appreciable fluid velocities, otherwise significant amounts of ore minerals could not be carried to the sites of deposition. A head gradient must exist for fluid movement to occur, implying an energy source. The permeability structure must also promote focusing of flow from the area of dissolution to the area of precipitation to concentrate the precipitated minerals.

Fluids may move to the site of eventual ore deposition by different means in different geologic settings. In the case of Mississippi Valley-type (MVT) lead-zinc deposits, the ores are stratabound, indicating the fluids moved through the matrix of the host rock. A **stratabound** ore deposit is one confined to a single stratigraphic unit. In the case of the massive sulfide deposits formed today near midocean spreading ridges (see section 13.5) and other ore deposits associated with extinct volcanic activity, fluids circulated through fractured volcanic rocks. The famous **Comstock Lode** near Virginia City in western Nevada is an example of mineralization strongly associated with faults, suggesting that faults provided the primary pathways for mineralizing fluids. The Comstock Lode was described by Criss and Champion (1991) as "the most spectacular epithermal bonanza ever discovered in the United States." From 1863 through

1880, the Comstock Lode produced 8 million ounces of gold and nearly 200 million ounces of silver. The Comstock Lode formed at about 13 Ma near volcanic flow deposits of the same age. Oxygen isotope analyses indicate that extensive hydrothermal flow occurred through an area of approximately 75 km^2. Ores are located near several plumelike areas of fluid upwelling as identified by oxygen isotope analyses. These include the steeply dipping Comstock normal fault. Flow was thus focused from a regional to a local scale along high-permeability pathways that allowed hot hydrothermal fluids at depth to reach the surface.

Generally speaking, an ore fluid must sample a relatively large volume of rock to acquire the ore minerals later precipitated. There must then be some focusing of flow so that ore minerals are concentrated, and the chemical and/or physical conditions must be right for ore precipitation. The simultaneous occurrence in the Earth's crust of all of these circumstances is relatively rare.

The depths and scales of the flow systems that are postulated to have formed hydrothermal ore deposits vary widely. Hydrothermal systems formed in the vicinity of magmatic intrusions are essentially close to the scale of the intrusion. Igneous intrusions may be only a few meters to tens of meters in diameter, or they may be regional in scale. A **batholith** is an igneous intrusion with an areal extent greater than 40 mi^2 (100 km^2) and an unknown (but great) thickness. Many epithermal deposits evidently involve the focusing of flow from an area of 10 to 100 km^2 to discrete, high-permeability pathways such as faults. Mississippi Valley-type lead-zinc ores are an example of an ore deposit that may have formed from a regional flow system extending over hundreds of kilometers.

Traditional thinking held that most hydrothermal ore deposits formed in the upper 5 km of the crust (Skinner, 1997). However, as our knowledge of fluids in the Earth's crust grows, it has become apparent that meteoric and evolved fluids circulate to depths as great as 15 km. It is thus conceivable that ore deposits may form throughout the entire upper crust. The German Continental Deep Drilling Program drilled a 9101-m-deep borehole into the crystalline rocks of the crust and measured in situ permeabilities in the range of 5×10^{-18} m^2 to 3×10^{-16} m^2 over the depth range of the borehole. In situ permeabilities tended to be three orders of magnitude higher than measurements on core samples. This was a surprising result, as it might have been expected that increasing effective stress with depth would close fractures and reduce the permeability developed through open fractures (Huenges et al., 1997). Indirect evidence for the circulation of meteoric waters to depths of at least 10 km comes from stable isotope studies of veins in the Canadian Cordillera (Nesbitt and Muehlenbachs, 1991, 1995). Quartz and carbonate veins found over an area greater than 100,000 km^2 had isotopic signatures indicating the fluids that formed the veins were derived from meteoric waters. Oxygen isotopes from these veins also exhibited a nearly homogeneous composition with little variation between sites. This can be interpreted as indicating a high degree of fluid-rock interaction, implying pervasive fluid circulation in the upper crust. A second characteristic of vein minerals from the Canadian Cordillera studied by Nesbitt and Muehlenbachs (1991, 1995) is the lack of intermediate isotopic values between those that characterize the source meteoric waters and the evolved fluids that formed the veins. Nesbitt and Muehlenbachs (1995) interpreted the lack of intermediate values to indicate that virtually all vein-forming fluids had equilibrated with their host rocks at temperatures exceeding 400°C. Lack of waters of an intermediate composition also implies that convecting meteoric fluids followed distinctly different up and down paths with little to no mixing. Overall, the picture that emerges is one of the pervasive and deep convection of meteoric water through the entire brittle section of the upper continental crust. Implied average permeabilities are of the order of 10^{-17} m^2, and fluid pressures must have been approximately hydrostatic for convection to occur. Nesbitt (1988) suggested that gold deposits formed from such convective circulation, with epithermal and mesothermal deposits representing different points on a continuum rather than distinct deposits formed through different mechanisms. The type and degree of fluid circulation in the continental crust do seem to depend, however, on the structural and tectonic regime. Extensional tectonics leads to high-angle faults creating pathways for fluids to enter and circulate through the crust to great depths. However, in areas dominated by compressional tectonics, the depths to which meteoric water can penetrate are more restricted. In these settings, the dominant fluid may be water released from dehydration reactions in rocks undergoing metamorphosis.

14.2. NEPTUNISTS AND PLUTONISTS.

The first modern theory of ore genesis of which we have a clear record was propounded by George Bauer (1494–1555), who published a book titled *De Re Metallica* (1556) under the latinized name Georgius Agricola (Agricola, 1950). Agricola proposed that ore deposits formed from surface waters entering the solid Earth through fissures. According to Agricola, groundwater circulating through the Earth became heated, dissolved minerals from rocks, and deposited them in veins. One of Agricola's important contributions was that he was the first to recognize that ore deposits postdated the age of the rocks in which they formed. Agricola was also the first to recognize that ore deposits are not entirely random phenomena but rather can be categorized and classified into types based on their form and origin.

As the young science of geology developed through the eighteenth century, a bitter scientific debate emerged over the origin of the fluids that were postulated to have formed ore deposits in the crust. **Neptunists** believed that ores formed as the result of surface waters percolating downward. The great advocate of the neptunist school of thought was the German professor Abraham Gottlob Werner (1749–1807). Werner not only believed that ore deposits formed by the downward percolation of a primeval universal ocean but also that all igneous and metamorphic rocks formed by precipitation from these same fluids. Werner was a gifted lecturer and was able to enthrall his students with imaginative extrapolations. His discourses included expositions of the economic uses of various minerals, how the nature of minerals and rocks within geographic regions controlled the composition of the soil, and in turn how the composition of the soil influenced the availability of resources and human civilization and wealth. Werner went so far as to claim that the direction of particular strata had determined the history of languages and human migrations. In discussing the nature of different types of building stones, he would digress into the entire history of architecture. Werner's eloquence and intellectual charm were unmatched. Men who had already achieved distinguished careers in science took on the task of learning German for the sole purpose of attending Werner's lectures (Lyell, 1867).

The best geologist is the person who has most the most experience in studying the Earth in its various habitats. Although evidently possessed of considerable intellect, Werner had derived his neptunist ideas entirely from the study of one small area in Germany. Through the power of his oratory, he convinced inumerable others that the entire world followed the pattern of this area. Lyell (1867) claimed that Werner even misinterpreted many of the few geologic outcrops he had studied in Germany and concluded that Werner's ideas had retarded the development of geology.

The most important spokesman for the opposing school of thought was the Scotch geologist James Hutton (1726–1797). Hutton was trained as a physician, but an inheritance allowed him the luxury of pursuing his scientific inclinations. A quiet man, Hutton chose to present his views in print. The first exposition of his theories appeared in 1788 in the *Transactions of the Royal Society of Edinburgh*. This was followed in 1795 by a book, *Theory of the Earth*. Hutton's writings were the first presentation of uniformitarianism. **Uniformitarianism** is the principle that past geologic events can be explained by the same physical laws and processes that govern present-day events. Arguably, uniformitarianism is the single most important concept in the history of geologic thought.

Cyrus Fisher Tolman: Teacher and Author

Cyrus Fisher Tolman (1873–1942) was born in Chicago on June 2, 1873. He graduated from the University of Chicago with a B.S. degree in 1896. He remained at Chicago as a graduate student until 1899, although his study was interrupted by service in the Spanish-American War.

Tolman's earliest work was in the geology of ore deposits. He was professor of geology and mining at the University of Arizona from 1905 to 1912 and was employed as Territorial Geologist from 1910 to 1912. In 1912, Tolman was lured away from Arizona by John Casper Branner to join the faculty of Stanford University in California. At Stanford, Tolman was one of the most beloved teachers and mentors in the department. He had a generous and genial nature, and took a personal interest in his students. He became affectionately known as "chief" by his students, due to his direction of the summer field camp in geology from 1913 to 1931. Under his direction, the summer courses in field geology at Stanford became national models. Tolman's syllabus of field methods became nationally famous and was adopted as the prototype for the manuals used by oil companies and other universities.

Tolman was unsurpassed in his contributions to the geological knowledge of the western United States. He made contributions to mining, petroleum, and groundwater geology. Tolman was active as a consultant in these areas and was employed by numerous large corporations and municipalities.

Cyrus Fisher Tolman (1873–1942).
Photograph reproduced courtesy of the GeoHistory archives at the School of Earth Sciences, Stanford University.

Hutton was not a particularly adept writer. Hallam (1983) characterized Hutton's writing as "verbose" and "obscure" and noted that it lacked a "coherently organized structure." The importance of Hutton's ideas would only become recognized when they became popularized by his colleague, John Playfair (1748–1819), in his book *Illustrations of the Huttonian Theory of the Earth* (1802) and later by Charles Lyell in his classic text *Principles of Geology*, first published in 1830.

Hutton was an advocate of the **plutonist** (also known as vulcanist) school of thought. The plutonist viewpoint was that igneous rocks formed from the cooling of molten rocks. One of Hutton's geologic epiphanies occurred on a field trip in 1785. In the mountains of Scotland, Hutton found an outcrop where veins of red granite branched out from a central mass and intruded into a schist and limestone. The granite-limestone contacts exhibited a clear metamorphic aureole, indicating the granite had intruded into the limestone as an igneous mass and then cooled. This was conclusive proof of the vulcanist viewpoint.

Plutonists also believed that ore deposits represented the cooled residue of molten material; ore minerals were not believed to be soluble in water.

For many years, Tolman taught a course in groundwater geology at Stanford. His course emphasized the necessity of combining geologic investigation with engineering analysis. Eventually, his classroom notes coalesced into the first English-language book on groundwater. Published in 1937, the book was titled simply, *Ground Water*. Showing a sense of prescience, in the preface of this book Tolman claimed:

> This book records the birth and describes the development of a new science—coordinated scientific data regarding the occurrence, motions, and activities of subsurface water, and the hydrologic properties of water-bearing materials—christened by O. E. Meinzer "Ground-water Hydrology."

A more somber tone is laid in the book's dedication:

> To my son and colleague, John van Steen Tolman, who, true to the highest traditions of his profession, gave his life in order to carry out faithfully the obligations imposed upon him as mine superintendent.

Tolman also established a scholarship in economic geology devoted to the memory of his son, John.

The following story is told in connection with the publication of Tolman's treatise on groundwater. After many years of development, Tolman misplaced the manuscript. He spent hours searching both his home and office, but to no avail. As the days and weeks passed, Tolman became testy and short-tempered. One day, Tolman stopped by the cigar store on University Avenue to pick up a box of his favorite Santa Fe cigars. "By the way, professor," the proprietor said, "the last time you were here, you left this bundle of papers." The "bundle of papers" was the mislaid manuscript. *Ground Water* went on to become the standard text on the subject for decades, being reprinted many times.

Through his teaching, C. F. Tolman brought to his students a realization that the science of hydrogeology can be both a fascinating study and an instrument for the betterment of mankind. Tolman's most famous student was Joseph F. Poland (1908–1991), who went on to a distinguished career at the U.S. Geological Survey. Poland became the world's leading expert on land subsidence. His research on the subject led to the saving of millions of dollars in construction costs through his advising on the redesign of irrigation canals, aqueducts, and roads. Poland determinined the cause of subsidence problems in Venice, Italy, and became known to Italians as the savior of that city.

FOR ADDITIONAL READING

Kildale, M. B. 1943. Cyrus Fisher Tolman, 1873–1942. *Economic Geology*, 38 (6): 541.

Both the neptunists and plutonists were partially correct. The plutonist theory of the origin of igneous rocks turned out to be correct. However, the idea that most ore deposits are the crystallized remnants of former magma was incorrect. Today, it is generally conceded that most ore deposits form by the precipitation of minerals from aqueous fluids.

14.3. FLUID INCLUSIONS.

Much of what we know concerning the nature of ore-forming fluids comes from the study of fluid inclusions. A **fluid inclusion** is a minute cavity in a crystal filled with fluid or gas. Fluid inclusions are believed to form when crystals grow, surround, and trap fluids. If a fluid inclusion is found in a crystal associated with an ore deposit, it may contain a sample of the fluid that formed the deposit. If the sealing of the inclusion occurs during the growth of the surrounding crystal, the inclusion is said to be a **primary fluid inclusion**. The two types of information most readily obtained from an analysis of fluid inclusions are the salinity of the trapped fluid and its temperature at the time it was trapped. The salinity is estimated by cooling the inclusion and noting the temperature at which the enclosed liquid freezes. The freezing temperature is a function of the total dissolved solids. The original temperature at the time of trapping is estimated by heating the inclusion until the separate gas and liquid phases disappear. The temperature at which this happens is the **homogenization temperature**. The operative assumption is that at the time the inclusion formed, a single homogeneous phase was present. The homogenization temperature is thus a minimum estimate of the temperature of the original hydrothermal fluid.

To infer the nature of ore-forming fluids from the analysis of fluid inclusions, certain assumptions are usually made. If one or more of these are assumptions are not correct, interpretations may be faulty (Roedder, 1979). One assumption is that the fluid trapped in the inclusion consisted of a single, homogeneous phase. This assumption cannot be shown to be correct when only a single inclusion is analyzed. However, when many inclusions in a single sample show a similar ratio of fluid to gas, the simplest interpretation is that a homogeneous fluid was trapped. Another assumption is that the cavity in which the fluid was trapped must not have changed in volume after sealing. Volume changes can occur by several means. Changes in the ambient pressure may cause the inclusion to shrink or expand, minerals may precipitate on the walls of the inclusion, and cooling may cause the inclusion cavity to contract. It is not certain, however, that these volume changes can lead to errors in interpretation. When the inclusion is re-heated to estimate the homogenization temperature, minerals that precipitated may again dissolve, and any contraction due to cooling will be reversed. Roedder (1979) concluded that volume changes due to pressure changes occur but are not important and can be ignored. A third assumption is that nothing has been added to or subtracted from an inclusion after sealing. Leakage is rare, except when rocks have been crushed or subject to other significant deformations. But a special type of leakage, termed necking, may be common and lead to significant errors in interpretation. **Necking** of a fluid inclusion is the formation of two or more **daughter inclusions** from a single original inclusion (Figure 14.2). The ratio of gas to fluid phases in daughter inclusions is not necessarily the same. The daughter inclusion with a higher ratio of liquid to gas homogenizes at a higher temperature, whereas the daughter inclusion with a lower ratio of liquid to gas homogenizes at a lower temperature. Necking and other changes in the shape of fluid inclusions are common and create much of the scatter found in homogenization temperatures (Roedder, 1979).

From the study of fluid inclusions, we know that the composition of ore-forming fluids is varied. However, all are saline and similar in composition to the fluids found in present-day geothermal systems. For many ore deposits, it is common to find that several types of fluids were involved in their creation. This finding leads to the interesting conclusion that although a fluid source is necessary for an ore deposit to form, the fluid itself is not the controlling factor (Skinner, 1997). The major dissolved constituents of aqueous ore fluids are sodium, potassium, calcium, and chlorine. It is usually quite difficult to discern from where a fluid obtained its dissolved constituents. Juvenile fluids from a cooling magma may be a source of some constituents, but most ore fluids (indeed, any moving fluid in the crust) evolve continually through matrix-fluid interactions along their flow path. There are "innumerable reactions" that may occur between a migrating ore fluid and the rocks it passes through (Skinner, 1997). Many, and possibly all, rocks may serve as sources of the metals found in hydrothermal ore deposits. It appears that the most important factor is the volume of rock sampled by moving fluids, not the rock type itself. If the ratio of fluid to rock is low enough, the moving fluid may become sufficiently enriched in metallic elements to have the potential to form an ore deposit.

14.4. Composition of Ore Fluids.

Fluids that form ore deposits may be oceanic, meteoric, or evolved (see section 4.5). With the

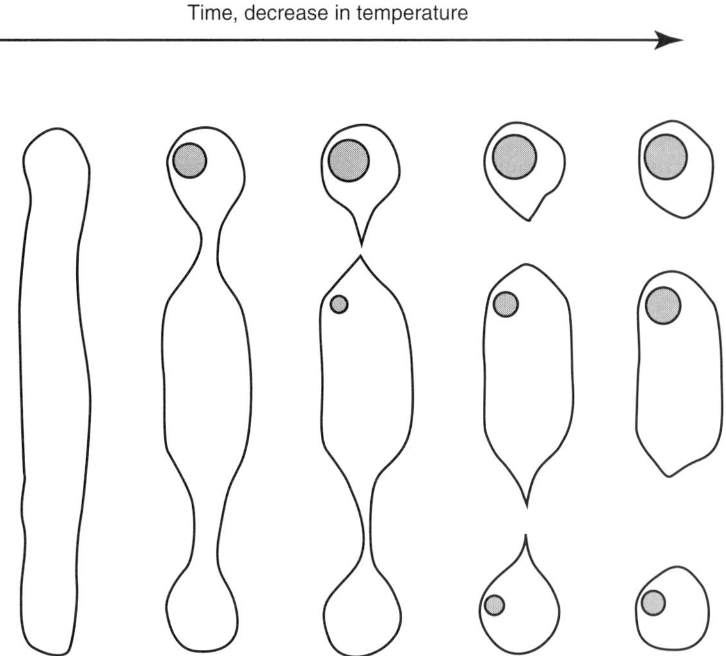

Figure 14.2 Necking down of a fluid inclusion. With increasing time and/or temperature one fluid inclusion may segregate into several daughter inclusions, each with a different ratio of gas to fluid. Analysis of necked inclusions may yield faulty conclusions concerning the temperature of ore-forming fluids.
(From Leach and Rowan, 1993, p. 968.)

exception of marine settings, nearly all of the aqueous fluids that form ores derive from a meteoric source and quickly evolve by interacting chemically with the rocks through which they are circulating. Juvenile fluids emanating from cooling igneous rocks are rare; this is true even for fluids circulating near magmatic bodies (Sharp and Kyle, 1988). Most common ore minerals are sulfides. A **sulfide mineral** is a mineral containing a metal and sulfur. Two common examples are galena (PbS) and pyrite (FeS_2). Most metals are minor constituents of ore-forming fluids, and most sulfide minerals have low solubilities in water. It is therefore difficult to explain how ore fluids can carry enough sulfides to form ore deposits. At the present time, we believe that most metals are carried by complex ions or ligands. A **ligand** is any molecule that binds to another. Ions that are thought to act as ligands in the transport of ore minerals include Cl^-, HS^-, and OH^-. Sulfide complexes such as $Zn(HS)_3$ or $HgS(H_2S)_2$ and chloride complexes such as $ZnCl_2$ or $CuCl_3$ are also potential ligands.

14.5. PRECIPITATION OF ORE MINERALS.

The mechanism by which ore minerals are precipitated from solution to form a deposit is usually uncertain. At least four mechanisms are potentially important (Skinner, 1997). (1) Temperature changes can cause precipitation of ore minerals by decreases in solubility or through more subtle and complex chemical changes, such as changes in the stability of the complex ions that are thought to transport ore metals. In most cases, a decrease in temperature implies a decrease in solubility, and ore minerals will precipitate from a saturated solution. One way to achieve a large change in temperature over a short distance is for a rising, hot ore-fluid to mix with a colder, near-surface fluid. (2) Solubility can change as a result of a change in fluid pressure, but the pressure change must be large to make a significant difference in solubility. The size of the pressure change needed is about 10^8 Pa (1000 bars), which is the hydrostatic pressure at the bottom of a water column 10 km high. A (partly) pressure-controlled phenomena that can lead to precipitation is boiling. Boiling increases the concentration of ore minerals in solution and selectively removes the more volatile constituents from solution, increasing the solution alkalinity and decreasing metal solubility. (3) Chemical reactions between an ore fluid and a rock are a major cause of precipitation. This conclusion is supported by the observation that certain ores are preferentially deposited on certain rocks. In general, three types of reactions are important. The removal of hydrogen ions from an ore solution decreases the acidity of the solution, leading to the precipitation of sulfides. Host rocks may also contribute a reduced sulfur species such as H_2S to the solution, again causing precipitation of sulfide minerals. The wallrock may change the oxidation state of the ore fluid, decreasing the stability of some complex ions. An example of this mechanism operating is the precipitation of precious metals, uranium, vanadium, and copper by the presence of organic matter causing a more reducing environment. (4) Chemical changes that take place when a hydrothermal fluid mixes with a different fluid may cause ore precipitation. If the mixed fluid had a different temperature, however, it is often difficult to conclude if the precipitation was caused by a chemical or a thermal change.

14.6. METALLOGENESIS.

Ore deposits are interesting in and of themselves, but in a larger sense, they are tracers for fluid flow events. The formation of an ore deposit may be a consequence of that fluid flow event, but it is not necessarily the most important consequence. Fluid flow may be both a cause and an effect. Flowing fluids may create phenomena such as ore deposits, but the flow system has an ultimate origin in some larger geologic event such as a plate collision or orogeny. **Metallogenesis** is the study of how and why ore deposits form through space and time, and the relationship of ore formation to large-scale geologic processes such as plate movements. Metallogenetic studies are motivated by the observation that specific types of ore deposits seem to be preferentially located with respect to space and time (Figure 14.3). Examples of ore deposits that are confined to certain characteristic regions cited by Skinner (1997) include magmatic-related deposits of the Andes, tin deposits of southeast Asia, and the lead-zinc deposits found in the Mississippi Valley of central North America. Other types of deposits are broadly distributed in a spatial sense but are temporally confined to certain geologic ages. Banded-iron formations are common in Precambrian rocks but are not found in any formations of Phanerozoic age. Similarly, the great coal measures of the Carboniferous age are unique in their size. Why the Carboniferous age alone gave rise to coal deposits of great age is not known, but the restriction of banded iron-formation to Precambrian time is believed to be related to changes in Earth's atmosphere. The primitive atmosphere is believed to have been oxygen-poor and reducing (Kasting, 1993), and the oceans are postulated to have contained a large reservoir of ferrous iron. As oxygen began to increase in the atmosphere, iron was precipitated wherever and whenever ferrous iron came into contact with atmospheric oxygen. The result was the formation of distinct banded-iron formations. The oceanic iron reservoir served as a sink for atmospheric oxygen until it was exhausted and the concentration of oxygen in the atmosphere began to approach today's levels. The history of banded-iron formation is thus believed to be an earmark for the evolution of Earth's atmosphere.

One hypothesis that seeks to explain systematic temporal variations in the distribution and style of ore deposits relates these trends to the cyclic aggregation and breakup of large continents (Barley

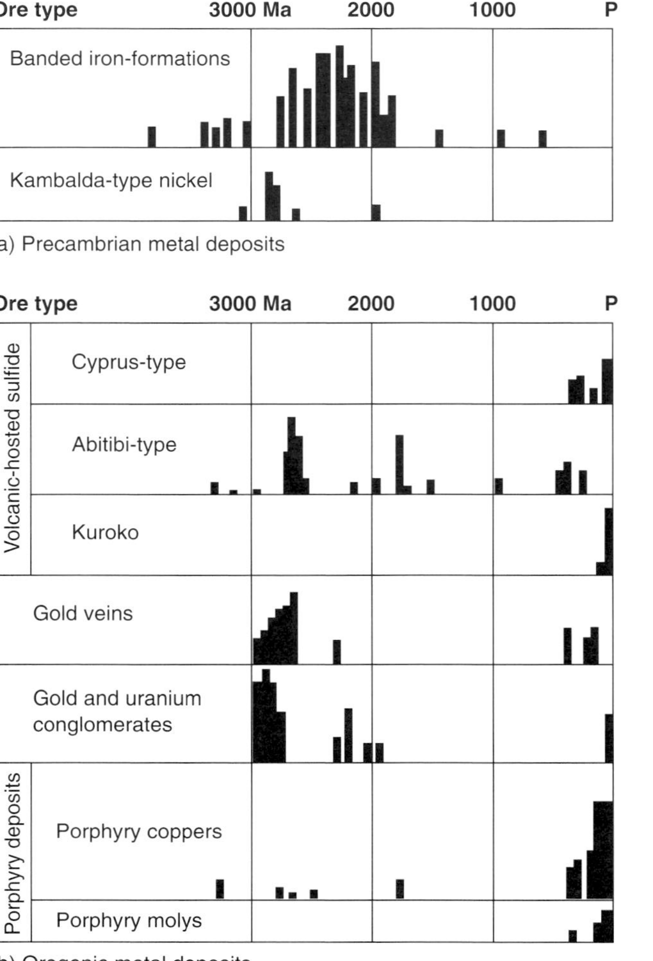

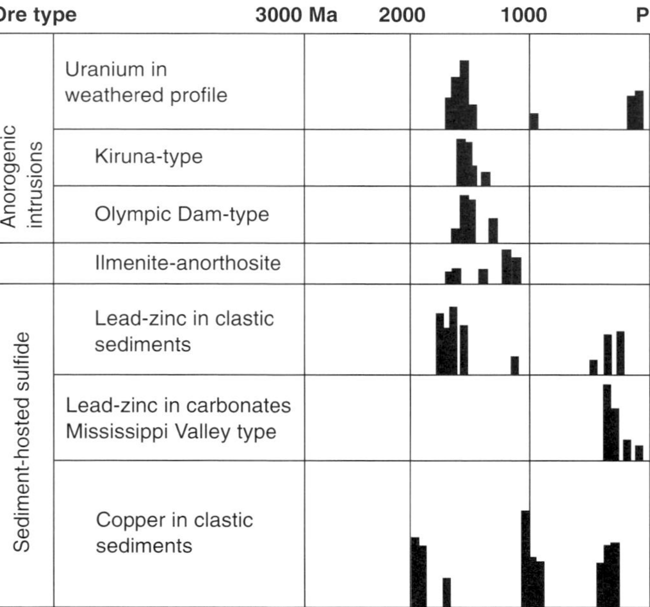

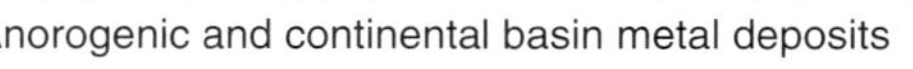

Figure 14.3 Styles of metal deposits through geologic time. The restriction of certain types ore deposits to distinct geologic periods suggests a relationship between ore formation and plate tectonic cycles.
(From Barley and Groves, 1992, p. 292.)

and Groves, 1992). The logic is that certain types of ore deposits are created in certain plate tectonic settings. These include mesothermal lode gold and volcanogenic massive sulfides in convergent margins, and metal deposits associated with anorogenic magmatism and/or continental sedimentation. The idea is that there is a so-called "supercontinent cycle" that repeatedly produces the type of plate tectonic settings that favor the formation of characteristic types of ore deposits. According to the hypothesis, fragments of continental crust move off upwelling domes and aggregate above convective downwellings to form supercontinents such as Pangea. These large continents inhibit subduction and mantle cooling, resulting in the buildup of heat and large-scale convective upwelling. The upwelling splits the supercontinent and the fragments continents drift apart, beginning another cycle. Thus, metal deposits that are anorogenic in origin should be most abundant toward the end of a period of aggregation, whereas those that are related to orogenesis and convergent margins should be most abundant during periods of aggregation.

14.7. Fluids from the Mantle?

Traditional theories of the formation of ore deposits in the Earth's crust have recently been challenged by Gold (1993; 1999). Gold suggests that in many instances, metallic ores may have been carried in solution by inorganic hydrocarbons moving more or less directly upward from the mantle.

Gold's ideas regarding the formation of metallic ore deposits are part of a larger theory whose key tenet is that most oil, gas, and coal deposits in the crust are not the remains of thermally altered organic matter but rather result from inorganic hydrocarbons streaming up from the mantle. Gold cites seven types of empirical evidence for his deep-Earth gas theory:

1. Petroleum reservoirs are often found in linear patterns extending for hundreds or thousands of kilometers that are more closely related to deep-seated and structural features of the crust than to sedimentary deposits.
2. **Koudryavtsev's Rule**: Hydrocarbon-rich areas tend to be enriched at all lower levels, regardless of geological age, extending down to and sometimes including the crystalline basement.
3. Methane is found in many areas where a biogenic explanation for its presence is improbable.
4. Hydrocarbon deposits of a large area often show common chemical features, regardless of the varied composition of the reservoirs in which they are found.
5. A number of hydrocarbon reservoirs appear to be refilling as they are emptied.
6. The constancy of the isotopic composition of carbonates throughout geologic time argues against the cumulative sequestration of organic carbon (which is enriched in the lighter carbon isotope, C-12) in the Earth's crust in the form of oil, gas, and coal deposits.
7. Hydrocarbons are often associated with the inert and inorganic gas helium.

There is no doubt that some methane upwells from the mantle. The presence of methane in midocean magmatism is prima facie evidence of abiogenic gas in the mantle. But it is generally not accepted that any more than a tiny fraction of the world's economic natural gas resources orginated abiogenically. Apps and van de Kamp (1993) wrote, "Overwhelming evidence supports the belief that the world's natural-gas resources come from the decomposition of organic matter in sedimentary rock formations." One line of evidence against the abiogenic theory is the carbon isotope ratios of commercial natural gases. Carbon contains two stable isotopes, a heavier C-13 and a lighter C-12. Terrestrial plants tend to be enriched in the lighter C-12 isotope. In contrast, magmatic methane, which is unquestionably of abiogenic origin, tends to contain more of the heavier C-13 isotope. A survey of commerically produced natural gases (Figure 14.4) shows that the isotopic composition of these gases tends to be similar to that produced artificially in the laboratory by heating plant-derived organic matter (Jenden et al., 1993). Gold (1999),

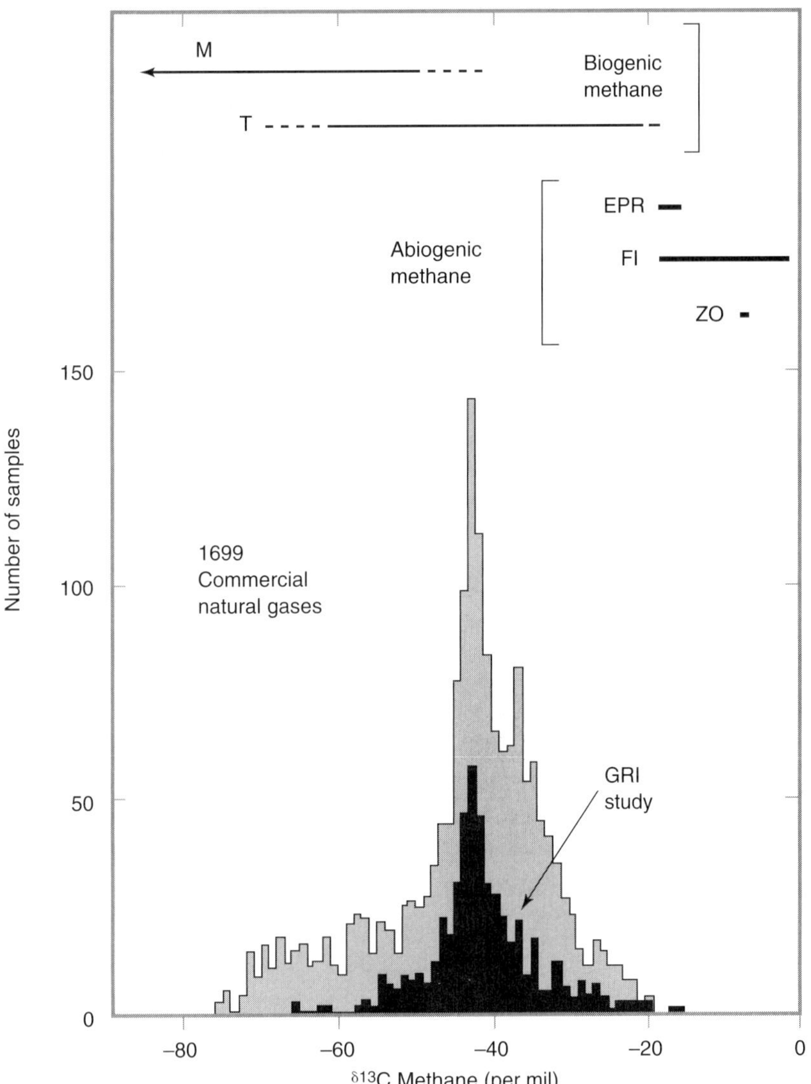

Figure 14.4 Distribution of carbon isotopes in commercially produced natural gases from throughout the world. Also shown are the carbon isotope ratios found in abiogenic and biogenic methane. Abiogenic sources are as follows, EPR = geothermal methane from the East Pacific Rise, FI = methane fluid inclusions in igneous rocks, ZO = methane from the Zambales Ophiolite. Biogenic sources are T = thermogenic methane produced by heating organic matter, and M = methane produced by microbial action on organic carbon in unconsolidated sediments. (From Jenden et al, 1993, p. 37.)

however, argues that magma-derived methane is atypical and that the methane found in commerical reservoirs has undergone fractionation by diffusion through the crust.

Gold (1999) claims two weaknesses in the hydrothermal theory of metallic-ore formation: (1) insolubility of metals in aqueous fluids, and (2) lack of suitable driving forces for fluid migration in the crust. The first argument (insolubility) can be countered by drawing on geologic time. Ore fluids percolating over thousands to millions of years can deposit large amounts of metallic ores, even if the concentration of the ore mineral in the ore fluid is low. The second argument is debatable, as a number of mechanisms exist for driving fluids through the continental crust. Notably, topographic gradients can create large-scale fluid movements that persist for tens of millions of years.

As support for his theory, Gold (1999) cites the example of diamonds. Diamonds are found primarily in kimberlite pipes. A **kimberlite pipe** is a pipe-shaped extinct volcanic orifice with a diameter of about 200 meters at the surface that narrows to a few meters at a depth of about 1 km. Kimberlite itself is a type of blue soil formed by the weathering of peridotite. Because diamonds are a unstable form of carbon at surface temperatures and pressures, they must have been brought up to the surface very quickly. Gold (1999) argues that kimberlite pipes formed through a high-pressure gas blowout from the mantle. As the gas enountered lower pressures in its upward journey, it underwent rapid adiabatic cooling with diamonds preserved by nearly instantaneous "supercooling."

The occurrence of hydrocarbons with metallic ore deposits provides some circumstantial evidence for Gold's theory. Gold (1999) notes that gold is often associated with carbon and relates that gold miners would often search for carbon trails to lead them to ore deposits. Other researchers have noted an association of organic matter with metallic ores. Kesler et al. (1994) wrote, "One of the enduring conundrums of Mississippi Valley-type (MVT) deposits [see section 14.8] is their relation to crude oil, natural gas, and other basinal hydrocarbons." Similarly, Anderson (1991) noted that "The relationship between organic material and Mississippi Valley-type deposits has been discussed many times, but it is fair to say that the relationship, if any, remains tentative and poorly understood."

A major problem in evaluating Gold's theory of ore depostion from inorganic hydrocarbons is our lack of knowledge of high-pressure chemistry. We do not know, for example, if hydrocarbons and metals form compounds that precipitate metals when moved from the high-pressure, high-temperature environment of the upper mantle to the upper crust. It is well accepted that some inorganic methane (CH_4) escapes from the mantle, and the formation of diamonds is evidence that carbon exists in the upper mantle. However, it is not accepted that a majority of the Earth's fossil fuels formed from mantle outgassing.

CASE STUDY 14.8 Case Study: Mississippi Valley-Type (MVT) Lead-Zinc Deposits.

The central part of the North American continent contains widespread sulfide mineralization. Most deposits are found in the drainage basin of the Mississippi River (Figure 14.5) and have been exploited as economic sources of lead and zinc since the middle of the nineteenth century (Ohle, 1991). Although MVT lead-zinc deposits have been studied intensively for more than 100 years, no consensus exists regarding the mechanism by which they formed.

14.8.1. CHARACTERISTICS OF MVT DEPOSITS.

Following Sverjensky (1986), MVT lead-zinc deposits have the following pertinent characteristics:

1. Ores occur primarily in carbonate rocks that form a thin cover over an igneous or highly metamorphosed Precambrian basement. The ores are clearly epigenetic (Anderson and Macqueen, 1988). An **epigenetic** ore is one that formed after the host rock was already in place.

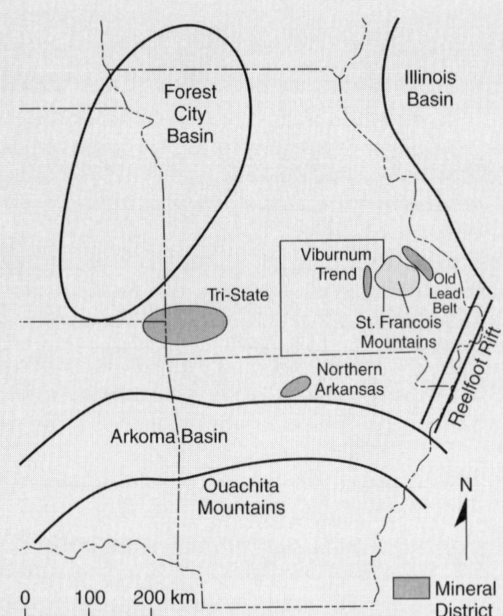

Figure 14.5 Location of Mississippi Valley-type lead-zinc deposits in northern Arkansas and south Missouri. The basinal-brine hypothesis postulates that these ore deposits formed when fluids were tectonically expelled from the Arkoma Basin during the Ouachita Orogeny in Late Pennsylvanian–Early Permian time.

From Shelton et al., 1992, p. 675.)

2. Ores are stratabound; that is, they are confined to particular geologic strata, especially areas of high permeability such as breccias and paleokarst.
3. The most common minerals found in MVT lead-zinc deposits are galena and sphalerite. Pyrite and marcasite are usually present in smaller quantities, and barite and fluorite are common in some districts (Anderson and Macqueen, 1988).
4. MVT deposits are not generally associated with igneous rocks, implying they are not epithermal deposits that formed as a result of fluid circulation induced by an intrusive heat source.
5. MVT ores are never found in basement rocks, but ores are often found in overlying sedimentary rocks near basement highs.
6. MVT ores are found at shallow depths, generally less than 600 m below the present day ground surface, and are believed to have never been buried more deeply than 1500 m.
7. Studies of fluid inclusions from MVT minerals reveal that the ore-forming fluid was highly saline (10 to 30% total dissolved solids) and hot (50 to 200°C), similar to the brines found in sedimentary basins today.
8. Reconstruction of the supposed thickness of the sedimentary cover at the time of ore formation shows that average geothermal gradients and surface temperatures are inadequate to explain the high-temperature fluids that apparently formed the ore deposits.
9. The deposition of sphalerite in the deposits of the Upper Mississippi Valley can be divided into three stages, each of which is composed of a number of distinctive color bands. Furthermore, these bands can be correlated over distances of hundreds of kilometers.
10. MVT deposits tend to be found on the margins of sedimentary basins.

14.8.2. The Basinal Brine Hypothesis.

The lack of any direct igneous heat source, as well as the apparent linkages with sedimentary basins, has led to the development of the **basinal brine hypothesis** for the formation of MVT lead-zinc deposits. The basinal brine hypothesis states that MVT ores formed when brines in sedimentary basins were driven to the margins of these basins, where one or more geochemical and/or physical mechanisms caused mineral precipitation. During the course of their travels, the brines picked up metals (primarily lead and zinc) that were subsequently deposited in highly permeable carbonate rocks. A tectonic event is commonly invoked to provide the driving force for brine migration. In the case of the MVT deposits found in Missouri and Arkansas, the brines were believed to have been expelled from the Arkoma Basin to the south during the Ouachita orogeny (Figure 14.5) in Late Pennsylvanian and Early Permian time (Leach and Rowan, 1986). The brines traveled northward through the Lamotte Sandstone until they reached the Ozarks in Arkansas. The Ozark uplift is an ancient feature believed to have been a topographic high as long ago

Case Study 14.8 *(continued)*

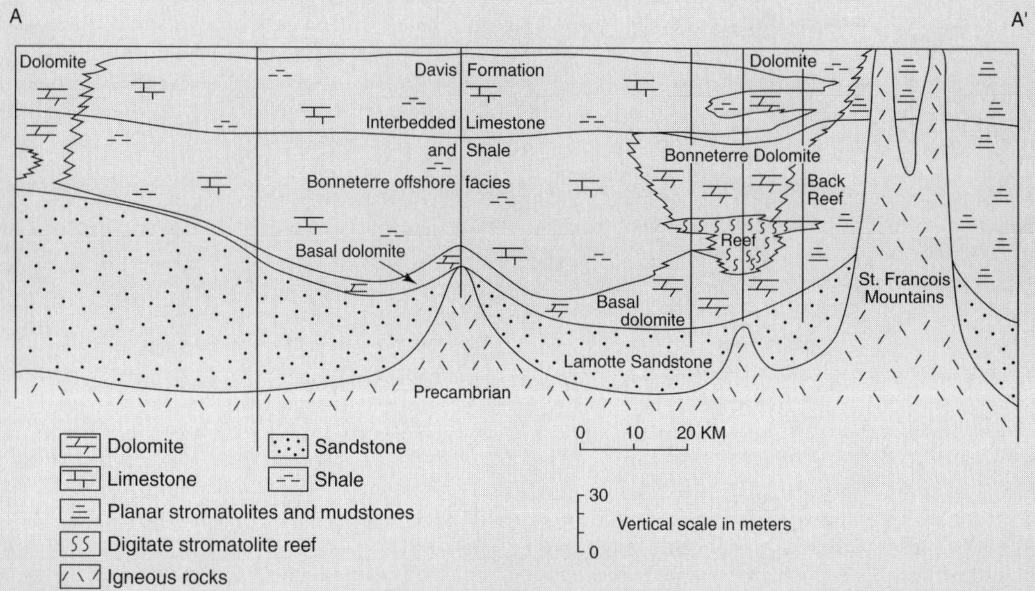

Figure 14.6 Geologic cross section through southeast Missouri, showing geologic relationships pertinent to the formation of Mississippi Valley-type lead-zinc deposits. Ore-forming fluids are believed to have migrated through the basal Lamotte Sandstone until forced up into overlying carbonate units where the Lamotte Sandstone pinched out against the basement rocks of the Ozark Uplift. Ores tend to be concentrated in areas of high permeability, such as paleo reefs.
(From Gregg and Shelton, 1989, p. 223.)

as the Pennsylvanian and before. Where the Lamotte Sandstone pinched out against the Precambrian-age crystalline rocks of the Ozarks, brines were forced upward into overlying carbonate units (Figure 14.6). One or more chemical and/or physical mechanisms then caused lead and zinc minerals to precipitate in carbonate rocks such as the Bonneterre Dolomite, with ores (and presumably fluid flow) being concentrated in the highly permeable sections of carbonate rocks, such as reefs.

Although the basinal-brine hypothesis is ascendant, considerable controversy remains concerning the mechanism that drove basinal brines to the site of ore deposition. In general, hypotheses are divided into three categories: (1) compaction-driven flow (Cathles, 1993b), (2) topography-driven flow (Garven, 1993), and (3) seismic pumping (Clendenin and Duane, 1990).

14.8.3. Compaction-Driven Flow.

The oldest theory of brine migration calls upon simple sediment compaction to provide the impelling force for brine migration. As the Ouachita orogeny proceeded in Late Pennsylvanian time, the Arkoma Basin formed as a foreland basin to the north of the uplifted region and received substantial amounts of sediment shed from the young mountains to the south. Although present-day sediment thickness in the Arkoma Basin is no more than 4 to 5 km, reconstructions of the sedimentary section suggest there has been 3 to 7 km of erosion (Lee et al., 1996), implying 7 to 12 km of sedimentary strata in late Pennsylvanian time. The potential thus existed for substantial amounts of pore fluids to have been expelled from the Arkoma Basin by the simple processes of burial and compaction.

The compaction hypothesis was dealt a severe blow by Cathles and Smith (1983), who first evaluated the hypothesis quantitatively by calculating the temperature of migrating brines. They found that the rates of fluid expulsion due to the process of sedimentation were likely to be 1000 times too low to explain the thermal anomalies associated with MVT deposits. As hot brines from the deep basin slowly migrate up dip to the basin margins, they cool off before they reach the sites of ore deposition. Cathles and Smith (1983) and Cathles (1993b) suggested that the hypothesis could, however, remain viable if hot brines were released episodically. Cathles (1993b, p. 109) also suggested that Arkoma Basin brines may have been forced from the basin by the generation of natural gas, as the Arkoma today is a "basin almost literally filled with gas."

The primary advantage of the compaction-driven brine hypothesis is that it starts with the right type of fluid: hot and salty. Another advantage is that the requirement of episodic release is strongly bolstered by the banded nature of the ore deposits, which apparently indicate that ore deposition was episodic instead of continuous. The compaction-driven flow hypothesis has two primary difficulties. The first is that there is a limited amount of fluid to work with; the second is the lack of a mechanism to store and release fluids. Cathles (1993b) estimated that if pore fluids expelled by compaction were responsible for the thermal anomalies associated with MVT deposits, they would have had to have been focused by a factor of 62 to achieve the flow velocities necessary to transport hot brines to the basin margins without cooling off. This focusing can be either temporal or spatial. There is no doubt that flow was spatially focused; otherwise, no ore deposits would have formed. However, ore deposits are also widespread. Leach (1979) pointed out that trace amounts of sphalerite are found throughout central and southern Missouri. The implication is that lead-zinc mineralization in the Ozark region was not restricted to the main ore districts but was regional in nature. Flow may not have been focused spatially but rather temporally, with distinct pulses of mineralizing fluids. The problem then becomes one of finding a mechanism to store and release fluids on a vast scale. No such mechanism has yet been proposed.

14.8.4. Topographic Flow.

Topographic flow is an appealing mechanism, as it is pervasive throughout the upper 5 to 15 km of the continental crust. Darcy's Law requires that flow must occur if a head gradient exists and the permeability is nonzero. These conditions are met virtually everywhere on the continents, although flow velocities are not necessarily significant in all regions. The topographic flow hypothesis has the inherent advantage of having unlimited amounts of fluid to work with, as basinal brines are continuously replenished by meteoric recharge. If permeability is high enough, flow velocities may be significantly higher than can be achieved in steady-state compaction-driven flow. The primary difficulty with invoking topography-driven flow as the mechanism that formed MVT lead-zinc deposits in central North America is that the fluid is of the wrong type. Although basinal brines are replenished as quickly as they flow to discharge areas, the recharging fluid is not hot and salty but cold and fresh. There is some question, therefore, as to whether or not topographic flow can satisfy the temperature and salinity constraints that have been imposed by our knowledge of ore-forming fluids gained through fluid inclusion analyses. Garven et al. (1993) modeled ore genesis in the Ozark region arising from fluid flow through the Arkoma Basin and were able to obtain steady-state temperatures in discharge regions in the neighborhood of 90 to 100°C at depths of 1 to 1.5 km. Garven et al. (1993) also found their computer models produced relatively short-lived pulses of fluid temperatures as high as120 to 130°C in areas of ore formation. However, Lin et al. (2000) later showed that the transient thermal pulses present in the Garven et al. (1993) models

Case Study 14.8 *(continued)*

were largely an artifact of improper boundary conditions and would not occur in nature. Computer simulations of heat transport by topographically driven groundwater flow therefore suggest that it is difficult to produce the high ore-formation temperatures (as high as 150°C) that are implied by the fluid inclusion data. Simulations of topographically driven flow may also be poorly or not at all constrained, often using optimistic or unrealistic estimates of permeability and background heat flow in order to match constraints imposed by fluid inclusions. Perhaps a more difficult problem for the topographic-flow hypothesis is the limited amount of dissolved solids available. Although the supply of water is essentially unlimited, throughgoing meteoric flow would soon flush out all original basinal brines. Even if substantial thicknesses of evaporites were present at some time in the Arkoma Basin, they would have been dissolved relatively quickly. Calculations by Deming and Nunn (1991) and Cathles (1993b) show that any conceivable supplies of solutes would be exhausted in 1 to 10 Ma, depending on flow rates and the degree to which flow was focused. The conundrum is that high flow velocities are required to satisfy the thermal constraints, yet the higher the flow velocity, the more quickly the basin's supply of salt is exhausted. Cathles (1993b) pointed out that the permeability of the rocks in which MVT deposits are found remains very high today. Once a topographic flow system is initiated by the uplift of a mountain range, it remains in place for millions of years. However, brines remain in the foreland of the Ouachita Mountains today. If a topographic flow system had existed there at some time in the past, these brines should have been flushed out.

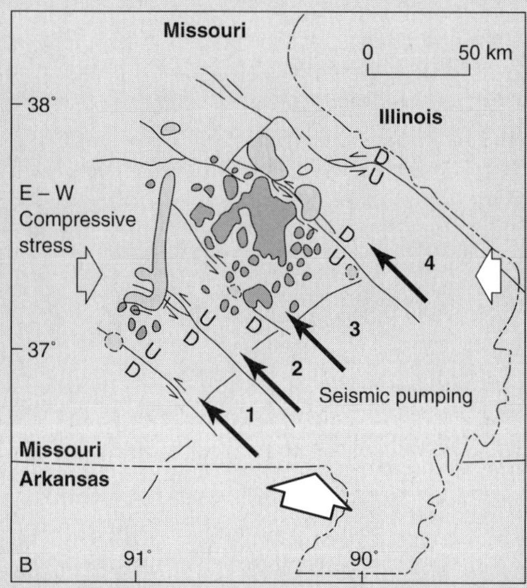

Figure 14.7 Schematic illustration of how seismic pumping may have formed Mississippi Valley-type lead-zinc deposits in southeast Missouri by focusing flow along four northwest-trending faults.
(From Clendenin and Duane, 1990, p. 118.)

14.8.5. Seismic Pumping.

Clendenin and Duane (1990) rejected the entire concept of a basinwide flow system driven either by sediment compaction or topography. Instead, they postulated that ore fluids were driven by seismic pumping along major northwest-trending faults to the ore districts in central Missouri (Figure 14.7).

Review Questions

1. Define the following terms in the context of hydrogeology:
   a. ore deposit
   b. hydrothermal ore deposit
   c. hypothermal
   d. mesothermal
   e. epithermal
   f. stratabound
   g. Comstock Lode
   h. batholith

They pointed out that the aquifer commonly invoked as the pathway for regional flow, the Cambrian Lamotte Sandstone, is not continuous from northern Arkansas into central Missouri. Evidence for fluid movement along faults includes (1) where faults intersect favorable host rocks, the ore zone is markedly enlarged and the ore is higher grade; (2) there was little to no movement of fluids across faults; and (3) acidic ore-fluids could not have traveled through carbonate aquifers because wherever acidic ore-fluids encountered carbonate rocks, they dissolved the rocks and produced cavities.

The fault-controlled seismic-pumping model of Clendenin and Duane (1990) was criticized by Leach and Rowan (1991), who pointed out that although there may be some association of ore minerals with faults, widespread low-grade mineralization argued for the existence of single, regional-scale flow system. Leach and Rowan (1991) also claimed that acidic ore-forming fluids could have traveled through carbonate rocks without dissolution if the brine was in equilibrium with the carbonate rocks as a result of a high CO_2 content.

14.8.6. CONCLUSIONS.

The analysis of MVT lead-zinc deposits is interesting, because it not only illustrates our ignorance of how and why fluid flow occurs in the Earth's crust but also demonstrates how theories follow evidence. In the case of MVT deposits, the best constraints we have on the ore-forming fluids are estimates of salinity and temperature from fluid-inclusion studies. Theories of fluid migration that invoke topographic or compaction-driven flow have concentrated on how to best satisfy those constraints. But where field relations involving faults have been invoked as the primary data source, the hypothesis that has emerged is seismic pumping.

The problem of how MVT deposits formed is complex. As the body of relevant data changes, so will our models. For example, some models of heat and fluid flow have relied on an apparent trend for homogenization temperatures to decrease to the north (Bethke et al., 1988). However, Shelton et al. (1992) re-examined these data and determined that no discernible thermal gradient exists, although this was disputed by Leach and Rowan (1993). Furthermore, trace and minor-element studies by Shelton et al. (1992) apparently documented that at least two fluids were involved in formation of the Missouri MVT deposits: one fluid that flowed northward from the Arkoma Basin and another that flowed westward from the Illinois Basin. Many models of ore formation have stressed matching what appear to be anomalously high fluid temperatures at relatively shallow depths. However, the actual depth of burial at the time of ore formation appears to be poorly constrained. References to original burial depths are often derivative and cannot be traced to the original source. Statements are usually made that the burial depth has been estimated by reconstructing the sedimentary section, but a totally eroded section cannot be reconstructed. If the burial depths at the time of ore genesis were greater than is thought, then what appears to be anomalously high temperatures from an invading hot fluid may be more simply explained as normal temperatures, and one of the more exacting constraints on ore formation may be disposed of.

i. neptunist
j. uniformitarianism
k. plutonist
l. fluid inclusion
m. primary fluid inclusion
n. homogenization temperature
o. necking
p. daughter inclusion
q. sulfide mineral
r. ligand
s. metallogenesis
t. Koudryavtsev's Rule
u. kimberlite pipe
v. epigenetic
w. basinal brine hypothesis

2. What is the most important class of both base and precious metal deposits?
3. Give two examples of how ore minerals can be concentrated by processes other than moving fluids.
4. What four factors are required for a hydrothermal ore deposit to form?
5. Do ore fluids move through rock matrices or faults? Explain.
6. At what depths in the Earth's crust do hydrothermal ore deposits form? What factors affect the depths at which hydrothermal ore deposits form?
7. Define and discuss the difference between neptunist and plutonist theories. Which theory proved to be correct?
8. What types of information concerning ore fluids are most commonly extracted from analysis of fluid inclusions?
9. List three assumptions that are usually employed in the analysis of fluid inclusions.
10. What process can create much of the scatter commonly seen in homogenization temperatures of fluid inclusions?
11. What four elements are the primary dissolved constituents of aqueous ore fluids?
12. Which is more important in providing a source for ore minerals: the volume or type of rock sampled by a moving fluid?
13. Are most ore fluids meteoric, juvenile, or evolved?
14. List four mechanisms that can cause ore minerals to precipitate from solution.
15. Why are banded-iron formations only found in Precambrian rocks?
16. Summarize arguments both against and for Gold's theory of metallic ore formation by inorganic mantle hydrocarbons.
17. What are the most common minerals found in Mississippi Valley-type lead-zinc deposits?
18. In what type of rocks are Mississippi Valley-type lead-zinc deposits found?
19. What was the temperature and salinity of the fluids that formed Mississippi Valley-type lead-zinc deposits in the North American midcontinent? How do we know?
20. What is the basinal brine hypothesis for the formation of Mississippi Valley-type lead-zinc deposits?
21. What three mechanisms have been proposed to move ore-forming fluids to the Mississippi Valley-type lead-zinc ore districts in south Missouri and north Arkansas? What are the advantages and difficulties of each mechanism?

Suggested Reading

Gold, T. 1999. *The Deep Hot Biosphere.* New York Copernicus, 235 pp.

Hallam, A. 1983. Neptunists, vulcanists, and plutonists. In *Great Geological Controversies*, pp. 1–28. Oxford University Press.

Skinner, B. J. 1997. Hydrothermal mineral deposits: What we do and don't know. In Barnes, H. L., ed. *Geochemistry of Hydrothermal Ore Deposits*, 3d ed., pp. 1–29. New York: John Wiley & Sons.

Glossary

A

abiogenic theories: theories which postulate that petroleum arises inorganically as a result of hydrocarbon outgassing from the mantle.

accretionary prism: a wedge-shaped accumulation of sediments and sedimentary rock that forms near a subduction zone.

adsorption: the process in which ions or molecules of a dissolved substance become attached to solid materials.

advection: the process by which a contaminant or any solute is carried by moving groundwater.

Anadarko Basin: sedimentary basin located in southwestern Oklahoma.

animal feeding operation (AFO): an agricultural enterprise where animals are kept in confined situations; feed is brought to the animals instead of the animal grazing or feeding in pastures.

anion: negatively charged ion.

anisotropic: an anisotropic material is one whose properties depend upon direction.

aquathermal pressuring: the increase in fluid pressure associated with the thermal expansion of heated pore fluid.

aqueduct: a pipe or channel designed to transport water from a remote source, usually by gravity.

aquiclude: a rock layer that completely excludes fluid flow through it. Geologic aquicludes are rare.

aquifer: a saturated geologic unit which is permeable enough to transmit significant quantities of water under ordinary hydraulic gradients (Freeze and Cherry, 1979).

aquitard: a stratum that can transmit quantities of water that are very significant for a variety of geologic problems, but is inadequate for supplying economic quantities of water to wells.

Archimedes principle: the principle which states that a solid denser than a fluid will, when submerged in that fluid, have its weight reduced by the weight of the fluid it displaces.

arithmetic average: average computed by summing terms equal to the product of the value of each component to be averaged and the weighting factor associated with that component. The sum of all weighting factors must equal one.

artesian aquifer: an aquifer in which the head of the groundwater is commonly found to be greater than the elevation of the land surface.

Athy's Law: observation that shale porosity tends to decrease exponentially with increasing depth.

atomic weight: the average mass of an atom of an element relative to a standard; since 1961 the standard unit of atomic mass has been 1/12 the mass of an atom of the isotope carbon-12.

auger: a tool for boring a hole in the Earth.

Avogadro's number: the number of atoms found in 12 grams of carbon-12 (6.0221367×10^{23}); named after the Italian physicist Amedeo Avogadro (1776–1856).

B

balneology: the practice of using mineral-rich spring water for the treatment of disease.

bar: unit of pressure equal to 10^5 Pascals. One bar also equals 14.5 lb-in^{-2} or 0.98692 atm, thus one bar is approximately equal to atmospheric pressure at the Earth's surface.

baseflow recession: the exponential decline of baseflow that occurs during the absence of precipitation.

baseflow: the component of stream flow which is longer in duration and declines more slowly than quickflow.

basinal brine hypothesis: theory which states that Mississippi Valley-type lead-zinc deposits formed when brines in sedimentary basins were driven to the margins of these basins where one or more geochemical and/or physical mechanisms caused mineral precipitation.

batholith: an igneous intrusion with an areal extent greater than 40 square miles (100 km^2) and an unknown (but great) thickness.

Bernoulli's Law: mathematical formula that relates pressure changes to changes in the velocity and potential energy of a fluid.

biodegradation: the decomposition of organic chemicals by living organisms.

biogenic theories: theories which postulate that petroleum forms from the transformation of organic material buried in sedimentary rocks as they are subjected to increasing pressure and temperature with increasing depth of burial.

black smoker: a hot spring located along eruptive fissures at the center of a spreading ridge which discharges hot water laden with sulfide minerals at temperatures near 350 °C.

Br-Cl-Na systematics: the study of similarities and differences amongst the composition and concentration of these ions (Bromine, Chlorine, and Sodium) in geologic fluids.

brittle-ductile transition: the zone where pressures and temperatures are sufficiently high that crustal rocks no longer deform by fracturing, instead they flow.

C

cable-tool drilling: a method of drilling in which a tool (commonly, a bit) suspended from a cable is alternately lowered and raised.

capillary force: the attractive force which draws water into a capillary tube or a rock pore and holds that water in place.

capillary fringe: a fully saturated layer containing water drawn up into the unsaturated zone by capillary forces.

capillary tube: small-diameter glass tube into which water rises due to a capillary force; rock pores behave like capillary tubes.

capillary-pressure head: the negative pressure produced by capillary forces in the unsaturated zone.

casing: a tube or pipe used to line wells.

catagenesis: the process in which kerogen is converted to oil and gas at elevated temperatures.

catchment: a **drainage basin.**

cation: positively charged ion.

cationic shift: the observation that as the salinity of groundwater increases, the relative proportion of Na decreases while the proportions of K, Mg, and Ca increase.

chemosynthesis: a process by which the base of food chain is created through chemical processes other than photosynthesis.

coesite: a type of quartz that forms at very high pressures.

compaction disequilibrium: the failure of sediments to reach equilibrium compaction conditions quickly enough; typically leads to overpressures occuring in areas undergoing high rates of sedimentation and low permeabilities.

Comstock lode: epithermal ore deposit near Virginia City, Nevada; an example of mineralization strongly associated with faults.

concentrated animal feeding operation (CAFO): an animal feeding operation which houses a large number of animals.

concretion: a hard feature formed by the concentration of certain substances.

conduction: the transport of heat through a substance by molecular collisions; the dominant mode of heat transport in the Earth's crust and lithosphere.

cone of depression: geometric form of the water table caused by drawdown.

confined aquifer: an aquifer that is bounded by less permeable layers (aquitards).

connate: water which was trapped in pore spaces at the time a rock was first formed; connate refers to pore fluid physically present at the time of sediment deposition.

convection: the transport of heat by the movement of a solid, liquid, or gas.

cost-benefit analysis: an analysis which attempts to quantitatively estimate both the cost and the benefit to be obtained from any action.

Coulomb wedge model: model of accretionary wedge deformation which is analogous to a bulldozer pushing a pile of sand, soil, or snow.

cracking: chemical process whereby larger, more complex molecules are broken down to smaller, simpler molecules. Oil is catagenically cracked to produce natural gas.

critical point: the pressure and temperature at which liquid and gaseous water can no longer be distinguished; below the critical point, liquid water and water vapor exist as distinct and separate phases, above the critical point, a single supercritical fluid exists at all temperature and pressure conditions.

critical taper: the shape for which an accretionary wedge is on the verge of failure under horizontal compression everywhere.

cryology: study of ice and snow.

D

Darcy velocity: specific discharge.

Darcy's Law: empirical relationship which states that the rate of fluid flow through a porous medium is proportional to the potential energy gradient within that fluid and a material property termed the hydraulic conductivity.

daughter inclusion: fluid inclusion formed when necking divides one fluid inclusion into two or more new inclusions.

daughter isotope: isotope formed by the radioactive decay of another isotope, the parent isotope.

decay constant: constant which describes the rate at which radioactive isotopes decay.

decollement: a detachment surface between two bodies of rock which have separate styles of deformation.

deduction: the process of reasoning in which a conclusion follows necessarily from the stated premises; it is inference by reasoning from the general to the specific.

dense hydrous magnesium silicate (DHMS): a mineral present in the upper mantle which may contain significant amounts of water.

diagenesis: any postdepositional change that takes place in a sediment or rock.

diamond-anvil cell: a device in which two gem-quality diamonds are used to apply extremely high pressures to samples.

diapir: a dome-shaped upwelling of a plastic material squeezed up through overlying strata.

diffuse source: a source of groundwater contamination which is due to the aggregate effect of widespread activity, such as septic tanks or agricultural chemicals.

diffusion coefficient: constant which describes the diffusion of a dissolved substance in a fluid.

diffusion equation: the mathematical formulation which describes the change of head with time.

diffusion: the transport of contaminants from regions of high concentration to regions of low concentration in the absence of fluid movement.

dilatancy hardening: an increase in the strength of a material which occurs in response to dilation.

dilatancy-diffusion model: model which theorizes that earthquakes are first delayed by dilatancy hardening, then initiated by diffusion of fluid into the dilated space near the fault.

dilatancy: a volume increase due to deformation.

dilatation: volumetric expansion.

dilatational jog: an area between two fault segments where extensional deformation may be concentrated during fault movement and an earthquake.

dispersion: mechanical mixing.

DNAPL (dense non-aqueous phase liquid): a NAPL whose density is higher than water.

dogma: a principle, belief, or statement of idea or opinion authoritatively considered to be absolute truth.

drainage basin: an area that has a common drain, usually a stream; also referred as a **watershed** or **catchment.**

drawdown: the decrease of head in an aquifer due to water withdrawal.

drill-stem test: procedure commonly used in the petroleum industry to estimate formation pressure and permeability.

dump: an uncovered landfill.

Dupuit assumptions: assumptions applied in a Dupuit analysis to calculate shallow flow in an unconfined aquifer.

dynamic school: school of thought which believes that abnormal fluid pressures in the Earth's crust are caused by a transient imbalance between pressure generation and dissipation inhibited by aquitards. The view of the dynamic school is that pressure seals do not exist.

dynamic viscosity: viscosity; the property of a fluid which offers resistance to flow, distinguished from kinematic viscosity.

E

effective diffusion coefficient: constant which describes the diffusion of a substance in a porous medium.

effluent: a stream that gains water from baseflow.

element: a substance that cannot be broken up into simpler substances by chemical means.

epicenter: the point on the ground surface directly above the hypocenter.

epigenetic: an ore that formed after the host rock was already in place.

epithermal: an ore deposit formed at relatively shallow depths at temperatures of 50 to 200 ºC.

equipotential: line connecting spatial locations which have the same value of head.

eustatic: a sea level change that is worldwide, not merely a local change due to land subsidence.

evaporation: the transformation of liquid water near and at the ground surface to water vapor in the atmosphere.

evaporite: a sediment or a rock deposited by evaporation of a solution such as seawater.

evapotranspiration: the sum of transpiration and evaporation.

evolved water: water which originated as either meteoric or oceanic water, but has subsequently changed its composition through chemical and/or physical processes.

expanding source area concept: the idea that during a precipitation event the water table in low-lying areas next to distributaries may rise to the surface and overland flow may occur. Following the end of the precipitation event, the water table drops and the "source areas" shrink in size.

F

fault: a fracture in the Earth's crust along which motion occurs; the motion must be in a direction parallel to the fault plane.

fault gouge: soft, fine-grained material that fills a fault, vein, or fracture, and presumably originates from the grinding motion that occurs during fault movement.

fault jog: offset in a fault plane.

fault-valve action: rupturing of a seal which maintains an area of fluid pressures above hydrostatic by faulting.

Fick's First Law: empirical observation that the flux of a dissolved substance by diffusion in a fluid is proportional to the concentration gradient.

finite difference method: numerical method of solving a differential equation which approximates derivatives by small, or finite, differences.

flank circulation: hydrothermal circulation through the oceanic crust which occurs away from the ridge axis, and is characterized by lower temperature fluids (< 200 ºC) and slower circulation (compared to ridge circulation).

flow line: streamline.

flow net: a plot of equipotential and streamlines; a flow net is a graphical solution to Laplace's equation.

flow tube: area between two streamlines.

fluid inclusion: a minute cavity in a crystal filled with fluid or gas.

flux: amount per unit area per unit time.

forearc basin: a sedimentary basin formed between a subduction-zone trench and a volcanic island arc.

foreland basin: a sedimentary trough which forms in the foreland of a fold and thrust belt, usually in an area characterized by a plate collision and compressional tectonics.

Fourier's Law of Heat Conduction: empirical observation that the amount of heat energy transported by conduction per unit time per unit area (the flux) is proportional to the thermal gradient and a material property termed the thermal conductivity.

fractionation: the separation of a mixture into fractions which possess different properties; commonly used in reference to mixtures of different isotopes.

funnel-and-gate system: a way of treating contaminated groundwater by funneling it through a gate which contains a reactive medium that neutralizes the contaminant.

geohydrology: hydrogeology.

geometric average: average computed by summing terms equal to the value of each component to be averaged raised to a power equal to the weighting factor associated with that component. The sum of all weighting factors must equal one.

geopressure: overpressure.

geothermal gradient: the rate at which temperature increases with depth.

greenhouse effect: the tendency of certain gasses in the atmosphere to retain heat and warm the Earth.

groundwater divide: a topographic ridge which separates areas of lesser elevations.

groundwater flow: the movement of groundwater below the water table.

groundwater: any subsurface aqueous fluid, either saline or fresh.

H

half-life: the time it takes for half of an initial amount of atoms of one element or isotope (the parent isotope) to change to atoms of another element or isotope (the daughter isotope).

harmonic average: average computed by summing the weighting factors associated with each component to be averaged, and then dividing this sum by the total sum of each weighting factor divided by its associated component value.

head: potentiometric surface or level to which a fluid will rise; if fluid density is constant, head is proportional to the potential energy per unit mass of fluid.

heat of fusion: the amount of heat or energy necessary to change a specified mass of a substance from a solid to a liquid while maintaining a constant temperature and pressure.

heat of vaporization: the amount of heat or energy necessary to change a specified mass of a substance from a liquid to a vapor while maintaining a constant temperature and pressure.

High Plains Aquifer: a water-bearing stratum underlying the central U.S. which consists primarily of sands and gravels deposited by streams draining the Rocky Mountains.

hinge line: an imaginary line in a topographically driven flow system which demarcates the region of downward moving flow from the region of upward moving flow.

homogenization temperature: the temperature at which the separate gas and liquid phases in a fluid inclusion disappear or homogenize.

hormesis: a positive response by an organism to the limited exposure of a toxic substance.

Hortonian overland flow: overland flow which occurs because rainfall is faster than infiltration.

hot spring: a spring whose water has a higher temperature than the human body (98.6 ºF or 37 ºC).

Hubbert-Rubey pore pressure hypothesis: the hypothesis that pore-fluid pressure reduces the normal stress across a fault and thus the force necessary for movement.

hydraulic conductivity: the property of a saturated porous medium which determines the rate at which a fluid moves through that medium in the presence of a potential energy gradient. The hydraulic conductivity depends on both the porous medium itself and the fluid which saturates it.

hydraulic diffusivity: ratio of hydraulic conductivity to specific storage.

hydraulic resistance: length divided by hydraulic conductivity.

hydrocarbon machine: all of the factors which affect the processes of hydrocarbon generation, migration, and accumulation; similar to "petroleum system".

hydrocompaction: compaction of soil in response to wetting.

hydrogeology: the branch of hydrology which studies underground fluids and their interaction with solid geologic materials.

hydrograph: a record of the volumetric discharge (volume per unit time) at a particular point in a stream over time.

hydrologic cycle: the circulation of water between the continents, the oceans, and the atmosphere.

hydrology: the study of terrestrial water.

hydroseism: a seismically induced water-level fluctuation in a well.

hydrostatic: a hydrostatic fluid is one in which the fluid pressure at any depth depends only on the weight of the overlying fluid.

hydrostatic: condition in which the fluid pressure at any depth is that due to the weight of the overlying fluid.

hydrothermal ore deposit: ore deposit formed by a hot fluid.

hypocenter: the three-dimensional point in the Earth's crust at which an earthquake originates.

hypothermal: an ore deposit formed at great depths at temperatures of 300 to 500 °C.

hypothesis: an explanation for a scientific observation.

hysteresis: the failure of a property that has been changed by an external agent to return to its original value when the cause of the change is removed; behavior which is not reversible.

I

illite: common clay mineral which can be formed diagenetically from smectite.

impoundment: a pond or lagoon used to hold liquid wastes.

induction: the process of inferring general relationships (physical laws) from specific instances.

infiltration capacity: maximum rate at which infiltration can take place.

infiltration: the entry of water into the ground surface and the subsequent flow of water away from the ground surface.

influent: a stream that loses water to the subsurface and thus recharges the water table.

insolation: solar radiation incident upon the surface of the Earth.

instream use: water used within a stream for purposes such as transportation, hydroelectric power, and recreation.

interception: precipitation caught by vegetation before it may reach the ground surface.

interflow: subsurface flow that takes place in the unsaturated or vadose zone between the ground surface and water table.

intracratonic basin: a sedimentary basin found in the central, tectonically stable interior of a continent.

ion exchange: the replacement of an ion attached to the surface of a solid by another ion in solution.

ion: a molecule with a net electrical charge.

isotope: a variation of an element which has the same number of protons in the nucleus, but a different number of neutrons.

isotropic: an isotropic material is one whose properties do not depend upon direction.

J

joint: a fracture along which no movement has occurred.

Joule: the standard metric unit of energy.

juvenile: water derived from the interior of the Earth that has not previously been present in the atmosphere or surface of the Earth.

K

karst: a type of topography which is formed in regions underlain by carbonate bedrock, characterized by the presence of sinkholes, caves, and other dissolution features.

kerogen: a generic petroleum precursor which can be divided into an inert component which has no potential to produce petroleum, a labile fraction which has the potential to chiefly generate oil, and a refractory component, which yields only gas.

kimberlite pipe: a pipe-shaped extinct volcanic orfice with a diameter of about 200 meters at the surface which narrows to a few meters at a depth of about 1 kilometer.

kinematic viscosity: dynamic viscosity divided by fluid density.

kinetic theory: the theory which states that the rate at which either oil or gas is formed catagenically is linearly dependent upon time, and exponentially dependent upon temperature.

Koudryavtsev's Rule: the observation that hydrocarbon-rich areas tend to be enriched at all lower levels, regardless of geological age, extending down to and sometimes including the crystalline basement.

L

Labile: that portion of kerogen which generates oil in the process of catagenesis.

laminar flow: fluid flow which occurs when slowly moving fluids are dominated by viscous forces, it is characterized by smooth motion in laminae or layers.

land subsidence: a gradual settling or sudden sinking of the Earth's surface due to subsurface movement of soil or rocks.

landfill: a hole in the ground that receives waste.

Laplace's equation: equation which states that the curvature of head is equal to zero.
Laplacian: the second spatial derivative, a scalar.
lateral dispersion: transverse dispersion.
law of unintended consequences: any action designed to improve environmental quality or reduce risk may have the opposite effect.
leachate: the liquid created by water infiltrating into landfills and leaching out substances from the wastes contained therein.
ligand: any molecule that binds to another.
limnology: the study of lakes.
linear velocity: the speed at which individual water molecules move.
liquefaction: the loss of shear strength a water-saturated sediment undergoes when shaken.
lithosphere: the outer, rigid section of the Earth which deforms elastically on a geologic time scale.
lithostatic: pressure due to the weight of the entire overburden (fluid plus matrix).
LNAPL (light non-aqueous phase liquid): a NAPL whose density is lower than water.
longitudinal coefficient of hydrodynamic dispersion: constant which is equal to the sum of the longitudinal dispersivity and the effective diffusion coefficient.
longitudinal dispersion: dispersion in the direction of a flow path.
longitudinal dispersivity: constant which describes dispersion in the direction of flow.

M

mass spectrometer: instrument used to measure the ratio of isotopes in an element by observing their deflection in an electrical or magnetic field.
matrix: the solid part of a rock.
megaplume: an unusually large plume.
membrane filtration: the filtering out of cations and anions by electrically polar clay molecules.
meniscus: a curved upper surface of a liquid in a container.
mesothermal: an ore deposit formed at intermediate depths at temperatures in the range of 200 to 300 °C.
metagenesis: the third stage of organic evolution. During metagenesis, residual kerogen which has not been converted to oil or gas becomes highly condensed and enriched in carbon. The ultimate product of metagenesis is pure carbon in the form of graphite.
metallogenesis: the study of how and why ore deposits form through space and time, and the relationship of ore formation to large-scale geologic processes such as plate movements.
meteoric water line: the trend line which shows the tendency for atmospheric water and precipitation to be increasingly depleted in the heavier isotopes with higher latitudes.
meteoric: water which occurs in, or originates from, the atmosphere.
method of multiple working hypotheses: an effort to bring up into view every rational explanation of the phenomenon in hand and to develop every tenable hypothesis relative to its nature, cause or origin, and to give all of these as impartially as possible a working form and a due place in the investigation (Chamberlin, 1897).
migration: the movement of petroleum in the subsurface; see **primary migration** and **secondary migration.**
mining groundwater: extraction of groundwater faster than it can be recharged.
miscible: miscible fluids are those that can be mixed in each other.
model: a representation of a geologic system or process.
molality: moles of solute per kilogram of solvent.
molarity: moles of solute per liter of solution.
mole: the amount of a substance that contains 6.02 x 10^{23} molecules (Avogadro's number).
molecular weight: a number equal to what a mole of a particular substance would weigh in grams.

N

NAPL (non-aqueous phase liquid): a liquid organic compound in the subsurface that has low solubility in water and therefore generally exists as a separate phase.
natural attenuation: any reduction of contaminant mass, toxicity, mobility, or concentration which occurs without human intervention.
necking: the formation of two or more daughter inclusions from a single original fluid inclusion.
Neptunist: school of geologic thought founded by the German professor, Abraham Gottlob Werner

(1749–1807), which believed that ore deposits as well as igneous and metamorphic rocks formed as the result of surface waters percolating downward from a primeval universal ocean,

Newton: the metric unit of force.

Newtonian fluid: a fluid in which shear stress is proportional to shear rate.

node: a location or point in space at which the solution to a differential equation is to be found.

normal fault: a fracture in the Earth's crust which is undergoing extension.

nucleus: the dense, central core of an atom containing neutrons and protons.

O

Occam's razor: the philosophical proposition that given two competing hypotheses equally supported by the evidence, the favored one is the simpler.

oceanic fluid cycle: the addition of fluids to the young rocks of the oceanic crust near the sites of their creation, and the return of the fluids to the surface at and near subduction zones where old oceanic crust is destroyed.

oceanic water: water which is found in the Earth's oceans, or water which is found elsewhere but has not significantly changed its composition from the time it left the oceans.

offstream use: surface water withdrawn from a lake, stream, or other body of water, or groundwater withdrawn from the solid Earth for human use.

oil: a complex liquid-phase mixture of organic compounds.

ophiolite: a piece of the oceanic crust (and possibly upper mantle) which escapes destruction through being thrust up into the continental crust in a plate collision.

ore deposit: a concentration of a mineral which is economically viable to mine.

osmosis: the tendency of water to diffuse through a semi-permeable membrane to make the concentration of water on one side of the membrane equal to that on the other side.

overland flow: the direct movement of water over the land surface to the nearest stream or body of water.

overpressure: fluid pressure in excess of hydrostatic.

overthrust fault: a reverse fault in which older strata have been superimposed above younger strata.

P

paradigm: a world-view or model of how nature works.

parent isotope: radioactive isotope which decays into another isotope, or daughter isotope.

parsimony: economy or simplicity of assumptions in logical formulation.

passive margin basin: an ocean basin that abuts a continental margin in the absence of orogenesis.

peat: a light, spongy material formed in temperate humid environments by the accumulation and partial decomposition of vegetable remains under conditions of poor drainage.

Peclet number: a dimensionless number which is a relative gauge of the magnitude of convective to conductive heat transport.

pedogenic: related to soil formation.

periodic table: a tabular arrangement of the elements according to their atomic number.

permeability: the capacity of a porous medium to transmit fluids in the presence of a potential energy gradient.

permeameter: a laboratory device used to measure permeability by applying Darcy's Law.

petroleum system: the essential geologic components and processes that may result in an economic petroleum accumulation; similar to "hydrocarbon machine".

petroleum: a natural subsurface hydrocarbon material, including oil and gas.

Phanerozoic: geologic time period consisting of the last 570 million years.

philosophy: the study of the fundamental nature of existence, of man, and of man's relationship to existence.

piston flow: flow in which the moving material does not diffuse; the advancing front is sharply and clearly demarcated.

plug flow: piston flow.

plume: a sudden release of hot water from an oceanic hydrothermal system. The released water is hotter and more buoyant than surrounding seawater, and thus rises above the discharge point until it dissipates and spreads laterally.

Plutonist: (also known as Vulcanist) school of thought which believed that igneous rocks and ore deposits formed from the cooling of molten rocks.

point source: a localized source of groundwater contamination, such as a single hazardous waste dump or landfill.

poise: an alternative viscosity unit; one poise = 0.1 $kg{\cdot}m^{-1}{\cdot}s^{-1}$.

polar coordinates: two-dimensional coordinate system based on angular and radial distances.

Popper's doctrine of falsifiability: the idea that it is impossible to prove that a hypothesis is true and that science advances by disproving hypotheses.

porosity: the fraction or percent void space in a porous medium such as a rock or sediment.

porous medium compressibility: ratio of the normalized volume change of a porous medium to the change in effective stress.

porous medium: a material with holes. In rocks, these holes may be present as fractures or as interstitial spaces between solid grains.

potamology: the study of surface streams.

potentiometric surface: head.

power: energy per unit time.

precipitation: water condensed from atmospheric water vapor which condenses and falls to earth.

pressure compartment: a three-dimensional hydraulically isolated volume of the Earth's crust which has a fluid pressure different from the ambient surroundings.

pressure seal: .a zone of rocks that prevent essentially all pore fluid movement over substantial intervals of geologic time; the term should not be applied to aquitards, and excludes capillary seals.

primary fluid inclusion: a fluid inclusion in which the sealing of the inclusion occurs during the growth of the surrounding crystal.

primary migration: the movement of petroleum out of the source rock in which it was generated to a reservoir.

principal directions of anisotropy: directions in which a material property has maximum values.

proton: a positively charged sub-atomic particle found in the nuclei of all atoms.

pump-and-treat: a method for remediating contaminated groundwater in which the affected water is pumped out of the ground and treated at the surface to remove the contaminant.

Q

quickflow: the component of stream flow which takes place coincident with, and shortly after, a precipitation event, it rapidly rises and then falls.

quicksand: a mixture of sand and water which has a low shear strength due to flowing water.

R

radiation: the transport of heat by the emission and absorption of electromagnetic energy.

radioactive isotopes: isotopes which undergo nuclear decay and change from one element into another; commonly used in geologic dating.

Rayleigh number: a dimensionless ratio which determines the onset of thermal free convection in an idealized, one-dimensional system.

refraction: bending.

refractory: that portion of kerogen which generates gas in the process of catagenesis.

relative permeability: the ratio of the effective permeability of a porous medium for a given fluid at a certain saturation level, to the intrinsic permeability of that same medium when 100 percent saturated with that same fluid.

remediation: any action that mitigates the undesirable effects of groundwater contamination.

repeat formation tester (RFT): a well-logging tool used in the petroleum industry which allows fluid-pressure measurements to be made repeatedly over short intervals in a borehole.

representative elementary volume (REV): the smallest volume that can successfully represent the heterogeneous microscopic domain of a porous medium by a fictitious, homogeneous continuum.

reservoir: a geologic unit saturated with oil or gas that is sufficiently porous so as to contain appreciable quantities of petroleum, and permeable enough (usually $\geq 10^{-14}$ m^2) to allow for large quantities of petroleum to be pumped out of the reservoir.

residence time: the average amount of time a water molecule spends in a reservoir.

residual saturation: the percent of void or pore space retained by the inhabiting fluid that cannot be replaced by the invading fluid; it is the saturation at which a fluid becomes discontinuous and immobile due to capillary forces.

resonance: the frequency at which forced oscillations have their maximum amplitude.

REV: see **representative elementary volume.**

reverse fault: a fracture in the Earth's crust which is undergoing compression and shortening.

reverse osmosis: the process of increasing the fluid pressure on the salty side of a semi-permeable membrane until water molecules move from the salty to the fresh side.

Reynolds Number: a dimensionless ratio of the inertial to the viscous forces in a fluid.

rheology: the study of how materials deform.

Richards Equation: equation which describes vertical fluid flow in the unsaturated zone.

ridge circulation: hydrothermal circulation through the oceanic crust which occurs in the immediate vicinity of spreading ridges, and is characterized by high temperature (350 °C), rapid circulation, and the presence of hydrothermal vents termed black and white smokers.

rift basin: a sedimentary basin formed by the extension of the lithosphere.

risk assessment: a way of ranking risks so that they may be avoided, reduced, or managed in the most efficient manner possible.

rotary drilling: a method of well-drilling where a cutting bit is attached to the lower end of a string of hollow drill pipe; rotation of the drill bit provides a cutting action. Drill cuttings are removed by circulation of a liquid (mud) down the inside of the drill pipe and back to the surface through the annulus between the drill pipe and surrounding rock.

S

salinity: total concentration of dissolved solids.

scalar: a quantity such as mass, length, temperature, or head, which does not depend upon direction.

scale analysis: a method of obtaining approximate answers to diffusion problems which utilizes the relationship between characteristic time and length, and provides insight into the physics of transient fluid flow into a porous medium.

science: a philosophy whose goal is the discovery of truth.

secondary migration: movement of petroleum within a reservoir.

sedimentary basin: a downwarped area of the Earth's crust where sediments accumulate and lithify to form sedimentary rocks over geologic time.

seismic pumping: fluid movement that occurs in response to a change in tectonic stress and/or strain.

seismic seiche: an oscillation in a body of surface water (e.g., a lake) caused by seismic waves.

semipermeable membrane: a material that neutral water molecules may move through, but charged ions such as Na^+ or Cl^- may not.

septic tank system: a sewage disposal system consisting of a settling tank and a drainage field which is used primarily for single residences.

serpentinization: conversion of magnesium- and iron-rich silicates found in the oceanic crust (e.g., olivine) to the mineral serpentine through hydration.

sinkhole: a depression in the land surface caused by collapse of an underground cavity.

slickenslide: a polished and striated surface which results from movement on a fault plane.

slug test: well test used to estimate permeability; fluid pressure within a well bore is instantaneously elevated by rapidly lowering a solid piece of pipe or slug into a well and permeability inferred from the resulting change of fluid pressure with time.

small comet hypothesis: the theory that Earth is currently accreting water from an influx of small, comet-like objects.

smectite: common clay mineral which undergoes a diagenetic conversion to the clay mineral illite with a release of water.

Smoluchowski's dilemma: the apparent mechanical impossibility of large overthrust faults.

SMOW: standard mean ocean water.

solute: substance dissolved in a solvent.

solution: the homogeneous mixture that results from a solute being dissolved in a solvent.

source rock: an organic-rich sedimentary rock that is capable of generating oil or gas.

spa: a fashionable hotel or resort that has a mineral spring; people visit spas for therapeutic reasons, as well as for relaxation and leisure.

sparging: the introduction of a gas into a liquid.

specific discharge: rate at which a volume of fluid which passes through a given area; specific discharge is a flux.

specific heat capacity: the amount of energy or heat necessary to raise the temperature of a specified mass of a substance by a specified amount.

specific storage: the normalized volume of fluid that is released from a porous medium in response to a change in head, it is a measure of an aquifer's ability to release or absorb groundwater.

specific yield: the volume of water that will drain from a saturated porous medium under the influence of gravity, divided by the total volume of the porous medium.

spring: a place where there is a natural flow of groundwater at the surface.

stable isotope: an isotope which is not radioactive.

standard mean ocean water (SMOW): composition of ocean water in terms of isotopes of hydrogen and oxygen used as a standard in isotopic studies.

static school: school of thought which maintains that abnormal fluid pressures in the Earth's crust, regardless of origin, are maintained by pressure seals.

statistical murder: the occurrence of unnecessary deaths because risks have not been reduced in the most efficient manner possible.

storativity: the product of aquifer thickness and specific storage; also, the volume of water released from (or taken into) storage per unit head drop (or increase) per unit surface area of a confined aquifer.

strain: a deformation; a change in length, area, or volume, that takes place in response to a change in stress.

stratabound: an ore deposit confined to a single stratigraphic unit.

streamline: line representing the direction of fluid flow.

stress: force per unit area; pressure.

strike-slip basin: a sedimentary basin which forms where the Earth's crust separates due to an offset in a strike-slip fault.

subsurface storm flow: lateral flow that occurs along the top of unsaturated zone when that area temporarily becomes saturated when precipitation is high.

sulfide mineral: a mineral containing a metal and sulfur.

surface tension: the strength of the attractive force between liquid molecules.

surfactant: a substance that, when added to a liquid, reduces its surface tension and increases its wetting properties; soaps and detergents are surfactants.

T

tailing: a decline in the rate of contaminant removal observed during pump-and-treat remediation.

TDS: total dissolved solids.

tensiometer: device used to measure the capillary-pressure head in the unsaturated zone.

theory of infiltration: the theory which states that the source of the water that discharges from springs and streams is precipitation which has infiltrated slowly into a porous Earth.

tortuosity: a dimensionless number defined as the flow path length for a water molecule divided by the straight-line distance.

tragedy of the commons: the tendency to deplete and ultimately destroy a resource which has a common ownership.

transmissivity: the product of aquifer thickness and hydraulic conductivity; also, the volumetric flux or flow rate per unit head gradient per unit aquifer width.

transpiration: the movement of water from the soil to the atmosphere by plants.

transverse dispersion: dispersion in a direction perpendicular to a flow path.

transverse dispersivity: constant which describes dispersion in a direction perpendicular to flow.

triple point: the pressure and temperature at which all three water phases (ice, liquid water, and steam) may coexist.

tritium unit (TU): manner of measuring tritium levels. One tritium unit is equal to a level of one tritium atom per 10^{18} hydrogen atoms; this is the level of tritium found in the natural environment before open-air testing of atomic bombs.

turbulent flow: fast-moving flow in which inertial forces dominate; it is characterized by random, three-dimensional motions of fluid molecules superimposed on the overall direction of flow.

U

unconfined aquifer: an aquifer that is not completely bounded by less permeable strata; the typical unconfined aquifer consists of near-surface unconsolidated sediments such as sands and gravels.

underpressure: fluid pressure below hydrostatic.

uniformitarianism: the principle that past geologic events can be explained by the same physical laws and processes which govern present day events.

unsaturated zone: the region between the ground surface and the water table where soil and/or rock pores are partially filled with air and partially filled with fluid.

upconing: the rise of a cone-shaped body of salt water which may occur when head is lowered in the overlying fresh-water strata.

V

vadose zone: the unsaturated zone.

vector: a quantity which depends upon direction, and can be specified completely only by quantitative representations of both magnitude and direction.

vein: a fracture or fault filled with ore minerals.

viscosity: property of a fluid which describes its resistance to flow.

volumetric moisture content: the dimensionless ratio of the water or fluid volume in a porous medium to the total volume of the porous medium.

W

wadsleyite: a dense hydrous magnesium silicate present in the upper mantle which may contain up to 3.3 percent water.

water table: the surface at which the solid Earth is saturated with fluid at atmospheric pressure.

watershed: a **drainage basin.**

Watt: a unit of power, or energy per unit time. One Watt is equal to one Joule per second.

weight percent: mass of solute in a solution expressed as a percentage of the solution's total mass.

wetting angle: the angle at which a meniscus intersects the container wall holding a liquid.

white smoker: a hydrothermal vent on the ocean floor which discharges fluids at temperatures in the range of 100-300 °C; the white color of the water discharging from these vents is due to white particles of silica, anhydrite, and barite that precipitate from solution when hot hydrothermal fluids encounter cold seawater.

References

Ackerman, D. J., 1996, Hydrology of the Mississippi River Valley Alluvial Aquifer, South-Central United States: U.S. Geological Survey Professional Paper 1416-D.

Ackermann, W. C., Colman, E. A., and Ogrosky, H. O., 1955, Where we get our water, From ocean to sky to land to ocean in, The Yearbook of Agriculture. 1955, U.S. Dept. of Agriculture, p. 41–51.

Agricola, D. R., 1950, De Re Metallica (translated from the Latin edition of 1556 by Herbert and Lovit Hoover): Dover, New York.

Ahrens, T. J., 1989, Water storage in the mantle: *Nature,* v. 342, p. 122–123.

Alley, W. M., Reilly, T. E., and Franke, O. L., 1999, Sustainability of ground-water resources: U.S. Geological Survey Circular 1186.

Alper, J., 1990, Archimedes, Plato make millions for big oil! *Science,* v. 248, no. 4960, p. 1190–1191.

Al-Shaieb, Z., Puckette, J. O., Abdalla, A. A., Tigert, V., and Ortoleva, P. J., 1994, The banded character of pressure seals in Ortoleva, P. J., ed., Basin Compartments and Seals, AAPG Memoir 61, p. 351–367.

Alt, J. C., 1995, Subseafloor processes in Mid-Ocean Ridge hydrothermal systems, in, Humphris, S. E., Zierenberg, R. A., Mullineaux, L. S., and Thomson, R. E., eds., Seafloor hydrothermal systems. American Geophysical Union Geophysical Monograph 91, American Geophysical Union, Washington, DC, pp. 85–114.

Ames, B. N., Magaw, R., and Gold, L. S., 1987, Ranking possible carcinogenic hazards: *Science.* v. 236, p. 271–277.

Anderson, G. M., 1991, Organic maturation and ore precipitation in Southeast Missouri: *Economic Geology.* v. 86, p. 909–926.

Anderson, G. M., and Macqueen, R. W., 1988, Mississippi Valley-type Lead-Zinc Deposits, in R. G. Roberts and P. A. Sheahan, eds., Ore Deposit Models. Geoscience Canada Reprint Series 3, p. 79–90.

Anderson, M. P., 1984, Movement of contaminants in groundwater: groundwater transport—advection and dispersion, in, Groundwater Contamination. p. 37–45, National Academny Press, Washington, DC.

Anderson, T. L., 1995, Water options for the blue planet, in, Bailey, R., ed., The True State of the Planet. The Free Press, New York, p. 268–294.

Angevine, C. L., Heller, P. C., and Paola, C., 1990, Quantitative sedimentary basin modeling. AAPG Continuing Education Course Notes Series 32, 133 pp.

Apps, J. A., and van de Kamp, P. C., 1993, Energy gasses of abiogenic origin in the Earth's crust, in, "The Future of Energy Gases." U. S. Geological Survey Professional Paper 1570, D. G. Howell, ed., p. 81–132.

Archie, G. E., 1950, Introduction to petrophysics of reservoir rocks. AAPG Bull., v. 34, p. 943–961.

ASCE (American Society of Civil Engineers), 1969, Saltwater intrusion in the United States: Proceedings of the American Society of Civil Engineers, Journal of the Hydraulics Division. v. 95, no. HY5, p. 1651–1669.

Athy, L. F., 1930, Density, porosity and compaction of sedimentary rocks. AAPG Bull., v. 34, p. 943–961.

Audet, D. M., 1995, Modelling of porosity evolution and mechanical compaction of calcareous sediments. *Sedimentology,* v. 42, p. 355–373.

Autin, W. J., Burns, S. F., Miller, B. J., Saucier, R. T., Snead, J. I., 1991, Quaternary geology of the Lower Mississippi Valley, in, Morrison, R. B., ed., Quaternary nonglacial geology. Conterminous U.S.: Boulder, Colorado, Geological Society of America, The Geology of North America, v. K-2, p. 547–581.

Baker, E. T., Lavelle, J. W., Feely, R. A., Massoth, G. J., and Walker, S. L., 1989, Episodic venting of hydrothermal fluids from the Juan de Fuca Ridge. *J. Geophys. Res.*, v. 94, p. 9237–9250.

Baker, E. T., German, C. R., and Elderfield, H., 1995, Hydrothermal plumes over spreading-center axes: global distributions and geological influences, in, Humphris, S. E., Zierenber, R. A., Mullineaux, L. S., and Thomson, R. E., eds., Seafloor hydrothermal systems. *American Geophysical Union Geophysical Monograph* 91.

Ballentine, C. J., O'Nions, R. K., Oxburgh, E. R., Horvath, F., and Deak, J., 1991, Rare gas constraints on hydrocarbon accumulation, crustal degassing and groundwater flow in the Pannonian Basin. *Earth and Planetary Science Letters,* v. 105, p. 229–246.

Banner, J. L., Wasserburg, G. J., Dobson, P. F., Carpenter, A. B., and Moore, C. H., 1989, Isotopic and trace element constraints on the origin and evolution of saline groundwaters from central Missouri. *Geochimica et Cosmochimica Acta,* v. 53, p. 383–398.

Barker, C., 1990, Calculated volume and pressure changes during the thermal cracking of oil to gas in reservoirs. *AAPG Bull.,* v. 74, p. 1254–1261.

Barley, M. E., and Groves, D. I., 1992, Supercontinent cycles and the distribution of metal deposits through time. *Geology,* v. 20, p. 291–294.

Bear, J., 1972, Dynamics of fluids in porous media. New York, Dover, 764 pp.

Beard, D. C., and Weyl, P. K., 1973, Influence of texture on porosity and permeability of unconsolidated sand. *AAPG Bull.,* v. 57, p. 349–369.

Beaumont, C., 1981, Foreland basins. *Geophysical Journal of the Royal Astronomical Society,* v. 65, p. 291–329.

Beaumont, P., 1973, A traditional method of ground-water utilisation in the Middle East. *Ground Water,* v. 11, no. 5, p. 23–30.

Bebout, G. E., 1995, The impact of subduction-zone metamorphism on mantle-ocean chemical cycling. *Chemical Geology,* v. 126, p. 191–218.

Belitz, K., and Bredehoeft, J. D., 1988, Hydrodynamics of Denver Basin: Explanation of subnormal fluid pressures. *American Association of Petroleum Geologists Bulletin,* v. 72, p. 1334–1359.

Belitz, K., and, Bredehoeft, J. D., 1990. Role of confining layers in controlling large-scale regional ground-water flow, in Neuman, S. P., and Neretnieks, I., eds. Hydrogeology of low-permeability environments — hydrogeology, selected papers, vol. 2. Verlag-Heinz, Hannover, Germany, p. 7–17.

Bennett, S. S., and Hanor, J. S., 1987, Dynamics of subsurface salt dissolution at the Welsh Dome, Louisiana Gulf Coast: in, Lerche, I., and O'Brien, J. J., eds. *Dynamical Geology of Salt and Related Structures.* Academic Press, Orlando, p. 653–677.

Bentley, H. W., Phillips, F. M., Davis, S. N., Habermehl, M. A., Airey, P. L., Calf, G. E., Elmore, D., Gove, H. E., and Torgersen, T., 1986, Chlorine 36 dating of very old groundwater 1. The Great Artesian Basin, Australia. *Water Resources Research,* v. 22, p. 1991–2001.

Berg, R. R., DeMis, W. D., and Mitsdarffer, A. R., 1994, Hydrodynamic effects on Mission Canyon (Mississippian) oil accumulations, Billings Nose area, North Dakota. *AAPG Bulletin,* v. 78, p. 501–518.

Berner, R. A., 1971, *Principles of Chemical Sedimentology.* McGraw-Hill, New York, 240 pp.

Berner, E. K., and Berner, R. A., 1996, *Global environment: water, air, and geochemical cycles.* Prentice Hall, Upper Saddle River, New Jersey, 376 pp.

Bernoulli, D., 1738, *Hydrodynamica, sive de viribus et motibus fluidorum commentarii. Opus academicum ab auctore, dum Petropoli ageret, congestum.* Argentorati, Sumptibus J. R. Dulseckeri, 304 pp.

Bethke, C. M., 1985, A numerical model of compaction-driven groundwater flow and heat transfer and its application to the paleohydrology of intracratonic basins. *Journal of Geophysical Research,* v. 90, p. 6817–6828.

Bethke, C. M., 1986, Inverse hydrologic analysis of the distribution and origin of Gulf Coast-type geopressured zones. *Journal of Geophysical Research,* v. 91, p. 6535–6545.

Bethke, C. M., 1988, Reply (to discussion on "A numerical model of compaction-driven groundwater flow and heat transfer and its application to the paleohydrology of intracratonic sedimentary basins"). *Journal of Geophysical Research,* v. 93, p. 3500–3504.

Bethke, C. M., Harrison, W. J., Upson, C., and Altaner, S. P., 1988, Supercomputer analysis of sedimentary basins. *Science,* v. 239, p. 261–267.

Bethke, C. M., 1989, Modeling subsurface flow in sedimentary basins. *Geologische Rundschau,* 78, p. 129–154.

Birkeland, P. W., and Larson, E. E., 1978, *Putnam's Geology* (third edtion). Oxford University Press, New York, 659 pp.

Biswas, A-K., 1970, A short history of hydrology, in, The Progress of Hydrology, Proceedings of the First International Seminar for Hydrology Professors, v. 2, p. 914–936.

Blanchard, D. C. and Tailleur, I. L., 1982, Temperatures and interval geothermal-gradient determinations from wells in National Petroleum Reserve in Alaska. US *Geological Survey Open File Report,* p. 82–391.

Blatt, H., 1970, Determination of Mean Sediment Thickness in the Crust: A Sedimentologic Model. *Geological Society of America Bulletin,* v. 81, p. 255–262.

Bodnar, R. J., and Costain, J. K., 1991, Effect of varying fluid composition on mass and energy transport in the Earth's crust. *Geophysical Research Letters,* v. 18, p. 983–986.

Bose, K., and Navrotsky, A., 1998, Thermochemistry and phase equilibria of hydrous phases in the system MgO-SiO_2-H_2O: implications for volatile transport to the mantle. *J. Geophys. Res.,* v. 103, p. 9713–9719.

Boswell, E. H., Cushing, E. M., and Hosman, R. L., 1968, Quaternary aquifers in the Mississippi Embayment. *U.S. Geological Survey Professional Paper 448–E.*

Bott, M. H. P., 1982, *The Interior of the Earth: its Structure, Constitution and Evolution.* Elsevier, London.

Bowman, I., 1911, Well-drilling methods: *U.S. Geological Survey Water Supply Paper 257.*

Brace, W. F., 1980, Permeability of crystalline and argillaceous rocks. *Int. J. Rock Mech. Min. Sci.,* v. 17, p. 241–245.

Brace, W. F., 1984, Permeability of crystalline rocks: new in situ measurements. *Journal of Geophysical Research,* v. 89, p. 4327–4330.

Brace, W. F., Walsh, J. B., and Frangos, W. T., 1968, Permeability of granite under high pressure. *J. Geophysical Research,* v. 73, p. 2225–2236.

Bradley, J. S., 1975, Abnormal formation pressure. *AAPG Bulletin,* v. 59, p. 957–973.

Bradley, J. S., and Powley, D. E., 1994, Pressure compartments in sedimentary basins: a review, in, Ortoleva, P. J., ed., Basin Compartments and Seals. *AAPG Memoir 61,* Am. Assoc. Petr. Geologists, Tulsa, p. 3–26.

Brantly, J. E., 1961, Percussion-drilling system, in *History of Petroleum Engineering,* p. 133–269, American Petroleum Institute, New York.

Bray, R. B., and Hanor, J. S., 1990, Spatial variations in subsurface pore fluid properties in a portion of southeast Louisiana: implications for regional fluid flow and solute transport. *Gulf Coast Association of Geological Societies Transactions,* v. 40, p. 53–64.

Bredehoeft, J. D., 1965, The drill-stem test: the petroleum industry's deep-well pumping test. *Ground Water,* v. 3, no. 3, p. 31–36.

Bredehoeft, J. D., 1992, Much contaminated ground water can't be cleaned up. *Ground Water,* v. 30, no. 6, p. 834–835.

Bredehoeft, J. D., and Pinder, G. F., 1973, Mass transport in flowing groundwater. *Water Resources Research,* v. 9, no. 1, p. 194–210.

Bredehoeft, J. D., and Papadopulos, I. S., 1965, Rates of vertical groundwater movement estimated from the Earth's thermal profile. *Water Resources Research,* v. 1, p. 325–328.

Bredehoeft, J. D., and Hanshaw, R. B., 1968, On the maintenance of anomalous fluid pressures: I. Thick Sedimentary Sequences. *Geological Society of America Bulletin,* v. 79, p. 1097–1106.

Bredehoeft, J. D., Back, W., and Hanshaw, B. B., 1982, Regional ground-water flow concepts in the United States: historical perspective, in Narasimhan, T. N., ed., *Recent trends in hydrogeology,* Geological Society of America Special Paper 189: Boulder, Colorado, p. 297–316.

Bredehoeft, J. D., Norton, D. L., Engelder, T., Nur, A. M., Oliver, J. E., Taylor, H. P., Jr., Titley, S. R., Vrolijk, P. J., Walther, J. V., and Wickham, S. M., 1990, Overview and Recommendations in, *The Role of Fluids in Crustal Processes.* National Academy Press, Washington, D.C., 170 pp.

Bredehoeft, J. D., Belitz, K. and Sharp-Hansen, S., 1992, The hydrodynamics of the Big Horn Basin: a study of the role of faults. *American Association of Petroleum Geologists Bulletin,* v. 76, p. 530–546.

Bredehoeft, J. D., Djevanshir, R. D., and Belitz, K. R., 1988, Lateral fluid flow in a compacting sand-shale sequence: South Caspian Basin. *American Association of Petroleum Geologists Bulletin,* v. 72, p. 416–424.

Bredehoeft, J. D., Neuzil, C. E., and Milly, P. C. D., 1983, Regional Flow in the Dakota Aquifer: a study of the role of confining layers. *U.S. Geological Survey Water-Supply Paper* 2237.

Bredehoeft, J. D., Djevanshir, R. D., and Belitz, K. R., 1988, Lateral fluid flow in a compacting sand-shale sequence: South Caspian Basin. *AAPG Bulletin,* v. 72, p. 416–424.

Bredehoeft, J. D., and Usselman, T. M., 1984, Groundwater contamination (letter). *Science,* v. 224, p. 1292.

Bredehoeft, J. D., Wesley, J. B., and Fouch, T. D., 1994, Simulations of the origin of fluid pressure, fracture generation, and the movement of fluids in the Uinta Basin, Utah. *AAPG Bull.,* v. 78, p. 1729–1747.

Brody, J. E., 1999, Communicating cancer risk in print journalism. *J. Natl. Cancer Inst.,* v. 25, p. 170–172.

Broecker, W. S., 1997. Will our ride into the greenhouse future be a smooth one? *GSA Today,* 7, no. 5: 1–7.

Bullard, E., 1954, The flow of heat through the floor of the Atlantic Ocean: Proceedings of the Royal Society of London, Series A, v. 222, p. 408–429.

Butler, M. A., 1933, Irrigation in Persia by Kanats. *Civil Engineering* (ASCE, New York), v. 3, no. 2, p. 69–73.

Byerlee, J., 1990, Friction, overpressure and fault normal compression. *Geophys. Res. Lett.,* v. 17, p. 2109–2112.

Cann, J. R., and Strens, M. R., 1989, Modeling periodic megaplume emission by black smoker systems. *J. Geophys. Res.,* v. 94, p. 12,227–12,237.

Carlston, C. W., 1943, Notes on the early history of water-well drilling in the United States. *Economic Geology,* v. 38, no. 2, p. 119–136.

Carlston, C. W., 1963, An early American statement of the Badon Ghyben-Herzberg principle of static fresh-water-salt-water balance. *Am. J. Sci.,* v. 261, p. 88–91.

Carslaw, H. S., 1921, *Introduction to the mathematical theory of the conduction of heat in solids.* London, Macmillan, 268 pp.

Carslaw, H. S., and Jaeger, J. C., 1959, *Conduction of heat in solids* (second edition). Oxford University Press, 510 pp.

Carson, R., 1962, *Silent spring.* Houghton Mifflin, Boston, 368 pp.

Cartwright, K., 1984, Shallow land burial of municipal wastes, in, *Groundwater Contamination,* National Academy Press, Washington, DC, p. 67–77.

Casagrande, A., 1937, Seepage through dams. *Journal of the New England Water Works Association,* v. 51, p. 131–172.

Case, L. C., 1945, Exceptional Silurian brine near Bay City, Michigan. *AAPG Bull.,* v. 29, p. 567–570.

Cathles, L. M. III, 1990, Scales and effects of fluid flow in the upper crust. *Science,* v. 248, p. 323–329.

Cathles, L. M., 1993a, A capless 350 C flow zone model to explain megaplumes, salinity variations, and high-temperature veins in ridge axis hydrothermal systems. *Economic Geology,* v. 88, p. 1977–1988.

Cathles, L. M., 1993b, A discussion of flow mechanisms responsible for alteration and mineralization in the Cambrian Aquifers of the Ouachita-Arkoma Basin-Ozark System: in, Horbury, A. D., and Robinson, A. G., eds., Diagenesis and basin development. *AAPG Studies in Geology* No. 36, p. 99–112.

Cathles, L. M., and Smith, A. T., 1983, Thermal constraints on the formation of Mississippi Valley-Type lead-zinc deposits and their implications for episodic basin dewatering and deposit genesis. *Economic Geology,* v. 78, p. 983–1002.

Cedergren, H. R., 1989, *Seepage, drainage, and flow nets* (third edition). New York, John Wiley.

Chamberlin, T. C., 1885, The Requisite and Qualifying Conditions of Artesian Wells. *Fifth Annual Report of the United States Geological Survey,* p. 131–173.

Chamberlin, T. C., 1897, The Method of Multiple Working Hypotheses. *J. of Geology,* v. 5, p. 81–85.

Chapman, D. S., Keho, T. H., Bauer, M. S., and Picard, M. D., 1984, Heat flow in the Uinta Basin determined from bottom hole temperature (BHT) data. *Geophysics,* v. 49, p. 453–466.

Chapelle, F. H., 1997, *The hidden sea: groundwater, springs, and wells.* Geoscience Press, Inc., Tucson, Arizona, 238 pp.

Cherry, J. A., 1983, Introduction to special issue on migration of contaminants in groundwater at a landfill: a case study. *J. of Hydrology,* v. 63, p. vii–ix.

Cherry, J. A., 1984a, Groundwater Contamination in, Environmental Geochemistry, Mineralogical Association of Canada Short Course Handbook, vol. 10, p. 269–306.

Cherry, J. A., 1984b, Contaminants in groundwater: chemical processes, in, Groundwater Contamination, p. 46–64, National Academy Press, Washington, DC.

Christensen, N. I., and Mooney, W. D., 1995, Seismic velocity structure and composition of the continental crust: a global view. *J. Geophys. Res.,* v. 100, p. 9761–9788.

Clark, G., 1944, Water in antiquity. *Antiquity,* v. 18, no. 69, p. 1–15.

Clark, L. C., Combs, G. F., and Turnbull, B. W., et al., 1996, Effects of selenium supplementation for cancer prevention in patients with carcinoma of the skin. *J. Am. Med. Assoc.,* v. 276, p. 1957–1963.

Clauser, C., 1992, Permeability of crystalline rocks. *EOS Transactions, American Geophysical Union,* v. 73, p. 233, 237–238.

Clauser, C., 1992, Scale effects of permeability and thermal methods as constraints for regional-scale averages, in, *Heat and Mass Transfer in Porous Media,* edited by M. Quintard and M. Todorovic, Elsevier, Amsterdam, p. 447–454.

Clayton, R. N., Friedman, I., Graf, D. L., Mayeda, T. K., Meents, W. F., and Shimp, N. F., 1966, The origin of saline formation waters: I. Isotopic composition. *J. Geophys. Res.,* v. 71, p. 3869–3882.

Clendenin, C. W., and Duane, M. J., 1990, Focused fluid flow and Ozark Mississippi Valley-type deposits. *Geology,* v. 18, p. 116–119.

Cooper, H. H., Jr., 1964, A hypothesis concerning the dynamic balance of fresh water and salt water in a coastal aquifer. *U. S. Geological Survey Water-Supply Paper 1613*–C, p. C1–C12.

Cooper, H. H., 1966, The equation of groundwater flow in fixed and deforming coordinates. *J. Geophys. Res.,* v. 71, p. 4785–4790.

Cooper, H. H., Jr., Bredehoeft, J. D., Papadopolous, I. S., and Bennett, R. R., 1965, The response of well-aquifer systems to seismic waves. *J. Geophys. Res.,* v. 70, p. 3915–3926.

Cooper, H. H., Jr., Bredehoeft, J. D., and Papadopulos, I. S., 1967, Response of a finite-diameter well to an instantaneous charge of water. *Wat. Res. Res.,* v. 3, p. 263–269.

Cooper, H. H., Bredehoeft, J. D., Papadopulous, I. S., and Bennett, R. R., 1968, The response of well-aquifer systems to seismic waves in, The Great Alaska Earthquake of 1964. *Hydrology,* p. 122–132.

Corbet, T. F., and Bethke, C. M., 1992, Disequilibrium fluid pressures and groundwater flow in the Western Canada Sedimentary Basin. *J. Geophys. Res.,* v. 97, p. 7203–7217.

Craig, H., 1961, Isotopic variations in meteoric waters. *Science,* v. 133, p. 1702–1703.

Craig, H., Boato, G., and White, D. E., 1956, Isotopic geochemistry of thermal waters: Proceedings, 2nd Conference on Nuclear Processes in Geologic Settings, National Research Council, Nuclear Science Series Report 19, p. 29–38.

Craun, G. F., 1979, Waterborne Disease—A Status Report Emphasizing Outbreaks in Ground-Water Systems. *Ground Water,* v. 17, no. 2., p. 183–191.

Criss, R. E., and Champion, D. E., 1991, Oxygen isotope study of the fossil hydrothermal system in the Comstock Lode mining district, Nevada, in, *Stable Isotope Geochemistry: A Tribute to Samuel Epstein,* H. P. Taylor, Jr., J. R. O'Neil, and I. R. Kaplan, eds., The Geochemical Society, Special Publication No. 3, pp. 437–447.

Criss, R. E., and Hofmeister, A. M., 1991, Application of fluid dynamics principles in tilted permeable media to terrestrial hydrothermal systems, Geophysical Research Letters, v. 18, p. 199–202.

Darcy, H., 1856, *Les fontaines publiques de la Ville de Dijon.* Victor Dalmont, Paris.

Darton, N. H., 1905, Preliminary report on the geology and underground water resources of the central Great Plains. *U.S. Geological Survey Professional Paper 32,* 433 pp.

Darton, N. H., 1909, Geology and underground waters of South Dakota. *U.S. Geological Survey Water Supply Paper 227.*

Davis, D., Suppe, J., and Dahlen, F. A., 1983, Mechanics of fold-and-thrust belts and accretionary wedges. *J. Geophys. Res.,* v. 88, p. 1153–1172.

Davis, E. E., and Becker, K., 1998, Borehole observatories record driving forces for hydrothermal circulation in young oceanic crust. *EOS,* v. 79, p. 369, 377–378.

Davis, E. E., and Chapman, D. S., 1996, Problems with imaging cellular hydrothermal convection in oceanic crust. *Geophys. Res. Lett.,* v. 23, p. 3551–3554.

Davis, E. E., Chapman, D. S., and Forster, C. B., 1996, Observations concerning the vigor of hydrothermal circulation in young oceanic crust. *J. Geophys. Res.,* v. 101, p. 2927–2942.

Davis, E. E., Wang, K., He, J., Chapman, D. S., Villinger, H., and Rosenberger, A., 1997, An unequivocal case for high Nusselt number hydrothermal convection in sediment-buried igneous oceanic crust. *Earth and Planetary Sci. Lett.,* v. 146, p. 137–150.

Davis, S. N., and DeWiest, R. J. M., 1966, *Hydrogeology.* John Wiley & Sons, New York, 463 pp.

de Marsily, G., 1986, *Quantitative hydrogeology, groundwater hydrology for engineers.* Academic Press, San Diego, 440 pp.

Deming, D., 1992, Catastrophic release of heat and fluid flow in the continental crust. *Geology,* v. 20, p. 83–86.

Deming, D., 1993, Regional Permeability Estimates from Investigations of Coupled Heat and Groundwater Flow, North Slope of Alaska. *Journal of Geophysical Research,* v. 98, p. 16,271–16,286.

Deming, D., 1994a, Factors necessary to define a pressure seal. *AAPG Bulletin,* v. 78, p. 1005–1009.

Deming, D., 1994b, Fluid Flow and Heat Transport in the Upper Continental Crust: in Parnell, J., ed., *Geofluids: Origin, Migration and Evolution of Fluids in Sedimentary Basins.* Geological Society Special Publication No. 78, p. 27–42.

Deming, D., 1994c, Overburden Rock, Temperature and Heat Flow — Essential Elements of the Petroleum System, in Magoon, L. B., and Dow, W. G., eds., The Petroleum System — From Source to Trap. *American Association of Petroleum Geologists Memoir 60,* p. 165–186.

Deming, D., and Nunn, J. A., 1991, Numerical Simulations of Brine Migration by Topographically-Driven Recharge. *Journal of Geophysical Research,* v. 96, p. 2485–2499.

Deming, D., Sass, J. H., Lachenbruch, A. H., and De Rito, R. F., 1992, Heat flow and subsurface temperature as evidence for basin-scale groundwater flow. *Geological Society of America Bulletin,* v. 104, p. 528–542.

DeSitter, L. U., 1947, Diagenesis of oil-field brines. *AAPG Bull.,* v. 31, p. 2030–2040.

Dessler, A. J., 1991, The small-comet hypothesis. *Rev. Geophys.,* v. 29, p. 355–382.

Dickey, P. A., 1969, Increasing concentration of subsurface brines with depth. *Chemical Geology,* v. 4, p. 361–370.

Dickey, P. A., 1975, Possible primary migration of oil from source rock in oil phase. *AAPG Bull.,* v. 75, p. 337–345.

Dickinson, G., 1953, Geological aspects of abnormal reservoir pressure in Gulf Coast, Louisiana. *AAPG Bull.,* v. 37, p. 410–432.

Dickinson, W. R., 2000, Effectively responding to the threat of global warming. *EOS, Transactions, American Geophysical Union,* v. 81, no. 9, p. 90.

Domenico, P. A., and Palciauskas, V. V., 1973, Theoretical analysis of forced convective heat transfer in regional ground-water flow. *Geological Society of America Bulletin,* v. 84, p. 3803–3814.

Domenico, P. A., and Palciauskas, V. V., 1979, Thermal expansion of fluids and fracture initiation in compacting sediments. *Geological Society of America Bulletin,* vol. 90, p. 953–979.

Domenico, P. A., and Schwartz, F. W., 1990, *Physical and Chemical Hydrogeology* (First Edition). Wiley, New York, 824 pp.

Domenico, P. A., and Schwartz, F. W., 1998, *Physical and Chemical Hydrogeology* (Second Edition). Wiley, New York, 506 pp.

Dow, W. G., 1974, Application of oil correlation and source-rock data to exploration in Williston Basin. *AAPG Bulletin,* v. 58, p. 1253–1262.

Downey, M. W., 1994, Hydrocarbon Seal Rocks, in Magoon, L. B., and Dow, W. G., eds., The Petroleum System — From Source to Trap. *American Association of Petroleum Geologists Memoir 60,* p. 159–164.

Duxbury, A. C., and Duxbury, A. B., 1994, *An introduction to the world's oceans* (fourth edition). Wm. C. Brown, Dubuque, Iowa, 472 pp.

Dupuit, A. J. E. J., 1863, *Études Théoriques et Pratiques sur le Mouvement des Eaux dans les Canaux Découverts et à Travers les Terrains Permeables* (Second Edition): Dunod, Paris.

Eddy, G. E., 1965, The effectiveness of Michigan's oil and gas conservation law in preventing pollution of the state's ground waters. *Ground Water,* v. 3, no. 2, p. 35–36.

Elder, J. W., 1965, Physical processes in geothermal areas, in, Lee, W. H. K., ed., *Terrestrial Heat Flow, American Geophysical Union Geophysical Monograph 8,* American Geophysical Union, Washington, DC, pp. 211–239.

EPA (US Environmental Protection Agency), 1973, Identification and Control of Pollution from Salt Water Intrusion. *EPA-430/9-73-013.*

EPA (US Environmental Protection Agency), 1977, Salt Water Intrusion in the United States. *EPA-600/8-77-011.*

EPA (US Environmental Protection Agency), 1983, Surface Impoundment Assessment National Report. *EPA 570/9-84-002.*

EPA (US Environmental Protection Agency), 1984, Protecting Ground Water, the Hidden Resource. *EPA 1.67/a:W 29.*

EPA (US Environmental Protection Agency), 1986, Septic systems and ground-water protection, an executive's guide. *EP 1.8 :Se 6.*

EPA (US Environmental Protection Agency), 1990a, National Water Quality Inventory, 1988 Report to Congress. *EPA-440-4-90-003,* 187 pp.

EPA (US Environmental Protection Agency), 1990b, Basics of Pump-and-Treat Ground-Water Remediation Technology. *EPA-600/8-90/003*

Fabryka-Martin, J., Davis, S. N., and Elmore, D., 1987, Applications of 129I and 36Cl in hydrology. *Nuclear Instruments and Methods in Physics Research,* v. B29, p. 361–371.

Faure, G., 1977, *Principles of isotope geology.* John Wiley, New York.

Fetter, C. W., 1994, *Applied Hydrogeology* (third edition). Macmillan, New York, 691 pp.

Fetter, C. W., 1999, *Contaminant hydrogeology* (Second Edition). Prentice-Hall, New Jersey, 500 pp.

Fetter, C. W., 2001, *Applied Hydrogeology* (fourth edition). Prentice-Hall, New Jersey, 598 pp.

Fisher, A. T., 1998, Permeability within basaltic oceanic crust. *Rev. Geophys.,* v. 36, p. 143–182.

Fisk, H. N., 1944, Geological investigation of the alluvial valley of the lower Mississippi River: U.S. Dept. Army, Mississippi River Commission, 78 pp.

Forcheimer, P., 1930, *Hydraulik* (third edition). B. G. Teubner, Leipzig and Berlin, 595 pp.

Fornari, D. J., and Embley, R. W., 1995, Tectonic and volcanic controls on hydrothermal processes at the mid-ocean ridge: an overview based on near-bottom and submersible studies, in, Humphris, S. E., Zierenber, R. A., Mullineaux, L. S., and Thomson, R. E., eds., *Seafloor hydrothermal systems, American Geophysical Union Geophysical Monograph 91:* American Geophysical Union, Washington, DC, pp. 1–46.

Fort, C. H., 1919, *Book of the damned.* Horace Liveright, New York.

Fowler, A. D., 1986, The role of regional fluid flow in the genesis of the Pine Point deposit, Western Canada Sedimentary Basin — A Discussion. *Economic Geology,* v. 81, p. 1014–1024.

Frank, B., 1955, Our Need for Water: the Story of Water as the Story of Man, in, *The Yearbook of Agriculture, 1955,* U.S. Dept. of Agriculture, p. 1–8.

Frank, L. A., 1990, *The big splash.* Carol Pub. Group, Seacaucus, New Jersey, 255 pp.

Frank, L. A., Sigwarth, J. B., and Craven, J. D., 1986, On the influx of small comets into the Earth's upper atmosphere II: interpretation. *Geophys. Res. Lett.,* v. 13, p. 307–310.

Frank, L. A., and Sigwarth, J. B., 1993, Atmospheric holes and small comets. *Rev. Geophys.,* v. 31, p. 1–28.

Frank, L. A., and Sigwarth, J. B., 1997a, Transient decreases of Earth's far-ultraviolet dayglow. *Geophys. Res. Lett.,* v. 24, p. 2423–2426.

Frank, L. A., and Sigwarth, J. B., 1997b, Simultaneous observations of transient decreases of Earth's far-ultraviolet dayglow with two cameras. *Geophys. Res. Lett.,* v. 24, p. 2427–2430.

Frank, L. A., and Sigwarth, J. B., 1997c, Detection of atomic oxygen trails of small comets in the vicinity of Earth. *Geophys. Res. Lett.,* v. 24, p. 2431–2434.

Frank, L. A., and Sigwarth, J. B., 1997d, Trails of OH emissions from small comets near Earth. *Geophys. Res. Lett.,* v. 24, p. 2435–2438.

Frank, L. A., and Sigwarth, J. B., 1999, Atmospheric holes: instrumental and geophysical effects. *J. Geophys. Res.,* v. 104, p. 115–141.

Frape, S. K. and Fritz, P., 1987, Geochemical trends for groundwaters from the Canadian Shield, in Fritz, P. and Frape, S. K. eds., Saline water and gases in crystalline rocks. *Geological Association of Canada, Special Paper, 33,* p. 19–38.

Freeze, R. A., 1974, Streamflow generation. *Rev. of Geophys.,* v. 12, p. 627–647.

Freeze, R. A., 1994, Henry Darcy and the Fountains of Dijon. *Groundwater,* v. 32, p. 23–30.

Freeze, R. A., and Cherry, J. A., 1979, *Groundwater.* Englewood Cliffs, New Jersey, Prentice-Hall, 604 pp.

Freeze, R. A., and Witherspoon, P. A., 1967, Theoretical analysis of regional groundwater flow. 2. Effect of Water-table configuration and subsurface permeability variation. *Water Resources Research,* v. 3, p. 623–634.

Fuchtbauer, H., 1967, Influence of different types of diagenesis on sandstone porosity. Proceedings of Seventh World Petroleum Congress, v. 2., p. 353–369.

Fuller, M. L., 1904, Contributions to the hydrology of eastern United States. *U. S. Geological Survey Water-Supply Paper 102.*

Furbish, D. J., 1997, *Fluid physics in geology, an introduction to fluid motions on Earth's surface and within its crust.* Oxford Univ. Press, New York.

Fyfe, W. S., 1997, Deep fluids and volatile recycling: crust to mantle. *Tectonophysics,* v. 275, p. 243–251.

Galloway, D., Jones, D. R., and Ingebritsen, S. E. (eds.), 1999, Land Subsidence in the United States. *U.S. Geological Survey Circular 1182.*

Galloway, W. E., 1974, Deposition and diagenetic alteration of sandstone in northeast Pacific are-related basins: Implications for graywacke genesis. *Geol. Soc. Am. Bull.,* v. 85, p. 379–390.

Garven, G., 1986, The role of regional fluid flow in the genesis of the Pine Point deposit, western Canadian Basin — A reply. *Economic Geology,* v. 81, p. 1015–1020.

Garven, G., and Freeze, R. A., 1984, Theoretical analysis of the role of groundwater flow in the genesis of stratabound ore deposits, 1, Mathematical and numerical model. *American Journal of Science,* v. 284, p. 1085–1124.

Garven, G., Ge, S., Person, M. A., and Sverjensky, D. A., 1993, Genesis of stratabound ore deposits in the mid-continent basins of North America. I. The role of regional groundwater flow. *Am. J. Sci.,* v. 293, p. 497–568.

Ge, S., and Garven, G., 1992, Hydromechanical modeling of tectonically driven groundwater flow with application to the Arkoma Foreland Basin. *J. Geophys. Res.,* v. 97, p. 9119–9144.

Gilbert, M. C., 1992, Speculations on the origin of the Anadarko Basin, in Mason, R., ed., *Basement tectonics 7 : proceedings of the Seventh International Conference on basement tectonics.* Boston, Kluwer Academic, p. 195–208.

Gillepsie, R. H., 1995a, Rise of the Texas oil industry: Part 1, Exploration at Spindletop. *The Leading Edge,* v. 14, no. 1, p. 22–24.

Gillepsie, R. H., 1995b, Rise of the Texas oil industry: Part 1, Spindletop changes the world. *The Leading Edge,* v. 14, no. 2, p. 113–115.

Gold, T., 1993, The origin of methane in the crust of the Earth, in, *The Future of Energy Gases.* U.S. Geological Survey Professional Paper 1570, D. G. Howell, ed., p. 57–80.

Gold, T., 1999, *The Deep Hot Biosphere.* Copernicus, New York, N.Y., 235 pp.

Goldman, M., 1996, Cancer risk of low-level exposure. *Science,* v. 271, p. 1821–1822.

Green, W. H., and Ampt, G. A., 1911. Studies on soil physics, 1: The flow of air and water through soils. *J. of Agricultural Science,* v. 4, p. 1–24.

Gregg, J. M., and Shelton, K. L., 1989, Minor- and trace-element distributions in the Bonneterre Dolomite (Cambrian), southeast Missouri: Evidence for possible multiple-basin fluid sources and pathways during lead-zinc mineralization. *Geol. Soc. Am. Bull.,* v. 101, p. 221–230.

Griswold, L. S., 1896, Notes on the geology of southern Florida. *Harvard Coll. Mus. Comp. Zoology Bull.,* v. 28, no. 2.

Gross, M. G., 1993, *Oceanography* (sixth edition). Prentice-Hall, 446 pp.

Hall, F. R., 1968, Base-Flow recessions—a review. *Water Resources Research,* v. 4, p. 973–983.

Hallam, A., 1983, *Great geological controversies.* Oxford Univ. Press, 182 pp.

Hannington, M. D., Jonasson, I. R., Herzig, P. M., and Petersen, S., 1995, Physical and chemical processes of seafloor mineralization at mid-ocean ridges, in, Humphris, S. E., Zierenber, R. A., Mullineaux, L. S., and Thomson, R. E., eds., *Seafloor hydrothermal systems.* American Geophysical Union Geophysical Monograph 91: American Geophysical Union, Washington, DC, pp. 115–157.

Hanor, J. S., 1979, Sedimentary genesis of hydrothermal fluids, in Barnes, H. L., ed., *Geochemistry of Hydrothermal Ore Deposits.* John Wiley, New York, p 137–168.

Hanor, J. S., 1987, History of thought on the origin of subsurface sedimentary brines, in, Landa, E. R., and Ince, S., eds, *The History of Hydrology.* American Geophysical Union History of Geophysics Series, v. 3, p. 81–91.

Hanor, J. S., 1993, Effective hydraulic conductivity of fractured clay beds at a hazardous waste landfill, Louisiana Gulf Coast. *Water Resources Research,* v. 29, p. 3691–3698.

Hanor, J. S., 1994a, Origin of saline fluids in sedimentary basins: in Parnell, J., ed., Geofluids: Origin, Migration and Evolution of Fluids in Sedimentary Basins. *Geological Society Special Publication No. 78,* p. 151–174.

Hanor, J. S., 1994b, Physical and chemical controls on the composition of waters in sedimentary basins. *Marine and Petroleum Geology,* v. 11, p. 31–45.

Hanor, J. S., and Sassen, R., 1990, Evidence for large-scale vertical and lateral migration of formation waters, dissolved salt, and crude oil in the Louisiana Gulf Coast, in Schumacher, D., and Perkins, B.F., eds., *Gulf Coast Oils and Gases: Their characteristics, origin, distribution, and exploration and production significance.* Proceedings 9th Annual Research Conference, Gulf Coast Section Society of Economic Paleontologists and Mineralogists , p. 283–296.

Hantush, M. S., 1960, Modification of the theory of leaky aquifers. *J. Geophys. Res.,* v. 65, p. 3713–3725.

Hardcastle, B. J., 1987, Wells ancient and modern—an historical review. *Quarterly Journal of Engineering Geology,* v. 20, p. 231–238.

Hardin, G., 1968, The Tragedy of the Commons. *Science,* v. 162, p. 1243–1248.

Harlan, R. L., Kolm, K. E., and Gutentag, E. D., 1989, *Water-well design and construction.* Elsevier, Amsterdam, 205 pp.

Healy, J. H., Rubey, W. W., Griggs, D. T., and Raleigh, C. B., 1968, The Denver Earthquakes. *Science,* v. 161, p. 1301–1310.

Hem, J. D., 1970, *Study and interpretation of the chemical characteristics of natural water.* USGS Water-Supply Paper 1473 (Second Edition).

Hessler, R. R., and Kaharl, V. A., 1995, The deep-sea hydrothermal vent community: an overview, in, Humphris, S. E., Zierenber, R. A., Mullineaux, L. S., and Thomson, R. E., eds., *Seafloor hydrothermal systems,* American Geophysical Union Geophysical Monograph 91. American Geophysical Union, Washington, DC, pp. 72–84.

Hitchon, B., 1969, Fluid flow in the western Canada sedimentary basin, 2, Effect of geology. *Water Resources Research,* v. 5, p. 460–469.

Hobba, W. A., Jr., Fisher, D. W., Pearson, F. J., Jr., and Chemerys, J. C., 1979, Hydrology and geochemistry of thermal springs of the Appalachians. *U.S. Geological Survey Professional Paper 1044–E.*

Hong, P., and Galen, M., 1992, The toxic mess called superfund. *Business Week,* no. 3265, May 11, 1992, p. 32.

Hornberger, G. M., Raffensperger, J. P., Wiberg, P. L., and Eshleman, K. N., 1998, *Elements of Physical Hydrology.* The Johns Hopkins Press, Baltimore, 302 pp.

Horton, R. E., 1933, The role of infiltration in the hydrologic cycle. *Transactions, American Geophysical Union,* v. 14, p. 446–460.

Hubbert, M. K., 1940, The theory of ground-water motion. *Journal of Geology,* v. 48, p. 785–944.

Hubbert, M. K., 1953, Entrapment of petroleum under hydrodynamic conditions. *American Association of Petroleum Geologists Bulletin,* v. 37, p. 1954–2026.

Hubbert, M. K., 1969, *The theory of ground-water motion and related papers.* Hafner, New York.

Hubbert, M. K., and Rubey, W. W., 1959, Role of fluid pressure in mechanics of overthrust faulting, I. Mechanism of fluid-filled porous solids and its application to overthrust faulting. *Geol. Soc. Am. Bull.,* v. 70, p. 115–166.

Huenges, E., Erzinger, J., Kück, J., Engeser, B., and Kessels, W., 1997, The permeable crust: geohydraulic properties down to 9101 m depth. *J. Geophys. Res.,* v. 102, p. 18,255–18,265.

Hunt, J. M., 1990, Generation and migration of petroleum from abnormally pressured fluid compartments. *AAPG Bulletin,* v. 72, p. 1–12.

Hutton, J., 1788, *Theory of the Earth; or an Investigation of the Laws Observable in the Composition, Dissolution, and Restoration of Land Upon the Globe.* Transactions, Royal Society of Edinburgh, v. 1, p. 217.

Hutton, J., 1795, *Theory of the earth, with proofs and illustrations.* In four parts: Edinburgh, W. Creech.

Ingerson, I. M., 1941, The hydrology of the southern San Joaquin Valley, California, and its relation to imported water supplies. *Transactions, American Geophysical Union,* v. 22, p. 20–45.

Ireland, R. L., 1986, Land subsidence in the San Joaquin Valley, California, as of 1983. *U. S. Geological Survey Water-Resources Investigations Report 85–4196.*

Jackson, S. A., and Beales, F. W., 1967, An aspect of sedimentary basin evolution: the concentration of Mississippi Valley-type ores during the late stages of diagenesis. *Canadian Society of Petroleum Geologists Bulletin:* v. 15, p. 393–433.

Jacob, C. E., 1939, Fluctuations in artesian pressure produced by passing railroad trains as shown in a well on Long Island, New York. *Transactions, Am. Geophysical Union,* v. 20, p. 666–674.

Jacob, C. E., 1940, On the flow of water in an elastic artesian aquifer. *Transactions, Am. Geophysical Union,* v. 22, p. 574–586.

Jeanloz, R., and Lay, T., 1993, The core-mantle boundary. *Scientific American,* v. 268, p. 48–55.

Jenden, P. D., Hilton, D. R., Kaplan, I. R., and Craig, H., 1993, Abiogenic hydrocarbons and mantle helium in oil and gas fields, in, *The Future of Energy Gases,* U. S. Geological Professional Paper 1570, D. G. Howell, ed., p. 31–56.

Jessop, A. M., 1990, *Thermal Geophysics.* Amsterdam, Elsevier.

Jonas, E. C., and McBride, E. F., 1977, *Diagenesis of sandstone and shale: Application to exploration for hydrocarbons.* Univ. of Texas at Austin, Continuing Ed. Program, Publ. 1, 165 pp.

Jorgensen, D. G., 1989, Paleohydrology of the Central United States. *U.S. Geological Survey Bulletin, Strategic and critical minerals in the midcontinent region, United States, Chapter D.*

Kappelmeyer, O., and Haenel, R., 1974, *Geothermics with special reference to application.* Berlin, Gebruder Borntraeger, 238 pp.

Kasting, J. F., 1993, Earth's early atmosphere. *Science,* v. 259, p. 920–926.

Kennicutt, M. C., II, McDonal, T. J., Comet, P. A., Denoux, G. J., and Brooks, J. M., 1992, The origins of petroleum in the northern Gulf of Mexico. *Geochimica Cosmochimica Acta,* v. 56, p. 1259–1280.

Keppler, H., 1996, Constraints from partitioning experiments on the composition of subduction-zone fluids. *Nature,* v. 380, p. 237–240.

Kesler, S. E., Jones, H. D., Furman, F. C., Sassen, R., Anderson, W. H., and Kyle, J. R., 1994, Role of crude oil in the genesis of Mississippi Valley-type deposits: Evidence from the Cincinnati arch. *Geology,* v. 22, p. 609–612.

Kharaka, Y. K., 1986, Origin and evolution of water and solutes in sedimentary basins, in Hitchon, B., Bachu, S., and Sauveplane, M., eds., *Proceedings, Third Canadian/American conference on hydrogeology, hydrogeology of sedimentary basins: application to exploration and exploitation.* National Water Well Association, Dublin, Ohio, p. 173–195.

Kiersch, G. A., 1965, Vaiont Reservoir Disaster. *Geotimes,* v. 9, no. 9, p. 9–12.

Kiersch, G. A., 1976, The Vaiont Reservoir Disaster, in, Tank, R., ed., *Focus on Environmental Geology, A Collection of Case Histories and Readings from Original Sources* (2nd Edition). Oxford Univ. Press, New York, p. 132–143.

Kiraly, L., 1975, Rapport sur l'état actuel des connaissances dans le domaine des caractères physiques des roches kartiques. *Internat. Union Geol. Sci.,* ser. B, no. 3, p. 53–67.

Kircher, A., 1678, *Mundus Subterraneus.* Amsterdam.

Klein, G. deV., 1991, Origin and evolution of North American cratonic basins. *South African Journal of Geology,* v. 94, p. 3–18.

Knauth, L. P., and Beeunas, M. A., 1986, Isotope geochemistry of fluid inclusions in Permian halite with implications for the isotopic history of ocean water and the origin of saline formation waters. *Geochimica Cosmochimica Acta,* v. 50, p. 419–433.

Knowles, D. B., 1965, Hydrologic aspects of the disposal of oil-field brines in Alabama. *Ground Water,* v. 3, no. 2., p. 22–27.

Kohout, F. A., 1964, The flow of fresh water and salt water in the Biscayne Aquifer of the Miami area, Florida in, *U.S. Geological Survey Water-Supply Paper 1613-C, p. C12–C35.*

Konikow, L. F., and Bredehoeft, J. D., 1992, Ground-water models cannot be validated. *Advances in Water Resources,* v. 15, p. 75–83.

Konikow, L. F., and Thompson, D. W., 1984, Groundwater contamination and aquifer reclamation at the Rocky Mountain Arsenal, Colorado, in, Groundwater Contamination, p. 93–103, National Academy Press, Washington, DC.

Kratz, G. P., 1998, Will corporate farms hog the market? *Deseret News,* September 16, 1998, p. B01.

Krauskopf, K. B., 1968, Preface, in, The Great Alaska Earthquake of 1964, *Hydrology,* National Academy of Science Publication 1603, Washington, DC, p. ix–xiii.

Kuhn, T. S., 1962, The structure of scientific revolutions. *Univ. of Chicago Press,* Chicago, 172 pp.

Lachenbruch, A. H., and Sass, J. H., 1977, Heat flow in the United States and the thermal regime of the crust, in Heacock, J. G., ed., The Earth's crust, its nature and physical properties. *Geophysical Monograph 20,* American Geophysical Union, Washington, D. C., p. 626–675.

Lachenbruch, A. H., Sass, J. H., Lawver, L. A., Brewer, M. C., Marshall, B. V., Munroe, R. J., Kennelly, J. P., Jr., Galanis, S. P., Jr., and Moses, T. H., Jr., 1987, Temperature and depth of permafrost on the Alaskan arctic slope, in Tailleur, I., and Weimer, P., eds., *Alaskan North Slope Geology,* v. 2: Bakersfield, California, Pacific section, Society of Economic Paleontologists and Mineralogists, p. 545–558.

Lachenbruch, A. H., Sass, J. H., Lawver, L. A., Brewer, M. C., Marshall, B. V., Munroe, R. J., Kennelly, J. P., Jr., Galanis, S. P., Jr., and Moses, T. H., Jr., 1988, Temperature and depth of permafrost on the Alaskan arctic slope, in Gryc, G., ed., Geology and exploration of the National Petroleum Reserve in Alaska, 1974 to 1982: *U.S. Geological Survey Professional Paper 1399,* p. 645–656.

Leach, D. L., 1979, Temperature and salinity of the fluids responsible for minor occurrences of sphalerite in the Ozark region of Missouri. *Economic Geology,* v. 74, p. 931–937.

Leach, D. L., and Rowan, E. L., 1986, Genetic link between Ouachita foldbelt tectonism and the Mississippi Valley-type lead-zinc deposits of the Ozarks. *Geology,* v. 14, p. 931–935.

Leach, D. L., and Rowan, E. L., 1991, Comment and Reply on "Focused fluid flow and Ozark Mississippi Valley-type deposits." *Geology,* v. 19, p. 190–191.

Leach, D. L., and Rowan, E. L., 1993, Fluid-inclusion studies of regionally extensive epigenetic dolomites, Bonneterre Dolomite (Cambrian), southeast Missouri: Evidence of multiple fluids during dolomitization and lead-zinc mineralization: alternative interpretation and reply. *Geol. Soc. Am. Bull.,* v. 105, p. 968–978.

Lee, Y., Deming, D., and Chen, K. F., 1996, Heat Flow and Heat Production in the Arkoma Basin and Oklahoma Platform, Southeastern Oklahoma. *J. Geophys. Res.,* v. 101, p. 25,387–25,401.

Lehr, J., Hurlburt, S., Gallagher, B., and Voytek, J., 1988, *Design and construction of water wells, a guide for engineers.* Van Nostrand Reinhold Company, New York, 229 pp.

Lin, G., Nunn, J. A., and Deming, D., 2000, Thermal buffering of sedimentary basins by basement rocks:

implications arising from numerical simulations. *Petroleum Geoscience,* v. 6, p. 299–307.

Lindley, T., 1999, The Panhandle: a Region Divided. *Daily Oklahoman,* October 17, 1999, p. 1–A.

Lindorff, D. E., and Cartwright, K., 1977, Ground-water contamination: problems and remedial action. *Environmental Geology Notes,* Illinois Geological Survey, no. 81, p. 1–30.

Lister, C. R. B., 1972, On the thermal balance of a mid-ocean ridge. *Geophys. J. Roy. Astr. Soc.,* v. 26, p. 515–535.

Lodemann, M., Fritz, P., Wolf, M., Ivanovich, M., Hansen, B. T., and Nolte, E., 1997, On the origin of saline fluids in the KTB (continental deep drilling project of Germany). *Applied Geochemistry,* v. 12, p. 831–849.

Lowell, R. P., Rona, P. A., and Von Herzen, R. P., 1995, Seafloor hydrothermal systems. *J. Geophys. Res.,* v. 100, p. 327–352.

Lucia, F. J., 1983, Petrophysical parameters estimated from visual descriptions of carbonate rocks — a field classification of carbonate pore space. *J. Petroleum Technology,* v. 35, p. 629–635.

Luckey, T. D., 1992, Hormesis and nurture with ionizing radiation, in, Ellsaesser, H. W., ed., *Global 2000 revisited, mankind's impact on spaceship Earth,* p. 189–251, Paragon House, New York.

Lyell, C., 1830, *Principles of geology; being an attempt to explain the former changes of the earth's surface, by reference to causes now in operation.* London, J. Murray.

Lyell, C., 1867, *Principles of Geology* (10th edition). J. Murray, London.

Mackay, D. M., and Cherry, J. A., 1989, Groundwater contamination: pump-and-treat remediation. *Environmental Science and Technology,* v. 23, no. 6, p. 630–636.

Mackenzie, A. S., and Quigley, T. M., 1988, Principles of geochemical prospect appraisal. *American Association of Petroleum Geologists Bulletin,* v. 72, p. 399–415.

Magara, K., 1980, Comparison of porosity-depth relationships of shale and sandstone. *J. Petroleum Geology,* v. 3, p. 175–185.

Magoon, L. B., and Dow, W. G. (eds.), 1994, The Petroleum System: from Source to Trap. *AAPG Memoir 60,* Am. Assoc. of Petroleum Geologists, Tulsa, Oklahoma, 655 pp.

Manning, C. E., and Ingebritsen, S. E., 1999, Permeability of the continental crust: implications of geothermal data and metamorphic systems. *Reviews of Geophysics,* v. 37, p. 127–150.

Marshall, E., 1991, A is for Apple, Alar, and....Alarmist? *Science,* v. 254, p. 20–22.

Martinsen, R. S., 1994, Summary of published literature on anomalous pressures: implications for the study of pressure compartments, in, Ortoleva, P. J., ed., *Basin Compartments and Seals.* AAPG Memoir 61, *Am. Assoc. Petr. Geologists,* Tulsa, p. 27–38.

Matthes, G. H., 1953, Quicksand. *Scientific American,* v. 188, no. 6, p. 97–102.

Mazor, E., 1991, *Applied Chemical and Isotopic Groundwater Hydrology.* Halstead Press, New York.

Mazzuchelli, G., 1737, Notizie istoriche e critiche intorno alla vita, alle invenzioni, ed agli scritti di Archimede Siracusano. *Brescia, Presso G. M. Rizzardi,* 128 pp.

McKenzie, D., 1978, Some remarks on the development of sedimentary basins. *Earth and Planetary Science Letters,* v. 40, p. 25–32.

McLaurin, J. J., 1896, Sketches in crude oil, published by the author at Harrisburg, Pennsylvania, 406 pp.

McMillion, L. G., 1965, Hydrologic aspects of disposal of oil-field brines in Texas. *Ground Water,* v. 3, no. 4, p. 36–42.

McNutt, M., 1992, Guymon OKs tax for jobs funds for pork plant win landslide. *Daily Oklahoman,* October 14, 1992.

McWilliams, J., 1972, Large saltwater-disposal systems of East Texas and Hastings Oil Fields, Texas, in, Cook, T. D., ed., *Underground Waste Management and Environmental Applications.* AAPG Memoir 18, *Am. Assoc. Petroleum Geologists,* Tulsa, Oklahoma.

Mead, W. J., 1925, The geologic role of dilatancy. *J. of Geology,* v. 33, p. 685–698.

Meade, C., and Jeanloz, R., 1991, Deep-focus earthquakes and recycling of water into the Earth's mantle. *Science,* v. 252, p. 68–72.

Meinzer, O. E., 1923, The occurrence of groundwater in the United States with a discussion of principles. *U.S. Geological Survey Water Supply Paper 489.*

Meinzer, O. E., 1923, Outline of ground-water hydrology, with definitions. *U.S. Geological Survey Water-Supply Paper 494,* 71 pp.

Meinzer, O. E., 1928, Compressibility and elasticity of artesian aquifers. *Economic Geology,* v. 23, p. 263–291.

Meinzer, O. E., 1934, The history and development of ground-water hydrology. *J. of the Washington Academy of Science,* v. 24, no. 1, p. 6.

Meinzer, O. E., 1942, Hydrology, in, Meinzer, O. E., ed., *Physics of the Earth—IX, Hydrology.* Dover Publications, New York, pp. 1–31.

Meissner, F. F., 1984a, Stratigraphic relationships and distribution of source rocks in the Greater Rocky Mountain Region, in, Woodward, J., Meissner, F. F., and Claypool, J. L., eds., *Hydrocarbon Source Rocks of the Greater Rocky Mountain Region: Rocky Mountain Association of Field Geologists 1984 Field Guide/Symposium Volume.* Rocky Mountain Association of Geologists, Denver, p. 1–34.

Meissner, F. F., 1984b, Petroleum Geology of the Bakken Formation, Williston Basin, North Dakota and Montana, in, Demaison, G., and Murris, R. J., eds., *Petroleum Geochemistry and Basin Evaluation, AAPG Memoir 35,* p. 159–179.

Meissner, F. F., 1991, Origin and migration of oil and gas, in, Gluskoter, H. J., Rice, D. D., and Taylor, R. B., eds., *Economic Geology, U.S.* Boulder, Colorado, Geological Society of America, The Geology of North America, Vol. P-2, p. 225–240.

Meissner, F. F., Woodward, J., and Clayton, J. L., 1984, Stratigraphic relationships and distribution of source rocks in the Greater Rocky Mountain Region, in, Woodward, J., Meissner, F. F., and Clayton, J. L., eds., *Hydrocarbon Source Rocks of the Greater Rocky Mountain Region.* Rocky Mountain Association of Geologists, Denver, Colorado, p. 1–34.

Mello, U. T., Karner, G. D., and Anderson, R. N., 1994, A physical explanation for the positioning of the depth to the top of overpressure in shale-dominated sequences in the Gulf Coast basin, United States. *J. Geophys. Res.,* v. 99, p. 2775–2789.

Mendeleev, D., 1877, L'origine du petrole. *Revue Scientifique,* 2e Ser., v. 8, p. 409–416.

Mendenhall, W. C., 1909, A phase of ground water problems in the West. *Economic Geology,* v. 4, p. 35–45.

Meyboom, P., 1961, Estimating ground-water recharge from stream hydrographs. *J. Geophys. Res.,* v. 66, p. 1203–1214.

Miller, D. W. (ed.), 1980, *Waste disposal effects on ground water.* Premier Press, Berkeley, California, 512 pp.

Mitra, S., 1988, Three-dimensional geometry and kinematic evolution of the Pine Mountain thrust system, southern Appalachians. *GSA Bulletin,* v. 100, p. 72–95.

Moore, J. C., and Vrolijk, P., 1992, Fluids in accretionary prisms. *Rev. Geophys.,* v. 30, p. 113–135.

Morrow, C. A., and Lockner, D. A., 1994, Permeability differences between surface-derived and deep drillhole core samples. *Geophysical Research Letters,* v. 21, p. 2151–2154.

Neglia, S., 1979, Migration of fluids in sedimentary basins. *AAPG Bulletin,* vol. 63, p. 573–579.

Nelson, P.H., 1994, Permeability-porosity relationships in sedimentary rocks. *The Log Analyst,* v. 35, no. 3, p. 38–62.

Nesbitt, B. E., 1988, Gold deposit continuum: a genetic model for lode Au mineralization in the continental crust. *Geology,* v. 16, p. 1044–1048.

Nesbitt, B. E., and Muehlenbachs, K., 1989, Origins and movement of fluids during deformation and metamorphism in the Canadian Cordillera. *Science,* v. 245, p. 733–736.

Nesbitt, B. E., and Muehlenbachs, K., 1991, Stable isotopic constraints on the nature of the syntectonic fluid regime of the Canadian Cordillera. *Geophysical Research Letters,* v. 18, p. 963–966.

Nesbitt, B. E., and Muehlenbachs, K., 1995, Geochemical studies of the origins and effects of synorogenic crustal fluids in the southern Omineca Belt of British Columbia, Canada. *GSA Bull.,* v. 107, p. 1033–1050.

Neuzil, C. E., 1986, Groundwater flow in low-permeability environments. *Water Resources Research,* 22, p. 1163–1195.

Neuzil, C. E., 1994, How permeable are clays and shales? *Water Resources Research,* v. 30, p. 145–150.

Neuzil, C. E., 1995, Abnormal pressures as hydrodynamic phenomena. *Am. J. Sci.,* v. 295, p. 742–786.

Neuzil, C. E., 2000, Osmotic generation of "anomalous" fluid pressures in geological environments. *Nature,* v. 403, p. 182–184.

Nielsen, D. R., Genuchten, M. Th. van, and Biggar, J. W., 1986, Water flow and solute transport processes in the unsaturated zone. *Water Resources Research,* v. 22, p. 89S–108S.

Nolet, G., and Zielhuis, A., 1994, Low S Velocities under the Tornquist-Teisseyre zone: evidence for water injection into the transition zone by subduction. *J. Geophys. Res.,* v. 99, p. 15,813–15,820.

North, F. K., 1985, *Petroleum Geology:* Allen & Unwin, Boston.

Norton, D. L., 1990, Pore fluid pressure near magma chambers, in: The role of fluids in crustal processes. Washington, D. C., National Academy Press, p. 42–49.

Norton, D., 1984, A theory of hydrothermal systems. *Annual Reviews of Earth and Planetary Sciences,* v. 12, p. 155–177.

Norton, D., and Knapp, R., 1977, Transport phenomena in hydrothermal systems: Cooling plutons. *American Journal of Science,* v. 277, p. 937–981.

NRC (National Research Council), 1984, Groundwater contamination. *National Academy Press,* Washington, DC, 179 pp.

NRC (National Reseach Council), 1994, Alternatives for ground water cleanup. *National Academy Press,* Washington, DC, 315 pp.

Nunn, J. A., 1994, Free thermal convection beneath intracratonic basins: thermal and subsidence effects. *Basin Research,* v. 6, p. 115–130.

Nunn, J. A., 1996, Buoyancy-driven propagation of isolated fluid-filled fractures: implications for fluid transport in Gulf of Mexico geopressured sediments. *J. Geophys. Res.,* v. 101, p. 2963–2970.

Nunn, J. A., and Sassen, R., 1986, The framework of hydrocarbon generation and migration, Gulf of Mexico continental slope. Transactions, Gulf Coast Association of Geological Societies, v. 36, p. 257–262.

Nur, A., 1972, Dilatancy, pore fluids, and premonitory variations of ts/tp travel times. *Seismological Society of America Bulletin,* v. 62, no. 5, p. 1217–1222.

Nur, A., and Walder, J., 1990, Time-dependent hydraulics of the Earth's crust in: The role of fluids in crustal processes. Washington, D.C., National Academy Press, p. 111–127.

O'Nions, R. K., and Oxburgh, E. R., 1983, Heat and helium in the Earth. *Nature,* v. 306, p. 429–431.

Ogata, A., 1970, Theory of dispersion in a granular medium. *U.S. Geol. Survey Professional Paper* 411-I.

Ohle, E. L., 1991, Lead and zinc deposits, in, Gluskoer, H. J., Rice, D. D., and Taylor, R. B., eds., *The Geology of North America,* vol. P-2, Economic Geology, U.S. Geol. Soc. Am., Boulder, Colorado, pp. 43–62.

Ohtake, M., 1974, Seismic activity induced by water injection at Matsushiro, Japan. *J. Phys. Earth,* v. 22, p. 163–176.

Oki, T., 1999, The global water cycle, in, Browning, K. A., and Gurney, R. J., eds., *Global energy and water cycles.* Cambridge Univ. Press, Cambridge.

Oliver, J., 1986, Fluids expelled tectonically from orogenic belts: their role in hydrocarbon migration and other geologic phenomena. *Geology,* v. 14, p. 99–102.

Oliver, J., 1992, The spots and stains of plate tectonics. *Earth Science Reviews,* v. 32, p. 77–106.

Ortoleva, P. J. (ed.), 1994a, Basin Compartments and Seals. *AAPG Memoir 61,* Am. Assoc. Petr. Geologists, Tulsa, 447 pp.

Orotleva, P. J., 1994b, Basin compartmentation: definitions and mechanisms, in, Ortoleva, P. J., ed., Basin Compartments and Seals. *AAPG Memoir 61,* Am. Assoc. Petr. Geologists, Tulsa, p. 39–51.

OTA (US Office of Technology Assessment), 1984, Protecting the Nation's Groundwater from Contamination, *OTA-O-233,* 244 pp.

Oxburgh, E. R., O'Nions, R. K., and Hill, R. I., 1986, Helium isotopes in sedimentary basins. *Nature,* v. 324, p. 632–635.

Parker, G. G., Ferguson, G. E., and Love, S. K., 1944, Interim report on the investigations of water resources in southeastern Florida with special reference to the Miami area in Dade County. *Report of Investigations, No. 4,* Florida Geological Survey, Tallahassee.

Parker, G. G., Ferguson, G. E., Love, S. K., and others, 1955, Water resources of southeastern Florida. *U.S. Geological Survey Water-Supply Paper 1255,* 963 pp.

Parker, B. L., Gillham, R. W., and Cherry, J. A., 1994, Diffusive disappearance of immiscible-phase organic liquids in fractured geologic media. *Ground Water,* v. 32, no. 5, p. 805–820.

Parks, G., Brittnacher, M., Chen, L. J., Elsen, R., McCarthy, M., Germany, G., and Spann, J., 1997, Does the UVI on Polar detect cosmic snowballs? *Geophys. Res. Lett.,* v. 24, p. 3109–3112.

Parks, G., Brittnacher, M., Elsen, R., McCarthy, M., O'Meara, J. M., Germany, G., and Spann, J., 1998, Comparison of dark pixels observed by VIS and UVI in dayglow images. *J. Geophys. Res.* (in press).

Parks, K. P., and Toth, J., 1995, Field evidence for erosion-induced underpressuring in upper Cretaceous and Tertiary strata, West Central Alberta, Canada. *Bull. of Canadian Petroleum Geology,* v. 43, p. 281–292.

Patchick, P. F., 1966, Quicksand and water wells. *Groundwater,* v. 4, no. 2, p. 32–46.

Patterson, J. W., 1989, Industrial wastes reduction. *Environmental Science and Technology,* v. 23, no. 9, p. 1032–1038.

Peacock, S. M., 1990, Fluid processes in subduction zones. *Science,* v. 248, p. 329–337.

Phillip, J. R., 1969, Theory of infiltration in *Advances in Hydroscience* v. 5, V. T. Chow, ed., p. 215–296.

Philip, J. R., 1995, Desperately seeking Darcy in Dijon. *Journal of the Soil Science Society of America,* v. 59, p. 319–324.

Phillips, F. M., Bentley, H. W., and Elmore, D., 1986, Chlorine-36 dating of old groundwater in sedimentary basins: in, Proceedings, Third Canadian/American

Conference on Hydrogeology: Hydrogeology of Sedimentary Basins: Application to Exploration and Exploitation, Banff, Alberta, Canada, June 22-26, 1986, edited by Hitchon, B., Bachu, S., and Sauveplane, C. M., National Water Well Association, Dublin, Ohio, p. 143–150.

Phillips, F. M., Tansey, M. K., Peeters, L. A., Cheng, S., and Long, A., 1989, An isotopic investigation of groundwater in the Central San Juan Basin, New Mexico: Carbon 14 dating as a basis for numerical flow modeling. *Water Resources Research,* v. 25, p. 2259–2273.

Phillips, S. L., Igbene, A., Fair, J. A., Ozbek, H., and Tavana, M., 1981, A technical handbook for geothermal energy utilization. *Lawrence Berkeley Lab. Rep. LBL-12810,* 46 pp.

Pickens, J. F., and Lennox, W. C., 1976, Numerical simulation of waste movement in steady groundwater flow systems. *Water Resources Research,* v. 12, p. 171–180.

Pinder, G. F., and Bredehoeft, J. D., 1968, Application of the digital computer for aquifer evaluation. *Water Resources Research,* v. 4, no. 5, p. 1069–1093.

Plank, T., 1996, The brine of the Earth. *Nature,* v. 380, p. 202–203.

Playfair, J., 1802, Illustrations of the Huttonian theory of the earth. Edinburgh,William Creech, 528 pp.

Poland, J. F., 1972, Subsidence and its control in *Underground Waste Management and Environmental Implications,* ed. by T. D. Cook, Am. Assoc. Petroleum Geologists Memoir 18, Tulsa, Oklahoma, p. 50–71.

Pollack, H. N., Hurter, S. J., Johnson, J. R., 1993, Heat flow from the Earth's interior: analysis of the global data set. *Reviews of Geophysics,* v. 31, p. 267–280.

Popper, K., 1959, *The logic of scientific discovery.* Hutchinson.

Powley, D. E., 1980, Pressures, normal and abnormal: AAPG Advanced Exploration Schools Unpublished Lecture Notes, 48 pp.

Press, F., and Siever, R., 1986, *Earth* (fourth edition). W. H. Freeman, New York.

Proshlyakov, B. K., 1960, Reservoir properties of rocks as a function of their depth and lithology. *Geol. Neft. Gaza,* v. 12, p. 24–29.

Quigley, T. M., Mackenzie, A. S., and Gray, J. R., 1987, Kinetic theory of petroleum generation, in Doligez, B., ed., *Migration of hydrocarbons in sedimentary basins.* Paris, Editions Technip, p. 649–665.

Raleigh, C. B., Healy, J. H., and Bredehoeft, J. D., 1976, An experiment in earthquake control at Rangely, Colorado. *Science,* v. 191, p. 1230–1237.

Raup, D. C., 1995, Historical Essay: The method by multiple working hypotheses by T. C. Chamberlin, with an introduction by David C. Raup. *J. of Geology,* v. 103, p. 349–354.

Revelle, R., and Maxwell, A. E., 1952, Heat flow through the floor of the eastern North Pacific Ocean. *Nature,* v. 170, p. 199–200.

Revil, A., Cathles, L. M. III, Shosa, J. D., Pezard, P. A., and de Larouziere, F. D., 1998, Capillary sealing in sedimentary basins: a clear field example. *Geophys. Res. Lett.,* v. 25, p. 389–392.

Richards, L. A., 1931, Capillary conduction of liquids through porous mediums. *Physics,* v. 1, p. 318–333.

Robertson, W. D., and Cherry, J. A., 1989, Tritium as an indicator of recharge and dispersion in a groundwater system in central Ontario. *Water Resources Research,* v. 25, p. 1097–1109.

Roedder, E., 1979, Fluid inclusions as samples of ore fluids in *Geochemistry of Hydrothermal Ore Deposits* (Second Edition), Barnes, H. L., ed., John Wiley, New York, pp. 684–737.

Roeloffs, E. A., 1988, Hydrologic precursors to earthquakes: a review: Pure and Applied Geophysics, v. 126, p. 177–209.

Rojstaczer, S., Wolf, S., and Michel, R., 1995, Permeability enhancement in the shallow crust as a cause of earthquake-induced hydrological changes. *Nature,* v. 373, p. 237–239.

Ross, M., and Skinner, H. C. W., 1994, Geology and health. *Geotimes,* v. 39, no. 1, p. 10–12.

Rouse, H., and Ince, S., 1957, *History of hydraulics: Iowa Institute of Hydraulic Research,* State University of Iowa, 269 pp.

Rubey, W. W., 1951, Geologic history of sea water, an attempt to state the problem. *Bull. Geol. Soc. Am.,* v. 62, p. 1111–1148.

Rusconi, G. A. (translator), 1592, *Della Architettvra.* Venice, Appresso i Gioliti, 143 pp.

Scholz, C. H., Sykes, L. R., and Aggarwal, Y. P., 1973, Earthquake prediction: a physical basis. *Science,* v. 181, p. 803–810.

Schwab, F. L., 1976, Modern and ancient sedimentary basins: comparative accumulation rates. *Geology,* v. 4, p. 723–727.

Schulze-Makuch, D., and Cherkauer, D. S., 1997, *Method developed for extrapolating scale behavior.*

EOS, Transactions American Geophysical Union, v. 78, p. 3.

Schulze-Makuch, D., Carlson, D. A., Cherkauer, D. S., and Malik, P., 1999, Scale dependency of hydraulic conductivity in heterogeneous media. *Groundwater,* v. 37, no. 6, p. 904–919.

Shaler, N. S., 1890, The topography of Florida. *Harvard Coll. Mus. Comp. Zoology Bull.,* v. 16, no. 7.

Sharp, J. M., and Domenico, P. A., 1976, Energy transport in thick sequences of compacting sediment. *Geological Society of America Bulletin,* v. 87, p. 390–400.

Sharp, J. M., Jr., and Kyle, J. R., 1988, The role of ground-water processes in the formation of ore deposits in Back, W., Rosenshein, J. S., and Seaber, P. R., eds., *Hydrogeology: The Geology of North America,* v. O-2: Boulder, Geological Soc. Am., pp. 461–483.

Shelton, K. L., Bauer, R. M., and Gregg, J. M., 1992, Fluid-inclusion studies of regionally extensive epigenetic dolomites, Bonneterre Dolomite (Cambrian), southeast Missouri: Evidence of multiple fluids during dolomitization and lead-zinc mineralization. *Geol. Soc. Am. Bull.,* v. 104, p. 675–683.

Sibson, R. H., 1987, Earthquake rupturing as a mineralizing agent in hydrothermal systems. *Geology,* v. 15, p. 701–704.

Sibson, R. H., 1994, Crustal stress, faulting and fluid flow: in Parnell, J., ed., Geofluids: Origin, Migration and Evolution of Fluids in Sedimentary Basins. *Geological Society Special Publication No. 78,* p. 69–84.

Sibson, R. H., Morre, J. McM., and Rankin, A. H., 1975, Seismic pumping—a hydrothermal fluid transport mechanism. *J. of the Geological Society,* London: v. 131, p. 653–659.

Silver, B. L., 1998, *The ascent of science.* Oxford Univ. Press, Oxford, 534 pp.

Simon, J. L. (ed.), 1995, *The State of Humanity.* Blackwell, Oxford, 694 pp.

Simpson, D. W., 1986, Triggered earthquakes. *Annual Review of Earth and Planetary Science,* v. 14, p. 21–42.

Skinner, B. J., 1997, Hydrothermal mineral deposits: what we do and don't know in *Geochemistry of Hydrothermal Ore Deposits,* Third Edition, H. L. Barnes, Ed., John Wiley & Sons, New York, pp. 1–29.

Sleep, N. H., 1971, Thermal effects of the formation of Atlantic continental margins by continental break up. *Geophysical Journal of the Royal Astronomical Society,* v. 24, p. 325–350.

Sleep, N. H., Nunn, J. A., and Chou, L., 1980, Platform Basins. *Annual review of Earth and Planetary Science,* v. 8, p. 17–34.

Slichter, C. S., 1899, Theoretical investigations of the motion of ground water. *U.S. Geological Ninteenth Annual Report,* part 2, p. 305–328.

Slichter, C. S., 1905, Field measurements of the rate of underground water. *U.S. Geological Survey Water Supply Paper 140,* p. 9–85.

Smith, A., 1953, *Blind white fish in Persia.* George Allen and Unwin Ltd., London.

Smoluchowski, M. S., 1909, Some remarks on the mechanics of overthrusts. *Geological Magazine,* v. 6, p. 204–205.

Smyth, J. R., 1994, A crystallographic model for hydrous wadsleyite (B-Mg_2SiO_4): an ocean in the Earth's interior? *American Mineralogist,* v. 79, p. 1021–1024.

Solley, W. B., Pierce, R. R., and Perlman, H. A., 2000, Estimated Use of Water in the United States in 1995: *U.S. Geological Survey Circular 1200.*

Starr, R. C., and Cherry, J. A., 1994, In situ remediation of contaminated ground water: the funnel-and-gate system. *Ground Water,* v. 32, no. 3, p. 465–476.

Stein, C., and Stein, S., 1992, A model for the global variation in oceanic depth and heat flow with lithospheric age. *Nature,* v. 359, p. 123–129.

Stein, C., and Stein, S., 1994, Constraints on hydrothermal heat flux through the oceanic lithosphere from global heat flow. *J. Geophys. Res.,* v. 99, p. 3081–3095.

Stipp, D., 1991, Throwing good money at bad water yields scant improvement. *Wall Street Journal,* v. 217, no. 95, May 15, 1991, p. A1.

Stueber, A. M., and Walter, L. M., 1991, Origin and chemical evolution of formation waters from Silurian-Devonian strata in the Illinois Basin, USA. *Geochimica et Cosmochimica Acta,* v. 55, p. 309–325.

Stueber, A. M., and Walter, L. M., 1994, Glacial recharge and paleohydrologic flow systems in the Illinois Basin: evidence from chemistry of Ordovician carbonate (Galena) formation waters. *GSA Bull.,* v. 106, p. 1430–1439.

Sverjensky, D. A., 1986, Genesis of Mississippi Valley-type lead-zinc deposits. *Annual Reviews of Earth & Planetary Science,* v. 14, p. 177–199.

Taylor, H. P., Jr., 1974, The application of oxygen and hydrogen isotope studies to problems of hydrothermal alteration and ore deposition. *Economic Geology,* v. 69, p. 843–883.

Terzaghi, K., 1925, *Erdbaumechanic auf Bodenphysikalischer Grundlage.* Franz Deuticke, Vienna.

Theim, G., 1906, *Hydrologische Methode.* Leipzig, Gebhardt.

Thelin, G. P., and Heimes, F. J., 1987, Mapping irrigated cropland from Landsat data for determination of water use from the High Plains Aquifer in parts of Colorado, Kansas, Nebraska, New Mexico, Oklahoma, South Dakota, Texas, and Wyoming: *U.S. Geological Survey Professional Paper 1400-C.*

Thompson, S. A., 1999, *Water use, management, and planning in the United States.* Academic Press, San Diego.

Tihansky, A. B., 1999, Sinkholes, West-Central Florida, in, Galloway, D., Jones, D. R., and Ingebritsen, S. E., eds. *U.S. Geological Circular 1182,* p. 121–140.

Titley, S. R., 1990, Evolution and style of fracture permeability in intrusion-centered hydrothermal systems in *The role of fluids in crustal processes.* Washington, D.C., National Academy Press, p. 50–63.

Tolman, C. F., 1937, *Ground water.* McGraw-Hill, New York.

Torgersen, T., 1980, Controls on pore-fluid concentration of ^{4}He and ^{222}Rn and the calculation of ^{4}He/^{222}Rn ages. *Journal of Geochemical Exploration,* v. 13, p. 57–75.

Torgersen, T., and Clarke, W. B., 1985, Helium accumulation in groundwater, 1. An evaluation of sources and the continental flux of crustal ^{4}He in the Great Artesian Basin, Australia. *Geochimica Cosmochimica Acta,* v. 49, p. 1211–1218.

Torgersen, T., Drenkard, S., Stute, M., Schlosser, P., and Shapiro, A., 1995, Mantle helium in ground waters of eastern North America: time and space constraints on sources. *Geology,* v. 23, p. 675–678.

Toth, J. A., 1962, A theory of ground-water motion in small drainage basins in central Alberta, Canada. *J. Geophys. Res.,* v. 67, p. 4375–4387.

Toth, J. A., 1963, A theoretical analysis of ground-water flow in small drainage basins. *J. Geophys. Res.,* v. 68, p. 4795–4811.

Toth, J., and Millar, R. F., 1983, Possible Effects of Erosional Changes of the Topographic Relief on Pore Pressures at Depth. *Water Resources Research,* v. 19, p. 1585–1597.

Toth, J., 1988, Ground water and hydrocarbon migration, in *The Geology of North America — Hydrogeology,* W. Back, J. S. Rosenshein, and P. R. Seaber, eds. Geological Soc. of America, v. O-2, p. 485–502.

Touloukian, Y. S., Saxena, S. C., and Hestermans, P., 1975, Viscosity, Thermophysical properties of matter. *TPRC Data Ser.,* v. 11, Plenum, New York, 1975.

Townend, J., and Zoback, M. D., 2000, How faulting keeps the crust strong. *Geology,* v. 28, no. 5, p. 399–402.

Turcotte, D. L., and Schubert, G., 1982, *Geodynamics applications of continuum physics to geological problems.* New York, John Wiley, 450 pp.

Twain, M., 1883, *Life on the Mississippi.* Boston, J. R. Osgood, 624 pp.

USGS (United States Geological Survey), 1999, The Quality of our Nation's Waters: Nutrients and Pesticides. *USGS Circular 1225.*

Vail, P. R., Mitchum, R. M., Jr., and Thompson, S., III, 1977. Seismic stratigraphy and global changes of sea level, Part 3: relative changes of sea level from coastal onlap. In Payton, C. E. (Editor), *Seismic stratigraphy — applications to hydrocarbon exploration.* AAPG Memoir 26. American Assoc. Petroleum Geologists, Tulsa, Oklahoma, pp. 83–97.

Viraraghavan, T., 1982, Effects of septic systems on environmental quality. *J. of Environmental Management,* v. 15, p. 63–70.

Vogel, H. U., 1993, The great well of China. *Scientific American,* v. 268, no. 6, p. 116–120.

Vorhis, R. C., 1968, Effects outside Alaska in *The Great Alaska Earthquake of 1964,* Hydrology, National Academy of Science Publication 1603, Washington, DC, p. 140–189.

Walker, T. R., 1961, Ground-water contamination in the Rocky Mountain Arsenal Area, Denver, Colorado. *Geological Soc. of American Bulletin,* v. 72, p. 489–494.

Waller, R. M., 1968, Water-sediment ejections in *The Great Alaska Earthquake of 1964,* Hydrology, National Academy of Science Publication 1603, Washington, DC, p. 97–116.

Walter, L. M., Stueber, A. M., and Huston, T. J., 1990, Br-Cl-Na systematics in Illinois Basin fluids: constraints on fluid origin and evolution. *Geology,* v. 18, p. 315–318.

Walther, J. V., 1990, Fluid dynamics during progressive regional metamorphism in *The role of fluids in crustal processes.* Washington, D.C., National Academy Press, p. 64–71.

Waples, D. W. , 1980, Time and temperature in petroleum exploration: application of Lopatin's method to petroleum exploration. *AAPG Bull.,* v. 64, p. 916–926.

Ward, J. C., 1964, Turbulent flow in porous media. *Proc. Amer. Soc. Civil Eng. No. HY5,* v. 90, p. 1–12.

Whelan, J. K., Kennicutt, M. C., Brooks, J. M., Schumacher, D., and Eglinton, L. B., 1994, Organic geochemical indicators of dynamic fluid flow processes in petroleum basins. *Organic Geochemistry,* v. 22, p. 587–615.

White, D. E., 1968, Environments of generation of some base-metal ore deposits. *Economic Geology,* v. 63, p. 301–335.

Wiedemeier, T. H., Rifai, H. S., Newell, C. J., and Wilson, J. T., 1999, *Natural attenuation of fuels and chlorinated solvents in the subsurface.* John Wiley & Sons, New York, 617 pp.

Williams, J. A., 1974, Characterization of oil types in the Williston Basin. *AAPG Bulletin,* v. 58, p. 1243–1252.

Williams, D. E., and Wilder, D. G., 1971, Gasoline pollution of a ground-water reservoir—a case history. *Ground Water,* v. 9, no. 6, p. 50–54.

Wilson, R., and Crouch, E. A. C., 1987, Risk assessment and comparisons: an introduction. *Science,* v. 236, p. 267–270.

Winograd, I. J., Landwehr, J. M., Ludwig, K. R., Coplen, T. B., and Riggs, A. C., 1997, Duration and structure of the past four interglaciations. *Quaternary Research,* v. 48, p. 141–154.

Winter, T. C., Harvey, J. W., Franke, O. L., and Alley, W. M., 1998, Ground water and surface water: a Single Resource. *U. S. Geological Survey Circular 1139.*

Wisler, C. O., 1959, *Hydrology* (2nd Edition). New York, Wiley.

Wood, R. M., 1994, Earthquakes, strain-cycling and the mobilization of fluids in *Geofluids: Origin, Migration and Evolution of Fluids in Sedimentary Basins,* J. Parnell, ed., Geological Society Special Publication No. 78, p. 85–98.

Wood, S. H., Wurts, C., Ballenger, N., Shaleen, M., and Totorica, D., 1985, The Borah Peak, Idaho, earthquake of October 28, 1983: hydrologic effects. *Earthquake Spectra,* v. 2, p. 125–150.

Wyllie, P. J., 1971, Experimental limits for melting in the Earth's crust and upper mantle in The structure and physical properties of the Earth's crust, J. G. Heacock, ed., *American Geophysical Union Monograph 14,* American Geophysical Union, Washington, D.C., p. 279–301.

Yates, M. V., 1985, Septic tank density and ground-water contamination. *Ground Water,* v. 23, no. 5, p. 586–591.

Zoback, M. D., Elders, W. A., Van Schmus, W. R., and Younker, L., 1988, The role of continental scientific drilling in modern earth sciences, scientific rationale and plan for the 1990's: a report to the interagency coordinating group for the continental scientific drilling program from the workshop on continental scientific drilling convened at Stanford University. August, 1988 (available from Mark Zoback, Stanford University).

INDEX